Heat Transfer Enhancement of Heat Exchangers

Dear MyCopy Customer,

This Springer book is a monochrome print version of the eBook to which your library gives you access via SpringerLink. It is available to you at a subsidized price since your library subscribes to at least one Springer eBook subject collection.

Please note that MyCopy books are only offered to library patrons with access to at least one Springer eBook subject collection. MyCopy books are strictly for individual use only.

You may cite this book by referencing the bibliographic data and/or the DOI (Digital Object Identifier) found in the front matter. This book is an exact but monochrome copy of the print version of the eBook on SpringerLink.

NATO ASI Series

Advanced Science Institute Series

A Series presenting the results of activities sponsored by the NATO Science Committee, which aims at the dissemination of advanced scientific and technological knowledge, with a view to strengthening links between scientific communities.

The Series is published by an international board of publishers in conjunction with the NATO Scientific Affairs Division

A Life Sciences	Plenum Publishing Corporation
B Physics	London and New York
C Mathematical and Physical Sciences	Kluwer Academic Publishers
D Behavioural and Social Sciences	Dordrecht, Boston and London
E Applied Sciences	
F Computer and Systems Sciences	Springer-Verlag
G Ecological Sciences	Berlin, Heidelberg, New York, London,
H Cell Biology	Paris and Tokyo
I Global Environment Change	

PARTNERSHIP SUB-SERIES

1. **Disarmament Technologies**	Kluwer Academic Publishers
2. **Environment**	Springer-Verlag / Kluwer Academic Publishers
3. **High Technology**	Kluwer Academic Publishers
4. **Science and Technology Policy**	Kluwer Academic Publishers
5. **Computer Networking**	Kluwer Academic Publishers

The Partnership Sub-Series incorporates activities undertaken in collaboration with NATO's Cooperation Partners, the countries of the CIS and Central and Eastern Europe, in Priority Areas of concern to those countries.

NATO-PCO-DATA BASE

The electronic index to the NATO ASI Series provides full bibliographical references (with keywords and/or abstracts) to about 50,000 contributions from international scientists published in all sections of the NATO ASI Series. Access to the NATO-PCO-DATA BASE is possible via a CD-ROM "NATO Science and Technology Disk" with user-friendly retrieval software in English, French, and German (©WTV GmbH and DATAWARE Technologies, Inc. 1989). The CD-ROM contains the AGARD Aerospace Database.

The CD-ROM can be ordered through any member of the Board of Publishers or through NATO-PCO, Overijse, Belgium.

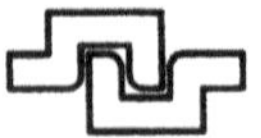

Series E: Applied Sciences – Vol. 355

Heat Transfer Enhancement of Heat Exchangers

edited by

S. Kakaç
University of Miami,
Coral Gables,
Florida, U.S.A.

A. E. Bergles
Rensselaer Polytechnic Institute,
Troy, NY, U.S.A.

F. Mayinger
Technical University of Munich,
Garching, Germany

and

H. Yüncü
Middle East Technical University,
Ankara, Turkey

Springer-Science+Business Media, B.V.

Published in cooperation with NATO Scientific Affairs Division

Proceedings of the NATO Advanced Study Institute on
Heat Transfer Enhancement of Heat Exchangers
Çeşme-Izmir, Turkey
May 25 – June 5, 1998

Library of Congress Cataloging-in-Publication Data

```
Heat transfer enhancement of heat exchangers / edited by S. Kakaç ...
  [et al.].
       p.   cm. -- (NATO science series. General sub-series E,
Applied sciences ; 355)

  1. Heat exchangers. 2. Heat--Transmission.  I. Kakaç, S. (Sadik)
II. Series.
TJ263.H425  1999
621.402'5--dc21                                          99-13875
```

DOI 10.1007/978-94-015-9159-1

Printed on acid-free paper

TABLE OF CONTENTS

Preface

This volume contains an archival record of the NATO Advanced study Institute on Heat Transfer Enhancement of Heat Exchangers held in Çeşme-İzmir Turkey, May 25 - June 5, 1998. The NATO ASIs are intended to be high level teaching activities in scientific and technical areas of current concern. Certainly, the subject of heat transfer enhancement needs no justification in this regard.

Energy and material-saving considerations, as well as economical and ecological incentives, have led to concerted efforts during the past 20-25 years to produce more efficient heat-exchange equipment. Because of the environmental considerations, design of new heat-exchange equipment for recycling is becoming more important. Therefore, if thermal energy can be conserved, the economical and ecological handling of thermal energy through heat exchangers will be possible. The usual goals are to reduce the size of a heat exchanger required for a specific heat duty and to upgrade the capacity of an existing heat exchange equipment. The present trend is moving toward compact heat exchangers, i.e., toward components with augmentation, enhancement or intensification. Therefore, the heat transfer enhancement in heat-exchange equipment becomes a challenge for the scientists and engineers in industry. World wide demand for efficient, reliable and economical heat-exchange equipment is accelerating rapidly, particularly in large-scale power and process industry, refrigeration and air-conditioning systems. Great improvements in energy conservation and the protection of the environment are possible if the fundamentals, methods of enhancement for heat transfer and design of efficient equipment with augmented surfaces were better understood.

During the ten working days of the Institute, the invited lecturers revive the current state-of-knowledge of heat transfer enhancement of heat exchangers. They discussed many techniques that have been developed to enhance heat transfer in single-phase forced convection, flow boiling, condensation, including modern advances in optical measuring techniques and tools to support energy conservation.

The sponsorship of the NATO Scientific Affairs Division is gratefully acknowledged; in person, we are very thankful to Dr. L.V. da Cunha, Director of ASI Programs who continuously supported and encouraged us at every phase of our activity. Our special gratitude goes to Drs. Egrican, M.Ünsal and T.Ayhan for their help in coordinating sessions during the Institute. We are very thankful to Liping Cao and A.Ümit Coşkun for their invaluable efforts in making the Institute a success. A word of appreciation is also due to the members of the sessions chairmen for their efforts in expediting the technical sessions. We are also grateful to the Kluwer Academic Publishers for their cooperation in preparing this archival record of the Institute, and F.Arınç, Secretary General of ICHMT for his guidance and help during the entire process of organization of this ASI.

Finally, our heartfelt thanks to all invited lecturers and authors, who provided the substance of the Institute, and to the participants for their attendance, questions, and comments.

Sadık Kakaç
Art E. Bergles
Franz Mayinger
Hafit Yüncü

INTRODUCTION TO HEAT TRANSFER ENHANCEMENT

-Preview of Contributions

SADIK KAKAÇ
Department of Mechanical Engineering
University of Miami
Coral Gables, FL 33124

Abstract. This lecture provides a general introduction to the subject of the Institute. Introductory remarks are made about heat transfer enhancements, surface augmentation or intensification in heat exchangers. The preview of lecture topics is introduced, indicating their contributions to the institute.

1. Introduction

Heat exchangers are devices that provide the flow of thermal energy between two or more fluids at different temperatures. These include power production, process, chemical and food industries, electronics, environmental engineering, waste heat recovery, manufacturing industry, and air-conditioning, refrigeration, and space applications.

In these applications, thermal-hydraulics and energy usage play dominant roles. Energy, space, and materials saving considerations, as well as the present-day global economics, have led to the expansion of efforts to produce more efficient heat exchange equipment for minimizing cost, which is to reduce the physical size of heat exchange equipment for a given heat duty. Therefore the main thermal-hydraulic objectives are to reduce the size of a heat exchanger required for a specific heat duty, to upgrade the capacity of an available heat exchanger and its operation with smaller approach temperature differences, or to reduce the pumping power.

There are various techniques used for heat transfer enhancement, which are segregated in two groups: (1) activate (2) passive techniques. The active techniques require external power to the surface(surface vibration, acoustic or electric fields). Passive techniques use specific surface geometries with surface augmentation.

1.1 BASICS OF HEAT TRANSFER ENHANCEMENT

In heat exchangers, depending on the applications, plain or enhanced/augmented heat transfer surfaces are used. Heat transfer between a wall and the fluid is given by

$$Q = hA(T_w - T_f) \tag{1}$$

or

S. Kakaç et al. (eds.), Heat Transfer Enhancement of Heat Exchangers, 1–11.
© *1999 Kluwer Academic Publishers.*

$$Q = (hA)_p (T_w - T_f) \tag{2}$$

The ratio of the (hA) of an augmented surface to that of a plain surface is defined as the enhancement ratio which is [1]

$$E = \frac{hA}{(hA)_p} \tag{3}$$

There are several methods to increase the hA value:
- a) Heat transfer coefficient can be increased without an appreciable increase in the surface area.
- b) Surface area can be increased without appreciable changes in heat transfer coefficient.
- c) Both the heat transfer coefficient and the surface area can be increased.

These methods will be discussed and examples will be given in the following chapters.

Heat transfer from a finned surface (augmented) is given by

$$Q = \eta hA(T_w - T_f) \tag{4}$$

or

$$Q = \eta hA\Delta T \tag{5}$$

where

$$\eta = [1 - \frac{A_f}{A}(1 - \eta_f)] \tag{6}$$

is called overall surface efficiency. ΔT is the temperature difference between the fluid stream and the wall temperature, and the total surface area on one side is $A = A_u + A_f$ (Figure 1)

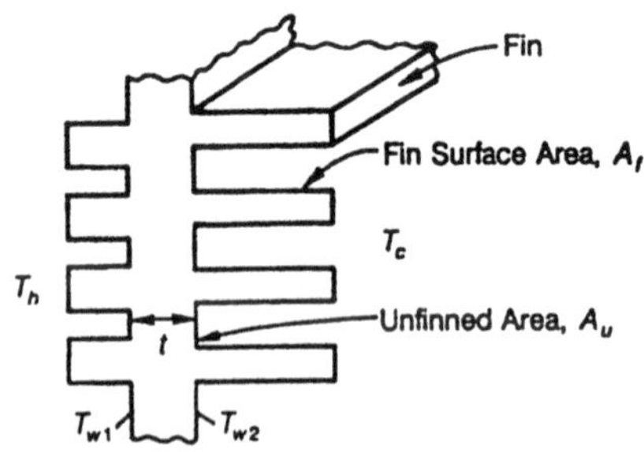

Figure 1. Finned Wall

Consider a two-fluid separated by a wall (heat exchanger). The total heat transfer rate, Q, between the fluids can be determined from the following equation:

$$Q = UA\Delta T_m \qquad (7)$$

The term *1/UA* is the overall thermal resistance which can be expressed as [2]:

$$R_t = \frac{1}{UA} = \frac{1}{U_o A_o} = \frac{1}{\eta_i h_i A_i} + \frac{R_{fi}}{\eta_i A_i} + R_w + \frac{R_{fo}}{\eta_o A_o} + \frac{1}{\eta_o h_o A_o} \qquad (8)$$

which includes fouling on both sides of the heat transfer surface. The fouling characteristics of an enhanced surface are important to be considered.

1.2 PERFORMANCE EVALUATION

The performance of the heat exchanger will be increased if the term (UA) is increased or the overall thermal resistance is reduced. The details of the principles of performance evaluation for single-phase and two-phase flows can be found in [1, 3].

The heat transfer enhancement increases with either or both of the (hA) terms, that is decreased convective resistances on both sides.

It can be seen from Eq. (7), the increased UA will contribute to

a) Reduction of the size of the heat exchanger: if the heat duty is kept constant, under the same temperature conditions, the length of the heat exchanger may be reduced. If the length and the inlet temperatures are kept constant, then the heat exchange rate will increase.

b) If Q and the total length (L) are kept constant, then ΔT_m may be reduced. This provides increased thermodynamic process efficiency (the second law efficiency) and yields lower system operational cost [4].

c) For fixed heat duty, Q, the pumping power can be reduced. However, this will dictate that the enhanced heat exchanger operates at a smaller velocity than the plain surface. This will require increased frontal area.

The selection one of the above improvements will depend on the objectives of designing the heat exchanger. The most important factor, in general, is the size of reduction, which results in cost reduction.

1.3 PUMPING POWER-SINGLE PHASE

Pumping power expenditure (pressure drop) is always to be considered by the designer.

$$P = \frac{\dot{m}\Delta p}{\eta_p \rho} \qquad (9)$$

In practice, the selected surface must satisfy the process specifications including pressure drop constraints. A surface that provides an enhancement level and satisfies all the specifications with the lowest possible pumping power expenditure is certainly preferred. The pressure drop limitation is a very important consideration of an enhanced surface versus a plain surface.

Let us consider a single-phase side passage (a duct) of a heat exchanger where the flow is turbulent and the surface is smooth. It can be shown that the pumping power expenditure per unit heat transfer area (W/m^2) can be expressed as [2]

$$\frac{P}{A} = C\frac{h^{3.5}\mu^{1.83}D_h^{1/2}}{k^{2.33}C_p^{1.17}\rho^2\eta_p} \tag{10}$$

where $C = 1.25 \times 10\ 4$

As can be seen from Eq. (10), the pumping power expenditure per heat transfer area depends strongly on fluid properties, as well as on the heat transfer coefficient, and hydraulic diameter of the flow passage.

Some important conclusions can be drawn from Eq. (10) (see Table 1):

1. With a high-density fluid such as a liquid, the heat exchanger surface can be operated at large values of h without excessive pumping power requirements.
2. A gas with its very low density results in high values of pumping power for even very moderate values of heat transfer coefficient.
3. A large value of viscosity causes friction power to be large even though density may be high. Thus, heat exchangers using oils must be designed for relatively low values of h to hold the pumping power within acceptable limits.
4. The thermal conductivity, k, also has a very strong influence; and therefore for liquid metals with very large values of thermal conductivity, the pumping power is seldom of significance.
5. Small values of hydraulic diameter, D_h, tend to minimize the pumping power.

TABLE 1 Pumping power expenditure for various fluid conditions

Fluid Conditions	Power Expenditure (W/m^2)
Water at 300 K	
h=3850 $W/m^2\cdot K$	3.85
Ammonia at 500 K, atm pressure	
h=100 $W/m^2\cdot K$	29.1
h=248 $W/m^2\cdot K$	697
Engine oil at 300 K	
h=250 $W/m^2\cdot K$	0.270×10^4
h=500 $W/m^2\cdot K$	3.06×10^4
h=1200 $W/m^2\cdot K$	65.5×10^4

Note: η_p=80%, D_p=0.0241 m.

Heat transfer coefficient of an enhanced surface which is given by the Colburn modulus, J, and friction factor of the surface, f, are usually presented as a function of Reynolds number ($D_h G/\mu$):

$$J = \frac{h}{Gc_p} Pr^{2/3} \tag{11}$$

The friction factor of an enhanced surface in a single-phase flow is higher than that of the smooth surface, when operated at the same Reynolds number. Therefore the enhanced surface would be allowed to operate at a higher pressure drop. The plain surface would give a higher heat transfer coefficient if it were also allowed to operate at a higher velocity, giving the same pressure drop as the enhanced surface. Therefore, as it can be seen from Eq. (10) and Table 1, the actual performance improvement cannot be obtained by calculating h/h_p and f/f_p at the same Reynolds number (at equal velocities).

Performance evaluation allows us to make a simple surface performance comparison and compare the thermal resistances of both fluid streams. The performance analyses allow the designer to quickly identify the most effective surface or other enhancement techniques of variable choices. These will be discussed in the various chapters of this volume.

If one of the fluids in the heat exchange undergoes a phase change, then this exchanger is called as a two-phase heat exchanger. Some of the examples are evaporators, boilers, reboilers and condensers.

In two-phase flow heat exchangers, pressure drop of phase changing fluid may affect the mean temperature difference in the heat exchanger, and the performance evaluation must take this effect into account. In a single-phase flow, the reduced pumping power can be considered as an objective function, which is not applicable to the two-phase flow situation. Detailed analysis of performance evaluation criteria for two-phase flow heat exchangers is presented in [1].

1.4 PUBLISHED LITERATURE ON HEAT TRANSFER ENHANCEMENT

Bergles et al. [5, 6, 7] have reported the journal publications on enhanced heat transfer. Bergles and Webb [8] summarize the information presented in the technical and patent literature. Figure 2, taken from Bergles et al. [7], shows the journal and conference publications per year between 1900 and 1990, which total 4345, as of December 1990. As August 1995, then total number of papers and reports increased to 5676 [10]. The period of rapid growth began about 1960. The patent literature is an equally important source of information on enhancement techniques. Figure 3 taken from Webb et al. [9] survey of the U.S. patent literature, shows 486 patent issued as of June 1982.

2. The Program of the Institute

During the 10 working days of this Institute, invited lecturers and papers as supplementary to the lectures will be presented. During the institute various heat enhancement/augmentation techniques and their performances will be introduced. This section introduces some of the topics that will be considered in detail in the following lectures.

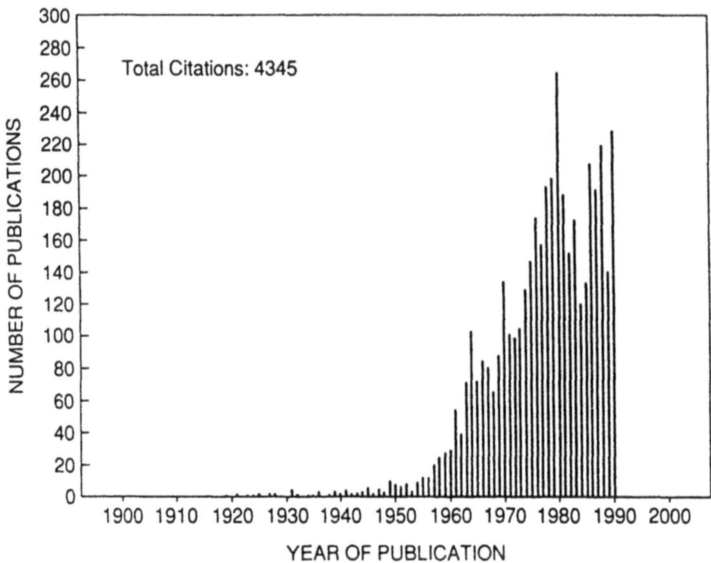

Figure 2. Citations on heat transfer augmentation versus year of publications.
Status as of December 1990 is illustrated (From Bergles et al. [6]).

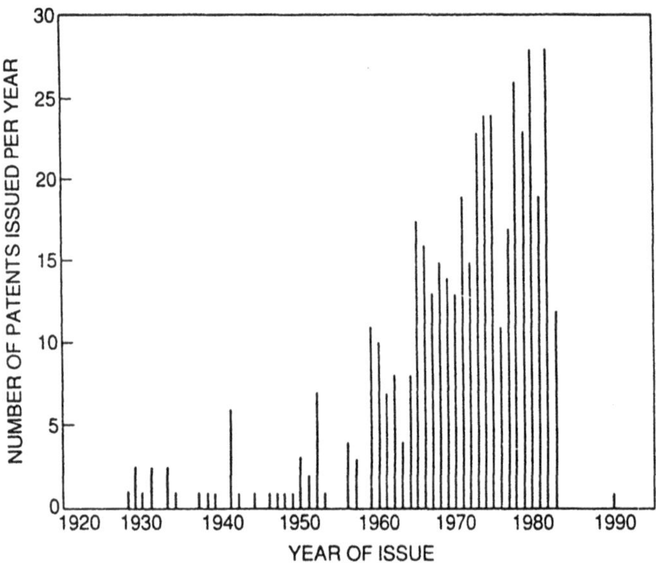

Figure 3. Patents on heat transfer augmentation versus year
of publication (From Webb et al. [9]).

Bergles presents **the imperative to enhance heat transfer** and considers many techniques that have been developed to enhance convective heat transfer, and examples are given for the application to various modes of heat transfer. This is an introduction to our entire program. Bergles then presents micro-fin tube technology where he outlines a comprehensive review of the enhanced "micro-fin" tubes that have been effectively applied to single-phase flows, flow boiling/evaporation and convective condensation. Microfin tubes are characterized by a large number of longitudinal or helical fins inside small diameter tubes with fin heights in the range of only 0.1-0.4 mm. They are widely used in air-conditioning and refrigeration. **The most recent experimental research on evaporation in microfin tubes** is reviewed and the latest prediction methods are critically surveyed by Thome.

Webb presents **a survey on the prediction of condensation and evaporation in microfin and micro-channel tubes**. There are several correlations to predict heat transfer of flow boiling (Shah, Kandlikar and Güngör and Winterton) and condensation (Traviss et al., Shah, Cavallini and Zecchin) in plain tubes. In this work, an improved predictive model based on the two-phase heat momentum transfer analogy is described which is called "equivalent Reynolds number model". For condensation in micro-fin tubes, both vapor shear and surface tension forces contribute to the condensing coefficient. It is shown that the vapor shear model is also applicable to evaporation inside tubes, and the nucleate boiling contribution is less than 15% and exists only at vapor qualities less than 50%.

In his second lecture, Webb decribes a series of studies to understand the mechanism of **boiling in structured surfaces** having sub-surface tunnels and surface pores.

Performance Enhancement of finned oval tubes with finned longitudinal vortex generators is introduced by Fiebig and Chen. As it is mentioned above, fins are used to reduce the thermal resistance on the gas side of a gas-liquid heat exchanger as in the case of a tube-fin heat exchanger. The heat transfer coefficients on the gas side is very small (30-300 $W/m^2 \cdot K$) compared with the liquid side heat transfer coefficient. To balance the resistance on both sides, a very large heat transfer area ratio of fin to tube would be needed; then the overall surface efficiency of the finned surface will decrease, Eq. (6), and the economy of fins decreases. Therefore heat transfer enhancement is necessary on the fins side. Fiebig and Chen introduce the delta Winglet pair (DWP) in a finned oval tube to enhance heat transfer on the gas side, and present flow structures and heat transfer analysis. Vortex geometries can be used as fins in plate heat exchangers, and as fin surface modifications in any finned heat exchangers.

There are a number of lectures on heat exchangers used in air-conditioning and refrigeration systems.

Successful design of evaporators and condensers in air-conditioning systems using the new refrigerants as working fluids would require heat transfer and pressure drop characteristics and heat transfer enhancement inside the tubes.

Ebisu and Torikoshi present **evaporation and condensation heat transfer characteristics of R-407C which is an alternative refrigeratnt for R-22**. They discuss a new heat transfer tube having a special inner surface which is named a "W-design heat transfer tube", and they provide information on the heat exchanger performance improvement by the use of the W-design heat transfer tubes in air-conditioning condensers and evaporators. Comparisons are made for the data of the W-design tube with those for the existing inner grooved tube.

8

The improvement of the heat transfer performance of the widely used air-cooled heat exchanger has always been great concern in the compact design of these heat exchangers. By the year 2030, the production of HCFC-22 will be phased-out. Ebisu and Torikoshi introduce **new concept air-cooled heat exchangers for energy savings in air-conditioning**. They developed a mesh finned tube heat exchangers to enhance the air-side heat transfer. They examine the heat transfer and flow friction characteristics of the mesh finned tubes, then they combine their available data for a mixture of HFC-32/HFC-134A to evaluate the mesh finned tube heat exchanger performance.

In the same category, Wang presents **energy savings in air-conditioning through heat transfer augmentation together with optimum design of air-cooled fin-tube heat exchangers-accounting for the effect of complex circuiting**. His lecture provides a methodology to model the air-cooled condenser and water heating coils. The calculated results are compared with experimental findings.

In his second lecture, Wang focuses on **the air-side performance of fin-and-tube heat exchangers**. Contents of this lecture include the reduction method of the air-side performance and the updated correlations for a typical fin-and-tube heat exchangers. The presented correlations for heat transfer and friction factors cover plain, wavy, louver, and covex-louver fin patterns. Friction factors include the entrance and exit loses as well. In addition, appropriate ε-NTU relationship are presented for the cross-flow configuration.

As mentioned by Ebisu and Torikoshi in their lecture, according to the Montreal Protocol, R-11, R-12, R-502 etc. will be replaced by more environmentally safe refrigerants such as R-134A, R-123, R-407, R-410A, ammonia, hydrocarbons, etc.

The lubricant used in the compressor circulates with the refrigerant through all components that make up a refrigeration system, including the evaporator and condenser. The circulating concentration of lubricant mixed with refrigerant probably varies from 0.2% to 10%, depending on the type of compressor (e.g. reciprocating, rotary, etc.) and on whether an oil separator exists. Thome presents a survey lecture under the title of **flow boiling of refrigerant-oil mixtures in plain and enhanced tubes (microfin tubes)**. He presents the results of experimental studies and new prediction models.

Performance enhancement of heat exchangers for semiconductor-chips manufacturing is presented by Yang. These heat exchangers can be classified, according to their function, into five major categories, which are all single-tube and single-fluid type in nature with heat transfer by conduction, convection, or radiation. These heat exchangers are for high-tech applications. Yang's lecture emphasizes how heat transfer performance can be enhanced in these heat exchangers for manufacturing semiconductor chips.

Kabov presents **cooling of local heat sources by subcooled liquid films** which provides state-of-the-art review of cooling of electronic and micro electronic equipment by moving liquid films and summarizes the experimental and modeling results at the Institute of Thermophysics of Russian Academy of Sciences. Experiments were carried out with a mixture of ethyl alcohol in water and with dielectric liquids up to 41 W/m^2 local heat sources. This lecture also discusses the enhancement of liquid film evaporation with a new method of heat transfer enhancement.

The second lecture by Kabov is **the enhancement of vapor condensation on horizontal tube with fins and spine-fins** in which he analyses the available experimental data and physical models on vapor condensation heat transfer on integral-fin tubes. He also discusses vapor condensation on high performance three-dimensional surfaces and spine-fin tubes. Then vapor condensations on tube banks are discussed. The available experimental data and physical models of condensation processes are analyzed and the map of condensate inundation effect is suggested.

The gasketed-plate heat exchangers (plate-and-frame) were introduced in the 1930s, mainly for the food industries because of the ease of cleaning; their design reached maturity in the 1960s with the development of more effective plate geometries, assemblies, and improved gasket materials. They are capable of meeting an extremely wide range of duties in many industries. They can be used as an alternative to shell-and-tube type of heat exchangers for low-and-medium pressure, liquid-to-liquid heat transfer applications. Design engineers require reliable and accurate predictive tools for their heat transfer and pressure drop characteristics, that would enable sizing, rating of such heat exchangers with some precision.

There are two lectures on plate-and-frame heat exchangers (PHE) that address to these issues. Manglik provides an exhaustive survey of the literature on PHEs, their thermal-hydraulic design consideration and applications for both single-phase and two-phase flows under the title of **enhanced heat transfer characteristics of plate heat exchangers for power and process industry applications**.

Sunden's lecture on **flow and heat transfer mechanisms in plate-and-frame heat exchangers** focuses attention on recent investigations that are aimed towards an improved understanding of the physical processes in plate-and-frame heat exchanger, and thus on development of reliable prediction and design methods.

Kumada and Himeji present **studies on a high-performance ceramic heat exchanger for ultra high temperatures**. A conventional metal heat exchanger has limits in its usage for the temperature limitations and for corrosive gas. A heat exchanger for high temperature is expected to recover heat from high energy gas. Ceramic heat exchangers may be the answer for high temperature and corrosive conditions. In this study, they address the development of the ceramic heat exchanger applying fluidized bed for generating high temperature gas; experiments are carried out with ultra high bed and recovered gas temperatures (>1000 °C). Smooth ceramic tubes, and ceramic tubes with transverse thin fins (first of the kind) were tested and their performance is evaluated. It is found that the heat transfer coefficient on the outside wall of the finned tube is almost 3 times as large as that of a smooth tube.

Fujita studies **boiling and evaporation of falling film on horizontal tubes and its enhancement on grooved tube**. A review of recent developments on evaporation and boiling heat transfer to falling films on horizontal tubes and an analysis of enhanced heat transfer in grooved or microfinned tubes is presented. Thin liquid films are utilized as an important component in various heat transfer processes because of their high heat transfer rate at low feed rates and small temperature differences.

10

Afgan and Carvalho review the growing concern of the scientific and political community of the world in the use of the available resources of our planet and in turn a sustainable energy development under the title of **sustainability for heat exchanger design**. Practical attention is given to the discussion of different aspects of sustainability in the present world, and a potential development in the energy development. In this lecture seven major areas with specific problems and their relevance to this sustainable energy development are listed. A special emphasis is devoted to the analysis of the current method for the condenser design and its comparison with the respective sustainability criterion.

Thermal energy is mostly provided via combustion processes. Therefore efficient burning processes are important factors in improving energy savings as heat recovery. Low combustion efficiency with non-negligible content of partially unburned components in the exhaust gas has a very strong negative effect on energy savings. Mayinger discusses **modern advances on optical measuring techniques-tools to support energy conservation** by which one can follow up the combustion process to provide, in detail, precise and instaneous information about the flame structure and provide very valuable hints, how one can influence the process to achieve high combustion efficiency.

The following and other presentation will supplement the lectures:

- **Enhancement of heat transfer with horizontal promoters**, by Onbasioglu and Egrican
- **Energy conservation in a hydraulic fueled diesel engine optimization of the mixture formation and combustion**, by Prechtl et al.
- **Advances in understanding of flame acceleration for improving combustion efficiency**, by Gerlach et al.
- **Enhancement of combined heat and mass transfer in rotary exchangers**, by Dinglreiter et al.
- **Heat and mass transfer with drying of water-based varnishes**, by Mintzlaff and Mayinger
- **Heat transfer enhancement in a plate heat exchangers with rib-roughened surfaces**, by Tauscher and Mayinger
- **Influence of the confinement on the FC-72 pool boiling from finned surface**, by Misale et al.
- **Enhancement of turbulent flow heat transfer in tubes by means of truncated hollow cane inserts**, by Ayhan et al.
- **Heat exchangers for thermoacoustic refrigerators: Heat transfer issues in oscillatory flows**, by Herman and Wetzel

Finally, suggestions for **further research on heat transfer enhancement** are presented in a session chaired by Bayazitoglu.

Nomenclature

A heat transfer surface area, m^2
A_f area of extended surface(fins), m^2
C_p specific heat, J/kg·K
D_h hydraulic diameter, m
E enhancement ratio

f fanning friction factor, $\Delta p = 4 f (L / D_h)(\rho u^2 / 2)$

h heat transfer coefficient, W/m$^2\cdot$K

k thermal conductivity, W/m$\cdot$K

m mass flow rate, kg/s

ΔT_m mean temperature difference, K

p pressure, Pa

P pumping power, W

Q heat transfer rate, W

R_f fouling resistance, m$^2\cdot$K/W

R_w wall resistance, K/W

T temperature, K

U overall heat transfer coefficient, W/m$^2\cdot$K

η_f fin efficiency

η overall surface efficiency

μ dynamic viscosity, kg/m$^2\cdot$s

ρ density, kg/m^3

References

1. Webb, R. L. (1994) *Principals of Enhanced Heat Transfer*, John Wiley, New York, NY.
2. Kakaç, S. and Liu, H. (1997) *Heat Exchangers Selection, Rating and Thermal Design*, CRC Press, Boca Raton, FL.
3. Webb R. L.(1987) Enhancement of single-phase heat transfer, Chapter 17 in S. Kakaç, R.K. Shah and W. Aung (eds.), *Handbook of Single-Phase Convective Heat Transfer*, John Wiley, New York, NY.
4. Bejan, A. (1995) *Energy Generation Minimization*, CRC Press, Boca Raton, Florida.
5. Bergles, A.E. (1980) Principles of heat transfer augmentation: single-phase and two-phase heat transfer, in S. Kakaç, A.E. Bergles, and F. Mayinger (eds.), *Heat Exchangers-Thermal-Hydraulic Fundamentals and Design*, Hemisphere, New York, NY.
6. Bergles, A.E., Jensen M.K., Somerscales, E.F.C. and Manglik, R.M. (1991) *Literature Review of Heat Enhancement Technology for Heat Exchangers in Gas Fired Applications*, GRI Report GRI 91-0146, Gas Research Institute, Chicago, IL.
7. Bergles, A.E., Nirmalan, V., Junkhan, G.H. and Webb, R.L. (1983) *Bibliography on Augmentation of Convective Heat and Mass Transfer II.* Heat Transfer Laboratory Report HTL-31, ISU-ERIAmes 84221, Iowa State University, December.
8. Bergles, A.E., Webb, R.L. (1985) A guide to the literature on convective heat transfer augmentation, in Shenkman, S.M., O'Brien, J.E., Habib I.S. and Kohler, J.A. (eds.), *Advances in Enhanced Heat Transfer -1985*, ASME Symposium, Vol. HTD-Vol.43, pp.81-90.
9. Webb, R.L., Bergles, A.E., and Junkhan, G.H. (1983) *Bibliography of US Patents on Augmentation of Convective Heat and Mass Transfer-II*, HTL-32, ISU-ERI-AMES 84257.
10. Bergles, A.E., Jensen, M.K. and Shome, B. (1995) *Bibliography on Enhancement of Convective Heat Transfer*, Heat Transfer Laboratory Report HTL-23, Rensselaer Polytechnic Institute, Troy, N.Y.

THE IMPERATIVE TO ENHANCE HEAT TRANSFER

Arthur E. Bergles

Department of Mechanical Engineering, Aeronautical Engineering and Mechanics
Rensselaer Polytechnic Institute
Troy, NY 12180-3590 USA

Abstract. This review considers the many techniques that have been developed to enhance convective heat transfer. There is the desire to encourage or accommodate high heat fluxes. After introducing the techniques, the applications to the various modes of heat transfer (single-phase free convection, single-phase forced convection [including performance evaluation criteria, active techniques, and compound techniques], pool boiling, convective boiling/evaporation, vapor-space condensation, and convective condensation) are described. Comments are offered regarding the generations of heat transfer technology; advanced enhancement represents 3rd generation heat transfer technology,.

1. Introduction

1.1 GENERAL CONSIDERATIONS

Energy and materials saving considerations, as well as economic incentives, have led to efforts to produce more efficient heat exchange equipment. Common thermal-hydraulic goals are to reduce the size of a heat exchanger required for a specified heat duty, to upgrade the capacity of an existing heat exchanger, to reduce the approach temperature difference for the process streams, or to reduce the pumping power. The first two objectives translate to an increase in the average heat flux of the heat exchanger, or the encouragement of high heat fluxes. In the case of systems with a specified heat dissipation, the goal is to cool the device, or accommodate a high heat flux, at moderate temperature difference. Implicit in these objectives, energy reduction (improvement of first law efficiency) or temperature difference reduction (improvement of second law efficiency) are important to efficient use of energy.

The study of improved heat transfer performance is referred to as heat transfer **enhancement, augmentation, or intensification**. In general, this means an increase in heat transfer coefficient. Attempts to increase "normal" heat transfer coefficients have been recorded for more than a century, and there is a large store of information. A recent survey by Bergles et al. [1] cites 5676 technical publications, excluding patents and manufacturers' literature. The rather recent growth of activity in this area is clearly evident from the yearly distribution of publications presented in Fig. 1.

13

S. Kakaç et al. (eds.), Heat Transfer Enhancement of Heat Exchangers, 13–29.
© 1999 *Kluwer Academic Publishers.*

14

Enhancement techniques can be classified either as *passive methods*, which require no direct application of external power (Fig. 2), or as *active methods*, which require external power. The effectiveness of both types of techniques is strongly dependent on the mode of heat transfer, which may range from single-phase free convection to dispersed-flow film boiling. Brief descriptions of these methods follow.

1.2 PASSIVE TECHNIQUES

Treated surfaces involve fine-scale alternation of the surface finish or coating (continuous or discontinuous). They are used for boiling and condensing; the roughness height is below that affecting single-phase heat transfer.

Rough surfaces are produced in many configurations ranging from random sand-grain type roughness to discrete protuberances. See Figs. 2(a).and 2(e). The configuration is generally chosen to disturb the viscous sublayer rather than to increase the heat transfer surface area. Application of rough surfaces is directed primarily toward single-phase flow.

Extended surfaces are routinely employed in many heat exchangers. See Figs. 2(b) and 2(d). Work of special interest to enhancement is directed toward improvement of heat transfer coefficients on extended surfaces by shaping or perforating the surfaces.

Displaced enhancement devices are inserted into the flow channel so as indirectly to improve energy transport at the heated surface. They are used with forced flow. See Figs. 2(c) and 2(f).

Swirl-flow devices include a number of geometric arrangements or tube inserts for forced flow that create rotating and/or secondary flow: coiled tubes, inlet vortex generators, twisted-tape inserts, and axial-core inserts with a screw-type winding.

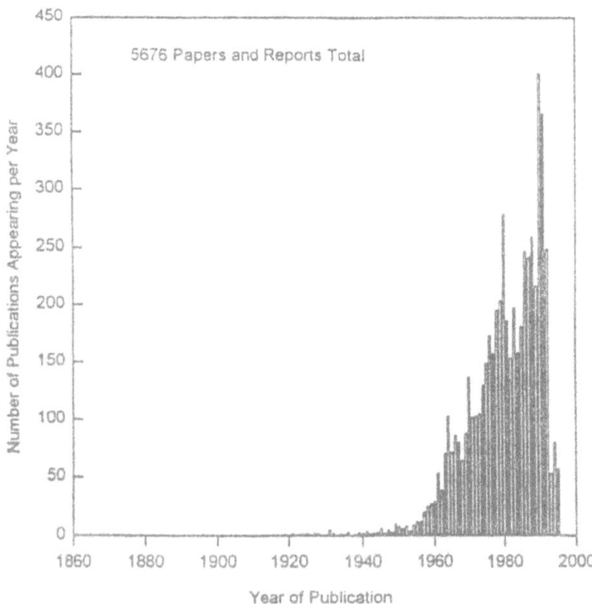

Figure 1. References on heat transfer enhancement versus year of publication , to mid 1995 (Bergles et al. [1]).

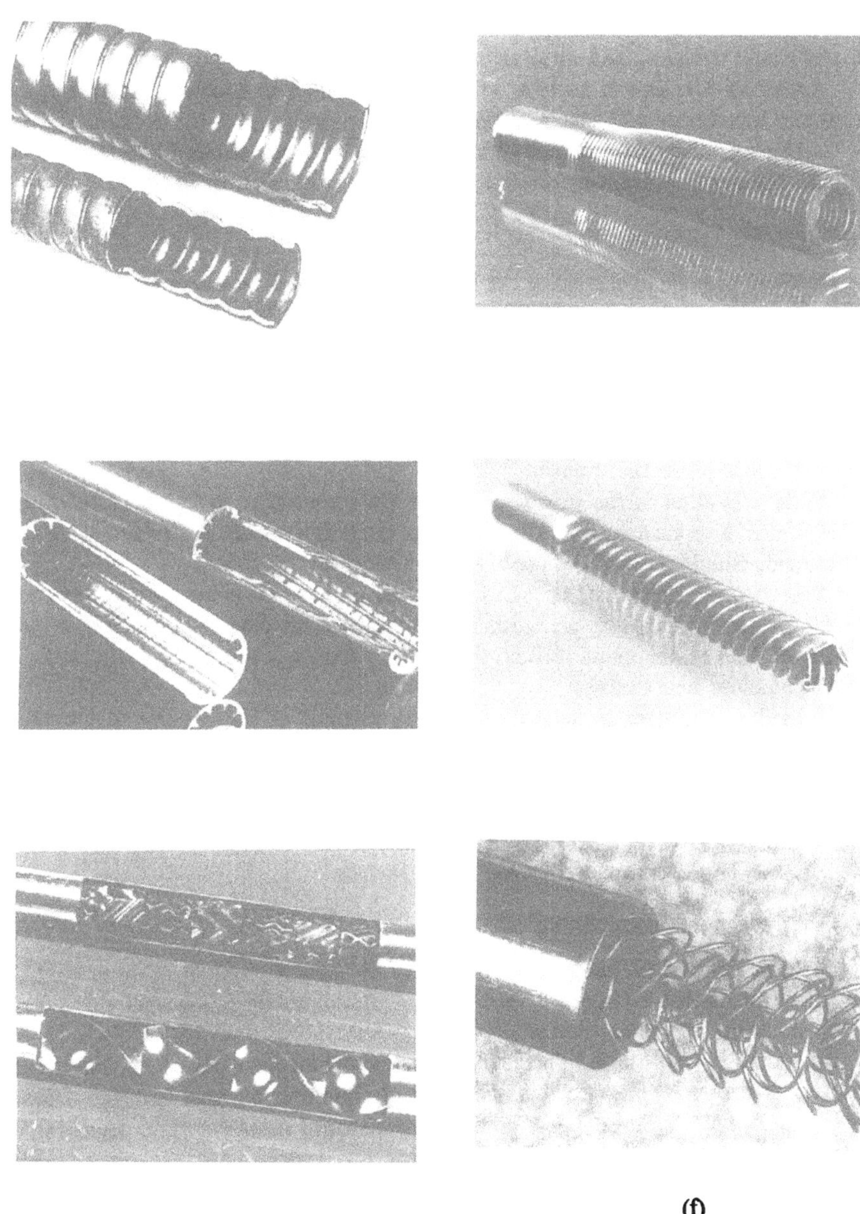

Figure 2. Tubes for enhancement of single-phase heat transfer: (a) corrugated or spirally indented tube with internal and external protuberances, (b) integral external fins, (c) integral internal fins, (d) deep spirally fined tube with internal and external protuberances, (e) static-mixer inserts, and (f) wire-wound insert.

Surface-tension devices consists of wicking or grooved surfaces to direct the flow of liquid in boiling and condensing.

Additives for liquids include solid particles and gas bubbles in single-phase flows and liquid trace additives for boiling systems.

Additives for gases are liquid droplets or solid particles, either dilute-phase (gas-solid suspensions) or dense-phase (fluidized beds).

1.3 ACTIVE TECHNIQUES

Mechanical aids involve stirring the fluid by mechanical means or by rotating the surface. Surface "scraping," widely used for batch processing of viscous liquids in the chemical process industry, is applied to the flow of such diverse fluids as high-viscosity plastics and air. Equipment with rotating heat exchanger ducts is found in commercial practice.

Surface vibration at either low or high frequency has been used primarily to improve single-phase heat transfer.

Fluid vibration is the practical type of vibration enhancement because of the mass of most heat exchangers. The vibrations range from pulsations of about 1 Hz to ultrasound. Single-phase fluids are of primary concern.

Electrostatic fields (DC or AC) are applied in many different ways to dielectric fluids. Generally speaking, electrostatic fields can be directed to cause greater bulk mixing of fluid or disruption of fluid flow in the vicinity of the heat transfer surface, which enhances heat transfer.

Injection is utilized by supplying gas to a stagnant or flowing liquid through a porous heat transfer surface or by injecting similar fluid upstream of the heat transfer section.   Surface degassing of liquids can produce enhancement similar to gas injection. Only single-phase flow is of interest.

Suction involves vapor removal, in nucleate or film boiling, or fluid withdrawal, in single-phase flow, through a porous heated surface.

1.4 FURTHER CONSIDERATIONS

Two or more of the above techniques may be utilized simultaneously to produce an enhancement that is larger than either of the techniques operating separately. This is termed *compound enhancement*.

It should be emphasized that one of the motivations for studying enhanced heat transfer is to assess the effect of an inherent condition on heat transfer. Some practical examples are: roughness produced by standard manufacturing, degassing of liquids with high gas content, rotating near exchanger ducts, surface vibration resulting from rotating machinery or flow oscillations, fluid vibration resulting from pumping pulsation, and electrical fields present in electrical equipment.

The surfaces in Fig. 2 have been used for both single-phase and two-phase heat transfer enhancement. The emphasis is on effective and cost-competitive (proved or potential) techniques that have made the transition from the laboratory to commercial heat exchangers. Broad reviews of developments in enhanced heat transfer are available [2-5].

2. Single-phase free convection

With the exception of the familiar technique of providing extended surfaces, the passive techniques have little to offer in the way of enhanced heat transfer for free convection. This is because the velocities are usually too low to cause flow separation or secondary flow. However, the restarting of thermal boundary layers in interrupted extended surfaces increases heat transfer so as to more than compensate for the lost area.

Mechanically aided heat transfer is a standard technique in the chemical and food industries when viscous liquids are involved. The predominant geometry for surface vibration has been the horizontal cylinder, vibrated either horizontally or vertically. Heat transfer coefficients can be increased 10-fold for both low-frequency/high-amplitude and high-frequency/low-amplitude situations. It is, of course, equally effective and more practical to provide steady forced flow. Furthermore, the mechanical designer is concerned that such intense vibrations could result in equipment failure.

Since it is usually difficult to apply surface vibrations to practical equipment, an alternative technique is utilized whereby vibrations are applied to the fluid and focused toward the heated surface. With proper transducer design, it is also possible to improve by several hundred percent.heat transfer to simple heaters immersed in gases or liquids

Electric fields are particularly effective in increasing heat transfer coefficients in free convection. Dielectrophoretic or electrophoretic (especially with ionization of gases) forces cause greater bulk mixing in the vicinity of the heat transfer surface. Heat transfer coefficients may be improved by as much as a factor of 40 with electrostatic fields up to 100,000 V. Again, the equivalent effect could be produced at lower capital cost and without the voltage hazard by simply providing forced convection with a blower or fan.

3. Single-phase forced convection

3.1 GENERAL CONSIDERATIONS

The present discussion emphasizes enhancement of heat transfer *inside* ducts that are primarily of circular cross section. Typical data for turbulence promoters inserted inside tubes are shown in Fig. 3. As shown in Fig. 3(a), the promoters produce a sizeable elevation in the Nusselt number, or heat transfer coefficient, at constant Reynolds number, or velocity. However, as shown in Fig. 3(b), this is accompanied by a large increase in the friction factor.

Surface roughness has been used extensively to enhance forced convection heat transfer. Integral roughness may be produced by the traditional manufacturing processes of machining, forming, casting, or welding. Various inserts can also provide surface protuberances. In view of the infinite number of possible geometric variations, it is not surprising that, even after more than 700 studies, no completely satisfactory unified treatment is available.

18

In general, the maximum enhancement of laminar flow with many of the techniques is the same order of magnitude, and seems to be independent of the wall boundary condition. The enhancement with some rough tubes, corrugated tubes, inner-fin tubes, various static mixers, and twisted-tape inserts is about 200 percent. The improvements in heat transfer coefficient with turbulent flow in rough tubes (based on nominal surface area) are as much as 250%. Analogy solutions for sand-grain-type roughness and for square-repeated-rib roughness have been proposed. A statistical correlation is also available for heat transfer coefficient and friction factor.

The following correlations are recommended for tubes with transverse or helical repeated ribs (Fig. 2a), with turbulent flow, (Ravigururajan and Bergles, [7]):

$$Nu_a \Big/ Nu_s = \left\{ 1 + \left[2.64\, Re^{0.036} (e/D)^{0.212} ((p/D)^{-0.21} (\alpha/90)^{0.29} (Pr)^{-0.024} \right]^7 \right\}^{1/7} \quad (1)$$

$$f_a / f_s = \left\{ 1 + \left[29.1\, Re^{\left(0.67 - 0.06\, p/D - 0.49\, \alpha/90 \right)} \right. \right.$$

$$\times (e/D)^{(0.37 - 0.157\, p/D)}$$

$$\times (p/D^{(-1.66 \times 10^{-6}\ Re - 0.33\, \alpha/90)} \qquad (2)$$

$$\times (\alpha/90^{(4.59 + 4.11 \times 10^{-6}\ Re - 0.15\, p/D)}$$

$$\times \left(1 + \frac{2.94}{n}\right) \sin\beta \big]^{15/16} \Big\}^{16/15}$$

Much work has been done to obtain the enhanced heat transfer of parallel angled ribs in short rectangular channels, simulating the interior of gas turbine blades. Jets are frequently used for heating, cooling, and drying, in a variety of industrial applications. A number of studies have reported that roughness elements of the

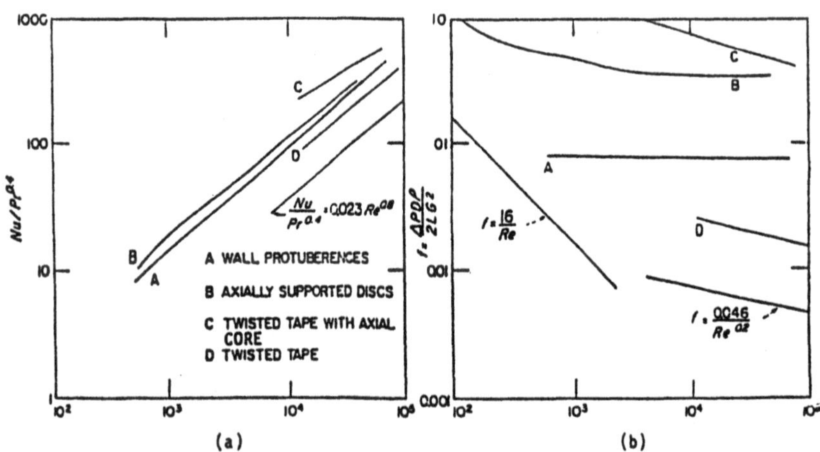

Figure 3. Typical data for turbulence promoters inserted inside tubes: (a) heat transfer data, (b) friction data [6].

transverse-repeated-rib type mitigate the deterioration in heat transfer downstream of stagnation.

Extended surfaces can be considered "old technology", as far as most applications are concerned. The real interest now is in increasing heat transfer coefficients on the extended surface. Compact heat exchangers of the plate-fin or tube-and-center variety use several enhancement techniques: offset strip fins, louvered fins, perforated fins, or corrugated fins. Coefficients are several hundred percent above the smooth-tube values; however, the pressure drop is also substantially increased, and there may be vibration and noise problems.

For the case of offset strip fins (Fig. 4a) the following correlations are recommended for calculating the j and f characteristics (Manglik and Bergles, [8])

$$j_h = 0.6522\, Re_h^{-0.5403}\, \alpha^{-0.1541}\, \delta^{0.1499}\, \gamma^{-0.0678}$$

$$\times \left[1 + 5.269 \times 10^{-5}\, Re_h^{1.340}\, \alpha^{0.504}\, \delta^{0.456}\, \gamma^{-1.055} \right]^{0.1} \tag{3}$$

$$f_h = 9.6243\, Re_h^{-0.7422}\, \alpha^{-0.1856}\, \delta^{0.3053}\, \gamma^{-0.2659}$$

$$\times \left[1 + 7.669 \times 10^{-8}\, Re_h\, 4.429\, \alpha^{0.920}\, \delta^{3.767}\, \gamma^{0.236} \right]^{0.1} \tag{4}$$

where j_h, and f_h, and Re_h are based on the hydraulic diameter given by

$$D_h = 4shl\, /\, [\,2(sl + hl + th) + ts\,] \tag{5}$$

These equations are based on experimental data for 18 different offset strip fin geometries, and they represent the data continuously in the laminar, transition, and turbulent flow regions (Fig. 4(b)).

Internally finned circular tubes are available in aluminum and copper (or copper alloys), Fig. 2(b). Correlations (for heat transfer coefficient and friction factor) are available for laminar flow, for both straight and spiral continuous fins.

Turbulent flow in tubes with straight or helical fins was correlated by Carnavos [9]:

$$Nu_h = 0.023\, Pr^{0.4}\, Re_h^{0.8} \left[\frac{A_F}{A_{Fi}} \right]^{0.1} \left[\frac{A_i}{A} \right]^{0.5} (\sec a)^3 \tag{6}$$

$$f_h = 0.046\, Re_h^{-0.2} \left[\frac{A_F}{A_{Fi}} \right]^{0.5} (\sec \alpha)^{0.75} \tag{7}$$

A numerical analysis of turbulent flow in tubes with idealized straight fins was reported. The necessary constant for the turbulence model was obtained from

20

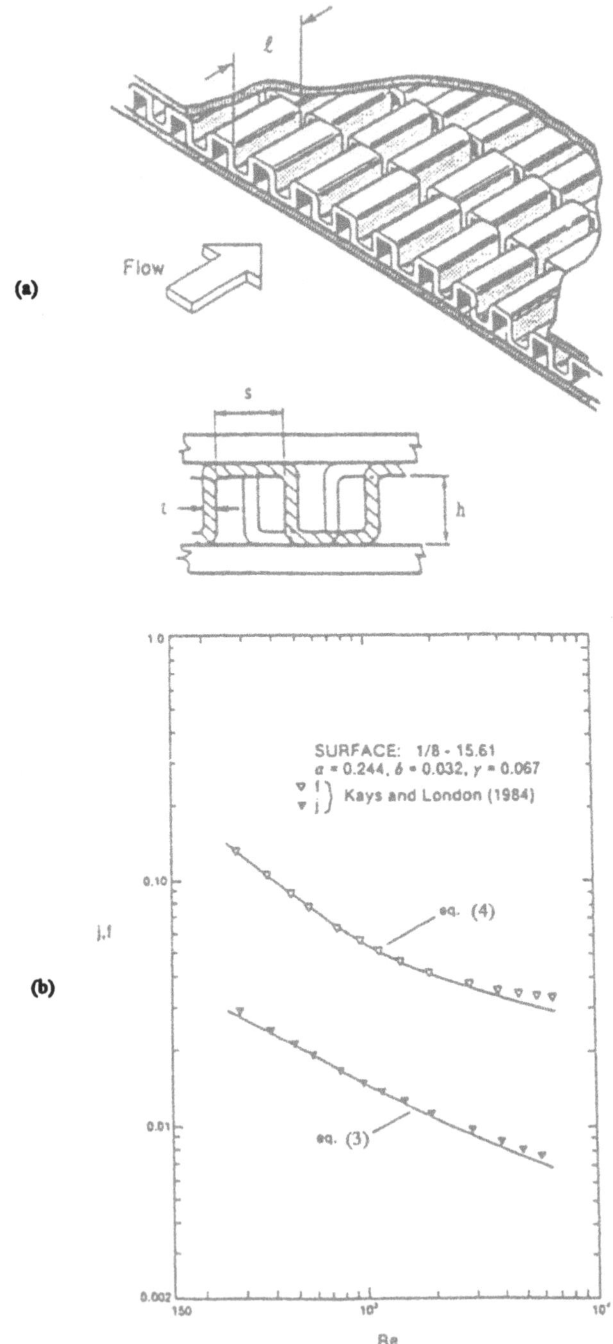

Figure 4. (a) Geometry of the offset strip fin; (b) comparison of the predictions with data from Kays and London (1984), (Manglik and Bergles, [8]).

experimental data for air. Further improvements in numerical techniques are expected, so that a wider range of geometries and fluids can be handled without resort to extensive experimental programs.

Many proprietary surface configurations have been produced by deforming the basic tube. The "convoluted, "corrugated," "spiral," or "spirally fluted" tubes (Figs. 2(a) and 2(e)) have multiple-start spiral corrugations, which add area, along the tube length. A systematic survey of the single-tube performance of condenser tubes indicates up to 400% increase in the nominal inside heat transfer coefficient (based on diameter of a smooth tube of the same maximum inside diameter); however, pressure drops on the water side are about 20 times higher.

Displaced enhancement devices are typically in the form of inserts, with the elements arranged to promote transverse mixing (static mixers, (Fig. 2(c)). They are used primarily for viscous liquids, to promote either heat transfer or mass transfer. Displaced promoters are also used to enhance the radiant heat transfer in high-temperature applications. In the flue-tube of a hot-gas-fired hot water heater, there is a trade-off between radiation and convection. Another type of displaced insert generates vortices, which enhance the downstream flow. Delta-wing and rectangular-wing promoters, both co-rotating and counter-rotating, have been studied. Wire-loop inserts (Fig. 2(f)) have also been used for enhancement of laminar and turbulent flow.

Twisted-tape inserts have been widely used to improve heat transfer in both laminar and turbulent flow. Correlations are available for laminar flow, for both uniform heat flux and uniform wall temperature conditions. Turbulent flow in tubes with twisted-tape inserts has also been correlated. Several studies have considered the heat transfer enhancement of a decaying swirl flow, generated, say, by a short twisted-tape insert.

3.2 PERFORMANCE EVALUATION CRITERIA FOR SINGLE-PHASE FORCED CONVECTION IN TUBES

Numerous, and sometimes conflicting, factors enter into the ultimate decision to use an enhancement technique: heat duty increase or area reduction that can be obtained , initial cost, pumping power or operating cost, maintenance cost (especially cleaning), safety, and reliability, among others. These factors are difficult to quantitize, and a generally acceptable selection criterion may not exist. It is possible, however, to suggest some performance criteria for preliminary design guidance. As an example, consider the basic geometry and the pumping power fixed, with the objective of increasing the heat transfer. The following ratio is then of interest

$$R_3 = \left(\frac{h_a}{h_o}\right)_{D_i,L,N,P,T_{in},\Delta T} = \frac{\left(Nu\,/\,Pr^{0.4}\right)_a}{\left(Nu\,/\,Pr^{0.4}\right)_s} = \frac{q_a}{q_s} \tag{8}$$

With the pumping power (neglecting entrance and exit losses) given as

$$P = NV\,A_F\,4\,f\,(L\,/\,D)\,\rho V^2\,/\,2 \tag{9}$$

and

$$f_s = 0.046 / Re_s^{0.2}, \tag{10}$$

$$A_{F,a} f_a Re_a^3 = 0.046\, A_{F,s} Re_s^{2.8} \tag{11}$$

The calculation best proceeds by picking Re_a, and reading $Nu_a / Pr^{0.4}$ and f_a. Re_s is then obtained from Eq. (11), and $Nu_a / Pr^{0.4}$ obtained from a conventional, empty-tube correlation. The desired ratio of Eq. (8) is then obtained. Typical results are presented in Fig. 5 for a repeated-rib roughness [10].

3.3 ACTIVE AND COMPOUND TECHNIQUES FOR SINGLE-PHASE FORCED CONVECTION

Under active techniques, mechanically aided heat transfer in the form of surface-scraping can increase forced convection heat transfer. Surface vibration has been demonstrated to improve heat transfer to both laminar and turbulent duct flow of liquids. Fluid vibration has been extensively studied for both air (loudspeakers and and sirens) and liquids (flow interruptors, pulsators, and ultrasonic transducers). Pulsations are relatively simple to apply to low-velocity liquid flows, and improvements of several hundred percent can be realized.

Some very impressive enhancements have been recorded with electrical fields, particularly in the laminar flow region. Improvements of at least 100% were obtained when voltages in the 10-kV range were applied to transformer oil. It is found that even with intense electrostatic fields, the heat transfer enhancement disappears as turbulent flow is approached in a circular tube with a concentric inner electrode.

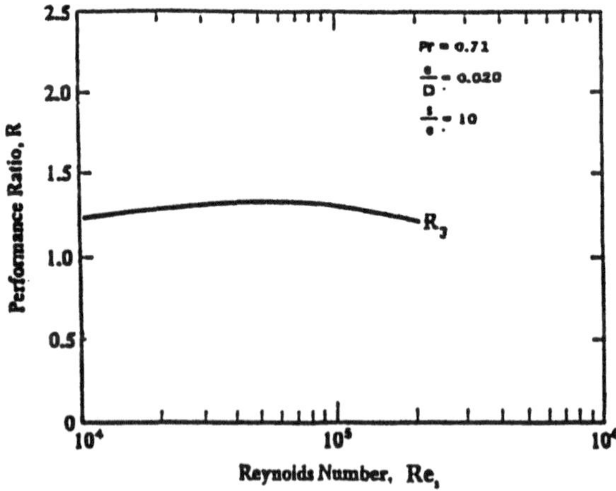

Figure 5. Constant-pumping-power performance criterion applied to repeated-rib roughness.

Compound techniques are a slowly emerging area of enhancement that holds promise for practical applications, since heat transfer coefficients can usually be increased above each of the several techniques acting alone. Some examples that have been studied are as follows: rough tube wall with twisted-tape inserts, rough cylinder with acoustic vibrations, internally finned tube with twisted-tape insert, finned tubes in fluidized beds, externally finned tubes subjected to vibrations, rib-roughened passage being rotated, gas-solid suspension with an electrical field, fluidized bed with pulsations of air, and a rib-roughened channel with longitudinal vortex generation.

4. Pool Boiling

Selected passive and active enhancement techniques have been shown to be effective for pool boiling and flow boiling/evaporation. Most techniques apply to nucleate boiling; however, some techniques are applicable to transition and film boiling.

It should be noted that phase-change heat transfer coefficients are relatively high. The main thermal resistance in a two-fluid heat exchanger often lies on the non-phase-change side. (Fouling of either side can, of course, represent the dominant thermal resistance.) For this reason, the emphasis is often on enhancement of single-phase flow. On the other hand, the overall thermal resistance may then be reduced to the point where significant improvement in the overall performance can be achieved by enhancing the two-phase flow. Two-phase enhancement would also be important in double-phase-change (boiling/condensing) heat exchangers.

As discussed elsewhere, surface material and finish have a strong effect on nucleate and transition pool boiling. However, reliable control of nucleation on plain surfaces is not easily accomplished. Accordingly, since the earliest days of boiling research, there have been attempts to relocate the boiling curve through use of relatively gross modification of the surface. For many years, this was accomplished simply by area increase in the form of low helical fins. The subsequent tendency was to structure surfaces to improve the nucleate boiling characteristics by a fundamental change in the boiling process. Many of these advanced surfaces are being used in commercial shell-and-tube boilers.

Several manufacturing processes have been employed: machining, forming, layering, and coating. In Fig.6(a), standard low-fin tubing is shown. Figure 6(c) depicts a tunnel-and-pore arrangement produced by rolling, upsetting, and brushing. An alternative modification of the low fins is shown in Fig. 6(d), where internal, Fig. 6(e), or external, Fig. 6(f), surface is also modified. Knurling and rolling are involved in producing the surface shown in Fig. 6(g). The earliest example of a commercial structured surface, shown in Fig. 6(b) is the porous metallic matrix produced by sintering or brazing small particles. Wall superheat reductions of up to a factor of ten are common with these surfaces. The advantage is not only high nucleate boiling heat transfer coefficients, but the fact that boiling can take place at very low temperature differences.

The structured boiling surfaces developed for refrigeration and process applications have been used as "heat sinks" for immersion-cooled microelectronic chips.

24

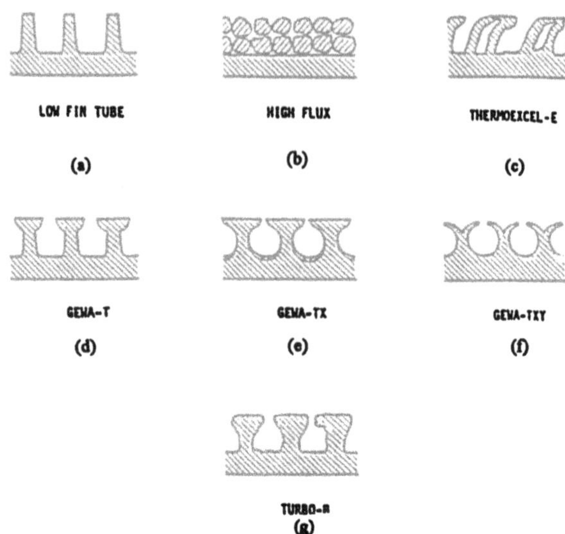

Figure 6. Examples of commercial structured boiling surfaces (Pate et al. [11]).

The behavior of tube bundles is often different with structured-surface tubes. The enhanced nucleate boiling dominates, and the convective boiling enhancement, found in plain tube bundles, does not occur.

Active enhancement techniques include heated surface rotation, surface wiping, surface vibration, fluid vibration, electrostatic fields, and suction at the heated surface. Although active techniques are effective in reducing the wall superheat and/or increasing the critical heat flux, the practical applications are very limited, largely because of the difficulty of reliably providing the mechanical or electrical effect.

Compound enhancement, which involves two or more techniques applied simultaneously, has also been studied. Electrohydrodynamic enhancement was applied to a finned tube bundle, resulting in nearly a 200 percent increase in the average boiling heat transfer coefficient of the bundle, with a small power consumption for the field.

5. Convective Boiling/Evaporation

The structured surfaces described in the previous section are generally not used for intube vaporization, due to the difficulty of manufacture. One notable exception is the High Flux surface in a vertical thermosiphon reboiler. The considerable increase in the low quality, nucleate boiling coefficient is desirable, but it is also important that more vapor is generated to promote circulation.

Helical repeated ribs and helically coiled wire inserts have been used to increase vaporization coefficients and the dryout heat flux in once-through boilers.

Numerous tubes with internal fins, either integral or attached, are available for refrigerant evaporators. Original configurations were tightly packed, copper, offset strip fin inserts soldered to the copper tube or aluminum, star-shaped inserts secured by drawing the tube over the insert. Examples are shown in Fig. 6. Average heat transfer coefficients (based on surface area of smooth tube of the same diameter) for typical evaporator conditions are increased by as much as 200 percent.

A cross-sectional view of a typical "micro-fin" tube, now widely used, is included in Fig. 6. The average evaporation boiling coefficient is increased 30-80 percent. The pressure drop penalties are less; that is, lower percentage increases in pressure drop are frequently observed. Twisted-tape inserts are generally used to increase the burnout heat flux for subcooled boiling at high imposed heat fluxes $\left(10^7 - 10^8 \ W/m^2\right)$, as might be encountered in the cooling of fusion reactor components. Increases in burnout heat flux of up to 200 percent have been reported at near-atmospheric pressure, but the enhancement drops off rapidly as the pressure is increased above 2 Mpa.

Since pressure drop is important to the stability of parallel cooling channels, recent work has emphasized both burnout and pressure drop [12].

6. Vapor-Space Condensation

Condensation can be either filmwise or dropwise. In a sense, dropwise condensation is enhancement of the normally occurring film condensation by surface treatment. The only real application is for steam condensers, because nonwetting coatings are not available for most other working fluids. Even after much study, little progress has been made in developing permanent hydrophobic coatings for practical steam condensers. The enhancement of dropwise condensation is pointless, because the heat transfer coefficients are already so high.

Surface extensions are widely employed for enhancement of condensation. The integral low fin tubing (Fig. 5(a)), used for kettle boilers, is also used for horizontal tube condensers. With proper spacing of the fins to provide adequate condensate drainage, the average coefficients can be several times those of a plain tube with the same base diameter. These fins are normally used with refrigerants and other organic fluids that have low condensing coefficients, but which drain effectively, because of low surface tension.

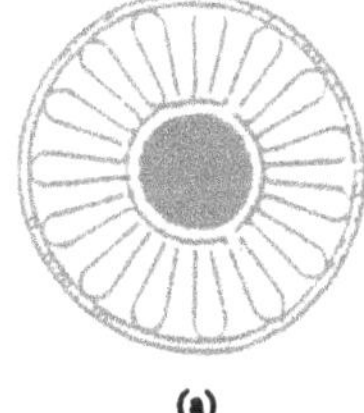

(a)   (b)   (c)

Figure 7. Inner-fin tubes for refrigerant evaporators.

Strip-fin insert	(a)
Star-shaped inserts	(b)
Micro-fin	(c)

The fin profile can be altered according to mathematical analysis to take full advantage of the Gregorig effect, whereby condensation occurs mainly at the tops of convex ridges. Surface- tension forces then pull the condensate into concave grooves, where it runs off. The average heat transfer coefficient is greater than that for an axially uniform film thickness. The initial application was for condensation of steam on vertical tubes used for reboilers and in desalination. According to numerical solutions, the optimum geometry is characterized by a sharp fin tip, gradually changing curvature of the fin surface from tip to root, wide grooves between fins to collect condensate, and periodic condensate strippers. Figure 8 schematically presents the configuration.

Recent interest has centered on three-dimensional surfaces for horizontal-tube condensers.   The considerable improvement relative to low fins or other two-dimensional profiles is apparently due to multidimensional drainage at the fin tips. Other three-dimensional shapes include circular pin fins, square pins, and small metal particles that are bonded randomly to the surface.

7. Convective Condensation

This final section on enhancement of the various modes of heat transfer focuses on in-tube condensation. The applications include horizontal kettle-type reboilers, moisture separator reheaters for nuclear power plants, and air-conditioner condensers.

Internally grooved or knurled tubes, deep spirally fluted tubes, random roughness, conventional inner-fin tubes have been shown to be effective for condensation of steam and other fluids.

The micro-fin tubes mentioned earlier have also been applied successfully to in-tube condensing. As in the case of evaporation, the substantial heat transfer improvement is achieved at the expense of a lesser percentage increase in pressure drop. By testing a wide variety of tubes, it has been possible to suggest some guidelines for the geometry, e.g., more fins, longer fins, and sharper tips; however, general correlations are not yet available. Fortunately for heat-pump operation, the tube that performs best for evaporation also performs best for condensation.

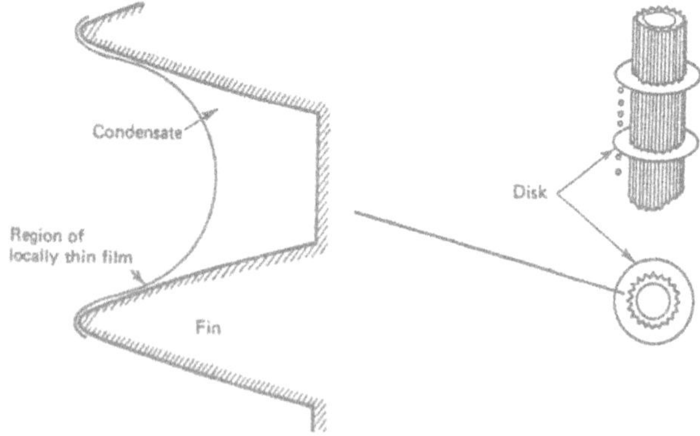

Figure 8. Recommended flute profile and schematic of condensate strippers[12].

Twisted-tape inserts result in rather modest increases in heat transfer coefficient for complete condensation of either steam or refrigerant. The pressure drop increases are large, due to the large wetted surface. Coiled tubular condensers provide a modest improvement in average heat transfer coefficient.

8. Concluding Remarks

Heat Transfer Enhancement - Second Generation Heat Transfer Technology - is well established. It is used routinely in the power and process industries particularly where phase change is involved. By way of explanation, first generation heat transfer enhancement is second generation heat transfer technology, and so on. What is now needed is Advanced Enhancement - Third Generation Heat Transfer Technology. Some examples are given in Table 1.

TABLE 1. The generations of heat transfer technology

	Tube-and-plate fins, single-phase
1st generation	bare tube
2nd generation	plain fins
3rd generation	longitudinal vortex generators on fins
	In channel, single-phase
1st generation	smooth channel
2nd generation	2-D roughness
3rd generation	3-D roughness
	Outside tubes, boiling
1st generation	smooth tube
2nd generation	2-D fins
3rd generation	3-D fins and metallic matrices
	In-tube, evaporation
1st generation	smooth tube
2nd generation	massive fins and inserts
3rd generation	micro-fins
	Outside tubes, condensing
1st generation	smooth tube
2nd generation	2-D fins
3rd generation	3-D fins and metallic matrices

We do not find a chronological relationship among the levels of enhancement. For example, the 3rd generation pool boiling surface High Flux was patented in 1968.

This review gives some indication why heat transfer enhancement is one of the fastest growing areas of heat transfer. Many techniques are available for improvement of the various modes of heat transfer. Fundamental understanding of the transport mechanism is growing, but, more importantly, design correlations are being established. New journals, e.g., *Journal of Enhanced Heat Transfer* and *International Journal of Heating, Ventilating, Air-Conditioning and Refrigerating Research*, feature this technology.

28

Nomenclature

A	heat transfer surface area, m^2
A_F	cross sectional flow area, m^2
c	constant pressure specific heat, $J/kg\,K$
D,d	diameter of plain tube, m
D_h	hydraulic diameter, m
e	protuberance height, m
f	Fanning friction factor, - ; $f_s = (1.58 \ln Re - 3.28)^{-2}$
G	mass flux, kg/m^2
h	heat transfer coefficient, $W/m^2 K$; strip fin height, m
j	$Nu/Re\,Pr^{1/3}$
k	thermal conductivity, W/mK
L	channel length, m
ℓ	length of one offset module of strip fins
Nu	Nusselt number hD/k, - ; $Nu_s = 0.5\,f\,RePr/(1 + 12.7(0.5f)^{0.5}$ $(Pr^{0.667} -1)$
n	number of sharp corners facing the flow, -
P	pumping power across heat exchanger, W
ΔP	pressure drop across heat exchanger, N/m^2
p	fin pitch; repeated-rib pitch, m
Pr	Prandtl number, $c\mu/k$
R	performance ratio, Eq. (8)
Re,Re_h	Reynolds number, $GD/\mu, GD_h/\mu, -$
s	lateral spacing of strip fins, m
$\dot{N}$	fin thickness at base, m
V	average flow velocity, m/s
α	spiral angle for helical ribs or fins, deg; aspect ratio s/h, -
β	contact angle of rib profile, deg
γ	ratio t/s, -
δ	ratio t/ℓ, -
μ	dynamic viscosity, Ns/m^2
ρ	density, kg/m^3

Subscripts

a	augmented or enhanced
h	based on hydraulic diameter
i	based on maximum inside (envelope) diameter
s	straight or plain; based on surface temperature

References

Bergles, A.E., Jensen, M.K., and Shome, B., Bibliography on Enhancement of Convective Heat and Mass Transfer, RPI Heat Transfer Laboratory Report HTL-23 (1995).

Bergles, A.E., Techniques to Augment Heat Transfer, *Handbook of Heat Transfer Applications* (Eds. W. M. Rohsenow, J. P. Hartnett, and E. N. Ganic) McGraw-Hill, New York, NY, 3–1—3–80, (1985).

Bergles, A.E., Heat Transfer Enhancement – The Encouragement and Accommodation of High Heat Fluxes, *Journal of Heat Transfer*, **119**, (1997), 8-19.

Thome, J.R., Enhanced Boiling Heat Transfer, Hemisphere, New York, NY, (1990).

Webb, R.L., Principles of Enhanced Heat Transfer, Wiley, New York, NY, (1994).

Bergles, A.E., Survey and Evaluation of Techniques to Augment Convective Heat and Mass Transfer, *Progress in Heat and Mass Transfer*, **1**, Pergamon, Oxford, England, (1969).

Bergles, A.E., Blumenkrantz, A.R., and Taborek, J., Performance Evaluation Criteria for Enhanced Heat Transfer Surfaces, *Heat Transfer 1974*, The Japan Society of Mechanical Engineers, Tokyo, II, (1974), 234-238.

Ravigururajan, T.S. and Bergles, A.E. General Correlations for Pressure Drop and Heat Transfer for Single-Phase Turbulent Flow in Internally Ribbed Tubes, *Augmentation of Heat Transfer in Energy Systems*, ASME, New York, HTD-52 (1985), 9-20.

Carnavos, T.C., Heat Transfer Performance of Internally Finned Tubes in Turbulent Flow, *Advances in Advanced Heat Transfer*, ASME, New York, (1979), 61-67 (1979).

Pate, M.B., Ayub, Z.H., and Kohler, J., Heat Exchangers for the Air-Conditioning and Refrigeration Industry: State-of-the-Art Design and Technology, *Compact Heat Exchangers*, Hemisphere, New York, (1990), 567-590.

Tong, W., Bergles, A.E., and Jensen, M.K., Critical Heat Flux and Pressure Drop of Subcooled Flow Boiling in Small-Diameter Tubes with Twisted-Tape Inserts, *Journal of Enhanced Heat Transfer*, **3**, (1981), 95-108.

Mori, Y., Hijikata, K., Hirasawa, S., and Nakayama, W., Optimized Performance of Condensers with Outside Condensing Surfaces, *Journal of Heat Transfer* **103** (1981), 96-102.

SUSTAINABILITY CRITERIA FOR HEAT EXCHANGER DESIGN

NAIM H.AFGAN AND MARIA G. CALVALHO
Instituto Superior Tecnico
Lisbon, Portugal

Abstract. The Lecture will review the growing concern of scientific and political community of the world has been expressed through the definition of sustainability as a parameter to measure our future behavior in the use of the available resources on our planet. Historically speaking, it was noticed in very early days of our past that human concern has been expressed in several documents which emphasize need for the rational use of global resources and the sustainability as a moral, cultural and philosophical pattern of our individual and collective behavior.

Special attention is devoted to the definition of sustainability and its generic meaning. In this respect a particular attention is given to the discussion of different aspects of sustainability in the present world. In order to present an engineering approach to the sustainable development, attention is focused to the review of sustainability criteria, as they have to be introduced in the future products

The main emphasis is given to review a potential development in the energy engineering science that may lead to a sustainable energy development. Seven major areas are listed with specific problems and their relevance to the sustainable energy development. This includes the following areas: energy resources and development; efficiency assessment; clean air technologies; information technologies; new and renewable energy resources; environment capacity; mitigation of nuclear power treat to the environment.

In order to quantify criteria for the sustainability assessment of any design of thermal system element indicators are defined to meet this requirement. In this respect, the efficiency of resources use and technology development is of fundamental importance

Special emphasize is devoted to the example related to the analysis of the current design method for the condenser design and its comparison with the respective sustainability criterion. In this respect the power plant condenser design is shown in the potential prospect for the new design criterion which might lead to the development of the new method for the condenser design or the better understanding for its potential improvement.

1. Introduction

Human society is becoming aware that future development of our civilization will strongly depend on the concern devoted to the understanding of global environmental, social, economic, cultural and religions problems. In this respect, coming tried millennium will be confronted with a number of treats resulting from growing fragility of the Earth's life support systems. The growing concern of scientific and political community of the world has been expressed through the definition of sustainability as a parameter to measure our future behavior in the use of the available resources on our planet. Historically speaking, it was noticed in very early days of our past that human concern has been expressed in several documents which emphasize need for the rational use of global resources and the sustainability as a moral, cultural and philosophical pattern of our individual and collective behavior. The sustainability as the principle of ethical approach in our life has been cited in number of ancient writing. Among those are:

a.) The Earth Chapter [1]
"The protection of the environment is essential for human well-being and the enjoyment of fundamental rights, and as such requires the exercise of corresponding fundamental duties "

31

S. Kakaç et al. (eds.), Heat Transfer Enhancement of Heat Exchangers, 31–47.
© *1999 Kluwer Academic Publishers.*

32

b.) Thomas Jefferson, Sept.6 1889 [2]
" Then I say the earth belongs to each generation during its course, fully and in its right no generation can contract debts greater than may be paid during the course of its existence "

Definition based on the religious believes playing the responsibility and duties toward the nature and Earth. In this respect it is of interest to enlighten that the Old Testimony in which the story of creation is told is a fundamental basis for Hebrew and Christian doctrine of the environment. In the world of Islam, nature is the basis for human consciousness. According to the Koran, while humankind is Gods vice-regent on Earth, God is the Creator and Owner of nature. But human beings are his trusted administrators, they ought to follow God instructions, that is, acquiesce to authority of Prophet and to the Koran regarding nature and natural resources.

Recently, several international forums have devoted the great concern and urgent warning to the government leaders of all nations that the fragility of Earth systems is reaching dangerous limits Resulting from that analysis a new definition of sustainability is generated. Among those are:

a.) The World Commission on Environment and Development (Brundtland Commission) [3]
"development that meets the needs of the present without compromising the ability of future generation to meet their own needs "
b.) the Agenda 21, Chapter 35 [4]
"development requires taking long-term perspectives, integrating local and regional effects of global change into the development process , and using the best scientific and traditional knowledge available "
c.) the Council of Academies of Engineering and Technological Sciences [5]
"It means the balancing of economic, social, environmental and technological consideration, as well as the incorporation of a set of ethic values "

Since the World Commission on Environment and Development has introduced notion of sustainability the debate has generated the tremendous interest and avalanche of publications. Even we still have not agreed definition; people tend to define sustainability in the ways that suit their particular application. Most sustainability definitions originate from the relation between humans and the resources they use. When introducing the goal of sustainability, it is not properly recognized that within complex hierarchical systems sustainability in a certain level may be in conflict with sustainability of systems in higher hierarchical levels. Very often sustainability is defined in the ways that suit particular applications. It should be also recognized that sustainability has become more of a political issue that the scientifically supported concept. In the scientific analysis of the sustainability concept it is either redundant, or ambiguous. For this reason some authors are inclined to talk about the paradox of sustainability [6,7]

Sustainable development focuses on the role and the use of sciences in supporting the prudent management of the environment and for the survival and future development of humanity [8,9,10,11]. It is recognized that scientific knowledge should be applied to articulate and support the goals of sustainable development, through scientific assessment of current conditions and future prospects for the Earth system. The program areas, which are in harmony with conclusions and recommendations of this International Conference on an Agenda of Science for Environment and Development into 21st Century, are:

a.) Strengthening the scientific bases for sustainable management;
b.) Enhancing scientific understanding;
c.) Improving long-term scientific assessment;
d.) Building up scientific capacity and capability.

It is essential for the implementation of this program that it be focused on the long-term perspectives and the global changes of life support systems. In particular, there is a need for a constant interaction with governmental, industrial, political, educational, cultural and spiritual authorities participating in the realization of the program. It is of crucial importance that, in the realization of the program, an active role be given to scientists from developing countries. Since the major part of the population increase is expected in the developing part of the world, the participation of scientists from developing countries will bridge deficiency in dealing with the problems which are immanent to their environment by an academic approach.

SUSTAINABLE ECONOMIC AND TECHNOLOGICAL DEVELOPMENT
Intergenerational and Intra-generational equity
Precautionary or risk aversion strategies
Biodiversity conservation
NONCONSUMPTIVE USE OF WORLD NATURAL RESOURCES
SUSTAINABLE WORLD
HUMAN RESOURCE DEVELOPMENT
Internalization of environmental costs
Enlightened institutions and infrastructure

Fig.1

UNSUSTAINABLE ECONOMIC AND TECHNOLOGICAL DEVELOPMENT
Resource shortages and Uneven distribution
Natural disasters
Health problems
Disadvantaged peoples
Deteriorating cities
Security issues
Poverty
LIFE SUPPORT RESOURCES:
WATER ENERGY ENVIRONMENT FOOD
GLOBAL CRISIS
HUMAN RESOURCE DEVELOPMENT
Global warming
Desrtification
Deforestation
Ozone depletion
Pollution

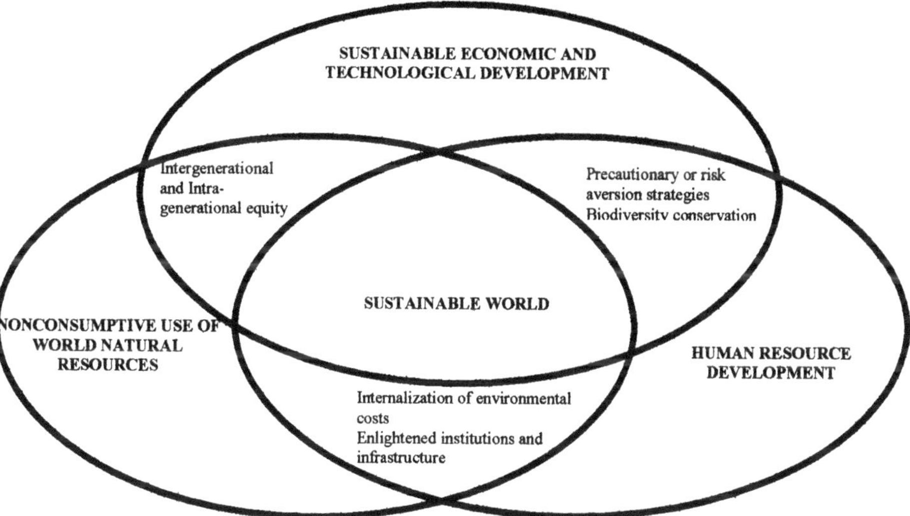

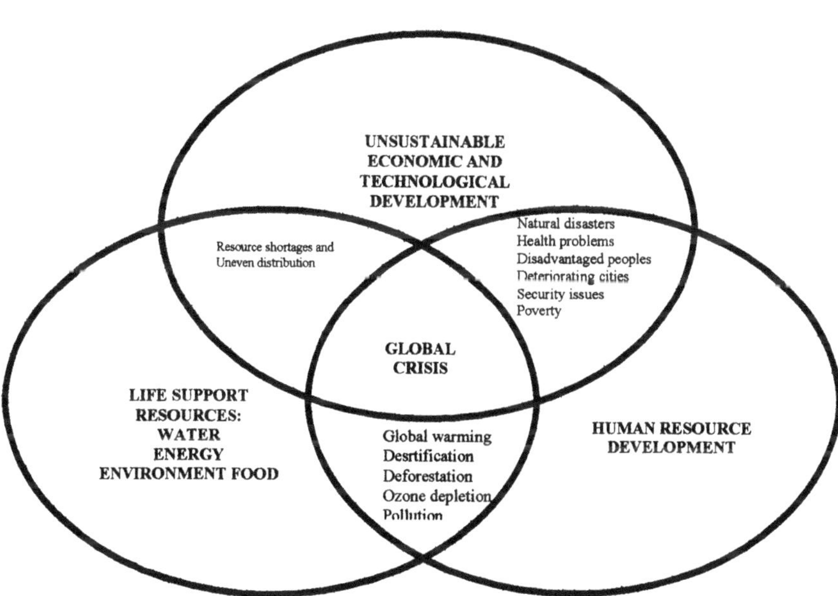

Fig.2

Even, sustainability development has become a political movement with a strong connotation related to the existing differences among the continents regions and countries, its strength should be seen in the promotion for the salvage of the planet, the only place for our human civilization. In this respect, the determination of the interested parties, including the United Nation Organizations, government organizations, non-government organizations and religious organizations, to recognize the sustainable development as the path for the creation of the future of new generations, is a guarantee for the economic prospective and social development.

There are several ways in which the ideas of sustainable development are presented and interpreted. Ecocentric interpretations indicate a conspicuous degree of reference to the ecosystems. Anthropocentric interpretation tends to put humans at the centre of the issues. Others such as biocentric interpretation focus on the protection of the elements of the biosphere. Our main concern in this general setting is energy and whatever we do for and with energy may be reflected in any of the above mentioned perspectives and interpretations. While this concern for sustainability in the energy sphere should include considerations at the global level, more importantly, there is the need to look at it as a regional issue in the overall global scenario. There are several perspectives on this.

Figures 1 and 2 attempt to distinguish sustainable and unsustainable   development .

2. Energy Resources And Its Consumption

Energy and matter constitute the earth's natural capital that is essential for human activities such as industry, amenities and services in our natural capital as the inhabitants of the planet earth may be classified as :

- Solar capital (provides 99% of the energy used on the Earth)
- Earth capital (life support resources and processes including human resources)

These, and other, natural resources and processes comprise what has become known as 'natural capital and it is this natural capital that many suggest is being rapidly degraded at

this time. Many also suggest that contemporary economic theory does not appreciate the significance of natural capital in techno-economic production.

All natural resources are, in theory, renewable but over widely different time scales. If the time period for renewal is small, they are said to be renewable. If the renewal takes place over a somewhat longer period of time that falls within the time frame of our lives, they are said to be potentially renewable. Since renewal of certain natural resources is only possible due to geological processes which take place on such a long time scale that for all our practical purposes, we should regard them as non-renewable. Our use of natural material resources is associated with no loss of matter as such. Basically all earth matter remains with the earth but in a form in which it can not be used easily. The quality or useful part of a given amount of energy is degraded invariably due to use and we say that entropy is increased.

The abundant energy resources at the early days of the industrial development of the modern society have imposed the development strategy of our civilisation to be based on the anticipated thinking that energy resources are unlimited and there is no other limitations which might affect human welfare development. It has been recognised that the pattern of the energy resource use has been strongly dependent on the technology development. In this respect it is instructive to observe the change in the consumption of different resources through the history of energy consumption. World-wide use of primary energy sources since 1850 is shown in Fig.3 [12] . F is the fraction of the market taken by each primary-energy source at a given time.

It could be noticed that two factors are affecting the energy pattern in the history. The first is related to the technology development and, the second, to availability of the respective energy resources. Obviously , this pattern of energy source use is developed under constraint immanent to the total level of energy resources consumption and reflects the existing social structure both in numbers and diversity [13,14,15]. The world energy consumption   is shown in Fig.4.[16]

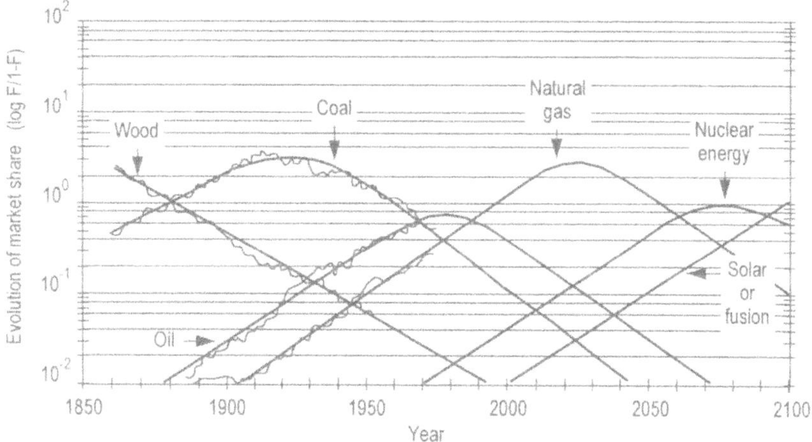

Fig.3

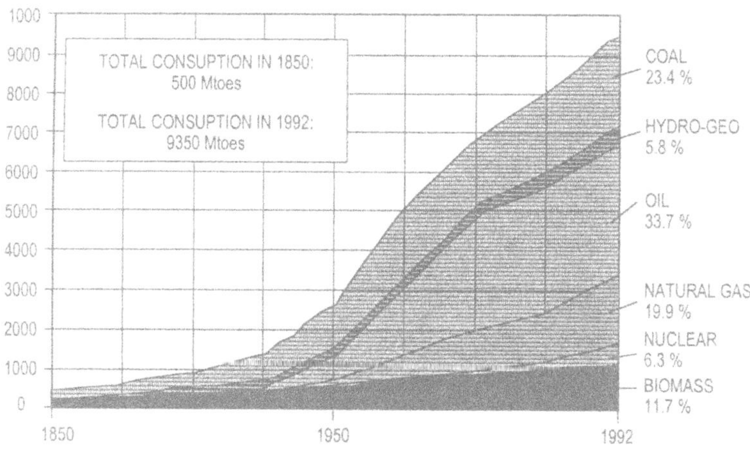

Fig.4

Looking at the present energy sources consumption pattern , it can be noticed that oil is a major contender, supplying about 40% of energy. Next, coal supply is around 30% , natural gas 20% and nuclear energy 6.5 %. This means that current fossil fuel supply is 90 % of the present energy use. In the last several decades our civilisation has witnessed changes which are questioning our long-term prospect. Fossil fuel, non-recyclable is an exhaustible natural resource that will be no more available one day . In this respect it is of common interest to learn how long fossil fuel resources will be available ,as they are the main source of energy for our civilisation. This question has attracted the attention of a number of distinguished authorities, trying to forecast the energy future of our planet. The Report of the Club of Rome " Limits to Growth" ,published in 1972 [17], was among the first ones which pointed to the finite nature of fossil fuel. After the first and second energy crisis the community at large has become aware of the possible the physical exhaustion of fossil fuels. The amount of fuel available is dependent on the cost involved .For oil it was estimated that proved amount of reserves has, over past twenty years , levelled off at 2.2 trillion of barrels produced under $ 20 per barrel. Over the last 150 years we have already used up one-third of that amount, or about 700 billion of barrels which leaves only a remaining of 1.5 trillion of barrels. If compared with the present consumption, it means that oil is available only for the next 40 years. Fig. 5 shows the ratio of the discovered resources to the yearly consumption for the fossil fuels [16].

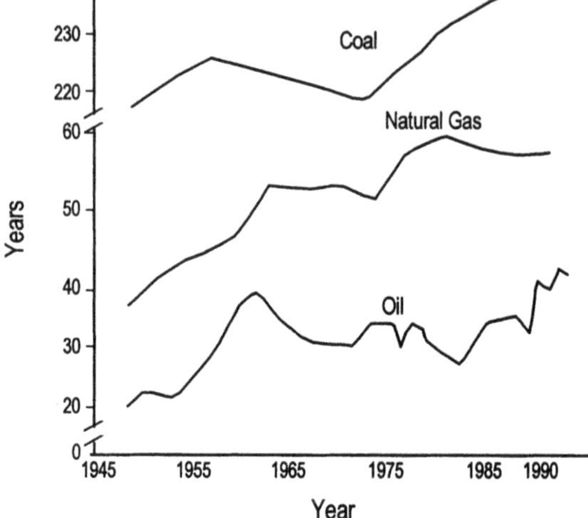

Fig.5

From this figure it can be noticed that coal is available for the next 250 years and gas for the next 50 years. Also , it is evident that as much as the fuel consumption is increasing, new technologies aimed to the   discovery of new resources are becoming available, leading to a slow increase of the time period for the exhausting of the available energy sources.

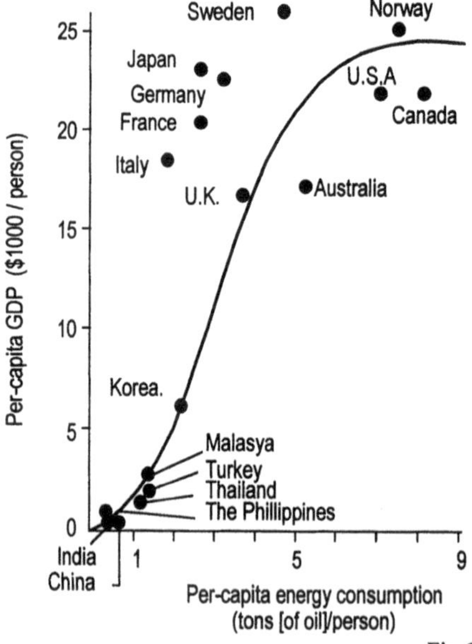

Fig.6

It is known that the energy consumption is dependent on two main parameters. Namely, the amount of energy consumed per capita and the growth of population. It has been proved that there is a strong correlation between the Gross Domestic Product and Energy consumption per capita. Fig.6 shows the economic growth and energy consumption for a number of countries, in 1990 [17].

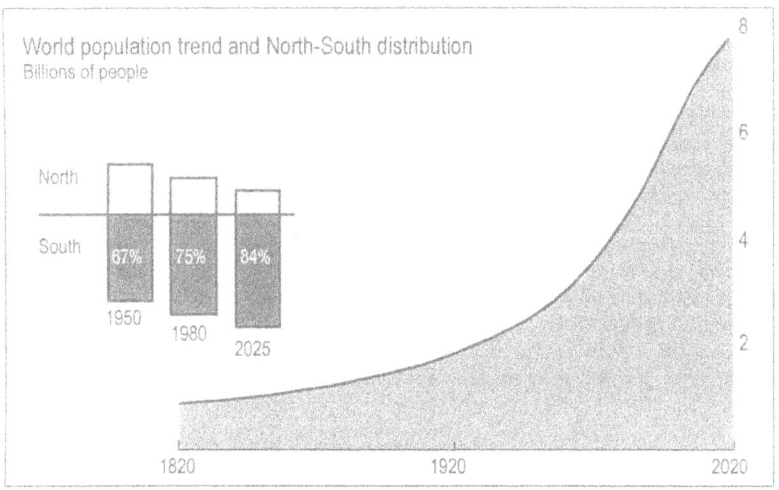

Fig.7

There is a number of scenarios which are used   for the forecast of the world economic development. With the assumption that the recent trend in the economic development will be conserved in the next 50 years and considering the demographic forecast in the increase of human population, as shown in Fig.7[18]

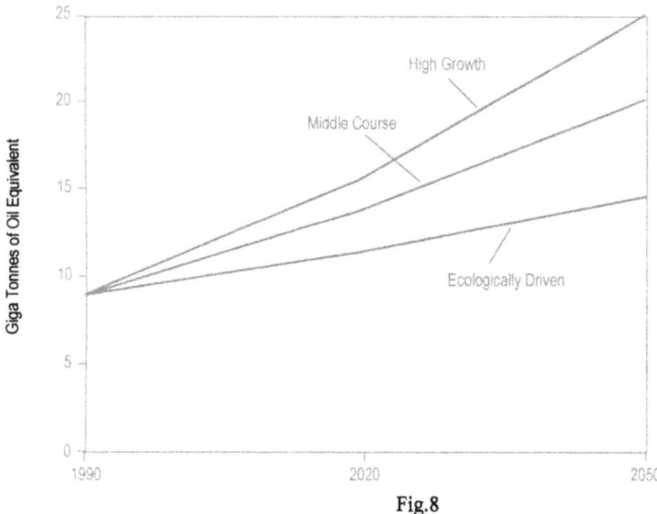

Fig.8

38

The future energy consumption could be calculated ,as shown in Fig.8 .[19] Compared with the available resources it is easily foreseen that the depletion of the energy resources is an immanent process which our civilisation will face in the near future. Nevertheless , whatever is the accuracy of our prediction methods and models, it is obvious that any inaccuracy in our calculation may affect only the time scale but not the essential understanding that the energy resources depletion process has begun and requires the human action before adverse effects may irreversibly enforce

Natural resources scarcity and economic growth are in fundamental opposition to each other . The study of the contemporary and historical beliefs showed [19] , that : (1) natural resources are economically scarce, and become increasingly so with the passage of time;(2) the scarcity of resources opposes economic growth. There are two basic versions of this doctrine. The first, the Malthusian, rests on the assumption that there are absolutely limits ; once these limits are reached the continuing population growth requires an increasing intensity of cultivation and, consequently, brings about diminishing returns per capita. The second , or Ricardian version, viewed the diminishing returns as current phenomena reflecting the decline in the quality of resources brought within the margin of a profitable cultivation. Besides these two models, there is also the so called "Utopian case' where there is no resources scarcity. There have been several attempts to apply these models to the energy resources in order to define the correlation between specific energy resources and economic growth. The substantial questions related to the scarcity , its measurement and growth are : (1) whether the scarcity of energy resources has been and/or will continue to be mitigated and (2) whether the scarcity has "de facto" impacted the economic growth. An analysis based on the relative energy prices and unit costs has been applied to natural gas , bitumen coal, anthracite coal and crude oil. The USA analysis in this respect can serve as the indication for the future trend in a world scale. Using two measures of scarcity - unit cost and relative resource price change in the trend of resource scarcity for natural gas, bitumen coal, anthracite coal and crude oil, over three decades are shown in Fig.9 [20].

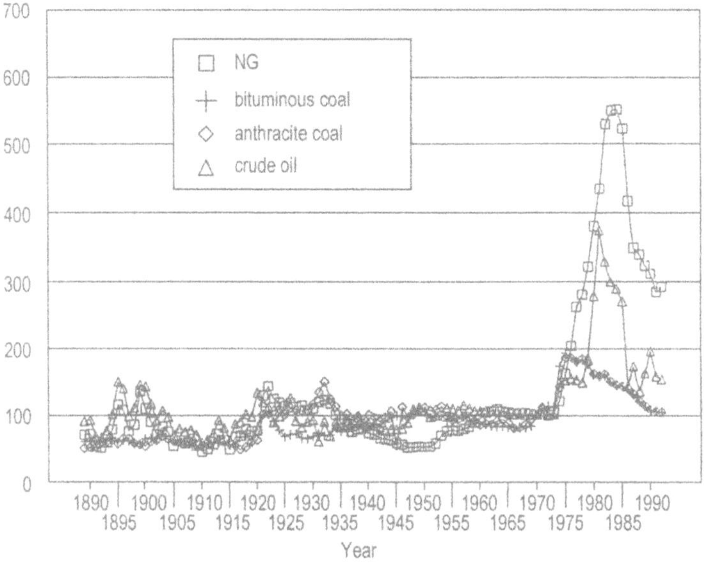

Fig.9

It can be noticed that each one of the energy resources has become significantly more scarce during the decade of the 1970s. The situation reversed itself during the 1980s. The change, that took place, has implications for the future economic growth to the extent resources scarcity and economic growth are interrelated, even if it was not proved that short term energy resources scarcity fluctuation has substantial implication on the long term economic growth. It has become obvious the need for   an active involvement in allocating scarce, non-renewable energy resources and its potential effect on the economic growth.

Since fossil fuels have demonstrated their economic superiority, more than 88 % of primary energy in the world in recent years has been generated from   fossil fuels. However   , the exhaust gases from combusted fuels have accumulated to an extent where a serious damage is being done to the world global environment. The accumulated amount of CO_2 in atmosphere is estimated at about 2.75 x 10^{12} t. The global warming trend from 1900 - 1990 is shown in Fig.10 [21].

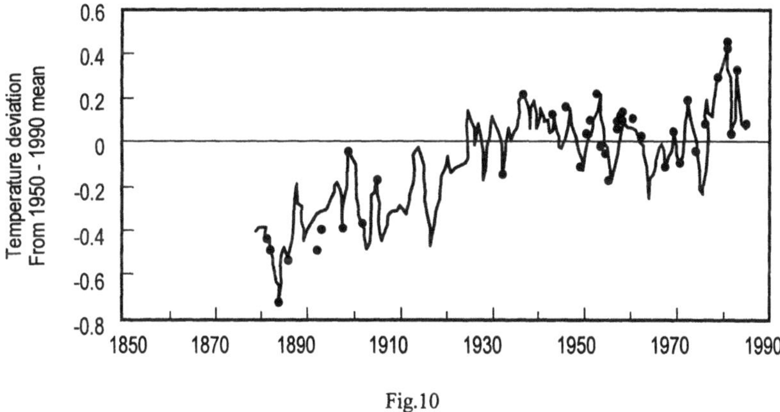

Fig.10

The future trend of the carbon dioxide concentration in the atmosphere can be seen from the Fig.11 [21].

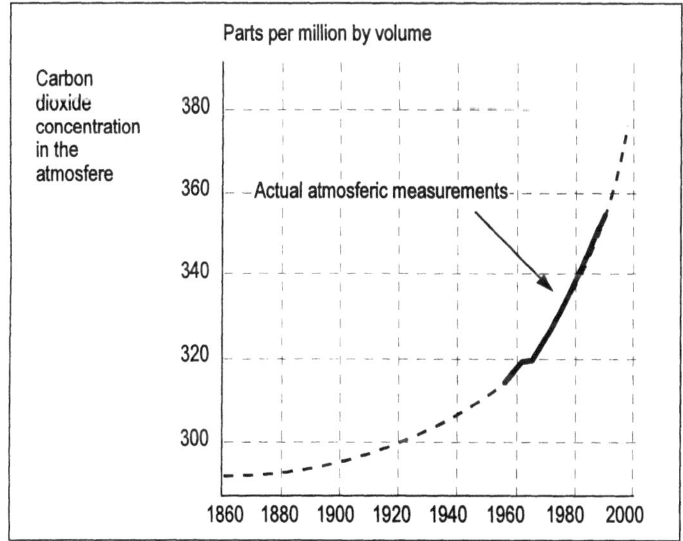

Fig.11

It is rather obvious that the further increase of the CO_2 will lead to disastrous effects to the environment. Also, the emission of SO_2 , NOx and suspended particulate matters will substantially contribute to exasperate the effect on the environment.

In a world scale, coal will continue to be a major source of fuel for the electric power generation. Many developing countries, such as China and India, will continue to use inexpensive ,abundant , indigenous coal to meet growing domestic needs [22,23,24]. This trend greatly increases the use of coal worldwide as economy in the other developing countries, continues to expand. In this respect the major long-term environmental concern about coal use has changed from acid rain to greenhouse gas emissions - primarily carbon dioxide from combustion .It is expected that coal will continue to dominate China's energy picture in the future. The share of coal, in primary energy consumption is forecast to be no less than 70 % during the period 1995 - 2010. In 1993 China has produced a total of 1.114 billion tons of coal, in 2000 it is planned 1.5 trillion and in 2010 it will be 2.0 . Since China is the third largest energy producer in world, after USA and Russia its contribution to the global accumulation of the CO_2 will be substantial if the respective mitigation strategies will not be adopted. The example of China is instructive in the assessment of the future development of developing countries and their need for   accelerated economic development.

3. Sustainability Criteria

There has been a number of attempts to define the criteria for the assessment of the sustainability of the market products. In this respect   the Working Group of UNEP on Sustainable Development has come out with the qualitative assessment criteria for the assessment of the product design .

Having those criteria as the bases , we would like to introduce them as the specific application in the energy system design. In this consideration energy system design is taken as the entity which should comply with the sustainability criteria.

Energy system design is defined as :

1. Strategic Design

The strategic design will require holistic planning that meets and considers all interrelated impacts e.g. logistic, space planning and resource planning. As regard the energy system, it may be interpreted as : mixed energy concept with optimization of local resources, urban and industrial planning with transport optimization, use of the renewable energy sources.

2. Optimized Design

The design optimization of the energy system means the selection of the structure and design parameters of a system to minimize energy cost under conditions associated with available material, financial resources, protection of the environment and government regulations, together with safety , reliability, availability and maintainability of the system.

3. Dematerialization of Design

This will imply that the energy system, plant and equipment are designed with optimal use of information technology in order to prevent duplication , prevent operational malfunction, assure rational maintenance scheduling. Dematerialization in the design may be seen as the introduction of knowledge based systems, use of virtual library, digitized video, use of on-line diagnostic systems, development of new sensor elements and development of new combustion technologies.

4. Longevity of Design

Complex energy system is commonly composed of different subsystems and individual equipment elements. It has been recognized that the life tie of the elements and subsystems is not equal. In this respect optimal selection of the life cycle for elements and subsystems may lead to the retrofitting procedure which will reflect need for the sustainable criteria application. Examples for this criterion can be seen as : modular design of the subsystems , standardization of the elements, lifetime monitoring and assessment, co-ordination of suppliers and buyers.

5. Life Cycle Design

This will mean that the energy system and its subsystems have to be designed to meet sustainability through every stage of the life cycle. It is known that the energy system is designed to work under different conditions in order to meet load change, environment change, social change ,etc. It is obvious that there will be different cycles for each of the mentioned time scale processes. In this respect the system has to fulfill its function without failing to meet sustainability requirements.

4. Sustainability Indicators

It is known that any numerical number, semantic expression or mathematical sign is information. Examples of these type of information can be fined in any reading of temperature or other time dependent variables describing state of the system. Information is also parameters used in the design of respective system. Also, positive or negative sign of the variable is also the information .

Collecting information and its processing will convert them in data. So , data represent agglomerated information, which are partially or finally processed. Examples of data can be find as a parameter, which describe evaluated information to be used for the specific purpose. In this respect the average inlet temperature of cooling water in the condenser is data obtained by the averaging procedure adapted for this purpose. Also. heat transfer coefficient used in the design of condenser is the data obtained by the experimental procedure for the heat transfer evaluation.

In order to use the data for the assessment of the respective system ,it is necessary to convert them into the indicator. So, the indicator represents the measuring parameter for the comparison between the different states or structure of the system. As example, the efficiency of the system is the indicator for the quality of energy use in respective system. Also, we can evaluate different structure of the systems by the indicator representing respective entities of the system. In this direction is the assessment of intelligence use in the improvement of the system compatibility with its surrounding measured by the respective indicators.

In order to quantify criterion's for the sustainability assessment of any design of thermal system element indicators are defined to meet this requirement. In this respect, the efficiency of resources use and the technology development is of the fundamental importance. It is obvious that the efficiency of the energy resource use is a short term approach which may give return benefit in the near future. As regard the technology development , the long term research and development is needed. In some cases it will require respective social adjustment in order to meet requirements of the new energy sources.

Potential improvement of the energy conversion process is a driving force for its development. In the assessment of the conversion process a promising tool is the energy analysis of the energy system. The energy analysis is based on the maximum potential availability and its use for the assessment of the conversion process . By definition the energy is a parameter for the validation of the efficiency of energy conversion process and system. Taking into account the law of thermodynamics the technology improvement appears as a significant factor responsible for an entropy change in energy system. The application of the principle of Carnet therefore allows to determine an absolute limit to any transformation of the deposit of free energy.

In order to facilitate the tool for sustainability assessment the indicators are used as parameters for the measurement of sustainability. In this respect ,attention is focused on the resource indicators and environment capacity indicators. Each of this entities has to be determined with specific parameters which can be used to determine the characteristic indicators for the assessment of the respective entities. The characteristic indicators are :

42

a. Possible Change Indicator

Possible change indicator is characterising changes, which are determined by the maximum change technically possible. It reflects the difference between state of the entity with maximum availability. If applied to the energy resource , it is a difference between known resources exploitable with known technology and current resource to be obtained with present technological capacity. As example, it can be shown that coal resource 1980 have been 2000 Motes and in 1985 2300Mtoes. Maximum potential resources to be used without effecting the available resources is 300 Mote. The indicator for the possible change is defined as the maximum potential resources and the average value for the period under consideration. So that the Possible Change Indicator (PCI) is 0.13. Possible change indicator is a function of the technology development in the resource discovery. This imply that there are two means for the increase of the resources, namely : the additional investment for the new discovery and the new technology for the resource exploration.

b. Current Change Indicator

Current change indicator is defined as the resource consumption change of the respective resource. It can be defined as the difference in current consumption change in the reference time period. Applying the measurement of this indicator to the period 1980 – 1982, it can be seen that Current Change Indicator (CCI) for coal is o.467. It is known that the current consumption of the energy resources strongly depend on the efficiency of energy use. Possible change in the efficiency of energy use includes the efficiency of the primary energy source conversion and second, the efficiency of the finale energy use.

c. Resource Indicator

By definition , the Resource Index (RI) is the ration between the Current Change Indicator and the Possible Change Indicator. In essence it will represent the qualitative measure of the resource depletion. Also, it may be interpreted as the resource sacristy change due to the resource depletion. For the case of coal resource the Resource Indicator is 0.065

d. Financial Effect of Resource Use

The resource indicator is reflecting material balance of the respective primary energy resource. In order to use it for the quality assessment of the primary energy use , it has to be connected with the financial effect resulting form its use.

4.1 DEFINITION OF RESOURCE INDICATOR FOR ENERGY SYSTEM ELEMENT

The resource indicator can be defined as the amount of material used to perform specific function . Material comprise, the construction material for specific equipment and material as the energy resource needed for respective function. Each element of thermal equipment represent the metallic construction build of the material by extraction from the available resources. The material is obtained by the processing of the respective material in the form defined by its function. In the processing of resource material adequate energy is needed to be spend for its finale form production. This energy is called production energy. Since, every element is designed to perform function reflecting specific operation in the system, the energy needed to perform respective function within the system.

In order to define indicator for resource use for the element of energy system let us represent them in following form

$$R_m = \frac{G_m}{E} \tag{1}$$

where
G_m - weight of material

E - energy transformation

In order to define its relative value which may be used for qualitative assessment of the specific element , we can write

$$\rho_m = \frac{R_m}{R_{m,opt}} \tag{2}$$

where
  $R_{m.opt}$ - is optimal resource consumption for specific element

In this case subscript m will reflect materials used for the production including metals, energy resource etc.

If it is assumed that the energy use for production of the respective element is

$$E_p = \left(E_m + E_{prod}\right)R_m \tag{3}$$

where
  E_m and E_{prod} - energy consumption for material production and material formation , respectively.

With assumption that the life time of element is τ_e , the energy used for element function performance is ξ , than the total energy consumption for the element function is

$$E_t = E_p + \xi * \tau_e \tag{4}$$

The energy resource consumption for the total energy consumption of the respective element is

$$R_e = \frac{E_e}{r_e} \tag{5}$$

where
  r_e - specific energy produced by respective energy resource

The relative value of energy resource is

$$\rho_e = \frac{R_e}{R_{e,opt}} \tag{6}$$

Following this methodology , it can be concluded that for the total resource consumption indicator is composed of the individual indicators reflecting respective materials used in the design of element and energy resource indicator.

4.2 ENVIRONMENT CAPACITY INDICATOR

The second criterion for sustainability assessment is based on the environment capacity consumption with the use of respective element design.

The optimal energy consumption is obtained by Carnot cycle with respective environment capacity consumption. If we divide obtained environment capacity with its optimal value , we will obtain

$$\varepsilon = \frac{C}{C_{opt}} \tag{7}$$

where

C and C_{opt} - environment capacity and optimal value, respectively

ε - environment capacity efficiency

It can be defined for every design of the energy system element the optimal environmental capacity value which will include all potential pollutants. Since, the environment capacity efficiency is function of number of parameters reflecting specific design it of special interest to the determine the value which correspond to the optimal option for the specific design. This will represent the environment capacity indicator for the assessment of the specific design.

5. Demonstration Of Sustainability Criterion For Heat Exchanger Design

For many years in power industry condensers are considered as the heat transfer elements which are representing the heat exchanger family. There are three numerical procedure for the calculation of heat transfer and fluid flow in heat exchanger. The first stage in the design of condenser is heat rating and heat transfer surface determination including sizing of tubes diameter and length. The second stage implies evaluation of two-dimensional study of two phase flow in shell side of the condenser. The threat stage comprise detailed three dimensional calculation of the velocity and temperature fields in the shell and tube side of heat exchanger.

For this exercise the first stage of the procedure is used and with the aim to define respective heat transfer surface sizing and pressure drop determination needed to attain stable function of the heat exchanger.

For the demonstration purpose the power plant condenser is used with the total heat rating Q = 700 MWt. Vapor flow G= 959 t/h with pressure p= 3.34 kPa and respective saturation temperature T= 26.8 °C. Water inlet temperature t = 16 °C and outlet temperature t = 24.8°C. Water mass flow rate G = 55 934 t/h.

TABLE 1

R	d [m]	Material
R = 0	0.028	90 Cu +10 Cu-Ni
	0.026	
R= TEMA	0.024	Carbon Steel

Table 1 shows. options taken into a consideration, namely; clean condenser without fouling margin, half fouling resistance case and total fouling resistance value as proposed by TEMA procedure. Also, change in the tube diameter and tube material is evaluated.

6. Sustainability Assessment Of Condenser Design

As it is indicated in the Table 1, the cases taken in the analysis of the sustainability assessment comprise the effect of following parameters, namely: fouling margin effect, change in the tube diameter and the use of different material for the condenser tubing. The aim of this assessment is to verify changes of sustainability indicators and show its effect on the global contribution of resource use and future development.

The fouling is an immanent process in the heat exchangers. In particular, fouling in condenser requires special attention in the condenser design. It is commonly adapted in the design procedure for the heat transfer rating and heat transfer surface determination that the fouling margin is expressed in form of fouling resistance in heat transfer calculation, resulting in oversizing of heat transfer surface. This approach in condenser design will result In the excessive use of the natural resources and at the same time contribute to the excess use of the environment capacity.

The respective indicators measure the assessment of fouling effect on the sustainability criteria. For this purpose a following indicators are used:

6.1 CONSTRUCTION MATERIAL INDICATORS

By definition the construction material indicator comprise the ratio between the difference in the weight of the material needed for the respective design of the condenser with and without fouling margin. Heat transfer surface sizing method is used to determine the difference in condenser weight. Attention is focused only on the tube material. As shown in Table 1 three tubes diameters are taken into a consideration and cooper alloy and carbon steel tube material is used as the tube material. For each of these combinations the clean and fouled surface size is calculated.

The results are presented in the form of the possible change indicator, current change indicator and resource change indicator for construction material.

Table 2 shows the data obtained for the cooper alloy tube and carbon steel tube material.

TABLE 2

Cooper alloy

D	G [t]		Δ G [t]	Δ G/E [t/Mwe]	PCI	CCI	RI
	R = 0	R = TEMA					
0.028	189,5	256	66.5	0.221	0.350	0.5	0.175
0.026	188.5	248	59.5	0.198	0.315	0.5	0.157
0.024	185	262	77	0.256	0.416	0.5	0.208

Carbon Steel

D	G [t]		Δ G [t]	Δ G/E [t/MWe]	PCI	CCI	RI
	R = 0	R = TEMA					
0.028	183	251	68	0.226	0.370	0.5	0.180
0.026	180	226	46	0.153	0.250	0.5	0.125
0.024	177	231	64	0.213	0.359	0.5	0.180

6.2 ENERGY RESOURCE INDICATOR

This indicator is reflecting the effect of the fouling on the energy resource consumption for the respective design of condenser. There are two contributions to this indicator. Namely, energy used in the conversion of the natural resource to the tube material production. In this execs the coal is used as the energy resource. The second contribution takes into account the energy needed for coolant pumping during condenser operation with assumption of the 30 years lifetime. The energy resource indicator is expressed in form of possible change indicator, current change of indicator and energy resource indicator if coal is the fuel under consideration.

The Table 3 shows the data obtained for the cooper and carbon steel tubing resource indicator.

TABLE 3

Material	$\Delta G_{coal}[10^6$ t]	(PCI)	CCI	RI
Cooper Alloy	15	0.57×10^{-3}	0.5	0.285×10^{-3}
Carbon Steel	12	0.43×10^{-3}	0.5	0.215×10^{-3}

6.3 ENVIROMENT CAPACITY INDICATOR

In this analysis only CO contribution due to the fouling effect is used for the assessment of the environment capacity indicator. As in the energy resource indicator the contribution to the environment capacity indicator is divided in two parts. One, representing contribution due to the material production process and second, resulting from the life time energy need for the operation of the system. The environment capacity indicator is expressed in form of possible change indicator, current change indicator and resource indicator for CO_2.

Table 4 shows the data obtained for the environment indicators for cooper and carbon steel condenser tube material.

TABLE 4

Material	$\Delta G_{CO2}[10^6 t]$	$(PCI)_{En}$	$(CCI)_{En}$	$(RI)_{En}$
Carbon Steel	6.0	0.57×10^{-3}	0.5	0.265×10^{-3}
Cooper Alloy	4.8	0.43×10^{-3}	0.5	0.215×10^{-3}

7. Discussion

The exercise presented in this paper is aimed only to indicate possibility to use the respective sustainability indicators as the tool for the assessment of the condenser design. The fouling effect is used as a possible cause for the adverse effect in the condenser operation. Since the condenser is the power plant element, which substantially affect the total power plant efficiency, it is of interest to show potential assessment of the adverse effect to the sustainability in general. Also, the effect of the fouling in the condenser on the sustainability indicators is not time dependent, so it gives possibility to use the available data obtained by the standard design procedure method. It should be emphasized that these analyses do not include the time change of the global resource indicators.

It could be noticed that the tube diameter has small effect on the sustainability indicators, but the change in the tube material is more pronounced. In particular this will be dominant if the total global resource indicator will be used in this evaluation. Since, the global steel resources are for several order of magnitudes larger than the cooper alloy resources it is obvious that the change to carbon steel condenser design may prove to be less sensitive to the fouling effect on the sustainability indicators.

Also, it could be noticed that the energy resource indicator for the carbon steel shows advantage of this material and will contribute to the lower use of the energy resource. In particular, it can be shown that with the use of oil as the energy resource this may be even more pronounced. If in this analysis will used the scarcity forecast of oil resources; the carbon steel may for the future power plant design.

The environment capacity is usually measured by the CO_2 contribution to the global atmosphere. In this respect the assessment of fouling effect is not the best case to represent the use of environment capacity indicator. It may be shown that in total CO_2 production, the part resulting from the material production is preceding the design of condenser and part resulting from operation will be spread over the time period the fouling process will take place. This may lead us to the conclusion that the carbon steel is more favorable material for the power plant condenser tubing from the sustainability point of view.

8. Conclusions

As the conclusion of this lecture a following should be emphasized:

1. The sustainability has become a notion, which imply human society activities, which are aimed to meet the essential of its definition and philosophy. In particular, it was emphasized that in the history of our society the main guidelines have been recognized by the religious, academic and scholar authorities.

2. Instrumentalization of the sustainability concept is defined with the respective indicators. In this respect the resource indicators are defined reflecting the material, energy resources and environment capacity.

3. The sustainability concept is demonstrated on the design of power plant condenser. Using the effect of fouling degradation of condenser, it was demonstrated the possibility for the different concept design assessment by the respective sustainability indicators.

4. The different condenser designs have been used to show the global effect of the potential influence of the new concept on the sustainability indicators. The sophisticated and intelligent design can be used in the future strategy of the energy systems.

References

1. The Earth Chapter : A Contribution Toward its Realization, Franciscan Center of Environment Studies, Roma,1995
2. C.S. Jenkinson, The Quality of Thomas Jefferson's Soul, White House Library
3. Report of The United Nation Conference on Environment and Development, Vol.1,Chapter 7,June,1992
4. Agenda 21, Chapter 35, Science for Sustainable Development, United Nations Conference on Environment and Development,1992
5. Declaration of the Council of Academies of Engineering and Technological Sciences
6. Norton G.B., A New Paradigm for Environment Management, Ecosystem Health : New Goals for Environment Management, Island Press, Washington DC, pp.23-41, 1992
7. Holdren J.P., Global Environmental Issues Related to Energy Supply: The Environmental Case for Increased Efficiency of Energy Use, Energy Conservation, Ed: Socolow R.H., Ross M., pp.975-992,1987
8. Curzio A.Q.,Zobali R., Cost of Sustainability , 50th Anniversary of UN , Rome, May,1996
9. Man and His Environment : Tropical Forest and the Conservation of Species, Pontifical Academia Scientiarum ,Documenta 26
10. Domingos J.J.D. O Desolvomento Sustentavel, Ingenium, Revista da Ordem dos Engenheiros, Julho/Agosto,1994,pp.95
11. Sustainable Concept as Applied within a Countries (Norway),1995
12. Marchetti C. , The " Historical Instant " of Fossil Fuel , Symptom of Sick World, Int. Journal Hydrogen Energy, 16, pp.563-575, 1991
13. Arnold M.St.J., Kersall G.J.,Nelson D.M., Clean Efficient Electric Generation for the Next Century: British Coal Toping Cycle, Combustion Technology for Clean Environment, Ed: M.G.Carvalho, W.A.Fineland, F.C. Lockwood, ,Ch. Papaloupolos
14. Mazzuracchio P,Raggi A, Barbiri G., New Method for Assessment the Global Quality of Energy System, Applied Energy 53(1996),pp.315-324
15. Noel D,A Recommendation of Effect of Energy Scarcity on Economic Growth, Energy, Vol.2,p 1-12,1995
16. Farinelli U. ,Alternative Energy Sources for the Third World : Perspective, Barriers, Opportunity, Pontifical Academy of Science, Plenary Session, Oct.25-29,1994
17. Medows D. Meadows H.,Randers D.L, Behrens J., , The Limits of Growth, Universe Book, 1972,New York
18. Keatiny M. Agenda for Change, Center for Our Common Future,1993
19. Barnett H.J.,Morse Ch., Scarcity and Growth, Resources for Future,Inc.,1963
20. Noel Uri, A Reconsideration of Effects of Energy Scarcity on Economic Growth, Energy, Vol.20, No1,pp.1-12.
21. Hought R.A. ,Woodwell G.M., Global Climatic Change, Scientific American, April issue, 1989
22. Wu K.,Li B., Energy Development in China, Energy Policy Vol.23,No.2,pp.167-178.
23. Painuly J.P.,Rao N.,Parickh J.,A Rural Energy-Agriculture Interaction Model Applied to Karnatak State, Energy ,Vol .20,No.3,pp.219-233
24. Hollander J.M.,Schnaider T.R., Energy Efficiency: Issues for the Decades, Energy,Vol.24,No.4, pp.273-287.

EXTENDED SURFACE HEAT TRANSFER IN HEAT EXCHANGERS AND PERFORMANCE MEASUREMENTS

P.J. HEGGS
Department of Chemical Engineering
UMIST, P.O. Box 88,
Manchester, M60 1QD, England.

Abstract. This discourse discusses the use of extended surface heat transfer in heat exchangers. The size and cost of an exchanger is dependent upon the heat duty, the allowable pressure drop, and the geometry of the heat transfer area. Often one heat transfer resistance within the overall heat transfer dominates and this determines the amount of area required for the duty. The dominance of this resistance is often broken by the use of extended surfaces (fins), which will result in a lower area and hence smaller exchanger. The extended surface performance can be predicted by a number of indicators: fin effectiveness, fin efficiency, enhancement and augmentation. This discourse will consider which, if any, of the above performance indicators are the best.

1. Introduction

Conservation of energy and extended surface heat transfer are inextricably linked. Two stream heat exchangers, Figure 1, for energy conservation implies transferring the maximum possible heat (energy/enthalpy) between the streams, whilst minimising the pressure drop of each stream, so that excessive pumping power is not consumed. The maximum heat that can be transferred between the two streams is directly related to the two inlet temperatures, T_{hl} and T_{cl} , and the flowing heat capacities of each stream, $\left(\dot{M}c_p\right)_h$ and $\left(\dot{M}c_p\right)_c$. The maximum heat that can be transferred is given by the following expression:

$$\dot{Q}_{max} = \left(\dot{M}c_p\right)_{min}\left(T_{hl} - T_{cl}\right) \tag{1}$$

where

$$\left(\dot{M}c_p\right)_{min} \text{ is the smaller of } \left(\dot{M}c_p\right)_h \text{ and } \left(\dot{M}c_p\right)_c \tag{2}$$

This is a hypothetical value, but its value is always worth knowing, because it is the upper limit of the heat transfer between the two streams. This hypothetical value would be obtained, when the outlet temperature of the $\left(\dot{M}c_p\right)_{min}$ stream is equal to the inlet

S. Kakaç et al. (eds.), Heat Transfer Enhancement of Heat Exchangers, 49–65.

temperature of the other stream and the heat exchanger must to be a 1:1 countercurrent configuration of infinite size.

More often than not, designers use the concept of logarithmic mean temperature difference in order to evaluate the actual heat transfer between the two streams and this requires the knowledge of both outlet temperatures for the enthalpy balances and the rate equation:

$$\dot{Q}_{act} = \left(\dot{M}\overline{c}_p\right)_h\left(T_{h1} - T_{h2}\right) \tag{3}$$

$$\dot{Q}_{act} = \left(\dot{M}\overline{c}_p\right)_c\left(T_{c2} - T_{c1}\right) \tag{4}$$

$$\dot{Q}_{act} = UA\,LMTD\,F_T \tag{5}$$

and
$$LMTD = \left(\Delta T_h - \Delta T_c\right)/\ln\left(\Delta T_h/\Delta T_c\right) \tag{6}$$

and F_T is a correction factor, which depends upon the exchanger configuration and flow arrangement, and accommodates the deviation of the mean temperature driving force in the exchanger to that occurring in a truly 1:1 counter-current exchanger[1]. The use of the equations (3) to (6) without knowledge of equation (1) does not provide the information with respect to how efficient is the exchange of heat between the two process streams. All these equations are dimensional and only relate to the amount of

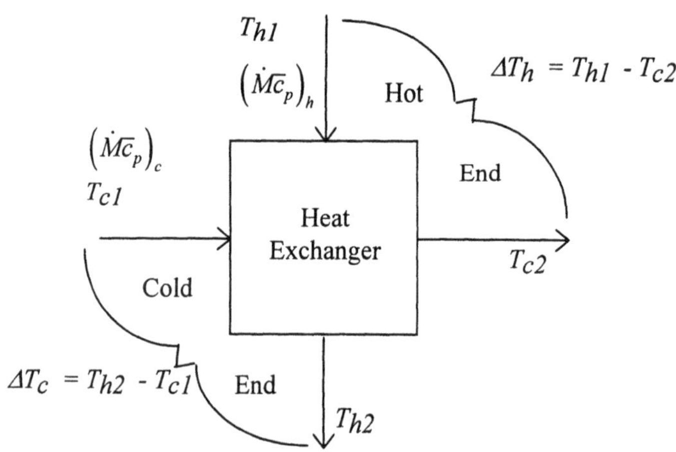

Figure 1. Schematic diagram of a two stream heat exchanger.

the heat duty, $\dot{Q}_{act}$. Any design and/or performance calculation should provide information relating to the efficiency of the heat duty with respect to the maximum possible heat duty between the two streams.

This information is provided automatically, if the thermal effectiveness - number of transfer units, $\varepsilon - N_{tu}$, methodology is used[1]. The definition of the thermal effectiveness is simply the ratio of either the equation (3) or (4) to equation (1):

$$\varepsilon = \dot{Q}_{act}/\dot{Q}_{max} \tag{7}$$

and the number of transfer units is defined as:

$$N_{tu} = UA\big/\left(\dot{M}\overline{c}_p\right)_{min} \tag{8}$$

It is also necessary to know the ratio of the two flowing heat capacities of the two streams;

$$C^* = \left(\dot{M}\overline{c}_p\right)_{min}\big/\left(\dot{M}\overline{c}_p\right)_{max} \tag{9}$$

and $\left(\dot{M}\overline{c}_p\right)_{max}$ is the larger of $\left(\dot{M}\overline{c}_p\right)_h$ and $\left(\dot{M}\overline{c}_p\right)_c$ $\tag{10}$

and the exchanger configuration and flow arrangement in order to proceed with calculations.

These three parameters: ε, N_{tu} and C^* are dimensionless and are bounded as:

$$0 \leq \varepsilon = \dot{Q}_{act} / \dot{Q}_{max} \leq 1.0 \tag{11}$$

$$0 \leq N_{tu} = UA / \left(\dot{M}\overline{c}_p\right)_{min} \leq \infty \tag{12}$$

and $$0 \leq C^* = \left(\dot{M}\overline{c}_p\right)_{min} / \left(\dot{M}\overline{c}_p\right)_{max} \leq 1.0 \tag{13}$$

Hence the conservation of energy within an exchanger occurs when the thermal effectiveness approaches unity. When the effectiveness is equal to unity, we have the hypothetical system, which was mentioned earlier, when there will be a zero temperature driving force at one end of the exchanger. This corresponds to a 1:1 countercurrent configuration of infinite size, such that the value of the number of transfer units is infinity.

Each exchanger configuration and flow arrangement has a unique relationship between the thermal effectiveness, the number of transfer units and the heat capacity ratio:-

$$\varepsilon = fn\{N_{tu}, C^*, configuration, flow\ arrangement\} \tag{14}$$

and these relationships can be transposed to give the value of UA in terms of the minimum heat capacity flowrate, the effectiveness and the heat capacity ratio :-

$$UA = \left(\dot{M}\overline{c}_p\right)_{min} \times fn\{\varepsilon, C^*, configuration, flow\ arrangement\} \tag{15}$$

Thus the value of UA depends upon the heat duty - through the value of the effectiveness, the ratio of the heat capacity flowrates and the stream with the smaller heat capacity flowrate. The physical size of the exchanger is directly related to the smaller heat capacity flowrate and how the design of a configuration to accommodate the value of UA.

It is important to note that each configuration and flow arrangement has a value of thermal effectiveness that cannot be bettered - ε_{max}. The first three exchangers and flow arrangements in Table 1 have values of ε_{max} equal to unity for all values of the heat capacity flowrate ratio. All other types have ε_{max} relationships dependent upon the heat capacity flowrate ratio - see in Table 1 for 1:1 co-current and the 1:2 configuration with both streams well -mixed. The value of ε_{max} for a particular exchanger and flow arrangement corresponds to a zero temperature difference driving force at one end of the exchanger - see Figure 1. These values are the thermodynamic limits for transfer of heat between the two streams in the single exchanger, often referred to as a single shell system. For values of the thermal effectiveness that exceed

ε_{max} for a particular exchanger and flow arrangement, then to fulfil the duty requires a train of exchangers arranged in series counterflow.

The most comprehensive set of $\varepsilon - N_{tu} - C^*$ relationships in analytic, algebraic and graphical forms have been published by ESDU[2].

TABLE 1.　ε_{max} values for some configurations and flow arrangements

TYPE	ε_{max}
1:1　Counterflow HX	1.0 for all values of C^*
1:1　Crossflow, both unmixed	1.0 for all values of C^*
1:1　Regenerators	1.0 for all values of C^*
All other types have ε_{max} relationships dependent upon the value of C^*	
1:1　Co-current HX	$1/(1 + C^*)$
1:2　Both streams mixed TEMA E Shell	$2\left/\left(1 + C^* + \sqrt{1 + C^{*2}}\right)\right.$ At $C^* = 1$　$\varepsilon_{max} = 0.586$

An important parameter for considering heat recovery between two process streams is the ineffectiveness of the heat duty, $(1-\varepsilon)$, which is the amount of heat that is not exchanged between the two streams. This is directly linked to the minimum temperature difference driving force, ΔT_{min} between the two streams and the difference between the two inlet temperatures[3]:-

$$1 - \varepsilon = \Delta T_{min} / \left(T_{h1} - T_{c1}\right) \qquad (16)$$

or

$$\varepsilon = 1 - \Delta T_{min} / \left(T_{h1} - T_{c1}\right) \qquad (17)$$

The minimum temperature ΔT_{min} is the smaller of ΔT_h and ΔT_c in Figure 1.

Pinch technology [4] is directly related to the value of ΔT_{min} for the total process and the values of the maximum energy recovery(*MER*) and the minimum energy requirements(*MER*). This is energy targetting, that is conservation by maximising the energy recovery between the process streams in the total system and the minimum use of utility sources. The User Guide on Process Integration in 1987 [5] recommended the following values of ΔT_{min} : "50°C for boiler house situations, 20°C for heavy chemicals industry and 5°C for refrigeration". Polley [6] has recommended the following values for particular exchangers: shell-and-tube 10°C, plate and frame 2°C and plate-fin 1°C respectively. All these values are very useful, but the value of thermal effectiveness for each enthalpy match in the network is not known until the matching design has been completed, that is the values of T_{h1} and T_{c1}, and C^* are known. The specification of small values of ΔT_{min} in process integration means large energy recovery, low energy requirements and highly efficient heat exchangers, because small values of ineffectiveness are obtained and, therefore, high thermal effectiveness. To accommodate the enthalpy matches in single shell (unit) exchangers requires configurations which have sufficient surface area (*A*) and geometrical features which result in the required UA – very often this requires the use of heat transfer augmentation.

2.   Heat Transfer Augmentation by Passive Devices – Extended Surfaces

Irrespective of which design methodology is employed, the designer must find the exchanger configuration and flow arrangement for a specified duty between two process streams, see eqn (13) for the ε-N_{tu} method and eqn (4) for the $LMTD$ technique. The reciprocal of the product UA is the total resistance to heat flow between the two streams and normally comprises five individual resistances as follows:-

$$\frac{1}{UA} = \frac{1}{(\alpha A)_h} + \frac{1}{(\alpha_{fo} A)_h} + R_w + \frac{1}{(\alpha_{fo} A)_c} + \frac{1}{(\alpha A)_c} \tag{18}$$

where α and α_{fo} are convective and fouling heat transfer coefficients respectively, and R_w is the conductive resistance of the wall between the two process streams. If the resistances on one side of the wall are much greater (at least an order of magnitude) than the summation of the remaining resistances, then it is worthwhile considering some form of augmentation on that side. This is often referred to the situation of controlling resistances. So if the hot side resistances are controlling, ie

$$\frac{1}{(\alpha A)_h} + \frac{1}{(\alpha_{fo} A)_h} \quad \gg \quad R_w + \frac{1}{(\alpha_{fo} A)_c} + \frac{1}{(\alpha A)_c} \tag{19}$$

then it is worth considering augmentation on the hot side. Note that the hot fouling coefficient has been used in eqn (19). If the process streams are non-fouling, then eqn (19) becomes

$$\frac{1}{(\alpha A)_h} \quad \gg \quad R_w + \frac{1}{(\alpha A)_c} \tag{20}$$

The above two expressions should be considered irrespective of the type of augmentation being considered – active or passive. This discourse will focus specifically on extended surfaces - passive augmentation.

3. Extended Surface (Fin) Heat Transfer

Extended surfaces come in many forms: clusters of cylindrical, rectangular, conical, concave parabolic and convex parabolic single fins attached to a planar wall; arrays of continuous rectangular, triangular, concave and parabolic fins on a planar wall; longitudinal rectangular fins on the outside of a tube arranged parallel to the axis of the tube; radial rectangular, triangular and hyperbolic fins on the outside of tubes; longitudinal fins on the inside of tubes; micro-sized fins on the inside of tubes, and various types of strip fins used in plate-fin exchangers. All the area of the extended surface is not fully utilised in the transfer of heat, because the temperature driving force from the fin surface to the surrounding fluid falls from the base of the fin along its length. If the fin is extremely long, then the temperature of the tip of the fin will become equal to that of the surrounding fluid. This is no good for heat transfer, but is perfect for measuring the temperature of a fluid by a thermocouple or thermometer. The evaluation of the temperature profile along a fin is normally obtained by solving a

54

second order ordinary differential equation subject to boundary conditions, which make it a boundary value problem. This equation is obtained after making a series of assumptions often referred to as the classical assumptions of Harper and Brown in 1922 [7]: material thermal conductivity is invariant, the film heat transfer coefficient is uniform, the heat flow is one-dimensional along the axis of the fin and perfect contact exists between the fin and the supporting primary surface. Coupled to these assumptions are that the fluid temperature is constant and the fin base temperature is a known value.

The general equation for the temperature distribution along a fin is

$$A_x \frac{d^2 T_f}{dn^2} + \frac{dA_x}{dn} \frac{dT_f}{dn} - \frac{\alpha A_s}{\lambda}\left(T_f - T_\infty\right) = 0 \tag{21}$$

where n is the direction normal to the surface, A_x is the cross-sectional area of the fin and A_S is the surface area per unit length of the fin. The boundary conditions are

at
$$n = 0, \quad T_f = T_B \tag{22}$$

and at
$$n = N, \quad -\lambda_f \frac{dT_f}{dn} = \alpha\left(T_f - T_\infty\right) \tag{23}$$

If the shape of the fin profile is such that the tip is a point, that is triangular, parabolic or hyperbolic, then the tip condition becomes

at
$$n = N, \quad T_f \neq T_\infty \tag{24}$$

If the fin is extremely long, then the tip condition becomes

at
$$n = \infty, \quad T_f = T_\infty \tag{25}$$

For long fins, it is often assumed that the fin tip area is small compared to the surface area of the fin, and the tip boundary condition is taken to be adiabatic, that is

$$\tag{26}$$
at
$$n = N, \quad \frac{dT_f}{dn} = 0$$

The original work of Harper and Brown [7] considered heat flow through a rectangular fin attached to a planar primary surface, see Figure 2. Let us revisit this problem making use of the generalised fin eqn (21) and appropriate boundary conditions. The fin is of thickness, $2t$, height, ℓ, and is relatively wide, so that the two end face areas are negligible. Hence the cross-sectional area of the fin profile is independent of fin length and is $2t$ x unit length, and the fin surface area per unit fin length is 2 x unit length, so that eqn (26) becomes

$$\frac{d^2 T_f}{dx^2} - \frac{\alpha}{\lambda_f t}\left(T_f - T_\infty\right) = 0 \tag{27}$$

with boundary conditions:

at
$$x = 0, \quad T_f = T_B \tag{28}$$

and at
$$x = \ell, \quad -\lambda_f \frac{dT_f}{dx} = \alpha\left(T_f - T_\infty\right) \tag{29}$$

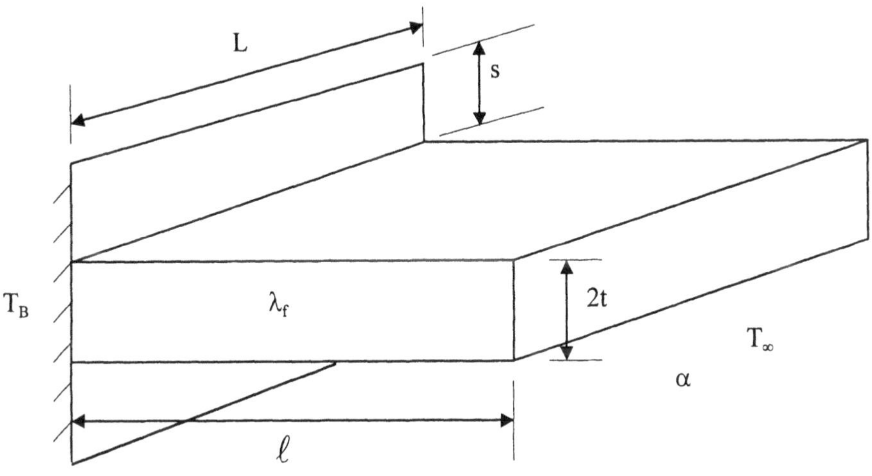

Figure 2. An isolated rectangular fin on a planar wall

The set of eqns (27) to (29) are made dimensionless by the introduction of the following normalised temperature and dimensionless fin length:-

$$\theta_f = \left(T_f - T_\infty\right)/\left(T_B - T_\infty\right) \tag{30}$$

$$y = x\sqrt{\frac{\alpha}{\lambda_f t}} \tag{31}$$

so that the set of eqns (27) to (29) become

$$\frac{d^2\theta_f}{dy^2} - \theta_f = 0 \tag{32}$$

at
$$y = 0, \quad \theta = 1 \tag{33}$$

and at
$$y = \ell\sqrt{\frac{\alpha}{\lambda_f t}}, \quad -\sqrt{\frac{\lambda_f}{\alpha t}}\frac{d\theta_f}{dy} = \theta_f \tag{34}$$

If the fin is very long, then the last boundary condition becomes

at
$$y = \infty, \quad \theta_f = 0 \tag{35}$$

The dimensionless temperature profile along the fin is given by the solution of the eqns (32) to (34) and is as follows:

$$\theta_{fa} = \cosh(y) - \sinh(y)\frac{\left[1 + \sqrt{\dfrac{\lambda_f}{\alpha t}}\,\tanh\left(\ell\sqrt{\dfrac{\alpha}{\lambda_f t}}\right)\right]}{\left[\sqrt{\dfrac{\lambda_f}{\alpha t}} + \tanh\left(\ell\sqrt{\dfrac{\alpha}{\lambda_f t}}\right)\right]} \tag{36}$$

and the temperature gradient at any point along the fin is

56

$$\frac{d\theta_{fa}}{dy} = sinh(y) - cosh(y) \frac{\left[1 + \sqrt{\frac{\lambda_f}{\alpha t}} \, tanh\left(\ell \sqrt{\frac{\alpha}{\lambda_f t}}\right)\right]}{\left[\sqrt{\frac{\lambda_f}{\alpha t}} + tanh\left(\ell \sqrt{\frac{\alpha}{\lambda_f t}}\right)\right]} \tag{37}$$

If the fin is extremely long, so that the tip boundary condition is eqn (35), then the temperature profile is given by

$$\theta_{fb} = exp(-y) \tag{38}$$

and the gradient is

$$\frac{d\theta_{fb}}{dy} = -exp(-y) \tag{39}$$

For long fins, where the tip area is small compared to the surface area of the fin, so that the tip boundary condition is eqn (26), the temperature profile is given by

$$\theta_{fc} = cosh(y) - sinh(y) \, tanh\left(\ell \sqrt{\frac{\alpha}{\lambda_f t}}\right) \tag{40}$$

and the gradient is

$$\frac{d\theta_{fc}}{dy} = sinh(y) - cosh(y) \, tanh\left(\ell \sqrt{\frac{\alpha}{\lambda_f t}}\right) \tag{41}$$

4. Performance Indicators

The heat flow from an isolated fin to the surrounding fluid is given by the flow of heat across the base of the fin and is given by the following expression:

$$\dot{Q}_f = -\lambda_f 2t \frac{dT_f}{dx}\bigg|_{x=0} = -2\sqrt{\alpha \lambda_f t}\,(T_B - T_\infty) \frac{d\theta_f}{dy}\bigg|_{y=0} \tag{42}$$

4.1 FIN EFFECTIVENESS, ζ_f

The fin effectiveness[8,9,10] is defined as the ratio of the heat flowing through the fin to that which would flow if the fin was not attached to the primary heat transfer surface:

$$\zeta_f = \dot{Q}_f / \alpha 2t(T_B - T_\infty) \tag{43}$$

For a fin of finite length, the gradient is given by eqn (37) with $y=0$ and the fin effectiveness is

$$\zeta_{fa} = \sqrt{\frac{\lambda_f}{\alpha t}} \sqrt{\frac{\left[1 + \sqrt{\frac{\lambda_f}{\alpha t}}\, tanh\left(\ell\sqrt{\frac{\alpha}{\lambda_f t}}\right)\right]}{\left[\sqrt{\frac{\lambda_f}{\alpha t}} + tanh\left(\ell\sqrt{\frac{\alpha}{\lambda_f t}}\right)\right]}} \tag{44}$$

Whereas for a fin of infinite length, the gradient is given by eqn (39) with $y=0$, so that the effectiveness becomes

$$\zeta_{fb} = \sqrt{\frac{\lambda_f}{\alpha t}} \tag{45}$$

If the fin is very long, so that the tip is adiabatic and the gradient is given by eqn (41) with $y=0$, then the effectiveness is

$$\zeta_{fc} = \sqrt{\frac{\lambda_f}{\alpha t}}\, tanh\left(\ell\sqrt{\frac{\alpha}{\lambda_f t}}\right) \tag{46}$$

The effectiveness for the fin of infinite length is the maximum possible and will be called **the maximum effectiveness, ζ_{max}**. Both the other expressions for the effectiveness eqns (44) and (46) become equal to the maximum effectiveness in the limit as the length of the fin tends to infinity. Hence the fin effectiveness is a function of two dimensionless groups: the maximum effectiveness eqn (45) and the aspect ratio, ℓ / t, because the argument for the hyperbolic terms can be rearranged as

$$\ell\sqrt{\frac{\alpha}{\lambda_f t}} = \frac{\ell}{t}\bigg/\sqrt{\frac{\lambda_f}{\alpha t}} \tag{47}$$

4.2 FIN EFFICIENCY, η_f

The commonly used fin efficiency was originally called the effectiveness by Harper and Brown[7], but was later called the efficiency by Gardner[11] and is the ratio of the heat flow through the fin to the heat flow if all the surface of the fin was at the base temperature:

$$\eta_f = \dot{Q}_f / 2\alpha(\ell + t)(T_B - T_\infty) \tag{48}$$

For a fin of finite length, the fin efficiency is obtained by using the gradient eqn (37) with $y=0$ in eqn (42) and substituting in the above eqn (48) to give

$$\eta_{fa} = \frac{t}{(\ell + t)}\sqrt{\frac{\lambda_f}{\alpha t}} \frac{\left[1 + \sqrt{\frac{\lambda_f}{\alpha t}}\, tanh\left(\ell\sqrt{\frac{\alpha}{\lambda_f t}}\right)\right]}{\left[\sqrt{\frac{\lambda_f}{\alpha t}} + tanh\left(\ell\sqrt{\frac{\alpha}{\lambda_f t}}\right)\right]} = \frac{t}{(\ell + t)}\zeta_{fa} \tag{49}$$

However, for a fin of infinite length the fin efficiency is zero. For a fin which is very long, then the fin efficiency is obtained from eqn (41) with $y=0$ in eqns (42) and (48) and is

$$\eta_{fc} = \frac{tanh\left(\ell\sqrt{\dfrac{\alpha}{\lambda_f t}}\right)}{\ell\sqrt{\dfrac{\alpha}{\lambda_f t}}} = \frac{t}{\ell}\zeta_{fc} \tag{50}$$

The fin efficiency, eqns (49) and (50), is simply the fin effectiveness divided by an appropriate aspect ratio of the base cross-sectional area to the surface area of the fin depending upon the boundary condition used at the tip of the fin. The fin efficiency given by eqn (50) is only a function of one dimensionless group, $\left(\ell\sqrt{\dfrac{\alpha}{\lambda_f t}}\right)$, and so has become very popular due to this particular property. It must always be noted when using this particular form of the efficiency, that there is no heat flow from the tip of the fin. It has been recommended that this form of the efficiency can be used to represent the more correct definition of the efficiency, eqn (49), by redefining the fin length in eqn (50) by $\ell' = \ell + t$.

4.3 ENHANCEMENT FACTOR, ε_f, OR SURFACE AUGMENTATION, Aug_s

Manzoor et al. [12] introduced an enhancement factor to account for the reciprocal of the total resistance on the fin side of an assembly to the resistance as if the extended surface was not present. This was an attempt to include the pitch of the fin into the consideration of the flow of heat from the primary surface with fins attached to it. This approach came from the analytical solution of the rigorous mathematical model of the one-dimensional representation of heat flow through an assembly with fins attached to one side of the primary surface [13]. The heat flow from the isolated fin and associated primary surface, which is the fin spacing, is given by

$$\dot{Q}_s = \left[2\alpha(p-t) - 2\sqrt{\alpha\lambda_f t}\,\frac{d\theta_f}{dy}\bigg|_{y=0}\right](T_B - T_\infty) \tag{51}$$

The enhancement factor was defined as

$$\varepsilon_f = 2\alpha p(T_B - T_\infty)/\dot{Q}_s \tag{52}$$

This enhancement factor can be readily represented in terms of either the effectiveness or the efficiency, and is given by the following formulae:

$$\varepsilon_{fa} = 1/\left[1 + \frac{t}{p}(\zeta_{fa} - 1)\right] = 1/\left[1 - \frac{t}{p} + \frac{(\ell+t)}{p}\eta_{fa}\right] \tag{53}$$

In hindsight it would have been more prudent to have introduced the reciprocal of the enhancement factor and have called it the surface augmentation factor, which is the ratio of the heat flow across the extended side of the assembly to that heat flowing across the bare surface.

$$Aug_s = \dot{Q}_s / 2\alpha p(T_B - T_\infty) = 1/\varepsilon_f \tag{54}$$

This surface augmentation factor accounts for the fins and the surface area not covered by the fins and is represented in terms of the fin effectiveness or efficiency as follows:

$$Aug_{sa} = 1 + \frac{t}{p}\left(\zeta_{fa} - 1\right) = 1 - \frac{t}{p} + \frac{(\ell + t)}{p}\eta_{fa} \tag{55}$$

This surface augmentation can now be used to give the heat flow on the fin side of the assembly by the following expression:

$$\dot{Q}_s = \alpha A_p Aug_{sa}\left(T_B - T_\infty\right) \tag{56}$$

Thus it is now possible to obtain the resistance of the extended side of the assembly from eqn (56):

$$R = \frac{1}{\alpha A_p Aug_{sa}} \tag{57}$$

The product $\alpha \times Aug_{sa}$ is identical to the weighted surface heat transfer coefficient, which is often quoted in the design of extended surface heat exchangers [15] and A_p is the area of the primary surface.

4.4 ASSEMBLY AUGMENTATION FACTOR, AUG_{ass}

The three performance factors: effectiveness, efficiency and surface augmentation do not provide any indication of the advantages to be obtained by using the extended surface assembly for the duty required between two process streams. The duty of an exchanger is given by

$$\dot{Q} = UA\Delta T_m \tag{58}$$

where the mean temperature difference over the exchanger, ΔT_m, is a function of the inlet and outlet temperatures of the two process streams, and the exchanger configuration and the flow arrangement within the exchanger. Hence if the exchanger configuration and the flow arrangement is kept the same and only the heat transfer surfaces are different: a surface without fins and one comprising a fin assembly, then the products of the overall heat transfer coefficient and transfer area are identical for both situations:

$$\dot{Q} / \Delta T_m = (UA)_p = (UA)_{ass} \tag{59}$$

the assembly augmentation factor is the ratio of the overall coefficient for the finned assembly to the overall coefficient for the plain primary surface

$$AUG_{ass} = \frac{U_{ass}}{U_p} = \frac{A_p}{A_{ass}} \tag{60}$$

This assembly augmentation factor immediately gives the magnitude of the size of reduction in the exchanger area and size, when an extended surface assembly is used instead of a plain primary surface. The original definition of the augmentation factor[14] was the ratio of the heat flow through the fin assembly (see Figure 3) to the heat flow across the plain primary surface and this is identical to eqn (60), but now has been developed from the rate equation for heat exchangers eqn (58).

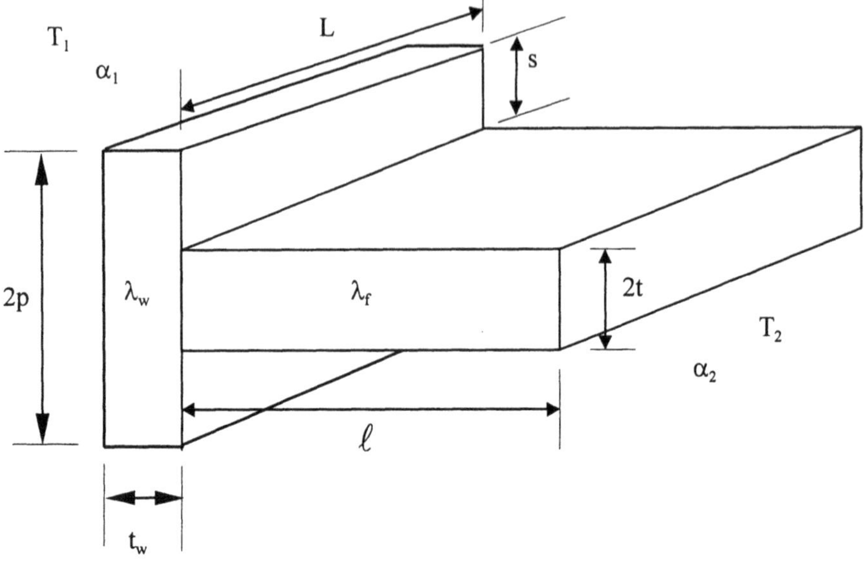

Figure 3. An extended surface assembly: planar wall with a rectangular fin

If the assembly has fins on only one side of the primary surface, Figure 3, then the assembly augmentation is given by

$$AUG_{ass} = \frac{\dfrac{1}{\alpha_1} + R_w + \dfrac{1}{\alpha_2}}{\dfrac{1}{\alpha_1} + R_w + \dfrac{1}{(\alpha Aug_{sa})_2}} \tag{61}$$

and if the assembly has fins on both sides of the primary surface, then the assembly augmentation is given by

$$AUG_{ass} = \frac{\dfrac{1}{\alpha_1} + R_w + \dfrac{1}{\alpha_2}}{\dfrac{1}{(\alpha Aug_{sa})_1} + R_w + \dfrac{1}{(\alpha Aug_{sa})_2}} \tag{62}$$

In a recent article, Wood et al. [16] discussed the various performance indicators after first setting up the one-dimensional mathematical model of the fin assembly with fins on one side of the primary surface. An analysis of the solution of the complete system was used to investigate the effects of the various parameters on the response of the performance indicators to the response to the heat flow through the assembly. The performance indicator change should correspond to the change in the heat flow. Unfortunately not all the indicators did so.

Table 2 lists the changes in assembly heat flow, fin effectiveness, fin efficiency, surface enhancement and augmentation, and finally, assembly augmentation for increases in the system parameters: fin length, thickness, spacing (with the pitch held constant), thermal conductivity and heat transfer coefficient, wall thickness and conductivity, and lastly, the coefficient on the unfinned side of the assembly. Only the assembly augmentation factor, AUG_{ass}, changes in a similar way to the overall heat flow for all the system parameters, except for the fin heat transfer coefficient. For this latter parameter, the assembly augmentation factor falls whereas the heat flow becomes larger. However this trend of the augmentation factor is correct, because it is a measure of the efficacy of the extended surface assembly with respect to a plain primary surface. If the heat transfer coefficient on the fin side increases, then the controlling resistance of the system becomes less dominant and this does not favour the use of the fins.

The fin effectiveness and efficiency cannot reflect changes in the system parameters unless they are related to the isolated fin, and the enhancement factor and surface augmentation do not alter with respect to the wall parameters and the unfinned side heat transfer coefficient. All these performance indicators are needed for the evaluation of the assembly augmentation factor and this is best performance indicator for the representation of heat flow through an extended surface assembly.

Table 2 Performance indicator response to increases in the magnitude of system parameters

Parameter	ℓ	t	s	λ_f	α_f	t_w	λ_w	α_{uf}
Q_{ass}	↑	↑	↓	↑	↑	↓	↑	↑
ζ_f	↑	↓	-	↑	↓	-	-	-
η_f	↓	↑	-	↑	↓	-	-	-
ε_f	↓	↓	↑	↓	↑	-	-	-
Aug_s	↑	↑	↓	↑	↓	-	-	-
AUG_{ass}	↑	↑	↓	↑	↓	↓	↑	↑

5. Application of the use of the performance indicators

The following problem is taken from the text book - Principles of Heat Transfer by F. Kreith, 2-52 on page 78 [17], however the units have been changed to SI:-

Heat is transferred from water through a brass wall (λ=78 W/m^2(K/m)). The addition of rectangular brass fins: 7.62×10^{-4} m thick ($2t$), 0.0254 m long (ℓ) and spaced 0.0125 m apart ($2s$), is contemplated. Assuming a water-side heat transfer coefficient of 170 W/m^{2}K and an air-side coefficient of 17 W/m^{2}K, compare the gain in heat transfer rate achieved by adding fins to (a) the water-side, (b) the air-side and (c) both sides. Neglect the temperature drop across the brass wall.

Solution:
The following dimensions and aspect ratios apply to this problem as follows: $t = 0.000381$ m, $p\ (= t+s) = 0.006631$ m, $\ell = 0.0254$ m, $\ell / t = 66.667$, $(\ell + t)/t = 67.667$, $t / \ell = 0.015$, $t / p = 0.05746$ and $(\ell + t)/p = 3.8880$.

Table 3 Evaluation of the fin effectiveness, efficiency and surface augmentation

Variable	Water-side	Air-side
$\zeta_{max}=\sqrt{\dfrac{\lambda_f}{\alpha t}}$ eqn (45)	34.70	109.74
$\dfrac{\ell}{t}\Big/\sqrt{\dfrac{\lambda_f}{\alpha t}}$ eqn (47)	1.9212	0.6075
ζ_{fa} eqn(44)	33.32	60.22
ζ_{fc} eqn(45)	33.24	59.52
η_{fa} eqn(49)	0.4925	0.8900
η_{fc} eqn(50)	0.4986	0.8927
Aug_{sa} eqn(52)	2.8572	4.4027

Table 4 Evaluation of the assembly augmentation for the various arrangements

Surface type	R_1	R_2	R_T	U	AUG_{ass}
-	m²K/W	m²K/W	m²K/W	W/m²K	-
Plain	0.05882	0.005882	0.06472	15.45	1.000
Fins water-side	0.05882	0.002059	0.06088	16.42	1.063
Fins air-side	0.01335	0.005882	0.01923	52.00	3.365
Fins both sides	0.01335	0.002059	0.01541	64.91	4.200

Table 3 contains the values of the maximum effectiveness, the argument for the hyperbolic functions in the various equations for the evaluation of the fin effectiveness and efficiency, and ultimately the surface augmentation for the water-side and air-side addition of the fins. The addition of fins to the air-side results in much larger values of the maximum effectiveness, fin effectiveness and efficiency, and surface augmentation. However, the fin effectiveness for the water-side is much closer to the maximum effectiveness for that side, so that longer fins on the water side would not provide much more heat flow. For the air-side, there would be advantages in using longer fins, because the fin effectiveness is considerably less than the maximum effectiveness. Also listed are the values of effectiveness and efficiency for the adiabatic fin tip boundary condition. These effectiveness values are lower than the respective values for the diabatic tip boundary condition, whereas the efficiency values show the opposite trend. The adiabatic tip condition values will under estimate the heat flow.

Table 4 lists the various resistances for the assembly configurations posed in the original problem and the corresponding overall heat transfer coefficients and assembly augmentation factors. For the plain primary surface, the resistance on the air-side is the controlling value and so the addition of fins to the water-side would not be recommended. This is borne out in the calculations: the air-side resistance becomes more dominant, but there is an increase in heat flow of 6.3% or the area could be reduced by 1.063. The addition of fins to the air-side dramatically increases the heat flow by 236.5% or the area could be reduced by 3.365. The addition of fins to the air-side has reduced the resistance of that side in the overall problem. Thus the addition of fins to both sides results an increase in heat flow of 320% or a reduction in area of 4.20.

This is an excellent problem for illustrating the effects of adding fins to a primary surface and how to make use of the various performance indicators.

6. Conclusions

The best performance indicator for extended surface heat transfer is the assembly augmentation factor.   The fin efficiency, which is the most commonly used indicator, does not provide the necessary information for the prediction of the heat flow and requires the evaluation of the total surface area in the subsequent calculation of the overall heat flow.   The fin effectiveness is a better indicator for the isolated fin and is much easier to use in the evaluation of the overall heat flow through the assembly.   The concept of the assembly augmentation factor needs to be taken up by the manufacturers and users of extended surfaces, and it would be advantageous if academics would teach this approach in their lectures on extended heat transfer.

7. Acknowledgements

The author is indebted to his academic colleagues: Professor D.B.Ingham, Drs. M.I.H. Bhatti, G.E.Tupholme and A.S.Wood, and his former research students, Drs. P.R.Stones, M.Manzoor, D.A.Gardner and J.M.Houghton, who have contributed to the development of the assembly augmentation factor concept.

8. Nomenclature

A	transfer area, m^2
A_x	cross-sectional area, m^2
A_s	surface area per unit length, m
AUG	assembly augmentation , dimensionless
Aug	surface augmentation, dimensionless
C^*	ratio of flowing heat capacities, dimensionless
c_p	heat capacity, J/kg K
F_T	correction factor
ℓ	fin length, m
$LMTD$	logarithmic temperature difference, K
$\dot{M}$	mass flow, kg/s
n	distance normal to surface, m
N	length, m
N_{tu}	number of transfer units
p	half fin pitch, m
$\dot{Q}$	heat flow (duty), W
R	thermal resistance, m^{2}K/W
s	half fin spacing, m
t	half fin thickness, m
T	temperature, K
U	overall heat transfer coefficient, W/m^{2}K
x	distance, m

y	dimensionless distance

Subscripts

act	actual
ass	assembly
B	base
c	cold
f	fin
fa	diabatic fin tip condition
fb	infinite fin
fc	adiabatic fin tip condition
fo	fouling
h	hot
m	mean
max	maximum
min	minimum
p	plain
s	surface
uf	unfinned
w	wall
∞	surrounding
1	inlet
2	outlet

Superscripts

$-$	mean

Greek

ΔT	temperature difference, K
α	heat transfer coefficient, W/m^2K
ε	thermal effectiveness
ε_f	surface enhancement
ζ_f	fin effectiveness
η_f	fin efficiency
θ	normalised temperature

9. References

1. Özisik, M.N. (1985) *Heat Transfer: A Basic Approach*, McGraw-Hill, New York.
2. ESDU (1998) *Design and Performance Evaluation of Heat Exchangers: The Effectiveness – N_{TU} Method, Parts 1-5*, ESDU Data Item Nos. 98003-98007, ESDU International plc, London.
3. Heggs, P.J. (1989) Minimum temperature difference approach concept in heat exchanger networks, *Heat Recovery Systems & CHP*, **9**, 367-375.
4. Linnhoff, B. (1993) Pinch analysis – a state-of-the-art review, *Trans. IChemE*, **71**, 503-522.

5. *User Guide on Process Integration for the Efficient Use of Energy* (1982), The Institution of Chemical Engineers, Rugby, UK.

6. Polley, G.T. (1995) Process Integration Lecture Notes, Department of Chemical Engineering, UMIST, Manchester, UK.

7. Harper, D.R. and Brown, W.B. (1922) Mathematical equations for heat conduction in the fins of air-cooled engines, NACA Report No. 158.

8. Gardner, K.A. (1942) Heat exchanger tube sheet temperatures, *Refiner Nat. Gas. Mann.*, **21**, 71-77.

9. Kern, D.Q. and Kraus, A.D. (1972) *Extended Surface Heat Transfer*, McGraw-Hill, New York.

10. Incropera, F.P. and DeWitt, D.P. (1990) *Fundamentals of Heat and Mass Transfer*, 3rd Edition, Wiley, New York.

11. Gardner, K.A. (1945) Efficiency of Extended Surfaces, *ASME Journal of Heat Transfer,* **67**, 621-631.

12. Manzoor, M., Ingham, D.B. and Heggs, P.J. (1983) The one-dimensional analysis of fin assembly heat transfer, *ASME Journal of Heat Transfer*, **105**, 646-651.

13. Wood, A.S., Tupholme, G.E., Bhatti, M.I.H. and Heggs, P.J. (1995) Steady-state heat transfer through extended plane surfaces, *Int. Comms. Heat Mass Transfer*, **22**, 99-109.

14. Heggs, P.J. and Stones, P.R. (1980) Improved design methods for finned tube heat exchangers, *Trans. IChemE*, **58**, 147-154.

15. Saunders, E.A.D. (1988) *Heat Exchangers: Selection, Design and Construction*, Longmann Scientific and Technical, UK.

16. Wood, A.S., Tupholme, G.E., Bhatti, M.I.H. and Heggs, P.J. (1996) Performance indicators for steady-state heat transfer through fin assemblies, *ASME Journal of Heat Transfer*, **118**, 310-316.

17. Kreith, F (1973) *Principles of Heat Transfer*, 3rd Edition, Harper and Row, New York.

MICROFIN TUBE TECHNOLOGY – THE EFFECTS OF SPIRAL ANGLE ON EVAPORATIVE HEAT TRANSFER ENHANCEMENT

S. -Y. OH and **A.E. BERGLES**
Rensselaer Polytechnic Institute
Department of Mechanical Engineering,
*   Aeronautical Engineering and Mechanics*
Troy, NY 12180-3590, U.S.A.

Abstract. The in-tube evaporation heat transfer of HFC-134a was examined for a smooth tube and five microfin tubes. The heat transfer coefficient in the smooth tube was compared with four correlations. The Yoshida correlation was found to be the best in annular flow, and the Kandlikar correlation (with $F_K= 2.33$) was the best in stratified flow, but the latter needs clarification for its applicability to the low quality range, because of the strong nucleate boiling effect cused by the large value of F_K. The effect of the spiral angle of microfin tubes was investigated to determine the optimal spiral angle. Spiral angles of 6, 12, 18, 25 and 44 degrees were tested at constant mean diameter, 8.71 mm (0.343 in.) and number of fins, 60. The optimal spiral angle was found to be mainly dependent on the mass flux, and was 18 degrees for a mass flux of $G=50$ kg/m^2 .s (36,865 lb$_m$/ft^2.h), 6 degrees for $G=100$ kg/m^2.s (73,730 lb$_m$/ft^2.h), and 6 or 18 degrees for $G=200$ kg/m^2.s (147,460 lb$_m$/ft^2.h).

1. Introduction

Because of the phaseout of chlorofluorocarbons imposed by the Montreal protocol, CFC-12, the widely used, but ozone-depleting refrigerant, has been replaced by the ozone-safe HFC-134a, in automotive air conditioning and refrigeration equipment. The energy efficiency of HFC-134a equipment is lower than that of CFC-12, because HFC-134a requires a more viscous lubricating oil.   However, energy efficiency standards have been tightened to reduce global warming due to the greenhouse effect. Microfin tubes have outstanding performance in enhancing heat transfer for both evaporation and condensation, and have been widely used to save energy in the air conditioning and refrigeration industries.   However, the effects of the various geometrical parameters of microfin tubes for HFC-134a, have not been studied, and have been only assumed to be the same as those for CFC-12. Spiral angle, one of the main geometrical parameters, is an important factor in determining the heat transfer enhancement, but its effect has not been thoroughly investigated for CFC-12, except by the patentees. Fujie et al.   [1]   claimed that a spiral angle of 7° enhanced heat

S. Kakaç et al. (eds.), Heat Transfer Enhancement of Heat Exchangers, 67–78.

transfer the best and specified the range of 4°-15°, and Shinohara et al. [?] claimed 7°-30°, in their patents. But a spiral angle of 18° has been accepted as the standard in industry, without further verification. In addition, for smooth tubes, only a few correlations for HFC-134a have been published, and the applicability of the existing correlations to HFC-134a has not been thoroughly investigated. In this paper, the in-tube evaporation heat transfer coefficient of HFC-134a was examined for a smooth tube and five microfin tubes. The heat transfer coefficients of the smooth tube were compared with four correlations: Shah [3] , Kandlikar [4], Watlett et al. [5] and Yoshida et al. [6]. Five microfin tubes were investigated to determine the optimal spiral angle, where the evaporation heat transfer coefficient becomes the maximum.

2. Literature survey

Ito and Kimura [7] studied evaporation heat transfer and pressure drop, and considered the effects of the spiral angle, the groove depth and pitch of the groove. They showed that the heat transfer coefficient reaches a maximum at 7 and 90 degrees of spiral angle for HCFC-22 and said that Goetler vortices and the high velocity of the liquid flow in the grooves caused the high heat transfer coefficients, but did not explain why 7 and 90 degrees were preferred. Kimura and Ito [8] reported for CFC-12 that the heat transfer coefficient is maximum at 15 degrees of spiral angle for annular flow, because of the secondary flow in the groove, and said that the coefficient increased with the spiral angle for the stratified flow region, because of the capillary phenomenon in the groove. Khanpara et al. [9] studied the geometrical parameters of the fins in microfin tubes for CFC-113 and reported that the higher the spiral angle is, the higher the heat transfer coefficient becomes. Yoshida et al. [10] observed the flow in a microfin tube by means of a fiberscope and, developed a model to predict the heat transfer coefficient. They assumed that the liquid flowed in the grooves, and the surplus liquid flowed at the bottom of the tube with a flat interface. The heat transfer coefficient, α_B, for the liquid at the bottom of the tube was evaluated from the correlation for a smooth tube, and the coefficient, α_G, for the liquid in the grooves was evaluated as heat transfer through the turbulent liquid film in the grooves. The circumferentially averaged heat transfer coefficient was calculated as an average of α_B and α_G with a weighting factor proportional to the occupied surface area. However, the detailed calculation procedure was not disclosed. Yoshida et al. [9] proposed a correlation predicting the evaporation heat transfer coefficient in a horizontal smooth tube, considering stratified and annular flows, separately. The correlation was said to be applicable to HFC-134a as well as CFC-12 and HCFC-22. Eckels and Pate [11] , 12]reported the average heat transfer coefficients for CFC-12 and HFC-134a, ignoring the effect of heat flux. Torikoshi et al. [13] reported the average evaporation and condensation heat transfer coefficients for HFC-134a and its oil mixtures, and noted that the microfin tube was more influenced by the presence of oil. Wattelet et al. [5] measured heat transfer coefficients for HFC-134a, CFC-12 and a mixture of HCFC-22/HCFC-124/HFC-152a, and proposed a correlation; but they did not compare in detail their data with the proposed correlation. Singh et al. [14] reported the flow boiling heat transfer coefficient of HFC-134a in a microfin tube, and showed that the heat transfer coefficient did not change with the quality at a mass flux of 50 kg/m^2·s

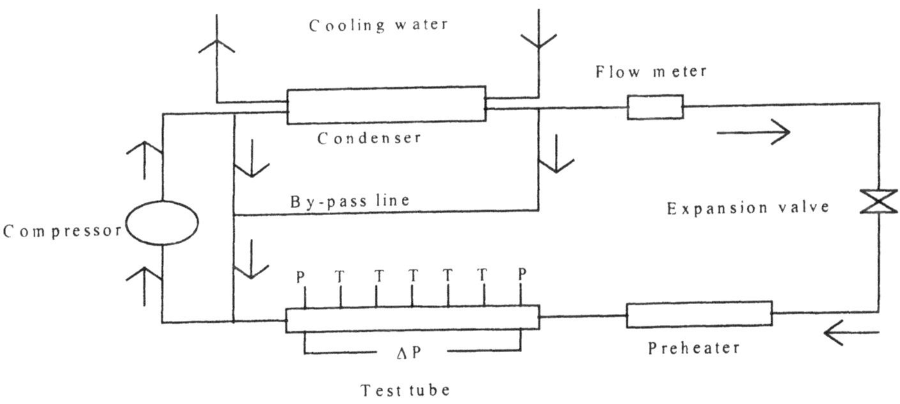

Figure 1 Schematic diagram of test loop

(36,865 $lb_m/ft^2 \cdot h$). The review carried out here established that there is considerable uncertainty on the effect of the spiral angle on the evaporation heat transfer coefficient for microfin tubes. The present study was developed to clarify this effect.

3. Experimental apparatus

A simplified schematic diagram of the test loop is shown in Figure 1. The test fluid was refrigerant HFC-134a, and the evaporation heat transfer coefficient was measured at 0.31 MPa (45.0 psi). The loop consisted of a 5-hp reciprocating compressor, water-cooled condenser, expansion valve, preheater, test tube, and hot gas and hot liquid by-passes. The flow rate was measured with a turbine-type flow meter. The copper test tube was 1.5-m (4.92-ft) long and 9.525-mm (0.375 in.) O.D., and was direct-heated by DC electricity. The inner diameter of the smooth tube was 8.21 mm (0.323 in.), and the mean diameter of the microfin tube was 8.71 mm (0.343 in.). The preheater was electrically heated using heating wire. The inlet quality at the entrance of the test tube was controlled by the amount of heat supplied to the preheater. The throttling process at the expansion valve was assumed as an adiabatic expansion. The test tube and the preheater were thermally insulated using fiber glass. A plastic fitting was used between the test tube and the preheater to isolate it electrically. The wall temperature of the test tube was measured using #30 BWG copper-constantan thermocouples attached on the outside of the tube. Temperatures were measured at five axial locations and four points at each location circumferentially. Each thermocouple bead was covered by a small layer of epoxy, and was electrically isolated from the test tube using thin plastic tape. No correction was made for the very small temperature gradient through the tube wall. The pressure at the inlet of the test tube was measured to establish the saturation temperature of the test fluid. The pressure drop was measured to account for the saturation temperature drop along the direction of the flow. The oil content of the test fluid was found to be less than 0.44%.

4. Data reduction

The local heat transfer coefficient was calculated by dividing the heat flux by the average temperature difference at each location. The heat transfer coefficients measured at both end of the test tube were not considered, because of the measurement error.

$$h_{local} = \frac{q}{T_{avg} - T_{sat}} \tag{1}$$

$$T_{avg} = \frac{1}{4}\left(T_{bottom} + T_{top} + T_{side1} + T_{side2}\right) \tag{2}$$

$$q = \frac{Q}{\pi DL} \tag{3}$$

The measurement uncertainty was estimated for the range of conditions encountered in this study. When the temperature difference was 2°C, typically encountered with microfin tubes at low heat flux and low mass flux, the uncertainty is about 5%. However, when the temperature difference is 0.3°C, typically encountered in microfin tubes at high mass flux and low heat flux, the uncertainty is about 30%. Since the latter is of the same order as the change in heat transfer coefficient due to spiral angle, caution must be exercised in interpreting the experimental results for these conditions. The effect of measurement uncertainty was, however, minimized by being very careful in installing temperature and pressure sensors, hooking up the instrumentation, making the measurements and reducing the data. The repetition of the data was found to be very good.

5. Results and discussion

5.1. SMOOTH TUBE

The evaporation heat transfer coefficients were measured and compared with four correlations: Shah [3], Kandlikar [4], Watlett et al. [5] and Yoshida et al. [6]. Both the Wattelet and Yoshida correlations were said to be applicable to HFC-134a, and the fluid specific constant (F_K) in the Kandlikar correlation was chosen as 1.63, as suggested by Eckels and Pate [12]. As shown in Fig.2, all the correlations except the Yoshida correlation tend to underpredict the heat transfer coefficient. The mean deviations are 27.0%, 20.0%, 26.8% and 19.1% for the Shah, Kandlikar, Wattelet, and Yoshida correlations, respectively. When the fluid specific constant (F_K) in the Kandlikar correlation is adjusted to 2.33, the mean deviation is reduced to 15.9%, as shown in Figure 3. In the stratified flow region (G=50 kg/m^2·s (36,865 lb$_m$/ft^2·h)), as observed in a sight section at the exit of the test tube, the heat transfer coefficients were compared with those of Wattelet et al. [5], and were found to be in good agreement, as shown in Figure 4(a). The test conditions were very similar; that is, Wattelet et al. [5] took data at a saturation temperature of 5°C (41°F) and tube

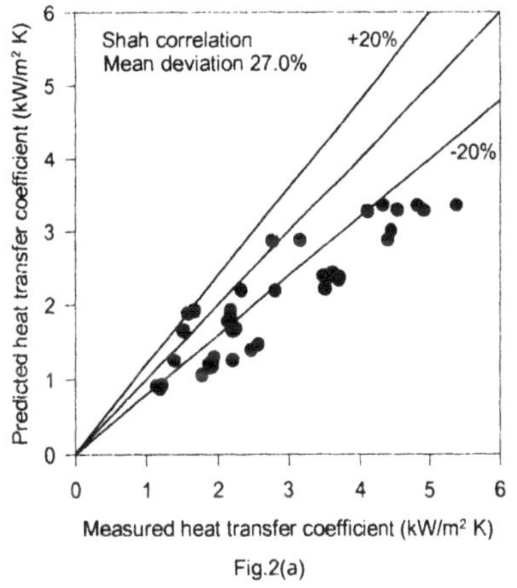

Fig.2(a)

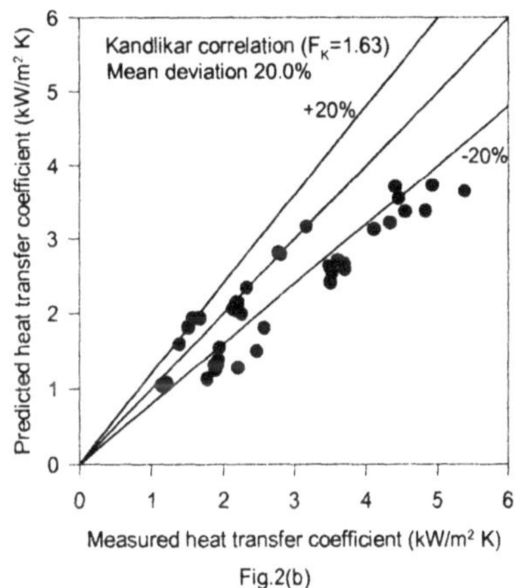

Fig.2(b)

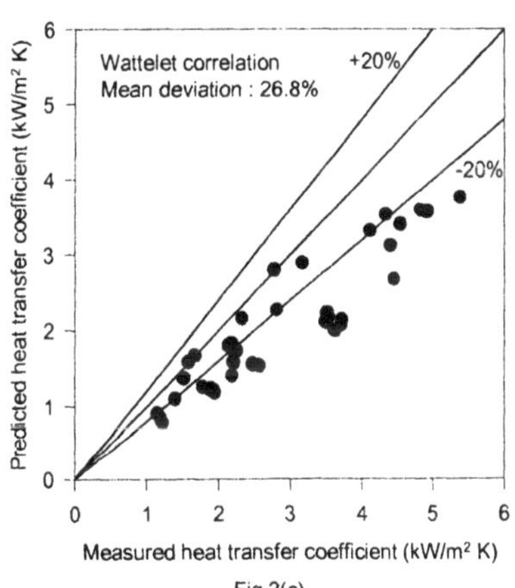

Fig.2(c)

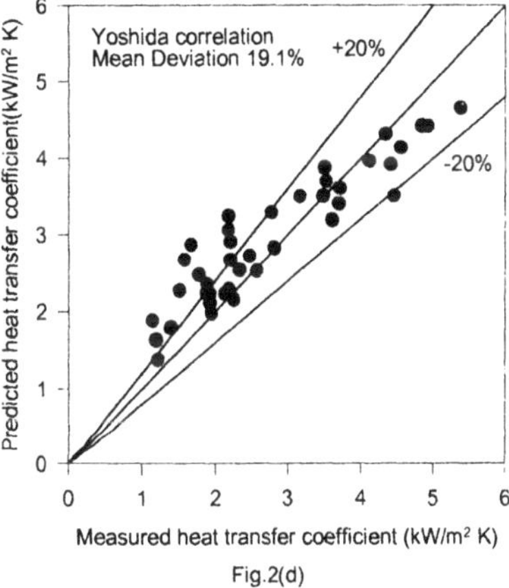

Fig.2(d)

Figure 2 Comparison of correlations for smooth tube

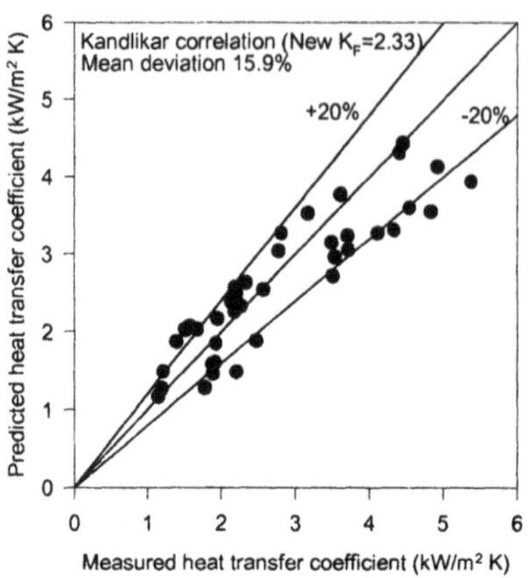

Figure 3 Kandlikar correlation with new F_k=2.33

diameter of 10.92 mm (0.430 in.), and in this study, two parameters were 1.62°C (34.9°F) and 9.525 mm (0.375 in.) O.D. Figures 4(a), (b), and (c) show that the heat transfer coefficient does not change greatly with quality. The Shah and Wattelet correlations predict this trend, but always underpredict. The prediction becomes worse as the heat flux increases, as can be seen in Figure 4 (c). The Yoshida and Kandlikar correlations fail to predict this trend, and are contradictory to each other. That is, the Kandlikar correlation predicts that the heat transfer coefficient decreases rapidly with quality and the Yoshida correlation predicts the opposite. The measured data fall between them, but do not confirm whether a nucleate boiling effect dominates in the lower quality region, as the Kandlikar correlation predicts, because of the lack of data in the lower quality range. Even though the Kandlikar correlation with F_K =2.33, shown in Figure 3, can predict the heat transfer coefficient with a reasonable mean deviation of 15.9%, it seems that the value of F_K=2.33 in the Kandlikar correlation is so large that the nucleate boiling effect is overemphasized, especially at a mass flux of G=50 kg/m²·s (36,865 lb$_m$/ft²·h)and heat flux of q=10 and 15 kW/m² (3,170 and 4,754 Btu/h·ft²). This correlation predicts that the heat transfer coefficient decreases with quality, but the measured data are almost constant with quality. Although the Kandlikar corelation with F_K =2.33 predicts the measured data better than the other correlations at G=50kg/m²·s (36,865 lb$_m$/ft²·h), it needs clarification for the effect of nucleate boiling in the lower quality range.

In the annular flow region (G=200 kg/m²·s (147,460 lb$_m$/ft²·h)), the heat transfer coefficient increases rapidly with quality as shown in Figures 4(d), (e), and (f), which is different from that in stratified flow. The Yoshida correlation outperforms the other correlations, with a mean deviation of 8.9% at G=200 kg/m²·s (147,460 lb$_m$/ft²·h). The Wattelet correlation predicts this trend, but underpredicts again with a mean deviation of 20.8% in this region. The Shah and Kandlikar correlations fail to predict this trend. The latter overemphasizes the nucleate boiling effect at lower quality range, and the former does not predict properly the convective effect at higher quality, as can be seen

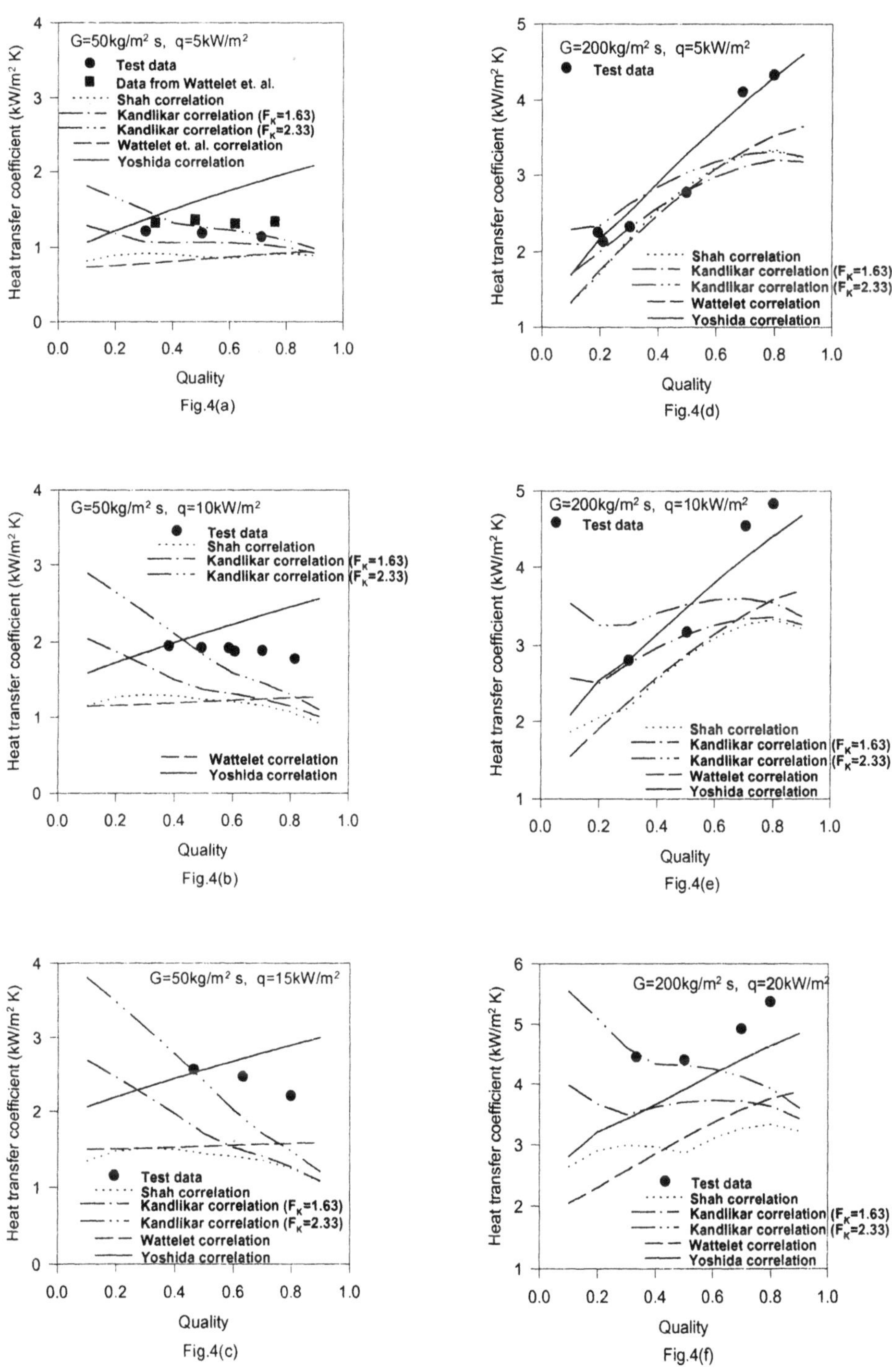

Figure 4 Evaporation heat transfer coefficient of HFC-134a in a smooth tube

74

in Figure 4(f). The Yoshida correlation seems to underestimate the nucleate boiling effect at a heat flux q=20 kW/m^2 (6,339 Btu/h·ft^2), where the measured heat transfer coefficient at quality of about x=0.3 seems to show this effect. It is very interesting that the Yoshida correlation considers stratified flow and annular flow separately, but it does not perform as well in the stratified flow region as in the annular flow region.

5.2. MICROFIN TUBE

Five microfin tubes were tested with spiral angles of 6, 12, 18, 25 and 44 degrees. The quality was essentially fixed at about 0.5, for this comparison, and various heat fluxes were considered, from 5 to 20 kW/m^2 (1,585 to 6,339 Btu/h·ft^2). The spiral angle at which the maximum heat transfer coefficient occurs is mainly dependent on the mass flux, G, as shown in Figures. 5(a), (b) and (c). At G=50 kg/m^2·s (36,865 lb$_m$/ft^2·h), the maximum occurs at a spiral angle of 18 degrees, and at G=100 kg/m^2·s (73,730 lb$_m$/ft^2·h), it occurs at 6 degrees. For both mass fluxes, the heat transfer coefficient reaches a minimum at a spiral angle of 25 degrees, and increases again at a spiral angle of 44 degrees. At G=200 kg/m^2·s (147,460 lb$_m$/ft^2·h), the maximum occurs at 6 degrees, but another peak occurs at 18 degrees. As the heat flux increases from 5 to 20 kW/m^2 (1,585 to 6,339 Btu/h·ft^2), the effect of the heat flux on the heat transfer coefficients at the two peaks becomes smaller. When spiral angle increases to 44 degrees, the heat transfer coefficient decreases further, regardless of the heat flux.

In Figure 8 in the paper of Kimura and Ito   [8]   the variation of the heat transfer coefficients with spiral angle, at G=45 kg/m^2·s (33,178 lb$_m$/ft^2·h) shows a different trend from that at G=65 kg/m^2·s (47,924 lb$_m$/ft^2·h), at the same q=8.5 kW/m^2 (2,694 Btu/h·ft^2) for CFC-12. That is, at G=45 kg/m^2·s (33,178 lb$_m$/ft^2·h), the heat transfer coefficient increases monotonically with spiral angle, but, at G=65 kg/m^2·s (47,924 lb$_m$/ft^2·h), the coefficient increases and then becomes almost constant when the spiral angle is 15 degrees and higher. In Figure 5(a) in the present study, the variation of the heat transfer coefficient with the spiral angle at G=50 kg/m^2·s (36,865 lb$_m$/ft^2·h), q=10 kW/m^2 (3,170 Btu/h·ft^2) shows the same trend as that at G= 65 kg/m^2·s (47,924 lb$_m$/ft^2·h), q=8.5 kW/m^2 (2,694 Btu/h·ft^2) in Kimura and Ito [8] , even though the mass flux is closer to G=45kg/m^2·s (33,178 lb$_m$/ft^2·h). Kimura and Ito [8] insisted that the heat transfer coefficient was proportional to sin(b)/q, in a stratified flow, where b is the spiral angle, and q is the heat flux. Although the flow was stratified at G=50kg/m^2·s (36,865 lb$_m$/ft^2·h) in the present study, Figure 5(a) shows that the heat transfer coefficient does not increase monotonically with the spiral angle. Also, Figure 5(a) does not show that the heat transfer coefficient is inversely proportional to the heat flux. The microfin tube of 6° spiral angle shows a very interesting feature in that the least enhancement of heat transfer is found in stratified flow and the greatest enhancement is found in annular flow. It seems that the secondary flow in the grooves prefers a certain spiral angle, depending on the mass flux and heat flux.

However, the optimal spiral angle does occur in the range which the patentees claim, i.e., 4°-30°. For CFC-12, the spiral angle at which the maximum heat transfer coefficient occurs, is 15 degrees as shown in Fig. 8 in Kimura and Ito   [8] , and for HCFC-22, it is 7 degrees as shown in Fig. 11 in Ito and Kimura [7] . For HFC-134a, the optimal angle was found to be dependent on the mass flux, 18 degrees for

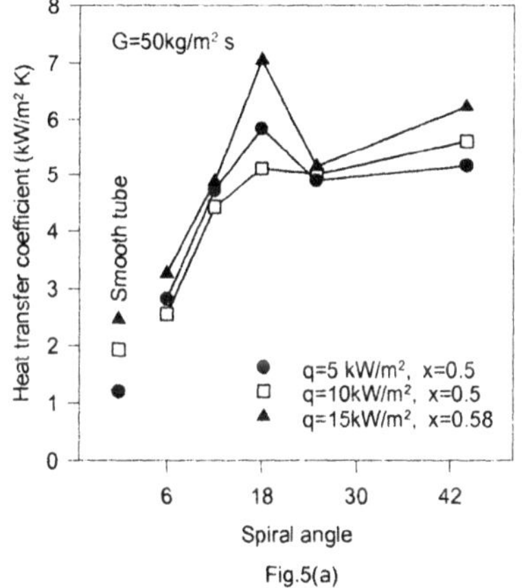

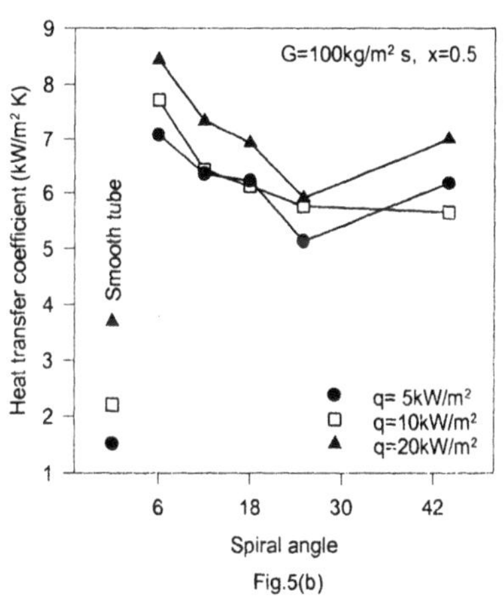

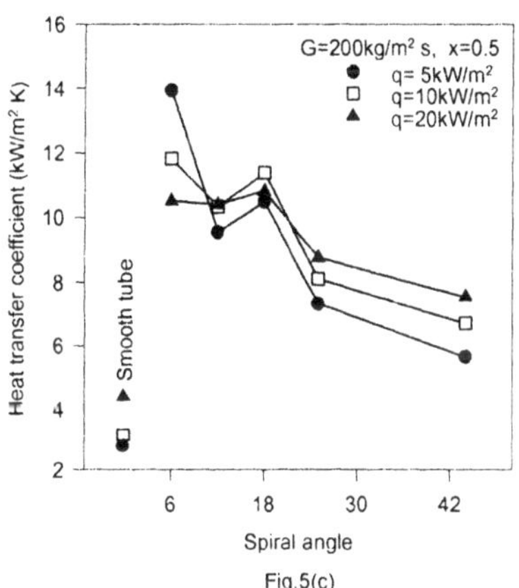

Figure 5 Effect of spiral angle on heat transfer coefficient at constant quality and constant mass flux, for various value of the heat flux

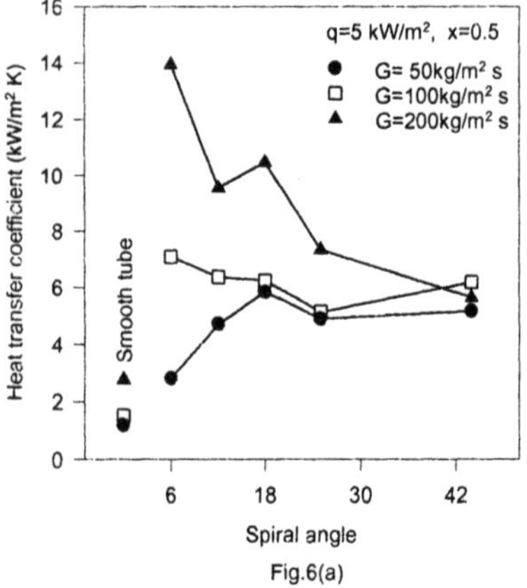

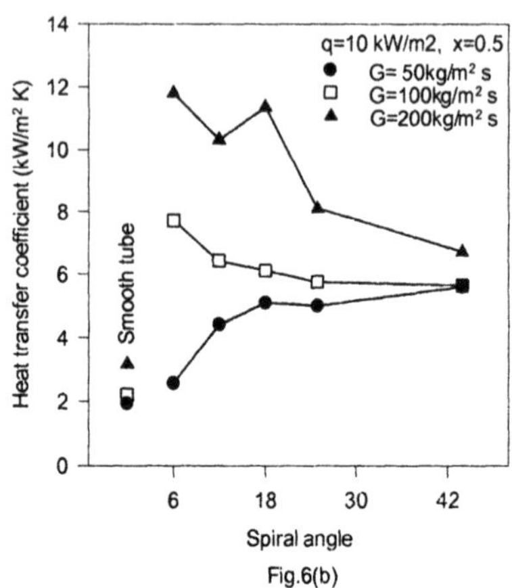

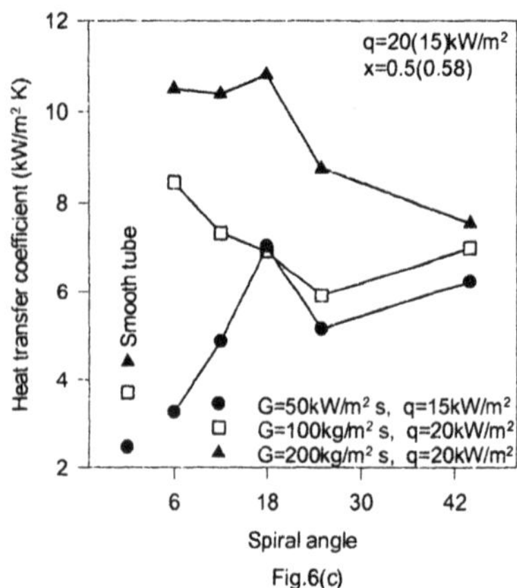

Figure 6 Effect of spiral angle on heat transfer coefficient at constant quality and constant heat flux, for various value of the mass flux

G=50kg/m^2·s (36,865 lb$_m$/ft^2·h), 6 degrees for G=100kg/m^2·s (73,730 lb$_m$/ft^2·h), and 6 or 18 degrees for 200kg/m^2·s (147,460 lb$_m$/ft^2·h). Thus, the spiral angle should be selected based on the intended mass flux to get the highest in-tube heat transfer coefficient for HFC-134a. Although the quality was fixed for this comparison at about 0.5, the same behavior was observed over the entire quality range of 0.3-0.8.

6. Conclusions

(1) For a smooth tube, the Yoshida correlation is the best in predicting the evaporation heat transfer coefficient in annular flow, and the Kandlikar correlation, with the fluid specific constant modified to F_K=2.33, is the best in stratified flow.
(2) The optimal spiral angle for HFC-134a in a microfin tube depends on the mass flux, and it is 18 degrees for G=50kg/m^2·s (36,865 lb$_m$/ft^2·h), 6 degrees for G=100kg/m^2·s (73,730 lb$_m$/ft^2·h) and 6 or 18 degrees for G=200kg/m^2·s (147,460 lb$_m$/ft^2·h).
(3) The optimal angle seems to depend on the working fluid, too, when the present study and previous works are considered.
(4) The comparison was made based on the limited data available and additional efforts are encouraged to confirm the findings in this paper.

Acknowledgments

The first author was financially supported by LG Electronics Inc., Seoul, Korea. The test tubes were provided by Mr. Petur Thors, Wolverine Tube, Inc., Decatur, Alabama. This research was partially supported by an ASHRAE Graduate Student Grant-In-Aid.

Nomenclature

D = inner diameter for smooth tube or
   mean diameter for the microfin tube (m)
h_{local} = local heat transfer coefficient (kW / m^2·°C)
L = heating lenth (m)
Q = heat supplied (kW)
q = heat flux (kW / m^2)
T_{avg} = average wall temperature (°C)
T_{bottom} = temperature at the bottom of the tube wall (°C)
T_{sat} = saturation temperature corresponding to the
   measured pressure (°C)
T_{top} = temperature at the top of the tube wall (°C)
$T_{side1,2}$ = temperatures at the tube wall 90° apart
   from the bottom (°C)

References

1. Fujie, K., Itoh, M., Innami, T., Kimura, H., Nakayama, W., and Yanagida, T.: Heat transfer pipe, U. S. Patent 4,044,797, 1977.
2. Shinohara, Y., Oizumi, K., Itoh, Y., and Hori, M.: Heat-transfer tubes with grooved inner surface, U. S. Patent 4,658,892, 1987.
3. Shah, M. M.: Chart correlation for saturated boiling heat transfer: Equations and further study, *ASHRAE Trans.* 88 (1) (1982), 185-196.
4. Kandlikar, S. G.: A general correlation for saturated two-phase flow boiling heat transfer inside horizontal and vertical tubes, *J. Heat Transfer* 112 (1990), 219-228.
5. Watlett, J. P., Chato, J. C., Souza, A. L., and Christoffersen, B. R.: Evaporative characteristics of R-12, R-134a, and a mixture at low mass fluxes, *ASHRAE Trans.*100(1) (1994), 603-615.
6. Yoshida, S., Mori, H., and Hong, H.: Flow boiling heat transfer to refrigerants and their mixtures inside horizontal tubes, Proceedings of International Seminar on Heat Transfer, Thermophysical Properties and Cycle Performance of Alternative Refrigerants. Japanese Association of Refrigeration, Tokyo, Japan, 1993, pp. 107-120.
7. Ito, M., and Kimura, H.: Boiling heat transfer and pressure drop in internal spiral-grooved tubes. *Bull. JSME* 22(171) (1979), 1251-1257.
8. Kimura, H., and Ito, M.: Evaporating heat transfer in horizontal spiral-grooved tubes in the region of low flow rates, *Bull. JSME* 24(195) (1981), 1602-1607.
9 Khanpara, J. C., Bergles, A. E., and Pate, M. B.: Augmentation of R-113 in-tube evaporation with micro-fin tubes, *ASHRAE Trans.* 92(2B) (1986), 506-524.
10. Yoshida, S., Matsunaga, T., Hong, H., and Nishikawa, K.: Heat transfer enhancement in horizontal, spirally grooved evaporator tubes, *JSME Int, J, Series II* 31 (1988), 505-512.
11. Eckels, S. J., and Pate, M. B.: Evaporation and condensation of HFC-134a and CFC-12 in a smooth tube and a micro-fin tube, *ASHRAE Trans.* 97(2) (1991), 71-81.
12. Eckels, S. J., and Pate, M. B.: An experimental comparison of evaporation and condensation heat transfer coefficient for HFC-134a and CFC-12, *Int. J. Refrigeration,* 14 (1991), 71-81.
13. Torikoshi, Kawabata, K., and Ebisu, T.: Heat transfer and pressure drop characteristics of HFC-134a in a horizontal heat transfer tube, Proceedings of the 1992 International Refrigeration Conference, Purdue University, West Lafayette, IN. 1992, pp. 167-176.
14. Singh, A., Ohadi, M. M., and Dessiatoun, S.: Flow boiling heat transfer coefficient of R-134a in a microfin tube, *J. Heat Transfer* 118 (1996), 497-499.

HEAT TRANSFER ENHANCEMENT BY WING-TYPE LONGITUDINAL VORTEX GENERATORS AND THEIR APPLICATION TO FINNED OVAL TUBE HEAT EXCHANGER ELEMENTS

MARTIN FIEBIG and YUWEN CHEN
Institut für Thermo- und Fluiddynamik
Ruhr-Universität Bochum
D-44780 Bochum, Germany

Abstract: Wing-type longitudinal vortex generators (LVGs) can be used as fins or fin modifications for heat transfer enhancement in compact heat exchangers. For the last 15 years a group at the Ruhr-University Bochum has studied systematically the characteristics of heat transfer surfaces with vortex generators. Recent results are reported here. Three heat transfer enhancement modes may be induced by LVGs: developing boundary layer, swirl (longitudinal vortices) and flow destabilization. First these enhancement modes were studied numerically and experimentally for a base configuration consisting of a rectangular winglet array in channel flow at a Reynolds number of 175. Then finned oval tube heat exchanger elements with and without punched delta winglets were studied numerically. Winglet number and arrangement on the fin were varied in the Re range from 100 to 500. Flow structure, temperature distribution, local and global heat transfer, and flow loss penalty were presented. Staggered winglet arrangements were superior to in-line arrangements. The fin heat transfer surface may be reduced by more than 50% compared to a plane fin for identical heat duty, pressure loss, mass flow rate and temperature difference of the hot and cold medium. Direction for further research is indicated.

1. Introduction

For energy conservation and for economic reasons it becomes increasingly important to enhance the effectiveness of liquid to gas and gas to gas heat transfer processes in numerous types of equipment. Many examples can be found in the process, power, automotive, and aerospace industries for heating, combustion, air conditioning, and refrigeration. The heat transfer performance is mostly limited by the poor gas side heat transfer. This motivates continuous efforts to improve gas side heat transfer in the interesting Reynolds number range from around a hundred to a few thousand.

Fins are widely used to reduce the gas side heat transfer resistance. Convective heat transfer can be further enhanced by a deliberate modification of the original heat

S. Kakaç et al. (eds.), Heat Transfer Enhancement of Heat Exchangers, 79–105.

80

transfer surface. Longitudinal vortex generators (LVGs) are surface modifications that generate vortex systems with dominant longitudinal or stream-wise vortices. The attractiveness of LVGs is that they enhance heat transfer coefficients at moderate flow losses or pressure loss penalties. Figure 1 shows wing-type LVGs in the form of thin slender plates of small aspect ratio that form an angle of attack with the main flow direction. They are called wing vortex generators when they are connected to the original heat transfer surface with their trailing edges, and winglet (half wing) vortex generators when their chords are connected with the original surface. In general LVGs can be attached to a surface, embossed or punched into a surface. An example of a longitudinal vortex generator manufactured by a combination of punching and embossing is shown in Fig. 2. In his thesis Behle [1] called this type of LVGs 'hutze'.

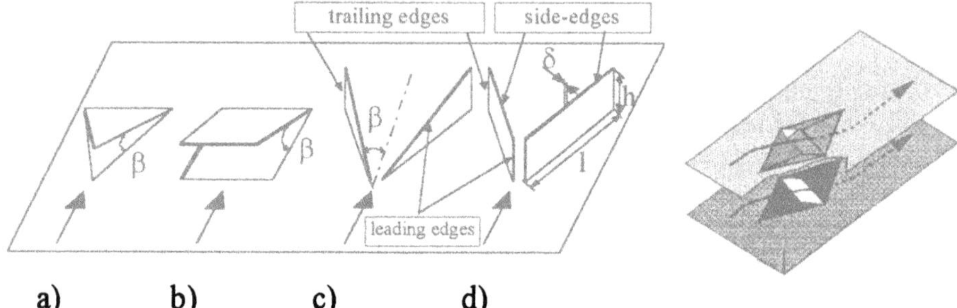

a)   b)   c)   d)

Figure 1. Basic Wing-type vortex generators: a) Delta wing;
b) Rectangular wing; c) Delta winglet pair; d) Rectangular winglet pair.

Figure 2. 'Hutze' -- LVG manufactured by punching and embossing.

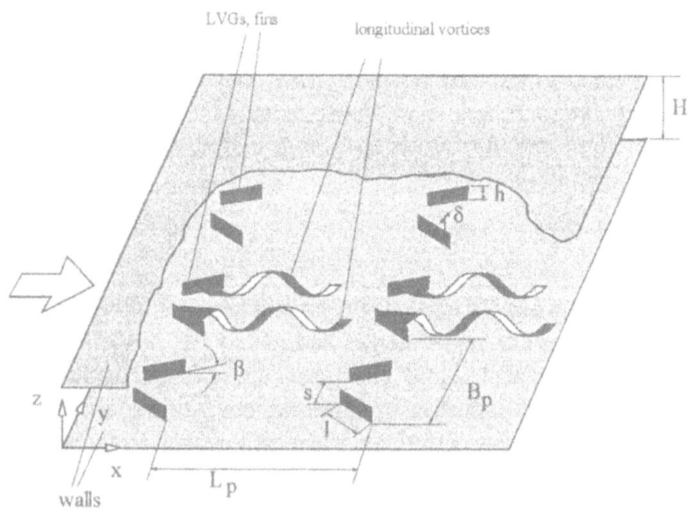

Figure 3. Channel with array of rectangular winglets at an angle of attack to the main stream and schematic of longitudinal vortices; h/H=0.5, L_p/h=10, B_p/h=8, l/h=4, δ/h=0.1, s/h=(1-sin β)·l/(2·h), β=45°.
All lengths are nondimensionalized by the channel height H*.

The vortex systems generated by LVGs and the associated heat transfer enhancement and pressure loss penalty are very complex. They depend nonlinearly on the dynamic and geometric parameters. Besides Reynolds and Prandtl number, the LVG form, arrangement and area relative to the original heat transfer surface are important factors. The many geometric parameters result in a high potential to improve the performance of heat transfer surfaces. But they also make it difficult to derive optimal configurations. The LVG configuration shown in Fig. 3 is governed by seven geometric parameters. It has been considered as a prototype by our group. It may be considered as a base configuration of the core of a plate fin heat exchanger element. The heat transfer enhancement and additional flow losses depend not only on the geometric parameters of LVGs but also on the heat transfer medium and on the on-coming flow situation, e.g. laminar or turbulent flow, entrance or fully developed flow. In Fig. 3, the primary longitudinal vortices generated along the upper edges of the winglets are indicated. They are generated by the pressure difference between the upstream and downstream sides of the winglets at an angle of attack β with respect to the main flow direction.

Publications on heat transfer enhancement by LVGs started to appear during 1969 and 1982. Since the middle eighties research in this field has been intensified. Surveys by Jacobi and Shah [2] and Fiebig [3-9] have discussed different aspects of heat transfer surfaces with vortex generators. During the last 15 years the research group of the first author has studied systematically the characteristics of surfaces with vortex generators for applications in compact heat exchangers. For that purpose, optical methods for determination of local heat transfer coefficients with high resolution, and numerical codes for simulation of three-dimensional unsteady flow with heat transfer for complex geometries have been developed.

In this lecture, parts of the recent results of our group are reported. In section 2 the heat transfer enhancement mechanism generated by LVGs in channel flow is discussed in detail. The geometry of Fig. 3 is used for a quantitative determination of the different heat transfer heat transfer enhancement modes.. A brief description of the numerical method is given in section 3. In section 4, first the state-of-the-art of heat exchanger elements with LVGs is summarized. Then the effects of punched delta-winglets on finned oval tube heat exchanger elements are reported for Reynolds numbers from 100 to 500. This range is of special interest for the automotive industry. Flow patterns, temperature fields, local and global heat transfer enhancement, pressure loss penalty, and performance evaluation are presented. Here we assume that the fins have no thermal resistance (constant wall temperature). In the companion paper [10] the effect of finite heat conduction of fins is discussed. In the concluding remarks, efforts for further research are also indicated.

2. Heat Transfer Enhancement Mechanism through LVGs

In single-phase flow three passive modes of heat transfer enhancement may be distinguished: (1) developing boundary layer, (2) swirl and (3) flow destabilization.

82

Depending on the Reynolds number all three modes may be induced by LVGs. They are coupled by the nonlinear fluid dynamics, and cannot be separated in general. The price to be paid for enhanced heat transfer is a pressure loss penalty. To illustrate the generation of the three modes by LVGs, the fully developed laminar channel flow (Re=175) over the periodic geometry of Fig. 3 is considered as a prototype. The LVGs are an array of rectangular winglet pairs attached to one channel wall with an angle of attack $\beta=45°$ to the main flow. Their length to height ratio is 4, their thickness to height ratio is 0.1, their longitudinal and lateral pitches are 10 and 8 respectively. Their tip separation is given by the condition that when rotated around their mid-point, the winglets form continuous cross ribs at $\beta=90°$. The total winglet area amounts to a little more than 10% of the plane wall area.

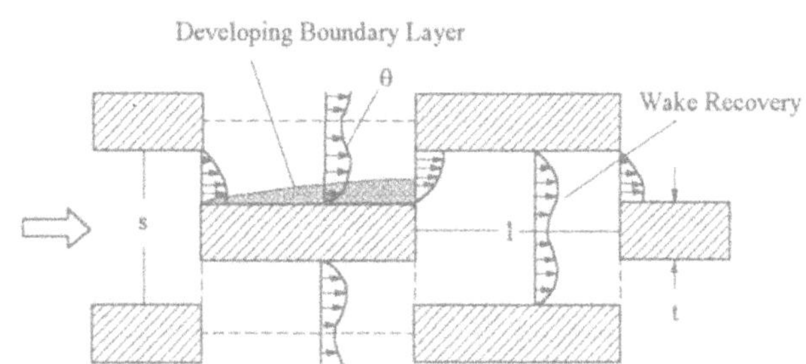

Figure 4. Stream-wise developing boundary layers.

The developing boundary layer mode is schematically shown in Fig. 4 for offset plates. Thermal boundary layers start at each offset plate with very high dimensionless wall temperature gradients, and correspondingly high heat transfer coefficients and Nusselt numbers. They decrease along each offset plate. The boundary layers separate at the trailing edges and form wakes. Behind each plate the wake recovers. Wake recovery is necessary so that the next plate may generate a new developing thermal boundary layer. When the offset plates are put in line with the other plates, no heat transfer enhancement would occur. The Nusselt number would continuously decrease towards the fully developed value. The heat transfer enhancement of the developing boundary layer mode results from the streamwise variations of the flow structure.

For the configuration of Fig. 3, boundary layers develop both on the up-stream and down-stream faces of each rectangular winglet. They start at the leading edges and separate at the trailing edges. Wake recovery takes place between the winglet rows. The flow is periodic in the flow direction. Figure 5 shows numerical results for the

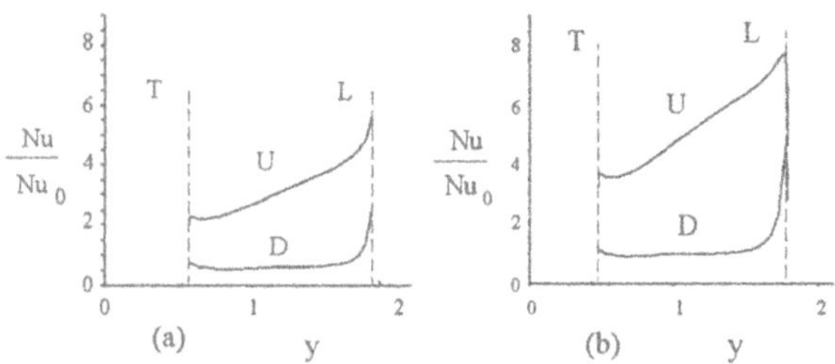

Figure 5. Height-averaged Nusselt number enhancement on the up- and downstream faces of a rectangular winglet shown in Fig. 3 at Re=175. (a) steady flow is realized by assuming symmetric lateral boundary conditions and (b) unsteady flow by periodic lateral boundary conditions.
Meanings of the symbols: T: trailing edge position; L: leading edge position; U: up-stream face of the winglet; D: down-stream face of the winglet. y coordinate as in Fig. 3 for the first winglet from the origin. Nu_0=3.77 for fully developed channel flow.

height-averaged heat transfer enhancement along the projection of the up-stream and down-stream faces of a winglet compared to the fully developed laminar plane channel flow with constant wall temperature (Nu_0=3.77). It was calculated by Müller [11] with the code FIVO for simulation of 3D unsteady flow [12]. For symmetric side boundary conditions the flow is steady at this Reynolds number and the Nusselt number enhancement is shown in Fig. 5(a). The down-stream face gives only strong heat transfer enhancement near the leading edge, while the upstream face generates heat transfer enhancement by more than a factor of 2 everywhere between the leading and trailing edge. Mean enhancement for both sides is a factor of 2.3. Part of the enhancement is caused however by the swirl generated by the winglets upstream because of the periodicity. In such cases it is impossible to separate the swirl effect from the developing boundary layer effect. Heat transfer near the juncture of the winglet and channel wall is very low because the velocities are small in that region. This reduces the height-averaged enhancement.

The swirl mode of heat transfer enhancement is sketched in Fig 6 for two pairs of counter-rotating longitudinal vortices that generate a corkscrew motion of a given wavelength in the lateral or span direction. The original dimensionless velocity and temperature profiles are shown in Fig. 6a. Note that the dimensionless temperature gradient at the wall is the dimensionless heat transfer coefficient or Nusselt number. In the induced down-wash regions, with velocities towards the wall, fluid with higher temperature is transported towards the wall into regions with lower temperatures, Fig. 6a-b. In the induced up-wash regions, with velocities away from the wall, fluid with lower temperature is transported away from the wall into regions with higher temperatures. For a sinusoidal spanwise variation of the temperature and velocity component perpendicular to the wall, the induced heat flux is always towards the wall, Fig. 6c-d. This

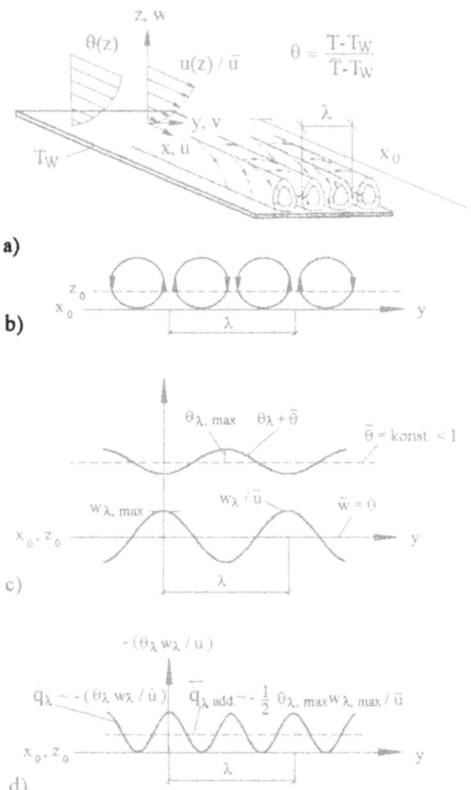

a)

b)

c)

d)

Figure 6. Spanwise fluctuations resulting from counter-rotating longitudinal vortices: θ is zero on the wall.
(a) Vortex structure;
(b) Cross section of the longitudinal vortex pairs at x_0;
(c) Induced and mean temperature and velocities at x_0,z_0;
(d) Convective heat flux profile towards the wall at x_0,z_0.

in turn increases the spanwise and span-averaged Nusselt number. The down- and up-wash regions induce however increased and decreased streamwise velocities, respectively. In turn they increase and decrease thermal energy transport in the main flow direction, respectively. As a consequence the heat transfer variations are not sinusoidal. Heat transfer enhancement is higher in the down-wash regions than in the up wash regions. The heat transfer enhancement of the swirl mode results from the spanwise variations of the flow structure.

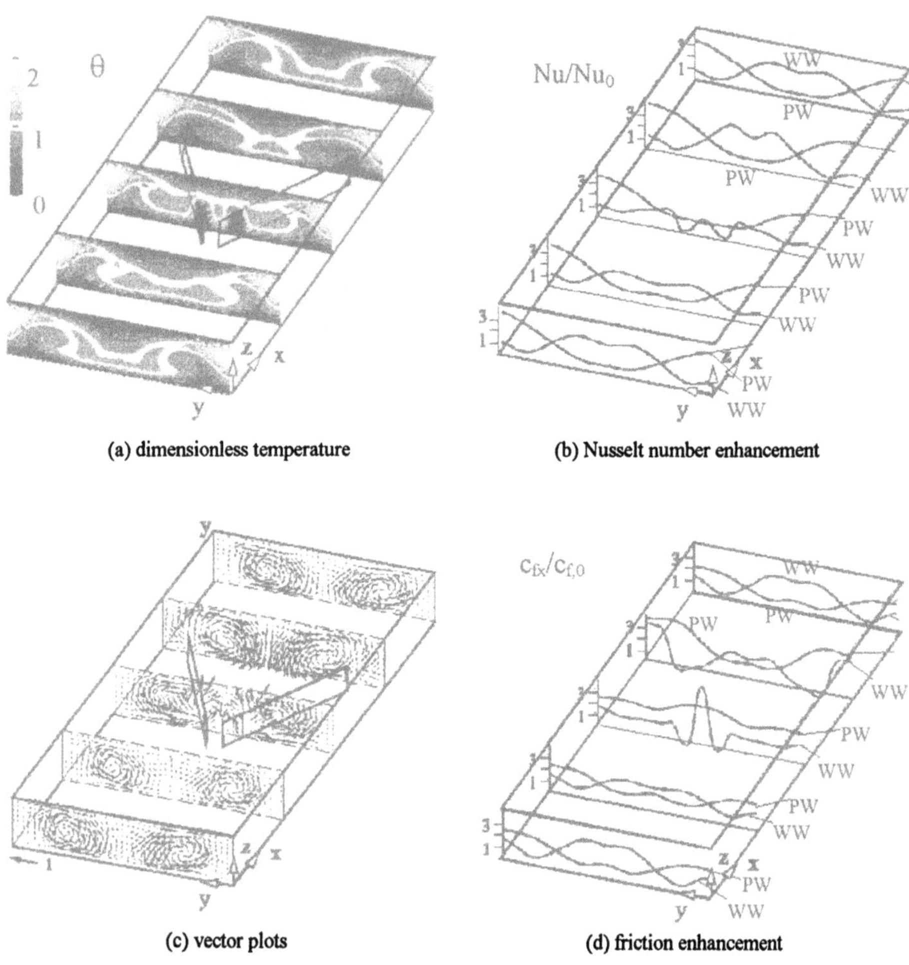

<table>
<tr><td>(a) dimensionless temperature</td><td>(b) Nusselt number enhancement</td></tr>
<tr><td>(c) vector plots</td><td>(d) friction enhancement</td></tr>
</table>

Figure 7 Steady thermal and flow characteristics for a periodic channel element with a rectangular winglet pair shown in Fig. 3 at Re=175. Steady flow is realized by using symmetric side boundary conditions.

a) dimensionless temperature distribution $\theta=(T-T_W)/(T_B-T_W)$ in five cross sections located at x=0, 1.25H, 2.5H, 3.75H, and 5.0H respectively;

b) corresponding local Nu enhancement Nu/Nu_0 on the plane wall (PW) and winglet wall (WW), $Nu_0=3.77$;

c) vector plots showing the counter-rotating vortices in the same five cross sections;

d) corresponding local streamwise friction coefficient enhancement $c_{fx}/c_{f,0}$ on the plane wall (PW) and winglet wall (WW), $c_{f0}=12\cdot Re^{-1}$.

The secondary flow induced by the winglets of Fig. 3 together with the corresponding temperature profiles, spanwise heat transfer distributions, and friction coefficients are shown in Fig. 7 for 5 cross sections of a periodic element. The rotational velocities are of the same order of magnitude as the mean flow velocity, see lower left corner of Fig. 7c for the measure. The dominance of the counter-rotating longitudinal vortices generated at the upper side edges of the winglets is apparent. But the vortex system generated by a winglet is complex. It consists of a primary longitudinal vortex, a horseshoe vortex which forms in front of the winglet and wraps around the winglet in the corner regions between winglet and channel wall, and leading edge and trailing

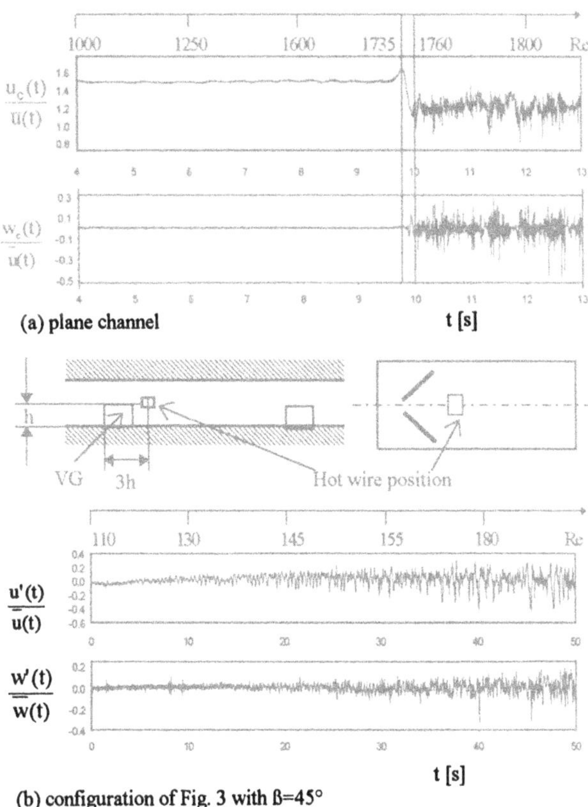

Figure 8 Transition in a slowly but continuously accelerated channel flow: u- and w-components at z/H=0.5 (a) plane channel, and (b) configuration of Fig. 3 with ß=45°. The hot-wire is located downstream of the eighth winglet-row. u_c: center-line velocity; $\overline{u}$: mean velocity; u', w': fluctuations.

edge vortices. The temperature boundary layers are thinner in the down-wash regions than in the up-wash regions, Fig. 7a, and consequently the spanwise heat transfer enhancement is higher in the down-wash regions than in the up-wash regions, Fig. 7b. But heat transfer enhancement may also occur in the up-wash region. The friction coefficient distributions in Fig.7d resemble qualitatively the corresponding Nusselt number distributions. This shows that the increased streamwise velocities in the down-wash areas are very important for the heat transfer enhancement, and the analogy between heat transfer and friction still holds qualitatively. But the energy needed for the generation of the vortices is mainly reflected in the drag of the winglets and causes more than 2/3 of the additional pressure loss. The average Nusselt number enhancement for both channel walls is 1.56. For the plane wall the heat transfer enhancement is 1.79, while for the winglet wall it is only 1.31 because of the low heat transfer close to the juncture of winglets and fins.

Flow destabilization leads to large amplitude self sustained oscillations or turbulence intensification. Heat transfer enhancement and flow loss penalty by turbulence are well recognized since Reynolds. The difference to the swirl mode is that heat transfer is enhanced not by variations of the flow in the lateral direction but by fast time variations of the flow. Destabilization of a flow depends on the Reynolds number and the velocity profile. Profiles with inflection points become unstable at smaller Reynolds numbers than profiles without inflection points. The critical Reynolds number decreases with increasing distance of the inflection point from the wall, see Drazin and Reid [13]. Longitudinal vortices change the velocities into profiles with inflection points and destabilize the flow. Figure 8 shows hot wire measurements by Weber [14] in fully developed plane channel flow and for

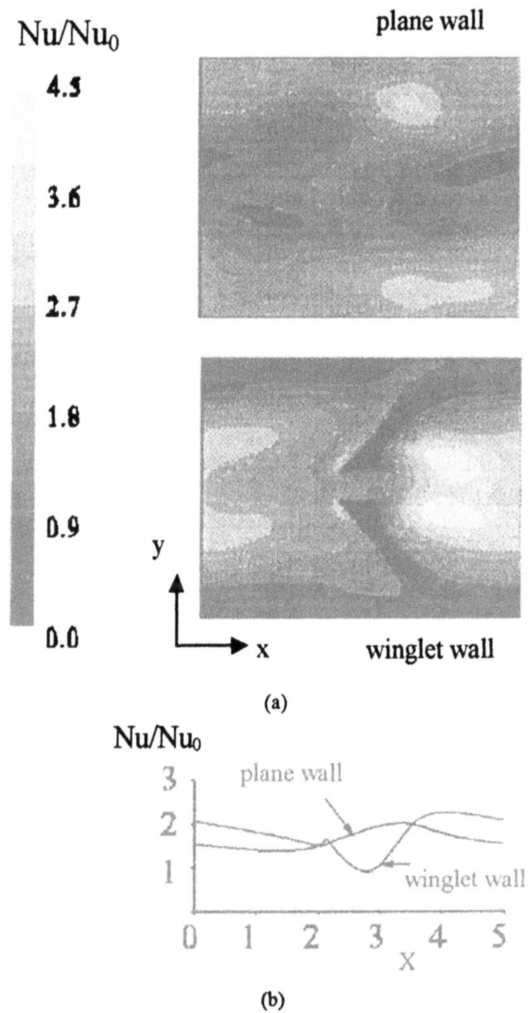

Figure 9. Nusselt number on the plane and winglet wall of Fig. 3. (a) instantaneous local and (b) time and span averaged values. Re=175 with periodic lateral boundary conditions, Nu$_0$=3.77.

the periodic configuration of Fig. 3. He increased the Reynolds number very slowly in both cases. For the plane channel, velocity fluctuations rose dramatically above the noise level of his wind tunnel at a Reynolds number of about 1735, and the centerline velocity dropped from the fully developed laminar value of 1.5 to a turbulence value of about 1.2 relative to the mean velocity, Fig. 8a. For the configuration of Fig. 3 the amplitudes of the fluctuations reached already values of more than 20% of the mean velocity at a Reynolds number of 150 which is an order of magnitude below the critical Reynolds number of plane channel flow, Fig. 8b.

But for the calculations with symmetric side boundary condition the flow was still steady at a Reynolds number of 175. When the side boundary condition was changed to the more realistic periodicity condition the flow became highly unsteady at the same Reynolds number. This allowed to study the effect of the destabilization mode on heat transfer separately from the two other modes at this Reynolds number of 175. As mentioned before the other two enhancement modes could not be isolated. Figure 9 shows local instantaneous, and time and span-averaged Nusselt number enhancement for the plane and winglet wall. The oscillations become apparent from the lateral asymmetry of the instantaneous heat transfer enhancement. The time and span averaged heat transfer values are now higher on the winglet wall than on the plane wall except in the area where the winglets are attached to the wall. For steady flow (symmetry side boundary conditions) the span averaged heat transfer on the plane wall was everywhere higher than on the winglet wall (see Fig. 7). For unsteady flow (periodic side boundary conditions) the average heat transfer enhancement for both walls was increased by roughly 10% to 1.71. For the winglet wall the increase was more than 30% to 1.79. For the plane wall a slight decrease to 1.64 occurred. On the winglets heat transfer also increased due to the fluctuations. The winglet-height-averaged Nusselt number enhancement is shown in Fig. 5(b) as a comparison to the steady values of Fig. 5(a). Heat transfer is increased everywhere. The mean value is increased by nearly 40% from 2.3 to 3.1. For the periodic element the heat transfer enhancement increased from 1.6 to 1.8 due to flow destabilization.

In general the three heat transfer enhancement modes are coupled by the nonlinear nature of fluid dynamics. Only the trick with the different side boundary conditions allowed us to separate the destabilization mode from the other two modes. And this worked only in a very limited Reynolds number range. At higher Reynolds numbers the symmetry side boundary conditions also lead to self-oscillating flows, and the destabilization mode cannot be separated from the other two modes. Oscillation amounts to about 13% of the total heat transfer at Re=175 for this configuration. For Reynolds numbers below 130 oscillation did not appear.

Generally all LVGs may induce the three heat transfer enhancement modes. Their coupling and relative importance depend on the specific geometry of the LVGs and the original flow situation without vortex generators. For low Reynolds numbers swirl and developing boundary layers are the dominant enhancement effects. Flow destabilization mode becomes active beyond a critical Reynolds number which depends strongly on the geometry.

3. Numerical Method

A computer design of compact heat exchanger surfaces with longitudinal vortex generators has been described by Fiebig and Mitra [9]. It models the geometry as channel elements with different boundary conditions. The computational scheme is based on a finite volume discretization with SIMPLEC pressure correction, see Van

Doormal and Raithby [15]. At present two versions of the program exist - one in body-fitted coordinates and the other in Cartesian coordinates. The latter is computationally less expensive than the former. For tube-fin heat exchanger elements the former version is used. The body-fitted grids are generated using algebraic or partial differential equation techniques. The longitudinal vortex generators in both versions of the program are simulated by imposing no-slip conditions on proper points in the three-dimensional grid.

The velocity, temperature, and pressure fields are determined by a time-marching solution of the unsteady continuity, Navier Stokes, and energy equations with constant properties. In a curvilinear co-ordinate system with Cartesian velocity components, the dimensionless forms of these equations in index notation are:

Continuity eq. $\quad \dfrac{\partial U_i}{\partial \xi_i} = 0$ $\hfill (1)$

Momentum eqs. $\quad J\dfrac{\partial u_k}{\partial t} + \dfrac{\partial}{\partial \xi_i}\left[U_i\, u_k - \dfrac{1}{Re}\dfrac{1}{J}\left(B_j^i\, \dfrac{\partial u_k}{\partial \xi_j} + \beta_j^i\, \omega_k^j \right) + p\,\beta_k^i \right] = 0$ $\hfill (2)$

Energy eq. $\quad J\dfrac{\partial T}{\partial t} + \dfrac{\partial}{\partial \xi_i}\left[U_i\, T - \dfrac{1}{Pe}\dfrac{1}{J}\left(B_j^i\, \dfrac{\partial T}{\partial \xi_j} \right) \right] = 0$ $\hfill (3)$

with $\quad U_i = u_j\,\beta_j^i\,,\quad B_j^i = \beta_k^i\,\beta_k^j\,,\quad \omega_j^i = \dfrac{\partial u_i}{\partial \xi_k}\,\beta_j^k$ $\hfill (4)$

and $\quad Re = \dfrac{u_0^* H^*}{v_0^*}\,,\quad Pr = \dfrac{v_0^*}{a_0^*}\,,\quad Pe = Re \cdot Pr$ $\hfill (5)$

where the Einstein summation convention applies, dissipation is neglected, and β_j^i is the cofactor of $\partial x_i / \partial \xi_j$ in the Jacobian J of the co-ordinate transformation $x_i = x_i(\xi_j)$. The Reynolds number Re, the Prandtl number Pr, and the Peclet number Pe are parameters for characterizing the velocity and temperature fields of a passage flow. In equations (1) to (4), the time is nondimensionalized by H^*/u_0^*, all lengths by H^*, and the velocity components by the average velocity u_0^*. The temperature is the difference between the local and the inlet temperature scaled by the difference of the tube temperature and the inlet temperature, $T = (T^* - T_0^*)/(T_T^* - T_0^*)$. The pressure is the difference between the local and a reference pressure divided by $\rho^* u_0^{*2}$,

$p = (p^* - p_r^*)/\rho^* u_0^{*2}$.

The computational domain is discretized into volume elements. All dependent variables are defined at the center of the elements (co-located grid). For the discretization the following difference formulas have been used:

time gradient: backward time difference of first-order accuracy
convective terms: flux blending of first-order upwind and central difference

diffusive terms: central difference
pressure gradient: central difference

The convective terms were treated with a 'deferred-correction approach' suggested by Khosla & Rubin [16], in which the up-wind part was calculated implicitly, while the difference between the central difference and up-wind part was calculated explicitly. The sets of algebraic equations were solved by the strongly implicit procedure of Stone [17]. More details of the computational techniques were given by Kost [18], Bai [19], Peric [20] and Chen [21]. Very good agreement between numerical and experimental results was shown in Chen [21].

4. Tube-Fin Elements With and Without Delta Winglets

For single, single row, and arrays of wing-type vortex generators in laminar air channel flow, experimental investigations by Fiebig et al. [22-23] and Tiggelbeck et al. [24], and numerical studies by Güntermann [25] show:
  (1) The two most important dimensionless parameters that control vortex structure and thereby heat transfer enhancement and flow loss penalty are ratio of LVG area to heat transfer area (or area ratio) and angle of attack of the LVGs;
  (2) Heat transfer enhancement increases with the area ratio and the angle of attack up to a maximum value which depends on LVG form and Reynolds number;
  (3) Pressure loss penalty increases with area ratio and angle of attack up to 90°;
  (4) Heat transfer enhancement and pressure loss penalty increase with Re;
  (5) Winglets are more effective than wings;
  (6) Delta forms are slightly more efficient than rectangular forms;
  (7) Punching gives slightly better performance than mounting;
  (8) Length to height ratio (or aspect ratio) of 2 is near an optimum for delta winglets.

For Reynolds numbers in the range of 600 to 3000, experimental results for finned circular tubes and finned flat tubes without and with winglets were reported by our group [26]. Local fin heat transfer data and global pressure losses were given for configuration with three tube-rows without and with winglets. But only one winglet pair per tube element was considered. Here numerical results of Chen et al. [21, 27-31] for finned oval tubes with up to four delta winglet pairs per tube element will be shown. Punched delta winglets of aspect ratio $\Lambda=2$ at an angle of attack $\beta=30°$ were selected in accordance with the results of early investigations, see points (5)-(8) above. The winglet height was fixed by the fin pitch so that the winglets acted also as pitch holders. The number of winglets and their position on the fin and relative to the tube were varied. The winglet to fin area ratios varied between 1.73% for one winglet pair and 6.92% for four winglet pairs per fin. No attempt at an optimization was made. The positioning of the winglets detailed in Fig. 12c and Table 1 was selected with the following background. Longitudinal vortices last over a long distance in the streamwise direction. Hence the first winglet was positioned close to the leading edge of the fin ($x_{pa}=0.2$ for the staggered and 0.6 for the in-line configurations), so that the

90

longitudinal vortex would act on a large part of the fins. Each winglet had a lateral extent of H. For the in-line configurations the winglets were roughly positioned in the middle between tube and the lateral edge of the fin. For the staggered configurations, the winglets generated counter-rotating vortices. The distance between the trailing edges of a staggered winglet pair is roughly the fin separation. It is expected that the staggered configurations will give higher heat transfer enhancement than the equivalent in-line configurations with the same winglet area and angle of attack because the vortices generated by the staggered winglet arrangement influence a larger part of the fin. The longitudinal pitch was 4.9 for the in-line configurations and 5.6 for the staggered configurations. The effectiveness of the second winglet row should be less than the first row because the vortices of the second winglet row act over a shorter length. All investigations were performed for air as the heat transfer medium with a Prandtl number of 0.7. The core of a finned oval tube heat exchanger is shown in Fig.10. A liquid flows through the tubes and air flows between the fins and around the tubes. The tube cross section is formed by smoothly connecting four arcs representing the front, sides and rear of the tube circumference, see Fig. 11.

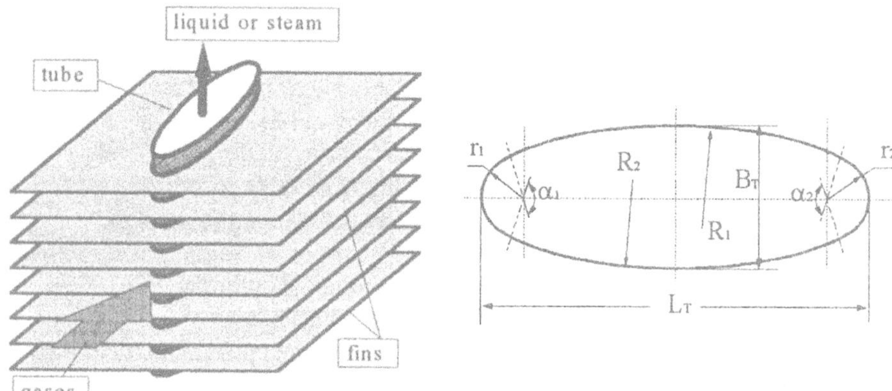

Figure 10. A tube-fin heat exchanger core.   *Figure 11.* The cross section of an oval tube.

4.1 GEOMETRY, COMPUTATIONAL DOMAIN, & BOUNDARY CONDITIONS

The tube-fin geometry with a fin-to-tube area ratio of 8.4 is kept the same for all configurations. This is a typical value for compact heat exchangers. But fin to tube area ratios of up to 30 may be used for very high conductivity fins and low Reynolds numbers.

Figure 12(a) shows a finned oval tube element with one delta-winglet pair. Two fins form a channel of height H, width $B(=9.1H)$ and length $L(=15.4H)$. Its cross section is obstructed by the oval tube and the punched delta winglets, the LVGs. The oval tube is positioned at the center of the fin with the dimension $L_T/B_T=5.5$, cross sectional area $A_{T,CS}=24.6H^2$, and $\alpha_1=\alpha_2=160°$, see Fig. 11. Winglets with h=H, $\beta=30°$ and $\Lambda=2$ are punched out of the fins. The thickness of the winglets is assumed to be zero. For small Reynolds numbers, the flow is steady and the computational domain can be reduced by using symmetry conditions on the mid-plane of the channel (y=B/2). The domain is extended by 6H over the rear edge of the fin in the *x*-direction to guarantee a recirculation-free flow at the exit. Figure 12(b) shows the computational domain with the employed boundary conditions.

Fig. 12(c) shows the configurations for which numerical results are given. They are not numbered sequentially, because results for the configurations with other numbers are not presented here. The reference configuration, config. 0 with plane fins, is not shown. The config. 3 has one delta winglet pair per tube element, configs. 6 and 11 have two and three winglet pairs respectively, and configs. 12 and 14 have two and four winglet pairs, respectively. The locations of the winglets are shown in Table 1.

Table 1 Locations of the winglets.

configs.		x_{PA}	y_{PA}	x_{PB}	y_{PB}	x_{PC}	y_{PC}
3		0.63	2.19	2.36	1.19	1.86	0.32
6	W1	0.63	2.19	2.36	1.19	1.86	0.32
	W2	5.55	2.19	7.28	1.19	6.78	0.32
11	W1	0.63	2.19	2.36	1.19	1.86	0.32
	W2	5.55	2.19	7.28	1.19	6.78	0.32
	W3	10.47	2.19	12.2	1.19	11.7	0.32
12	WA	0.2	3.185	1.93	2.185	1.43	1.319
	WB	1.46	0.25	3.19	1.25	2.69	2.116
14	WA1	0.2	3.185	1.93	2.185	1.43	1.319
	WB1	1.46	0.25	3.19	1.25	2.69	2.116
	WA2	5.8	3.185	7.53	2.185	7.03	1.319
	WB2	7.85	0.25	9.58	1.25	9.08	2.116

W: winglet; Definitions of (x_{PA}, y_{PA}) ..., see Fig. 12(a)

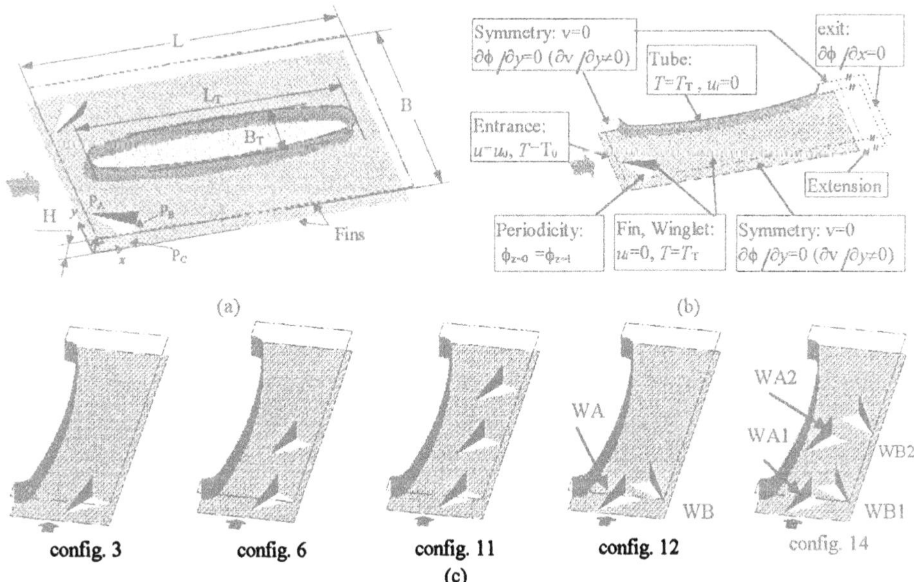

Figure 12 (a) Geometrical model, (b) computational domain with boundary conditions, and (c) the investigated configurations. In (a) and (b) only one delta-winglet pair is shown. T_0=0, T_T=1, u_0=1.

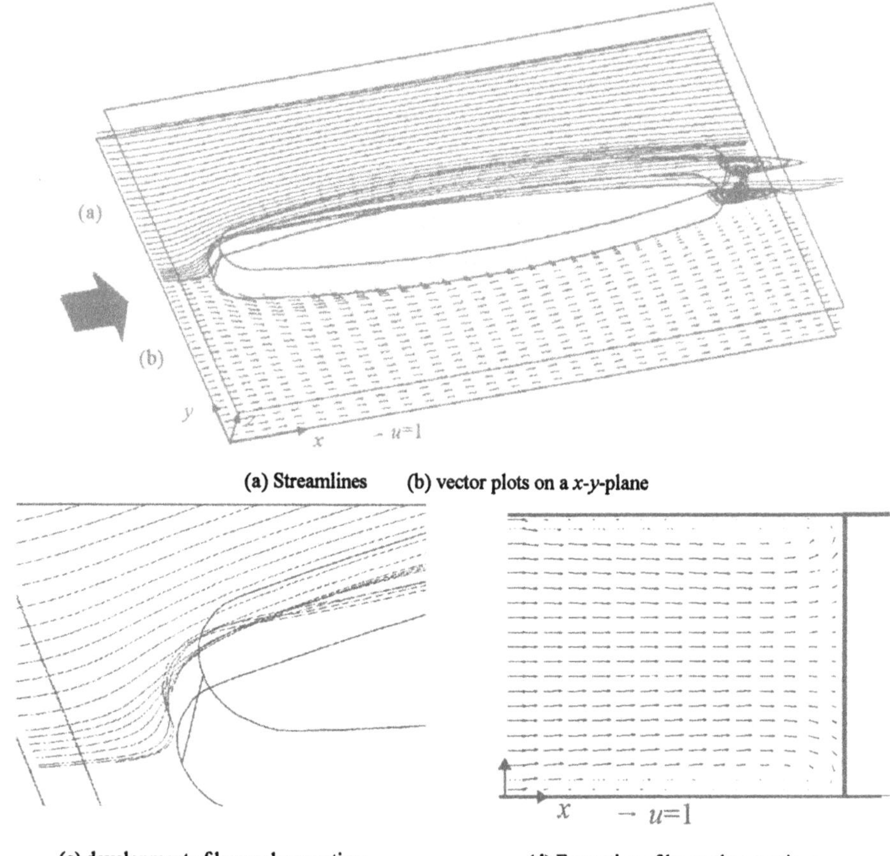

(a) Streamlines   (b) vector plots on a *x-y*-plane

(c) development of horse shoe vortices   (d) Formation of horse shoe vortices

Figure 13 Selected velocity vectors and streamlines in a finned oval-tube element at Re=300.
   (a) Streamlines starting in the entrance plane at z/H=0.1;
   (b) Vector Plots of the velocity components on a x-y plane of z=0.1H;
   (c) Enlargement of (a) near the front stagnation point of the tube, where horse-shoe vortex develops;
   (d) Velocity vectors in the x-z-plane of the stagnation point showing the formation of the horse shoe vortices near the fins;

4.2 OVAL TUBE WITH PLANE FINS

4.2.1 Flow Pattern

Main features of the flow between the fins of the finned oval tube element are:
- a developing flow near the fin leading edges,
- a pair of weak horse-shoe vortices developing in the stagnation region of the tube and wrapping round the tube in the juncture regions between tube and fins,
- a developing flow along the circumference of the tube with very low velocities in the juncture region between tube and fins,

- flow separation from the tube near the juncture of the side arcs and the rear arc of the oval tube,
- a small recirculation zone behind the tube,
- a helical vertical vortex in the tube wake that results from the coupling of recirculation and vertical pressure gradient.

Figure 13 shows vector plots of the velocity components in a x-y plane of z=0.1(H) and streamlines stemming from the same plane at the entrance for Re=300. The oval tube induces a convergent-divergent flow. This together with the horse-shoe vortices result in a spirally flow in the passage. The streamlines of the secondary flow

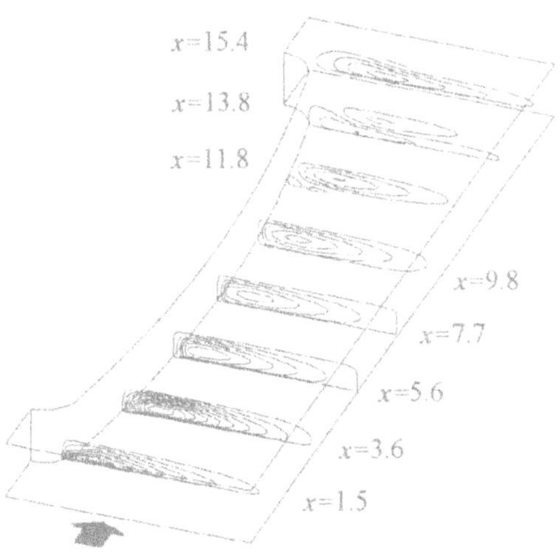

Figure 14. Streamlines of the secondary flow in eight cross-sections of the duct at Re=300. A counter-rotating spirally flow is observed in the passage. Only the streamlines in the lower half are shown.

are shown in Fig. 14 for the lower half of eight cross sections. The streamlines in the upper half is symmetric to those in the lower half.

4.2.2 Temperature distribution and Heat Transfer

Figures 15 and 16 show the local dimensionless mean fluid temperatures and Nusselt numbers on the fin for Re=100 and 300. The Nusselt number is the dimensionless temperature gradient at the wall scaled by the difference between fin and bulk temperatures (T_f-T_B(x)). Four regions can be distinguished:

- The entry region from the entrance to a short distance upstream of the tube: Here the thermal boundary layers develop as in the entrance of a flat duct.
- The near tube region: Here a thermal boundary develops along the tube circumference. The isotherms run almost parallel to the tube wall, i.e., the mean temperature gradient in the spanwise direction dominates.
- The outer region away from the tube: Here the temperature field in the main flow direction also develops as in a flat duct.
- The separation or wake region: Here the fluid temperatures are high because of the recirculation and the temperature gradients are small.

Generally, fin heat transfer is high near the leading and is low in the tube wake, see Fig. 16. Laterally from the tube, Nu increases from a minimum at the juncture with the tube to a maximum and then it decreases somewhat to the side edge. The low velocities in the juncture region are responsible for the weak convective thermal transport. The maxima of Nu concur with the velocity maxima. The local maximum Nu in the front stagnation region of the tube is caused by the horse-shoe vortices. For x > 6 with

94

Re=100, the Nusselt numbers on the fin tend to the fully developed channel value of 3.77. Close to the tube and in the wake, Nu becomes smaller than the value of fully developed channel flow. The small velocities in the juncture regions and wake results in high mean temperatures and small temperature gradients on the wall. If the local bulk temperatures instead of the cross sectional ones were used, higher Nu in the vicinity of the tube would have results.

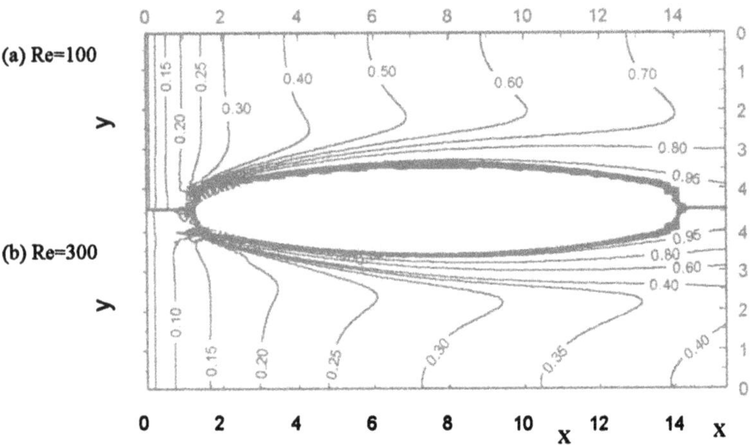

Figure 15 Local dimensionless mean fluid temperatures at Re=100 and 300. The dimensionless temperature is zero at the entrance and one on the tube.

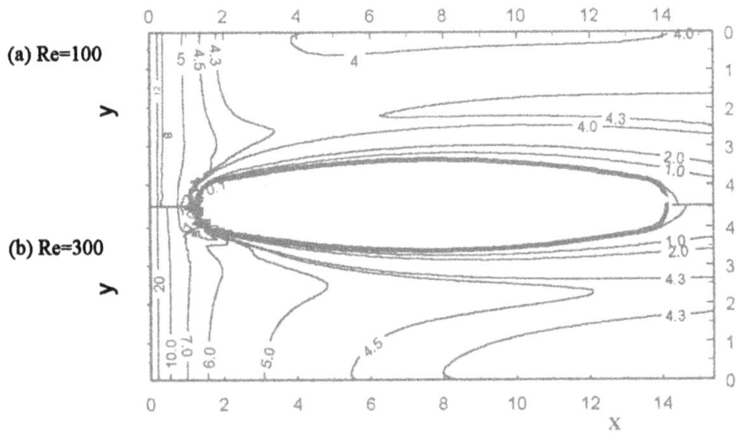

Figure 16. Isolines of Nusselt number distribution on the fin at Re=100 and 300.

Figure 17 shows the span averaged Nusselt number distributions for Re=100, 300 and 500 respectively. Very large gradients of Nusselt number appear in the leading edge of the fin. The disturbances of the tube with the formation of horse-shoe vortices slow down this tendency. Small local peaks of Nu_{sp} between $x=1$ and $x=2$ result from the horse-shoe vortices. Heat flux decreases slowly in the streamwise direction after that. The Nusselt number for Re=100 is almost constant in the range of the tube. A further decrease of the heat transfer occurs in the wake of the tube, where fluid recirculates and thus heat transfer is very weak. Because of the very small

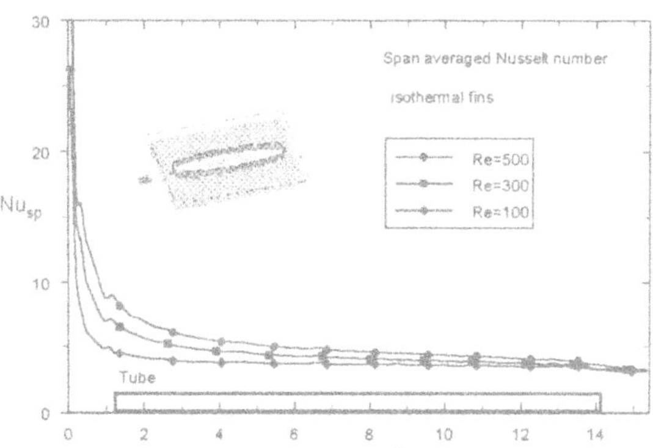

Figure 17. Span averaged Nusselt number distributions for Re=100, 300 and 500 respectively.

'dead water zone', decrease in the span-averaged value is not evident.

4.3 OVAL TUBE WITH WINGLET-FINS

4.3.1 Flow Pattern

For in-line winglet configurations, essentially one primary longitudinal vortex is generated by the winglets, see Figs. 18(a) and (c). Only immediately behind the second winglet, two closely spaced co-rotating vortical flows can be observed. They soon coalesce to one stronger vortex. From Fig. 18(c) we notice that the core of the vortex from the first winglet passes around the trailing edge of the second winglet, and then wraps the vortex core from the second winglet further downstream. For staggered winglets as in config. 12, two longitudinal vortex systems are observed, see Figs. 18(b) and (d). The induced vortices near the tube and to the lateral side of the fin are much weaker than the primary winglet vortices. The swirl generated by winglet B (the one away from the tube) is stronger than that generated by winglet A (the one near the tube). This results from the higher velocities encountered by the winglet further away from the tube and further downstream from the leading edge.

4.3.2 Temperature Distribution and Heat Transfer

The swirl and the developing boundary layers at the punched edges result in changes of the temperature distribution. For configuration 12 the temperature field in the flow passage is shown in Fig. 19. In regions of flow towards the wall (down-wash) the temperature boundary layers are thinner than in regions of flow away from the wall (up-wash). In Fig. 20, velocity vectors and Nusselt numbers (Nu) are shown in a cross section behind the winglets. The strong primary vortices, the weaker induced vortices and the secondary flow caused by the tube can be observed. On each side of the fin, three local maxima and minima of the Nu occur. They correspond to the down-wash and up-wash regions of the swirling flow. The average Nu of both sides of the fin is everywhere larger than that of a fin without winglets. But locally the minima may be lower than the Nu for the oval tube with plane fins. Nu is zero at the tube-fin juncture.

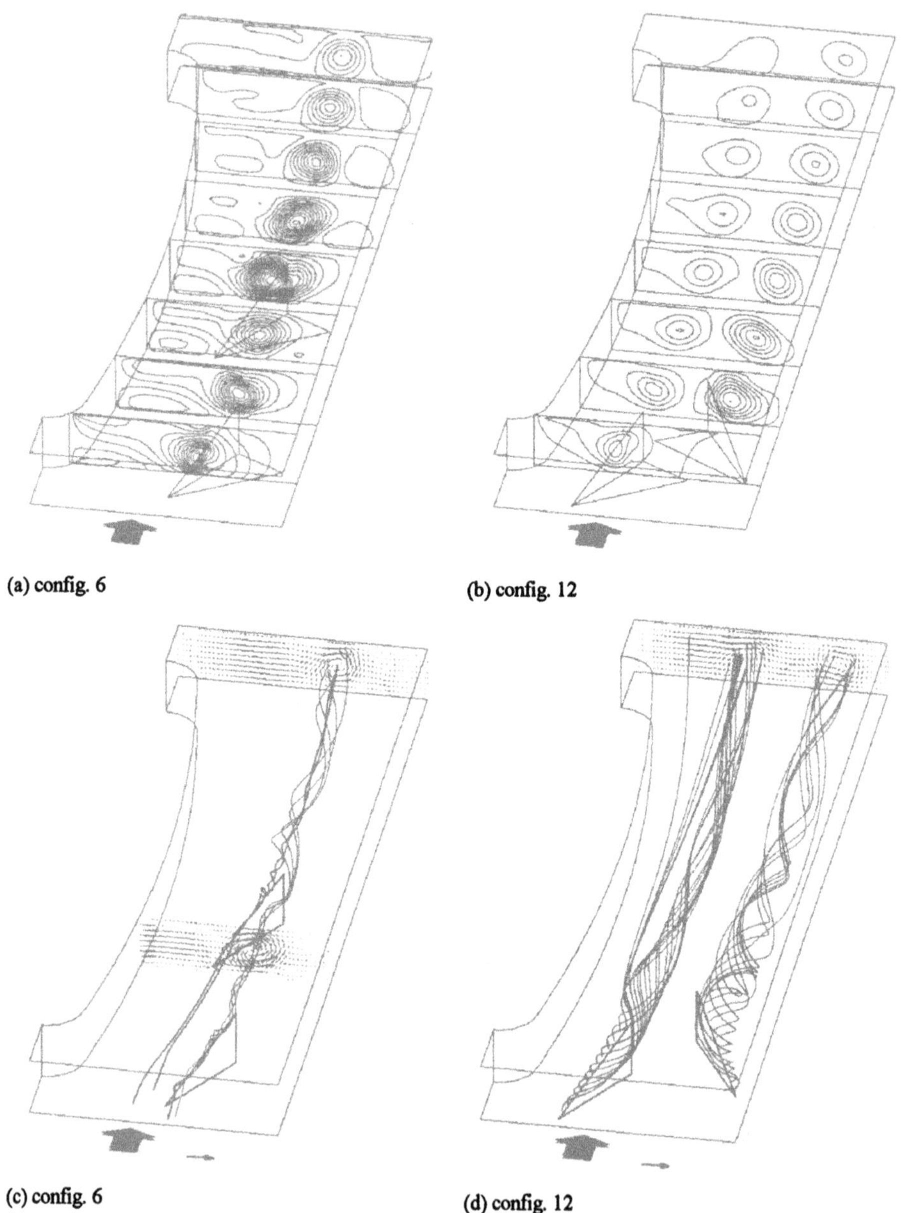

(a) config. 6

(b) config. 12

(c) config. 6

(d) config. 12

Figure 18. Vortex systems in configs. 6 and 12 at Re=300.
    (a)–(b): Streamlines of the secondary flow in eight cross sections located at 1.5, 3.6, 5.6, 7.7, 9.7, 11.8 13.8, and 15.4 respectively.
    (c): Selected streamlines of the vortex core in the config. 6.
    (d): Selected streamlines from the leading edges of the winglets in config. 12.

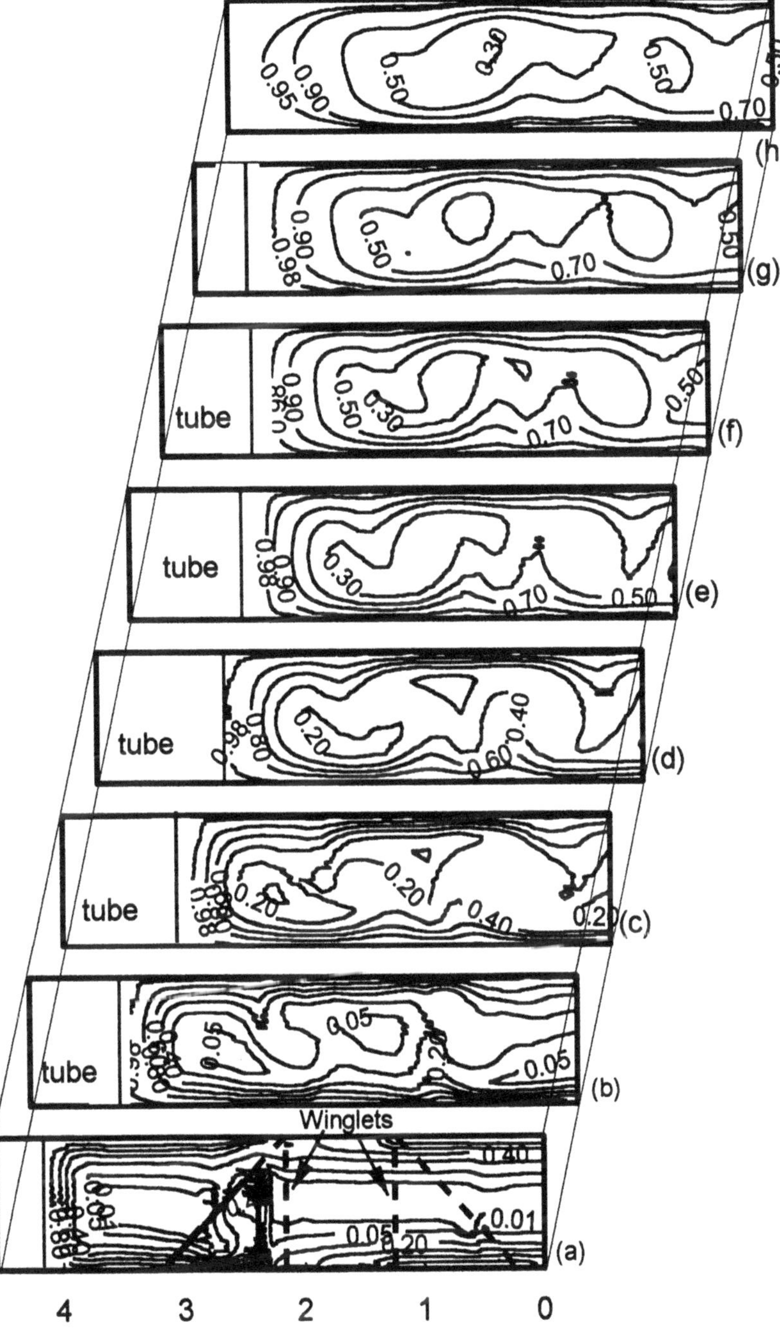

Figure 19. Fluid temperature distribution in eight cross sections of the duct in config. 12 at Re=300. The distorted isothermal lines show the mixing of the fluid by the swirling flow. From (a) to (h), x is located at 1.5, 3.6, 5.6, 7.7, 9.7, 11.8, 13.8, and 15.4 respectively. Solid or dash-lines show winglet parts upstream or downstream of the section in which winglets are drawn.

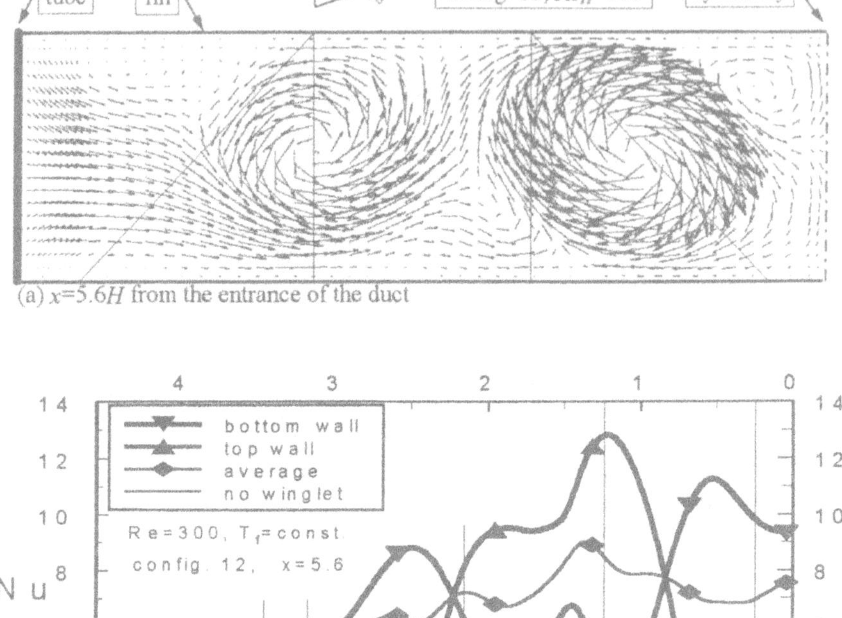

Figure 20. (a) Velocity vectors in the section of x=5.6 in config. 12 at Re=300 show: 1) the down-wash and up-wash regions of the primary vortices; 2) the induced vortices on the up-wash sides; and 3) the secondary vortices near the tube. (b) fin Nusselt number distributions in the same section as in (a). Fin Nusselt numbers become zero at the juncture with the tube.

Figure 21 shows heat fluxes in three sections for config. 11 with three winglets in-line and config. 0 with plane fins. The fin heat fluxes are the sum of both fin sides. It is the product of Nu of both sides and the difference between the fin and bulk temperature. The cross sections are indicated in Fig. 21a. They intersect the punched winglets. For the heat flux presentation, the winglets are assumed to be placed back into the punched areas. For the first section at x=1.8, the heat flux distribution is characterized as follows: Starting from the lateral edge, no enhancement over the plane fin value occurs as long as the fin is plane, y≤0.32. Between 0.32<y≤1.5, the heat flux is strongly enhanced up to a factor of about two through the developing boundary layers on both sides of the winglet. For y>1.5, the enhancement is due to the swirl. At first, the heat flux drops only weakly and then more strongly till it merges with the curve for the plane fin (config. 0) near y≈3. The local maximum near the tube at y≈3.8 is caused by the horse-shoe vortices generated by the tube. In the second and third sections x=6.7 and 11.6, the influence of the horse-shoe vortices cannot be detected any more. The second and third winglets have booster effects for the vortices. The local heat fluxes relative to the plane fin configuration are further enhanced up to values of about 2.5 at x=6.7 and up to values of about 2.3 for x=11.6. The decrease of heat flux enhancement in the third cross section relative to the second section is caused by the higher mean temperature relative to that of the reference configuration. In the second and third section, the distributions are similar, but enhancement occurs also for y<0.32 because the primary vortex from the winglet has spread laterally. The enhancement on the winglets and downstream of the punched edges is now not only caused by developing boundary layers but also by the swirl generated by the winglet upstream.

The span-averaged Nusselt number enhancements with respect to the oval tube with plane fins are shown in Fig. 22. In the regions where winglets are punched local maxima occurred followed by a decrease to a level higher than that upstream of the winglets. These maxima result from the combination of swirl effect on the fins and developing boundary layers on both sides of the winglets and the punched edges. On the winglets themselves, swirl effect also occurred if swirling flow was generated upstream. The two effects can not be separated. But the booster effect of the second row of winglet becomes clear. For config. 14, the local maximum of the second staggered winglet row reaches a value of 6.8 times the value of the plane fin, while it is about four for the first staggered row. The mean Nusselt number enhancements for configs. 3, 6, 12, and 14 are 16%, 32%, 66% and 121% of the Nusselt number of config. 0. The Nusselt number enhancement is related to the heat transfer enhancement by the factor of $(T_f - T_B)_0 / (T_f - T_B)$, where the subscript "0" stands for the value of config. 0.

The span-averaged pressures in Fig. 23 show that the form drag of the winglets is mainly responsible for the pressure loss penalty. Except for the punched winglets region all curves run essentially parallel. This was also the case in the experimental results of [22]. The pressure loss penalty for configs. 3, 6, 12 and 14 are 10 %, 29 %, 30% and 70%, respectively. Figure 24 shows the streamwise integrated heat fluxes for the same configurations and Re. The heat transfer enhancement for configs. 3, 6, 12 and 14 are 15%, 27%, 49% and 74%, respectively. Except for configuration 12, the heat transfer enhancement differs only slightly from the pressure drop penalty. For configs. 3 and 12 the heat transfer enhancement is about 50% higher than the pressure loss penalty. The two staggered winglets in config. 12 result in a much higher heat transfer enhancement with about the same pressure loss penalty as the two in-line winglets in config. 6. The second row of winglets in configurations 6 and 14 caused about 50% higher pressure losses than the first row but resulted in only about 60% and 50% additional heat transfer enhancement. The higher pressure losses of the second row are caused by the higher local mean velocities because of the displacement effect of the tube. The smaller global heat transfer enhancement by the second row is due to the lower basic heat transfer level at their streamwise position and the fact that the generated vortices act only over a shorter streamwise distance.

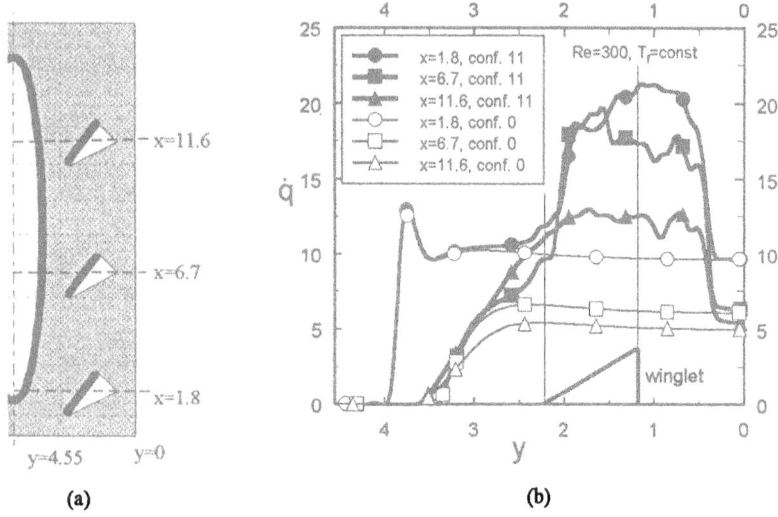

Figure 21. Heat fluxes in three sections of configs. 11 and 0 at Re=300. (a) locations, (b) distributions.

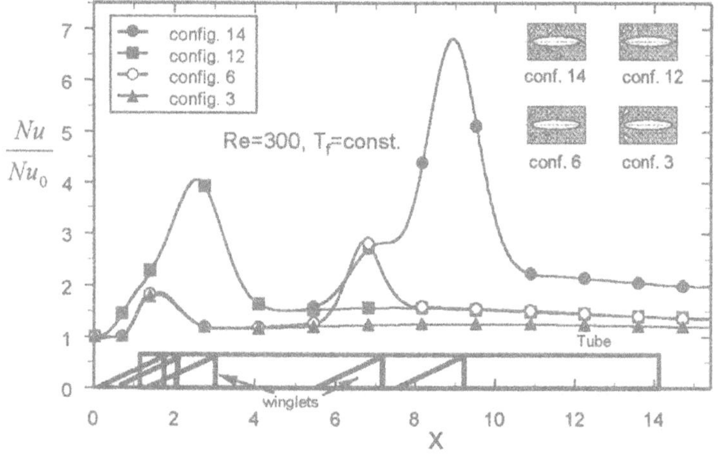

Figure 22. Span-averaged Nusselt number enhancement with respect to config. 0 at Re=300.

The variable geometry criterion, the VG1-criterion of Webb [32-33], is often used to evaluate the effectiveness of a heat transfer surface based on a comparison with a reference surface A_0. Conditions for the comparison are identical heat rate, pumping power, mass flow, and mean temperature difference between the two transport media of the heat exchanger. A smaller area ratio (A/A_0) leads to a smaller heat exchanger and correspondingly larger savings in material and manufacturing costs. Figure 25 shows the area ratios for configs. 3, 6, 12, and 14 with Re_0, the reference Reynolds number of the reference configuration, ranging from 100 to 500. (A/A_0) decreases with increasing

Re. The area saving remains within 20% for config. 3 and within 30% for config. 6. The staggered winglet arrangement of config. 12 provides better performance than the in-line winglet arrangement of config. 6, though both have two winglet pairs per tube element. Config. 14 provides the best performance. The area saving is more than 50% even at Re=100. This means that less than 50% of fin area is needed..

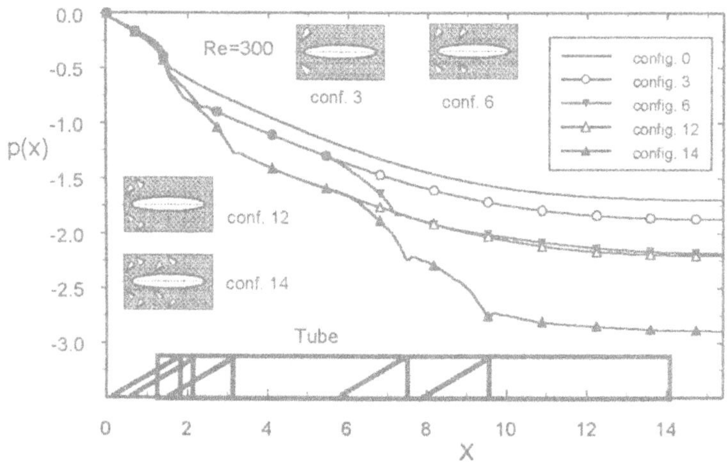

Figure 23. Span-averaged pressure distributions for five tube-fin configurations.

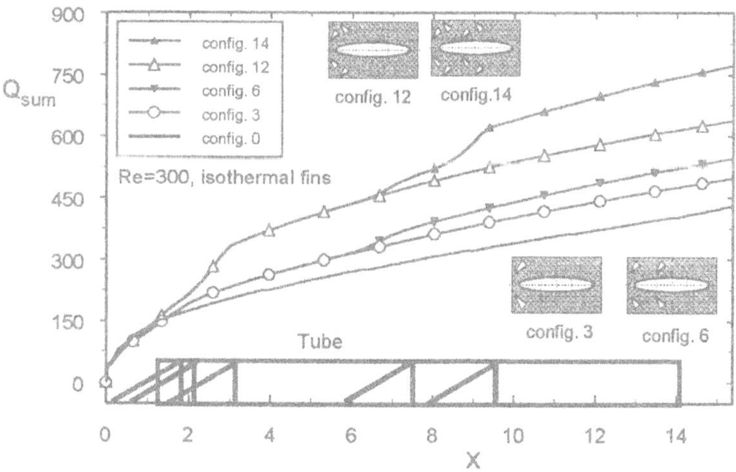

Figure 24. Streamwise integrated heat fluxes in five tube-fin configurations.

5. Concluding Remarks

For a plate fin heat exchanger element with fins consisting of an array of periodic rectangular winglet vortex generators , it was shown that these LVG may enhance heat transfer by creating developing boundary layers, swirl and large amplitude oscillations with large amplitude at Re=175. At this low Re, flow oscillations increased heat transfer by roughly 15%, while boundary layer development and swirl increased heat transfer by about 60%. Developing boundary layers and swirl lead to heat transfer enhancement at all Reynolds numbers. Flow oscillations and heat transfer enhancement induced by flow destabilization occur only when the Reynolds number is above the critical value, which depends strongly on geometry. For the investigated finned oval tube configurations with delta winglets as LVGs, the flow was steady in the considered Reynolds number range up to 500.

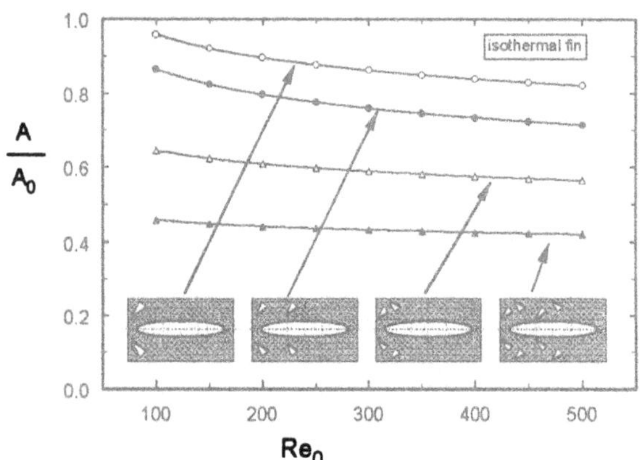

Figure 25. Ratios of heat transfer surfaces calculated with the VG1 criterion.

For the oval tube with plane fins, the reference configuration, the flow structure in the passage between the fins around the oval tube was complex due to the horse shoe vortices around the front region of the tube, and the flow separation in the rear part of the tube with the recirculation and helical vortex behind the tube. The corresponding temperature distribution was also complex and resulted in strongly varying fin heat transfer in the streamwise and lateral directions and very low fin heat transfer behind the tube and in the juncture region of fins and tube.

Finned oval tubes with five different delta winglet configurations were numerically studied for Re ranging from 100 to 500. The additional pressure drop was mainly caused by the form drag of the winglets. It was proportional to the exposed winglet area and the local mean velocity squared. The differences in pressure drop between staggered and in-line configurations were small when the projection of the total winglet area was the same. The heat transfer enhancement for staggered winglets was considerably higher than for in-line configurations for identical exposed total winglet area. The transport of fluid from the wall to the core, or vice versa, was more effective by two counter-rotating longitudinal vortices generated by the staggered arrangement of winglets than by essentially one vortex generated by in-line arrangement of winglets. For the staggered winglet arrangements, the outer winglets (those away from the tube) were more effective for heat transfer enhancement than the inner ones because the local mean velocity and the temperature difference between fin and fluid were larger.

The variation of the pressure loss and heat transfer with Reynolds number was similar for different configurations. Heat transfer enhancement and pressure loss penalty increased with increasing Re. The configuration which gave the best performance at one Reynolds number was also superior for the other Reynolds numbers. According to the VG1-criterion, more than 50% of the area could be saved for the staggered configuration with four pairs of delta winglets per tube element in the considered Re range. The winglets amounted to only 6.92% of the fin area. This demonstrated the effectiveness of LVGs for heat transfer enhancement, especially for low Reynolds numbers.

The influence of the winglet height has not been studied because here the winglets also functioned as pitch holders. Earlier investigations have shown that favorable winglet heights decrease with increasing Reynolds number.

The non-linear hydrodynamics does not allow a performance prediction when the fins are modified by differently arranged delta winglets.

This study did not attempt to optimize longitudinal vortex generator configurations for fin-tube or plate fin compact heat exchangers. But the presented results together with previous investigations indicated how should further studies might be pursued. Manufacturing possibilities and fin efficiency should always be kept in mind. But it has been shown by Chen & Fiebig [10] that the best configuration for fins with infinite conductivity was also the best for fins with finite conductivity. First the best winglet form could be established for a single row of winglets for the most compact lateral packing. Next the winglet height, angle of attack and arrangement on both sides of the fin could be varied till a relative performance maximum is reached. Then the longitudinal pitch could be varied to establish the best spacing in the streamwise direction. It might be advantageous to have higher LVG packing in the front part of the fin than in the rear part. Finally the effect of finite fin conductivity should be evaluated. When the fin efficiency becomes to low, the fin to tube area may be reduced.

Nomenclatures

Letters

A	area
a	thermal diffusivity
B	width
c_f	friction coefficient
H	channel height
h	winglet height
J	Jacobian
k	thermal conductivity
L	length
l	winglet length
p	pressure
$\dot{Q}_{sum}$	streamwise summed heat flux
T	temperature
t	time
u, v, w	velocity components
x, y, z	Cartesian coordinates

Subscripts:

0	reference
B	Bulk
cs	cross-sectional averaged
f	fin
i, j, k	indices
p	period
sp	span-averaged
T	tube

Superscripts

*	dimensional quantities

Dimensionless parameters

Nu	Nusselt number, $(\partial T/\partial z)_{z=\{0;1\}} / (T_f\text{-}T_B)$
Pr	Prandtl number, (v^*/a^*)
$\dot{q}$	heat flux, $((\partial T/\partial z)_{z=1}\text{-}(\partial T/\partial z)_{z=0})$
Re	Reynolds number, $(u_0^* H^*/v^*)$
T_B	bulk temperature, $T_B(x)=\iint |u|T dxdy/(|u|dxdy)$

Greek Symbols :

θ	dimensionless temp. in periodic channel flow	ϕ	general dependent variable
Λ	aspect ratio of delta winglets, l/h	ν	kinematic viscosity
β	angle of attack	ρ	density
δ	thickness	ξ, η, ζ	general coordinates

References

1. Behle, M. (1997) Wärme- und strömungstechnische Untersuchung stanzgeprägter Hutzen als Längswirbelerzeuger, Cuvillier Verlag, Göttingen.
2. Jacobi, A. M., Shah, R. K.,1995. Heat Transfer Surface Enhancement through the Use of Longitudinal Vortices: A Review of Recent Progress. Exp. Therm. Fl. Sci., 1995, 11:295-309.
3. Fiebig, M. (1995) Vortex Generators for Compact Heat Exchangers, Int. J. of enhanced Heat Transfer, Vol. 2, No. 1-2, pp. 43-61.
4. Fiebig, M. (1995) Embedded Vortices in Internal Flow: Heat Transfer and Pressure Loss Enhancement, Int. J. of Heat and Fluid Flow, Vol. 16, No. 5, pp. 376-388.
5. Fiebig, M. (1996) Vortices: Tools to Influence Heat Transfer--Recent Development, *Procs. of the 2nd European Thermal Sciences and 14th UIT Heat Transfer Conference*, Rome, Italy, 29-31 May.
6. Fiebig, M. (1997) Vortices and Heat Transfer, ZAMM, Z. angew. Math. Mech. Vol. 77, No. 1, pp. 3-18.
7. Fiebig, M. (1997) Wing-Type Vortex Generators Heat Transfer Enhancement Mechanisms and Potential for Heat Transfer Surfaces and Heat Exchangers, Procs. 15th UIT National Heat Transfer Conference, Vol. 1, pp. 3-22, Torino, 19-20 June 1997.
8. Fiebig, M. (1997) Vortices, Generators and Heat Transfer, Trans. IChemE, vol. 76, Part A, pp. 108-123, Feb. 1998..
9. Fiebig, M. & Mitra, N.K. (1998) Experimental and Numerical investigation of Heat Transfer Enhancement with Wing-Type Vortex Generators, in International Series on Developments in Heat Transfer: Computer Simulations in Compact Heat Exchangers, (eds. B. Sunden & M. Faghri), Computational Mechanics Publications, Southampton, UK and Boston, USA, Vol. 1, pp. 227-254.
10. Chen, Y., Fiebig, M. (1998) Effect of Fin Heat Conduction on the Performance of Punched Winglets in finned Oval Tubes. this book.
11. Müller, U. (1994) Wärmeübergang und Druckverlust bei Hochleistungswärmeübertragungsflächen mit periodischen Wirbelerzeugern. Bedeutung von Anstellsinkel und Randbedingungen, Diplomarbeit Nr. 94-08, Institut für Thermo- und Fluiddynamik, Ruhr-Universität Bochum.
12. Große-Gorgemann, A. (1996) Numerische Untrsuchung der laminaren oszillierenden Strömung und des Wärmeüberganges in Kanälen mit rippenförmigen Einbauten, VDI-Verlag GmbH, Düsseldorf.
13. Drazin, P.G. & Reid, W.H. (1984) Hydrodynamic Stability, Cambridge University Press.
14. Weber, D. (1996) *Experimente zu selsterregt instationären Spaltströmungen mit Wirbelerzeugern und Wärmeübertragung*, Cuvillier Verlag, Göttingen.
15. Van Doormaal, J. P., Raithby, G. D. (1984) Enhancements of the SIMPLE Method for Predicting Incompressible Fluid Flows, Numerical Heat Transfer, Vol. 7, pp.147-163.

16. Khosla, P.K., Rubin, S.G. (1974) A Diagonally Dominant Second-Order Accurate Implicit Scheme, Computer & Fluid, 2, pp. 207-209.
17. Stone, H. L. (1968) Iterative Solution of Implicit Approximations of Multidimensional Partial Differential Equations, SIAM J. Numer. Anal., Vol. 5, No. 3, pp. 530-558.
18. Kost, A. (1993) Strömungsstruktur und Drehmomentübertragung in hydrodynamischen Kupplung, VDI-Verlag GmbH, Düsseldorf.
19. Bai, L. (1995) Numerische Untersuchung von turbulenten Strömungen in hydrodynamischen Kupplungen, VDI-Verlag GmbH, Düsseldorf.
20. Peric, M. (1985) A finite Volume Method for the Prediction of Three-Dimensional Fluid Flow in Complex Ducts, Dissertation, University of London.
21. Chen, Y. (1998) *Leistungssteigerung von Wärmeübertragern mit berippten Ovalrohren durch Längswirbelerzeuger--Numerische Simulation von Strömung und konjugiertem Wärmeübergang*, Shaker Verlag, Aachen, Germany & Maastricht, Holland.
22. Fiebig, M., Kallweit, P., Mitra, N. K. (1986) Wing Type Vortex Generators for Heat Transfer Enhancement, *Proc. 8th Int. Heat Transfer Conf.*, vol.6, pp. 2909-2913.
23. Fiebig, M., Kallweit, P., Mitra, N. K., Tiggelbeck, S. (1991) Heat Transfer Enhancement and Drag by Longitudinal Vortex Generators in Channel Flow, *Experimental Thermal and Fluid Science*, Vol. 4, pp. 103-114.
24. Tiggelbeck, S., Mitra, N.K., Fiebig, M. (1994) Comparison of Wing-Type Vortex Generators for Heat Transfer Enhancement in Channel Flows, *Journal of Heat Transfer*, Vol. 116, pp. 880-885.
25. Güntermann, T. (1992) Dreidimensionale stationäre und selbsterregtschwingende Strömungs- und Temperaturfelder in Hochleistungswärmeübertragern mit Wirbelerzeugern, VDI-Verlag GmbH, Düsseldorf.
26. Fiebig, M., Valencia, A., Mitra, N.K. (1994) Local Heat Transfer and Flow Losses in Fin-and-Tube Heat Exchangers with Vortex Generators, a Comparison of Round and Flat Tubes, *Experimental Thermal and Fluid Sciences*, Vol. 8, pp. 35-45.
27. Chen, Y., Fiebig, M., Mitra, N. K. (1998). Heat Transfer Enhancement of a Finned Oval Tube with Punched Wing-Type Longitudinal Vortex Generator, accepted as general paper for presentation at and for publication in the *Proc. of the 11th IHTC*, 23-28 August 1998, Kyongju, Korea.
28. Chen, Y., Fiebig, M., Mitra, N. K., (1998) Conjugate Heat Transfer of a Finned Oval Tube, Part A: Flow Patterns, *Numerical Heat Transfer, in press.*
29. Chen, Y., Fiebig, M., Mitra, N. K. (1998). Conjugate Heat Transfer of a Finned Oval Tube, Part B: Heat Transfer Behavior, *Numerical Heat Transfer,* in press
30. Chen, Y., Fiebig, M., Mitra, N. K. (1998) Conjugate Heat Transfer of a Finned Oval Tube with a Punched Longitudinal Vortex Generator in Form of a Delta Winglet — Parametric Investigations of the Winglet, *Int. J. Heat Mass Transfer,* in press .
31. Chen, Y., Fiebig, M., Mitra, N. K. (1998) Heat Transfer Enhancement of a Finned Oval Tube with Punched Longitudinal Vortex Generator In-Line, *Int. J. Heat Mass Transfer,* in press .
32. Webb, R. L. (1981) Performance Evaluation Criteria for Use of Enhanced Heat Transfer Surfaces in Heat Exchanger Design, *Int. J. Heat Mass Transfer,* Vol. 24, No. 4, pp. 715-726.
33. Webb, R. L. (1982) Performance Evaluation Criteria for Air-Cooled Finned Tube Heat Exchanger Geometries, *Int. J. Heat Mass Transfer,* Vol. 25, No. 11, pp. 1770.

EFFECT OF FIN HEAT CONDUCTION ON THE
PERFORMANCE OF PUNCHED WINGLETS IN FINNED OVAL TUBES

YUWEN CHEN and MARTIN FIEBIG

Institut für Thermo- und Fluiddynamik
Ruhr-Universität Bochum, 44780 Bochum, Germany

Abstract: Wing-type longitudinal vortex generators were employed for performance enhancement of finned tube heat exchanger elements. The effect of finite fin heat conduction on the performance of fins and longitudinal vortex generators was considered. Three-dimensional developing laminar flows and conjugate heat transfer in high performance finned oval tube elements, with and without punched delta-winglets, were studied numerically with a Finite-Volume Method. Body-fitted grids were used to satisfy exactly the thermal and hydrodynamic boundary conditions. The conjugate heat transfer was realized by iterations of convection in the flow field and conduction in the fin. Reynolds numbers of $100<Re<500$, and fin conduction parameters of $100<Fi<1000$ were varied. Temperature fields, local fin heat transfers and fin efficiencies were presented. Performance of finned oval tube elements with up to four in-line or staggered winglets was evaluated. For the investigated base configuration, performance enhancement through winglets increased with increasing Reynolds number and fin conductivity. For performance comparisons of different configurations, results of isothermal fins could be transferred to non-isothermal fins. The performance enhancement by winglets for fins with finite conduction may be reduced by about 20% in practical applications compared to that for isothermal fins.

1. Introduction

Finned tube heat exchangers (Fig. 1) find wide applications in automobiles, air-conditioners and chemical industry. Fins are used to reduce the thermal resistance by enlarging the transfer surface on the gas side where heat transfer is normally very poor due to the thermophysical properties of gases. The differences of heat transfer coefficients on the gas and the liquid sides may be so large that a very large fin-to-tube area ratio (up to two orders of magnitude) is needed to balance the resistance on both sides. But the efficiency and thus the economy of fins decreases quickly with increasing area ratio. Heat transfer enhancement on fins is therefore necessary. Longitudinal

S. Kakaç et al. (eds.), Heat Transfer Enhancement of Heat Exchangers, 107–122.

108

vortex generators (LVGs) are an effective modification for heat transfer enhancement. Mechanisms of heat transfer enhancement through LVGs -- developing boundary layers, swirl and flow destabilization -- are described in [1] and numerous references are cited there. Punched winglets are an effective kind of wing-type LVGs. Characteristics of heat transfer enhancement and pressure loss penalty in heat exchanger elements with LVGs are complex due to the nonlinear fluid dynamics and the various geometric parameters that affect the performance of LVGs. The characteristics are further complicated by the finite fin heat conduction, which results in non-homogeneous fin temperatures in finned tubes, and strongly non-homogeneous fin temperatures if winglets as those shown in Fig. 2 are punched out of the fin. On the punched boundaries and on the edges of winglets, local heat transfer maxima occur due to the developing boundary layers and swirl. As a consequence local fin temperature minima appear in those areas (assuming the tube as the heat source). Weakened heat transfer enhancement is the consequence. This contradiction occurs for heat transfer enhancement in extended surfaces. Here we consider this problem in finned oval tube with punched delta winglets as an example.

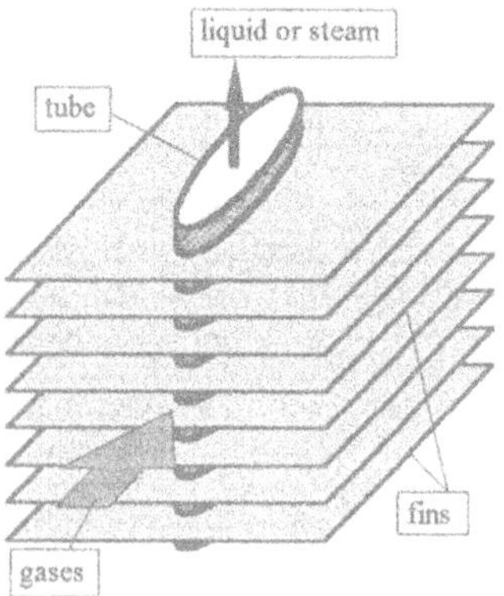

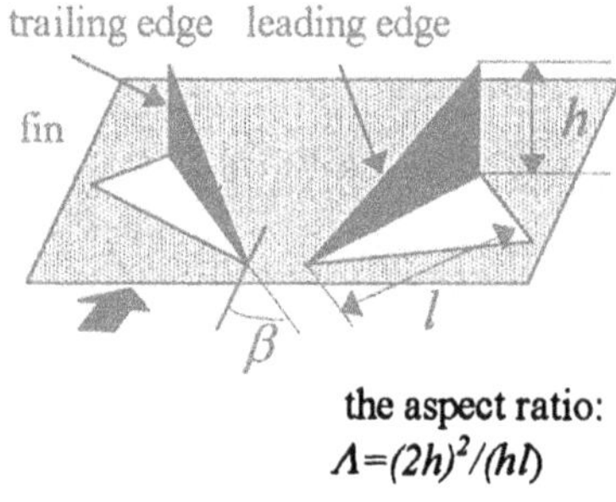

Figure 1 Finned oval tube heat exchanger core. The tube is the heat source in this work.

Figure 2 Schematic of a punched delta winglet pair. β: angle of attack, Λ: aspect ratio, h: winglet height, l: winglet length.

The fin heat fluxes and fin temperatures influence each other. It is called conjugate heat transfer of convection and conduction. Experimentally, detailed investigation of the conjugate heat transfer in heat exchanger elements with complex geometry is very difficult. For simple configurations with known or assumed flow structure and heat transfer coefficient, conjugate heat transfer can be approximated analytically [2, 3]. Early investigations of conjugate heat transfer in finned tube assumed known heat transfer coefficients or Nusselt numbers, and solved the heat conduction by different methods [4-6]. Moffat [7] concluded that Nusselt numbers measured with one set of boundary conditions can not be simply transferred to situations with other boundary conditions. The fin heat transfer in a finned tube varies strongly due to the complex flow structures between the fins and around the tube, e.g., the horse-shoe vortex around the front of the tube, flow separation on the tube and recirculation with a helical vortex in the tube wake [1, 8-9]. The interaction of these structures with the longitudinal vortices generated by the LVGs (here the winglets) results in an even more complex flow structure, and corresponding heat transfer distribution on the fins [1, 10-11].

Investigations with conjugate heat transfer will improve the performance evaluation of fins and LVGs in finned tubes. For numerical simulation, the conjugate heat transfer can be described by the Navier-Stokes equations with the energy equation for heat convection in the flow field and the heat conduction equation for the fin. For steady flows with small Biot number the influence of the fin material and thickness can be combined into a fin parameter Fi which is the product of the ratio of fin thickness to fin separation times the ration of fin thermal conductivity to fluid thermal conductivity [12-13]. For $Fi\rightarrow0$, no conduction in fins occurs and the fin efficiency becomes zero. For $Fi\rightarrow\infty$, the fin temperature is equal to the tube temperature and the fin efficiency becomes 100%. Generally Fi ranges from 10^2 to 10^3. For example, if air (thermal conductivity $k^*=0.026$ (W/mK) at 20°C) is used as fluid and δ^*/H^* is assumed to be 0.1, fins of aluminium (thermal conductivity $k^*=221$ (W/mK) at 20°C) have $Fi=850$, fins of steel with 0.6% of carbon content (thermal conductivity $k^*=50$ (W/mK) at 20°C) have $Fi=192$; and fins of stainless steel ($k^*\approx15$ (W/mK) at 20°C) have $Fi=58$.

The objective of this paper was the detailed investigation of temperature fields and heat transfer distributions in finned oval tubes with and without punched winglets to clarify the local and global differences resulted from the finite fin conduction. Of most importance was the study of transferability from results of isothermal fins to results of non-isothermal fins regarding the performance comparison of different configurations. This paper will present results of five configurations of finned oval tubes with and without punched winglets for Reynolds numbers ranging from 100 to 500 and for Fi parameters ranging from 100 to 1000. Velocity and temperature fields were simulated numerically. Performances of the configurations were evaluated.

2. Theoretical Formulation and Solution Procedure

2.1 THEORETICAL FORMULATION

Figure 3(a) shows a finned oval tube element with a delta winglet pair. Two fins of thickness δ form a channel of height H, width $B(=9.1H)$ and length $L(=15.4H)$. An oval tube ($L_T/B_T=5.5$, cross sectional area $A_{T,CS}=24.6H^2$) is located at the center of the fin. The area ratio of fin to tube is 8.4. Delta winglet pairs with $\beta=30°$ and $\Lambda=2$ are punched out of the fin. The height of the winglets (h) is equal to the channel height H, so that the winglet can also function as a spacer of fins. The thickness of the winglet is assumed to be zero. For small Reynolds numbers, the flow is steady and the computational domain can be reduced by using symmetry conditions on the mid-plane of the channel ($y=B/2$). The domain is extended by $6H$ over the rear edge of the fin in

x-direction to guarantee a recirculation-free flow at the exit. Figure 3(b) shows the computational domain with the employed boundary conditions.

Fig. 3(c) shows the investigated configurations (they are not numbered sequentially). There is no winglets in config. 0 (not shown), one delta winglet pair in config. 3, there are two in-line delta winglet pairs in config. 6, two staggered delta winglet pairs in config. 12 and four staggered delta winglet pairs in config. 14. The locations of the winglets are shown in [1].

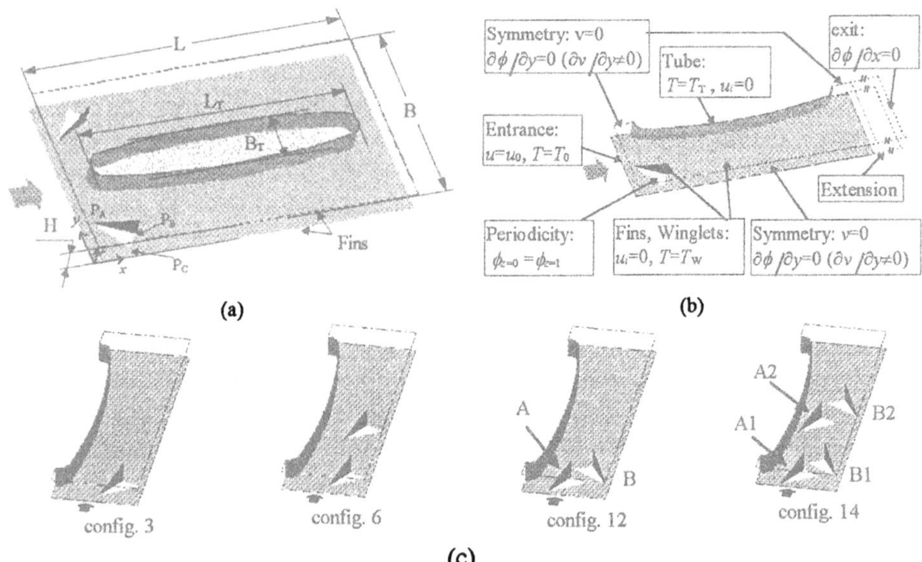

Figure 3 (a) Geometrical model, (b) computational domain with boundary conditions, and (c) the investigated configurations. In (a) and (b) only one DWP is shown. $T_0=0$, $u_0=1$, $T_T=1$.

The velocity and temperature fields in the channel were calculated by solving the three dimensional Navier-Stokes and energy equations for an incompressible fluid with constant properties. The temperature field in the fin was obtained by solving the conduction equation. In curvilinear co-ordinate system with Cartesian velocity components, these equations in dimensionless form and in index notation are:

$$\frac{\partial U_i}{\partial \xi_i}=0 \tag{1}$$

$$J\frac{\partial u_k}{\partial t}+\frac{\partial}{\partial \xi_i}\left[U_i\,u_k-\frac{1}{Re}\frac{1}{J}\left(B_j^i\frac{\partial u_k}{\partial \xi_j}+\beta_j^i\,\omega_k^j\right)+p\,\beta_k^i\right]=0 \tag{2}$$

$$J\frac{\partial T}{\partial t}+\frac{\partial}{\partial \xi_i}\left[U_i\,T-\frac{1}{Pe}\frac{1}{J}\left(B_j^i\frac{\partial T}{\partial \xi_j}\right)\right]=0 \tag{3}$$

$$\frac{1}{Ft}J\frac{\partial T_f}{\partial t}-\frac{\partial}{\partial \xi_i}\left[\frac{1}{J}\left(B_j^i\frac{\partial T_f}{\partial \xi_j}\right)\right]=\frac{1}{Fi}J\,\dot{q} \tag{4}$$

with $\quad U_i = u_j \beta^i_j, \qquad B^i_j = \beta^i_k \beta^j_k, \qquad \omega^i_j = \dfrac{\partial u_i}{\partial \xi_k} \beta^k_j$

and $\quad Re = \dfrac{u^*_0 H^*}{v^*}, \qquad Pr = \dfrac{v^*}{a^*}, \qquad Pe = Re \cdot Pr, \qquad Fi = \dfrac{\delta^* \, k^*_f}{H^* \, k^*}, \qquad Ft = \dfrac{1}{Pe} \dfrac{a^*_f}{a^*}$

where the Einstein summation convention applied and dissipation is neglected. The coefficients β^i_j are the cofactors of $\partial x_i / \partial \xi_j$ in the Jacobian J of the co-ordinate transformation $x_i = x_i(\xi_j)$. The conduction in the fin is considered as two-dimensional because the Biot number on the fin was very small. In eqs. (1) to (4), the time is nondimensionalized by H^*/u^*_0, all lengths were by H, and the velocity components are by the average velocity u^*_0 at the entrance. The temperature is the difference between the local and the inlet temperature scaled by the difference of the tube temperature and the inlet temperature $(T^* - T^*_0)/(T^*_T - T^*_0)$. The pressure is the difference between the local and a reference pressure divided by $\rho^* u^{*2}_0$.

The dimensionless groups Re, Pr, and Pe are the Reynolds number, the Prandtl number and the Peclet number, respectively, which determine the flow and heat transfer in a duct. The conjugate heat transfer is further described by two parameter Fi and Ft. The convective heat removal to both sides of the fin $\dot{q}$ couples the convective heat transfer to the fins and conductive heat transfer in the fins. It is defined as:

$$\dot{q} = \dot{q}\big|_{z=0} - \dot{q}\big|_{z=1}$$

with

$$\dot{q}\big|_{z=\{0;1\}} = \frac{1}{J}\left(\frac{\partial T}{\partial \xi}\beta^1_3 + \frac{\partial T}{\partial \eta}\beta^2_3 + \frac{\partial T}{\partial \zeta}\beta^3_3 \right)_{z=\{0;1\}} \tag{5}$$

and it is nondimensionalized by dividing by $(T^*_T - T^*_0)k^* / H^*$.

2.2 NUMERICAL SOLUTION

A grid generation technique was used to discretize the computational domain into a number of control volumes by solving the Poisson equation [14]. The differential eqs. (1-4) were discretized by a finite-volume method [15]. The SIMPLEC- algorithm [16] was used for the pressure-velocity correction. To avoid decoupling of the velocity and pressure field caused by the co-located arrangement of the variables, a momentum interpolation was applied [17]. The algebraic equation system was solved by the strongly implicit procedure of Stone [18]. Details of the solution procedure and validation can be found in [19-23]. Very good agreement of the numerical and experimental Nusselt number distributions is noted in [20]. The numerical solution of the velocities and temperature equations for one time interval was performed by the following steps:

112

1. The momentum equations were solved with the best available initial values;
2. A pressure-velocity correction based on SIMPLEC was performed;
3. Steps 1 and 2 were repeated until a divergence-free velocity field were obtained;
4. The energy equation was solved with the known velocity field and a best available temperature field on the fin and in the fluid;
5. The local heat flux to the fin $\dot{q}\,(x,y)$ was calculated;
6. The heat conduction equation was solved with the local heat flux as heat sources.
7. Steps (4) to (6) were repeated until a converged solution was obtained. The convergence criterion was set to be 10^{-5}.

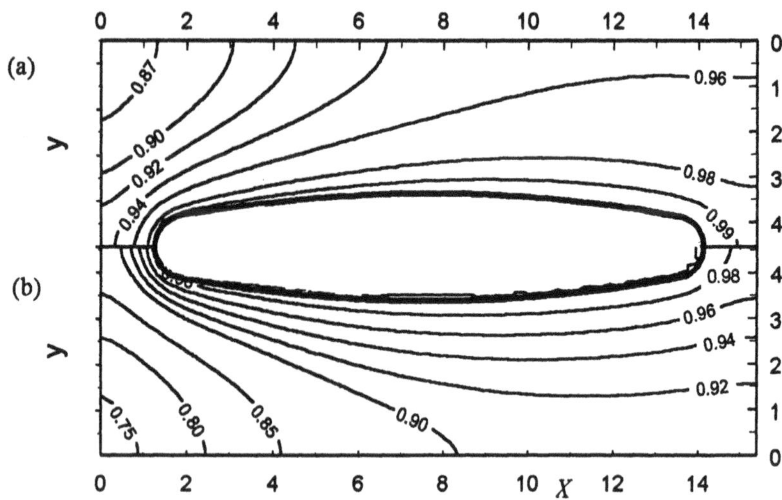

Figure 4. Fin temperatures in a finned oval tube with plain fins at (a) *Re*=100 and (b) *Re*=300 with *Fi*=500.

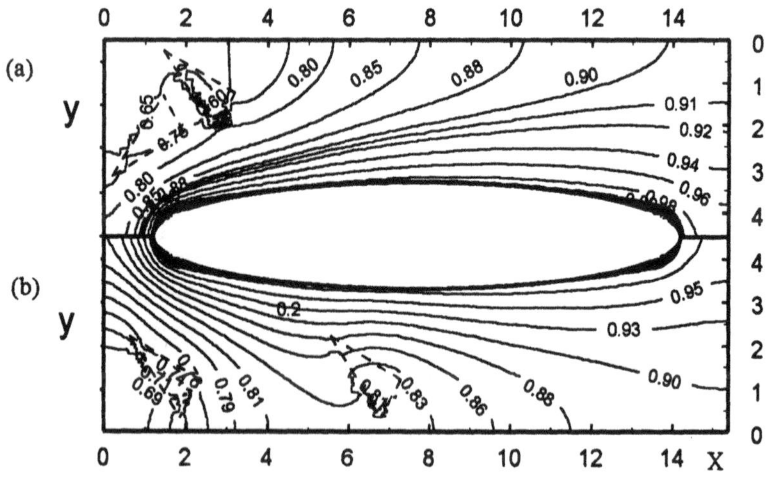

Figure 5. Fin temperatures in (a) config. 12 and (b) config. 6 at *Re*=300 and *Fi*=500.

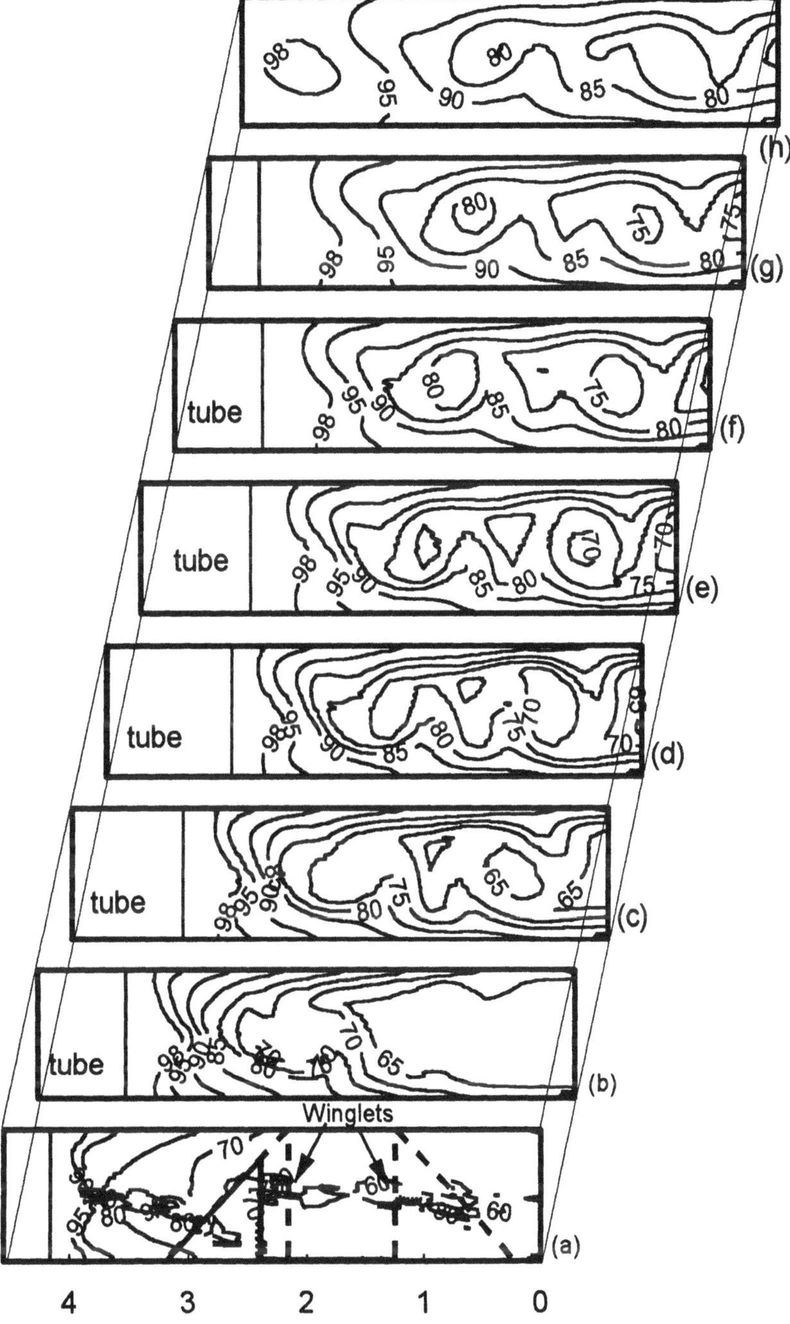

Figure 6. Ratios of the local fluid temperatures with Fi=500 to those with *Fi* → ∞ at *Re*=300 in config. 12. Solid or dashed lines indicate that the parts of the winglets are upstream or downstream of the respective section. The sections from (a) to (h) are located at *x*=1.5, 3.7, 5.6, 7.7, 9.8, 11.8, 13.8, and 15.4 respectively.

114

3. Results and Discussion

3.1 TEMPERATURE FIELDS

The flow patterns and vortex formations in finned oval tubes without and with punched winglets are presented in [1]. To demonstrate the effect of fin heat conduction on the fin heat transfer, selected results for finned oval tube with and without punched winglets were compared. Figure 4 shows the fin temperatures in config. 0 at $Re=100$ and 300 with $Fi=500$. Generally the difference of tube and fin temperatures remained small in the investigated range of Reynolds numbers and Fi values. The minimum temperatures appear around the corner of the leading and side edge. Maximum temperatures appear near the tube and in the tube wake. The fin temperature distributions for $Re=100$ and $Re=300$ are quite similar. Fin temperatures decrease with increasing Reynolds numbers. Increasing Re from 100 to 300, the minimum fin temperatures decrease by more than 10% of the tube temperature ($T_T=1$), while the fin temperatures in the tube wake differ only about 1% to 2% of the tube temperature from each other. For fins with punched winglets, the local fin temperatures around the winglets were lowered because of the enhanced heat transfer resulting from developing boundary layers on the winglet faces and the swirl of the longitudinal vortices, (see Fig. 5). In addition, punching interrupts the heat conduction. Near the leading edge, fin temperatures in config. 12, in which two winglets near the leading edge of the fin are punched, were lower than those in config. 6, in which one winglet near the fin leading edge is punched. Comparing the fin temperatures near the leading edges in Fig. 4(b) and Fig. 5(a), we noticed that local fin temperature was reduced by about 20% near the winglet. Near the side edge, fin temperatures in config. 6 were lower than those in config. 12. In the vicinity of the tube and in the tube wake, fin temperatures in both configurations were about the same, which means that heat transfers in that area are not much affected by the winglets. Because the dimensionless fin temperature of a isothermal fin is unity, the presented dimensionless fin temperatures can be regarded as ratios of the fin temperatures with finite conduction to those with $Fi \rightarrow \infty$.

Except for the lowered values, the fluid temperature distributions of non-isothermal fins are similar to those of isothermal fins [1]. To study the local differences of the temperature field, ratios of the local fluid temperatures of non-isothermal fins and isothermal fins $(T(x,y,z)|_{Fi} \,/\, T(x,y,z)|_{Fi \rightarrow \infty})$ in percentage were calculated. Figure 6 shows the ratios of fluid temperatures with $Fi=500$ to those with $Fi \rightarrow \infty$ at $Re=300$ in eight sections of config. 12. Higher percentage of the temperature ratio indicates smaller temperature difference between the isothermal fins and non-isothermal fins; and vice versa. Generally speaking, temperature differences near the lateral plane of the fins, upstream of the passage, and in the core of the flow are larger than those near the tube, downstream of the channel, and near the fins respectively. Minimum temperature differences appeared near the tube because the tube temperature is constant. Maximum differences occur near the entrance of the passage due to the leading edge effect. Local maxima occurs near the vortex-cores. The gradual decrease of the temperature difference in the streamwise direction suggests that heat transfer in the rear part of the fins with $Fi=500$ may be larger than that with isothermal fins. This phenomenon is called "heat transfer excess" which will be discussed later.

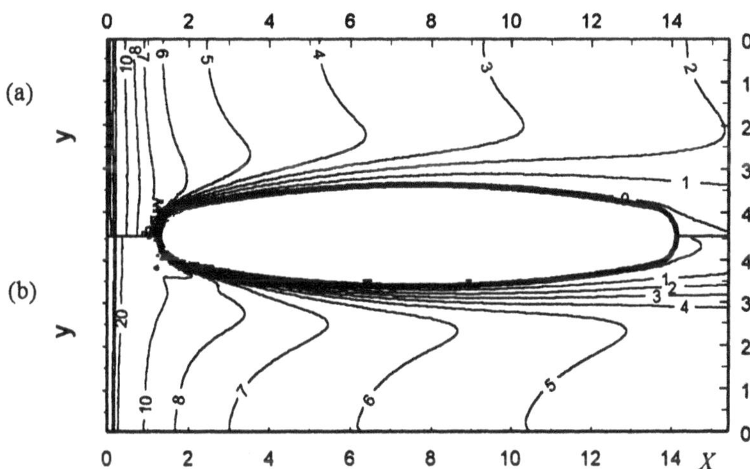

Figure 7. Fin heat fluxes in a finned oval tube with plain fins at (a) $Re=100$ and (b) $Re=300$ with $Fi=500$.

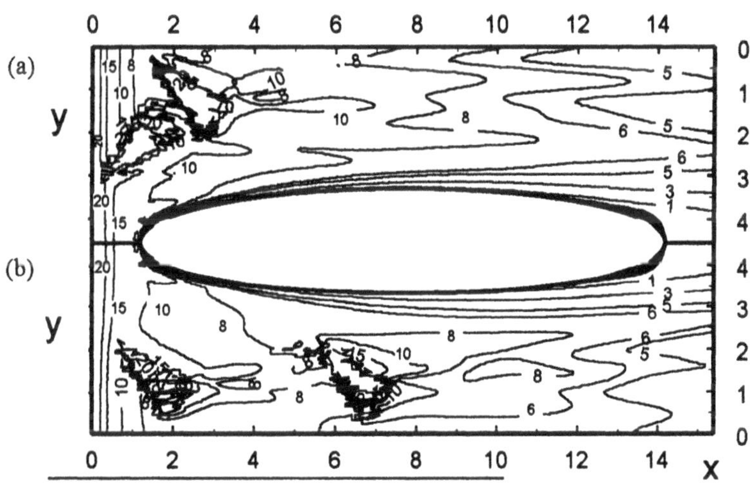

Figure 8. Fin heat fluxes in (a) config. 12 and (b) config. 6 at $Re=300$ and $Fi=500$.

3.2 FIN HEAT FLUXES AND FIN EFFICIENCIES

Figure 7 shows the fin heat fluxes $\dot{q}(x,y)$ for config. 0 at $Re=100$ and 300 with $Fi=500$. Near the fin leading edge $\dot{q}(x,y)$ are high. They decrease strongly in the main flow direction. The local maxima in the area of $2<y<3$ concurs with the local maxima of the streamwise velocity [1]. For $Re=300$, heat fluxes around $x=10$ are about the same

116

values as those around x=2 for Re=100. In the tube wake and in the vicinity of the tube except near the front stagnation area, heat fluxes are very small. The punched winglets changed the values and the distribution of fin heat flux behind the winglets (Fig. 8). Beginning with the winglets, the effects of developing boundary layers and swirl enhanced the heat transfer. The distributions of heat fluxes were complicated compared to those in config. 0. The vortices in config. 12 covers a larger fin area than those in config. 6. The zigzag curves behind the winglets resulted from the down-wash and up-wash effects of the vortices [1]. Heat fluxes on the winglet faces are high and varied strongly. This made it difficult to present them clearly. Heat transfers near the tube and in the tube wake were hardly affected by the vortices.

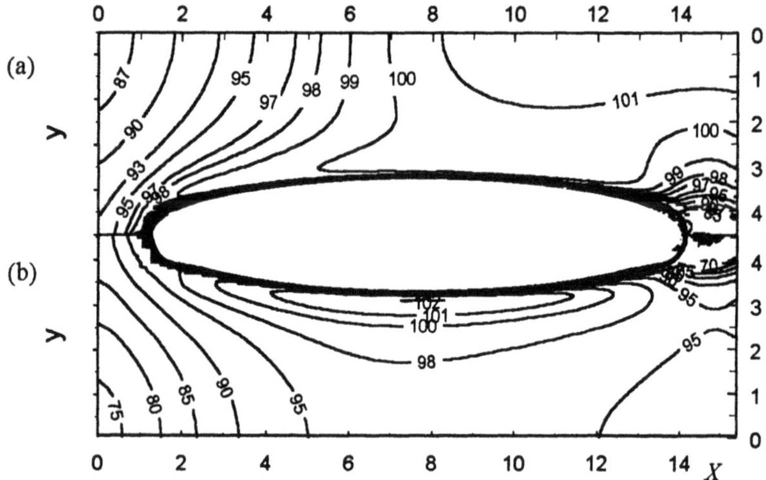

Figure 9. Local fin efficiencies in a finned oval tube with plain fins at (a) Re=100 and (b) Re=300 with Fi=500.

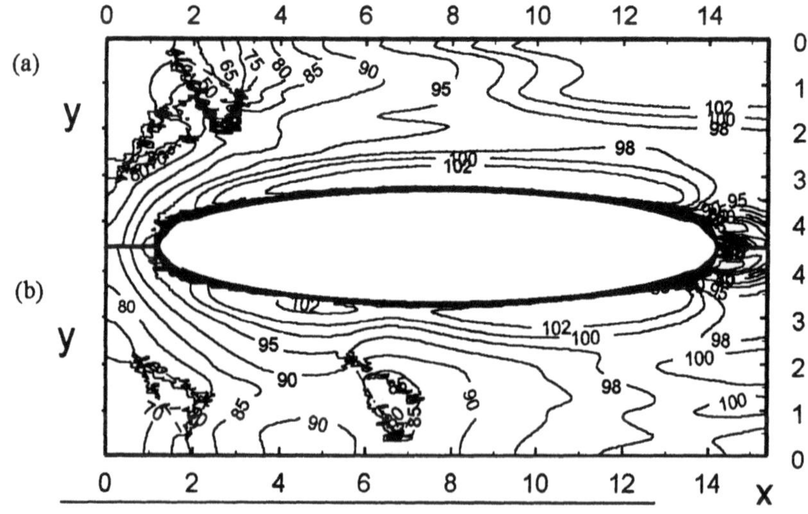

Figure 10. Local fin efficiencies in (a) config. 12 and (b) config. 6 at Re=300 and Fi=500.

The fin efficiency η is a measure of the fin thermal performance. It is defined as the ratio of the heat rate of a real fin to that of an isothermal fin having the base temperature (here the tube temperature) and thus the maximum driving temperature difference. The fin efficiency depends on the conduction in the fins, the convection to the fins and the shape and measure of the fins. Due to the fact that higher and lower global fin efficiencies do not direct related to higher or lower heat transfer, but only indicate the effect of fin temperatures on heat transfer, we calculated only the local fin efficiencies to demonstrate the local effect of the fin heat conduction. Figures 9(a-b) show the local fin efficiencies of config. 0 at Re=100 and 300 with Fi=500. Fin efficiencies near the leading edge, especially near the corner of the leading and side edge, are low due to the strong heat transfer, the long distance to the tube, and thus the low fin temperature there (see Figs. 4 and 7). Near this corner, the fin efficiency for Re=300 is about 10% lower than that for Re=100 with the same Fi value. Lower fin efficiencies also appear in the tube wake, though heat fluxes are small and fin temperatures are high there. This results from the recirculation with a helical vortex in that area [1]. Note that the fin efficiencies may become larger than 100%, locally in the downstream part of the fins. This phenomenon was called "heat transfer excess" in [8]. It results from the strong streamwise variation of the convection together with the finite fin conduction. For isothermal fins, the high heat transfer near the leading edge together with the large driving temperature difference results in higher bulk temperature downstream than fins with finite conduction. The fin temperatures in the downstream part are quasi-isothermal in the investigated range of Fi and Re, while the fluid temperature is lower for fins with finite conduction. This lead to higher driving temperature differences in the downstream part of the fins with finite conduction than isothermal fins. The heat transfer coefficients are about the same in that region for both cases. "Heat transfer excess" is the consequence. For Re=300, the area with η>100% appears only in the vicinity of the tube, while for Re=100 it covers about half the fin area. Figures 10(a-b) show the local fin efficiencies in configs. 12 and 6 at Re=300 and Fi=500. The fin efficiencies around the winglets are lower than those in config. 0. For config. 6, fin efficiencies around the first winglet ranges from 70% to 80%, and around the second winglet from 80% to 85%. For config. 12, fin efficiencies near the winglets is as low as 50%. The area of "heat transfer excess" becomes larger than that in config. 0. For config. 12, heat transfer excess occurs near the tube as well as near the corner of the side and trailing edge of the fin. For config. 6, the area of heat transfer excess near the tube is affected by the vortices of the second winglet.

3.3 PERFORMANCE COMPARISON

To compare the performance of the investigated heat transfer surfaces, the VG1-criterion by Webb [24-25] was employed. Conditions for the comparison in VG1-criterion are identical heat rate, pumping power, mass flow and mean temperature difference between the transport media. The criterion evaluates the effectiveness of a heat transfer surface by calculating the area ratio (A/A_0), where A_0 is a reference surface (Here A_0 refers to the area with config. 0). A smaller area ratio means a larger area saving. To calculate (A/A_0), correlations of $j(Re)$ and $f(Re)$ for each configuration were developed, where j was the Colburn factor and f the friction factor. They are defined as:

$$j = \frac{Nu_m}{RePr^{1/3}} \qquad , \qquad f_{app} = 2\Delta p \frac{A_{frt}}{A_{ht}}$$

where Nu_m is the global mean Nusselt number, Δp the pressure drop over an element, A_{frt} the frontal area and A_{ht} the heat transfer area.

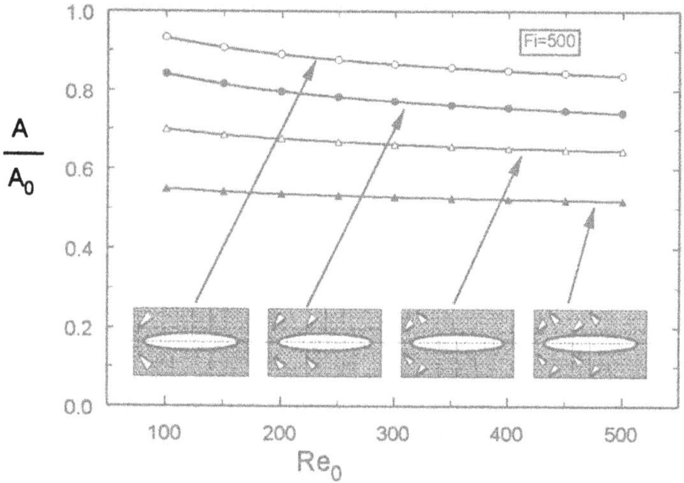

Figure 11. Area ratios for configs. 3, 6, 12 and 14 with respect to config. 0 according to the VG1-criterion .

Figure 11 shows the dependence of area ratios on Re for configs. 3, 6, 12, and 14 in $100 < Re_0 < 500$. The values of (A/A_0) decreases with increasing Re, i. e., the performance of winglets increases with increasing Re. The dependence of (A/A_0) on Re decreases with increasing number of winglets. The area saving remains within 20% for config. 3 and within 30% for config. 6. The two staggered winglets in config. 12 provides better performance than the two in-line winglets in config. 6. For $Re=500$ and $Fi=500$, (A/A_0) is 0.645 and 0.741 for config. 12 and config. 6, respectively. The pressure differences for both configurations are almost the same [1]. Config. 14 with four staggered delta winglet pairs provides the best performance. For $Re=500$, the area saving is 48.1%. Compared to the results in [1], we notice that a configuration with better performance for isothermal fins keeps its better performance for non-isothermal fins. Figure 12 shows the dependence of area ratios on Fi, the conduction capacity of the fin. The curves have the same tendency on Re. The area ratios decrease with increasing Fi. The distance between the curves increases slightly with increasing Re. For better comparison, the area savings $(1-A/A_0)$ at several Re and Fi are listed in Table 1. For $Re=500$, the area saving for non-isothermal fins with $Fi=100$, 500 and

1000 is reduced by 23.0%, 18.8%, and 15.7% respectively compared to the value with isothermal fins.

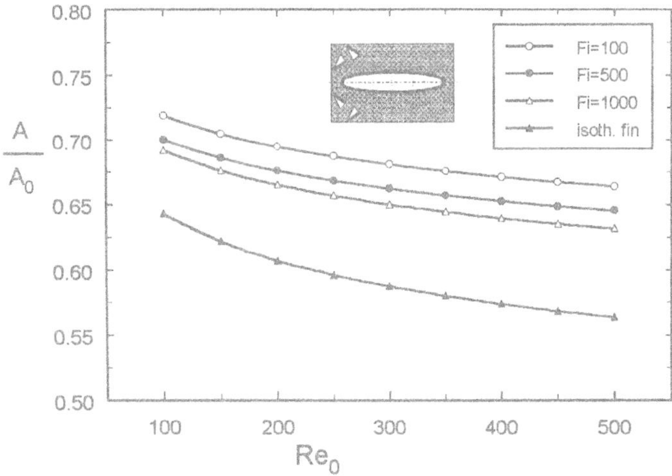

Figure 12. Area ratios for config. 12 at different *Fi* values with respect to config. 0 according to the VG1-criterion .

Table 1   Area savings ($1-A/A_0$) at several *Re* and *Fi* for config. 12.

Re \\ Fi	100	500	1000	∞
100	0.2810	0.2997	0.3080	0.3568
300	0.3191	0.3376	0.3496	0.4124
500	0.3361	0.3546	0.3682	0.4366

4. Concluding Remarks

Three-dimensional flow and conjugate heat transfer of finned oval tubes with a fin-to-tube area ratio of 8.4 were calculated for plain fins and fins with punched winglets in staggered or in-line arrangement. Reynolds numbers *Re* in the range of 100 and 500 as well as fin conduction parameters *Fi* in the range of 100 and 1000 were varied. Performances for different configurations and different parameters were evaluated.

- The performance enhancement by winglets increased with increasing Reynolds number *Re* and increasing *Fi*. The dependence of performance enhancement on *Re* decreased with increasing number of winglets.

- With respect to a finned oval tube without winglets (config. 0), about 50% of the heat transfer area could be saved by employing four staggered winglets (config. 14) at Fi=500.
- Regarding the performance comparison of different configurations, results of isothermal fins could be transferred to non-isothermal fins.
- The area saving for non-isothermal fins may be reduced by about 20% in practical applications compared to that for isothermal fins.
- In the vicinity of punched winglets, local minina of fin temperatures and local maxima of heat fluxes occurred due to the heat transfer enhancement by developing boundary layers and swirl.
- In the rear part of the fin, heat fluxes for non-isothermal fins might be larger than those for isothermal fins.
- The fluid temperature difference between non-isothermal fins and isothermal fins decreased in the streamwise direction.

Nomenclature

Letters

A	area
a	thermal diffusivity
B	width
H	channel height
h	winglet height
J	Jacobian
k	thermal conductivity
L	length
l	winglet length
p	pressure
T	temperature
t	time
u, v, w	velocity components
x, y, z	Cartesian coordinates

Greek Symbols :

Λ	aspect ratio of delta winglets, l/h
β	angle of attack
δ	thickness
ν	kinematic viscosity
ρ	density
$\xi_{i,j,k}$	general coordinates
ϕ	general dependent variable

Subscripts:

0	reference
B	Bulk
f	fin
frt	frontal
ht	heat transfer
i, j, k	indices
m	mean
T	tube
w	wall

Superscripts

*	dimensional quantities

Dimensionless parameters

Fi	fin parameter, $(\delta^*/H^*)\cdot(k_f^*/k^*)$
Ft	fin parameter, $(1/Pe)\cdot(a_f^*/a^*)$
Nu	Nusselt number, $(\partial T/\partial z)_{z=\{0,1\}} / (T_f-T_B)$
Pr	Prandtl number, (ν^*/a^*)
$\dot{q}$	heat flux, $((\partial T/\partial z)_{z=1}-(\partial T/\partial z)_{z=0})$
Re	Reynolds number, $(u_0^* H^*/\nu^*)$
$T_B(x)$	bulk temperature, $\iint |u|T\,dx\,dy/(|u|\,dx\,dy)$
$\eta(x,y)$	fin efficiency, $\dot{q}(x,y)/\dot{q}(x,y)_{Fi\to\infty}$

References

1. Fiebig, M. & Chen, Y. (1998) Heat Transfer enhancement by Wing-Type Longitudinal Vortex Generators and Their Application to Finned Oval Tube Heat Exchanger Elements, this book.

2. Harper, D.R. & Brown, W.B. (1923) Mathematical Equation for Heat Conduction in the Fins of Air-cooled Engines, NACA Report No. 158.

3. Luikov, A.V. (1974) Conjugate Convective Heat Transfer Problem, *Int. J Heat Mass Transfer*, Vol. 17, pp. 257-265.

4. Sparrow, E.M. & Lin, S.H. (1964) Heat Transfer Characteristics of Polygonal and Plate Fins, *Int. J. Heat Mass Transfer*, Vol. 7. p.951-953.

5. Rosman, E.C., Karajilevscov, P., & Saboya, F.E.M. (1984) Performance of One- and Two-Row Tube and Plate Fin Heat Exchangers, *J. Heat Transfer*, Vol. 106, pp. 627-632.

6. Baehr, H.D. & Schubert, F. (1959) Die Bestimmung des Wirkungsgrades quadrstischer Scheibenrippen mit Hilfe eines elektrisches Analogie-Verfahrens, *Kältetechnik*, 11 Jahrgang, Heft 10, pp. 320-325.

7. Moffat, R.B. (1990) Experimental Heat Transfer, *Procs. of the 9th IHTC*, Jerusalem, Vol. 1, pp. 187-205.

8. Chen, Y., Fiebig, M., & Mitra, N.K. (1997) Numerical Investigation of Fin Efficiencies in a Finned Oval Tube, *Procs. 15th IMACS World Congress*, 24-29 August, Berlin, pp. 761-766.

9. Chen, Y., Fiebig, M., & Mitra, N.K. (1997) Numerical Investigation of Flow and Heat Transfer in a Finned Oval Tube, *Procs. of the 3rd ISHMT-ASME Heat and Mass Transfer Conference*, pp.735-741, IIT Kanpur, India, Narosa Press.

10. Chen, Y., Fiebig, M., & Mitra, N.K. (1997) Influence of the Angle of Attack of a Wing-Type VG on the Heat Transfer and Flow Loss of a Finned Oval Tube, *Procs. Eurotherm seminar #55, "Heat Transfer in single phase flows 5"*, 18-19 September Santorini, Greece.

11. Chen, Y., Fiebig, M., & Mitra, N.K. (1998) A Numerical Study of Finned Oval Tube Heat Exchanger Elements with Winglet Vortex Generators, *Procs. of the 1998 International Conference on Energy & Environment*, 4-6 May, Shanghai, China.

12. Fiebig, M., Grosse-Gorgemann, A., Chen, Y., & Mitra, N.K. (1995) Conjugate Heat Transfer of a Finned tube. Part A: Heat Transfer behavior and Occurrence of Heat Transfer Reversal, *Numerical Heat Transfer, Part A*, Vol. 28, pp. 133-146.

13. Fiebig, M., Chen, Y., Grosse-Gorgemann, A., & Mitra, N.K. (1995) Conjugate Heat Transfer of a Finned tube. Part B: Heat Transfer Augmentation and

avoidance of Heat Transfer Reversal by LVG, *Numerical Heat Transfer, Part A*, Vol. 28, pp. 147-155.

14. Hilgenstock, A. (1988) A Fast Method for Elliptic Generation of 3D Grids with Full Boundary Control, in: *Numerical Grid Generation in Computational Fluid Mechanics*, ed. by S. Sengupta et al., Swansea, Pineridge, pp. 137-146.

15. Patankar, S.V. (1980) *Numerical Heat Transfer and Fluid Flow*, Hemisphere Publishing Corporation.

16. Van Doormaal, J. P., & Raithby, G. D. (1984) Enhancements of the SIMPLE Method for Predicting Incompressible Fluid Flows, *Numerical Heat Transfer*, Vol. 7, pp.147-163.

17. Rhie, C.M., & Chow, W.L. (1983) Numerical Study of the Turbulent Flow Past an Airfoil With Trailing Edge Separation, *AIAA J.*, Vol. 21, No. 11, pp. 1525-1532.

18. Stone, H.L. (1968) Iterative Solution of Implicit Approximations of Multidimensional Partial Differential Equations, *SIAM J. Numer. Anal.*, Vol. 5, No. 3, pp. 530-558.

19. Peric, M. (1985) A Finite Volume Method for the Prediction of Three-Dimensional Fluid Flow in Complex Ducts, Dissertation, University of London.

20. Chen, Y. (1998) *Leistungssteigerung von Wärmeübertragern mit berippten Ovalrohren durch Längswirbelerzeuger--Numerische Simulation von Strömung und konjugiertem Wärmeübergang*. Shaker Verlag, Aachen, Germany.

21. Kost, A. (1993) Strömungsstruktur und Drehmomentübertragung in hydrodynamischen Kupplung, VDI-Verlag GmbH, Düsseldorf.

22. Bai, L. (1995) Numerische Untersuchung von turbulenten Strömungen in hydrodynamischen Kupplungen, VDI-Verlag GmbH, Düsseldorf.

23. Große-Gorgemann, A. (1996) Numerische Untrsuchung der laminaren oszillierenden Strömung und des Wärmeüberganges in Kanälen mit rippenförmigen Einbauten, VDI-Verlag GmbH, Düsseldorf.

HEAT TRANSFER AND FLUID FLOW IN
RIB-ROUGHENED RECTANGULAR DUCTS

BENGT SUNDÉN
Division of Heat Transfer
Lund Institute of Technology
Box 118, 221 00 Lund, Sweden

Abstract. This lecture considers flow fields, friction factors, and local and average heat transfer coefficients in rib-roughened ducts for applications in compact heat exchangers and cooling systems in gas turbine systems. Details of the flow pattern and the influence of rib configuration, rib angle, rib pitch and height are discussed, and physical interpretations of the results are provided.

1. Introduction

Improvement of the heat transfer process is desired in heat exchangers to enable reductions in weight and size, to increase the heat transfer rate or to diminish the mean temperature difference between the fluids and thus to improve the overall process efficiency. In gas turbine cycles the turbine inlet temperature becomes higher and higher to improve the cycle efficiency. Efficient cooling of the first rows of guiding vanes and blades is therefore very important. Internal cooling by air in duct flow is now very common. However, only a limited amount of the air flow is available for such cooling, and thus efficient heat transfer is required. Similarly in combustor wall cooling, the trend is to apply pure convective cooling or combined convective and impingement cooling methods in ducts. Here also, a limited amount of air flow is available, and the pressure drop has to be low. This calls for a very effective heat transfer process.

The methods to improve or enhance the heat transfer usually involve extended surfaces or some form of surface modification, destabilisation of the flow field and/or generation of secondary flow. The methods being used for heat transfer enhancement are referred to as active or passive. The active methods require external power such as surface vibration, electric or acoustic fields. The passive methods use surface modifications or specific types of surface geometries and in some cases fluid additives are used to establish the enhancement. In [1], Webb has presented an overview of heat transfer enhancement in general. In this lecture rib-roughened ducts are considered.

123

S. Kakaç et al. (eds.), Heat Transfer Enhancement of Heat Exchangers, 123–140.
© 1999 *Kluwer Academic Publishers.*

2. Problem Statement

Figure 1 shows a sketch of a rectangular duct with transverse ribs, inclined ribs and V-shaped ribs. In addition to these geometries, a rectangular duct with multiple V-shaped ribs will be considered here.

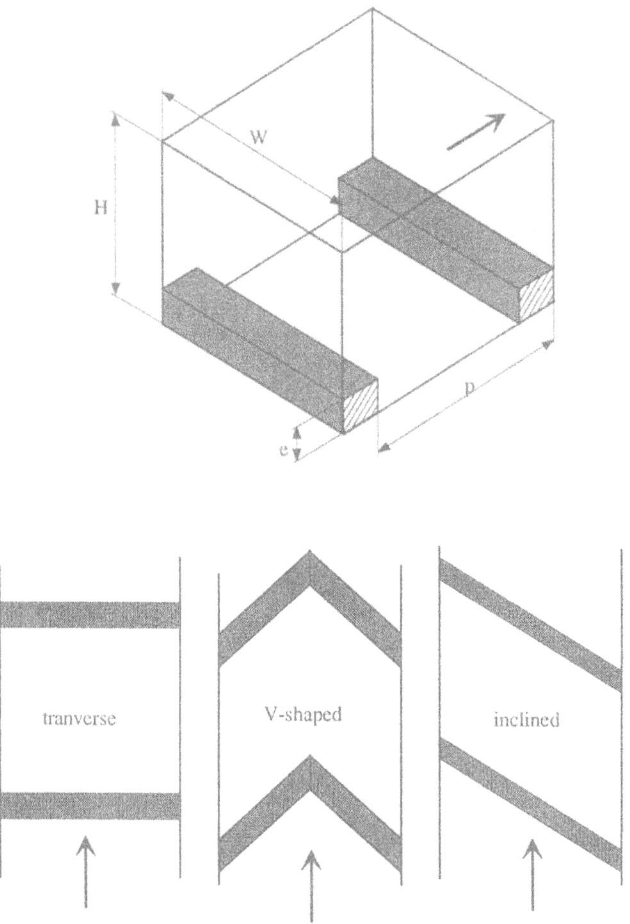

Figure 1. Principal sketch of a rectangular duct with various ribs.

In Figure 2, two real rib-roughened radiator tubes with inclined ribs are shown. The outside top surfaces are highlighted. The aspect ratios, H/W, are 1:8.4 and 1:13.1. The rib height to hydraulic diameter ratios are 0.058 and 0.065. The pitch to height ratios are 22.2 and 20.

Figure 2. Outside top surfaces of two rib-roughened radiator tubes/ducts.

3. Experimental Investigations

A variety of investigations concerning rib-roughened walls in square and rectangular ducts exist. However, depending on the application, the design of the ribs will be different. In radiator tubes or ducts the manufacturing process will result in ribs with a radius of curvature of the same order of magnitude as the rib height. For internal cooling ducts of gas turbine blades and wall cooling ducts of combustors, the ribs may have sharp corners, i.e., the cross section of the ribs will be like a square or a rectangle. If the ribs are perpendicular to the main flow, the shape of the ribs will have a significant influence on the pressure drop and heat transfer, but if the ribs are inclined and a secondary flow is set up, the actual rib shape might be less important.

Investigations in the past mainly concerned the overall pressure drop or friction factor and the average heat transfer coefficient or Nusselt number, while more recent studies involve advanced experimental techniques, e.g., LDV and liquid crystal thermography to study the details of the flow pattern and temperature field. Improvements on the understanding of the physical processes have been made.

3.1 REVIEW OF PUBLISHED WORKS

In [2] Han et al. presented an investigation of the effects of rib shape, angle of attack and pitch to height ratio on the friction factor and heat transfer coefficient in a parallel plate duct experiment at Reynolds numbers in the range of 3 000-30 000. They found that a symmetrical rib arrangement gave the same results as a staggered rib arrangement, and that the rib shape influenced the friction factor but had only a modest effect on the heat transfer coefficient. In contrast, a numerical investigation by Arman and Rabas [3], and an experimental investigation by Liou and Hwang [4], show, however, a significant influence of the rib shape on the heat transfer. Han et al. found that the highest heat transfer for a given friction power was achieved with ribs with a 45° angle of attack. Metzger et al. [5] investigated the effects of rib angle and orientation on local heat transfer in a square channel. They expected different large-scale secondary flows, depending on the orientation of the ribs. A two-cell pattern was expected when the ribs are placed in parallel planes, and a one-cell pattern was expected when the ribs are placed in intersecting planes on the opposite walls. It was found that 60° angled ribs giving a two-cell flow pattern provided the best heat transfer performance. In a similar investigation Han et al. [6] also included V-shaped ribs with 45° and 60° half-angle, and the ribs could point in the main flow direction or in the opposite direction. The ribbed walls were in-line and the Reynolds number covered the range of 15 000 to 90 000. The V-shaped ribs directed opposite to the main flow provided the highest heat transfer enhancement while the V-shaped ribs directed with the main flow generated the largest pressure drop penalty. It was conjectured that differences in the secondary flow were the explanation for the observed results. Figures 3 and 4 show results for the friction factor and Nusselt number from Han et al. [6].

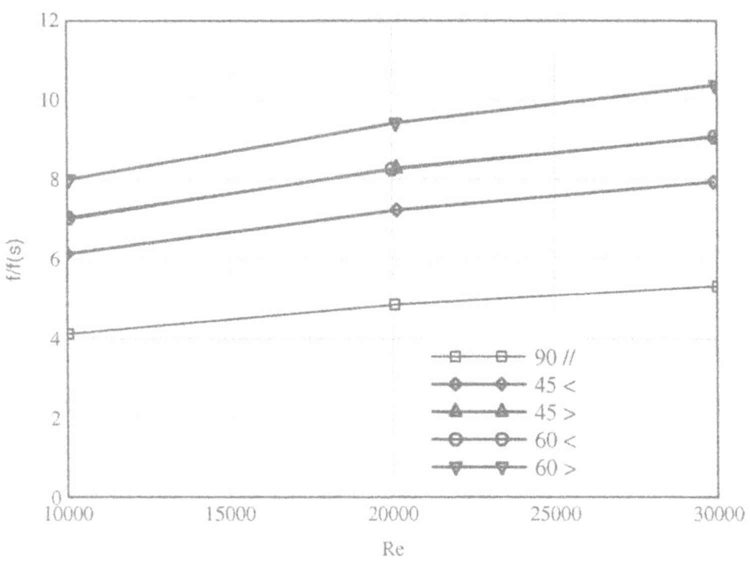

Figure 3. Friction factor vs Reynolds number for different rib orientations.
W/H = 1, e/D_h = 0.0625, P/e = 10. f (s) = f for corresponding smooth surface.

Han and co-workers [7] investigated the pressure drop and heat transfer in a square duct with broken ribs. As in previous investigations two of the walls were ribbed and e/D_h = 0.0625 and P/e = 10, similar to investigation [6]. It was concluded that the broken rib pattern provided somewhat higher heat transfer, but the pressure drop was almost equal to that of the corresponding continuous ribs.

Zhang et al. [8] proposed that grooves placed between adjacent ribs could improve the overall heat transfer by creation of vortices, which would increase the exchange between the fluid and the wall. An investigation for a rectangular duct with aspect ratio W/H = 10, two ribbed walls with e/D_h = 0.028 and P/e = 10 was carried out. Grooves were created midway between two adjacent ribs. Figure 5 depicts a sketch of the ribbed-grooved duct of Zhang et al.

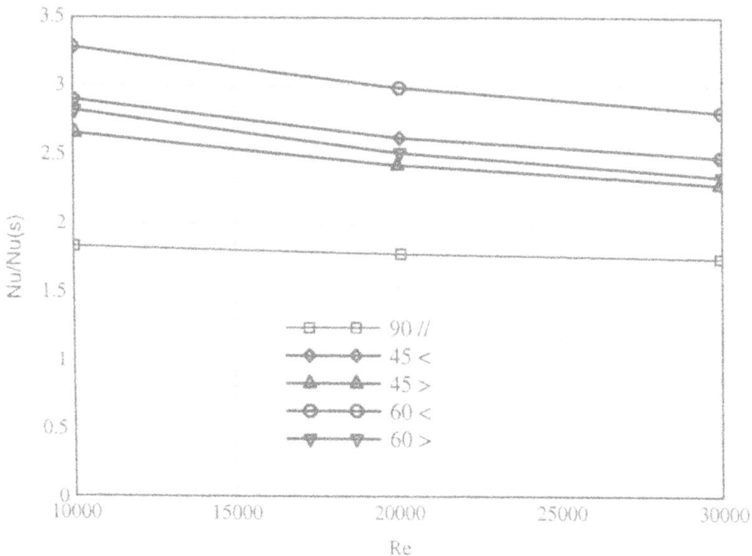

Figure 4. Nusselt number vs Reynolds number for different rib orientations. W/H = 1, e/D$_h$ =0.0625, P/e = 10. Nu(s) = Nu for corresponding smooth surface.

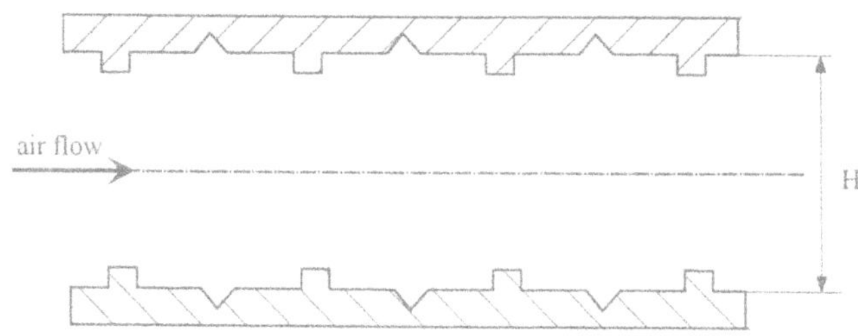

Figure 5. Ribbed-grooved duct investigated by Zhang et al. [8].

Another attempt to further improve the heat transfer from ribbed walls was presented by Hwang and Liou [9, 10]. They investigated the performance of perforated 90° ribs. A rectangular duct with W/H = 4, e/D$_h$ = 0.081 and P/e = 10 was considered. The holes are penetrated by the fluid, hence the wake or separated region downstream a rib is eliminated. The heat transfer coefficient downstream of the rib is then increased. It was found that the overall heat transfer coefficient was higher than that for solid ribs and the pressure drop penalty was slightly less. Liou et al. [11] investigated the periodic heat transfer in a channel with ribs detached from one wall, and they found an optimum clearance ratio.

Taslim et al. [12] performed an experimental investigation using liquid crystal thermography to determine the local Nusselt numbers in a square duct with angled, V-shaped and discrete ribs on two opposite walls. Only one wall was heated with a constant heat flux, while the other walls were adiabatic. Their results indicated that the

128

heat transfer was highest for V-shaped ribs pointing downstream contrary, to the results of Han et al. [6].

As mentioned previously, overall friction factors and average heat transfer coefficients have been the major interest in most of the published works and thus investigations of the accompanied flow fields are rare. As the results are interpreted and the physical mechanisms have to be revealed such investigations are necessary. Liou et al. [13] presented LDV-measurements of the flow field in a rectangular duct with transverse ribs at a Reynolds number of 33 000. Hirota et al. [14] investigated the flow and temperature fields in a square duct with ribs only on one of the walls at Re = 65 000. Both investigations suggested that the secondary flow is a very important mechanism for the redistribution of heat and momentum.

In vehicle heat exchangers, e.g., radiators, rib-roughened and dimpled surfaces are employed to enhance the heat transfer of the inside duct wall. Investigations of the overall thermal and hydraulic performance of such real ducts have been carried out by Farell et al. [15] and Olsson and Sundén [16]. For radiators W/H is between 5 to 20 and the rib-roughening is rolled in the sheet-metal as the ducts are formed. Since the hydraulic diameter is about 3 mm and the duct wall thickness is 0.1 to 0.5 mm it is not possible to have sharp-edged roughness elements. In section 3.2 of this chapter such rib-roughening will be considered.

3.1.1 *Wall functions for rib-roughened surfaces.*
Based on the data by Han and co-workers [6, 7, 17] wall functions for the velocity and temperature fields have been established. The equations, given below, are limited to the so-called fully rough flow regime. This is attained if the dimensionless rib height $e^+ >$ 50. The wall functions read:

$$u^+ = 2.5\ln\frac{y}{e} + R\!\left(e^+\right) \tag{1}$$

$$T^+ = 2.5\ln\frac{y}{e} + G\!\left(e^+\right) \tag{2}$$

The functions $R\!\left(e^+\right)$ and $G\!\left(e^+\right)$ are given as

$$R\!\left(e^+\right) = a\cdot\left(e^+\right)^b \tag{3}$$

$$G\!\left(e^+\right) = c\cdot\left(e^+\right)^d \tag{4}$$

The dimensionless rib height e^+ is defined as

$$e^+ = \frac{u^* e}{\upsilon} \tag{5}$$

The constants a, b, c, d in equations (3) and (4) depend on P/e and the rib orientation, see [6, 7, 17].
The range of validity of equations (1) and (2) is restricted to: $1 < W/H < 4$, two ribbed walls, $0.021 < e/D_h < 0.078$, $10 < P/e < 20$, $8\,000 < Re < 80\,000$ and $Pr = 0.704$.

The advantage of using wall functions in numerical analysis of flow and heat transfer in rib-roughened ducts will be presented in a later section.

3.1.2 *Correlations for f and Nu*
Correlations for the overall friction factor and the average Nusselt number are available in the literature [17]. Due to their complexity they are not provided here.

3.2 RECENT WORK

In a recent investigation [18, 19, 20], Olsson and Sundén have performed detailed experimental studies of the flow fields and the establishment of large scale secondary flows in rectangular ducts with ribbed walls. Concurrent information of the flow structure, pressure drop and heat transfer, as well as the importance of rib configuration and its layout, should be useful for design engineers and researchers who are interested in heat transfer enhancement to enable significant improvements of heat exchangers.

Figure 6 presents a description of some rib configurations considered in [18, 19]. In Figure 6, all the ducts have identical hydraulic diameter, rib height and rib pitch. Thus as the results are compared the importance of the rib configuration is revealed. Smoke wire visualisations and LDV-measurements were carried out in separate test rigs where the ducts could be upscaled about 6 times and the aspect ratio W/H is about 8. The rib dimensions are specified as: $e/D_h = 0.06$, $P/e = 21$ (37, 46 for multiple V-shaped ribs). The pressure drop and heat transfer data were taken in another two separate test rigs where the ducts were upscaled about twice. For details of the data evaluation and estimation of uncertainties, the reader is referred to [18, 19].

In [20] a particular investigation of the performance of a rectangular duct with multiple V-shaped ribs was carried out. The duct had an aspect ratio $W/H = 8$, the rib height to diameter ratio was varied as $0.056 < e/D_h < 0.112$, the rib angle was in the range of $\phi = 15\text{-}45°$ while the pitch to height ratio of the ribs was $P/e > 15$. Figure 7 depicts the duct with multiple V-shaped ribs.

The experimental procedure, data evaluation and estimation of uncertainties follow those in [18, 19].

3.2.1 *Flow Distributions*
Figure 8 presents some flow results for the cross rib-roughened duct in Figure 6. It may be expected that a one cell secondary flow is set up, since the ribs turn the flow upwards along the right wall and downwards along the left wall. The second panel in Figure 8 shows the smoke visualisation results, which at the top and bottom of the duct indicate a motion similar to the conjectured one. Further away from these walls the smoke visualisation indicates fluid motions in opposite direction across the mid-plane. At this specific instant the smoke traces seem to have been influenced by several vortex-like structures and not a stable one cell pattern. The LDV vector plot of the secondary mean flow velocity (panel 3) shows a flow pattern that is consistent with the schematic pattern. The motion is mainly directed along the walls according to the rib direction.

130

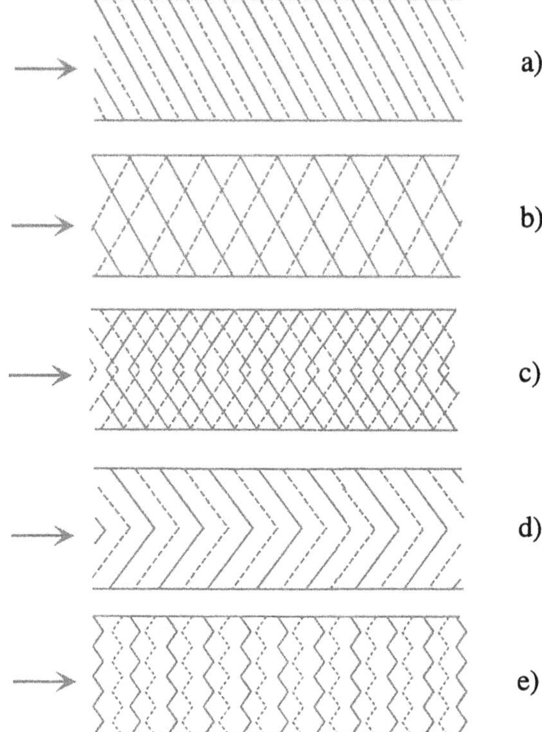

Figure 6. Sketch of rib configurations in [18, 19]. (Dashed lines on bottom side wall, solid line on top side wall). a) Parallel rib-roughened, b) cross rib-roughened, c) cross V-rib-roughened, d) parallel V-rib-roughened, e) multiple V-rib-roughened .

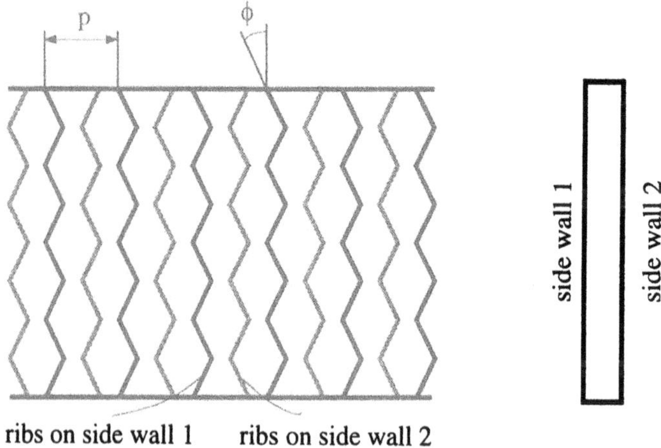

Figure 7. Duct with multiple V-shaped ribs. In the cross-sectional plane the ribs are not shown.

The consequences of this motion are seen in the contour plots of the axial velocity and the r.m.s. velocity fluctuations (panels 4 and 5). The secondary flow over the ribs forces low momentum and high turbulent kinetic energy fluid from the wall towards the right hand side of the duct. Thus, the finite height of the ribs causes the real secondary flow to differ from a simple one cell pattern.

In Figure 9 corresponding flow pictures are shown for the cross V-rib-roughened duct (see Figure 6). The smoke wire visualisation picture was found to be very stable at Re = 1100, in contrast to what was found for the other ducts. The secondary velocity vectors and the smoke pattern agree well with the conjectured pattern. The axial flow is separated into symmetric parts, i.e., one high momentum region on each side of the horizontal centre line, separated by a region where the flow is from left to right. Regions of strong velocity fluctuations coincide with regions where the secondary flow is pushing fluid away from a wall.

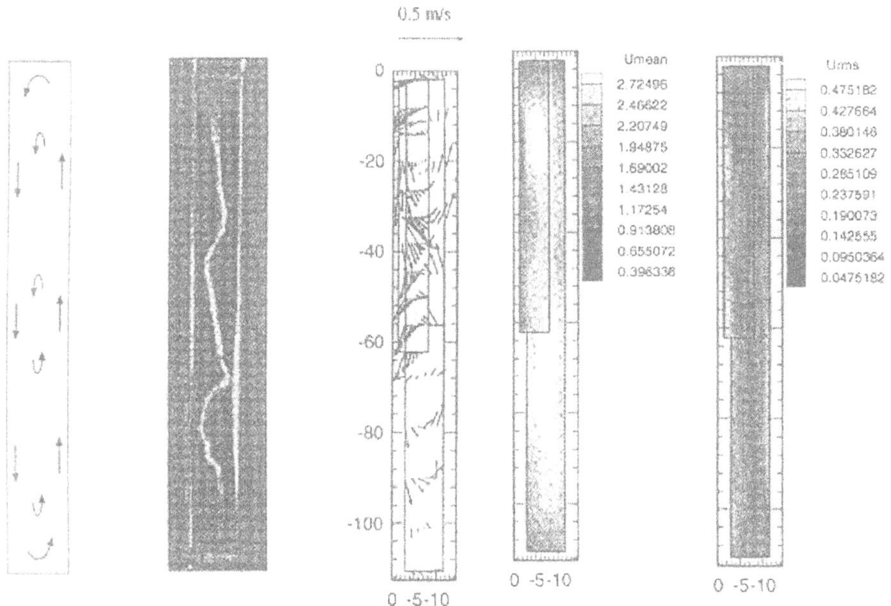

Figure 8. Secondary flow and axial mean and r.m.s. velocity contours for a cross rib-roughened duct. Re = 1 000 in smoke visualisation (second panel) and Re = 3 000 in LDV measurements.

Figure 10 provides flow pictures for the duct with the multiple V-shaped ribs. This duct is also called a swirl flow duct (tube) since the purpose is to create swirling motions in cells, with an aspect ratio of unity, by a proper arrangement of the ribs. The conjectured secondary flow pattern consists of eight longitudinal vortices. The smoke visualisation picture supports this. Panel 4 reveals that high momentum fluid is convected either to the right or to the left by the secondary flow. In regions of low momentum fluid, the velocity fluctuations are approximately twice those in the high momentum regions. The magnitude of the secondary velocities in this duct is smaller than for the ducts in Figures 8 and 9. However, the length scale for the secondary motion is smaller.

132

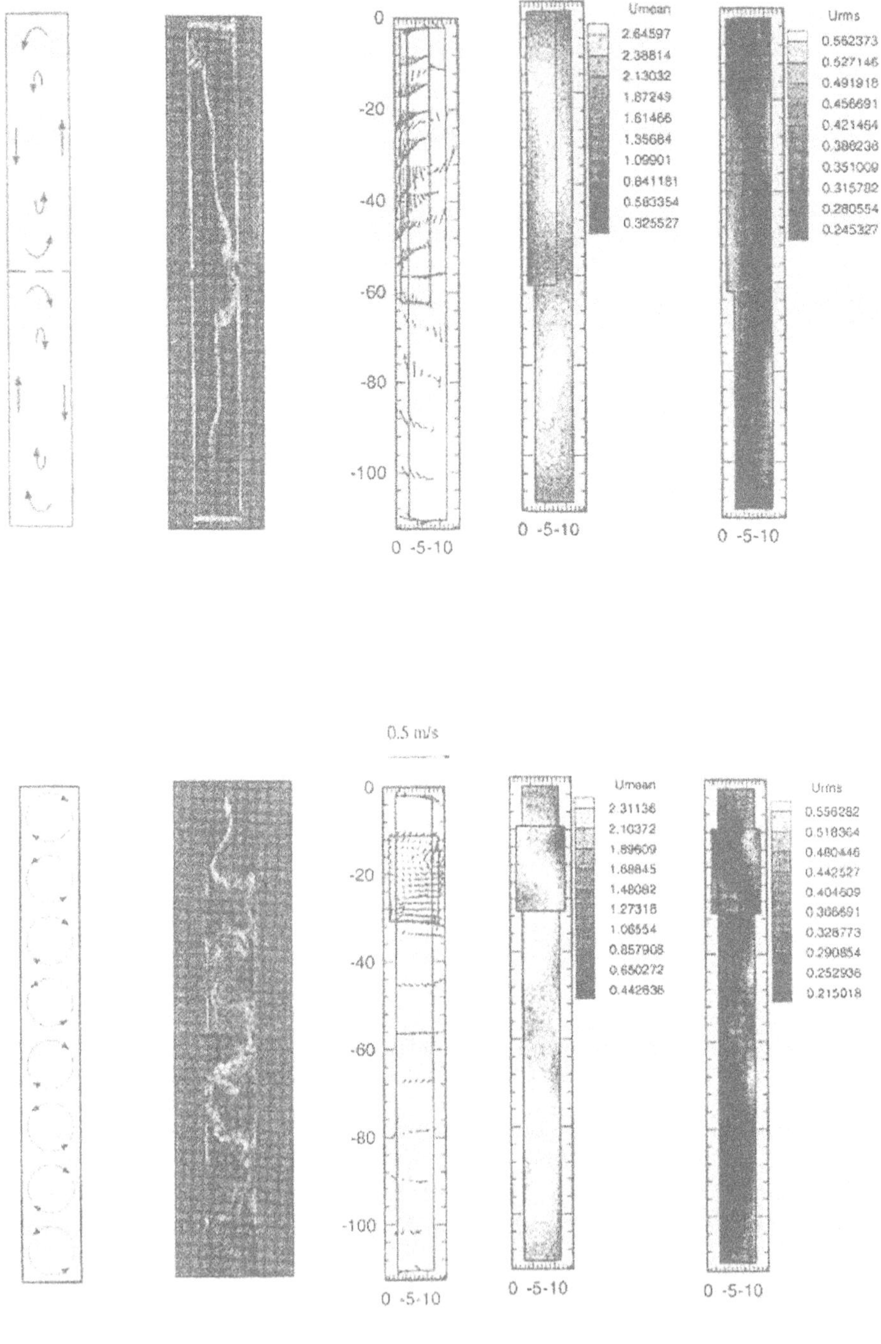

Figure 10. Secondary flow and axial mean and r.m.s. velocity contours for a duct with multiple V-shaped ribs. Re = 1 100 in smoke visualisation (second panel), Re = 3 000 in LDV-measurements.

It should be noted that the LDV-measurements were carried out at only one position for each duct. At similar positions upstream and downstream, the time-averaged flow pattern should be identical due to the periodicity of the geometry and thus the flow. In between such positions the flow pattern will differ somewhat but it is believed that the presented results will reveal some practically features of relevant flow fields.

3.2.2 *Performance Comparison*

In order to compare the thermal-hydraulic performance of the ducts in Figure 6, the so-called flow area goodness factor (j/f_{app} vs Re) and the volume goodness factor (heat transfer coefficient versus pumping power per unit heat transfer area). Figure 11 shows the flow area goodness factor. As expected j/f_{app} is less than 0.5, which would be the highest obtainable value for boundary layer type flow over a flat plate.

The highest values of j/f_{app} is attained by a corresponding smooth duct. However, this smooth duct was only about 56 hydraulic diameters in length and thus the influence of entrance region appears on both j and f_{app}. For Re > 3 000 all curves of the rib-roughened ducts are enclosed by the curves for the parallel V-ribbed ducts, and the highest j/f_{app} appears with the flow in the direction opposite the V-ribs (<<). The lowest j/f_{app} is achieved if the flow is in the same direction as the V-ribs (>>). The cross-ribbed and parallel-ribbed ducts provide higher j/f_{app} than the swirl flow tube (multiple V-shaped ribs) and the cross V-ribbed duct. The swirl flow tube or the duct with multiple V-ribs provides in fact the highest j-factor for Re < 2 000, but requires an increased pumping power and thus j/f_{app} becomes modest.

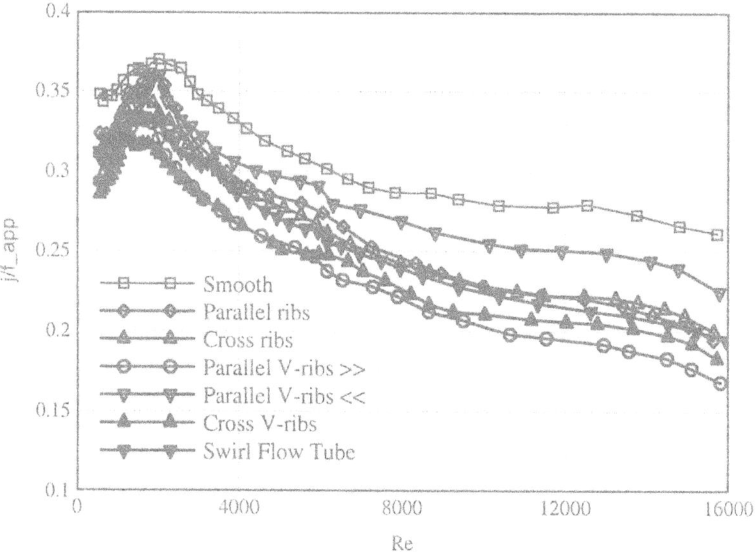

Figure 11. Flow area goodness factor j/f_{app} vs Reynolds number.

Figure 12 shows the volume goodness factor, i.e., the heat transfer coefficient versus the pumping power per unit heat transfer area. Also this figure indicates that the parallel V-

134

ribbed duct with flow in opposite direction (<<) provides the highest position, and thus should be regarded as the most efficient one.

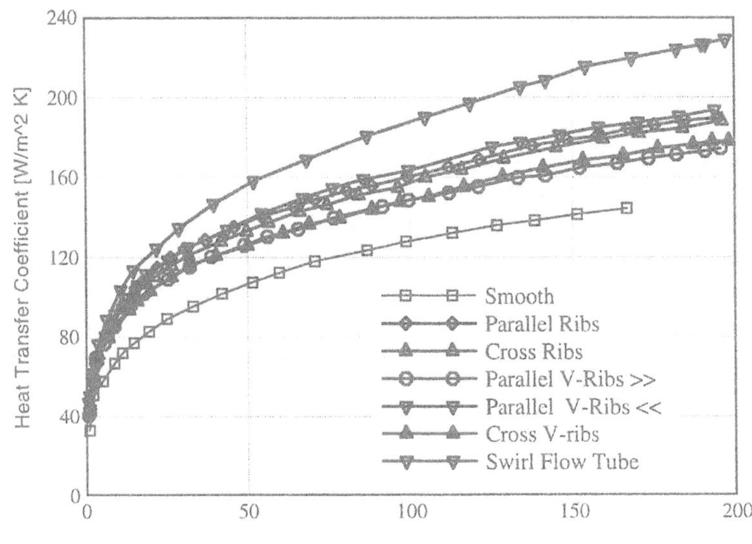

Figure 12. Volume goodness factor chart.

3.2.3 *Multiple V-shaped ribs*

For the duct type shown in Figure 7, the rib dimensions were varied. The following correlations for the j- and f-factors were established (by least squares fits to data), see also [20]:

$500 < \text{Re} < 1\ 500$

$$j = 2.313 \cdot \text{Re}^{-0.4948} \cdot \left(\frac{e}{H}\right)^{0.5275} \cdot \left(\frac{P}{H}\right)^{-0.2387} \cdot \left(\frac{\phi}{90}\right)^{0.2389} \tag{6}$$

$$f = 20.82 \cdot \text{Re}^{-0.5306} \cdot \left(\frac{e}{H}\right)^{0.9212} \cdot \left(\frac{P}{H}\right)^{-0.3513} \cdot \left(\frac{\phi}{90}\right)^{0.1886} \tag{7}$$

$2\ 000 < \text{Re} < 15\ 000$

$$j = 0.3221 \cdot \text{Re}^{-0.2457} \cdot \left(\frac{e}{H}\right)^{0.5160} \cdot \left(\frac{P}{H}\right)^{-0.3143} \tag{8}$$

(independent of ϕ)

$$f = 1.028 \cdot \left(\frac{e}{H}\right)^{1.2982} \cdot \left(\frac{P}{H}\right)^{-0.6108} \cdot \left(\frac{\phi}{90}\right)^{-0.0987} \tag{9}$$

(independent of Re)

In equations (7) and (9), the friction factor is the Fanning friction factor corresponding to fully developed periodic flow. The correlations do not cover the range $1\ 500 < \text{Re} < 2\ 000$, which may be referred to as a transitional regime. If one assumes that the Nusselt number is proportinal to $\text{Pr}^{1/3}$ then the heat transfer results (obtained for air flow) are also applicable to other fluids if scaling by the Prandtl number ratio to the power one third is carried out.

3.2.4   Mechanisms

It was shown that a secondary flow is set up in the rib-roughened ducts. The aspect ratio of the swirling motion differs between the ducts. The main objective of creating a secondary flow is to establish or improve the exchange of heat between the core region and wall regions. In a rectangular duct, the largest temperature gradients are found in the direction normal to the wide walls. A convective motion in this direction is thus expected to be beneficial for heat transfer enhancement. The duct with multiple V-ribs and creating several swirling cells was found to provide the highest j-factor for Re < 2 000.

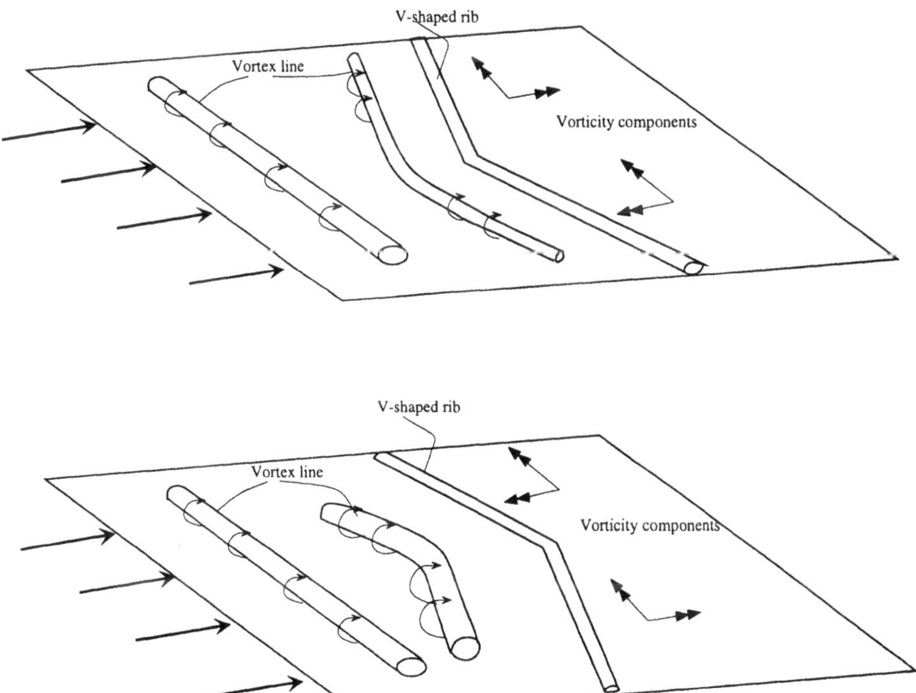

Figure 13. Influence on wall vorticity by V-shaped ribs. a) <<, b) >>.

The duct with V-ribs pointing upstream or opposite to the main flow direction (<<) and the duct with V-ribs pointing downstream (>>) were found to enclose the other ducts in Figures 11 and 12. To obtain an understanding of this a vortex line close to a wall is studied (see Figure 13).

The vortex line is bent to a V-shape similar to the ribs if the vortex line is close to the wall. The vortex line will be stretched and the vorticity amplified by the velocity gradients in the vicinity of the ribs. Consequently, the vortex line now has both axial and spanwise components. The axial component is associated with the secondary flow, while the spanwise component has similar behaviour as the original vortex line. In Figure 13a, the axial vorticity components will act as an inflow pair of vorticies resulting in thinning of the boundary layer, and consequently the heat transfer is enhanced.

Self-propagation of the axial vorticity components is directed towards the wall, this which tends to keep the strong vorticity close to the wall. It seems likely that this vortex interaction increases the local heat transfer and shear stress. In Figure 13b, vortex stretching also may occur, but as the vortex line is bent, the axial vorticity components act differently compared to Figure 13a. An outflow vortex pair now occurs and the boundary layer thickness is increased. Self-propagation of the axial vorticity components is directed away from the wall, and thereby the local heat transfer and shear stress are decreased.

4. Numerical Investigations

The complex flow and heat transfer processes in enhanced geometries, like rib-roughened ducts, are very difficult to handle, as the geometries are more or less complex, turbulence is present. There is separation and reattachment, etc. In the past some numerical investigations for ribbed ducts have been performed. Fodemski and Collins [21] predicted heat transfer and friction in a rectangular duct equipped with square ribs spaced at $P/e = 6$. They used the standard k-ε model with wall functions as the Reynolds number was high. The predictions were found to qualitatively agree with other results. Reasonable agreement with temperature distributions measured by holographic interferometry was reported. Acharya et al. [22] predicted flow and heat transfer in a ribbed rectangular duct ($W/H = 4.9$, $e/H = 0.1$, $P/e = 20$) using the non-linear and standard k-ε models. Wall functions were applied. Both models performed poorly in the separated region just downstream the ribs. The local Nusselt numbers were underpredicted compared to experimental data.

At relatively low Reynolds numbers that may occur in heat exchangers, the standard wall function approach for the wall boundary conditions is not appropriate. Low Reynolds number turbulence models or so-called two-layer models are needed to resolve the flow and temperature fields close to the ribs and in between the ribs. The former models have been used by Chang and Mills [23], Arman and Rabas [24], Fusegi [25], Iacovides and Raisee [26] and Abdon and Sundén [27]. The two-layer model approach has been applied by Arman and Rabas [24] and Iacovides and Raisee [26]. In [27] ribbed ducts with $W/H = 12$, $P/e = 10$ and $e/D_h = 0.032\text{-}0.102$ were investigated.

Distributions of velocity vectors and local Nusselt numbers, from [27], are provided in Figure 14.

The recirculation zones upstream and downstream of a rib are clearly visible. The reattachment region between two ribs is also evident, and the flow effect on the local Nusselt number is shown. It has been found that for small e/D_h the friction factor and the average Nusselt number are predicted to be in reasonable agreement with corresponding experimental data [2, 28]. For large e/D_h the predictive ability was not satisfactory. This finding agrees with that in [29].

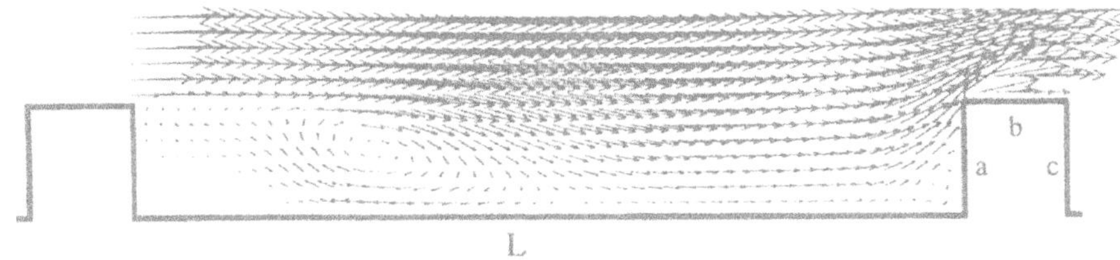

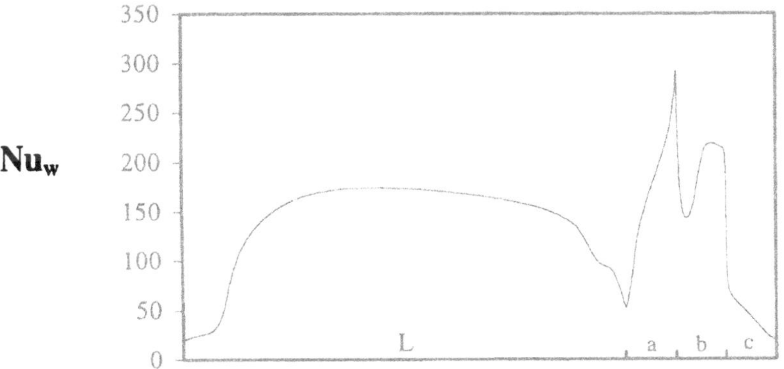

Figure 14. Local Nusselt number and the flow velocity for $e/D_h = 0.056$, $Re \cong 20\,200$.

The disadvantage with the low Reynolds number model is that it requires a great number of grid points to resolve the flow and temperature fields, and thus the computational effort may be considerable. A way to circumvent this problem is to consider the enhanced surface as a rough one. The detailed modelling of the enhancement promotors, i.e., the ribs, is then avoided. The key here is to use wall functions valid for ribbed surfaces. Such wall functions were presented in equations (1)

138

and (2). Simulations using these wall functions have been presented in [30-32]. In [32] this approach was applied to maximise the heat transfer for a given pressure drop and fixed mass flow rate. It was found that V-shaped or broken V-shaped ribs could enhance the heat transfer significantly, up to 80 %, compared to a smooth duct. The accuracy of these calculations was established by comparison with experimental data [2, 28].

5. Concluding Remarks

The general conclusions of the information presented on rib-roughened ducts can be summarised as follows:
- For solid ribs orientated perpendicular to the main flow, P/e = 10 is appropriate for good heat transfer.
- The rib height is important.
- The shape of the ribs affects the pressure drop more than the heat transfer.
- Angled ribs may be favourable as both heat transfer and pressure drop are considered.
- V-shaped ribs pointing in the upstream direction (<<) were found to be superior if the volume goodness factor and the flow area goodness factor were considered.
- Multiple V-shaped ribs performed extremely well at low Reynolds number.
- Perforated ribs may improve the overall performance.
- To reveal the mechanisms for the momentum and heat transfer processes, detailed measurements of flow and temperature fields are desirable.
- Numerical investigations are partially successful, but the predictive ability is low for large rib heights.
- More advanced turbulence models need to be tested in this problem.

6. Acknowledgements

Financial support from NUTEK (Swedish Board for Industrial and Technical Development), Valeo Engine Cooling Systems AB, AB Volvo, Swedish Gas Turbine Center (GTC) is kindly acknowledged. The word-processing of this manuscript was skilfully and efficiently carried out by Ms Gunvi Andersson.

Nomenclature

a	duct height, m
a,b,c,d	constants, -
c_p	specific heat, J/kgK
D_h	hydraulic diameter, m
e	rib height, m
e^+	dimensionless rib height $\left(= u^* y / \upsilon\right)$, -
f	Fanning friction factor, -
f_{app}	apparent friction factor, -
G	additional function, -

H	duct height, m
j	Colburn factor, -
Nu	Nusselt number, -
P	rib pitch, m
q_w	wall heat flux, $W/m^2 K$
R	additional function, -
Re	Reynolds number, -
T	temperature, K
T_w	wall temperature, K
T^+	dimensionless temperature $\left(= \rho c_p u^* \left(T_w - T \right) / q_w \right)$, -
U	local mean axial velocity, m/s
U_m	cross-sectional average axial velocity, m/s
u^*	friction velocity, m/s
u^+	dimensionless velocity $\left(= U / u^* \right)$, -
u'_m	r.m.s. axial velocity fluctuation, m/s
W	duct width, m
x, y, z	Cartesian coordinate system, m

Greek symbols

α	heat transfer coefficient, $W/m^2 K$
υ	kinematic viscosity, m^2/s
ρ	density, kg/m^3

References

1. Webb, R.L. (1994) *Principles of Enhanced Heat Transfer*, Wiley-Interscience, New York.
2. Han, J.C., Glicksman, L.R. and Rohsenow, W.M. (1978) An investigation of heat transfer and friction for rib-roughened surfaces, *Int. J. Heat Mass Transfer*, 21, 1143-1156.
3. Arman, B. and Rabas, T.J. (1992) Disruption shape effects on the performance of enhanced tubes with the separation and reattachment mechanism, *ASME HTD-Vol. 202*, 67-75, ASME, New York.
4. Liou, T.M. and Hwang, J.J. (1993) Effect of ridge shapes on turbulent heat transfer and friction in a rectangular channel, *Int. J. Heat Mass Transfer*, 36, 931-940.
5. Metzger, D.E., Fan, C.S. and Yu, Y. (1990) Effects of rib angle and orientation on local heat transfer in square channels with angled roughness ribs, in R.K. Shah. A.D. Kraus and D.E. Metzger, *Compact Heat Exchangers*, Hemisphere Publishing Corporation, New York, pp. 151-167.
6. Han, J.C., Zhang, Y.M. and Lee, C.P. (1991) Augmented heat transfer in square channels with parallel, crossed and V-shaped angled ribs, *ASME J. Heat Transfer*, 113, 590-596.
7. Han, J.C., Zhang, Y.M. and Lee, C.P. (1992) High performance heat transfer ducts with parallel broken and V-shaped broken ribs, *Int. J. Heat Mass Transfer*, 35, 513-523.
8. Zhang, Y.M., Gu, W.Z. and Han, J.C. (1994) Heat transfer and friction in rectangular channels with ribbed or ribbed-grooved walls, *ASME J. Heat Transfer*, 116, 58-65.
9. Hwang, J.J. and Liou, T.M. (1995) Heat Transfer and friction in a low aspect ratio rectangular channel with staggered perforated ribs on two opposite walls, *ASME J. Heat Transfer*, 117, 843-850.
10. Hwang, J.J. and Liou, T.M. (1995) Heat transfer and friction in a rectangular channel with perforated turbulence promotors using holographic interferometry measurement, *Int. J. Heat Mass Transfer*, 38, 3197-3208.
11. Liou, T.M. Wang, W.B. and Chang, Y.J. (1995) Holographic interferometry study of spatially periodic heat transfer in a channel with ribs detached from one wall, *ASME J. Heat Transfer*, 117, 32-39.
12. Taslim, M.E., Li, T. and Kercher, D.M. (1996) Experimental heat transfer and friction in channels roughened with angled, V-shaped and discrete ribs on two opposite walls, *ASME J. Turbomachinery*, 118, 20-28.

140

13. Liou, T.M., Wu, Y.Y. and Chang, Y. 1(1993) LDV measurements of periodic fully developed main and secondary flows in a channel with rib-disturbed walls, *ASME J. Fluids Engineering*, 115, 109-114.

14. Hirota, M., Fujita, H. and Yokosawa, H. (1994) Experimental study on convective heat transfer for turbulent flow in a square duct with a ribbed rough wall (characteristics of mean temperature field), *ASME J. Heat Transfer*, 116, 332-340.

15. Farell, P., Wert, K. and Webb, R.L. (1991) Heat transfer and friction characteristics of turbulator radiator tubes, *SAE Paper 910197, SAE Transactions*, 100, 218-230.

16. Olsson, C.O. and Sundén, B. (1996) Heat transfer and pressure drop characteristics of ten radiator tubes, *Int. J. Heat Mass Transfer*, 39, 3211-3220.

17. Han, J.C. (1988) Heat transfer and friction characteristics in rectangular channels with rib turbulators, *ASME J. Heat Transfer*, 110, 321-328.

18. Olsson, C.O. and Sundén, B. (1997) Fluid flow and heat transfer in rib-roughened tubes, in M. Giot et al. (eds.), *Experimental Heat Transfer, Fluid Mechanics and Thermodynamics 1997*, Edizioni ETS, Pisa, Vol. 3, pp. 1655-1662.

19. Olsson, C.O. and Sundén, B. (1998) Experimental study of fluid flow and heat transfer in rib-roughened channels, *Experimental Thermal and Fluid Science*, 16, 349-365.

20. Olsson, C.O. and Sundén, B. (1998) Thermal and hydraulic performance of a rectangular tube with multiple V-shaped ribs, *ASME J. Heat Transfer*, accepted for publication.

21. Fodemski, T.R. and Collins, M.W. (1988) Flow and heat transfer simulations for two- and three-dimensional smooth and ribbed channels, in *Proceedings of the 2nd UK National Conference on Heat Transfer*, ImechE, London, UK, Vol. 1, pp. 845-860.

22. Acharya, S., Dutta, S., Myrum, T.A. and Baker, R.S. (1993) Periodically developed flow and heat transfer in a ribbed duct, *Int. J. Heat Mass Transfer*, 36, 2069-2082.

23. Chang, B.H. and Mills, A.F. (1993) Turbulent flow in a channel with transverse rib heat transfer augmentation, *Int. J. Heat Mass Transfer*, 36, 1459-1469.

24. Arman, B. and Rabas, T.J. (1994) Two-layer model predictions of heat transfer inside enhanced tubes, *Numerical Heat Transfer A*, 25, 721-724.

25. Fusegi, T. (1995) Turbulent flow calculations of mixed convection in a periodically-ribbed channel, *J. Enhanced Heat Transfer*, 2, 295-305.

26. Iacovides, H. and Raisee, M. (1997) Computation of flow and heat transfer in 2-D rib-roughened passages, in K. Hanjalic and T.W.J. Peters (eds.), *2nd Int. Symposium on Turbulence, Heat and Mass Transfer*.

27. Abdon, A. and Sundén, B. (1998) Investigation of a turbulence model for wall cooling of combustion chambers, *ASME Paper 98-GT-540*.

28. Han. J.C., Glicksman, L.R. and Rohsenow, W.M. (1979) Correction to an investigation of heat transfer and friction for rib-roughened surfaces, *Int. J. Heat Mass Transfer*, 22, 1587-1588.

29. Saidi, A. and Sundén, B. (1998) Calculation of convective heat transfer in square-sectioned gas turbine blade cooling channels, *ASME Paper 98-GT-204*.

30. Cunha, F.J. (1992) Turbulent flow and heat transfer in gas turbine blade cooling passages, *ASME Paper 92-GT-239*.

31. Youn, B., Yuen, C. and Mills, A.F. (1994) Friction factor for flow in rectangular ducts with one side rib-roughened, *ASME J Fluids Engineering*, 116, 488-493.

32. Abdon, A. and Sundén, B. (1998) A numerical method to calculate the cooling performance of a rib-roughened duct, in A.J. Nowak et al. (eds.), *Advanced Computational Methods in Heat Transfer V*, Computational Mechanics Publications, Southampton, pp. 381-391.

ON THE AIRSIDE PERFORMANCE OF FIN-AND-TUBE HEAT EXCHANGERS

CHI-CHUAN WANG
Energy & Resources Laboratories,
Industrial Technology Research Institute,
D500 ERL/ITRI
Bldg. 64, 195~6 Section 4, Chung Hsing Rd.,
Chutung, Hsinchu, Taiwan 310

E-mail:ccwang@erl.itri.org.tw

Abstract. The present study focuses on the airside performance of fin-and-tube heat exchangers. Contents of this lecture include the data reduction method of the airside performance and the updated correlations for the typical fin-and-tube heat exchangers which including plain, wavy, louver, and convex-louver fin patterns. It is strongly recommended that the investigators should clearly address the details of their data reduction methods. In addition, appropriate ε-*NTU* relationships should be carefully examined before applying them to the determination of the heat transfer coefficients. It is also suggested that the entrance and exit loss should be included in the estimation of friction factors.

1. Introduction

Air-cooled heat exchangers are very common in the process and HVAC&R industries. For HVAC&R applications, the air-cooled heat exchangers generally consist of equally spaced parallel plates and an array of tubes which pass perpendicularly through the plates. In practical application, the dominant resistance is usually on the airside. Therefore enhanced fin patterns have often been used to effectively improve the airside performance. *Figure 1* shows the common plate fin-and-tube heat exchangers, including plain, wavy, louver, and convex-louver. For space-limited application, the use of enhanced fins are especially helpful since considerable size reduction can be achieved. However, the airside performances are generally considered as proprietary. As a result, it is often very difficult to accurately size or rate a heat exchanger owing to the lack of reliable airside performance data.

In addition, the open literature on the airside performance may not be consistent. *Figure 2* shows the reduced heat transfer performance by McQuiston, [1], Seshimo and

141

S. Kakaç et al. (eds.), Heat Transfer Enhancement of Heat Exchangers, 141–162.
© 1999 *Kluwer Academic Publishers.*

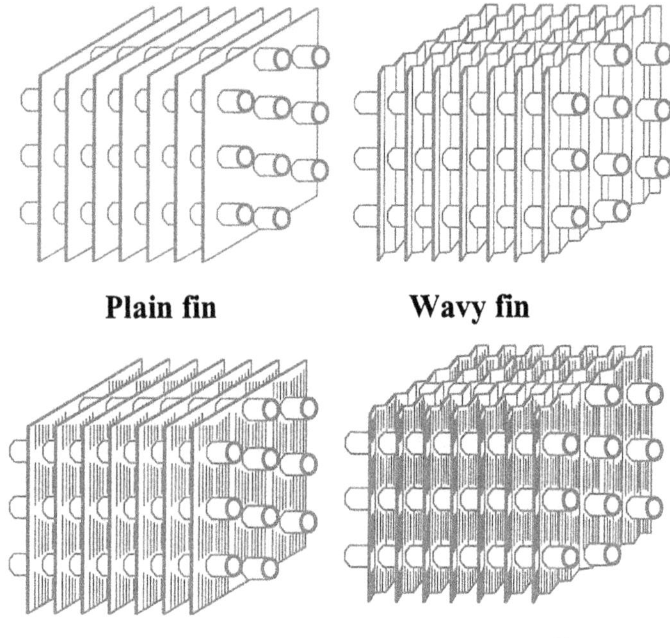

Plain fin   **Wavy fin**

Louver   **Convex-Louver fin**

Figure 1. Schematic of various fin patterns for fin-and-tube heat exchangers.

Fujii [2], Kayansayan [3], and Wang et al. [4]. Note that the fin patterns tested by these investigators were all plain fins. In addition, the test samples by the above-mentioned investigators consisted of D_o=9.52 mm (before expansion), P_t=25.4 mm, P_l=22 mm, and N=4. Though their fin pitch is not the same, however, as pointed out by Wang et al. [4,5] and Rich [6], the effect of fin pitch on the heat transfer performance is quite small for plain fin-and-tube heat exchangers having 4-row configuration. Therefore it is expected that the airside performance for the preceding samples may not deviate so much. Nonetheless, as shown in the figure, the test results differ as much as 100%. Possible explanations about the deviations in the test results include (1) contact resistance; (2) data reduction method; and (3) experimental uncertainties. Most of the investigators claimed a negligible contact resistance and acceptable uncertainties in their investigations. However, their arguments are questionable, being especially doubtful for those samples that were not mechanically expanded.

As seen, even for a similar fin geometry, significant differences in the airside performance are seen; therefore the purpose of this lecture is to clarify the differences. In addition, updated correlations for various fin patterns based on reliable and consistent data are proposed in this study.

2. Determination of Heat Transfer Coefficients

The airside heat transfer coefficients were usually obtained from the overall thermal resistance relations:

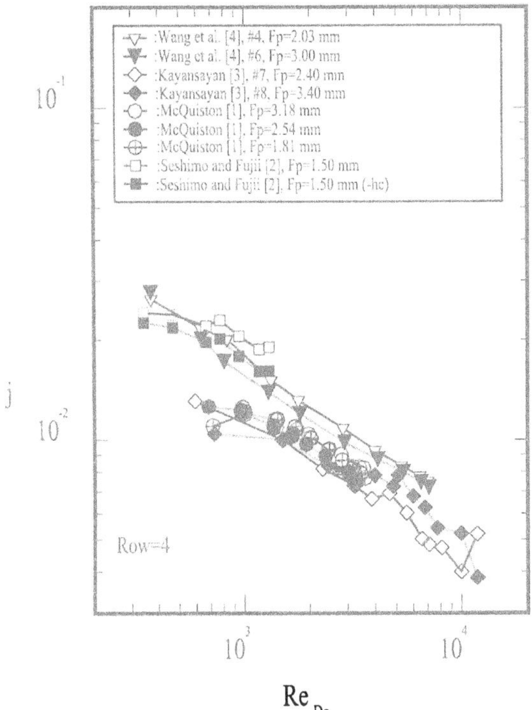

Figure 2. Comparison of j values from McQuiston [1], Seshimo and Fujii [2]
Kayansayan [3], and Wang et al. [4].

$$\frac{1}{UA} = \frac{1}{\eta_o h_o A_o} + \frac{\delta_w}{k_w A_w} + \frac{1}{h_i A_i} \tag{1}$$

where h_i is the heat transfer coefficient of tube-side, and h_o is the heat transfer coefficient of the airside. The wall resistance, $\dfrac{\delta_w}{k_w A_w}$, usually accounts less than 1% of the total resistance, and was sometimes neglected by investigators (Seshimo, [7], Seshimo and Fujii, [2]). The airside heat transfer coefficient can be obtained by subtracting the tube-side resistance and wall resistance from the overall resistance, i.e.

$$h_o = \frac{1}{\eta_o} \left(\frac{1}{U} - \frac{\delta_w A_o}{k_w A_w} - \frac{A_o}{h_i A_i} \right)^{-1} \tag{2}$$

For easier obtaining the airside performance, water was usually adopted as the working fluid in the tube side since its heat transfer coefficients can be easily evaluated

144

from the available correlations. For single-phase fluid within a smooth tube, h_i can be calculated quite accurately from the Gnielinski [8] semi-empirical correlation:

$$h_i = \left(\frac{k}{D}\right)_i \frac{(\mathrm{Re}_i - 1000)\,\mathrm{Pr}(f_i/2)}{1 + 12.7\sqrt{f_i/2}\,(\mathrm{Pr}^{2/3} - 1)} \qquad (3)$$

where

$$f_i = \left(1.58\ln(\mathrm{Re}_i) - 3.28\right)^{-2} \qquad (4)$$

$$\mathrm{Re}_i = \rho V D_i / \mu_i \qquad (5)$$

The η_o in Eq. (1) is the surface efficiency, and is related to the fin surface area, total surface area, and fin efficiency:

$$\eta_o = 1 - \frac{A_f}{A_o}(1 - \eta) \qquad (6)$$

The determination of fin efficiency, η, would involve evaluation of modified Bessel function. For circular fin geometry, the fin efficiency is given as

$$\eta_{circular} = \frac{2r_i}{M(r_o^2 - r_i^2)}\left[\frac{K_1(Mr_i)I_1(Mr_o) - K_1(Mr_o)I_1(Mr_i)}{K_1(Mr_o)I_0(Mr_i) + K_0(Mr_i)I_1(Mr_o)}\right] \qquad (7)$$

where

$$M = \sqrt{\frac{h_o P}{kA}} \qquad (8)$$

Schmidt [9] proposed an approximation of Eq. (7) which simplifies the calculation of fin efficiency.

$$\eta = \frac{\tanh(mr\phi)}{mr\phi} \qquad (9)$$

where

$$m = \sqrt{\frac{2h_o}{k_f \delta_f}} \qquad (10)$$

$$\phi = \left(\frac{R_{eq}}{r} - 1\right)\left[1 + 0.35\ln\left(R_{eq}/r\right)\right] \qquad (11)$$

For a staggered layout,

$$\frac{R_{eq}}{r} = 1.27\frac{X_M}{r}\left(\frac{X_L}{X_M} - 0.3\right)^{1/2} \qquad (12)$$

For a one-row coil or inline layout,

$$\frac{R_{eq}}{r} = 1.28 \frac{X_M}{r} \left(\frac{X_L}{X_M} - 0.2 \right)^{1/2} \tag{13a}$$

Hong and Webb [10] show that the Schmidt method would have an error higher than 5% for $R_{eq}/r > 3$ and $m(R_{eq}-r) > 2.0$. However, it is found that most practical applications would involve $R_{eq}/r < 3$ and $m(R_{eq}-r) < 2.0$. Therefore, the Schmidt approximation of the fin efficiency is often used in the data reduction. The investigations by McQuiston, [1], Kayansayan [3], and Wang et al. [4] all adopted this fin efficiency formula. Seshimo and Fujii [2], however, used a different fin efficiency formula in their reductions, i.e.

$$\eta = \left\{ 1 + h_o \frac{\left(D_f - D_c \right)^2}{6 k_f \delta_f} \left(\frac{D_f}{D_c} \right)^{0.5} \right\}^{-1} \tag{13b}$$

Seshimo and Fujii [2] claimed this formula was derived based on the assumption of the annular fins with uniform heat transfer coefficient. It is not clear about the derivation of the fin efficiency formula. However, examination of Eqs. (9) and Eq. (13b) for the above-mentioned test samples shows negligible difference. Nonetheless, it should be pointed out that Seshimo [2,7] had used different approaches to reduce the airside heat transfer coefficients. In a previous paper by Seshimo [7], he had used the identical method cited in the previous section. *Figure 3* presents the comparison

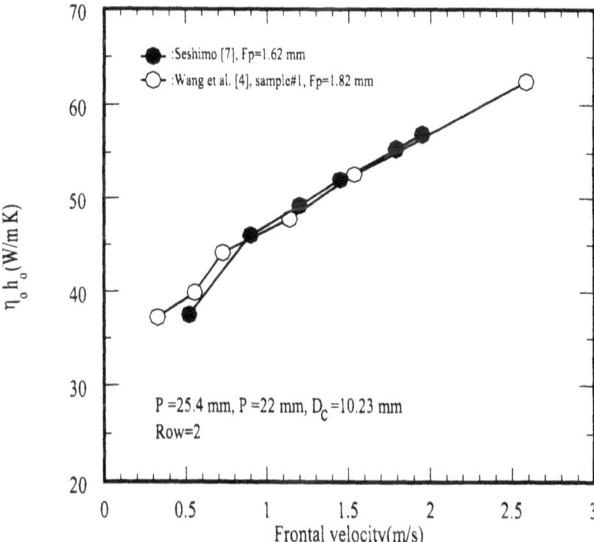

Figure 3. Comparison of the $\eta_o h_o$ values from Seshimo [7] and Wang et al. [4].

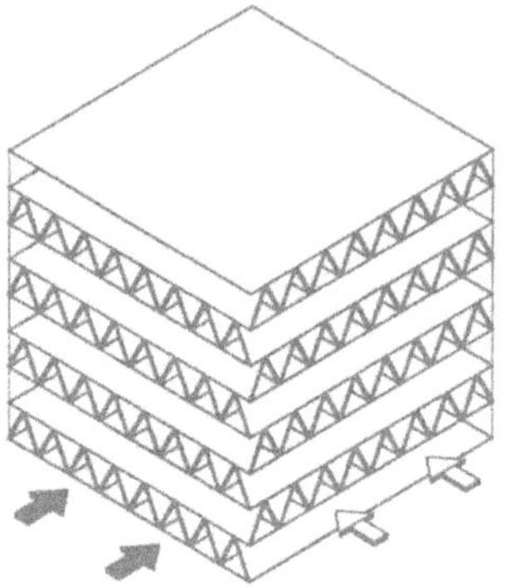

(a) Pure Cross Flow, N ⟶ ∞

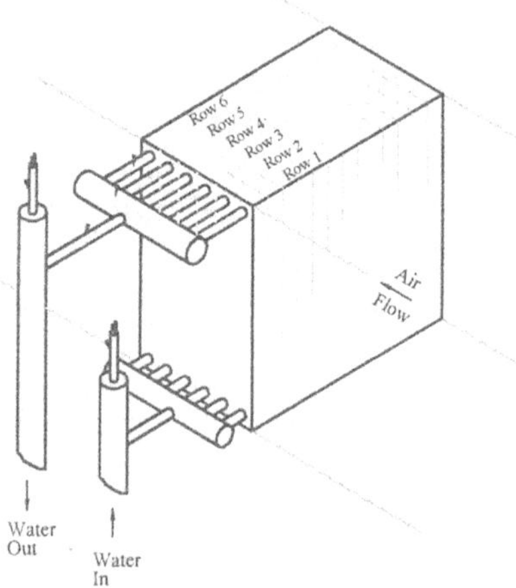

(b) Fin-and-tube Heat Exchanger with Finite Row

Figure 4.   Schematic of the heat exchanger having cross-flow
configuration; (a) infinite passage; and (b) finite number
of tube row for the present fin-and-tube heat exchangers.

between his test results and those by Wang et al. [4]. Notice that the heat exchangers
compared in this figure had the same plain fin pattern and identical longitudinal tube
pitch, transverse tube pitch, and tube diameter but a slightly different fin pitch. As
shown in the figure, the test results by Seshimo [7] and Wang et al. [4] are almost
identical. In a paper by Seshimo and Fujii [2], they had obtained the heat transfer
coefficients by subtracting an additional resistance, contact resistance, from the overall
resistance, i.e.

$$h_o = \frac{1}{\eta_o} \left(\frac{1}{U} - \frac{A_o}{h_i A_i} - \frac{A_o}{h_c A_t} \right)^{-1} \qquad (14)$$

The values of the contact conductance, h_c, is calculated based on the correlation derived by Naito [11], i.e.

$$h_c = \left(20000 \cdot \left[D_{o,after} - D_{o,before}\right] + 2.5\right) \cdot \delta_f \cdot 10^7 \qquad (15)$$

where $D_{o,after}$ and $D_{o,before}$ is the tube diameter after and before expansion, respectively. Seshimo [12] pointed out the values are about 10,000~14,000 W/m²K for their test samples. For a similar fin geometry (P_t=25.4 mm, P_l=22 mm, D=9.52 mm, and full collar), Sheffield et al. [13] reported that h_c ranged from 10,607 to 30,828 W/m²K. In practice, it is very hard to accurately predict the contact conductance, and may be very difficult to differentiate it from the overall conductance. Hence, most of the published works on the airside performance had absorbed the contact resistance into the airside performance. In this connection, we have combined the contact resistance which was calculated by Eq. (15) and the "pure" airside heat transfer coefficients from Seshimo and Fujii [2]. As seen in *Figure 2*, the test results by Wang et al. [4] agree favorably with those of Seshimo and Fujii [2]. In comparison to the test results of Seshimo and Fujii [2] and Wang et al. [4], the test results by McQuiston [1] and Kayansayan [3] were

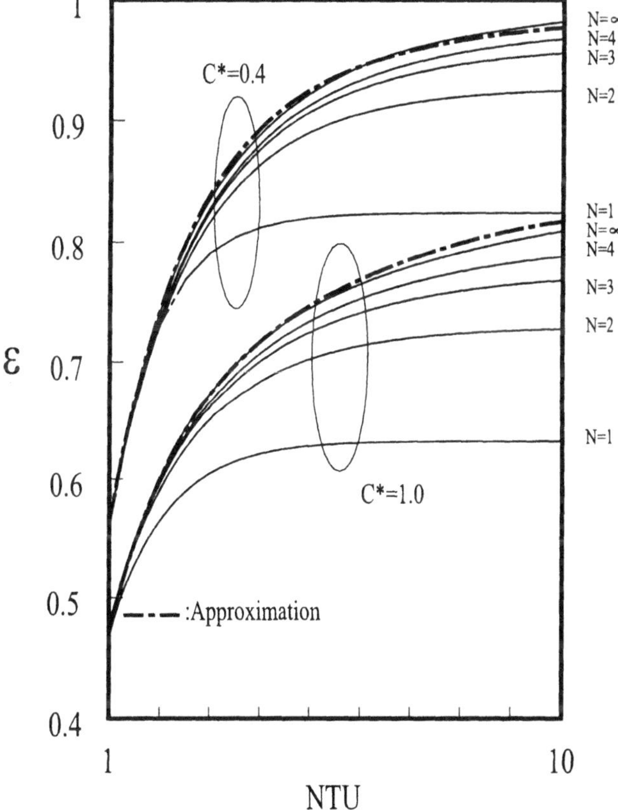

Figure 5. The ε-NTU relationship accounting the number of tube rows.

148

about 20~30% lower. There are several possible explanations for their test results. The first possible reason may be due to contact resistance. Kayansayan [3] claimed a contact conductance of 26,320 W/m^{2}K; hence it is neglected in their data reduction. However, as seen in *Figure 2*, the test results by Kayansayan [3] are quite scatter, and their heat transfer performance vs. fin pitch is not consistent. Therefore, it is likely that Kayansayan [3] may have underestimated the contact resistance in his test samples.

Table 1 ε-NTU relationship for cross-flow configuration.

N Row	Side of C_{min}	Formula
1	Air	$\varepsilon = \dfrac{1}{C^*}\left[1-e^{-C^*\left(1-e^{-NTU}\right)}\right]$
	Tube	$\varepsilon = 1 - e^{-\dfrac{\left(1-e^{-NTU\cdot C^*}\right)}{C^*}}$
2	Air	$\varepsilon = \dfrac{1}{C^*}\left[1-e^{-2KC^*\left(1+C^*K^2\right)}\right],\ K = 1 - e^{-NTU/2}$
	Tube	$\varepsilon = 1 - e^{-2K/C^*\left(1+\dfrac{K^2}{C^*}\right)},\ K = 1 - e^{-NTU\cdot C^*/2}$
3	Air	$\varepsilon = \dfrac{1}{C^*}\left[1-e^{-3KC^*\left(1+C^*K^2(3-K)+\dfrac{3\left(C^*\right)^2K^4}{2}\right)}\right],\ K = 1 - e^{-NTU/3}$
	Tube	$\varepsilon = 1 - e^{-3K/C^*\left(1+\dfrac{K^2(3-K)}{C^*}+\dfrac{3K^4}{2\left(C^*\right)^2}\right)},\ K = 1 - e^{-NTU\cdot C^*/3}$
4	Air	$\varepsilon = \dfrac{1}{C^*}\left[1-e^{-4KC^*\left(1+C^*K^2\left(6-4K+K^2\right)+4\left(C^*\right)^2K^4\left(2-K\right)+\dfrac{8\left(C^*\right)^3K^6}{3}\right)}\right]$ $K = 1 - e^{-NTU/4}$
	Tube	$\varepsilon = 1 - e^{-4K/C^*\left(1+\dfrac{K^2\left(6-4K+K^2\right)}{C^*}+\dfrac{4K^4\left(2-K\right)}{\left(C^*\right)^2}+\dfrac{8K^6}{3\left(C^*\right)^3}\right)}$ $K = 1 - e^{-NTU\cdot C^*/4}$
∞		$\varepsilon = 1 - \exp\left[NTU^{0.22}\cdot\left\{\exp\left(-C^*\cdot NTU^{0.78}\right)-1\right\}/C^*\right]$

Note: Unmixed-Unmixed Formula

In addition, the overall heat conductance, UA, is needed in the reduction of airside heat transfer coefficient from Eq. (1). Since the effectiveness ε and capacity ratio C^* can be obtained from the experimental condition. Therefore, the overall conductance can be evaluated from the arrangement of heat exchanger. For a pure unmixed/unmixed cross-flow configuration having an infinite number of passages (*Figure 4a*), the corresponding ε-NTU relationship [14] is

$$\varepsilon = 1 - \exp\left[NTU^{0.22} \cdot \left\{ \exp\left(-C^* \cdot NTU^{0.78} \right) - 1 \right\} \Big/ C^* \right] \qquad (16)$$

Notice that this relation is an approximation of a cross-flow configuration having both side unmixed, and is valid for an infinite number of tube rows. Typical fin-and-tube heat exchangers (*Figure 4b*) would involve finite number of tube rows. Hence the effect of the number of tubes row should be taken into account in the ε-NTU relationships (N is finite). The corresponding ε-NTU relationships, taken from the ESDU [15], for 1 to 4 row configuration are tabulated in Table 1. For the number of tube row greater than 4, ESDU [15] suggests an unmixed/unmixed flow arrangement having N = ∞ as the approximation. *Figure 5* shows a plot of ε vs. NTU for $C^* = 0.4$ and 1.0. As seen in the figure, the approximation by Eq. (16) gives very good agreement with those of N = ∞. For NTU < 2, the differences in effectiveness for N > 1 are negligible. However, for NTU > 3, the difference in effectiveness is greater than 0.1 between N = 1 and N = ∞. Therefore unacceptable results of heat transfer coefficients may occur when applying Eq. (16) to reduce the heat transfer coefficients. It should be further emphasized here that the use of correct ε-NTU relationships should be carefully examined before applying the present heat transfer correlations to size or rate a heat exchanger.

The following is a detailed description of the calculation procedures for a typical rating process.

(1) Depending on the fin patterns, obtain the heat transfer coefficients, h_o, from the correlations proposed in the present study.

(2) Calculate the fin efficiency using the Schmidt approximation [9]

(3) Calculate the surface efficiency, η_o, from fin efficiency η.

(4) Obtain the in-tube heat transfer coefficients, h_i, from the specific correlations. For single-phase fluid within a smooth tube, h_i is evaluated from the Gnielinski [8] semi-empirical correlation.

(5) Calculate the overall heat transfer resistance from the relationship of the overall heat transfer resistance from Eq. (1),

(6) Obtain the NTU from $NTU \equiv UA/C_{min}$

(7) Use the appropriate ε-NTU relationships to calculate the effectiveness ε. For pure cross-flow, use the relationships summarized in Table 1.

(8) Obtain the heat transfer rate by $\dot{Q} = \varepsilon \dot{Q}_{max}$.

3. Reductions of Friction Factors

In the reduction of friction factors for plate fin heat exchangers as shown in *Figure 4a*, entrance and exit loss were generally subtracted. Hence the friction factors would simply represent the friction caused by the fins. The core friction of the heat exchanger is calculated from the pressure drop equation proposed by Kays and London [16], i.e.

$$f = \frac{A_c \, \rho_m}{A_o \, \rho_1}\left[\frac{2\rho_1 \Delta P}{G_c^2} - \left(K_c + 1 - \sigma^2\right) - 2\left(\frac{\rho_1}{\rho_2} - 1\right) + \left(1 - \sigma^2 - K_e\right)\frac{\rho_1}{\rho_2}\right] \quad (17)$$

where A_o and A_c stand for the total surface area and the flow cross-sectional area, respectively. The term, σ, is the ratio of the minimum flow area to frontal area. Depending on the core configuration, the entrance and exit loss coefficients K_c and K_e can be found from Kays and London [16]. For the present fin-and-tube heat exchangers, periodic entrance and exit occurs as air flows across the heat exchangers. Thus it is very difficult to differentiate the fin friction loss from the total pressure drop. Therefore it may be more appropriate to include the entrance and exit loss in the friction factors, i.e.

$$f = \frac{A_c \, \rho_1}{A_o \, \rho_m}\left[\frac{2\Delta P}{G_c^2 \rho_1} - \left(1 + \sigma^2\right)\left(\frac{\rho_1}{\rho_2} - 1\right)\right] \quad (18)$$

In the present study, definition (18) is used to reduce all the test data.

4. Correlations for Various Fin Patterns

It is obvious that no single curve can be expected to describe the complex behaviors for both j and f factors. Therefore attempts are made to correlate the test results by using a multiple regression technique. The basic forms of the correlations are:

$$j = C_1 \, \mathrm{Re}_{Dc}^{C_2} \quad (19)$$

$$f = C_3 \, \mathrm{Re}_{Dc}^{C_4} \quad (20)$$

It is assumed that C_1, C_2, C_3, and C_4 are dependent on the physical dimensions of the heat exchanger. A separate multiple linear regression was performed to determine the exponents, C_2 and C_4 of the heat exchangers. The determinations of C_1 and C_3 are analogous to C_2 and C_4. After a trial-and-error process, the final correlations for the j and f factors for various fin patterns are summarized as follows:

4.1. Correlations for Plain Fins

The first successful heat transfer and friction correlations for plain fin geometry having staggered layout was proposed by McQuiston [17]. However, the predictive ability of the friction factors, as pointed out by Gray and Webb [18], was quite poor. Gray and Webb [18] had significantly improved the predictive capability of friction factors, and the heat transfer correlation by Gray and Webb [18] is comparable to that of McQuiston [17]. A recent investigation by Wang et al. [5,19] indicated that the Gray and Webb correlation [18] considerably underpredicts the heat transfer data for those coils having smaller diameter tubes. The mean deviations of the Gray and Webb correlation can be over 25%. In addition, the deviations of the predictive values for friction factor by Gray and Webb correlation is even higher. Therefore, this has motivated the present investigator to develop an updated correlation for the plain fin pattern [20].

The database used by Wang and Chi. [20] are from Rich [6,21], Seshimo and Fujii [2], Wang et al. [4,5], and newly data reported by Wang and Chi [20]. A total of 74 samples were compiled in developing the correlation. Notice that the original test results of Seshimo and Fujii [2] had excluded the contact resistance. In the present study, their original airside resistance was combined with the contact resistance using Eq. (15). After a detailed evaluation of the database, the recommended correlation for heat transfer performance are given as follows:

For $N = 1$,

$$j = 0.173 \, \mathrm{Re}_{D_c}^{-0.346} \left(\frac{P_t}{P_l} \right)^{P1} \left(\frac{D_c}{F_p} \right)^{1.161} \left(\frac{D_h}{F_p} \right)^{1.035} \left(\frac{F_p}{P_t} \right)^{P2} \tag{21}$$

where

$$P1 = -0.22 \ln\left(\mathrm{Re}_{Dc} \right) + 1.88 \tag{22}$$

$$P2 = 0.106 \ln\left(\mathrm{Re}_{Dc} \right) \tag{23}$$

For $N \geq 2$,

$$j = 0.078 \, \mathrm{Re}_{D_c}^{P3} \, N^{P4} \left(\frac{F_p}{D_h} \right)^{P5} \left(\frac{P_t}{F_p} \right)^{1.026} \tag{24}$$

where

$$P3 = 0.16 \ln\left(N \left(\frac{F_p}{D_c} \right)^{0.42} \right) - 0.349 \tag{25}$$

$$P4 = \frac{-0.094 \left(\dfrac{P_l}{D_h} \right)^{1.38}}{\ln\left(\mathrm{Re}_{Dc} \right)} - 1.405 \tag{26}$$

$$P5 = 1.263 \ln\left(\frac{\mathrm{Re}_{Dc}}{N} \right) - 5.97 \tag{27}$$

152

$$D_h = \frac{4 A_c L}{A_o} \tag{28}$$

The friction factors is given as:

$$f = 0.0146\, \mathrm{Re}_{D_c}^{P6} \left(\frac{P_t}{P_l}\right)^{1.959} \left(\frac{F_p}{D_c}\right)^{P7} N^{0.021} \tag{29}$$

where

$$P6 = -0.0535 + \frac{0.01166}{\ln\left(\dfrac{P_t}{P_l}\right)} + 0.123\left(\frac{F_p}{D_c}\right) \tag{30}$$

$$P7 = 2.319 - \frac{19.59}{\ln\left(\mathrm{Re}_{Dc}\right)} \tag{31}$$

As illustrated from Table 2, Eqs. (21,24) can describe 87.4% of the j factors within 15% and Eq. (29) can correlate 84.2% of the friction factors within 15%. In addition to the correlation proposed in this paper, several other correlations were tested against the data. These correlations include the McQuiston [17], and Gray and Webb [18]. The results of the comparison are shown in Table 2. As seen, the mean deviation of heat transfer correlation for the present correlation, the McQuiston correlation, and the Gray and Webb correlation are 7.64%, 34.4%, and 15.8%, respectively. On the other hand, the mean deviation for friction factors are 8.97%, 40.6%, and 22.4%, respectively.

Table 2 Comparisons of the proposed plain-fin correlation with other correlations

Deviation	Present Correlation		McQuiston (1978b)		Gray and Webb (1986)	
	j	f	j	f	j	f
±10%	76.3	67.4	34.7	18.7	38.9	22.3
±15%	87.4	84.2	46.7	25.8	51.6	33.9
±20%	92.8	92.5	56.3	29.2	62.8	45.5
Average Deviation (%)	0.59	0.76	-20.4	-12.2	-12.4	-20.5
Mean Deviation (%)	7.64	8.97	34.4	40.6	15.8	22.4

$$\text{Average Deviation} = \frac{1}{K}\left(\sum_{1}^{K} \frac{j(f)_{pred} - j(f)_{exp}}{j(f)_{exp}}\right) \times 100\%$$

$$\text{Mean Deviation} = \frac{1}{K}\left(\sum_{1}^{K} \frac{\left|j(f)_{pred} - j(f)_{exp}\right|}{j(f)_{exp}}\right) \times 100\%$$

K: Number of data points

4.2. Correlations for Louver fin Pattern

The database for the louver fin-and-tube heat exchangers consists six different louver fin patterns. Detailed description of the present louver fins and the terminology of

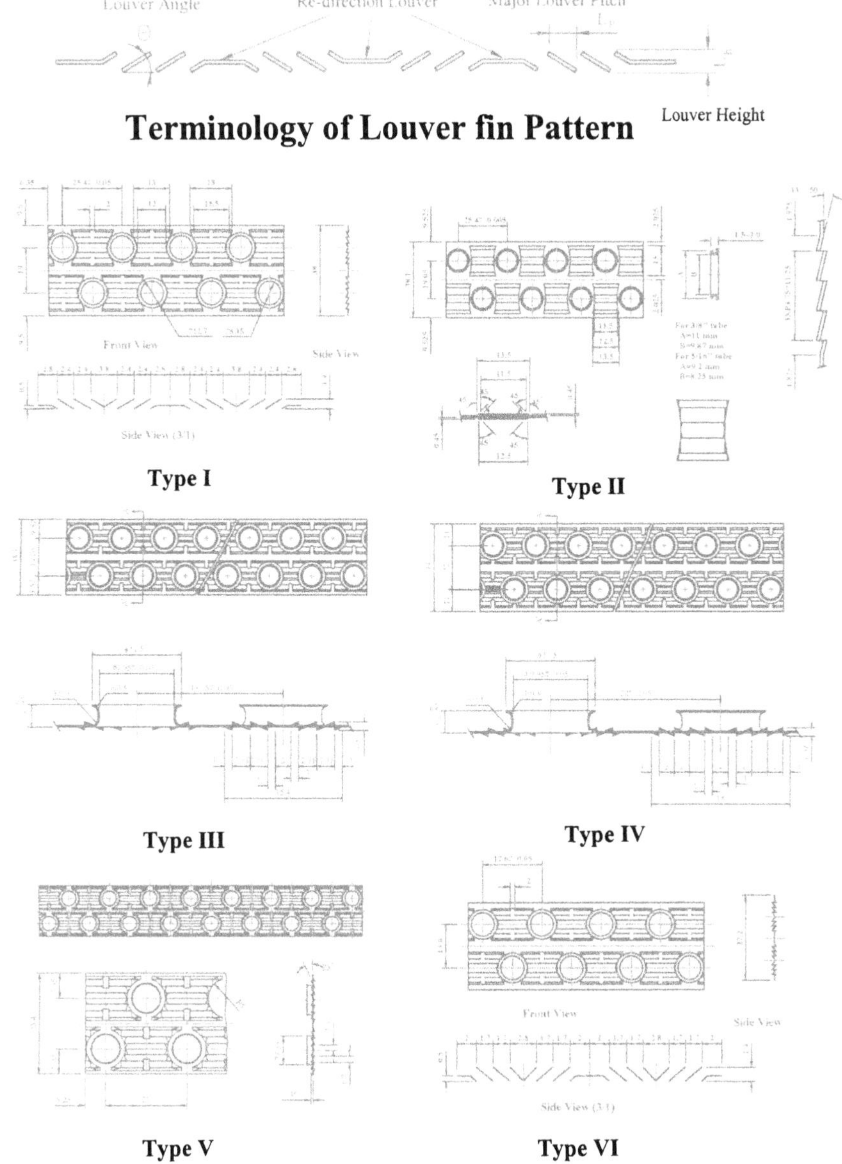

Figure 6. Detailed louver-fin patterns for developing the correlation.

154

louver fin is shown in *Figure 6*. A total of 49 samples are used in the development of correlations. The data are from Wang *et al.* [22] (17 samples, *Figure 6*, Type I), Wang *et al.* [23] (14 sample, *Figure 6*, Type II). Chi *et al.* [24] (6 samples, *Figure 6*, Type III), Wang et al. [25] (4 samples, *Figure 6*, Type IV), Chi *et al.* [24] (4 sample, *Figure 6*, Type V), and Wang [26] (4 samples, *Figure 6*, Type VI).

The final heat transfer and friction correlations are given as follows:

$$j = 0.0501\,\mathrm{Re}_{D_c}^{0.094}\left(\frac{P_l}{P_t}\right)^{1.44}\left(\frac{Fp}{Dc}\right)^{0.322} N^{-0.809}\left(\frac{L_p}{Fp}\right)^{-0.177}\left(\frac{L_h}{L_p}\right)^{-0.254} \quad (32)$$

and

$$j = 0.962\,\mathrm{Re}_{D_c}^{L1}\left(\frac{P_l}{P_t}\right)^{-1.51}\left(\frac{Fp}{Dc}\right)^{-0.107} N^{-1.06}\left(\frac{L_h}{L_p}\right)^{-0.433} \quad (33)$$

where

$$L1 = \left[-0.501\left(\frac{P_l}{P_t}\right)^{-0.424} N^{-0.253}\left(\frac{L_h}{L_p}\right)^{-0.127}\right] \quad (34)$$

Notice that Eq. (32) is valid for $\mathrm{Re}_{Dc} < 1{,}000$, and the louver fins with re-direction louvers (Type I, III, IV, V, and VI). In addition, the ranges of applicability of Eq. (32) are (1) $N \geq 4$, Fp/Dc ≥ 0.14; or (2) $N = 2$ or 3, Fp/Dc ≤ 0.14. Eq. (33) is valid for (1) type II fin pattern; (2) louver fins with re-direction louver and $N = 1$; and (3) louver fins with re-direction louver having $\mathrm{Re}_{Dc} > 1{,}000$.

The correlation of friction factor is given as:

$$f = 0.72\,\mathrm{Re}_{Dc}^{Z1}\left(\frac{F_p}{D_c}\right)^{Z2} N^{0.322}\left(\frac{P_l}{P_t}\right)^{-2.494}\left(\frac{L_h}{L_p}\right)^{1.34}\left(\frac{L_p}{F_p}\right)^{1.69} \quad (35)$$

$$Z1 = -0.392\left(\frac{P_l}{P^\cdot}\right)^{-0.726} N^{0.0782}\left(\frac{L_h}{L_p}\right)^{0.352}\left(\frac{L_p}{F_p}\right)^{0.63} \quad (36)$$

$$Z2 = -3.3 + 3.72\left(\frac{P_l}{P_t}\right)^{-1.0} - 1.29\left(\frac{P_l}{P_t}\right)^{-2} \quad (37)$$

Table 3 Comparison of the proposed correlation with the experimental data

Deviation	±10%	±15%	±20%	±25%	Mean Deviation	Average Deviation
j	79.7%	93.2%	96.4%	99.1%	6.83%	0.97%
f	59%	75.3%	86.9%	95.9%	10.1%	1.17%

Eqs. (32,33) can describe 93.2% of the j factors within ±15% and Eq. (35) can correlate 75.3% of the friction factors within ±15%. The results of the comparisons with all the test data are tabulated in Table 3. As seen, the proposed heat transfer correlation gives a mean deviation of 6.83% while the proposed friction correlation shows a 10.1% mean deviation.

4.3. Correlations for Wavy Fins

Figure 7 presents the terminology of the wavy fin geometry which includes the corrugation angle θ, waffle height P_d, and projected wavy length X_f. The proposed correlation for the wavy fin geometry is given as follows (based on Wang et al. [27]):

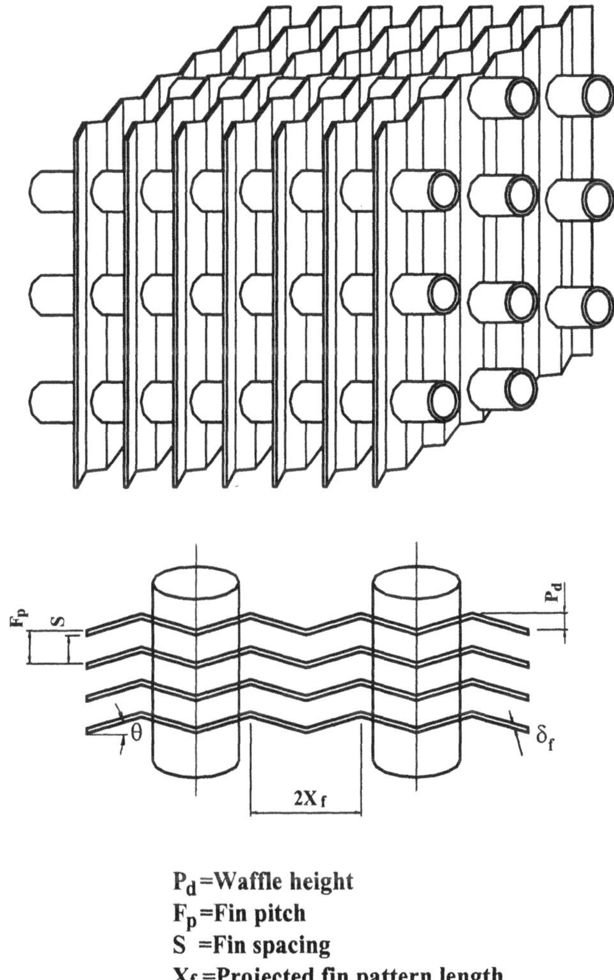

P_d =Waffle height
F_p =Fin pitch
S =Fin spacing
X_f =Projected fin pattern length
δ_f =Fin thickness
θ =Corrugation Angle

Figure 7. Schematic of a typical wavy fin-and-tube heat exchanger.

156

$$j = 1.79097\,\mathrm{Re}_{Dc}^{-0.1707-1.374\left(\frac{P_l}{\delta_f}\right)^{-0.493}\left(\frac{F_p}{D_c}\right)^{-0.886}N^{-0.143}\left(\frac{P_d}{X_f}\right)^{-0.0296}}\left(\frac{P_l}{\delta_f}\right)^{-0.456}N^{-0.27}\left(\frac{F_p}{D_c}\right)^{-1.343}\left(\frac{P_d}{X_f}\right)^{0.317}$$

(38)

$$f = 0.05273\,\mathrm{Re}_{Dc}^{f1}\left(\frac{P_d}{X_f}\right)^{f2}\left(\frac{F_p}{P_t}\right)^{f3}\left(\log_e\left(\frac{A_o}{A_t}\right)\right)^{-2.726}\left(\frac{D_h}{D_c}\right)^{0.1325}N^{0.02305}$$

(39)

$$f1 = 0.1714 - 0.07372\left(\frac{F_p}{P_l}\right)^{0.25}\log_e\left(\frac{A_o}{A_t}\right)\left(\frac{P_d}{X_f}\right)^{-0.2}$$

(40)

$$f2 = 0.426\left(\frac{F_p}{P_t}\right)^{0.3}\log_e\left(\frac{A_o}{A_t}\right)$$

(41)

$$f3 = \frac{-10.2192}{\log_e(\mathrm{Re}_{DC})}$$

(42)

The above-mentioned equations are constructed based on the test results of wavy fin pattern having larger tube diameter, larger transverse tube pitch, and longitudinal tube pitch (as shown in Table 4).

Table 4 Geometric dimensions of the wavy fin-and-tube heat exchangers

Type	Nominal Tube Diameter (mm)	Dc (mm)	Fin Pitch (mm)	P_t (mm)	P_l (mm)	Row No.
I	12.7	13.62	3.0~6.3	31.75	27.5	1~6
II	15.88	16.71	3.0~6.3	38.1	33	1~4

Equation (38) can describe 93.02% of the experimental data within 10% while Eq. (39) can describe 91.8% of the experimental data within 10%.

For wavy fin-and-tube heat exchangers having smaller diameter tube (Dc=8.5~10.4 mm, P_t=25.4 mm, and P_l=19.05~25.4 mm), Wang et al. [28] proposed the following correlations which were based the test results of 27 samples [29,30]:

$$j = 0.324\,\mathrm{Re}_{D_c}^{J1}\left(\frac{F_p}{P_l}\right)^{J2}(\tan\theta)^{J3}\left(\frac{P_l}{P_t}\right)^{J4}N^{0.428}$$

(43)

where

$$J1 = -0.229 + 0.115\left(\frac{F_p}{D_c}\right)^{0.6}\left(\frac{P_l}{D_h}\right)^{0.54} N^{-0.284}\ln(0.5\tan\theta) \quad (44)$$

$$J2 = -0.251 + \frac{0.232 N^{1.37}}{\left(\ln(\mathrm{Re}_{Dc}) - 2.303\right)} \quad (45)$$

$$J3 = -0.439\left(\frac{F_p}{D_h}\right)^{0.09}\left(\frac{P_l}{P_t}\right)^{-1.75} N^{-0.93} \quad (46)$$

$$J4 = 0.502\left(\log_e(\mathrm{Re}_{Dc}) - 2.54\right) \quad (47)$$

The correlation for friction factor is given as:

$$f = 0.01915\,\mathrm{Re}_{Dc}^{F1}(\tan\theta)^{F2}\left(\frac{F_p}{P_l}\right)^{F3}\left[\ln\left(\frac{A_o}{A_t}\right)\right]^{-5.35}\left(\frac{D_h}{D_c}\right)^{1.3796} N^{-0.0916} \quad (48)$$

where

$$F1 = 0.4604 - 0.01336\left(\frac{F_p}{P_l}\right)^{0.58}\ln\left(\frac{A_o}{A_t}\right)(\tan\theta)^{-1.5} \quad (49)$$

$$F2 = 3.247\left(\frac{F_p}{P_t}\right)^{1.4}\ln\left(\frac{A_o}{A_t}\right) \quad (50)$$

$$F3 = \frac{-20.113}{\ln(\mathrm{Re}_{Dc})} \quad (51)$$

The above-mentioned heat transfer correlation gives a mean deviation of 6.44% while the proposed friction correlation shows a 5.01% mean deviation. Detailed comparisons of the present correlation with the experimental data are given in Table 5.

Table 5 Comparison of the proposed correlation with the experimental data

Deviation	±10%	±15%	±20%	±25%	Mean Deviation	Average Deviation
j	79.6%	95.1%	97.4%	98.5%	6.44%	0.35%
f	84.9%	97.3%	99.6%	100%	5.01%	0.24%

4.4. Correlations for Convex-louver fin Pattern

The database for convex-louver are based on the test results by Wang et al. [30,31]. Notice that the following correlations are developed using only one fin pattern. Therefore readers should carefully examine their convex-louver fin pattern before using the following correlations.

$$j = 16.06 \, \mathrm{Re}_{Dc}^{-1.02(\frac{P_f}{D_c})-0.256} \left(\frac{A_o}{A_t} \right)^{-0.601} N^{-0.069} \left(\frac{P_f}{D_c} \right)^{0.84} \tag{52}$$

The recommend correlation for friction factors is:

For $\mathrm{Re}_{Dc} \leq 1000$

$$f = 0.264 \left[0.105 + 0.708 e^{\frac{-\mathrm{Re}_{Dc}}{225}} \right] \mathrm{Re}_{Dc}^{-0.637} \left(\frac{A_o}{A_t} \right)^{0.263} \left(\frac{F_p}{D_c} \right)^{-0.317} \tag{53}$$

For $\mathrm{Re}_{Dc} > 1000$

$$f = 0.768 \left[0.0494 + 0.142 e^{\frac{-\mathrm{Re}_{Dc}}{1180}} \right] \left(\frac{A_o}{A_t} \right)^{0.0195} \left(\frac{F_p}{D_c} \right)^{-0.121} \tag{54}$$

Eq. (52) can describe 95.9% of the experimental data within 15% while Eqs. (53,54) can describe 90.7% of the experimental data within 15%. Note that the j factor correlation over-predicts the test results of the 4-row and F_p=1.21 mm. However, in practical applications, the convex-louver fin-and-tube heat exchangers are employed with 1 or 2 row configuration.

5. Conclusions

The present lecture focus on the air-side performance of typical fin-and-tube heat exchangers. It is strongly recommended that investigators should clearly address the details of their data reduction methods. In addition, before use of certain correlations, designers should examine their detailed tube layout (use the correct ε-NTU relationship) and the applicability of the correlations. In addition to the data reduction method, updated correlations, which were developed based on consistent data reduction method for fin-and-tube heat exchangers are reported. The proposed correlations include plain, wavy, louver, and convex-louver fins.

6. Acknowledgement

The author would like to express gratitude for the Energy R&D foundation funding from the Energy Commission of the Ministry of Economic Affairs, Taiwan, for providing financial support of this study.

Nomenclature

A area, m²

A_c minimum flow area, m²

A_o total surface area, m²

A_t external tube surface area, m²

C heat capacity rate, W/K

c_p specific heat at constant pressure, J/(kg·K)

D_c fin collar outside diameter, m

D_i inside tube diameter, m

D_f equivalent fin diameter, m

f Fanning friction factor, dimensionless

F_p fin pitch, mm

G_c mass flux of the air based on the minimum flow area, kg/m²·s

h_o air-side heat transfer coefficient, W/m²·K

h_c contact conductance, W/m²·K

h_i tube-side heat transfer coefficient, W/m²·K

I_0 modified Bessel function solution of the first kind, order 0

I_1 modified Bessel function solution of the first kind, order 1

j $Nu/RePr^{1/3}$, the Colburn factor, dimensionless

k fluid thermal conductivity, W/m·K

K_0 modified Bessel function solution of the second kind, order 0

K_1 modified Bessel function solution of the second kind, order 1

K_c abrupt contraction pressure-loss coefficient

K_e abrupt expansion pressure-loss coefficient

L_h louver height, mm

L_p louver pitch, mm

$\dot{m}$ mass flow rate, kg/s

N Number of longitudinal tube rows, dimensionless

NTU $U \cdot A/C_{min}$, Number of transfer unit, dimensionless

ΔP pressure drop, Pa

F_p fin pitch, mm

P circumference of the fin, m

P_d waffle height, mm

P_l longitudinal tube pitch, mm

Pr   Prandtl number, dimensionless

P_t transverse tube pitch, mm

$\dot{Q}$ heat transfer rate, W

$\dot{Q}_{max}$ $C_{min}(T_{water,in} - T_{air,in})$, the maximum possible heat transfer rate, W

Re_{Dc} $\rho V D_c/\mu$, Reynolds number based on tube outside diameter, including collar, dimensionless

r   radius of the tube diameter, including collar fin thickness, mm

R_{eq} equilibrium radius for circular fin, mm

160

S	fin spacing, $F_p - \delta_f$, mm
T	temperature, °C
U	overall heat transfer coefficient, W/m²·K

$$X_L \quad \left\{ \begin{array}{l} \sqrt{(P_t/2)^2 + P_l^2}\,\big/2, \text{ for staggered layout} \\[2mm] P_l\big/2 \text{ , for inline and 1 - row coil} \end{array} \right\} , \text{ geometric parameter, mm}$$

X_M	$P_t/2$, geometric parameter, mm
X_f	projected wavy length, mm

Greeks Symbols

δ_f	fin thickness, mm
δ_w	thickness of tube wall, mm
ε	$\dot{Q}_{ave}\big/\dot{Q}_{max}$, heat exchanger effectiveness, dimensionless
θ	corrugation angle or louver angles, degrees
η	fin efficiency, dimensionless
η_0	surface efficiency, dimensionless
μ	dynamic viscosity of fluid, Pa·s
ρ	mass density of fluid, kg/m³
σ	contraction ratio of cross-sectional area, dimensionless

Subscripts

1	air side inlet
2	air side outlet
air	air side
avg	average value
i	tube side
in	inlet
f	fin surface
m	mean value of the inlet and outlet conditions
min	minimum value
max	maximum value
o	total surface
water	water side
w	wall of the tube

References

1. McQuiston, F.C. (1978) Heat, mass and momentum transfer data for five plate-fin-tube heat transfer surfaces, *ASHRAE Transaction*, **84**(1), 266-293.

2. Seshimo, Y. and Fujii, M. (1991) An experimental study of the performance of plate fin and tube heat exchangers at low Reynolds number, Proceeding of the *3rd ASME/JSME Thermal Engineering Joint Conference,* **4**, 449-454.
3. Kayansayan, N. (1993) Heat transfer characterization of flat plain fins and round tube heat exchangers, *Experimental Thermal and Fluid Science,* **6**, 263-272.
4. Wang, C.C., Hsieh, Y.C., Chang, Y.J. and Lin, Y.T. (1996) Sensible heat and friction characteristics of plate fin-and-tube heat exchangers having plane fins, *Int. J. of Refrigeration,* **19**(4), 223-230.
5. Wang, C.C., Lee, W.S., and Chang, C.T. (1997) Sensible heat transfer characteristics of plate fin-and-tube exchangers having 7-mm tubes, *AIChE Symposium Series,* **93**(314), 211-216.
6. Rich, D.G. (1973) The effect of fin spacing on the heat transfer and friction performance of multi-row, smooth plate fin-and-tube heat exchangers, *ASHARE Transaction,* **79**(2), 135-145.
7. Seshimo, Y. (1988) Effectiveness and fin efficiency of plate-fin and tube heat exchangers, *Transactions of the JAR,* **5**(2), 133-141, in Japanese.
8. Gnielinski, V. (1976) New equation for heat and mass transfer in turbulent pipe and channel flow, *Int. Chem. Engng.,* **16**, 359-368.
9. Schmidt, Th.E. (1949) Heat transfer calculations for extended surfaces, *Refrigerating Engineering,* April, 351-357.
10. Hong, T.K. and Webb, R L. (1996) Calculation of fin efficiency for wet and dry fins, *Int. J. HVAC&R Research,* **2**(1), 27-41.
11. Naito, N. (1970) *SHASE Transactions* (The Society of Heating, Air-conditioning and Sanitary Engineers of Japan), **44**(5), pp. 1-, in Japanese, quoted from Seshimo (1988).
12. Seshimo, Y. (1998) private communication.
13. Sheffild, J.W. and Sauer, H.J. Jr. (1989) Experimental investigation of thermal conductance of finned tube contacts, *Exp. Thermal Fluid Sci.,* **1**, 107-121.
14. McQuiston, F.C. and Parker, J.D (1994) *Heating, Ventilating, and Air-conditioning,* 4th ed. Chapter 14, John Wiely and Sons, Singapore, p. 571.
15. ESDU 86018 (1991) Engineering Science Data Unit 86018 with amendment A, July 1991, 92-107, ESDU International plc, London.
16. Kays, W.M. and London, A.L. (1984) *Compact Heat Exchangers,* 3rd ed. McGraw-Hill, New York.
17. McQuiston, F.C. (1978) Correlation of heat, mass and momentum transport coefficients for plate-fin-tube heat transfer surface, *ASHRAE Transactions,* **84**(1), 294-308.
18. Gray, D.L. and Webb, R.L. (1986) Heat transfer and friction correlations for plate finned-tube heat exchangers having plain fins, *Proc. 8th. Heat Transfer Conference,* pp. 2745-2750.
19. Wang, C.C. and Chang, C.T. (1998) Heat and mass transfer for plate fin-and-tube heat exchangers; with and without hydrophilic coating, *Int. J. Heat and Mass Transfer,* **41**, 3109-3120.
20. Wang, C.C. and Chi, K.U. (1998) Heat transfer and friction characteristics of plain fin-and-tube heat exchangers: new experimental data and new correlation, paper in review.

162

21. Rich, D.G. (1975) The effect of the number of tubes rows on heat transfer performance of smooth plate fin-and-tube heat exchangers, *ASHRAE Transactions*, **81**(1), 307-317.
22. Wang, C.C., Chang, Y.P., Chi, K.U. and Chang, Y.J. (1998) An experimental study of heat transfer and friction characteristics of typical louver fin and tube heat exchangers, *Int. J. of Heat and Mass Transfer*, **41**, no. 4-5, 817-822.
23. Wang, C.C., Chang, Y.P., Chi, K.U. and Chang, Y.J. (1998) A study of non-redirection louver fin-and-tube heat exchangers, *Proceeding of Institute of Mechanical Engineering, Part C, Journal of Mechanical Engineering Science*, **212**, 1-14.
24. Chi, K.U., Wang, C.C., Chang, Y.J. and Chang, Y.P. (1998) A comparison study of compact plate fin-and-tube exchanger, *ASHRAE Transactions*, in press.
25. Wang, C.C., Lee, C. J. and Chang, C.T. (1998) Some aspects of the fin-and-tube heat exchangers: with and without louvers, *Journal of Enhanced Heat Transfer*, in press.
26. Wang, C.C. (1998) ERL louver fin test data, unpublished data for four louver fins.
27. Wang, C.C., Lee, C. J., Lin, Y. T. and Chang, Y.J. (1998) An investigation of wavy fin-and-tube heat exchangers; a contribution to databank, accepted for publication in *Experimental Heat Transfer*.
28. Wang, C.C. Jang, J.Y. and Chiou, N.F. (1998) A heat transfer and friction correlation for wavy fin-and-tube heat exchangers, paper in review.
29. Wang, C.C., Jang, J.Y. and Chiou, N.F. (1998) Effect of waffle height on the air-side performance of wavy fin-and-tube heat exchangers, accepted for publication in *Heat Transfer Engineering*.
30. Wang, C.C., Tsi, Y.M. and Lu, D.C. (1998) Comprehensive study of convex-louver and wavy fin-and-tube heat exchangers, *AIAA J. of Thermophysics and Heat Transfer*, **12**, 423-430.
31. Wang, C.C., Chen, P.Y. and Jang, J.Y. (1996) Heat transfer and friction characteristics of convex-louver fin-and-tube heat Exchangers, *Experimental Heat Transfer*, **9**, 61-78.

OPTIMUM DESIGN OF AIR-COOLED FIN-AND-TUBE HEAT EXCHANGERS:ACCOUNTING FOR THE EFFECT OF COMPLEX CIRCUITING

CHI-CHUAN WANG
Energy & Resources Laboratories,
Industrial Technology Research Institute,
D500 ERL/ITRI
Bldg. 64, 195~6 Section 4, Chung Hsing Rd.,
Chutung, Hsinchu, Taiwan 310

E-mail:ccwang@erl.itri.org.tw

Abstract. This lecture presents a rationally based methodology to model air-cooled condenser and water heating coils. The effect of complex circuiting can be taken into account by the proposed indexing technique. The calculated results agree well with experimental data. It is found that the mal-distribution of refrigerant flow is strongly related to thermal-hydraulic interactions between air and refrigerant. In addition to the model, preliminary test results of a 1-circuit wavy finned condenser having different arrangements are also reported.

1. Introduction

Air-cooled heat exchangers used for air-conditioning and refrigeration applications generally consist of plate fin-and-tube heat exchangers. The plate fin-and-tube heat exchangers are usually composed of mechanically or hydraulically expanded round tubes into parallel continuous fins. Depending on the application, the heat exchangers can be produced in one or more rows.

To effectively improve the performance of air-cooled heat exchangers, passive enhancement techniques are often employed. The methods are (1) by using enhanced fins/tubes surfaces; (2) by increasing total surface area; (3) by increasing the effective mean temperature difference between the air flow and refrigerant flow. The use of enhanced fin/tube may be very effective if the enhancement is employed on the dominant resistance side. This also leads to possible increase of initial cost and some potential pitfalls (e.g., fouling and higher pumping power). Increase of surface area may be again effective. However, in space-limited application, use of larger heat exchangers may not be acceptable. Enlargement of the effective mean temperature may be economical but the degree of augmentation may not be very large. In practice, the most common method to enlarge temperature difference is via circuitry. The main objective of the present study is focused on this effect, as applied to refrigerant condensers.

163

S. Kakaç et al. (eds.), Heat Transfer Enhancement of Heat Exchangers, 163–184.
© 1999 *Kluwer Academic Publishers.*

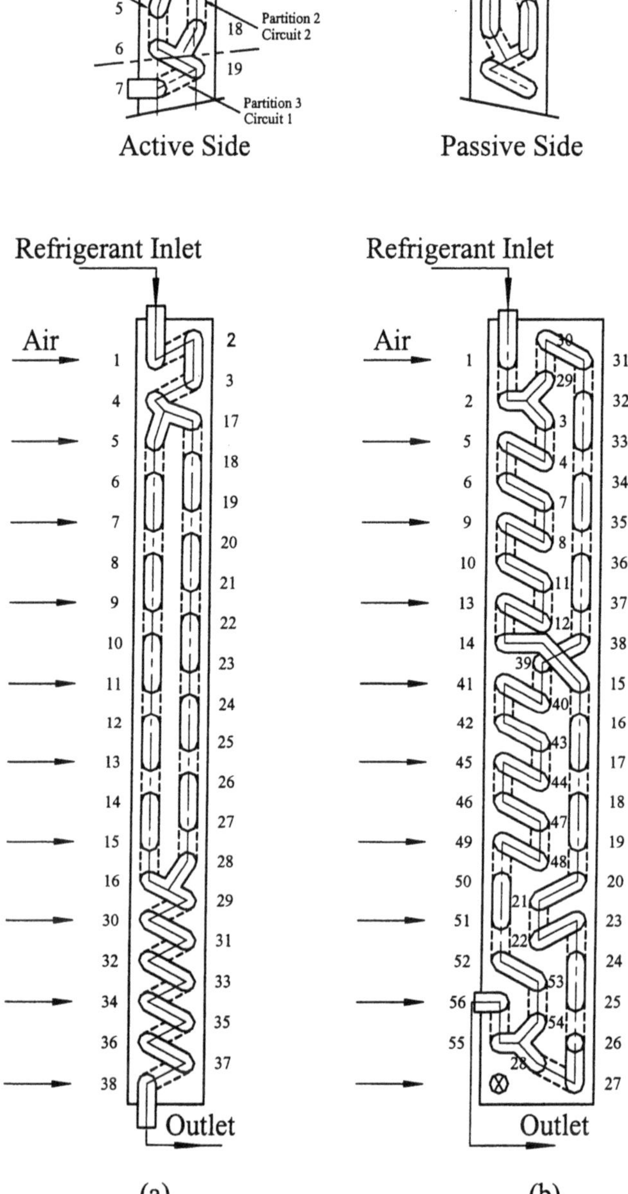

Figure 1. Schematic of typical circutry arrangement.

In practical use, circuiting such as shown in *Figure 1* is often encountered. The number of methods to circuit refrigerant in a heat exchanger are numberless. Fortunately, the arrangements are limited by the manufacturing constraints. As seen in *Figure 1*, the inlet and outlet tubing occur on one side of the heat exchanger-denoted here as the active side of the coil. All solid bends or hairpins are on the opposite or passive side of the coil. A hairpin is the 180-degree bend created by bending a length of copper tubing into a "U" shaped configuration.

The purposes of the circuiting are (1) provide better refrigerant-side distribution; and (2) increase the mean temperature difference between the air and refrigerant; hence higher heat transfer rates can be achieved. The selection of circuiting is important subject of air-cooled condensers having multiple-row configurations. As shown in *Figure 2*, typical in-tube condensation process may involve superheated single-phase heat transfer (region I), superheated condensation (region II), saturated condensation (region III), and subcooled single-phase heat transfer (region IV). In region I, the wall temperature is greater than that of the vapor saturation temperature, in which heat transfer takes place via single phase forced convection. In region II, the bulk temperature of the refrigerant vapor is still superheated, while the wall temperature is lower than the vapor saturation; as a result, the condensation takes place. In region III, the refrigerant flow is saturated, which results in saturated condensation. In region IV, the refrigerant is completely condensed, and hence heat transfer mechanism is the same as region I.

As illustrated in *Figure 2*, the heat transfer characteristics are significantly different along the condenser. Furthermore, the corresponding mean temperature difference

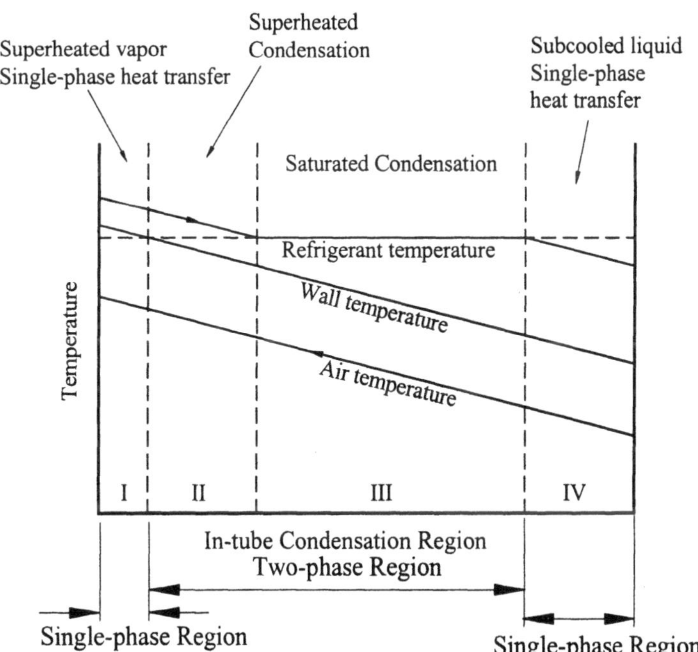

Figure 2. Schematic of various heat transfer regime for in-tube condensation.

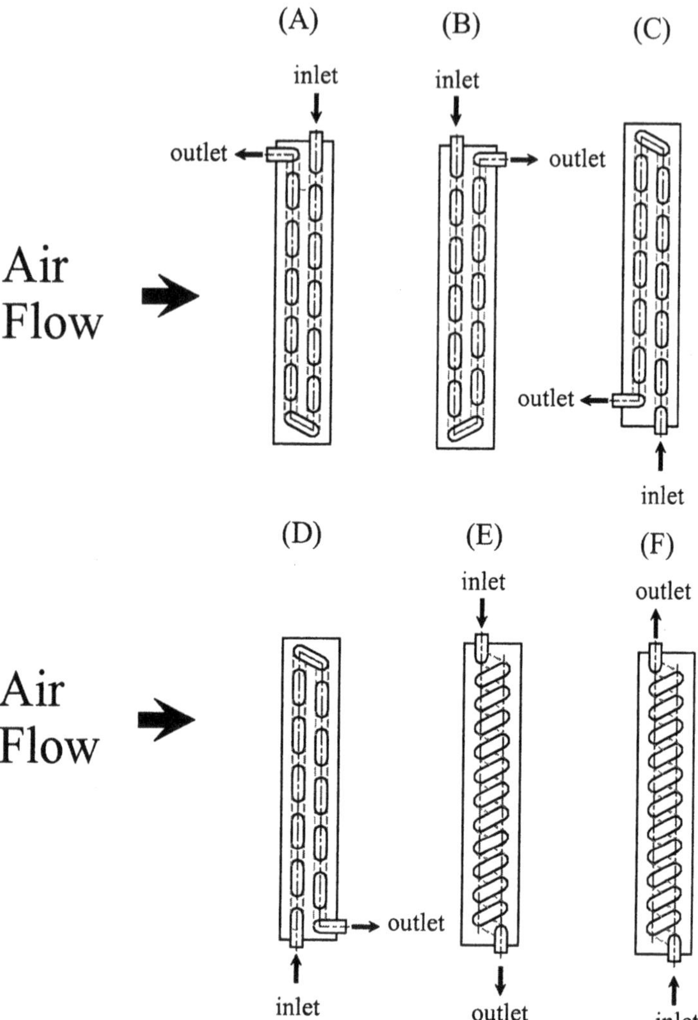

Figure 3. Schematic of the 1-circuit arrangement tested in this study.

between the air-flow and refrigerant changes significantly along the condenser. Therefore, to effectively improve the heat transfer performance while limiting the pressure drop of the refrigerant flow, the use of multiple refrigerant circuits is very popular. Usually, for maximum performance, the maximum number of circuits are at the inlet superheated vapor region, and the minimum number of circuits exist at the subcooled outlet.

The design of circuitry is heavily influenced by practical experience. Several rules were suggested by Hogan [1]. For instance, an active side of the coil should contain connections of the inlet and outlet tubes, while a passive side should contain only tube bends. Hogan [1] also suggested that the heat exchanger should be divided into unique sub-heat exchangers which contain equal refrigerant flowrate. Though there are some basics guidelines for arranging circuits [1], there is still no general algorithm that can take into account of the effect of complex circuiting under practical conditions. In addition, experimental information regarding the circuitry is very rare. Therefore, the main purpose of this lecture is two-folds. Firstly, preliminary data that relates to a simple circuitry is presented. In addition, a rationally based model is proposed to calculate the detailed variations of relevant circuiting to assist the design of condensing coils and water heating coils.

2. Experimental Examination of the Performance for 1-circuit Heat Exchangers

A preliminary experiment was performed to examine the performance of circuitry. Six samples of fin-and-tube heat exchangers were made and tested. The fin pattern is wavy fin, and the corresponding geometry is given as follows:

Frontal area of the heat exchanger (W×H): 595×305 mm.

Waffle height (P_d): 1.18 mm.

Fin pitch (F_p): 1.7 mm.

Number of tube row (N): 2.

Notice there is only one circuit for the test sample, and the detailed arrangements are given in *Figure 3*. Arrangements (A) and (C) are counter-cross flow, arrangements (B) and (D) are parallel-cross; and arrangements (E) and (F) are z-shape cross flow. Note that the inlets for arrangements (A), (B), and (E) are located at the upper portion while the inlets of arrangements (C), (D), and (F) are located at the lower portion of the test samples.

Experiments were conducted in an environmental chamber. The test apparatus is based on the air-enthalpy method proposed by an ANSI/ASHRAE standard [2]. Cooling capacity was measured from the enthalpy difference of the air flowrate across the test sample. The air flow measuring apparatus is constructed based on ASHRAE 41.2 standard [3]. Refrigerant R-22 was used as the working fluid. The test conditions are given as follows:

Ambient inlet air temperature: 25±0.3 °C

Saturation refrigerant temperature at the inlet of condenser: 45±0.2 °C

Inlet superheat of the refrigerant flow: 6±1 °C

Table 1. Refrigerant-side pressure drop for arrangement (E) and (F).

Frontal Velocity	Δp (kPa) $G=100$ kg/m²s		Δp (kPa) $G=300$ kg/m²s	
(m/s)	(E)	(F)	(E)	(F)
0.7	7.1	7.7	33.3	32.3
1.0	6.5	7.8	28.9	27.4
1.5	6.1	7.8	27.9	26.5
2.0	5.9	7.9	25.8	25.3

Figure 4 shows the exit quality vs. frontal velocity for arrangement (A) to (F). For a given arrangement (counter-cross arrangement, parallel-cross arrangement, and z-shape arrangement), the exit quality is relatively independent of the location of inlet position (upper or lower portion). However, it should be pointed out that the refrigerant-side

pressure drops may be different due to the effect of gravity. Table 1 presents the pressure drops for arrangements (E) and (F) having G=100 and 300 kg/m² s. For G=100 kg/m² s, it is seen that the pressure drops for arrangement (F) are considerably higher than those of arrangement (E). For G=300 kg/m² s, the pressure drops for arrangement (E) and (F) are comparable.

The exit quality for the counter-cross arrangements (A and C) shows a lower value than those of parallel-cross flow arrangements (B and D). However, the z-shape arrangements (E) and (F) give the lowest exit quality. The results seem confusing since one would expect counter-cross arrangement would have a better performance than other arrangements.

To explain this phenomenon, one may have to examine the detailed variations of the wall temperature along the condenser as shown in *Figure 5* (arrangement A, C and E). For G=100 kg/m² s, one can clearly see that the temperature decrease for arrangement (A) occurs sooner than for arrangement (E). For instance, the wall temperature for tube #7 is approximately 7 °C lower than that of arrangement (E). The result substantiates that the counter-cross flow may have a higher heat transfer performance than the z-shape arrangement. However, the wall temperature for arrangement (A) does not consistently decrease along the condenser. Actually, the temperature may even increase when the tube number is higher than #15 (G=100 kg/m² s). Notice that the z-shape

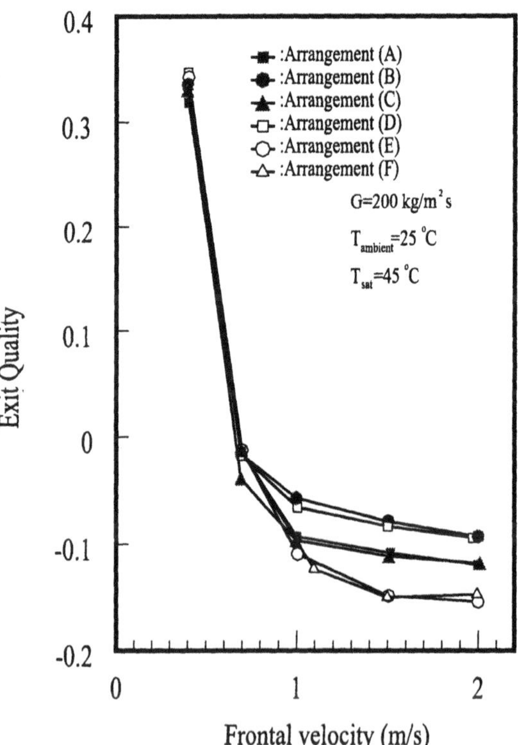

Figure 4. Exit quality for arrangements (A)..(F).

arrangement (E) does not show this phenomenon. For G=200 or 300 kg/m^{2}s, the result is similar; wall temperature for arrangement (A) drops more rapidly than for arrangement (E) but reaches an asymptotic value that is higher than that of arrangement (E) at the exit.

The explanation of this phenomenon for arrangement (A), is because the temperature at the inlet portion (row 2, upper side) is much higher than that at the exit (row 1, upper side). As a result, the heat conduction along the fin transfers energy back to the refrigerant. For the z-shaped arrangement (E), the temperature difference between the neighboring tube is relatively small, therefore reversed heat transfer by conduction is negligible. The test results suggest that the counter-cross flow arrangement may not achieve a highest subcooling temperature. This may significantly offset the system performance. To eliminate the reversed conduction effect and improve the performance of the counter-cross flow arrangement, possible solutions for improving the performance of the counter-cross arrangement can be seen from *Figure 6*. *Figure 6a* has removed several redundant tubes at the exit. This may be very helpful from both heat transfer and economical point of view. *Figure 6b* has used two separate 1-row to form a 2-row condenser. The tiny gap between these two row may inhibit the heat conduction from the hot inlet portion back to the cold exit portion.

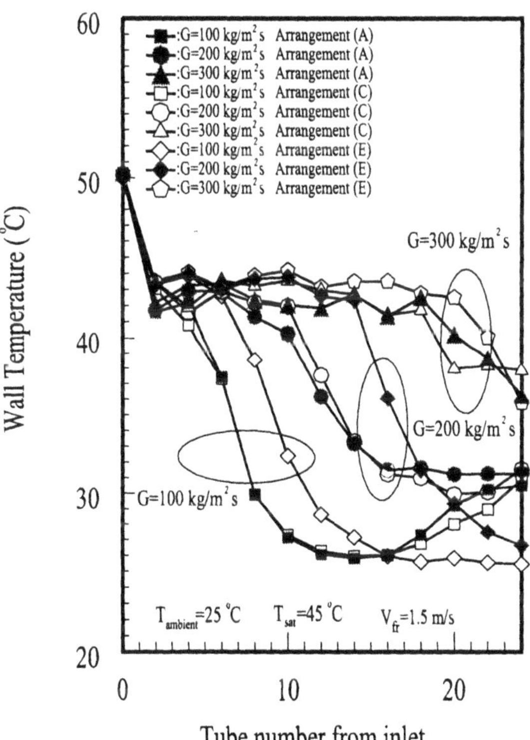

Figure 5. Variations of wall temperature for arrangements (A), (C), and (E)

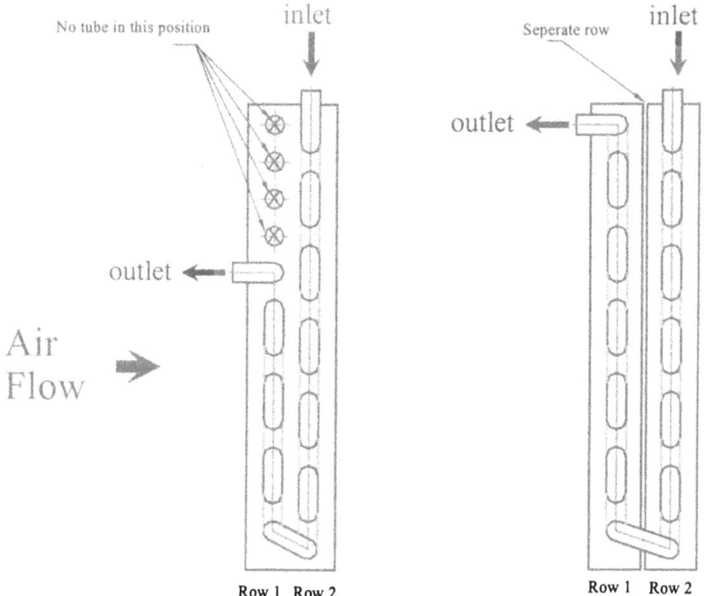

Figure 6. Possible improvements for the counter-cross flow arrangement.

3. The Circuit Model

The previous test results show that the effect of circuitry is very complicated, even for a simple arrangement. It is very difficult to examine all the possible circuitry through experiments. In this lecture, a rationally-based model which can take into account the complex circuitry is proposed as follows. The major limitations of the present modeling are:

1.   The heat exchanger may be divided into a number of unique sub-heat exchangers. No further splitting/combining within the sub-heat exchanger is allowed. Interactions between the sub-heat exchangers are also not allowed.

2.   All hairpin bends occur on the passive side of the coil, where complex junctions are made on the active side of the coil.

3.1 Tube Index Technique

As shown in *Figure 1*, practical design of an air-cooled condenser may involve quite complex circuiting. Therefore, to keep track of the refrigerant flow within the tube, it is necessary to record the tube pass within the condenser. The proposed model can simulate the frequently encountered arrangements. The terminology of the present study is as follows:

Partition: The term partition is defined as the tube portion of refrigerant flow following the split/dividing and preceding the combination of tubes. For instance, as shown in *Figure 1a*, the refrigerant enters the coil at tube 1, and passes through tube 2,

3, and 4 to split into two circuits. Therefore tubes 1~4 are regarded as Partition 1. Partition 2 is for tubes 6~28, and Partition 3 contains tubes 29~38.

Circuit: the number of circuit is defined as the number of refrigerant branches after splitting or dividing. For instance, as given in *Figure 1a*, partition 1 contains 1 refrigerant circuit, partition 2 contains 2 refrigerant circuit, and partition 3 contains 1 refrigerant circuit. The numbering technique proposed to keep track of the refrigerant flow should follow the "Partition first, and then the circuit" rule within each partition. The number of the tube row, partition, circuit, tubes, tubes in each partition/circuit, and their corresponding indexes are illustrated in Table 2a. Table 2b depicts the example of the corresponding tube indexes for *Figure 1b*.

Table 2a. Partition/Circuit Index for *Figures 1a* and *1b* Arrangements

Coil	Row No.	Partition	Partition Index	Circuit	Tubes Per Circuit	Tubes Per Partition	Circuit Index	Starting Tube Index	Ending Tube Index
Figure 1a	2	3	1	1	4	4	1	1	4
			2	2	12	24	1	5	16
							2	7	28
			3	1	10	10	1	29	38
Figure 1b	3	3	1	1	2	2	1	1	2
			2	2	26	52	1	3	28
							2	29	54
			3	1	2	2	1	55	56

Table 2b. Relationship of the Tube Indexes and Their Corresponding row and column Indexes

(a)				(b)		
Column Index	Row 1	Index 2		Column Index	Row 1	Index 2	Index 3

Column Index	Row (1)	Index (2)	Column Index	Row (1)	Index (2)	Index (3)
1	1	2	1	1	30	3
2	4	3	2	2	29	32
3	5	17	3	5	3	33
4	6	18	4	3	4	34
5	7	19	5	9	7	35
6	8	20	6	10	8	36
7	9	21	7	13	11	37
8	10	22	8	14	12	38
9	11	23	9	41	39	15
10	12	24	10	42	40	16
11	13	25	11	45	43	17
12	14	26	12	46	44	18
13	15	27	13	49	47	19
14	16	28	14	50	48	20
15	30	29	15	51	21	23
16	32	34	16	52	22	24
17	34	36	17	56	53	25
18	36	37	18	55	54	26
19	37	38	19	×	28	27

For coil (b) at Row Index 1, Column Index 19, "× " means no tube in this location.

172

3.2 Solution Algorithm

In analyzing the heat exchanger, the tube can be further divided into small elements. The conventional effectiveness-NTU method is employed in each small element. The iteration process is described as follows:

1. Guess an initial refrigerant flow distribution if any splitting is specified.
2. For each partition, use the conservation relations to solve the refrigerant distribution (if any splitting occurs) and the heat exchanges between the air and tube side. The conservation relations include
   (a) Energy conservation between the air and tube side. This is written for each small element as
   $$\Delta \dot{Q}_{air} = \dot{m}_r (i_{in} - i_{out}) = \dot{m}_{air} c_{p,air} \Delta T_{air} = \varepsilon \dot{Q}_{max}$$
   (b) Momentum balance among the circuits within each partition
   $$\Delta p_1 = \Delta p_2 = \ldots = \Delta p_N$$
   where N is the number of circuit in each partition.
3. Continue the prescribed solving methodology to the next partition, and then to the final refrigerant outlet.
4. Check the overall energy balance between air-side and tube-side, if this is not met, then repeat process 1 to 4. The iteration is terminated when the difference of heat transfer rate between air-side and tube-side is less than 0.1%.

In order to fulfill the above-mentioned solution strategy, additional constituti relationships are needed for the closure of the model.

3.3 Constitutive Relationships

Note that the overall heat transfer resistance and the corresponding pressure drops along the each small tube element are given as

$$\Delta p_r = \Delta p_a + \Delta p_b + \Delta p_f + \Delta p_g \tag{1}$$

$$\frac{1}{UA} = \frac{1}{\eta_o h_o A_o} + \frac{\delta_w}{k_w A_w} + \frac{1}{h_i A_i} \tag{2}$$

where Δp_a, Δp_b, Δp_f, and Δp_g denote the acceleration pressure drop, pressure drops in the bends, frictional pressure drop, and gravitational pressure drop. The bend pressure drop exists only when the refrigerant migrates from one tube to another (eg. connects by bends), and the gravitational pressure drop exists when the level changes caused by the connecting bends. To calculate the pressure drops and the overall heat transfer coefficients, knowledge of the friction factors and corresponding heat transfer coefficients for the air-side and tube-side are needed. They are described as follows:

1. For single phase fluids, the pressure drops are evaluated as

$$\Delta p_a \approx 0 \tag{3}$$

$$\Delta p_f = \frac{4L}{D_i} f \frac{G^2}{2\rho} \tag{4}$$

$$\Delta p_g = \rho g \Delta h \tag{5}$$

$$\Delta p_b = \frac{4 f G^2}{2\rho}\left[\frac{z}{D}\right] \tag{6}$$

where the Fanning friction factor, f, can be computed from

$$f = 0.079\,\text{Re}^{-0.25}, \quad 4000<\text{Re}<30000 \tag{7}$$

$$f = 0.046\,\text{Re}^{-0.2}, \quad \text{Re}>30000 \tag{8}$$

Single-phase pressure drop attributable to a bend is expressed as an equivalent pipe length/diameter ratio, detailed description of the relation between equivalent length, z/D, vs. r/D can be found in the monograph by Chisholm [4].

2. Pressure gradients for two-phase condensing flow, as depicted by Collier and Thome [5]:

$$\left(dp/dz\right)_a = G^2 \frac{d}{dz}\left[\frac{x^2}{\alpha \rho_g} + \frac{(1-x)^2}{(1-\alpha)\rho_f}\right] \tag{9}$$

$$\left(dp/dz\right)_g = g \sin\theta\left[\alpha\rho_g + (1-\alpha)\rho_f\right] \tag{10}$$

$$\Delta p_b = \Delta p_{ce} + \Delta p_{ef} \tag{11}$$

where the void fraction, α, is taken from the Baroczy correlation [6]

$$\alpha = \left[1+\left(\frac{1-x}{x}\right)^{0.74}\left(\frac{\rho_g}{\rho_f}\right)^{0.65}\left(\frac{\mu_f}{\mu_g}\right)^{0.13}\right]^{-1} \tag{12}$$

In addition, the frictional pressure gradient is calculated from

$$\left(dp/dz\right)_{lo} = \phi_{lo}^2\left(dp/dz\right)_{lo} \tag{13}$$

where the two-phase multiplier, ϕ_{lo}^2, is obtained from the Friedel correlation [7].

Invoking these conditions and integrating from the inlet of two-phase region to exit yields the total pressure drop of the two-phase region.

3. The single-phase heat transfer coefficients for regions 1 and 4 can be obtained from the Gnielinski correlation [8],

$$Nu = \frac{(\text{Re}_{D_i}-1000)\,\text{Pr}(f_i/2)}{1+12.7\sqrt{f_i/2}\,(\text{Pr}^{2/3}-1)} \tag{14}$$

where

174

$$h_i = Nuk/D_i \tag{15}$$

$$f_i = \left(1.58\ln(\mathrm{Re}_{D_i}) - 3.28\right)^{-2} \tag{16}$$

4. For the saturated condensation region (region III),

The heat transfer coefficients in the saturated region is taken from the Shah [9] correlation:

$$h_i = h_f \left[\left(1-x\right)^{0.8} + \frac{3.8x^{0.76}(1-x)^{0.04}}{P_r^{0.38}} \right] \tag{17}$$

where P_r is the reduced pressure and x is the vapor quality and h_f is the single-phase forced convective heat transfer coefficients, which can be evaluated using the Gnielinski correlation [8].

5. Superheated condensation region,

Lee et al. [10] proposed a superposition model that combine the saturated condensation heat transfer and sensible forced convection heat transfer coefficient, that is given as

$$h_{\mathrm{sup}} = \frac{T_g - T_{sat}}{T_g - T_{sat}} h_{fc} + \frac{T_{sat} - T_w}{T_g - T_w} h_{sat} \tag{18}$$

where h_{fc} and h_{sat} are forced convection heat transfer coefficient and saturated condensation heat transfer coefficient, respectively. In the present study, the Gnielinski correlation [8] is used for the forced convection heat transfer coefficient and the saturated heat transfer coefficient is calculated by Eq. (17).

6. Enhancement factors for micro-fin tubes for the single-phase and two-phase region.

For single-phase superheated vapor and subcooled flow, general heat transfer and friction correlations by Wang et al. [11] are adopted. For condensation, the enhancement factors were calculated, as suggested by Cavallini et al. [12].

7. The air-side heat transfer coefficients, h_o,

Note that h_o is highly dependent on the fin pattern. This value can be directly obtained from the test results provided by the manufacturer if it is available. Otherwise, there are several references for correlation and test data of various fin patterns, namely (1) plain fin geometry (Rich [13,14], Gray and Webb [15], and Wang et al. [16,17]); (2) wavy fin pattern, (Kim et al. [18], Wang et al. [19,20]; (3) louver fin pattern (Chang et al. [21], Wang et al. [22,23]; and (4) convex-louver fin pattern (Wang et al. [24,25]).

8. In the ε-NTU method, the number of heat transfer units (NTU) is defined as

$$NTU \equiv UA/C_{\min} \tag{19}$$

The UA product was calculated using the ε-NTU relation. For single-phase fluid, the relation of unmixed-unmixed for cross-flow is adopted. For the two-phase region, the ε-

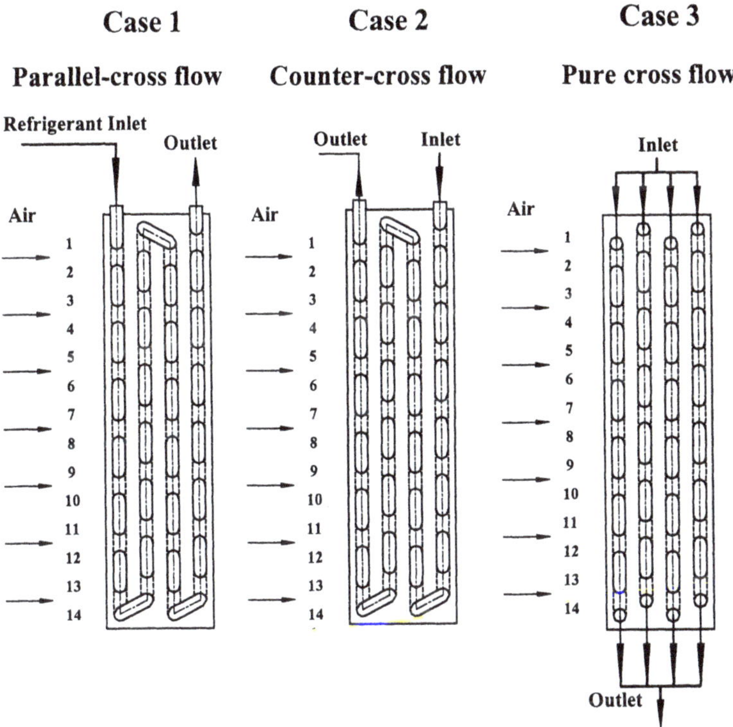

Figure 7. Schematic of the circuitry arrangement for the wter coils.

NTU relationship may become counter-flow relation if the refrigerant-side pressure drop is negligible, i.e.

$$\varepsilon = 1 - e^{-NTU} \tag{20}$$

where

$$\varepsilon \equiv \frac{\dot{Q}}{\dot{Q}_{max}} \tag{21}$$

Notice that Equation (20) does not account for the pressure drop along the condenser. If the two-phase refrigerant pressure drop cannot be neglected, an effective heat capacity rate of the refrigerant side can be calculated as follows:

$$C_{eff} = \frac{C_{air}\Delta T_{air}}{\Delta T_r} \tag{22}$$

where ΔT_r is the refrigerant temperature drop by corresponding to the refrigerant pressure drop. In this connection, the ε-*NTU* relation for single-phase fluids is

$$C^* = \frac{C_{air}}{C_{eff}} \tag{23}$$

9. Transport and thermodynamic properties of the liquid phase,

In the superheated condensation and saturated condensation region, the properties of the liquid phase are assumed to be at saturation. Therefore, the corresponding transport and thermodynamic properties are related to the local saturation temperatures. At the subcooled region, the transport properties were related to the local liquid temperature while the thermodynamic properties can be approximated by the corresponding saturated property.

10. Transport and thermodynamic properties of the vapor phase,

The transport properties of the vapor phase are assumed to be associated with their corresponding vapor temperature while the thermodynamic properties are related to their local pressure and vapor temperature.

The relations of thermodynamic and transport properties equations are constructed based on the ASHRAE Handbook [26] and the JSME Thermophysical Properties of Fluids Data Book [27], using the least-square fit method.

4. Results and Discussion

The validation of the present model was carried out for a 4-row water heating coil. The flow arrangement of the test coil includes cross-parallel, cross-counter, and pure cross flow arrangements as is shown in *Figure 3*. The test 4-row coil had a plain fin configuration (D_o=10.0 mm, F_p=2.31 mm, P_t=25.4 mm, P_l=19.05 mm). The length, width, and height of the sample are 76.2 mm, 595 mm, and 355.6 mm, respectively. The air-side heat transfer coefficients, h_o, were obtained from the pure cross flow arrangement (*Figure 7c*). Detailed description of the experimental facility and data reduction method can be found in previous studies conducted by Wang et al. [16,19]. Experimental uncertainties associated with the heat transfer rate, following the single-sample analysis proposed by Moffat [28], are less than 3.3% throughout the test range. The mean heat transfer coefficients obtained in the experiment were then implemented to the present model to simulate the heat transfer performance of cross-parallel and cross-counter circuitry. *Figure 8* presents the overall heat transfer rate vs. frontal velocity. As seen in the figure, the heat transfer rate is sensitive to the change of refrigerant circuiting, and the proposed model gives satisfactory agreements with the test data. The overall cross-counter flow arrangement shows a 15.6% higher heat transfer rate at V_{fr}=6.4 m/s comparing to that of the overall cross-parallel flow arrangement. The results show that the heat transfer coefficients are independent of flow arrangement. The heat transfer coefficients should be independent of the path arrangement. Different flow arrangement results in the difference in their corresponding correction factor F (for those using UA-LMTD-F method), or the corresponding ε-NTU relationship (for those using ε-NTU method). The heat transfer coefficient should not be affected by the pass arrangement. Seshimo [29] has experimentally and theoretically proved the results. Note that in practical applications, the correction factor, F, or the corresponding ε-NTU relationship could be too complicated to obtained due to complex circuitry. Therefore, most of the calculations were carried out based on a pure cross-flow arrangement, and as a result, unacceptable errors may occur.

As mentioned in a previous section, the present model is valid for air-cooled condenser. For validation of the applicability of the present model, it is compared with test results by Kaga et al. [30]. Their fin-and-tube heat exchanger consists two separate refrigerant circuits as depicted in *Figure 9a*. Detailed test conditions were

1   Refrigerant: R-22
2   Condensing pressure: 1.81~1.825 MPa
3   Inlet temperature of refrigerant: 70.4~72.3 °C
4   Refrigerant flow Rate: 0.0088~0.01 kg/s
5   Inlet air temperature: 20 °C
6   Frontal velocity across test coil: 1.5 m/s
7   Fin pitch: 1.5 mm

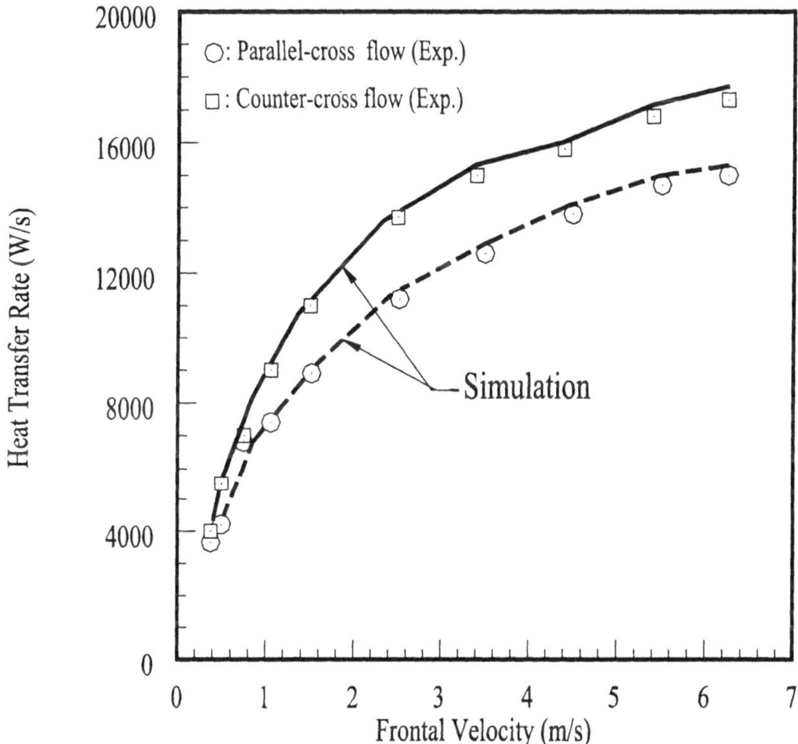

Figure 8. Comparisons of heat transfer rate between the test data and the predictive results for cross-parallel and counter-cross arrangement.

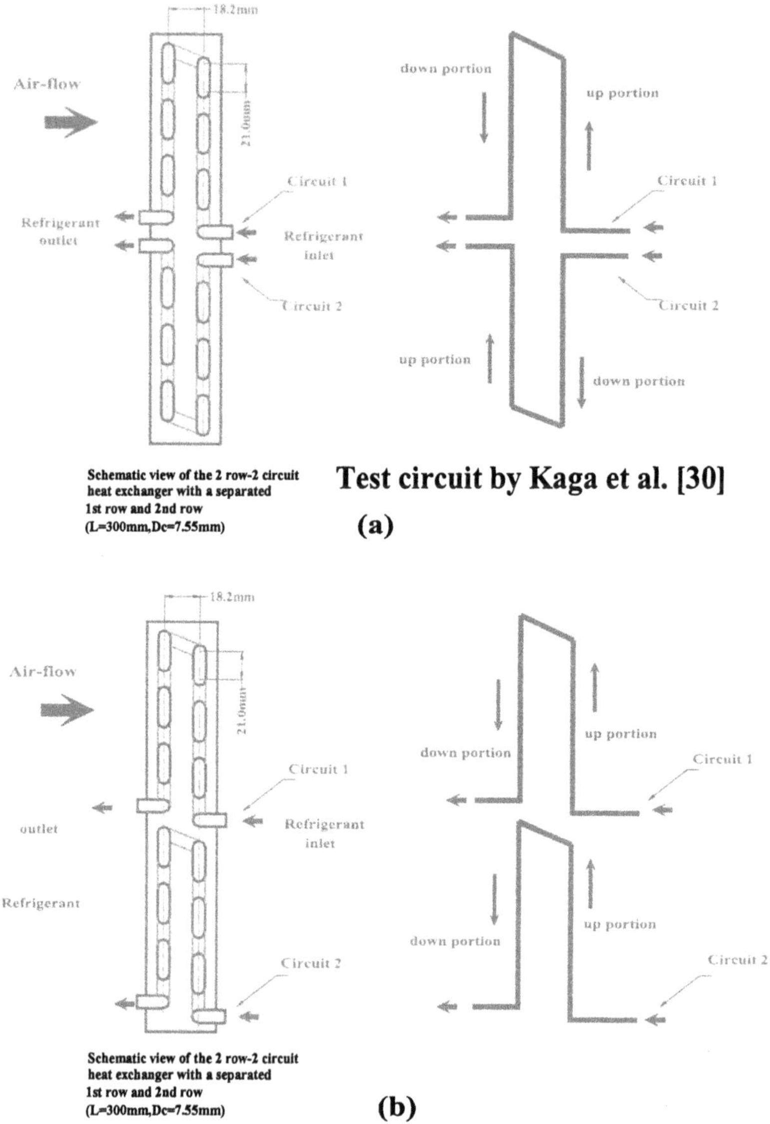

Schematic view of the 2 row-2 circuit
heat exchanger with a separated
1st row and 2nd row
(L=300mm,Dc=7.55mm)

Test circuit by Kaga et al. [30]

(a)

Schematic view of the 2 row-2 circuit
heat exchanger with a separated
1st row and 2nd row
(L=300mm,Dc=7.55mm)

(b)

Figure 9. Schematic of circuitry arrangement by Kaga et al. [30].

Figure 10 shows the distributions of refrigerant temperature variations for each circuit along the condenser having $\dot{m}_r = 0.01$ kg/s $(G \approx 135$ $kg/m^2 s)$ It is seen that the present calculations agree with the trend of the experimental data. The predicted heat transfer rate, $\dot{Q}_{cal} = 2110$ W, is very close to the test result of $\dot{Q}_{exp} = 2043$ W. It is interesting to know that a detectable mal-distribution of refrigerant flowrate exists along the condenser. The flowrate inside refrigerant circuit 1 is approximately 54.3% of the total flowrate which is about 8.6% higher than that of circuit 2. Note that the tube length for each circuit is the same, and the arrangement of these two refrigerant circuits is symmetrical. As depicted in the previous section, the total pressure drop of the refrigerant circuit is given as $\Delta p_r = \Delta p_a + \Delta p_b + \Delta p_f + \Delta p_g$. Therefore, for adiabatic or single-phase test conditions, it is expected that the refrigerant flow in each circuit should be the same. Moreover, for the identical circuits arrangement as seen in *Figure 9b*, it is obvious the refrigerant flowrate for both circuit is the same. However, for the symmetrical arrangements as shown in *Figure 9a*, the predictions and the experimental data reveal a significant mal-distribution of refrigerant flowrate. The difference in flowrate may be attributed to different total gravity forces. As seen in Equation (10), the gravitational pressure gradient is $\left(dp/dz\right)_g = g \sin\theta \left[\alpha\rho_g + (1-\alpha)\rho_f\right]$ which indicates that the gravitational pressure gradient is composed of a vapor-phase contribution, $g\sin\theta\alpha\rho_g$, and a liquid-phase contribution, $g\sin\theta(1-\alpha)\rho_f$. For an air-cooled refrigerant condenser, the major flow pattern is annular at the initial stage of condensation (see the two-phase visualization of flow pattern by Wang et al. [31]). As a result, the void fraction is large at this stage, therefore the gravity term is comparatively small compared to the total pressure gradient. Actually, in the "up-portion" of circuit 1, the gravitational pressure drop is only 3.9% of the total pressure drop. However, in the "down-portion" of circuit 1, the contribution by gravitation is reversed. Further, since the quantity of the liquid condensate in this portion is much larger than that of the "up-portion", hence the "pressure-gain" in this portion is very pronounced. A close examination of the "pressure-gain" caused by the gravitational gradient in the "down-portion" of circuit 1 is approximately 27.7% of the total gradient. In addition, contrary to the refrigerant circuit 1, the sign of the gravitational gradient in the "up-portion" is different from circuit 1 and is regarded as "pressure loss". Accordingly, this pressure gradient yields the mal-distribution of the refrigerant flow. Note that this study classified the condensation into "superheated condensation" and "saturated condensation". As given in *Figure 10*, inclusion of superheated condensation would give better predictions of the distribution of the refrigerant temperature compared to those without inclusion of superheated condensation.

This mal-distribution of refrigerant flowrate may become more pronounced as the refrigerant flowrate decreases. The comparison of the calculated results and the test results by Kaga et al. [30] for $\dot{m}_r = 0.0088 kg/s$ $(G \approx 120\ kg/m^2 s)$ confirmed this.

Actually, the calculated results indicated that the flowrate inside refrigerant circuit 1 is approximately 56% of the total flowrate, which is about 12% higher than that of circuit 2. Notice that this phenomenon is strongly related to the inlet mass flowrate. For a higher inlet mass flowrate, $\dot{m}_r = 0.02 kg/s$ $(G \approx 270\ kg/m^2 s)$, the difference in flowrate for circuit 1 and circuit 2 is reduced to 2.6%. Eventually, the degree of mal-distribution is somewhat smaller than that at lower inlet mass velocities. Notice that this phenomenon cannot be modeled without the inclusion of two-phase pressure gradient, especially the gravitational pressure gradient.

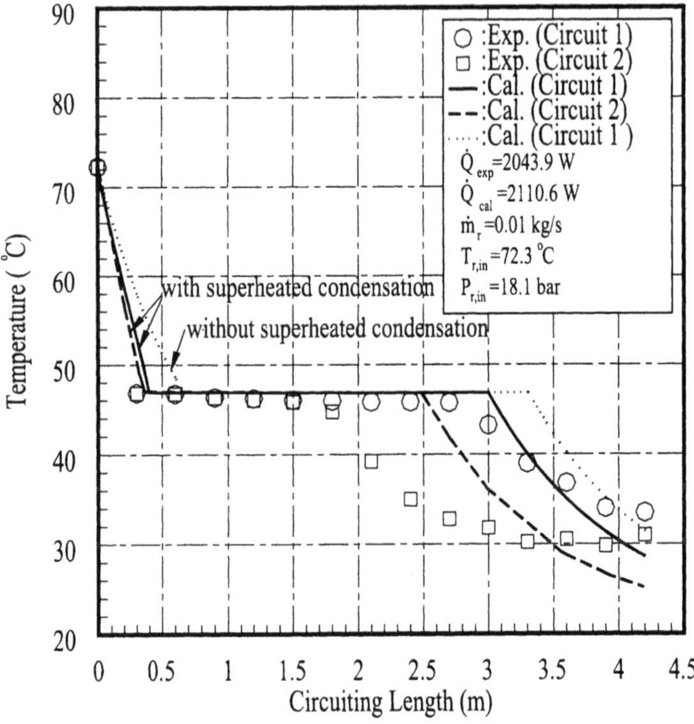

Figure 10. Comparison of the refrigerant temperature distribution between the test data of Kaga et al. [30] and the present preditive results (m =0.01 kg/s)

5. Conclusions

The primary purpose of this study is to present a rationally based model to handle the complex circuitry within air-cooled condenser and water heating coils via an indexing technique. The calculated results for the water heating coils agree well with experimental data, and the calculated results of two-phase condensing flow also agree with the basic trend of air-cooled condensers. The mal-distribution of the refrigerant flow is strongly related to the contribution of the gravitational pressure gradient. In addition, experimental data related to 1-circuit heat exchangers having counter-cross, parallel-cross, and z-shaped arrangements are reported. Compared with the z-shaped arrangement, the counter-cross flow arrangement shows a quicker drop of refrigerant temperature initially. However, due to the contribution of heat conduction, the refrigerant temperature may rise in the downstream of the condenser.

6. Acknowledgement

The author would like to express gratitude for the Energy R&D foundation funding from the Energy Commission of the Ministry of Economic Affairs, Taiwan, for providing financial support of this study.

Nomenclature

C_p — specific heat, J/kg·K
C — heat capacity rate, W/K
A — total surface area, m^2
$dP_{f,l}$ — frictional pressure gradient for liquid flowing alone, Pa
$dP_{f,v}$ — frictional pressure gradient for gas flowing alone, Pa
$dP_{f,lo}$ — frictional pressure gradient for total mixture flowing as liquid, Pa
D_i — inner diameter of tube, m
D_o — outer diameter of tube, m
f — friction factor, dimensionless
F — correction factor used by UA-LMTD-F approach, dimensionless
F_p — fin pitch, mm
g — acceleration due to gravitation, m/s^2
G — mass velocity, kg/m^2·s
H — height of the heat exchanger, m
h — heat transfer coefficient, W/m^2·K
h_{fc} — single-phase forced convection heat transfer coefficient, W/m^2·K
h_{sat} — saturation condensation heat transfer coefficient, W/m^2·K
h_{sat} — superheated condensation heat transfer coefficient, W/m^2·K
$i_{r,in}$ — refrigerant enthalpy at the inlet of refrigerant temperature, kJ/kg
$i_{r,out}$ — refrigerant enthalpy at the outlet of refrigerant temperature, kJ/kg
k — thermal conductivity, W/m·K
$\dot{m}_r$ — refrigerant mass flowrate, kg/s
Nu — Nusselt number, dimensionless
NTU — Number of transfer unit, dimensionless
p — pressure, MPa
P_d — waffle height of the wavy fin pattern, mm
P_l — longitudinal tube pitch, mm
P_r — reduced pressure, dimensionless
P_t — transverse tube pitch, mm
Pr — Prandtl number, dimensionless
$\dot{Q}$ — heat transfer rate, W
r/D — radius of curvature to tube diameter ratio for tube bend, dimensionless
Re — Reynolds number, dimensionless
U — overall heat transfer coefficient, W/m^2·K
V_{fr} — frontal velocity across coil, m/s
W — width of the heat exchanger, m
T — temperature, °C

182

T_{sat} saturation temperature, °C
Δh level change of height, m
ΔT_m mean temperature difference, °C
Δp_a acceleration pressure drop, Pa
Δp_f frictional pressure drop, Pa
Δp_g gravitational pressure drop, Pa
Δp_b pressure drop by bends, Pa
x vapor quality, dimensionless
z/D   equivalent pipe/length ratio for tube bend, dimensionless

Greek letters
α void fraction, dimensionless
ε heat exchanger effectiveness, dimensionless
ρ density, kg/m^3
δ_w thickness of tube wall, mm
μ viscosity, N·s/m^2
ϕ_{lo} two-phase frictional multiplier (for liquid only), dimensionless

Subscripts
air   air side
cal   calculated results
eff   effective value
exp   experimental data
f   fluid phase
g   gas phase
max   maximum value
min   minimum value
r   refrigerant side

References

1. Hogan, M.R. (1980) The development of a low-temperature heat pump grain dryer, PhD thesis, Purdue university, pp. 116-127.
2. ANSI/ASHRAE standard 37 (1988) Methods of testing for rating unitary air-conditioning and heat pump equipment.
3. ASHRAE standard 41.2 (1987) Standard Methods for Laboratory Air-flow Measurement, American Society of Heating, Refrigerating and Air-conditioning Engineers, Inc., Atlanta.
4. Chishom, D. (1983) *Two-phase Flow in Pipelines and Heat Exchangers,* Pitman Press Ltd, Bath, pp. 154~166.
5. Collier, J.G., and Thome. J.R. (1994) *Convective Boiling and Condensation,* 3rd Ed., Oxford Science Publication, p. 48.
6. Baroczy, C.J. (1965) A systematic correlation for two-phase pressure drop, AIChE reprint 37 presented at *8th National Heat Transfer Conf.,* Los Angeles, August.

7. Friedel, L. (1979) Improved friction pressure drop correlations for horizontal and vertical two-phase pipe flow, European Two-Phase Flow Group Meeting, Paper E2, June.

8. Gnielinski, V. (1976) New equation for heat and mass transfer in turbulent pipe and channel flow, *Int. Chem. Engng.*, **16**, 359-368.

9. Shah, M.M. (1979) A general correlation for heat transfer during film condensation inside pipes, *Int. J. of Heat Mass Transfer*, **22**, 547-556.

10. Lee, C.C., Teng, Y.T., Lu, D.C. and Fang, L.J. (1991) An investigation of condensation heat transfer of superheated R22 vapor in a horizontal tube, *Proceedings of the 2nd World Conference on Experimental Heat Transfer, Fluid Mechanics, and Thermodynamics*, 1051-1057.

11. Wang, C.C., Chiou, C.C. and Lu. D.C. (1996). Single-phase heat transfer and flow friction correlations for microfin tubes, *Int. J. Heat and Fluid Flow*, **17**, 500-508.

12. Cavallni, A., Doretti, L., Klammsteiner, N., Longo, G.A. and Rossetto. L. (1995) Condensation of new refrigerants inside smooth and enhanced tubes, *Proc. of the 19th Int. Congress of Refrigeration*, **IVa**, 105-114.

13. Rich, D.G. (1973) The effect of fin spacing on the heat transfer and friction performance of multi-row, plate fin-and-tube heat exchangers, *ASHRAE Transactions*, **79(2),** 137-145.

14. Rich, D.G. (1975) The effect of the number of tube rows on heat transfer performance of smooth plate fin-and-tube heat exchanger, *ASHRAE Transactions*, **81(1),** 307-317.

15. Gray, D.L. and Webb, R.L. (1986) Heat transfer and friction correlations for plate finned-tube heat exchangers having plain fins *Proc. 8th. Heat Transfer Conference*, 2745-2750.

16. Wang, C.C., Hsieh, Y.C., Chang, Y.J. and Lin, Y.T. (1996) Sensible heat and friction characteristics of plate fin-and-tube heat exchangers having plane dins, *Int. J. of Refrigeration*, **19**(4), 223-230.

17. Wang, C.C., Lee, W.S. and Chang, C.T. (1997) Sensible heat transfer characteristics of platc fin-and-tube exchangers having 7-mm tubes, *AIChE Symposium Series*, **93**(314), 211-216.

18. Kim, N.H., Youn, J.H. and Webb, R.L. (1997) Heat transfer and friction correlations for wavy plate fin-and-tube heat exchangers, *ASME Journal of Heat Transfer*, **119(3),** 560-567.

19. Wang, C.C., Fu, W.L. and Chang, C.T. (1997) Heat transfer and friction characteristics of typical wavy fin-and-tube heat exchangers, *Experimental Thermal and Fluid Science*, **14(2),** 174-186.

20. Wang, C. C., Jang, J. Y. and Chiou, N. F. (1998) Effect of waffle height on the air-side performance of wavy fin-and-tube heat exchangers, accepted for publication in *Heat Transfer Engineering*.

21. Chang, W.R., Wang, C.C., Tsi, W.C, and Shyu, R.J. (1995) Air side performance of louver fin heat exchanger, *Proc. of the 4th ASME/JSME Thermal Engineering Joint Conference*, **4**, 367-372.

22. Wang, C.C., Chang, Y.P., Chi, K.U. and Chang, Y.J. (1998) An experimental study of heat transfer and friction characteristics of typical louver fin-and-tube heat exchangers, *Int. J. of Heat and Mass Transfer*, **41**, no. 4-5, 817-822.

184

23. Wang, C.C. Chang, Y.P., Chi, K.U. and Chang, Y.J. (1998) A study of non-redirection louver fin-and-tube heat exchangers, *Proceeding of Institute of Mechanical Engineering, Part C, Journal of Mechanical Engineering Science*, **212**, 1-14.
24. Wang, C.C, Chen, P.Y. and Jang, J.Y. (1996) Heat transfer and friction characteristics of convex-louver fin-and-tube heat exchangers, *Experimental Heat Transfer*, **9**, 61-78.
25. Wang, C.C., Tsi, Y.M. and Lu, D.C. (1998) Comprehensive study of convex-louver and wavy fin-and-tube heat exchangers, *AIAA J. of Thermophysics and Heat Transfer*, **12**, 423-430.
26. ASHRAE, ASHRAE Handbook Fundamentals Volume (1997) Atlanta.
27. JSME Data Book (1983) *Thermalphysical Properties of Fluids*, JSME.
28. Moffat, R.J. (1988) Describing the uncertainties in experimental results, *Experimental Thermal and Fluid Scienc*, **1**, 3-17.
29. Seshimo, Y. (1988) Effectiveness and fin efficiency of plate-fin and tube heat exchangers, *Transactions of the JAR*, **5**(2), 133-141, in Japanese.
30. Kaga, K., Yamada, K., Kotoh, S., Takeshita, M., Yamanaka, G. (1994) Improvement of a capacity of plate fin tubed heat exchanger by thermal analysis considering thermal interaction between pipes by heat conduction in fins, *Proc. 10th Int. Heat Transfer Conf.*, Brayton. UK, the industrial Sections Paper, paper no. I/2-CHE-8, pp. 99-104.
31. Wang, C.C., Chiang, C.S. and Lu, D.C. (1997) Visual observation of flow pattern of R-22, R-134a, and R-407C in a 6.5 mm smooth tube, *Experimental Thermal and Fluid Science*, **15**, No. 4, 395-405.

FLOW AND HEAT TRANSFER MECHANISMS IN PLATE-AND-FRAME HEAT EXCHANGERS

BENGT SUNDÉN
Division of Heat Transfer
Lund Institute of Technology
Box 118, 221 00 Lund, Sweden

Abstract. This lecture focus attention on the transport of heat and momentum in plate-and-frame heat exchangers. In particular numerical procedures are considered and comparisions with experimental data are provided. A brief review of previous investigations is also given.

1. Introduction

Plate-and-frame heat exchangers (PHE) are used in many different processes in a broad range of temperatures with a variety of substances. PHEs are used frequently but to a lesser extent than shell-and-tube heat exchangers. Relative to the latter, PHEs often have higher overall heat transfer coefficient, higher compactness (m^2/m^3) and are easier to clean. However, they present limitations in pressure and temperature levels. Various types of PHEs exist. They are all assembled of plates with an embossed surface area enhancement pattern. Today, the most common pattern in use is the chevron or herringbone design. In the past a wash-board like pattern was common. The embossed pattern results in a larger heat transfer surface than a flat plate, improves the plate stiffness and assures the channel gap. The geometry of the corrugation is very important for the thermal-hydraulic performance of the heat exchanger. Each plate has four corner ports which, in pairs, establish access to the narrow flow passages on either side of the plate. Usually also there are distribution areas between the ports and the major embossed plate area.

The different channels are commonly separated by a sealing gasket. Development of new technologies has resulted in brazing or welding plates together. Welded PHE:s are classified as fully-welded or semi-welded. The former ones do not use any elastomer sealings between the plates, while in the latter ones pairs of plates are welded together. For the semi-welded type, every pair of plates makes up a cassette in which one fluid flows. The other fluid flows between neighbouring cassettes which are sealed with gaskets.

Figure 1 shows a plate with a chevron pattern. Plates are usually stacked together in a symmetric or mixed arrangement as indicated in Figure 2.

The heat transfer enhancement in PHE:s is related to the plate characteristics and may be attributed to increased effective heat transfer area, small hydraulic diameter

185

S. Kakaç et al. (eds.), Heat Transfer Enhancement of Heat Exchangers, 185–206.
© 1999 *Kluwer Academic Publishers.*

flow channels, separation and reattachment of boundary layers, and vortex or swirl flow generation. The inclination angle, corrugation amplitude, corrugation wavelength and the profile of the waviness are important for the local and overall heat transfer coefficients as well as for the pressure losses. As plate-and-frame heat exchangers are considered, the term thermal length is often used. It is defined as the temperature

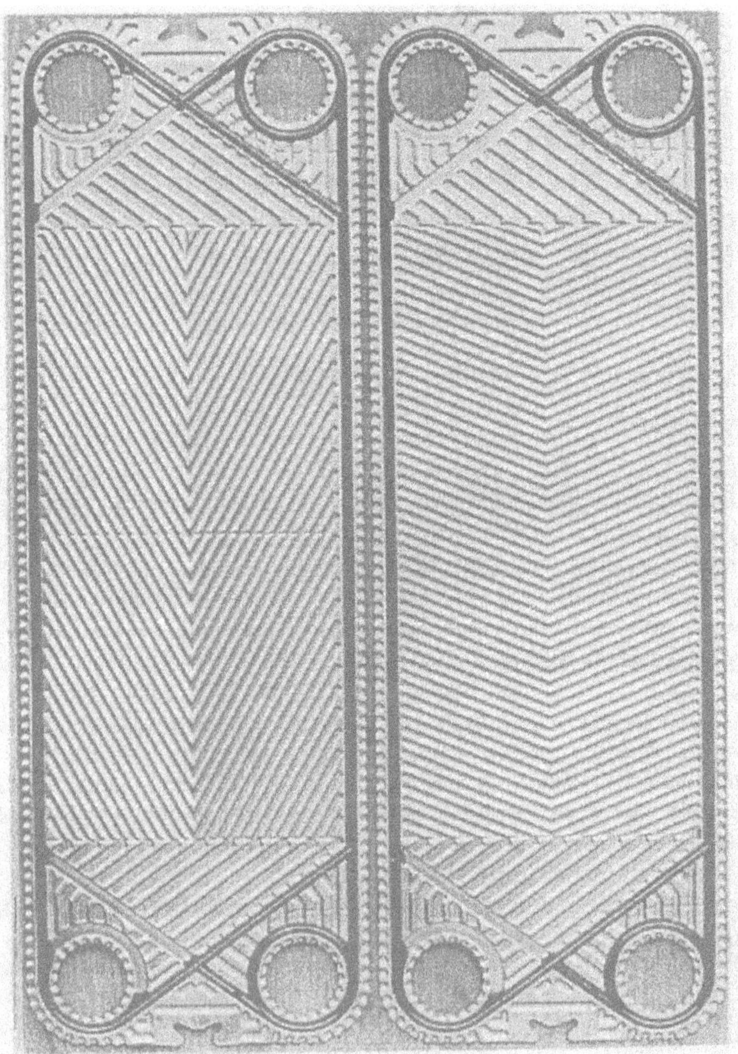

Figure 1. Plate with chevron pattern.

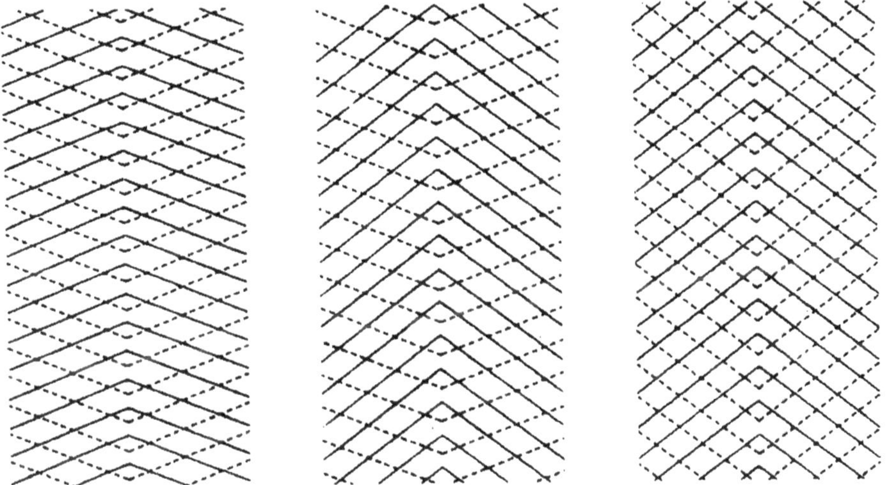

Figure 2. Sketch of plate arrangement.

difference of one of the fluids divided by the logarithmic mean temperature difference across the heat exchanger (corresponding to pure counter flow). Plates with a dense pattern have a large pressure drop but also a high heat transfer coefficient. A thermally long channel then prevails. On the other hand if the plates have a more open pattern the pressure drop and the heat transfer coefficient will be low. One then has a thermally short channel. The plates can be assembled in several ways. The package of plates may be splitted up in a number of streaks for the fluids. Figure 3 illustrates one streak and two streaks couplings.

Overall pressure drop and heat transfer characteristics have been presented by Okada et al. [1], Marriot [2], Focke et al. [3], Muley and Manglik [4], Thonon et al. [5] and Muley and Manglik [6]. These investigations are for single-phase fluids. Manglik [7] presented recently a summary of available correlations for the friction factor and Nusselt number of chevron plates. An investigation of the frictional pressure drop for air-water two-phase flow in PHEs was presented by Kreissig and Müller-Steinhagen in [8]. They gave recommendations for a calculation procedure based on their measurements. An investigation of the performance of some PHEs as evaporators in refrigeration units was recently presented in [9]. The constant C in the Chisholm equation for the two-phase multiplier, see [10], was found to be much larger than reported for tubular flow. Measurements of local transfer coefficients in single-phase flow have been carried out only to a very limited extent. Refs. [11-13] belong to such investigations.

Investigations of the flow field in PHEs have been carried out only to a limited extent mainly because of the complex geometry and narrow passages with a small hydraulic diameter. The conjectured flow patterns in the washboard type corrugated plate and the herringbone type plate are shown in Figures 4a and 4b, respectively.

188

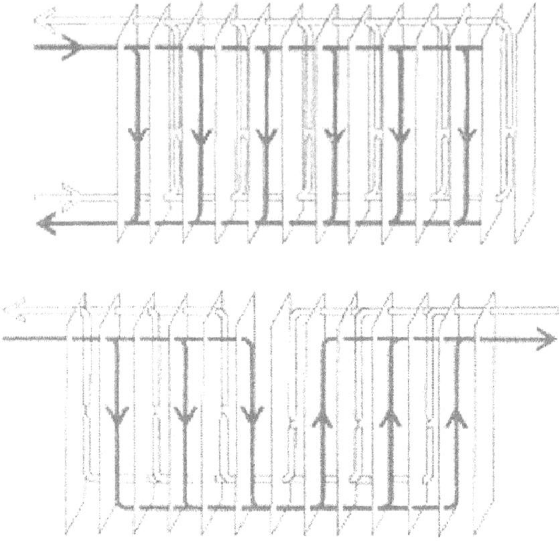

Figure 3. Various couplings in PHE:s.

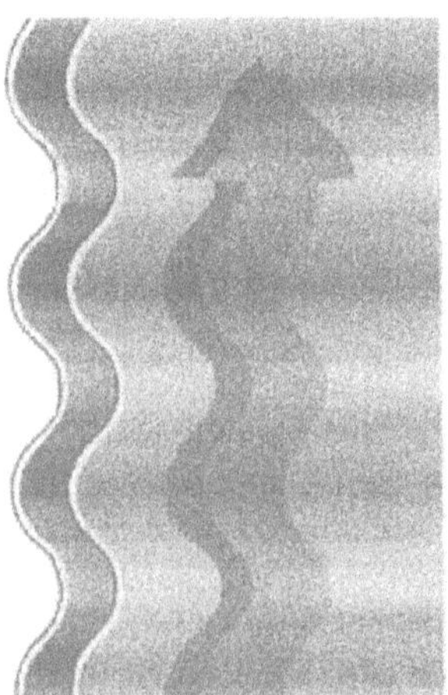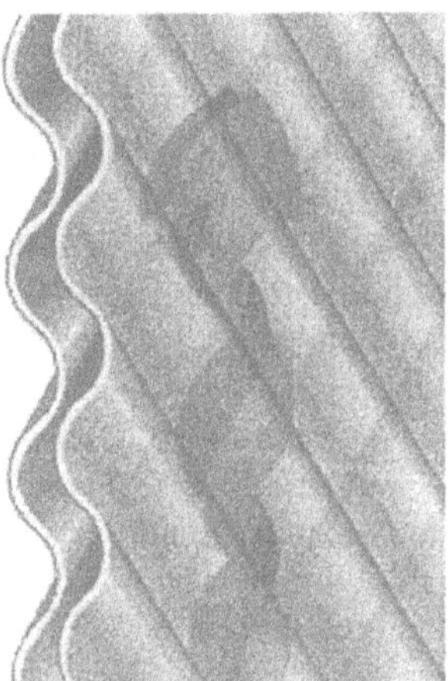

Figure 4. a) Washboard pattern plates. b) Herringbone pattern plates.

For plates with a washboard corrugation pattern it is believed that turbulence is promoted by the continuously change in flow direction and velocity. Plates with herringbone corrugation pattern are usually assembled with the pattern pointing in opposite directions on two adjacent plates. The created flow channels are believed to promote a swirling motion of the fluid. In [14] flow visualisations for a few brazed PHE geometries were presented. Tracing particles were used, and a top transparent plate was necessary. A CCD-camera with a acquisition system was used. The local velocity vector did not show any predominant flow direction.

With the rapid development of methods for numerical solution of non-linear partial differential equations and advanced computers, predictions of temperature and flow fields and associated heat fluxes and stresses, respectively, may be possible to some extent. However, numerical investigations of the flow and temperature fields in PHE:s are rare but recently some publications have been presented. Ciofalo et al. [15, 16] found that the standard k-ε model with wall functions gave acceptable results at high Reynolds numbers (Re) but failed at lower values of Re. Laminar flow results were acceptable at low Re number and for moderate angles between the plates. They found best overall agreement with experimental data if a low Reynolds number k-ε model or large eddy simulations (LES) were used. It should be noted that only a so-called unit cell with periodic conditions was considered and not the entire plate surfaces. A similar numerical investigation for PHEs with three different corrugation inclination angles was presented by Mehrabian et al. [17]. Only laminar cases were considered. The calculated friction factors were much higher than corresponding experimental values. No calculated Nusselt numbers were given.

The present lecture will present some recent numerical investigations, and the background of the numerical modelling approach, of flow and heat transfer in some PHEs. However, the analysis will be restricted to the chevron corrugation pattern of the plates and not the ports and the distribution areas (see Figure 1).

2. Problem Statement

The geometries investigated are consistent with typical Alfa Laval Thermal AB plate-and-frame heat exchangers. Figure 5 shows a segment of the geometries of the

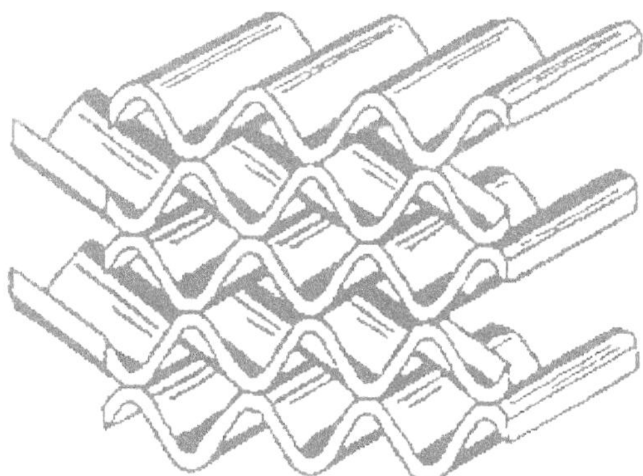

Figure 5. Segment of the geometries of the corrugated plates.

corrugated plates. In the numerical solution approach, the smallest volume, for which periodic conditions can be realised, is attempted. The whole plate areas will consist of many such cells. Figure 6 depicts the unit or elementary cell considered in the present analysis. In Figure 6 the most important dimensions are indicated, namely the pitch p, height b, angle 2β and plate thickness t.

A top view of the unit cell is shown in Figure 7. This view can also be considered as the plane of contact points. Figure 8 shows a side view of the unit cell, which can be considered as the plane of geometrical symmetry.

The main purpose of this work is to present methods to determine the flow and heat transfer characteristics and to provide friction factors and Nusselt numbers for typical PHEs. The mass conservation, momentum and energy equations, including turbulence models, are solved numerically. In this particular lecture a low Reynolds number k-ε model and the RNG-k-ε model with wall functions are used to model the turbulence field.

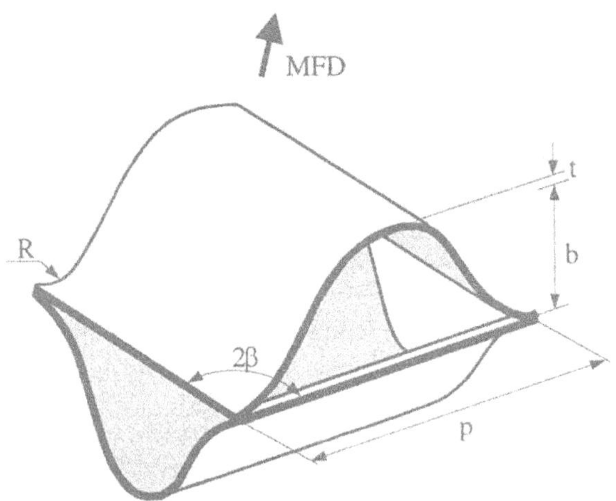

Figure 6. Unit cell considered in the numerical solution approach.

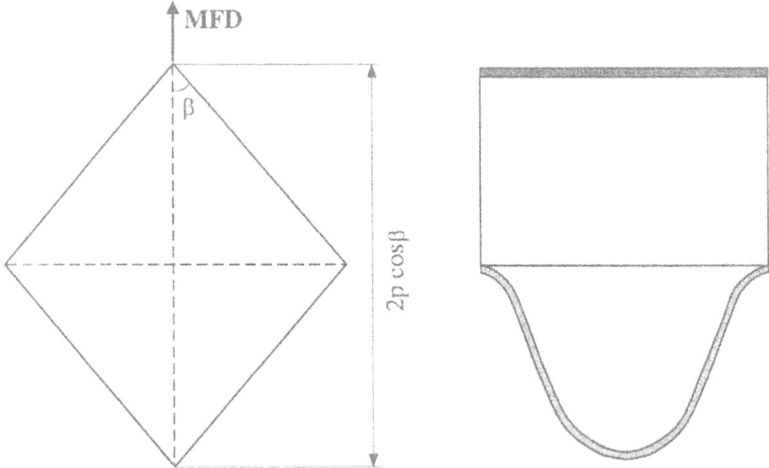

Figure 7. Top view of the unit cell.   Figure 8. Side view of the unit cell.

3. Governing Equations

The governing equations are the continuity, momentum and energy equations. Fully developed periodic laminar and/or turbulent flow is prescribed. Steady-state and constant thermophysical properties are assumed. One then has in a Cartesian coordinate system:

$$\frac{\partial}{\partial x_j}\left(\rho U_j\right)=0 \tag{1}$$

$$\frac{\partial}{\partial x_j}\left(\rho U_i U_j\right)=-\frac{\partial p}{\partial x_i}+\frac{\partial}{\partial x_j}\left\{\mu\left(\frac{\partial U_i}{\partial x_j}+\frac{\partial U_j}{\partial x_i}\right)\right\}+\frac{\partial}{\partial x_j}\left(-\rho\overline{u_i u_j}\right) \tag{2}$$

$$\frac{\partial}{\partial x_j}\left(\rho U_j T\right)=\frac{\partial}{\partial x_j}\left(\frac{\mu}{\mathrm{Pr}}\frac{\partial T}{\partial x_j}+\left(-\rho\overline{u_j t}\right)\right) \tag{3}$$

The turbulent stresses $-\rho\overline{u_i u_j}$ and the turbulent heat fluxes $\rho c_p \overline{u_j t}$ are modelled as described in the following sections.

3.1   TURBULENCE MODELS

To close the system of the mean flow equations (1) – (3) a so-called turbulence model is needed. Such modelling is very important to enable reasonable and successful predictions of turbulent convective heat transfer and fluid flow. Models based on the time-averaged equations are classified as: a) zero-equation models, i.e., mixing length models, b) one-equation models, c) two-equation models, d) Reynolds stress models and e) algebraic stress models. The models considered here belong to category c. Besides these model types, so-called large eddy simulations (LES) based on space-filtered equations are becoming more frequent. Also there are the so-called direct numerical simulations (DNS), where the unsteady forms of the governing equations are solved directly. However, the last two approaches are still at the research level. Further information on turbulence modelling in general can be found in [18-20].

3.1.1   *RNG k-ε Model*
The basic idea of the dynamic Re-Normalization Group (RNG) in turbulent flow is to systematically remove the small scales of turbulence to a point where large scales are resolvable. The mathematical derivation is quite complex, and is not reported here. In a rather recent development of this model, Yakhot et al. [21] expressed the RNG-model as shown below. It can be used with wall functions and compared to the standard k-ε model, and it involves a modification of the ε-equation.
The following equations are used:

192

$$\frac{\partial}{\partial x_j}\left(\rho U_j k\right) = \frac{\partial}{\partial x_j}\left\{\sigma_k\left(\mu+\mu_\tau\right)\frac{\partial k}{\partial x_j}\right\} + P_k - \rho\varepsilon \qquad (4)$$

$$\frac{\partial}{\partial x_j}\left(\rho U_j \varepsilon\right) = \frac{\partial}{\partial x_j}\left\{\sigma_\varepsilon\left(\mu+\mu_\tau\right)\frac{\partial \varepsilon}{\partial x_j}\right\} + C_{\varepsilon 1}\cdot\frac{\varepsilon}{k}\cdot P_k - C_{\varepsilon 2}\rho\frac{\varepsilon^2}{k} - R \qquad (5)$$

where $\sigma_k = \sigma_\varepsilon = 1.39$, $C_{\varepsilon 1} = 1.42$ and $C_{\varepsilon 2} = 1.68$. The production term P_k is found by

$$P_k = -\rho\overline{u_i u_j}S_{ij} \qquad (6)$$

with

$$S_{ij} = \frac{1}{2}\left(\frac{\partial U_i}{\partial x_j} + \frac{\partial U_j}{\partial x_i}\right) \qquad (7)$$

The remaining term R is defined as

$$R = \rho\frac{C_\mu\eta^3\left(1-\eta/\eta_o\right)}{1+\beta_{RNG}\eta^3}\cdot\frac{\varepsilon^2}{k} \qquad (8)$$

where

$$\eta = \frac{k}{\varepsilon}\sqrt{2S_{ij}S_{ij}} \qquad (9)$$

In equation (8), $\eta_o = 4.377$ and $\beta_{RNG} = 0.012$. The turbulent viscosity μ_τ is found from

$$\mu_\tau = \rho C_\mu\frac{k^2}{\varepsilon} \qquad (10)$$

with $C_\mu = 0.845$.

This version of the RNG k-ε model should not be used with damping functions (to get a low Reynolds number model) since the small scales already have been removed. It should be noted that the RNG k-ε model contains a strain-dependent correction term through the variable R, which may avoid the limitations of the standard k-ε model in complex and recirculating flows.

3.1.2   A Low Reynolds Number k-ε Model

At low Reynolds numbers the so-called log-law is not valid, and modifications of the standard k-ε model are needed to enable it to cope with low Reynolds number flow. To ensure that the viscous stresses dominate over the turbulent stresses at low Reynolds numbers and in the viscous sub-layer adjacent to solid walls, damping functions are introduced. The equation for k is identical to equation (4) while the equation for the dissipation $\tilde{\varepsilon}$ is given by

$$\frac{\partial}{\partial x_j}\left(\rho U_j \tilde{\varepsilon}\right) = \frac{\partial}{\partial x_j}\left\{\left(\mu + \frac{\mu_\tau}{\sigma_\varepsilon}\right)\frac{\partial \tilde{\varepsilon}}{\partial x_j}\right\} + C_{1\varepsilon}f_1\frac{\tilde{\varepsilon}}{k}P_k - C_{2\varepsilon}f_2\rho\frac{\tilde{\varepsilon}^2}{k} + \rho E \quad (11)$$

The turbulent viscosity μ_τ is determined by

$$\mu_\tau = \rho C_\mu f_\mu \frac{k^2}{\tilde{\varepsilon}} \quad (12)$$

For the damping functions f_μ, f_1 and f_2, various expressions have been suggested in the literature. Here the expressions by Launder and Sharma [22] are employed. Thus one has

$$f_\mu = \exp\left\{\frac{-3.4}{\left(1 + R_t/50\right)^2}\right\} \quad (13)$$

with

$$R_t = \frac{\rho k^2}{\mu \cdot \tilde{\varepsilon}} \quad (14)$$

$$f_1 = 1 \quad (15)$$

$$f_2 = 1 - 0.3\exp\left(-R_t^2\right) \quad (16)$$

The relation between ε and $\tilde{\varepsilon}$ is

$$\varepsilon = \tilde{\varepsilon} + 2\frac{\mu}{\rho}\cdot\left(\frac{\partial}{\partial x_j}\left(\sqrt{k}\right)\right)^2 \quad (17)$$

The damping function E is given by

$$E = 2\cdot\frac{\mu\mu_\tau}{\rho^2}\left\{\frac{\partial^2 U_i}{\partial x_j^2}\right\}^2 \quad (18)$$

194

In this low-Reynolds-number turbulence model, the constants are set to: $\sigma_k = 1.0$, $\sigma_\varepsilon = 1.217$, $C_\mu = 0.09$, $C_{1\varepsilon} = 1.44$ and $C_{2\varepsilon} = 1.92$.

3.1.3 *Turbulent Stresses and Turbulent Heat Fluxes*

The models presented in sections 3.1.1 and 3.1.2 are usually applied as linear eddy viscosity models to express the turbulent stresses and the turbulent heat fluxes in equations (2) and (3), respectively. One then has

$$- \rho \overline{u_i u_j} = 2\mu_\tau S_{ij} - \frac{2}{3} \rho k \delta_{ij} \tag{19}$$

$$\overline{\rho u_j t} = -\frac{\mu_\tau}{\sigma_T} \cdot \frac{\partial T}{\partial x_j} \tag{20}$$

The turbulent Prandtl number was set here to $\sigma_T = 0.9$. $\qquad$ (21)

4. Periodic Conditions

The method of treating periodic conditions follows that outlined by Patankar et al. [23]. As is evident from Figures 6-8, the upper and lower parts of the unit cell may be regarded as separate channels and in the main flow directions the pressure is written as

$$p = -\gamma x_i + p^* \tag{22}$$

where γ is the pressure drop per unit length.

A constant wall heat flux q_w is imposed on all walls. To create a periodic temperature field a source term containing the mean temperature gradient has to be included in the energy equation. This source term may be written as

$$S_T = -u \frac{\Delta T}{L_{eff}} \tag{23}$$

where

$$\frac{\Delta T}{L_{eff}} = \frac{1}{L_{eff} \cdot \dot{m} \cdot \rho \cdot c_p} \int\limits_{A_w} q_w dA_w \tag{24}$$

L_{eff} is an effective length of the cell in the main flow direction that will be defined in the next section.

5. Additional Relations

To calculate some of the properties in the source terms and to find the friction factor
and Nusselt number, additional relations are needed.

The rate of mass flow is calculated according to

$$\dot{m} = \int_{A_{in}} \vec{U}_{in} \cdot d\vec{A}_{in} \tag{25}$$

To achieve a certain Reynolds number the following procedure is adopted.

$$\Delta p^{new} = (1-\omega)\Delta p^{old} + \omega \cdot \Delta p^{old} \cdot \frac{\mathrm{Re}^{new}}{\mathrm{Re}^{old}} \tag{26}$$

where ω is a relaxation parameter. Re^{old} is the actual Reynolds number calculated as

$$\mathrm{Re}^{old} = \frac{\rho U_{eff} \cdot D_h}{\mu} \tag{27}$$

In equation (27) the U_{eff} and D_h are:

$$U_{eff} = \frac{\dot{m}}{\rho A_c} \tag{28}$$

$$D_h = \frac{4 \cdot V}{A_w} \tag{29}$$

The cross-sectional area A_c is calculated according to the relation

$$A_c = 2b \cdot p \cdot \frac{\sin(\beta)}{\sin(2\beta)} \tag{30}$$

The friction factor f is calculated as

$$f = \frac{2 \cdot \Delta p \cdot D_h}{\rho L_{eff} \cdot U_{eff}^2} \tag{31}$$

where the effective channel length is

$$L_{eff} = \frac{V}{A_c} \tag{32}$$

The Nusselt number is given by

196

$$Nu = \frac{q_w \cdot D_h}{\lambda \cdot (T_w - T_b)} \tag{33}$$

This is the local Nu number that is calculated at every wall grid point. The average value is found from

$$\overline{Nu} = \frac{D_h \displaystyle\int_{A_w} q_w dA_w}{\lambda \displaystyle\int_{A_w} (T_w - T_B) dA_w} \tag{34}$$

The bulk temperature T_B is calculated by

$$T_B = \frac{\displaystyle\int U \cdot T dA}{\displaystyle\int U \cdot dA} \tag{35}$$

6. Numerical Solution Procedure

To deal with complex geometries, a body or boundary-fitted coordinate method is applied. In principle, the complex domain in physical space is mapped into a rectangular domain in computational space by using a curvilinear coordinate transformation. This means that the Cartesian coordinate system x_i in the physical domain is replaced by a general coordinate system ξ_i.

The governing partial differential equations are transformed to algebraic equations by a general finite-volume technique. A non-staggered mesh arrangement is applied, and the velocity components are kept in the fixed Cartesian coordinate directions. The convective terms are treated by the QUICK-scheme, except for the k, ε, $\tilde{\varepsilon}$ equations for which the hybrid scheme is used. The pressure velocity coupling is handled by the SIMPLEC-algorithm. In the iterative solution procedure, underrelaxation factors were chosen as: U and V velocities: 0.4-0.45, W velocity: 0.3-0.45, k and ε $(\tilde{\varepsilon})$: 0.3-0.5, enthalpy or temperature: 0.4-0.8. For the pressure, the underrelaxation could be chosen close to unity in most cases but a value of 0.35 was chosen in some cases.

As the RNG k-ε model was used, wall functions were applied to create the wall boundary conditions.

Nowadays several commercial computer codes are available and such codes are being more frequently applied in industry. The results to be presented in this article were in fact obtained by applying the code CFX (previously CFDS-FLOW3D), although access to an in-house code is available.

6.1 THE MESH

A multiblock approach was applied for generating the mesh or grid. Each duct of the unit cell in Figure 6 may be covered with a three-dimensional array of hexahedral control volumes with a structured and body-fitted grid. Advantage is taken of the fact that the unit cell can be mapped smoothly into a cuboid having one inlet and one outlet face (see Figure 9). To avoid singularities in the mapping, the line contact between the upper and lower plate has to be replaced by a surface contact. A generic control volume of one of the ducts can be identified by the three indices i, j, k (inner co-ordinates).

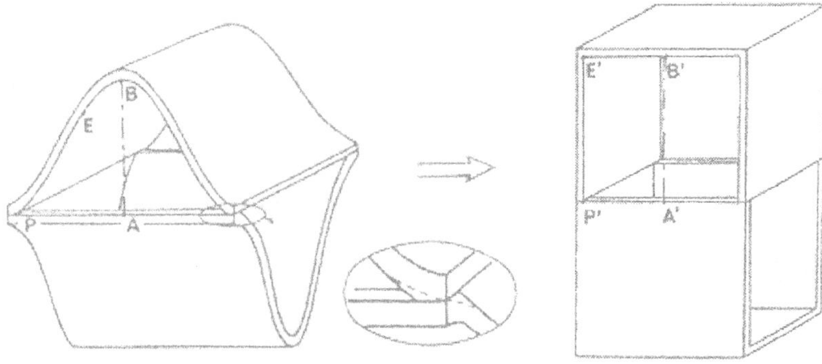

Figure 9. The unit cell and its mapping into a cuboid.

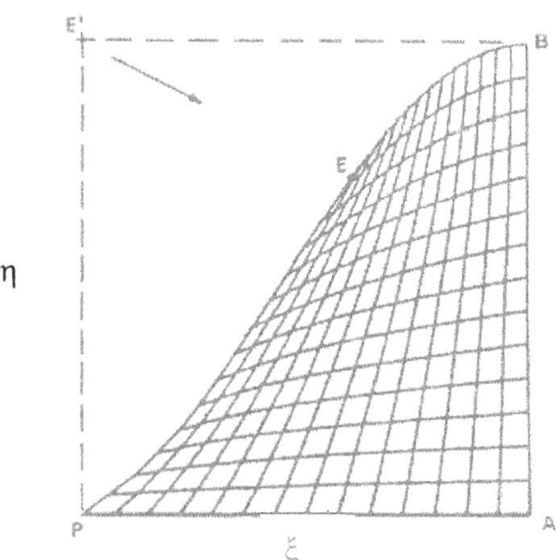

Figure 10. Auxiliary grid.

As the grid is generated, a two-dimensional grid is first build in the region PEBA in Figure 9 by creating a rectangular grid in the mapped region P'E'B'A' in Figure 9, and distorting it as shown in Figure 10. Local co-ordinates ξ and η are introduced in the

198

plane of the section normal to the furrow. Point E, which is mapping the corner E' of the corresponding cube, is arbitrarily located along the curve PB. The final positions of the grid points lying on the boundary PEBAP are given by subdividing the two lines PA and EB (lines j = constant) into N_ξ equal increments, and the two lines PE and AB (lines k = constant) are divided into N_η equal increments.

The co-ordinates of the inner grid points are computed by solving the elliptic Laplacian equations

$$\nabla^2 \xi = 0 \qquad (36)$$

$$\nabla^2 \eta = 0 \qquad (37)$$

The discretized form of these equations reads:

$$\xi_{k,j} = \frac{1}{8} \sum_{nb} \xi \qquad (38)$$

$$\eta_{k,j} = \frac{1}{8} \sum_{nb} \eta \qquad (39)$$

The notation nb in the summation means that the sum is over the eight neighbouring grid points. The co-ordinates of the grid points at each iteration are the average of those of the eight neighbours at the previous iteration. This method has been used successfully in other research papers, e.g., Ciafalo et al. [15] and Amsden and Hirt [24]. The method converges quickly to a smooth body-fitted grid like that shown in Figure 10. The method converges fast, typically within 100 iterations. The resulting two-dimensional grid is mirror-reflected around the line AB in Figure 10, and projected in the inlet plane of each duct. Then it is extruded in the furrow axis direction.

6.2 SAMPLE CALCULATIONS

Two geometries of the plate corrugation pattern in PHEs were considered, and results will be presented. For geometry 1 the following dimensions are valid: $p = 10 \cdot 10^{-3}$ m, $b = 3.6 \cdot 10^{-3}$ m and $R = 1.2 \cdot 10^{-3}$ m. The angle β is varied as: $\beta = 0°$, $\beta = 22.5°$ and $\beta = 45°$. The number of grid points was 20×20×20 or 25×25×25 for each block (2). The Reynolds number was in the range of 400-10 000.

Geometry 2 has the dimensions: $b = 3.05 \cdot 10^{-3}$ m, $p = 1.07 \cdot 10^{-2}$ m and $R = 2.3 \cdot 10^{-3}$ m. For this geometry 32×16×32 control volumes were used in each block. The Reynolds number range was 900-20 000. For this geometry experimental results were available for the average Nusselt number and the friction factor. These results, delivered by Alfa Laval Thermal AB, Lund, were deduced from overall performance tests. The chosen numbers of grid points were found appropriate as a balance between computational times and sufficient numerical accuracy was obtained.

For geometry 1 the low Reynolds number k-ε model was used while for geometry 2 the RNG k-ε model was applied. In both geometries water was the flowing fluid. For geometry 1 the Prandtl number was set 12.8, while for geometry 2 the Prandtl number was set to 4.33.

7. Results and Discussion

7.1 GEOMETRY 1

Figure 11 shows the velocity field in the mid plane corresponding to the top view in Figure 7. In this case the Reynolds number is 6042 and the angle β is 45°.
Two symmetric regions are observed and the influence of the contact points between the upper and lower plate is clearly visible.

Figure 12 depicts the flow pattern at a mid plane parallel with the lower furrow axis. It is obvious that a secondary flow is set up on the flow along a corrugation when its path is crossed by streams of fluid flowing along the corrugations on the opposite wall. This secondary flow has a swirling character which may play an important role in promoting the heat transfer. It has been noted by comparing the results for various angles β, that the secondary flow seems to be weaker and less complex as β is decreased.

Figures 13 and 14 present the average Nusselt numbers and friction factors, respectively, for three values of the angle β. The case β = 0 corresponds to a straight duct.

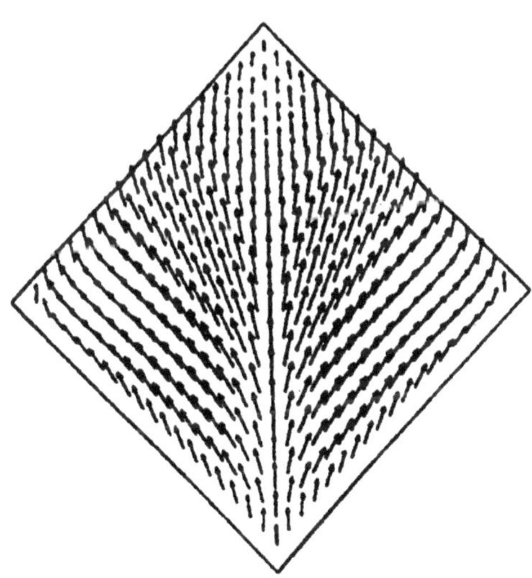

Figure 11. Velocity vectors in a mid plane (see Figure 7), β =45°, Re = 6042.

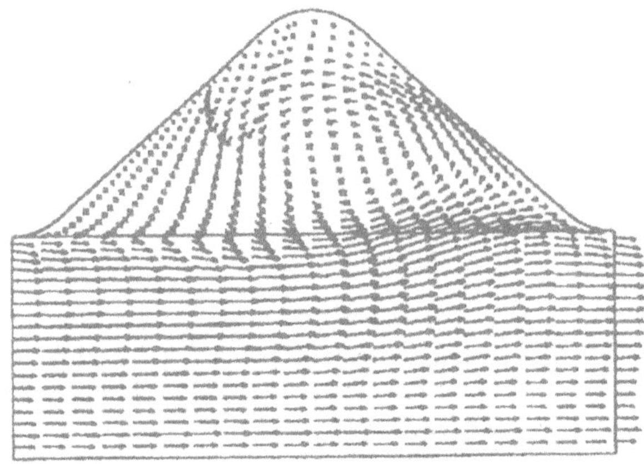

Figure 12. Velocity field in a vertical plane parallel with the lower furrow axis. $\beta = 45°$, Re = 6042.

For $\beta = 0$ the values are close to those of a straight circular duct. It is obvious from Figure 13 that with $\beta = 45°$, the highest Nu numbers are achieved. However, the friction factors are also high. The geometry with $\beta = 22.5°$ presents high Nu numbers but a relatively modest friction factor increase compared to the case $\beta = 0$. Thus with $\beta = 22.5°$ a decent heat exchanger geometry has been identified. Further details may be found in [24]. The results also agree with those of Ciofalo et al. [15, 16] for different, but similar geometries.

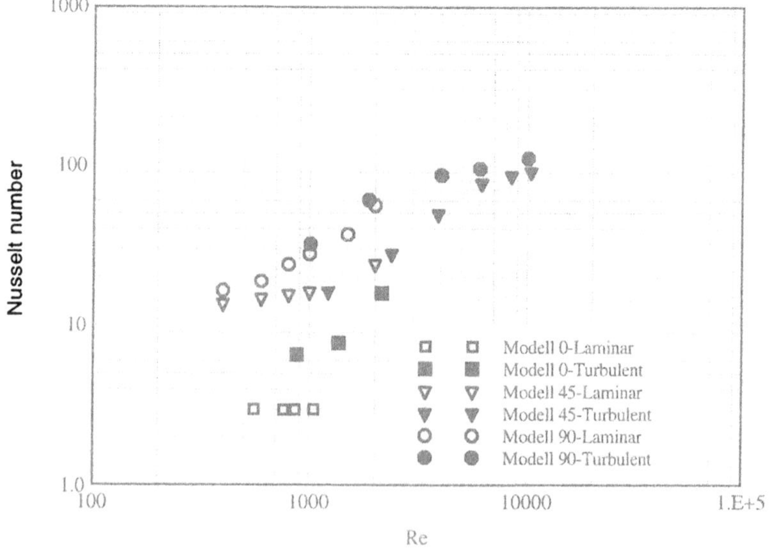

Figure 13. Nusselt number vs Reynolds number for $\beta = 45$, 22.5 and 0.

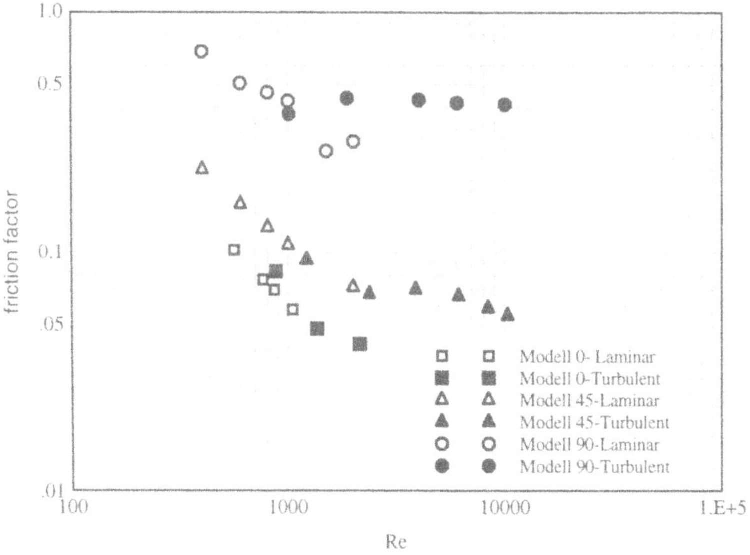

Figure 14. Friction factor vs Reynolds number for β = 45, 22.5 and 0.

7.2 GEOMETRY 2

Figure 15 shows the flow pattern, at mid plane parallel with the lower furrow axis, for this geometry. It is comparable to Figure 12.

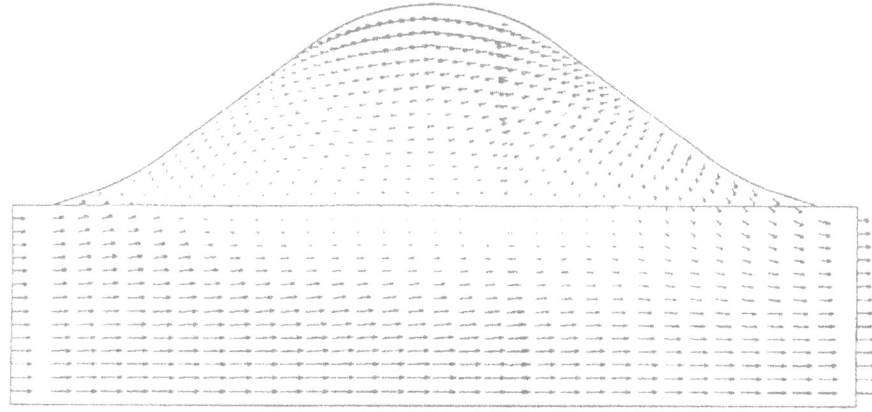

Figure 15. Velocity field in a vertical plane parallel with the lower furrow axis. Geometry 2.

Figures 16 and 17 depict the friction factor and the average Nusselt number results, respectively. The function g(Pr) used in Figure 17 is g(Pr) = $Pr^{0.42}$. In general, the numerical values are lower than the experimental ones. The trend is that the differences increase by increasing the Reynolds number. The maximum underprediction of the Nusselt number is 25 %, while the underprediction of the friction factor is between 17-40 %. However, it should be noted that the experimental friction factor is independent of the Reynolds number which may imply that the plate surfaces in the experiments, are rough in some way. This is not included in the numerical calculations.

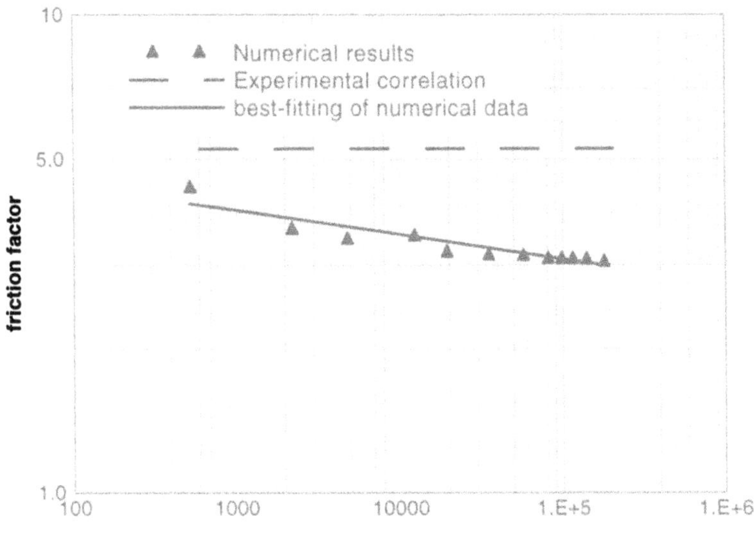

Figure 16. Friction factor vs Reynolds number. Geometry 2.

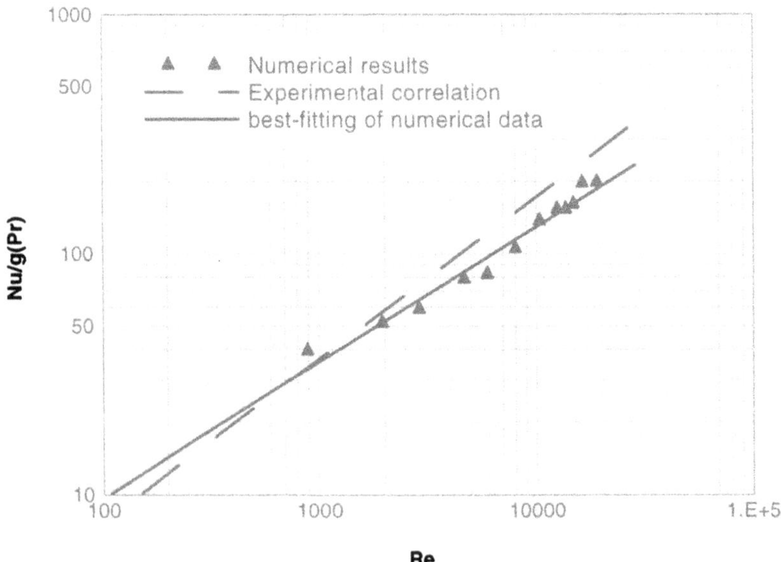

Figure 17. Nusselt number vs Reynolds number. Geometry 2.

It should also be noted that the experimental values for the unit cell were achieved from overall data from which the inlet and outlet regions, and the flow distribution areas were subtracted. A certain additional degree of uncertainty is thus present in the experimental values presented in Figures 16 and 17. Nevertheless, the computed results are promising and encouraging, and in general the agreement with experiment is better than that presented in Mehrabian et al. [17].

Based on the numerical calculations, the following correlations for f and Nu for the unit cell of geometry 2 were established:

$$f = 7.912 \, \mathrm{Re}^{-0.102} \tag{40}$$

$$Nu = 0.693 \, \mathrm{Re}^{0.568} \, \mathrm{Pr}^{0.42} \tag{41}$$

It should be noted that the correlations (40) and (41) are only valid for geometry 2. Further information on the case of geometry 2 can be found in [25, 26].

8. Concluding Remarks

Convective flow and heat transfer in plate-and-frame heat exchangers were presented. A review, followed by a numerical investigation and presentation of results, was provided. In particular two geometries were considered. For the first one the corrugation angle β was varied while for the fixed dimensions of geometry 2 comparison with experimental data was given. The following conclusions are drawn:

- The numerical solutions revealed the existence of a secondary flow on the main stream along a corrugation when its path is crossed by the fluid flow along the corrugations on the opposite walls.
- This secondary flow is believed to affect the heat transfer process and result in enhancement.
- Increasing the angle β increased the Nusselt number and also the friction factor.
- For the case with available experimental data, the Nusselt number was underpredicted by at most 25 % while the corresponding friction factor was underpredicted by 17-40 %.
- It is suggested that more advanced turbulence models be applied.
- Additional experimental investigations of the flow and temperature fields in a unit cell are needed for further confirmation of the numerical modelling approach.

9. Acknowledgements

Financial support was received from Alfa Laval Thermal AB, Lund, Sweden, and some experimental data were provided by this company. The co-operation with M.Sc. Leif Hallgren is kindly acknowledged. Ms Gunvi Andersson wordprocessed this manuscript very efficiently and accurately.

Nomenclature

A_c	cross sectional area, m^2
A_w	heat transfer wall area, m^2
b	internal height, m
c_p	specific heat, J/kgK
$C_\mu, C_{\varepsilon 1}, C_{\varepsilon 2}$	constants in turbulence model, -
D_h	hydraulic diameter $\left(4V / A_w\right)$, m
E	damping term, m^2/s^4
f	friction factor, -
f_1, f_2, f_μ	damping functions, -
k	turbulent kinetic energy, m^2/s^2
L_{eff}	effective duct length, m
$\dot{m}$	rate of mass flow, kg/s
Nu	Nusselt number, -
p	pressure, Pa
$p*$	periodic pressure, Pa
p	pitch, m
Pr	Prandtl number, -
q_w	wall heat flux, W/m^2
Re	Reynolds number, -
R_t	turbulent Reynolds number, -
R	radius, m
S_{ij}	mean rate of strain tensor, $1/s$
S_T	source term temperature field, K/s
T	time-averaged temperature, K
T_w	wall temperature, K
T_B	bulk temperature, K
t	fluctuating temperature, K
t	plate thickness, m
U_i	Cartesian velocity vector, m/s
V	volume, m^3
x_i	Cartesian co-ordinates, m

Greek symbols

β_{RNG}	parameter in RNG-k-ε model, -
β	chevron angle, $degree$
γ	pressure gradient, Pa/m
δ_{ij}	Kronecker delta, -
$\varepsilon, \tilde{\varepsilon}$	turbulent dissipation, m^2/s^3
η	coordinate in grid generation, m
η	variable in extra term in the ε-equation, -
η_o	fixed value, -
λ	thermal conductivity, W/mK
μ	molecular dynamic viscosity, kg/ms
μ_τ	turbulent viscosity, kg/ms
ξ	coordinate in grid generation, m
ρ	density, kg/m^3

$-\rho\overline{u_i u_j}$	turbulent stresses, N/m^2
$\rho c_p \overline{u_j t}$	turbulent heat flux, W/m^2
σ_k	turbulent Pr-number, k-equation, -
σ_ε	turbulent Pr-number, ε-equation, -
σ_T	turbulent Pr-number, T-equation, -
ω	relaxation parameter, -

References

1. Okada, K., Ono, M., Tominura, T., Okuma, T., Konno, H. and Ohtani, S. (1972) Design and heat transfer characteristics of a new plate heat exchanger, *Heat Transfer-Japanese Research*, 1, 90-95.
2. Marriot, J. (1977) Performance of an Alfaflex plate heat exchanger, *Chemical Engineering Progress*, 73, 73-78.
3. Focke, W.W., Zachariades, J. and Olivier, I. (1985) The effect of the corrugation angle on the thermohydraulic performance of plate heat exchangers, *Int. J. Heat Mass Transfer*, 28, 1469-1479.
4. Muley, A. and Manglik, R.M. (1995) Experimental investigation of heat transfer enhancement in a PHE with $\beta = 60°$ chevron plates, *in Heat and Mass Transfer 95*, Tata McGraw-Hill, New Delhi, pp. 737-744.
5. Thonon, B., Vidil, R. and Marvillet, C. (1995) Recent research and developments in plate heat exchangers, *J. Enhanced Heat Transfer*, 2, 149-155.
6. Muley, A. and Manglik, R.M. (1997) Experimental study of turbulent flow heat transfer and pressure drop in a plate heat exchanger with chevron plates, in P.H. Oosthuizen et al. (eds.), *ASME HTD-Vol. 346*, ASME, New York, pp. 69-76.
7. Manglik, R.M. (1996) Plate heat exchangers for process industry applications: Enhanced thermal-hydraulic characteristics of chevron plates, in R.M. Manglik and A.D. Kraus (eds.), *Process, Enhanced and Multiphase Heat Transfer*, Begell House Inc., pp. 267-276.
8. Kreissig, G. and Müller-Steinhagen, H.M. (1992) Frictional pressure drop for gas/liquid two-phase flow in plate heat exchangers, *Heat Transfer Engineering*, 13, 42-52.
9. Sterner, D. and Sundén, B. (1997) Performance of plate-and-frame heat exchangers as evaporator in a refrigeration system, in *Proc. 5th UK National Conference on Heat Transfer*, IChemE, London, UK, Industrial session II.
10. Collier, J.G. and Thome, J.R. (1994) Convective boiling and condensation, 3rd edition, Oxford University Press.
11. Rosenblad, G. and Kullendorff, A. (1975) Estimating heat transfer rates from mass transfer studies on plate heat exchanger surfaces, *Wärme- und Stoffübertragung*, 8, 187-191.
12. Gaiser, G. and Kotte, V. (1990) Effects of corrugation parameters on local and integral heat transfer in plate heat exchangers and regenerators, in *Proc. 9th Int. Heat Transfer Conference*, Vol. 5, pp. 85-90.
13. Heggs, P.J., Sandham, P., Hallam, R.A. and Walton, C. (1997) Local transfer coefficients in corrugated plate heat exchanger channels, in *Proc. 5th UK National Conference on Heat Transfer,* IChemE, London, UK, Session A.
14. Hessami, M.A. and Alguine, V. (1995) Optical measurement of fluid flow in compact brazed-plate heat exchangers, in V. Sernas et al. (eds), *ASME HTD-Vol. 314*, ASME, New York, pp. 57-67.
15. Ciofalo, M., Collins, M.W. and Stasiek, J. (1996) Investigation of flow and heat transfer in corrugated passages-Part II: Numerical simulations, *Int. J. Heat Mass Transfer*, 39, 165-192.
16. Ciofalo, M., Collins, M.W. and Stasiek, J.A. (1998) Flow and heat transfer predictions in flow passages of air preheater: Assessment of alternate modeling approaches, in B. Sundén and M. Faghri (eds), *Computer Simulations in Compact Heat Exchangers*, Computational Mechanics Publications, Southampton, UK, pp. 169-225.
17. Mehrabian, M.A., Quarini, G.L., Poulter R. and Tierney, M.J. (1997) Effect of herringbone angle on the performance of plate heat exchangers, in *Proc. 5th UK National Conference on Heat Transfer*, IChemE, London, UK, Industrial session II.

18. Launder, B.E. (1988) On the computation of convective heat transfer in complex turbulent flows, *ASME J. Heat Transfer*, 110, 1112-1128.

19. Abbot, M.B. and Basco, D.R. (1989) *Computational fluid dynamics – An introduction for engineers*, Longman Scientific & Technical, Harlow, UK.

20. Wilcox, D.C. (1993) *Turbulence modelling for CFD*, DCW Industries Inc., LaCanada, USA.

21. Yakhot, V., Orzag, S.A., Thangam, S., Gatski, T.B. and Speziale, C.G. (1992) Development of turbulence models for shear flows by a double expansion technique, *Phys. Fluids*, A4, 1510-1520.

22. Launder, B.E. and Sharma, B.I. (1974) Application of the energy-dissipation model of turbulence to the calculation near a spinning disc, *Letters in Heat Mass Transfer*, 1, 131-138.

23. Patankar, S.V., Liu, C.H. and Sparrow, E.M.. (1977) Fully developed flow and heat transfer in ducts having streamwise-periodic variations of cross-sectional areas, *ASME J. Heat Transfer*, 99, 180-186.

24. Claesson, J. (1996) Numerical simulations of heat transfer and fluid flow in plate-and-frame heat exchangers, M.Sc. thesis 96/2, Division of Heat Transfer, Lund Institute of Technology, Lund, Sweden (in Swedish).

25. Di Piazza, I. (1997) A numerical investigation of convective heat transfer and fluid flow in plate-and-frame heat exchangers, Report March 1997, Division of Heat Transfer, Lund Institute of Technology, Lund, Sweden.

26. Sundén, B. and Di Piazza, I. (1998) Numerical analysis of fluid flow and heat transfer in plate-and-frame heat exchangers, to be presented at ASME IMECE 1998, Anaheim, California.

HEAT TRANSFER ENHANCEMENT IN A PLATE HEAT EXCHANGER WITH RIB-ROUGHENED SURFACES

R. TAUSCHER, F. MAYINGER
Lehrstuhl A für Thermodynamik
Technische Universität München
85747 Garching
Germany

Abstract. Experimental and numerical investigations of the forced convection heat transfer in flat channels with rectangular cross section are presented in this paper. The heat transfer is enhanced by rib-roughened surfaces applied to the wider walls of the duct. The flow rates have been varied between the Reynolds numbers 500 and 10.000, covering the range from laminar to low turbulent flow. Various configurations (rib: shape, size, spacing, angle of attack, arrangement, duct: width, height, wall temperature) have been investigated. The local heat transfer coefficients have been obtained by holographic interferometry. Local hydrodynamic parameters (with special interest to turbulent kinetic energy and Reynolds shear stresses) were measured by a laser Doppler velocimeter (LDV). According to the configurations examined experimentally, numerical CFD-calculations also have been performed. To increase the accuracy of the numerical calculations for problems with simultaneous laminar and turbulent flow regimes, an empirical procedure is proposed.

1. Introduction

The efficiency of compact heat exchangers can be improved for example by means of boundary layer modification and active surface enlargement. The knowledge about the thermo-hydrodynamic parameters of the flow (e.g., temperature and velocity fields, size of recirculation areas) in the boundary layer of heated walls is important to purposefully improve the heat transfer. The high surface area densities of compact heat exchangers lead to ducts with small aspect ratios. To avoid high pressure drop and pumping power for the fluid, low velocity flow is often applied. Ribs can be used to induce turbulence and thus enhance the heat transfer. The optimization of heat exchangers therefore always has to be aimed at an increase of the heat transfer simultaneously with a minimum increase of pressure drop.

S. Kakaç et al. (eds.), Heat Transfer Enhancement of Heat Exchangers, 207–221.

208

Heat transfer enhancement by means of various techniques is an important task for research, which is also paid tribute to by the huge and growing number of publications on this subject, especially in recent times. A bibliography on enhancement of convective heat transfer is given by Bergles et al. [1].

2. Experimental Setup

Schematics of the experimental apparatus and the modular test section are shown in figure 1. Ambient air from the climate-controlled laboratory is sucked into the inlet section of the heat exchanger by a compressor equipped with a flow regulating throttle. The adiabatic inlet section leads to an almost hydrodynamically developed flow, with a thermally developing flow regime in the heated test section. The test section consists of a rectangular duct (length L = 300 mm, width W = 30-150 mm, height H = 0-32 mm) with isothermically heated upper and lower aluminium walls (T_{Wall} = 20 - 90°C). The aluminium plates are heated by thermostats, with water as the working fluid to ensure a uniform and constant wall temperature. Measurements and numerical calculations prove that even at the highest heat transfer rates occuring during the experiments, the temperature difference between rib base and top is neglectable. The side walls are made of glass to allow optical access; they can be considered as nearly adiabatic. The test section is connected to a smooth 700 mm inlet duct (adiabatic) and a 150 mm outlet duct with the same aspect ratio. The upper and lower test section walls (i.e., heat transfering walls) have rib-roughened surfaces. The modular design of the test section allows a quick variation of a large number of geometric parameters. Global data for heat transfer and pressure drop have been obtained by conventional measuring techniques such as thermocouples and an inclined tube manometer, respectively.

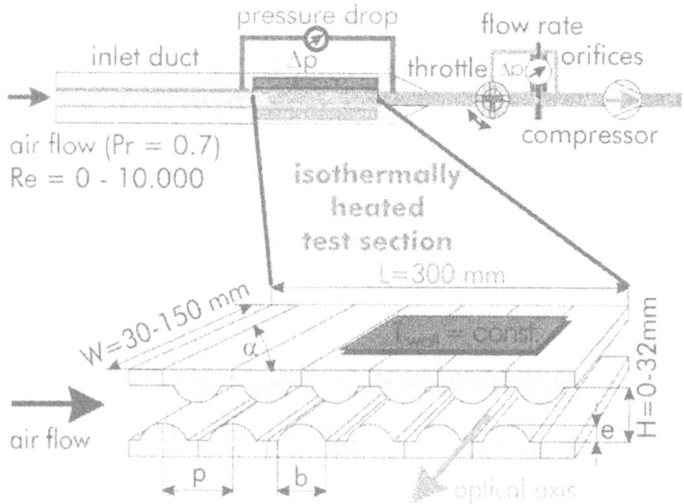

Figure 1: Experimental apparatus - modular test section

2.1 VARIATION OF EXPERIMENTAL PARAMETERS

The following parameter variations have been investigated experimentally so far (figure 2):

- smooth duct / rib roughened surfaces
- rib shape: circular, rectangular, triangular, wing shape, grooves
- rib width: b = 5, 10, 15, 20 mm
- rib height: e = 1.5, 3 mm
- groove angle: γ = 60°, 90°
- rib spacing: p/e = 6.67, 10, 13.33, 15, 16.67, 20, 26.67
- rib angle: α = 15°, 45°, 75° 90°
- rib arrangement: inline, staggered (symmetric, asymmetric), crossed
- duct width: W = 30, 60, 90, 120, 150 mm
- duct height: H = 4, 5, 6, 8 ,9, 10, 12, 14, 22, 32 mm
- wall temperature: T_w = 30, 50, 70, 90°C

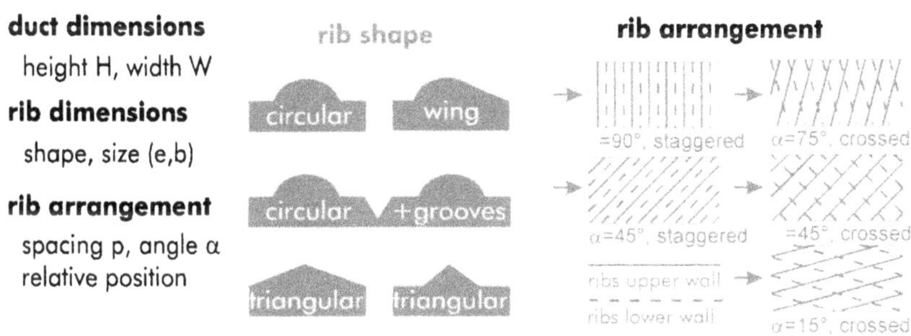

Figure 2: Shape, size and arrangement of turbulence promotors

2.2 SETUP FOR OPTICAL MEASUREMENTS

The experimental setup for the optical measurement techniques is depicted in figure 3. Two-dimensional temperature fields, and thus the local heat transfer in the duct, were examined by holographic interferometry (real-time method). This technique allows the continuous and online observation of the temperature field in the duct, avoiding any influence on the fluid and therefore the effect to be measured. It shows the thermo-hydrodynamic flow pattern within the duct in real time. Any optical accessible configuration (in general 2-dimensional problems) can be investigated by means of this technique. Detailed information on this technique and its applications are given by Mayinger [2], [3], [4], Tauscher [5] and other authors (e.g., Klas [6], Herman [7], Chen [8], Panknin [9]). For the principal understanding of the interferograms shown in this report (figures 5 and 6), it is only important to know that the interference lines (i.e., black and white fringes) of the images are approximatively equal to the isotherms of the flow (infinite fringe method). The distance of the isotherms along lines perpendicular to the heat-transferring wall has to be measured, and will deliver the temperature gradient at the wall and thus the local heat transfer coefficient at a position x in the duct:

$$h_x = \frac{-k(\partial T/\partial y)_w}{(T_w - T_b)_x} \tag{1}$$

The local fluid bulk temperature $T_{b,x}$ is defined for incompressible flow as

$$T_{b,x} = \frac{1}{A_c\, u_m} \int_{A_c} u\, T\, dA_c \tag{2}$$

(Kakac [10]). Using the LDV data together with the local temperature information from holographic interferometry, the correct bulk temperature can be calculated.

This time-consuming procedure is supported by a digital image processing system developed for this purpose. Velocities and turbulence intensities were measured by a single-frequency laser Doppler velocimetry system in the free cross section areas of the duct, and especially in the recirculation and reattachment zones between the turbulence promotors.

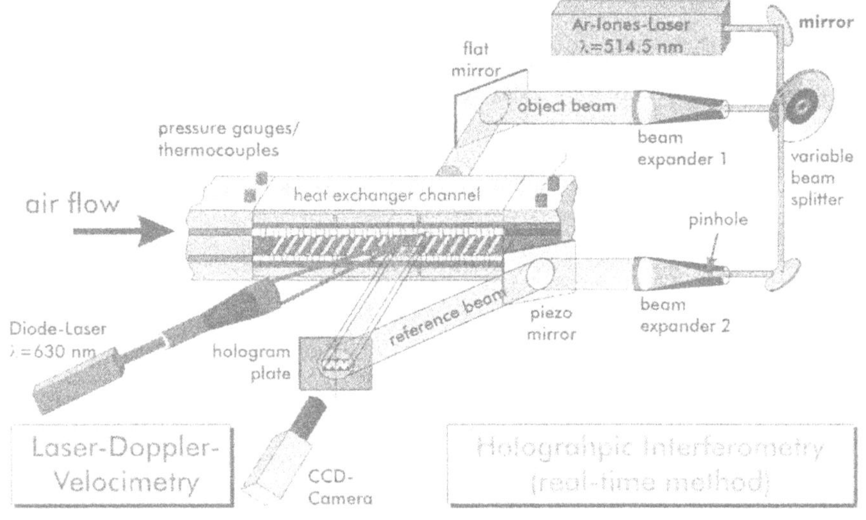

Figure 3: Setup for optical measuring techniques

3.   Experimental Results

Due to limited space only a selection of results will be presented in this paper. Further results and possibilties of the measuring techniques are presented by Tauscher et al. [11],[5] Mayinger [2], [3], [4] and Klas [6].

3.1 VERIFICATION OF MEASUREMENTS

At the beginning of the experimental work, a comprehensive verification of the measurements had to be performed to ensure the reliability and accuracy of the applied techniques. Three main verification procedures have been performed:

1. Comparison of the global measurements (heat transfer and pressure drop) for a smooth duct with data from literature (e.g. Shah [12], Kays [13], Zukauskas [14] and others).
2. Check of the integrated local heat transfer with the global calorimetric measurements: The local heat transfer measurements (holographic interferometry) show differences of less than 8 % compared to the calorimetric measurements, over the whole range of Reynolds numbers for all experimental configurations.
3. Comparision of theoretical predictions for dimensionless velocity (u^+) and temperature (T^+) wall functions (figure 4).

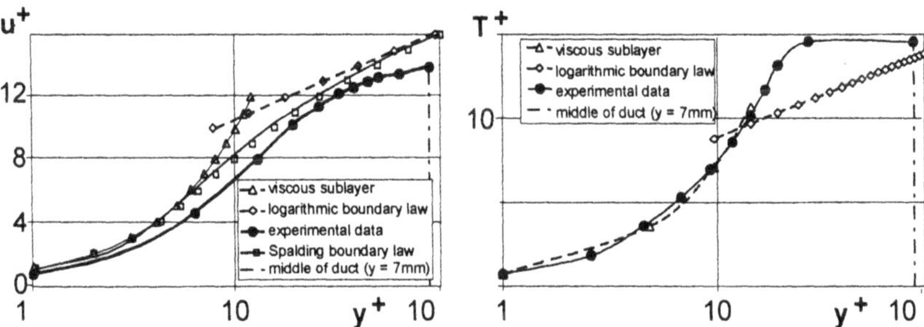

Figure 4: Dimensionless velocity (u^+) and temperature (T^+) in the boundary layer

3.2 HOLOGRAPHIC INTERFEROMETRY

As an expamples figure 5 shows the temperature field in a duct with triangular turbulence promotors for different Reynolds numbers. One can easily detect the areas with good heat transfer (dense fringe pattern at the upstream sides of the ribs) and the ones with bad heat transfer (downstream sides, recirculation zones) and the flow reattachment points between the ribs. The effect of the ribs to induce turbulence can be seen especially in the interferograms for Re = 1500. While the flow is still laminar at the beginning of the test section, it becomes turbulent approximatively after one third of the length of the flow path and is fully turbulent at the end. The evaluation of these interferograms provides the local heat transfer. In figure 6 the development of the local Nusselt number (for various Reynolds numbers) in a region almost at the end of the heated test section is depicted. It can be seen that the local heat transfer always has maxima located just before the top of a rib, minima just behind the ribs and, depending on Reynolds number, further local maxima in the spacing between the ribs. This is due to the fact that at a certain velocity the flow separates at the rib top, and areas in the downstream region of the rib develop where the fluid is recirculating. The amount of flow which is bound in these vortices can be quantified by LDV measurements.

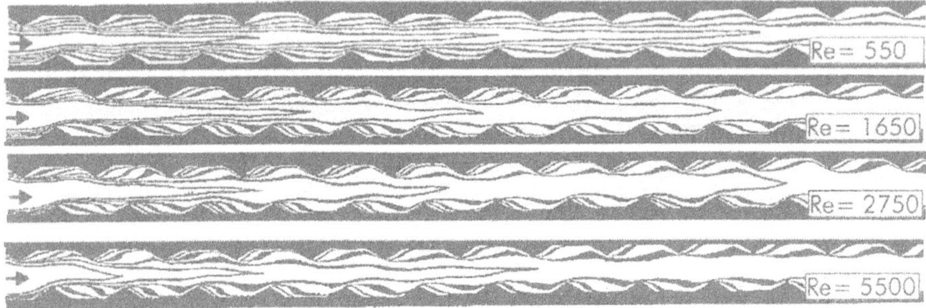

Figure 5: Temperature field (holographic interferograms after binarizaton with the digital image processing system) of the complete heat exchanger duct with triangular turbulence promotors (p/e = 10, e/b = 0.3, e/H = 0.214, $\alpha = 90°$)

The local maxima in the spacing between the ribs result from reattaching flow. The behaviour described above is very similar for different rib shapes, but other parameters, such as rib spacing, have much more effect on the flow (Tauscher [11]).

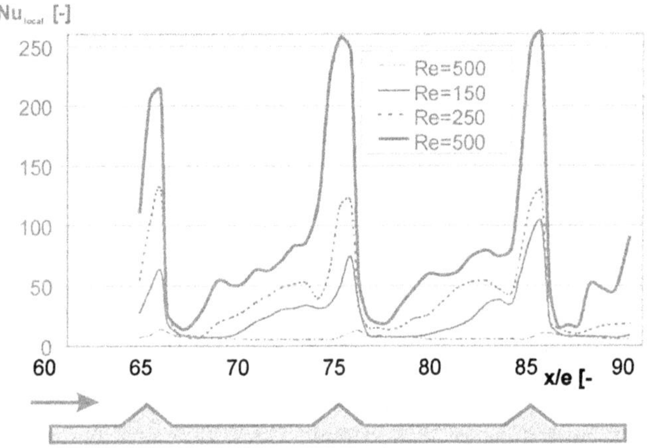

Figure 6: Local Nusselt number in heat exchanger duct with triangular turbulence promotors (p/e = 10, e/b = 0.3, e/H = 0.214, $\alpha = 90°$)

One possibility to enhance heat transfer with a moderate increase in pressure drop is to apply grooves in the spacing between the ribs (e.g., Zhang et al. [15]). In figure 7 interferograms for ducts with circular ribs and for the same configuration but with additional grooves in the rib spacings are presented. One can see how the grooves influence the flow and shorten the recirculation areas compared to the configuration with a flat spacing. The interferograms give no information on the influence on the pressure drop. Therefore global (and local) pressure drop measurements had to be performed for an estimation of the performance of the configurations.

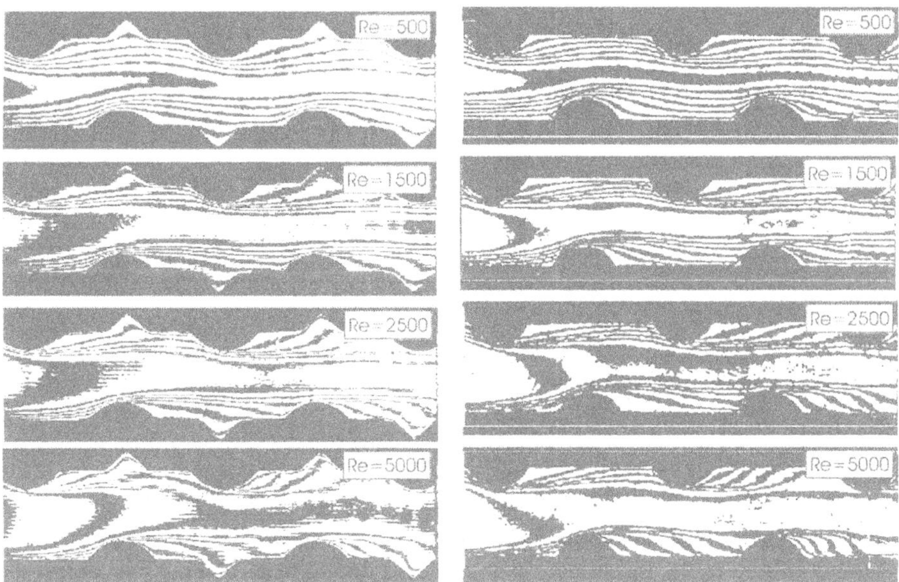

Figure 7: Temperature field in duct with circular turbulence promotors and grooves in rib spacing and in duct with circular turbulence promotors only
(p/e = 10, e/b = 0.3, e/H = 0.214, α = 90°)

The local Nusselt numbers for the grooved configuration is presented in figure 8. The comparison with figure 6 shows that further local maxima in the rib spacing have been created.

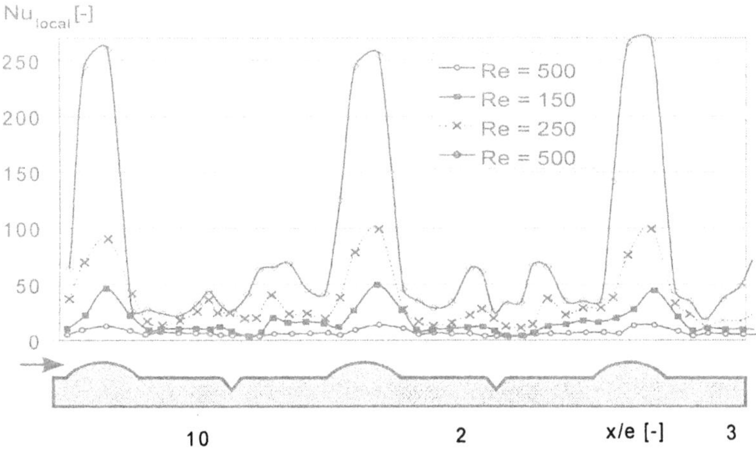

Figure 8: Local Nusselt number in heat exchanger duct with circular turbulence promotors and additional grooves in the spacings (p/e = 10, e/b = 0.3, e/H = 0.214, α = 90°)

214

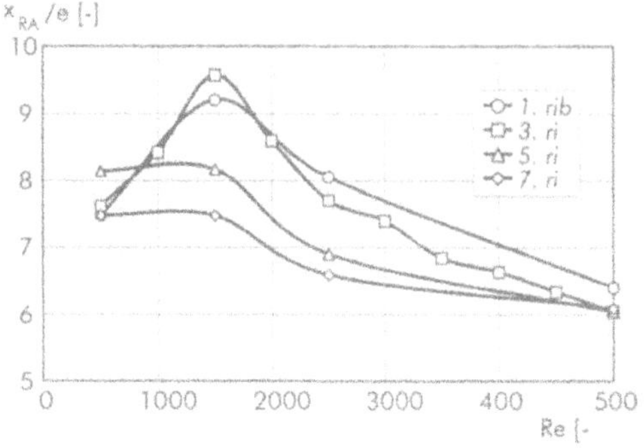

Figure 9: Flow reattachment points (circular ribs, p/e = 10, e/b = 0.3, e/H = 0.214, α = 90°)

The positions of the reattachment points between the ribs vary with rib number and Reynolds number. The graphs in figure 8 show the reattachment positions for a duct arrangement with circular ribs. In laminar flow the distance of the reattachment points, measured from the upstream rib edge, is about 7.5 e– 8 e, and is constant with the duct length. In the transition region it moves downstream and reaches almost the pitch length of 10 e in the inlet region at Re = 1500. With higher flow rates it moves back upstream.

3.3   LASER DOPPLER VELOCIMETRY

Input and verification data for CFD-calculations were obtained by a laser Doppler velocimeter. Special emphasis was given to the measurement of turbulent kinetic energy (figure 11) and Reynolds shear stresses. But also the size and location of recirculation areas (causing poor heat transfer) between the ribs (figure 10) was examined.

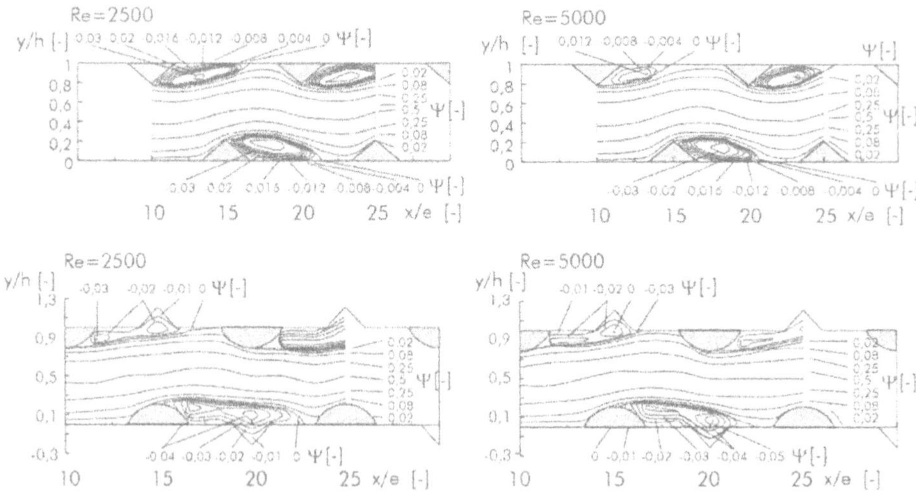

Figure 10: Streamlines of air flow in heat exchanger duct

Generally the part of the air flow which is bound in recirculation zones behind the ribs is small compared to the total flow rate (figure 10). Depending on configuration, duct position and Reynolds number, up to 6% of the flow can be bound in such vortices, which have only a little exchange with the main flow, and therefore cause poor heat transfer in these areas. This areas prevent cold flow from the inner regions of the duct to get in contact with heat-transferring walls, too. Therefore it is important to keep these areas as small as possible and allow the flow, separated at the rib tops, to reattach the walls. Grooves between the ribs can promote this behaviour (see also figure 7).

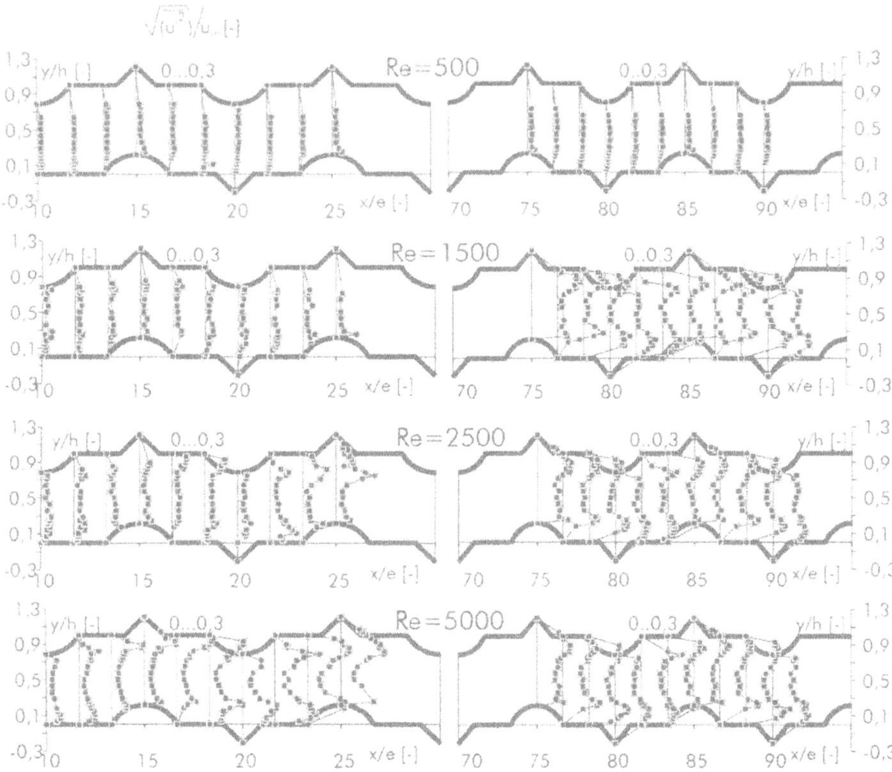

Figure 11: Turbulence intensity profiles at the inlet and outlet of the test section

The turbulence-inducing effect of the ribs can be seen in figure 11. The turbulence intensity increases with the number of ribs passed by the air flow. While the turbulence intensity cannot be increased very much and sustained at low Reynolds numbers (e.g., Re = 500) due to viscous damping effects, it can be increased at Re = 1500 and thus cause the transiton from laminar to turbulent flow approximatively at one third of the duct length (which is also confirmed by the interferograms from figure 5). At higher flow rates flow disturbance, only by few ribs, will be sufficient to cause the transition.

3.4   PERFORMANCE OF SELECTED RIB ARRANGEMENTS

There exist numerous ways to describe and judge the performance of a heat transferring system. One may choose different criteria depending on given requirements when designing a heat exchanger. One criterion to judge the performance of a configuration is to evaluate the increase in Nusselt number of the rib-roughened channel compared to a smooth channel (Nu/Nu_0) in relation to the ratio of friction factor between a rough and a smooth channel $(\xi/\xi_0)^{1/3}$, which is used in this paper. By means of this kind of comparison, all the data refer to the same pumping power for the fluid $(Nu/Nu_0)/(\xi/\xi_0)^{1/3}$). Other possibilities for a performance analysis would be to plot the Colburn factor $j = St\, Pr^{2/3}$ $[=(h/Gc_p)(c_p\mu/k)^{2/3}]$ vs. Re (e.g. Kakac et al. [10]), the flow area goodness factor (j/ξ), the volume goodness factor E (= P/A) etc. .

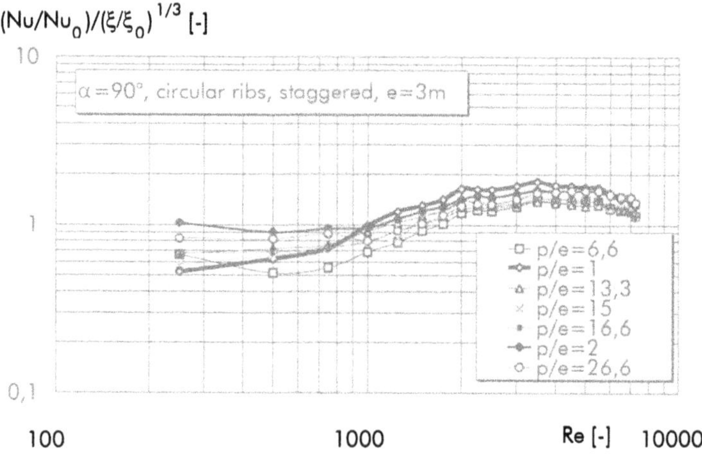

Figure 12: Influence of rib spacing on heat transfer and friction factor
(p/e = 6.67 - 26.67, e/b = 0.3, e/H = 0.214, α = 90°)

The influence of the rib spacing on heat transfer and friction factor is presented in figure 12. To judge the performance of the different configurations we have to examine the laminar and the turbulent region separately. In the laminar regions the performance is worse for all ribbed configurations compared to a smooth channel. In the transition region where the ribs can induce turbulence the performance of the ribbed surfaces becomes better. A spacing of p/e = 10 works best in turbulent flow when one considers both heat transfer and friction factor. The diagram also indicates the start of the laminar-turbulent transition, depending on Reynolds number and rib spacing.

The application of grooves between the ribs in the reattachment and recirculation areas can shift the transition to lower Reynolds numbers, and therefore improve the performance of the arrangement compared to the not-grooved configurations (figure 13). In the full turbulent region their performance is similar to the good performance of the circular ribs.

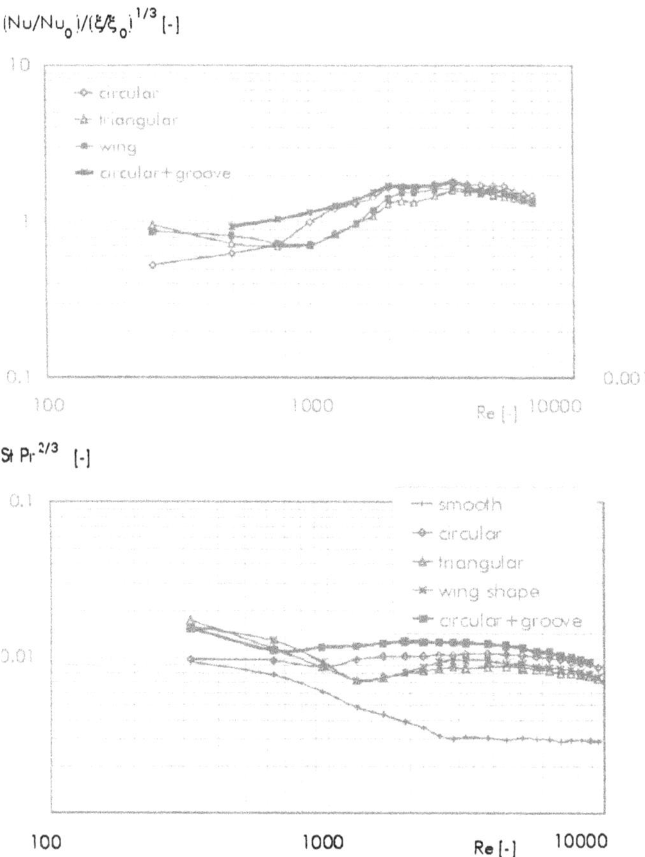

Figure 13: Influence of rib shape on performance
(p/e = 10, e/b = 0.3, e/H = 0.214, α = 90°)

4. Numerical Calculations

The numerical simulation of the heated duct with rib-roughened surfaces was performed with the commercial CFD-codes CFX and TASCFlow3D. These codes work with a finite-volume method based on finite elements. The advantage of the finite-volume method lies in its inherent conservatism, i.e., the conservation equations are fulfilled automatically. Influences of turbulence are considered with different turbulence models (differential-stress-model etc.). In order to reduce computational time the calculation grid was reduced to a 2-dimensional problem whenever possible (e.g., rib vs. main flow angle α = 90°). After a verification of the numerical results by experimental data (figure 14), a parameter variation by computer only shall be possible.

The main problem when using commercial CFD-codes is caused by the fact that the codes cannot decide whether to compute a laminar or turbulent flow, or even detect a transition between these flow patterns.

218

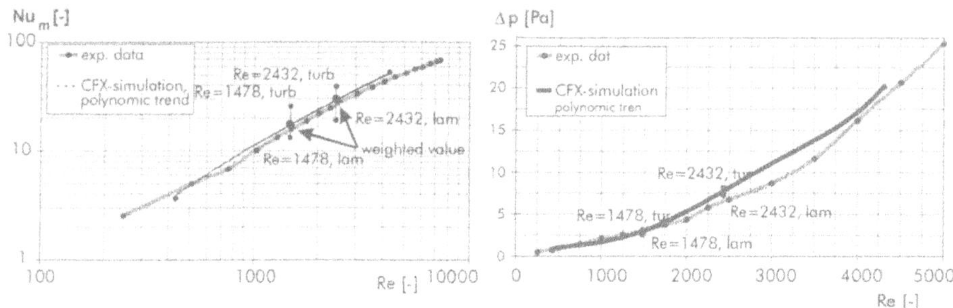

Figure 14: Mean Nusselt number and pressure drop, numerical calculation and experimental data (circular ribs, p/e = 10, e/b = 0.3, e/H = 0.214, α = 90°)

If one compares the numerically calculated velocity profiles with the experimental data, one can notice, that the differences increase as soon as there is only a little turbulence in the duct (figure 15). It may seem astonishing, but even in laminar flow the codes have problems to compute the recirculation areas correctly. The uncertainties are even increased as soon as transition from laminar to turbulent flow occurs. A practical way to overcome this problem is to "split" the duct in a laminar, a transition and a turbulent region for the calculation (which is done in figure 14 by a weighting of the laminar and turbulent calculated values with the real laminar and turbulent flow lenghts). So far, one can only locate these regions from experimental data (e.g., LDV).

Unfortunately the turbulence promotors have the best effect in the transition region, thus the main interest in the present investigations concentrated on this region. The analysis of the possibilities for heat transfer enhancement shows that the localization of the laminar-turbulent transition region is essential for an estimation of the optimal effect of turbulence promotors. The modelling of turbulence in a simulation involves the problem that one has to indicate either turbulent or laminar flow for the entire calculation. Therefore a detection of the transition is not possible, although such detection would be necessary to calculate heat transfer and pressure drop correctly.

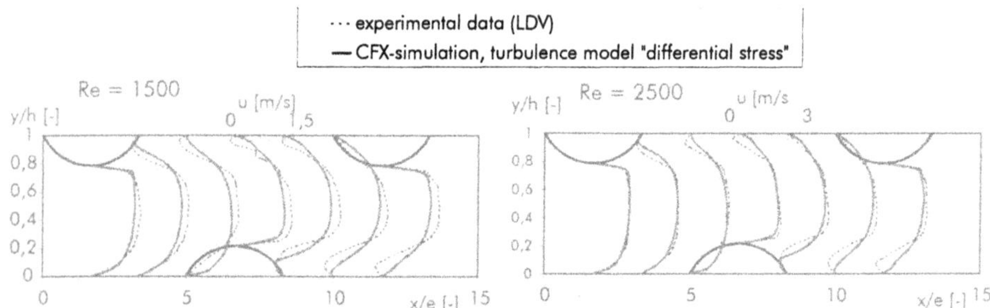

Figure 15: Numerically calculated velocity profiles
(circular ribs, p/e = 10, e/b = 0.3, e/H = 0.214, α = 90°)

For the localization of the transition region a possible emperical procedure is discussed in the following. This procedure uses an effect, which was observed when comparing the velocity profiles, calculated with different models. Figure 16 shows the relative divergence of the calculated u-velocity components in the laminar and turbulent model (differential-stress-model). The continuous line (at 5%) indicates the value which was used as limit deviation to judge the start of the transition. One has to be aware that this value is highly influenced by the duct and rib configuration (e.g., the ratio of rib height to duct height e/H). The results show that the transition at a flow with Re = 2500 starts between the third and fourth rib. The transition criterion for Re = 1500 is only fulfilled behind the sixth rib.

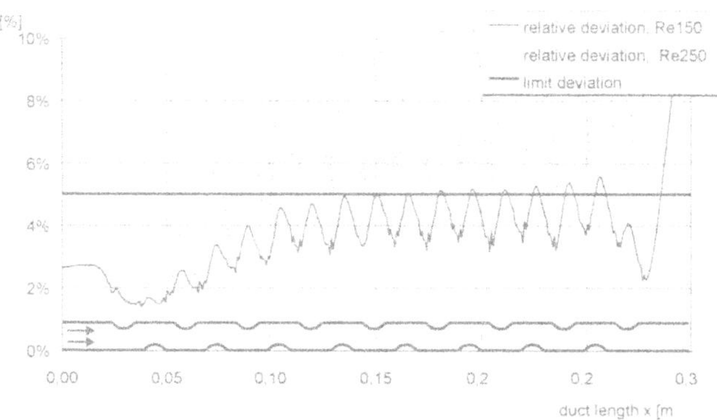

Figure 16: Relative deviation of u-velocity component (laminar and turbulent model)
(circular ribs, p/e = 10, e/b = 0.3, e/H = 0.214, α = 90°)

With the knowledge of the position of the transition point, the codes can be specified to calculate laminar upstream and turbulent downstream. Of course there is still the problem of defining the size of the transition region. A comparison of numerical calculation and experimental data for a wide range of parameter variations will be necessary to prove the general applicability of the procedure described above.

5.   Concluding Remarks

Turbulence promotors (ribs) applied to the heat transferring surfaces of heat exchanger ducts can induce turbulence, and will cause a transition from laminar to turbulent flow at lower Reynolds numbers compared to a smooth duct. These ribs show their best effect in regions where they can induce turbulence, i.e., for Reynolds numbers that are a little bit smaller than the critical Reynolds number. The most effective rib spacing is at a rib-pitch-to-height ratio of p/e = 10. In general the turbulence promotors show their best effect in the transition region from laminar to turbulent flow. Keeping in mind that the enhancement of heat transfer has to be paid by a higher pressure drop the application of grooves in the spacing of the ribs was invesitigated and showed an improvement, especially by an earlier laminar-turbulent transition.

Although the numerical simulations show quite good results, there is still need for experiments as data base for verification and improvement of the codes, considering especially models for flows near or in the transition region from laminar to turbulent flow. Compared to other turbulence models, the so called differential stress model gives the best results and is used throughout the turbulent calculations presented in this report. One possible way to increase the accuracy of numerical calculation in the transition region or in ducts with both laminar and turbulent regions will be the splitting of the calculation regions. An automatic prediction of the transition location by computer only is not yet possible . There is still need for a great number of experimental data to proove the reliability of the proposed empirical method and to generalize the transition criterion for different duct and rib arrangements.

This paper and other publications are also available at the authors' internet homepage (http://www.thermo-a.mw.tu-muenchen.de/~tauscher/).

Nomenclature

A	total heat transfer area, m^2
A_c	flow cross sectional area, m^2
A_{ff}	free flow cross sectional area, m^2
b	rib width at base, m
c_p	specific heat capacity, J/kgK
D_{hyd}	hydraulic diameter, m, $D_{hyd} = 4A/P$
e	rib height, m
G	mass velocity, kg m^2/s, $G = \dot{m}/A_{min}$
h	convection heat transfer coefficient, W/m^2K
H	duct height, m
k	thermal conductivity, W/mK
L	duct length, m
$\dot{m}$	mass flow rate, kg/s
Nu	Nusselt number $Nu = h\,D_{hyd}/k$
P	perimeter, m
p	rib spacing, m, pressure, Pa
Pr	Prandtl number, $Pr = \mu/(\alpha\rho)$
Re	Reynolds number, $Re = \rho w D_{hyd}/\mu$
St	Stanton number, $St = Nu/Pe = Nu/(RePr) = h\alpha/(kw_m) = h/(G\,c_p)$
T	temperature, K, °C
Tu	turbulence intensity
u, v, w	velocity components, m/s
u', v', w'	turbulent fluctuation velocity, m/s
W	duct width, m
w_m	mean velocity, m/s
x,y,z	duct coordinates, m
Ψ	2-dimensional stream function
α	thermal diffusivity, m^2/s, rib angle of inclination vs. main flow, °

μ dynamic viscosity, $kg/m^2\,s$
ρ density, kg/m^3
ξ friction factor
Subscripts
b   bulk
m   mean
w   wall

REFERENCES

1. Bergles, A.E., Jensen, M.K., Shome, B. (1995) *Bibliography on Enhancement of Convective Heat and Mass Transfer*, Heat Transfer Laboratory Report HTL-23, Rensselaer Polytechnic Institute, Troy, USA
2. Mayinger, F. (Ed..) (1994) *Optical Measurements*, Techniques and Applications, Springer-Verlag, Berlin, Germany
3. Mayinger, F., Klas, J. (1992) *Compact Heat Exchangers*, - In: Heat Transfer. Vol. 1. (IChemE Symposium Series No. 129). Ed.: Institution of Chem. Engineers, Rugby, UK. New York: Hemisphere, pp. 36-49.
4. Mayinger, F., Panknin, W. (1994) *Holography in Heat and Mass Transfer*, Proc. of the 5th Int. Heat Transfer Conf., Tokyo, Japan
5. Tauscher, R., Mayinger, F.(1997) *Advances in Heat Transfer by Optical Techniques*, Proc. of the 2nd Int. Symposium on Heat Transfer and Energy Conservation, Guangzhou, PR China
6. Klas, J. (1993) *Wärmeübergang in Strömungskanälen ohne und mit Turbulenzpromotoren*, Diss. TU-München, Germany
7. Herman, C. (1992) *Wärmeübergang in einzelnen und kommunizierenden parallelen Kanälen mit genuteten Oberflächen*, Diss. TU-Hannover, Germany
8. Chen, Y.-M. (1985) *Wärmeübergang an der Phasengrenze kondensierender Blasen*, Diss. TU-München, Germany
9. Panknin, W., (1977) *Eine holographische Zweiwellenlängen-Interferometrie zur Messung überlagerter Temperatur- und Konzentrationsgrenzschichten*, Diss. TU-Hannover, Germany
10. Kakac, S., Liu, H. (1997) *Heat Exchanger Selection, Rating, and Thermal Design*, CRC Press, Boca Raton, USA
11. Tauscher, R., Mayinger, F. (1997) *Enhancement of Heat Transfer in a Plate Heat Exchanger by Turbulence Promotors*, Proc. of the Int. Conf. on Compact Heat Exchangers for the Process Industries, Snowbird, Begell House, Inc., N.Y., USA
12. Shah, R.K., London, A.L. (1978) *Laminar Flow Forced Convection in Ducts*, Academic Press, New York, USA
13. Kays, W.M., London, A.L. (1973), *Hochleistungswärmeübertrager*, Akademie-Verlag, Berlin, Germany
14. Zukauskas, A. (1989) *High-Performance Single Phase Heat Exchangers*, Hemisphere Publishing Corperation, New York, USA
15. Zhang, Y. M.., Gu, W. Z., Han, J. C. (1994) *Heat Transfer and Friction in Rectangular Channels with Ribbed or Ribbed-Grooved Walls*, Transactions of the ASME, Vol. 116, pp. 58-65

HEAT TRANSFER AUGMENTATION IN CHANNELS WITH POROUS COPPER INSERTS*

TUNCER M. KUZAY AND JEFFREY T. COLLINS
Experimental Facilities Division, Advanced Photon Source
9700 South Cass Avenue
Argonne National Laboratory
Argonne, Illinois 60439
USA

Abstract. Copper mesh, compressed and formed into porous matrices of various shapes and sizes, has been routinely used in high heat load/flux component cooling at the Advanced Photon Source (APS) to significantly enhance the heat transfer performance. Substantial research has been performed at the APS over the last eight years in order to better quantify and optimize the selection of the mesh copper matrix attributes (porosity, wire size, core size, bonding technique, etc.) for various water-cooled component applications in single phase. The same mesh configuration can also be applied in cryogenic cooling of optical components, such as monochromators, mirrors and multilayers, with liquid nitrogen. This paper reviews the experimental data and the analytical calculations, compares the data with existing single- and two-phase correlations, and interprets the results for a cryogenically cooled monochromator.

1. Introduction

Porous inserts have been used for enhancing heat transfer in cooling channels for high heat flux applications [1-6]. At the Advanced Photon Source facility (APS) at Argonne National Laboratory (ANL), the heat flux from the powerful APS x-ray beam that impinges upon the x-ray beamline components is of the order of 600 W/mm^2 in normal incidence. Through innovative grazing-angle engineering, the actual heat flux impinging on the beam-interacting components is reduced to under 30 W/mm^2, which is manageable using high temperature materials and advanced cooling concepts. Deionized water in single phase is normally used as the cooling medium for beamline components. Safety requires strict equipment and personnel protection system interlocking. Interlocked flow/pressure set points are carefully chosen so as to compromise the high reliability and availability of this complex user facility, which normally operates around the clock. Therefore, two-phase cooling is avoided for assured reliability and mitigation of vibrations. The high heat flux and heat load levels handled by the x-ray components require highly enhanced cooling methods wherever appropriate to assure long-term structural integrity of the components as well as to minimize the use of expensive deionized cooling water in the facility. For high heat transfer enhancement, we have chosen the use of rolled and compressed copper mesh inserts in the cooling channels of a variety of components [7-8]. Extensive research to optimize the cooling enhancement, and to minimize the pressure drop and deionized water use resulted in the selection of an enhanced heat transfer coefficient

* This work is supported by the U.S. Department of Energy, BES-ER, under contract no. W-31-109-ENG-38.

S. Kakaç et al. (eds.), Heat Transfer Enhancement of Heat Exchangers, 223–247.
© 1999 *Kluwer Academic Publishers.*

224

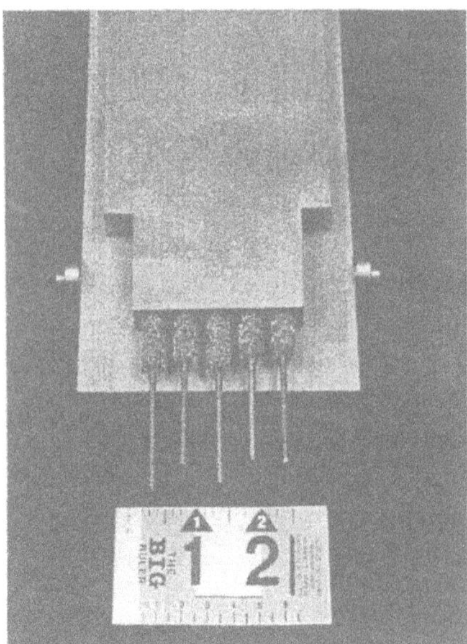

Figure 1. Use of copper mesh inserts at the Advanced Photon Source x-ray beam photon shutters. The five centeral channels have porous copper inserts. The two side channels are inlets for deionized cooling water.

on the order of 2.5 to 3.0 W/cm^{2}K. Compared to plain channels of the same size, this enhancement represent, on an average, a 6-7 fold increase in the value of the heat transfer coefficient, while there is similar level of reduction in the required deionized water usage. Figure 1 (above) shows a typical use of the porous copper inserts at the APS. Shown is a photon shutter blade.

The photon shutter is used to stop the x-ray beam on demand by the beamline user. In this case the five central channels, which cover the area on which the beam can impinge, are equipped with porous copper inserts that are subsequently brazed to the channel walls for intimate thermal contact. The two side channels are used as inlets for the deionized cooling water, which returns through the five central channels in parallel flow. Due to the high pressure drop caused by the porous inserts, typical parallel channel flow instability concerns are totally eliminated.

While the enhanced cooling of the mechanical components on the beamlines is satisfied with the above scheme, it has been discovered that cryogenic cooling of the optical components with similar conductive mesh inserts also offers distinct advantages [9-10]. Because the optical components exhibit very low tolerance to flow-induced vibrations and jitter, the use of porous media in the flow channels of such optical elements was suggested. The low flows sufficient for an efficient porous media heat transfer operation, coupled with the intrinsic turbulence-reducing characteristics of such media, achieve highly quiet but very effective cooling in these components. Optical components, exemplified by the monochromator crystals and mirrors, typically use materials such as single-crystal silicon or diamonds, natural or man-made, and specialty ceramics and semiconductors. It is widely known that silicon and especially diamond exhibit very desirable thermo-physical properties at cryogenic temperatures, particularly with liquid nitrogen (LN2). At cryogenic temperatures, the inherent high thermal conductivity of the optical materials as well as the low thermal expansion coefficient (nearly zero for diamond) result in very low thermal distortions in the components, which is very desirable for optical performance [11-13]. In cooling of optical components in high heat flux applications, liquid nitrogen, as a coolant, is an attractive economic choice. On the negative side, cryogenics, in general, will reach boiling at a lower heat flux. They exhibit an undesirable rapid and uncontrollable vapor blanketing upon boiling due to small saturation temperature differences. Porous inserts are known to offset some of the negative aspects of two-phase heat transfer in cryogenically cooled plain-tube channels. The inserts are known to cause larger saturation temperature differences in the cooled channel hence boiling is harder to start in a channel with a porous insert. On the other hand, the bulk temperature difference in the fluid cross section is smaller in a porous medium particularly for metallic inserts, compared to a plain channel. Hence while the cryogenically cooled plain channels can operate in a subcooled boiling mode, the porous-insert channels are known to transform rapidly from onset of boiling (OOB) to rapid full voiding. Hence the classic onset of nucleate boiling (ONB) followed by a steady nucleate boiling does not generally exist in channels with porous inserts. As such, in the case of porous-channel heat transfer phenomena, we will use the terminology OOB throughout in place of the conventional ONB. Only if acoustically benign and maintainable in a steady fashion subcooled convective boiling permissible is in operations. Otherwise, single-phase cryogenic cooling is the preferred method. Thus one needs to know precisely the initiation of boiling and transition through the subcooled boiling with liquid nitrogen.

The limits of subcooled nucleate boiling in porous media are quite complicated for any modeling or extraction from existing boiling heat transfer correlations. In addition to the complex structure of the medium, sufficient theory is not available to describe a number of the physical aspects of the problem. For instance, what are the criteria for bubble formation? What controls the bubble size and whether it will collapse, stick to the surface or the fibers, or whether it will detach from the surface and move? How much superheat is needed locally for the onset of nucleate boiling? Moreover, nucleation is a statistical process, and the complex structure of the porous matrix makes the problem even more complicated. Due to a lack of theories to characterize these processes on a macroscopic scale, theoretical analysis by numerical simulation seems to be unattainable at this time. Therefore an experimental program was undertaken to generate information that can be used in practice to estimate heat transfer and to predict pressure drop.

2. Experimental Program

The test tubes, schematically shown in Fig. 2, are made of oxygen-free high conductivity (OFHC) copper. All test tubes are the same size with 12.7-mm (1/2 -in) OD and 9.53-mm (3/8-in) ID. Two Kapton-encapsulated thermofoil heaters are continuously wrapped around the tube with about 1.6-mm spacing in between the wraps. Each heater has been rated at 250 W heating capacity. Miniature copper-constantan thermocouples (TC) with 0.005-mil wire diameters are attached to the copper surface at six locations using an indium-eutectic solder.

Rolled and uniformly compressed copper mesh is used in the flow channel to enhance heat transfer. These copper mesh inserts were made from what is called 8 X 8 sieve (8x8 openings per inch square) with 0.32-mm-diameter wire, woven fabric. After rolling a cut piece of the copper fabric around a 4-40 threaded stainless steel rod, the mesh is uniformly compressed to shape in an in-house designed and automated mesh-making machine. The initial mesh fabric size determines the desired porosity. Practically thousands of such mesh inserts have been made to size to satisfy the needs of the APS project and for use by other high heat load applications. The mesh-making machine can duplicate the desired finished product shape to within a mil (.025 mm). We are able to produce any mesh from a tight 60 percent porosity to the loosest 90 percent porosity in any increment desired. In the extensive tests, porosity was varied from 70 to 90 in increments of two percent, and the mass flow rate was varied in the range zero to 15 kg/min. For better clarity, however, in the summary plot only the test data for porosities from 70 to 90 percent in increments of four percent are shown. At the APS, the high heat flux components use $\varepsilon = 76$ percent porous mesh in the optimized design for single-phase de-ionized water cooling. In the subsequent cryogenic cooling test, three copper mesh porosities at 76, 82 and 88 percent are utilized with two tubes in each case, one with brazed and another with non-brazed (mechanical wall contact) mesh. In all cases, tubes were attached to the flexible flow lines via synthetic couplings to reduce axial heat losses.

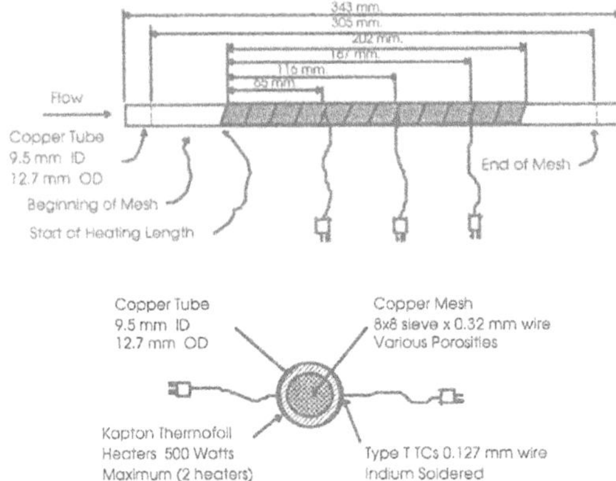

Figure. 2. Schematic of the instrumented test tube.

Rolled and uniformly compressed copper mesh is used in the flow channel to enhance heat transfer. These copper mesh inserts were made from what is called 8 X 8 sieve (8x8 openings per inch square) with 0.32-mm-diameter wire, woven fabric. After rolling a cut piece of the copper fabric around a 4-40 threaded stainless steel rod, the mesh is uniformly compressed to shape in an in-house designed and automated mesh-making machine. The initial mesh fabric size determines the desired porosity. Practically thousands of such mesh inserts have been made to size to satisfy the needs of the APS project and for use by other high heat load applications. The mesh-making machine can duplicate the desired finished product shape to within a mil (.025 mm). We are able to produce any mesh from a tight 60 percent porosity to the loosest 90 percent porosity in any increment desired. In the extensive tests, porosity was varied from 70 to 90 in increments of two percent, and the mass flow rate was varied in the range zero to 15 kg/min. For better clarity, however, in the summary plot only the test data for porosities from 70 to 90 percent in increments of four percent are shown. At the APS, the high heat flux components use $\varepsilon = 76$ percent porous mesh in the optimized design for single-phase de-ionized water cooling. In the subsequent cryogenic cooling test, three copper mesh porosities at 76, 82 and 88 percent are utilized with two tubes in each case, one with brazed and another with non-brazed (mechanical wall contact) mesh. In all cases, tubes were attached to the flexible flow lines via synthetic couplings to reduce axial heat losses.

Each test tube is approximately 343 mm in total length; the copper mesh inside is 305 mm long. The Ohmic heated length of the tube is 203 mm. Therefore an unheated porous flow entry length of about 50.8 mm exists in each test tube followed by the heated section, ending with another unheated porous section, 50.8 mm long. The hydraulic diameter length (L/Dh) of the test tubes is approximately 32, and the heated length is about 21. The uniform ohmic heat was applied via a 2 kW capacity-regulated DC power supply.

The test section instrumentation includes instruments to measure mixed mean inlet and outlet temperatures and the inlet-to-outlet differential pressure drop; a precision turbine flow meter in water tests (a precision Coriolis-effect mass flow meter was used in cryogenic tests); and the TC instrumentation. Power input was measured using a combination of a precision shunt for current measurement and precision resistors for the voltage measurements. The TCs were usually placed as an averaging group of two at each measurement point. On all the tubes, the three TC groups were placed at 65 mm (2.56-in); 116 mm (4.62-in); and, 167 mm (6.54-in) from the start of the heated section (x=0 position). These axial TC locations correspond to a flow hydraulic length of (from start of the mesh) 12.7, 17.5 and 26.5 diameters, respectively. In most cases, the heat transfer coefficient extracted at the last TC location is regarded as the representative value. The inside wall temperatures needed to calculate the local heat transfer coefficient at the thermocouple locations were obtained in the conventional manner by subtracting the calculated copper wall temperature drop from the measured outer wall temperatures under the applied uniform heat flux.

3. Experimental Results with Deionized Water

3.1 PRESSURE DROP WITH WATER

Pressure drop data for the 9.5-mm test tubes used in the experiments are shown in Figs. 3 and 4 (below) for the single-phase water flow in dimensional and non-dimensional forms respectively. Porosities were varied from 70 to 90 in increments of two percent. However for better clarity in Fig. 3, the data in increments of four percent porosity variation are shown against the square of the mass flow rate.

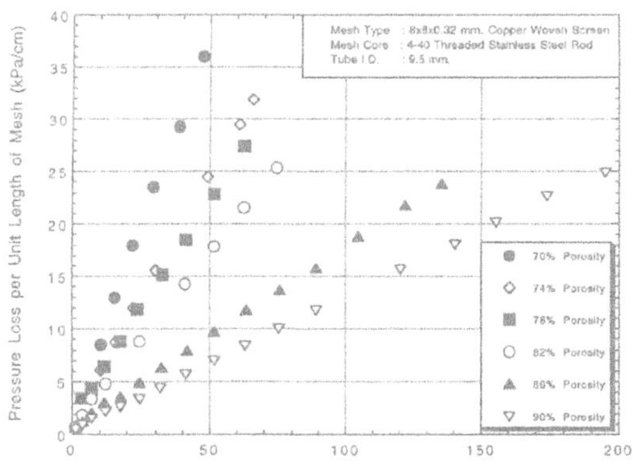

Figure 3. Pressure loss vs. mass flow rate squared for tubes containing mesh inserts of various porosities and with water as the coolant.

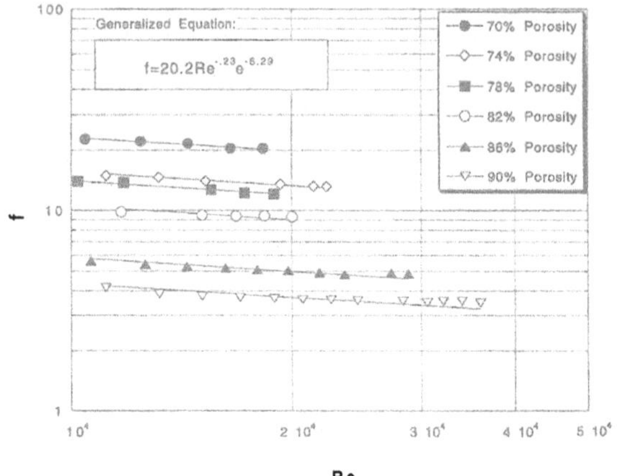

Figure 4. Friction factor vs. Reynolds number correlation in tubes with inserts of various porosities and with water as the coolant.

The pressure drop with porous inserts is large and follows a linear variation with the mass flow rate squared. As anticipated also, the higher porosities cause less pressure drop. Pressure drop measurements are done with good accuracy and show excellent repeatability with negligible scatter.

Non-dimensional pressure drop data, shown as friction coefficient versus Reynolds number (Re), are presented in Fig. 4. Data considered in the plot are all taken at a Re number above 10,000 to be sure that fully developed flow conditions are being treated. The complete pressure drop test data for the brazed insert tubes have been correlated well in the following form:

$$f = 20.2 Re^{-0.23} \varepsilon^{-6.29} \qquad (1)$$

This correlation will be revisited later in connection with the cryogenic tests with liquid nitrogen.

3.2   HEAT TRANSFER DATA WITH WATER

In Fig. 5, the reduced Nusselt (Nu) number is shown versus the Re number for brazed tubes having porosities from 70 to 90 percent in four percent increments. Again, in the experimental program, data were generated in two percent porosity increments and data follow smoothly the trends shown in Fig. 5 in a regular fashion. A Re number range from 3000 to almost 35,000 is attained using the approach velocity and the tube ID. The solid line is the Dittus-Boelter correlation with cross symbols indicating the experimentally established plain tube data. The slight slope change between the Dittus-Boelter correlation and the plain tube data is assumed to be due to the limited length of the experimental tube ($L/D_h \cong 36$). However, for the porous-insert-tubes, this limitation is not significant as fully developed flow conditions are attained quickly in the tube.

An examination of Fig. 5 indicates that brazed porous insert channels provide excellent enhancement in heat transfer. Unlike ribbed or finned tubes, the heat transfer enhancement with the porous tubes is effective starting at low Re numbers and diminishes only gradually as the Re number increases. Enhancement appears to be a weak function of porosity in the data range. With 70 percent porosity, enhancement at Re= 10^4 appears to be about six fold.

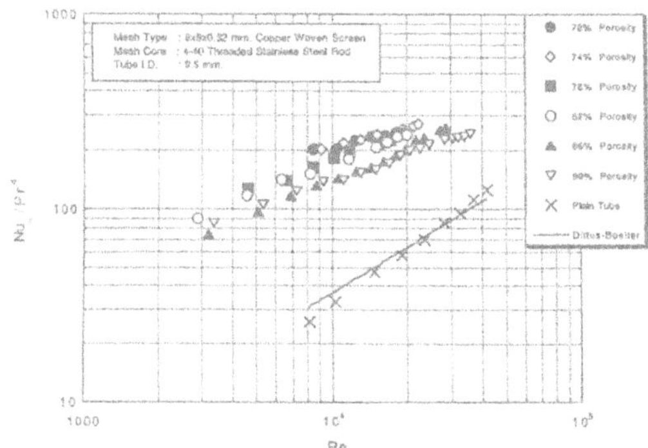

Figure 5. Correlation of Nusselt number versus Reynols number for tubes containing inserts with various porosities and comparison with the values for plain tubes.

One experimentally established strong effect is whether or not the copper insert is brazed to the inner channel wall. In the non-brazed case, the mechanical fit is a sliding fit. Porous inserts are manufactured within a 25-μm tolerance in diameter. A typical performance difference between the brazed and non-brazed insert cases is illustrated in Figs. 6 and 7 for the 76 percent porosity case (in our optimized use, the ε=76 percent). Figure 6 shows that the pressure drop is about the same with both tubes. It was observed that, if a sliding fit is achieved and the brazing is done carefully, pressure drop with the non-brazed tube will only be about a couple percent lower than the brazed ones. Figure 7 is a plot of the "average heat transfer coefficient" in the test tubes. As expected the brazed tube exhibits a

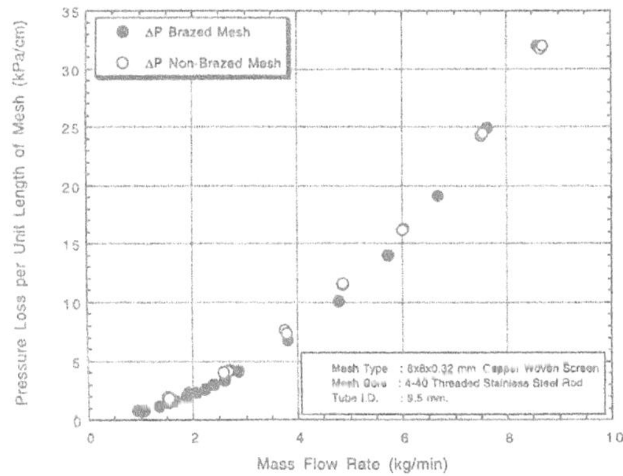

Figure 6. Pressue drop versus mass flow rate for a tube containing either brazed or non-brazed 76% porous mesh.

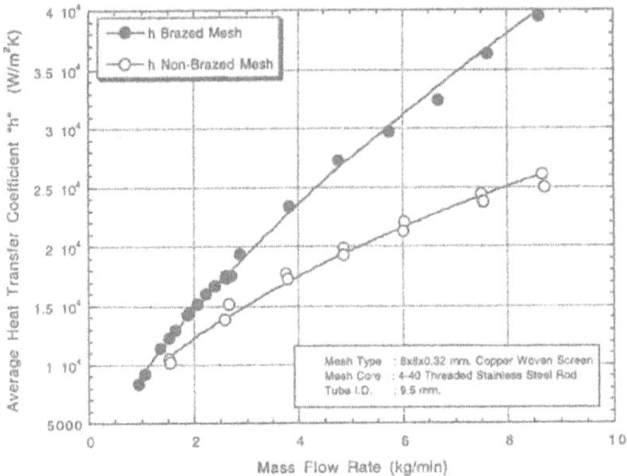

Figure 7. Average heat transfer coefficient data vs. mass flow rate for the 9.5-mm-diameter tube containing either brazed or non-brazed 76% porous mesh.

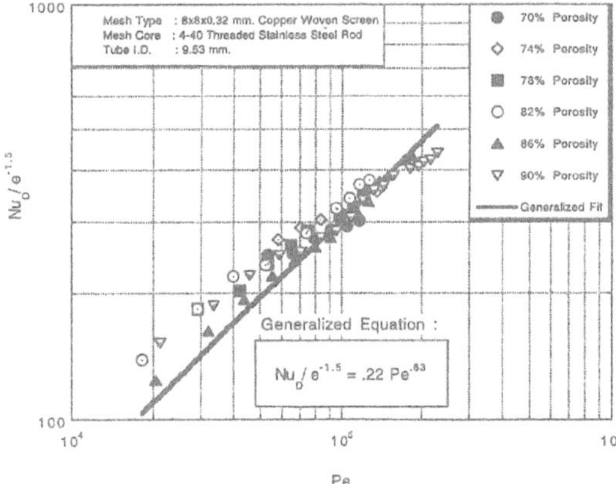

Figure 8. Heat transfer correlation of Nusselt number with Peclet number in water for tubes containing brazed porous inserts at porosities from 70 to 90 percent.

superior heat transfer enhancement compared to the non-brazed tube. We find that data is repeatable in other porosities except that with increasing porosity the difference between the two becomes less pronounced. This is also intuitively correct. A general correlation is offered in Fig. 8 (above) for heat transfer with water for tubes with brazed porous inserts. This correlation applies to all the tubes we tested from 70 to 90 percent porosity. The correlation form is:

$$Nu_D/\varepsilon^{-1.5} = 0.22\,Pe^{\,0.63} \tag{2}$$

Goodness of the fit is 93 % RMS which is very satisfactory. This correlation form will be discussed again later in the study when correlation of heat transfer with cryogenically-cooled porous tubes is examined.

4. Mathematical Model and Analysis of Single-Phase Flow and Heat Transfer in Porous Cooling Channels

4.1 MOMENTUM TRANSPORT

For the momentum transfer in the rolled and compressed porous matrix under study, the best fit was found to be a formulation of a mix of the Brinkman [14] extension and the Cai model [15] with three terms expressing different viscous and inertia effects in the porous matrix as compared those in a packed bed.

$$-\frac{dP}{dx} = C_L \frac{\mu\,u}{d_f^2} \frac{(1-\varepsilon)^2}{\varepsilon^3} + C_T \frac{\rho\,u^2}{d_f} \frac{(1-\varepsilon)}{\varepsilon^3} - \frac{\mu}{\varepsilon}\frac{1}{r}\frac{d}{dr}\left(r\frac{du}{dr}\right) \tag{3}$$

Under incompressible and fully developed flow assumptions, the Brinkman extension is expressed by the third term on the right-hand side in Eqn. (3) to satisfy the no-slip boundary condition at the wall. The two empirical constants C_L and C_T in Eqn (3) were determined from a curve fit to the experimental pressure drop data with the porous matrix in which the working fluid was water [16]. The values of C_L and C_T were found to be 177.2 and 0.4, respectively, for the porous copper matrix with 75 to 85 percent porosity and a mesh size of 8x8x0.32 mm and 8x8x0.20 mm These values were considered to be sufficiently accurate for water, and for liquid nitrogen.

4.2 ENERGY TRANSPORT

Energy transfer was modeled based on the assumption of local thermal equilibrium between the solid and fluid phases in the porous matrix, i.e., by a one-equation model. The transient energy equation was written as:

$$\left(\varepsilon\rho_f c_{p_f} + (1-\varepsilon)\rho_s c_{p_s}\right)\frac{\partial T}{\partial t} + \rho_f c_{p_f} u\frac{\partial T}{\partial x} =$$
$$\frac{\partial}{\partial x}\left(k_{eff_x}\frac{\partial T}{\partial x}\right) + \frac{1}{r}\frac{\partial}{\partial r}\left(k_{eff_r}\,r\frac{\partial T}{\partial r}\right) + \frac{1}{r^2}\frac{\partial}{\partial\theta}\left(k_{eff_\theta}\frac{\partial T}{\partial\theta}\right) \tag{4}$$

where subscripts f and s denote the fluid and solid phases, respectively. It should be noted that the first term goes to zero at steady state. However, the solution procedure was based on an explicit finite difference numerical scheme, and therefore, the transient term was kept. The effective thermal conductivity of the porous medium was based on the work of Koh and Fortini [17] and is expressed as:

$$\frac{k_{eff_i}}{k_s} = \left(\frac{1-\varepsilon}{1+n\varepsilon^2}\right) \tag{5}$$

232

where n is a constant whose recommended value for felt metals is 10, and i can be x, or r direction. The boundary conditions for the problem were as follows:

$$\text{at} \quad r = R \quad u = 0 \quad \text{no} - \text{slip boundary conditions} \qquad (6)$$

$$\text{at} \quad r = 0 \quad u = u_0$$

where u_o is the core velocity that can be obtained from the quadratic equation, which results when the Brinkman term is dropped from the momentum equation. The boundary conditions for the energy equation include the inlet fluid condition as well as the prescribed wall heat flux condition. These boundary conditions are expressed as:

$$\text{at} \quad x = 0 \quad T = T_{in} \qquad \qquad \text{inlet condition}$$

$$\text{at} \quad r = 0 \quad \left.\frac{\partial T}{\partial r}\right| = 0 \qquad \qquad \text{symmetry condition}$$

$$(7)$$

$$\text{at} \quad r = R \quad q''_w(q) = k_{eff}\left.\frac{\partial T}{\partial r}\right|_{r=R} \qquad \text{wall heat flux condition}$$

Finite difference methods were used in obtaining the solutions for the above equations. Details have been provided previously [16]. Some sample results are presented here to show the success of the

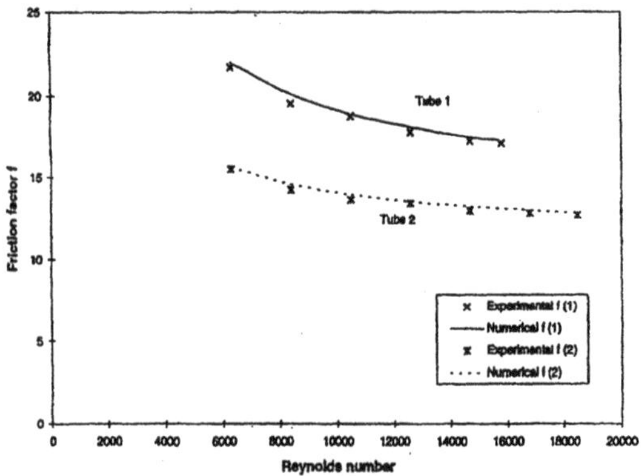

Figure 9. Analytical prediction of the friction factor versus Reynolds number for two test tubes. Tube 1 has a 75% porosity; tube 2 has an 85% porosity insert. Both inserts are brazed.

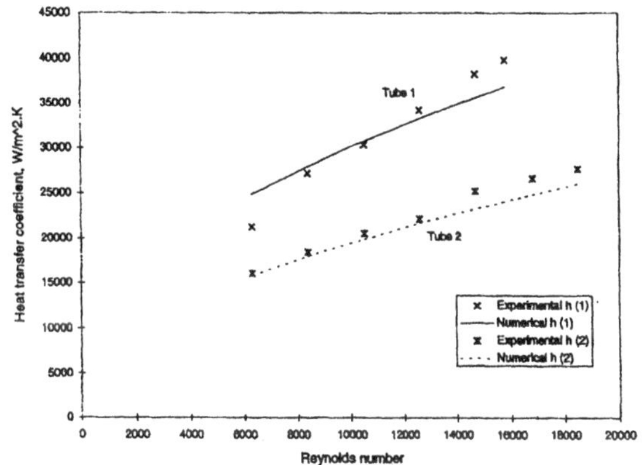

Figure 10. Analytical prediction of the average heat transfer coefficient versusReynolds number for the two test tubes described in Fig. 9.

analytical predictions with the earlier test data with water as the coolant. In Fig. 9, experimental and numerical results are provided for the friction coefficient for two test tubes. These tubes have brazed copper mesh; the first one having 75 percent porosity and the second 85 percent porosity. Agreement between the numerical predictions and the experimentally established values is very good.

In Fig. 10 (above), the average heat transfer coefficient along the total tube length is shown against the Re number. Both the calculated values and the experimentally determined values are shown for the two test tubes above. Agreement is satisfactory given the uncertainties involved in establishing the average heat transfer coefficient in the experiments and the analytical modeling of the complex problem. Other analytical predictions have been reported in [16].

The above examples show that analytical prediction of the heat transfer problem with water as the coolant was successful. Application of the analytical techniques to predict the heat transfer with liquid nitrogen cooling is to be tried in future studies.

5.   Cryo-Cooling Experiments with Liquid Nitrogen

Test tubes for the cryogenic tests with liquid nitrogen were made and instrumented similar to those for the deionized water test tubes (Fig. 2). The test section is in a vacuum tank that is kept normally at 10^{-2} Torr. The inlet pressure to the liquid nitrogen pump is adjustable up to 1000 kPa. The closed-loop, pressurized subcooled LN2 pumped through the test section returns to the dewar vessel where it circulates through the heat exchanger submerged in a bath of atmospheric LN2. The atmospheric LN2 boils off, removing heat from the pumped liquid, thereby maintaining a constant fluid inlet temperature to the test section of about 80 K. Liquid nitrogen is introduced into the test section with no sub-cooling. The head pressure is regulated to be at a preset value between 550 and 585 kPa in all experiments. Hence the saturation temperature of LN2 was about 97 K or so, depending on the head pressure. The total heat load applied to the test section was uniformly set at five values: 100, 200, 300, 400 and 500 Watts.

Experiments measured cryogenic heat transfer using LN2 in round copper tubes. All test tubes were placed in a horizontal configuration in this phase of the experiments. Three copper mesh porosities of 76, 82 and 88 percent were utilized with two tubes in each case, one with brazed and another with non-brazed (mechanical wall contact) mesh. Further details of the test section and the results will be published [18].

Tubes were tested in a progressive fashion from single-phase heat transfer to the onset of boiling and beyond. A sample case of heat transfer data pertaining to the phenomena described above is shown

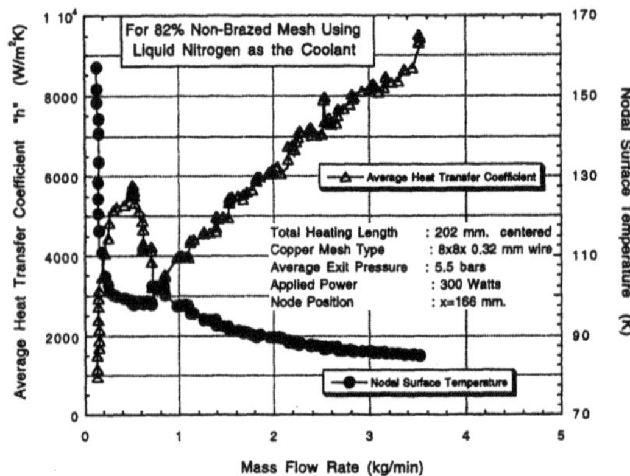

Figure 11. A sample case of experimental data for the surface temperature and the average heat transfer coefficient. The tube contained 82% non-brazed porous mesh. The total heat load was 300 W. The right vertical scale indicates the temperature reading measured by the end thermocouple (Node #3).

in Fig.11 (above) for a test tube containing 82 percent porous non-brazed mesh subjected to a 300 W total heat load.

The temperature reading indicated by the end thermocouple (Node #3) is shown, on the right vertical scale, against the stepwise reduced mass flow rate in the test section. Starting at about 3.5 kg/min mass flow rate, the flow was reduced and shows single-phase heat transfer until 0.8 kg/min mass flow is reached.

At this mass flow point, onset of boiling (OOB) is reached as indicated by a sudden drop in the wall temperature. A narrow temperature plateau is traced with further reduction in the mass flow rate. This is the subcooled/nucleate boiling region that, with the stepwise reduction in the mass flow rate, degenerates into rapid voiding and hence a wall temperature runaway condition. At this point the experiment is terminated by manually increasing the mass flow rate to save the test section from a potential burnout.

The average heat transfer coefficient at the same node corresponding to the measured temperature is shown in Fig. 11 on the left vertical scale. It should be noted that, during the single-phase heat transfer tests up to the OOB point, all the thermocouple and the flowmeter data are meaningful. Hence the inlet and outlet TCs along with the flowmeter reading provide an independent check on the heat input (enthalpy rise) into the test section. Up to the OOB point, heat transfer calculations are based upon the known power input and the measured inlet temperature of the single-phase LN2. Bulk fluid temperatures and inner wall temperatures at each node are calculated on this basis. Beyond the OOB point, where two-phase LN2 is present, the enthalpy rise of the fluid as determined by the inlet and outlet TCs along with the mass flowmeter reading does not match the heat input into the test section. An independent measure of fluid quality is required to account for a proper heat balance. Unfortunately, a means of measuring fluid quality was not available for these experiments. Beyond the OOB point, the measured single-phase enthalpy rise along with the inlet temperature is used to determine a conservative estimate of the heat transfer. Without knowledge of the fluid quality, the known power input will lead to erroneous and sometimes negative results for the heat transfer coefficient if used in these calculations. Consequently, although plotted as read, the quantitative data from the test section should be viewed as suspect beyond the OOB point and can be useful only for trend information.

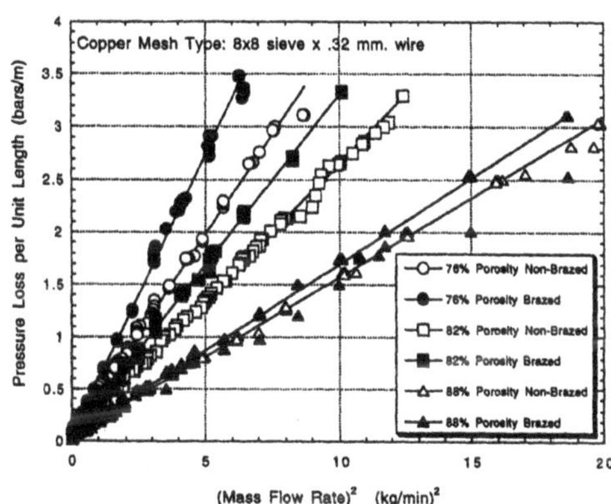

Figure 12. Pressure loss versus square of the mass flow rate in tubes with inserts of various porosites and with liquid nitrogen as the coolant.

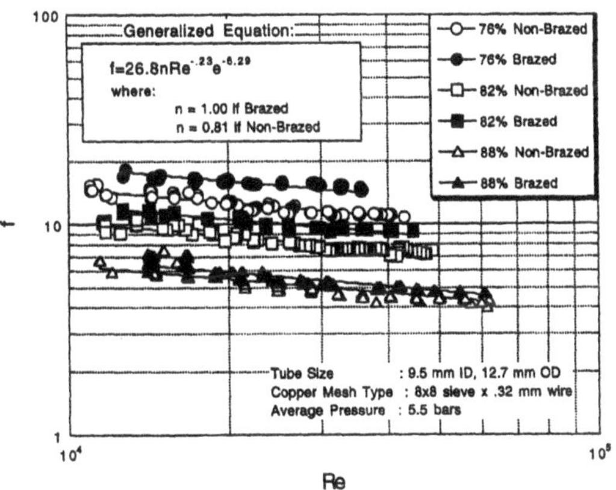

Figure 13. Friction factor versus Reynolds number correlation in tubes with inserts of various porosities and with liquid nitrogen as the coolant.

5.1   PRESSURE DROP

Pressure drop data for the 9.5-mm test tubes used in the experiments are shown in Fig. 12 (above) for the single-phase LN2 flow. The pressure drop in the porous tubes correlates with the square of the flow rate, both for the brazed and the non-brazed inserts. However, the pressure drop in the brazed tube, as expected, is always higher than that in the non-brazed tube at the same flow rate.

Non-dimensionalized pressure drop data, as friction coefficient versus Reynolds number, are shown in Fig. 13. Both the brazed and non-brazed data have the same slope. The complete pressure drop test data have been correlated as follows:

$$f = C\ 26.8\ Re^{-0.23}\ \varepsilon^{-6.29} \qquad \text{where,} \qquad (8)$$

$$C = 1.0 \text{ for brazed inserts}$$

$$C = 0.81 \text{ for non-brazed inserts}$$

The correlation form is identical to the one used for the water tests, however, the coefficient for the cryogenic case appears to be about 20 percent higher for the brazed inserts than for the non-brazed inserts. We do not have an explanation for this. In the non-brazed tube tests, same heater tube has been used with different porous inserts. The nature of the fit is a "snug fit" in all cases. If another fit quality is used, the correlation may not be as accurate.

5.2   HEAT TRANSFER

Heat transfer data have been analyzed separately for the plain tubes and the tubes with porous conductive inserts. The plain tube single-phase heat transfer has been measured against the traditional Sieder-Tate correlation to verify the experimental procedures for a variety of heater powers from 100 to 500 W (Fig. 14). The solid line is the Sieder-Tate heat transfer correlation for circular plain tubes. The single-phase heat transfer is consistent with this correlation, with the data scatter within ± 5 percent. In Fig. 14, one can follow the progression of the averaged effective heat transfer coefficient, "h_{eff}", on the tube wall.

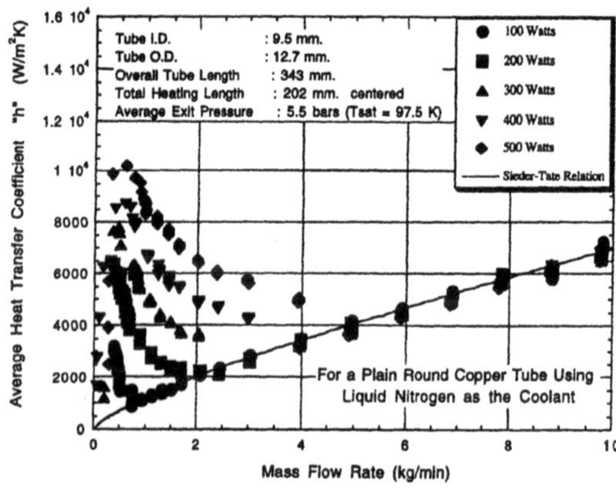

Figure 14. Average heat transfer data vs. liquid nitrogen mass flow rate for a 9.5-mm plain test tube at various total power levels.

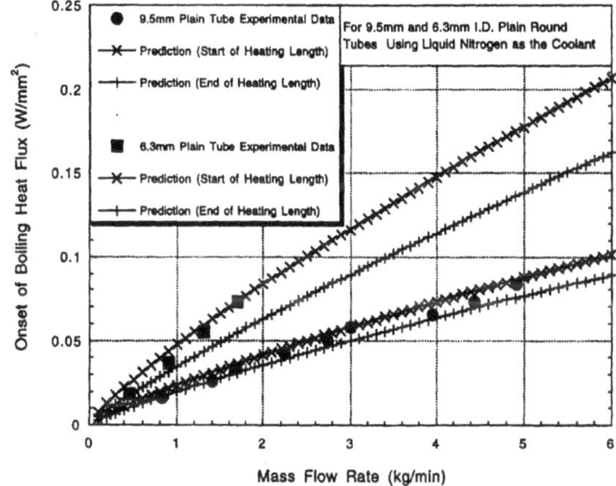

Figure 15. Onset of boiling heat flux data vs. liquid nitrogen mass flow rate for the 9.5-mm-and 6.3-mm-diameter plain test tubes and comparison with predictive correlations.

The progression of forced convection up to the ONB point is a classic case as covered in the foregoing. At and after ONB, the reduced mass flow rate enhances the heat transfer rates from the wall under the combination effect of two-phase forced convection. Hence the averaged effective heat transfer coefficient continually rises until such point where attainment of critical heat flux causes a sudden drop in the effective h_{eff}, where tests were immediately terminated to prevent tube burn-out. Figure 15 (above) presents the onset of boiling data for the cases of the 9.5-mm diameter (3/8 inch) and the 6.4-mm-diameter (1/4 inch) plain tubes. For the 9.5-mm tube, data are shown for all the uniform heat flux test cases at various mass flow rates from less than 0.5 to 5 kg/min. Diversion points from the Seider-Tate curve in Fig.14 are taken to be onset of nucleate boiling (ONB) in a plain tube. At the lowest heater power of 100 W, this occurs at about 0.8 kg/min. At the highest heater power test of 500 W it occurs at about 5 kg/min. The plain tube tests with the 9.5 mm tube is tested in 50 W total power increments. Hence the nine data points for the 9.5-mm tubes plotted in Fig. 15 are taken at heater powers from 500 to 100 W, in 50 W intervals. Because the exact location of the ONB could not be determined in the tests, the correlation was tested for the two ends of the heater length. The two solid lines show the ONB heat flux using the beginning and the end of the heated length based on correlations [19 and 20]. Data are bracketed by the resulting correlation band in close agreement in magnitude and trend. At mass flow rates above the ONB, heat transfer is single-phase convection. Under the ONB flow rate it is two-phase. The ONB heat flux indicated in the vertical ordinate is heat flux at the inner tube wall calculated from the measured total heater power applied via the heaters on the outside tube surface.

Some limited heat transfer measurements were also made with a 6-mm-diameter (1/4 inch) diameter test tube (see Fig. 15). The four data points plotted are taken at 400, 300, 200 and 100 W heater power. At 400 W, the onset of boiling occurred at about 1.8 kg/min and at 100 W at 0.5 kg/min mass flow rate of the liquid nitrogen. Again data are bracketed by the ONB correlation curves using the two ends of the heater length.

Focusing on Fig. 15, one can deduce that, to support a wall heat flux level of 0.1 W/mm² without boiling, one has to pass more than 6 kg/min LN2 in a 9.5-mm-diameter tube. At somewhat reduced head pressure this is shown to be reduced significantly to about 3 kg/min for a 6.4 mm tube. Existing plain tube correlations can be used confidently to determine any parameter needed for a given set of conditions.

6.   Heat Transfer in Tubes with Porous Conductive Inserts

For the 9.5-mm-diameter (3/8 inch) tube, three porosities of 76, 82 and 88 percent have been studied. Figure 16 is an illustrative composite plot of the nodal temperature against the varying mass flow rate at a particular case of 300 W total heater power. The plain tube nodal temperature has also been added to provide a basis for comparative discussions.

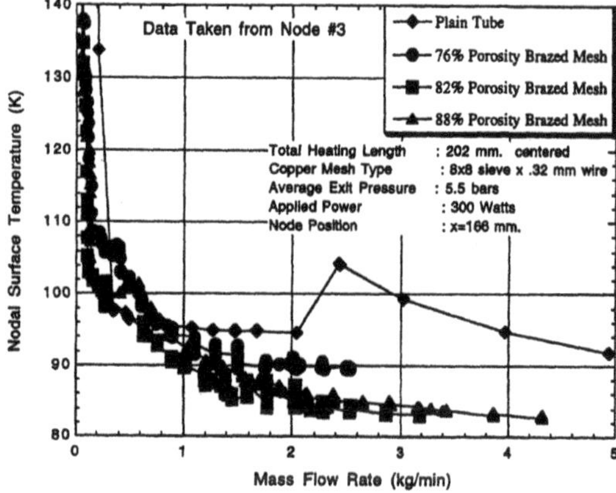

Figure 16. Sample surface temperature data vs. liquid nitrogen mass flow rate for a 9.5-mm plain test tube and for tubes with porous mesh inserts at 300 W input power.

Focusing on the particular case of 76 percent porosity in Fig. 16, one can see that, as the mass flow rate is continuously reduced from 2.6 kg/min the nodal wall temperature steadily increases as expected. At about 0.4 kg/min mass flow rate, a sudden drop in the wall temperature occurs terminating with a couple of prominent narrow peaks followed by a very steep runaway temperature condition. The peaks indicate instability caused by the onset of boiling at this particular node location. The run-away temperature condition is the very rapid voiding condition in the porous tube after the OOB. Compared to the case of the plain tube provided in the plot, there is no prolonged convection and nucleate boiling regime plateau after the OOB. For the case of the tube with 88 percent porous mesh, the phenomenon is similar except that the single OOB peak is well demarcated and the follow-up boiling regime is still short but a little more lingering than for the case of the tube with 76 percent porous mesh. Hence there appears to be some porosity effect on the apparent width of the boiling plateau.

In addition, the effective heat transfer coefficient, h_{eff}, has been plotted against the mass flow rate. To illustrate the behavior, a composite plot is presented in Fig. 17 for the 76, 82 and 88 percent porosity conditions, for the tubes with both the brazed and non-brazed mesh. Also shown in the same plot are the plain tube test data and the classic Sieder-Tate correlation for a plain tube. Data plots are only shown for the 300 W total heater power case, but trends are similar for all the power test cases run. As the mass flow rate of LN2 through the tube is reduced in a continuous manner at the same heater power and pressure, the heat transfer coefficient is reduced commensurate with some form of the Reynolds number dependence. At a certain point, which is approximately 0.6 kg/min mass flow rate for this chosen case, OOB takes place, signaled by a sharp rise in the value of h_{eff}. However as noted earlier a boiling instability sets in quickly as

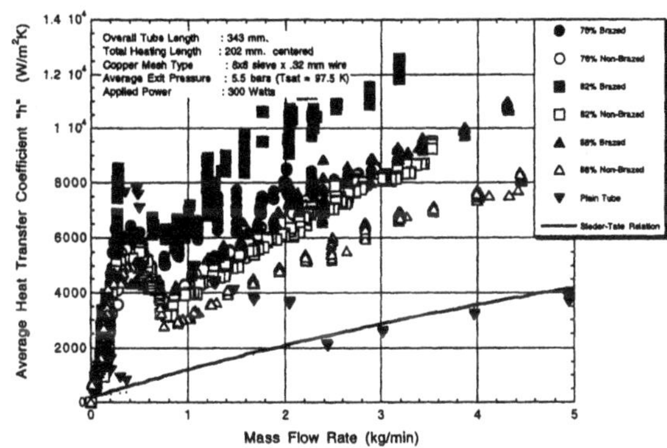

Figure 17. Sample case of the average heat transfer coefficient data vs. liquid nitrogen mass flow rate for a 9.5-mm tube with porous mesh inserts at 300 W power levels.

demarcated by a sharp dome. Subsequently, with further reduction in the mass flow rate, the critical heat flux condition is reached as indicated by where a very sudden drop in the value of h_{eff}. This drop corresponds to the runaway wall temperature condition, at which point the test was terminated. Under the test conditions applied, ONB for the plain tube is seen to take place at about 2.4 kg/min mass flow rate. The critical heat condition is reached near 0.8 kg/min mass flow rate. At about 0.5 kg/min mass flow, the wall temperature runaway condition is experienced. In the mass flow rate range from 2.4 to 0.8 kg/min, the plain tube is stable in the convective nucleate boiling regime. This is beneficial if one is interested to stably cool a heated object with LN2.

Both the tubes with brazed and non-brazed porous mesh have superior heat transfer performance in single-phase LN2 cooling relative to that of the plain tube. Again the brazed porous mesh is superior, in all the cases, to the non-brazed porous mesh. To give some quantitative values, let us focus on the 3.0 kg/min LN2 flow in Fig. 18. The plain tube in single-phase yields an h_{eff} of about 2,800 W/m² K. For the tube with the brazed mesh, h_{eff} is in the range 8,000 to 12,000 W/m² K for the three porosities tested. For the tube with the non-brazed mesh this range is lower, about 6,000 to 8,000 W/m² K. Relative to a plain tube with boiling the heat transfer enhancement attained with the porous mesh is significant, i.e., up to a 4-fold increase.

Still another observation from Fig. 18 is that while the plot of the data for the tubes with the brazed mesh exhibits wider scatter, the plot of the data for the tubes with the non-brazed porous mesh is relatively smooth at all power levels. This is not unexpected since brazing of a porous mesh inside a tube, no matter how skillfully done, is difficult to control for uniformity especially at the wall interface. Hence a near-wall effect is introduced.

6.1 POROUS MEDIUM BOILING HEAT TRANSFER CORRELATIONS

There are no well-established heat transfer correlations in the literature. for porous medium Several tentative heat transfer correlations have been offered in the form of Nusselt vs. Peclet numbers with or without the explicit term involving porosity [21, 22 and 23]. For forms involving an explicit porosity term, Nu and Pe are based on the hydraulic diameter of the flow tube. For non-explicit forms, Nu and Pe are

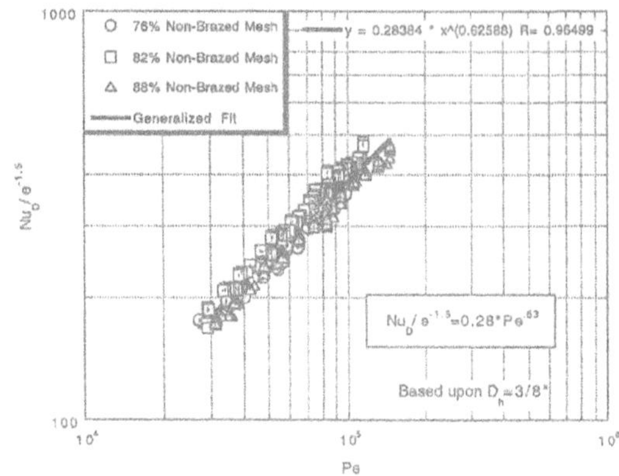

Figure 18. Correlation of reduced Nusselt number vs. Peclet number for liquid nitrogen heat transfer in tubes with mesh inserts of various porosities (based on tube hydraulic diameter).

defined by a characteristic pore size. Thus the porosity effect is included in the definition of Nu and Pe. AS indicated below, Gortyshov and coworkers [21 and 22] offered two correlations explicitly involving the porosity effect.

With porous ceramic inserts:

$$Nu_D = 0.325\, Re^{0.65}\, Pr^{0.56}\, \epsilon^{-5.6} / Pr_w^{0.14}, \tag{9}$$

and, with metallic foam inserts:

$$Nu_d = 0.606\, Pe^{0.56}\, \epsilon^{-5.2} \quad \text{(claimed to be within 12\% error RMS).} \tag{10}$$

Of these two correlations, the more relevant latter form has been checked with our experimental data. The characteristic pore size was conveniently determined from a consideration of the cross-sectional geometry of the mesh fabric rolled and pressed into the final insert to achieve the desired porosity. The resulting pore dimensions are indicated in the figures. Repeated data fits strongly suggested that the best fit was obtained in the above form as long as the explicit porosity term was totally omitted. The porosity effect is already folded into the relationship through using characteristic pore size in defining Nu and Re. If, however, the flow tube hydraulic diameter is used instead, then the explicit porosity term as suggested by Gorytshov was needed. The results are shown in Figs. 19 and 20 for the tubes with the non-brazed porous insert using both correlative forms, respectively:

$$Nu_D/\epsilon^{-1.5} = 0.28\, Pe^{0.63} \qquad \text{and,} \tag{11}$$

$$Nu_d = 0.135\, Pe^{0.62} \tag{12}$$

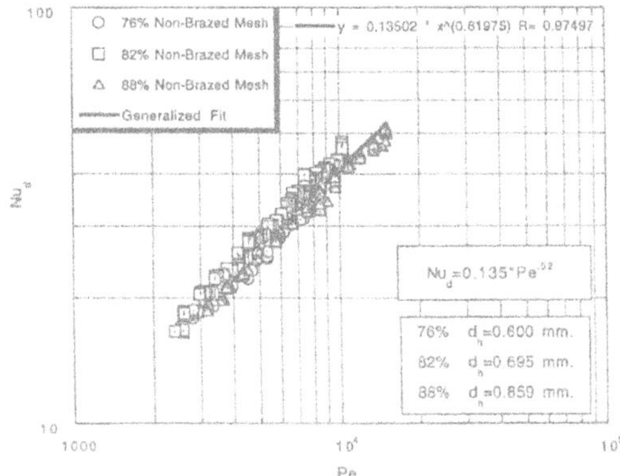

Figure 19. Correlation of Nusselt number vs. Peclet number for liquid nitrogen heat transfer in tubes with mesh inserts of various porosities (based on pore hydraulic diameter).

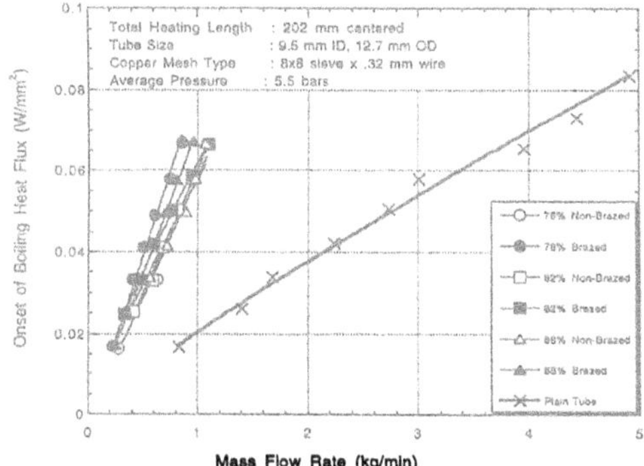

Figure 20. Onset of boiling heat flux vs. liquid nitrogen mass flow rate in tubes with inserts of various porosities.

In either form the fit is good within R=97 percent RMS. But the data fit form is not as good for the tubes with the brazed inserts. The best fit to the data is represented by:

$$Nu_d = 0.727 \; Pe^{0.467} \tag{13}$$

with a fit goodness factor of R= 87 percent RMS. The larger deviation to the fit should be expected because of the much larger data scatter in brazed tube tests as opposed to the data with the non-brazed tube. It is suspected that the previously alluded near-wall effects are responsible for the increased data scatter.

7.   Heat Transfer Enhancement Using Porous Mesh

Porous mesh inserts in flow channels result in large heat transfer enhancement, which has been very well established. However in cryogenic applications, as in optical crystal cooling, emphasis has been shifted from enhancement level to maximum practicable flux level such that cryogenic cooling is viable without going into full boiling due to the low heat flux levels associated with cryogenics at the onset of boiling.

An unpublished literature review [24] revealed a very limited number of investigations that analyzed heat transfer in forced convective boiling flow in tubes/channels with porous inserts of different configurations. In chronological order, these are the investigations reported by Megerlin *et al.* [1], Gortyshov *et al.* [22] and Kuznetsov and Chikov [23]. Megerlin and coworkers investigated single-phase flow as well as forced convective boiling flow heat transfer in stainless steel tubes with stainless steel mesh and brush inserts. In single-phase, they reported heat transfer rates increased up to nine times in convective heat transfer in tubes with mesh inserts as compared to the case of plain tubes. They also reported that the burnout heat fluxes were 2 to 3 times the values for plain tubes for the same mass fluxes. Roizen *et al.* [25] reported work in which the porous inserts were made of highly permeable porous cellular porcelain. They too found that the critical heat fluxes accompanying boiling in a channel with porous inserts was 2.5 to 3 times higher than the values in channels with no inserts for the same mass fluxes. And finally, Kuznetsov and Chikov [23] also stated that " . . . the critical heat flux in a porous channel is several times higher than that in a smooth one when dealing with high velocity flows in short channels." It should be emphasized that no such comparisons are available regarding the onset of subcooled nucleate boiling. Here emphasis is on the ONB point because, once this is reached, there is a very short operational span before the tube will be substantially voided. Hence, with tubes having porous inserts in forced convection, the farthest point one can comfortably operate is the subcooled nucleate boiling point.

Results from the test data on the OOB point heat flux versus the mass flow rate are plotted in Fig. 20 for all test tubes. The plot suggests that, all the OOB data for the tubes with non-brazed porous inserts are almost independent of porosity. Hence reviewing a case at 1 kg/min mass flow rate, the maximum boiling heat flux is about 0.06 W/mm^2. The OOB for the plain tube occurs at 0.02 W/mm^2. Thus we see a factor of three enhancement in heat transfer over that for a plain tube, in agreement with the conclusion reached by other researchers cited above. A stronger porosity effect was found for tubes with brazed porous inserts. Heat transfer enhancement at OOB is certainly larger than 3-fold relative to a plain tube, and, for the tubes with 76% porosity, it is nearly a 4-fold enhancement. The OOB heat flux for the tube with the 76% porous insert is about 0.08 W/mm^2.

A recent study examined a LN2 cooled silicon crystal using our 6.4-mm (0.25-inch) porous copper inserts [26]. Seven such porous inserts mechanically fitted into the cooling channels, 60 mm long, were used to cool the crystal during an x-ray beam experiment. The data showed that, by using LN2 flow in the range 10.4 to 13.1 l/min for the crystal, up to 1800 W total power from the x-ray beam could be supported before ONB was detected. (the monochromator absorbed at least 80 percent of the incident power.) The average heat flux on the surface under the beam was as high as 19 W/ mm^2. The input data translate into an average coolant channel wall heat flux of about 1.44 W/mm^2 and a heat transfer coefficient of about 0.26 W/cm^{2}K estimated from the Dittus-Boelter correlation. Obviously, a significant enhancement of the heat transfer coefficient was achieved using the inserts in the 6.4-mm-diameter channels to support the heat flux levels cited in the study without the OOB. The study concluded that, for the same flow rate, a cryo-cooled silicon crystal with 76% porous inserts would support a 3-fold higher total heat load compared to a crystal with plain channels before OOB was encountered (1800 W versus 600 W). In an unpublished work [24] based on the existing literature, the authors also concluded that, with the 76% porous 9.5-mm channels, an average channel wall heat flux of about 1.42 W/ mm^2 could be supported in the subcooled boiling regime with LN2 at a 13.5 l/min flow rate (9.54 kg/min). For high-precision optical work, this is of course operationally unacceptable because of the inevitable flow jitter at such high flow speeds (2.7 m/s).

However, it may be acceptable in applications that are not sensitive to flow-induced vibrations. For enhanced heat transfer with porous inserts in a 9.5-mm-diameter channel, one may suggest a practical operational range of 3.2 to 4.7 kg/min LN2 flow per channel, which should support 0.8 to 1.0 W/ mm^2 coolant wall heat flux without the voiding concerns.

In conclusion, the tubes with non-brazed porous inserts require much higher boiling heat flux relative to a plain tube under the same operational conditions to initiate boiling. Tubes with brazed porous inserts require an even higher, up to 4-fold, increase in heat flux to initiate boiling.

8. Heat Transfer Enhancement Versus Pressure Drop Considerations

No heat transfer enhancement is complete without discussing the attendant pressure drop considerations, which is addressed in the following. In single-phase water cooling, our engineering applications have been optimized using porous copper mesh with 76% porosity. The typical 9-mm flow channel we use has a rather larger pressure drop, about 7 kPa/cm at about 0.1 kg/s mass flow of deionized water (DIW). This yields significant enhancement of the convective heat transfer coefficient at about 2.8 W/cm^{2}K. Our optimization has been driven to achieve the required heat transfer enhancement with minimal deionized water use rather than for minimization of the pressure drop (pumping power). Outside this specific choice for our

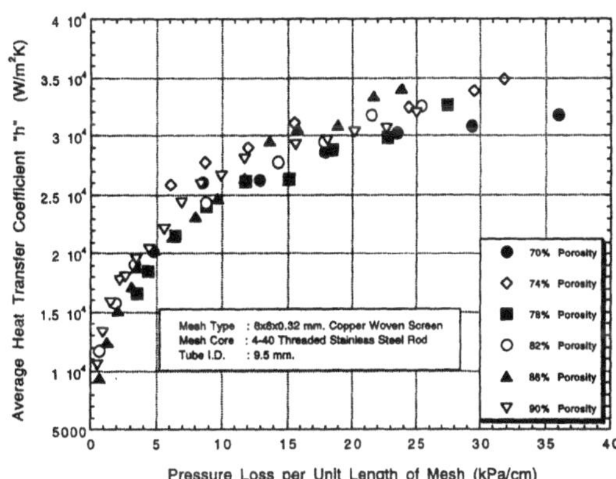

Figure 21. Average heat transfer coefficient with water vs. pressure loss per unit length in tubes with mesh inserts of various porosities from 70 to 90 percent.

applications, the heat transfer coefficient versus the pressure drop relationship is presented in Fig.21 for all the test cases in the study with water. Data in Fig.21 suggests that effect of porosity variation in the range from 70 to 90 percent on enhancement of heat transfer is rather limited. In this range, the maximum "Δh" increment achievable appears to be within 500 W/m^{2}K at all specific pressure drop levels.

In the optical element cooling case using a cryogen like LN2, the inherent poor heat transfer capacity and low boiling heat flux of LN2 are the primary concerns. Again maximum heat transfer enhancement is the prime consideration rather than optimization of the pumping power. Figure 22 has been plotted with this optimization criterion in mind and shows the heat transfer coefficient versus the pressure drop in a typical 9.5-mm flow channel. First, we observe that, for the same pressure drop, the tubes with brazed porous inserts will have a higher heat transfer coefficient relative to the tubes with non-brazed

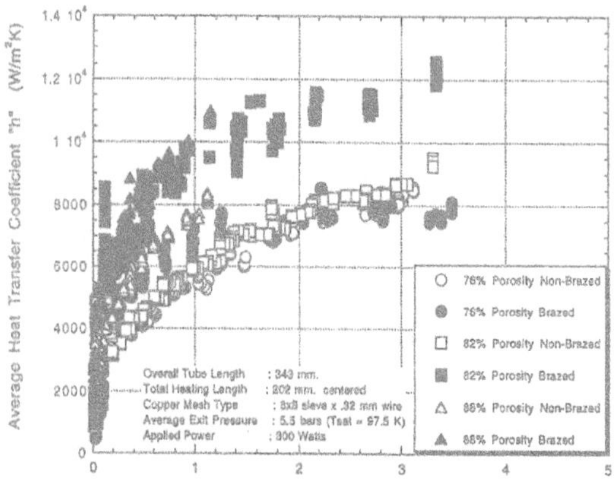

Figure 22. Average heat transfer coefficient with liquid nitrogen vs. pressure loss per unit length in tubes with mesh inserts of various porosities from 76 to 88 percent.

porous inserts. This difference is as much as 50% if curve fits were passed through the test data, which exhibit sizable scatter, particularly with the brazed tubes. The relative increase in the magnitude of the heat transfer coefficient levels off at about the 200 kPa/m pressure drop level. While this level of pressure drop appears to be very large in applications with long cooling channels and almost not sustainable for even subcooled LN2 operations, in real life applications, it is, in fact, very usable. To give an example, in a typical application of optics cooling, a monochromator crystal will be of the order of 15 cm long. To achieve a high effective heat transfer coefficient, say, above 1 W/cm^2 K, the pressure drop through the optics will be of the order of 30 kPa.

9. Conclusions

Conductive porous inserts in flow channels achieve superior heat transfer enhancement in special applications. We have examined heat transfer with deionized water and liquid nitrogen in tubes that have brazed or non-brazed porous copper matrix inserts with varying porosities. The intent here has been to understand the mechanics of heat transfer and pressure drop in porous tubes with convective cooling. Pressure drop and heat transfer correlations have been generated for a set of porosity values in a 9.5-mm ID tube (also a 6-mm ID tube in cryogenic studies).

Heat transfer experiments prove that the insertion of porous copper mesh into plain tubes enhances the convective heat transfer coefficient significantly. With water cooling, the enhancement achievable in single-phase heat transfer is virtually unlimited and totally a function of the pressure drop supportable in a particular application. In our particular application, an optimal enhancement level of about 6-7 fold has been achieved while accepting a tolerable pressure drop.

However in the case of cryogenic cooling, the achievable heat transfer enhancement is bound by the onset of boiling (ONB) in porous tubes. Unlike the case of the plain tube, once the boiling is initiated, such tubes tend to void very rapidly with practicably no sustainable margin in the nucleate boiling regime, for example. In boiling, with tubes in which the porous insert is brazed to the tube wall for the best thermal contact, the heat transfer enhancement is found to be on the order of four-fold relative to a plain tube. This is of the same order of magnitude, albeit somewhat higher, that the few studies in this area had shown previously. Certainly this enhancement and the accompanying pressure drop are related to the porosity. While the lower porosity has the highest specific heat transfer enhancement, optimized against the specific pressure drop, the higher porosity tube will give a better heat transfer coefficient enhancement.

The onset of nucleate boiling (ONB) in plain tubes is found to be very predictable with the existing correlations. No such correlations exist for the tubes with porous inserts.

Porous matrix inserts offer a significant advantage in cooling, providing a jitter-free operation and a much higher effective heat transfer coefficient, at grossly reduced flow rates relative to plain tubes and finned tubes. The jitter-free characteristic of fluid flow with porous insert makes them eminently useful where flow-induced vibration is a concern in the performance of devices such as high-precision optical devices like mirrors and monochromators.

Acknowledgments

Use of the Advanced Photon Source was supported by the U.S. Department of Energy, Basic Energy Sciences, Office of Energy Research, under Contract No. W-31-109-Eng-38.

Nomenclature

c_p	specific heat, (J/kg K)		t	time, (s)
C_L, C_T	coefficients in Eq. (3)		T	temperature, (^oC)
d	pore size, (m)		u	superficial velocity, (m/s)
d_f	fiber diameter, (m)		u_o	superficial porous core velocity, (m/s)
f	friction coefficient			
G	mass flow rate, (kg/min)		*Greek letters*	
h	heat transfer coefficient, (W/m^2K)		ε	porosity
k	thermal conductivity, $(W/m\ K)$			
$LN2$	Liquid nitrogen		*Subscripts*	
l	length of a tube, (m)		eff	effective
Nu	Nusselt number		f	fluid phase
p	pressure, (Pa)		i	inlet
Pe	Peclet number		l	saturated liquid
q_w''	wall heat flux, (W/m^2)		m	mean (bulk)
Re	Reynolds number		s	solid phase
			sat	saturation
			w	inner wall

References

1. Megerlin, F. E. Murphy, R.W. and Bergles, A.E., (1974) "Augmentation of Heat Transfer in Tubes by Use of Mesh and Brush Inserts," Trans. ASME J. Heat Transfer, Vol.96, pp.145-151
2. Apollonov, V.V. et al., (1978) "Increase in the Thresholds for Optical Failure of Metallic Mirror Surfaces during Cooling through Structure with Open Porosity," Pis'ma Zh. Tekh. Fiz., 4, No.19, pp.1193-2297

3.	Polyaev, V. M. and Maiorov, V. A., (1982) "Heat Exchange in a Channel with Porous Inserts in a Forced Cooling System," Izv. Vyssh. Uchebn. Zaved., Mashinostr., No.7, pp.51-55

4.	Mairov,V.A. et al., (1984) "Intensification of Convective Heat Exchange in Channels with Porous High-Thermal-Conductivity Filler. I. Heat Exchange with Local Thermal Equlibrium inside the Permeable Matrix," Inzhenerno-Fizicheskii Zhurnal, Vol. 47(1), pp.13-24

5.	Mairov, V.A. et al., (1984) "Intensification of Convective Heat Exchange in Channels with Porous High-Thermal-Conductivity Filler. II. Forced Heat Transfer Regime," Inzhenerno-Fizicheskii Zhurnal, Vol. 47(1), pp.199-205

6.	Apollonov, V.V. et al., (1988) "Thermophysical Principles of Cooled Laser Optics Based on New-Type Penetrable Structures," Experimental Heat Transfer, Fluid Mechanics and Thermodynamics (Eds. R.K.Shah, E. N. Ganic and K.T.Yang), Elsevier Science Publishing Co.

7.	T. M. Kuzay (1990) "Fixed Mask Assembly Research for Advanced Photon Source Insertion Devices," Argonne National Laboratory Report ANL-90/20

8.	Kuzay, T.M., Collins, J.T., Khounsary, A.M., and Morales Gilberto, (1991) "Enhanced Heat Transfer With Wool-Filled Tubes," ASME /JSME 3rd Joint Heat Conference, Book No. I0309E-1991, Reno, Nevada, pp. 145-151

9.	<u>Process of Making Cryogenically Cooled High Thermal Performance Crystal Optics</u>, U.S. Patent 5,123,982, 1992

10.	T. M. Kuzay, (1992) "Cryogenic Cooling of X-ray Crystals Using A Porous Matrix", Rev. Sci. Instrum. 63 (1), 468-472

11.	Lin Zhang, (1993) "Cryogenic Cooled Silicon-Based X-ray Optical Elements- Heat Transfer Limit", SPIE Vol. 1997 High Heat Flux Engineering II, pp. 223-229

12.	B. X. Yang et al., (1993) "Performance analysis of cryogenic silicon Laue monochromators at APS undulators", SPIE Vol. 1997 High Heat Flux Engineering II, pp. 302-315

13.	Z. Wang et al., (1995) "Thermal and deformation analysis of a cryogenically cooled silicon monochromator for high-heat-flux synchrotron sources", Rev. Sci. Instrum. Vol. 66 (2), pp. 2267-2269

14.	Brinkman, H. C., (1947) "A Calculation of the Viscous Force Exerted by a Flowing Fluid on a Dense Swarm of Particles", App. Sci. Res., vol. A1, pp. 27-34

15.	Cai, Z., (1993) "Evaluation of the Non-Darcy Flow through Fibrous Media," J. Material Processing and Manufacturing Science, Vol. 2, pp. 19-39

16.	M. Sözen, T. M. Kuzay, (1996) "Enhanced Heat Transfer in Round Tubes with Porous Insert," Int. J. Heat and Fluid Flow, Vol. 17, 124-129

17.	Koh, J. C. Y. and Fortini, A., (1971) "Thermal Conductivity and Electrical Resistivity of Porous Material," Topical Report, NASA CR-120854

18.	T.M. Kuzay, J.T. Collins and J. Koons, "Boiling Heat Transfer in Channels with Porous Copper Inserts," accepted for publication in Int. J. Heat & Mass Trans.

19.	Van P. Carey, (1992) *Liquid-vapor Phase-change Phenomena*, Hemisphere Publishing, pp.491-506.

20.	Collier, John. G., (1972) *Convective Boiling and Condensation*, McGraw Hill

21.	Yu.F. Gortyshov, G.B.Murav'ev, and I.N.Nadyrov, (1987) "Experimental Study of Flow and Heat Exchange in Highly Porous Structures," Engng -Phys. Journal, 53, 357

22.	Gortyshov, Y. F., Nadyrov, I. N., Ashikmin, S. R. and Kunevich, A. P., (1991) "Heat Transfer in the Flow of a Single-Phase and Boiling Coolant in a Channel with a Porous Insert," J. Engineering Physics and Thermophysics (trans. from Russian), vol. 60(2), pp. 202-207

23.	Kuznetsov, V. V. and Chikov, S. B., (1993) "Heat Transfer under Conditions of Flow along a Fuel Element Placed in a Porous Medium," High Temperature (trans. from Russian), vol. 31(2), pp. 242-246

24.	M. Sözen and T. M. Kuzay, (1995) "A study of cryo-cooling of high heat flux components by the use of cooling channels with porous inserts," Unpublished
25.	Roizen, L. I., Rachitskii, D. G., Rubin, I. R., Vertogradskaya, L. M., Yudina, L. A. and Pypkina, M. B., (1982) "Heat Transfer with Boiling of Nitrogen and Freon-113 on Porous Metallic Coatings," High Temperature (trans. from Russian) Vol. 20(2), pp. 264-270
26.	C. S. Rogers, D. M. Mills, and W. -K. Lee, (1995) "Performance of a liquid-nitrogen-cooled, thin silicon crystal monochromator on a high-power, focused wigler synchrotron beam", Rev. Sci. Instrum, 66(6), 3494-3499

BOILING ON STRUCTURED SURFACES

RALPH L. WEBB
Penn State University
University Park, PA, U.S.

LIANG-HAN CHIEN
Chang-Gung University
Kwei-Shan, Taiwan ROC

Abstract. This paper describes a series of studies to understand the mechanism of boiling in structured surfaces having sub-surface tunnels, and surface pores. Innovative visualization that allowed observation within the tunnels conclusively shows that in saturated boiling, the tunnel is vapor filled, except for thin liquid films on the tunnel walls and menisci in the corners. Evaporation on liquid menisci in the tunnel corners is the principal boiling mechanism for the structured surfaces. Experiments were performed to define the effect of pore diameter, pore pitch, and tunnel size on performance in structured boiling surfaces. The Dry-Out Heat Flux increases with the increase of total open area. At a certain reduced heat flux, part of the tunnel will become flooded and the performance will be reduced. Smaller pore size will inhibit flooding at reduced heat flux. Visualization experiments were performed to determine the bubble departure diameter, bubble frequency, waiting and growth periods, and nucleation site density. These data provided the basis for models to predict these parameters as a function of surface geometry and fluid properties. A mechanistically based model was developed to predict the boiling heat flux as a function of $(T_w - T_s)$, pore and tunnel dimensions, and fluid properties. The model predicted the heat transfer data for R-11, R-123, R-134a, and R-22 within ±33% (0.20 MSD).

1.   Introduction

Perhaps the most significant advances in enhanced heat transfer technology have been made in special surface geometries that promote high-performance nucleate boiling. These special surface geometries provide an increased heat transfer coefficient. Although the surface area may be increased, it is common to define the boiling coefficient in terms of the projected, or flat plate, surface area. The fact that "roughness" can improve nucleate boiling performance has been known for 60 years. For nearly 35 years this was regarded as an interesting, but commercially unusable

S. Kakaç et al. (eds.), Heat Transfer Enhancement of Heat Exchangers, 249–284.
© 1999 *Kluwer Academic Publishers.*

concept because the heat transfer improvement lasted for only a matter of hours before "aging" caused the performance to decay to the plain surface value. In the period 1955-1965, fundamental advances were made in understanding the character of nucleation sites and the shape necessary to form stable vapor traps. This understanding provided a basis for industrial research [1960-1975] that led to the development of commercially viable enhanced surface geometries. The first of these high performance geometries was patented in 1968. In 1980 six nucleate boiling surface geometries were commercially available. These have generally been used in the refrigeration and process industries.

Two basic types of enhanced boiling surfaces have been developed. These are porous and structured surfaces. High boiling performance can be obtained using either approach. However, the structured surfaces have found the greatest commercial application, principally because the cost is lower. Structured surfaces on tubes use cold metal working to form a radial finned tube having a high area density of reentrant nucleation sites, which are interconnected below the surface. Typically, these are reentrant grooves or tunnels, rather than discrete cavities. Webb [1] and Thome [2] survey the techniques used to develop enhanced boiling surfaces, which include the Wieland GEWA-TW tube (Fig. 1a), Hitachi Thermoexel-E (Fig. 1b) surface, Wolverine Turbo-B surface(Fig. 1c),and the Trane "bent fin" surface (Fig. 1e). All of these surfaces have interconnected sub-surface tunnels and surface pores (or narrow gaps at the surface). The author proposes that the Thermoexel-E, Turbo-B, and Turbo-B II geometries operate by the same mechanism. It is probable that the mechanism of the GEWA-TW and bent fin geometries having narrow fin gap (s_g) operates by a closely related mechanism. Two key geometric characteristics of the surfaces are the (1) Subsurface tunnels and (2) Surface pores or fin gaps. These key features may be further defined by the following geometric parameters:

1. Subsurface Tunnels: Tunnel pitch (P_t), tunnel height (H_t), tunnel width (W_t), and tunnel base radius (R_b), tunnel shape.
2. Surface Pores (or Gaps): Pore diameter (d_p), and pore pitch (P_p).

Thus, a geometry having surface pores has six geometrical dimensions, plus the tunnel shape. Figure 1b illustrates a generic surface having subsurface tunnels and surface pores, which may be defined by the above dimensions. The commercial enhanced tubes do not have such precisely defined enhancement dimensions. Figure 2 shows an enlarged photograph of the Wolverine Turbo-B II tube. This figure shows that the surface pores are of an elliptical shape. One possibility for measurement of the pore diameter is to use the minor diameter of the ellipse. Experimental data have been taken on flat plate geometries having subsurface tunnels and surface pores. These include Nakayama et al. [3,4] who tested a Fig. 1b type surface which was made by soldering a thin copper sheet onto a horizontal finned plate. Pores were made in the copper sheet above the rectangular tunnels. The tunnel depth in their tests is 0.4-to-0.6 mm. Water, R-11 and liquid nitrogen were used as working fluids. Nakayama et al. [4] compared surfaces having pores of uniform pore size and mixed pore size (d_p = 0.05-to-0.15 mm). They found that the surface pores having a uniform pore diameter (d_p = 0.1 mm) had the best boiling performance for R-11. The effect of pore pitch, tunnel size and tunnel shapes were not investigated.

Arshad and Thome [5] tested a planar surface similar to the Nakayama et al. [3] surface, for which they also varied the tunnel cross section shape. They soldered micro-drilled cover plates on top of grooved surfaces and tested circular, rectangular, and triangular shape tunnels. They found the surfaces having 0.25 mm pore diameter had the best boiling performance. Circular tunnels generally required a larger incipient superheat. Triangular tunnels had high performance at low heat flux, but they dried out at intermediate heat flux.

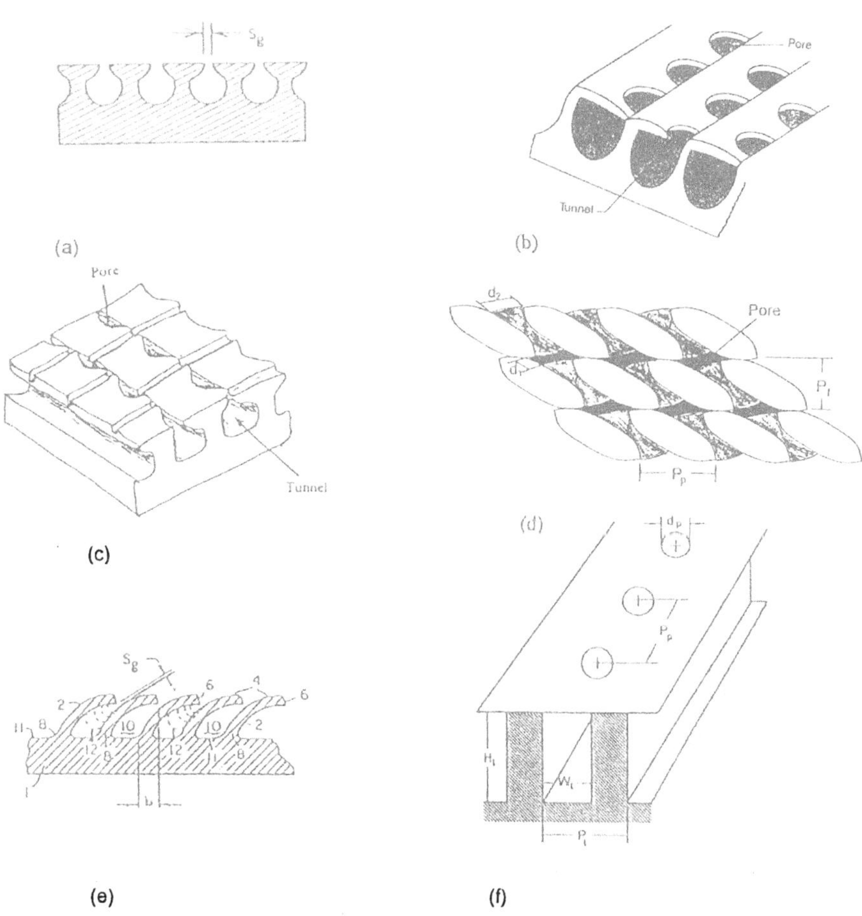

Fig. 1   Illustration of enhanced nucleate boiling surfaces: (a) Wieland GEWA-TW, (b) Hitachi Thermoexcel-E, (c) Wolverine Turbo-B, (d) Wolverine Turbo-B II, (e)Trane Bent fin, (f) Surface geometry studied (from [17] Figure 1.1)

252

Perspective side view

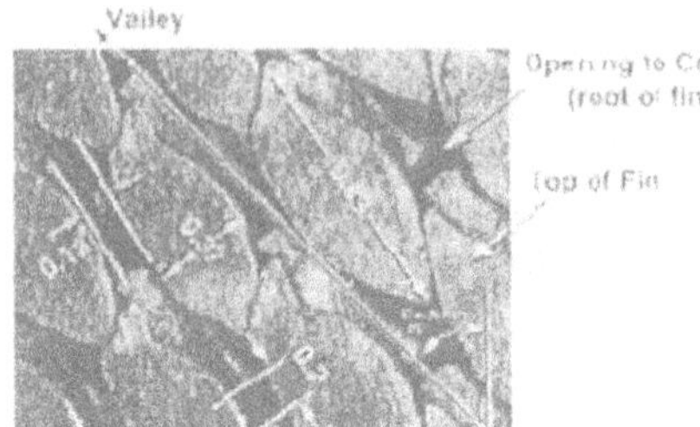

Top view

Fig. 2 An enlarged photograph of the Wolverine Turbo-B II tube.

Chien and Webb [6, 7, 8, 9, 10] have recently published a series of papers that provide new understanding of the boiling mechanism in the Fig. 1b tunneled surface, and of the effect of the pore and tunnel dimensions, and the tunnel shape on the surface performance. These results are summarized and discussed in this paper.

Nakayama [11] proposed that three possible boiling mechanisms may exist. These are the "flooded mode," the "suction-evaporation mode," and the "dried-up mode." Nakayama et al. [3] proposed a "suction-evaporation mode" model. This model assumed evaporation on menisci in the corners of the tunnel. Haider [12] and Webb and Haider [13] developed an analytical model of boiling in tunneled surfaces, based on the "flooded mode," which assumes alternate zones of liquid slugs and vapor plugs in the sub-surface tunnels. Chien and Webb [10] developed an improved semi-analytical model, which applied the Webb and Haider [13] model to account for the dynamic behavior of nucleate boiling and assumed suction-evaporation mode inside the subsurface tunnels.

2. Visualization Experiments and Boiling Mechanism

Understanding of the boiling mechanism must result from visualization experiments of the boiling process. This is very difficult to do, because the small surface pores block view of the subsurface tunnels, and because the pores and tunnels are so small.

Nakayama et al. [3] performed the first visualization study to observe the liquid-vapor conditions in the tunnel. Their boiling apparatus consisted of a single tunnel having a thin cover sheet, in which pores were made. This consisted of a base block, two glass side plates, end plates, and a 0.05-mm thick metal lid with a series of pores drilled along its center. These parts formed a rectangular (1 mm × 1 mm) cross-section tunnel immersed in a liquid pool of R-11. They observed the liquid-vapor condition in the tunnel through the glass side walls. Liquid and vapor regions existed in the tunnel before heat was applied. Boiling activation was recorded by a high-speed cine camera, or a still camera oriented for a side view of the tunnel. Tunnel heights varied between 0.5 and 1.0 mm, and the width was held fixed at 1.0 mm. The pore diameters were uniform, ranging from 0.05 to 0.5 mm in different lids tested. They found that the liquid was vaporized and driven from the tunnel through the pores leaving liquid menisci remaining in the corners when heat was applied. They concluded that the flooded mode does not exist when nucleate boiling exists at the pores. These characteristics were observed for all test geometries except for the lid having the largest pores (0.5 mm). In this case, expulsion of the liquid from the tunnel was never complete, and liquid could be seen sloshing around in the tunnel while occupying roughly 10% to 50% of the tunnel volume. Their observations indicate that nucleation will not occur for the flooded mode if the pore diameter is less than 0.5 mm. Based on this study, Nakayama [4] proposed that three possible boiling mechanisms may exist. These are the "flooded mode," the "suction-evaporation mode," and the "dried-up mode." Nakayama et al. [11] concluded that the suction-evaporation mode existed under boiling conditions. They reported no quantitative results on heat flux vs. boiling ΔT_{ws}, because significant heat losses occurred from the glass side walls of the tunnel. Their apparatus contained only one tunnel, the side and top walls were not heat transfer surfaces, and significant heat loss occurred from the tunnel side walls. Their experiment does not simulate the interaction of bubbles that exists in a boiling surface having multiple tunnels. Hence, one cannot conclusively state that their single tunnel observations are applicable to multiple-tunnel surfaces having metal side and top walls in thermal communication.

Arshad and Thome [5] conducted a visualization experiment using an apparatus similar to the one used by Nakayama et al. [11], for which they view the ends of the tunnel. Their apparatus consisted of a brass block having grooves on its top surface. Triangular, rectangular, and circular cross-sectional grooves were tested. Copper cover plates were soldered on top of the grooves and pores were made on the cover plates. They found nucleation always started from one corner of triangular or rectangular tunnels. They conclude evaporation on the thin liquid film is the principal boiling mechanism for the structured surfaces.

2.1 CHIEN AND WEBB VISUALIZATION EXPERIMENTS

It is desirable to visualize boiling on a surface having multiple tunnels, and copper side walls. Chien and Webb [8] performed visualization experiments on finned tube geometry that closely simulates the actual tube geometry. The experiments viewed a 19.5 mm diameter, copper, integral-fin tube having 0.8 mm fin height, and 1575 fins/m. A 0.02 mm thick, transparent polypropylene tape was wrapped over the fins. Adhesive backing on the tape provided adherence to the fin tips. Pores of a fixed diameter were made in transparent plastic (polypropylene) tape using a heated needle attached to the tip of a hot soldering iron. The heated needle melted the plastic material and created

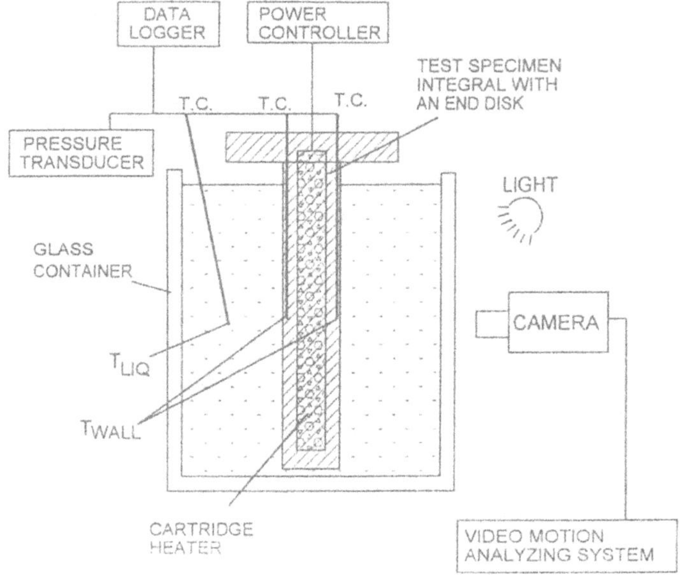

Fig. 3 Apparatus for vertical tube visualization experiments.

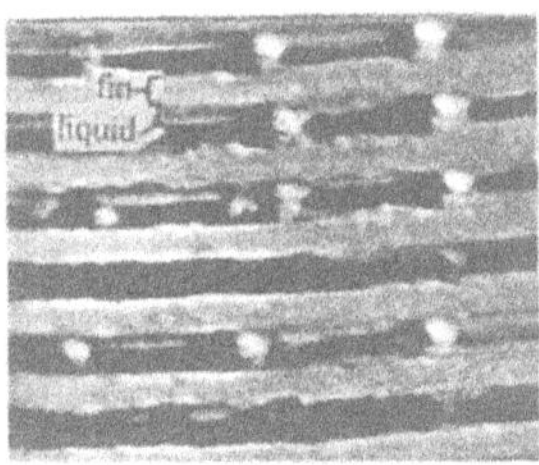

enlarged section

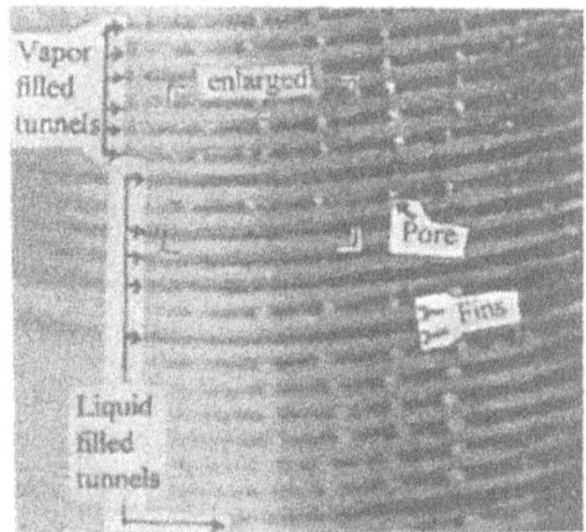

Fig. 4 Saturated boiling on a vertical tube at 1 kW/m^2.

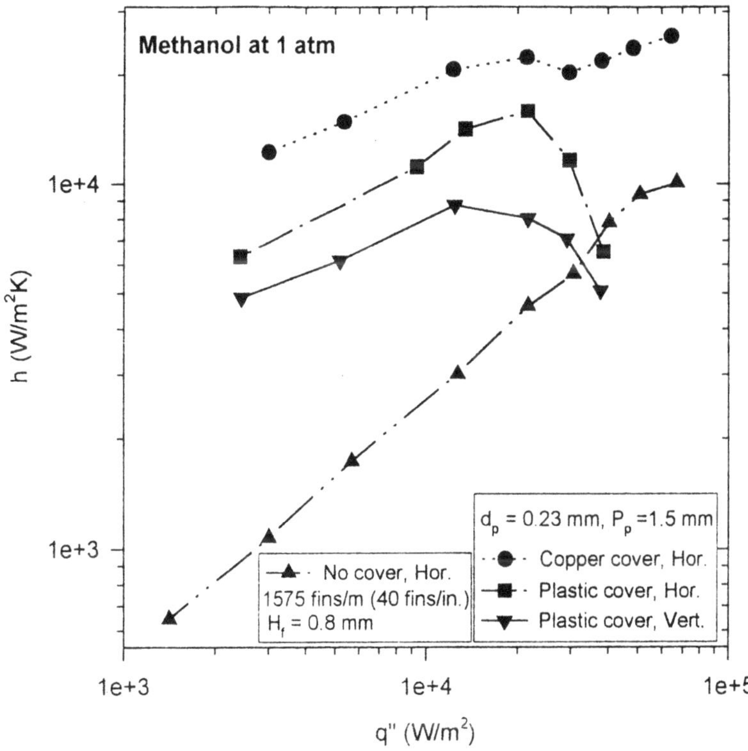

Fig. 5 Boiling performance for plastic and copper foil covers.

one permanent pores of diameter $(d_p) = 0.23$ mm and pore pitch $(P_p) = 1.5$ mm. In case, a 0.05 mm copper foil was soldered on the fin tips. Pores of the same diameter and pitch were also formed in the copper foil wrapped tube. Methanol at 1.0 atm $(T_{sat} = 64.5$ °C) was used as the working fluid. This specially made tube had 9.52 mm inside diameter, in which a 500 W electric cartridge heater 129 mm long was inserted. The tube was tested in either the vertical (or horizontal) orientation using the Fig. 3 test cell described by Chien and Webb [9]. The instrumentation used is described in Section 3.

A Kodak EktaPro 1000 high speed photography system was used to photograph the boiling phenomena in the tunnels. It was also used to measure the bubble formation data. This high speed photography system includes a video camera and amotion analyzing system. In the present experiments, frame speeds between 125 and 1000 frames/sec were used. A Nikon 60 mm micro lens with extension tubes was used on the video camera to facilitate close-up photography. The high speed images were recorded on a special tape, which facilitated play back at slow motion (1-to-30 frames/ sec) by using the motion analyzing system on a CRT to obtain approximately 20-to-30 power of magnification. The high magnification allowed detailed observation of the liquid-vapor activity inside the tunnels. The same lens (Nikon 60 mm micro) was used to take still pictures.

The transparent plastic tape allowed observation of the liquid vapor conditions inside the tunnels, as shown in Figure 4. The vapor and liquid were distinguished by the difference of reflection and the movement of liquid meniscuses. The vapor filled

256

tunnels are brighter than the liquid filled tunnels as shown on Fig. 4, because the reflection from liquid is different than for vapor filled region. For a vapor filled tunnel, the surface pores appear as shining white spots. However, liquid filled tunnels are much darker than those on a vapor filled tunnel, and the surface pores are difficult to see.

Figure 5 shows the boiling performance of the tube for copper foil and the transparent covers having the same pore diameter and pitch. For the same orientation (horizontal), the copper cover provides moderately higher enhancement than the plastic cover. However, the plastic cover provides considerable enhancement compared to the finned tube without a pored cover foil. A key factor causing these differences is that the thermal conductivity of copper foil is much larger than that of the polypropylene film. An additional factor is that the copper foil was deformed into the tunnel and formed re-entrant shape cavities. Because the boiling curves show the same enhancement trend for $q'' < 20$ kW/m^2, their boiling mechanism are expected to be the same. When the tube was tested in the vertical orientation, boiling coefficient is approximately 10-to-20% lower than that of the horizontal tube.

At the low heat flux (1.0 kW/m^2) of Fig. 4, six tunnels are liquid filled and the remaining tunnels were vapor filled. Figure 6 shows boiling at three different heat fluxes, 3, 5, and 10 kW/m^2. Bubble diameters and frequencies were also measured and are reported by Chien and Webb [9]. Figure 6 shows that all tunnels were vapor filled except for the liquid menisci in corners. For decreasing heat flux, the number of active pores and the bubble frequency decreased. At $q'' = 3$ kW/m^2, only one pore was active (Fig. 5c), and all of the tunnels remained vapor filled. When the heat flux was reduced to $q'' < 1$ kW/m^2, all pores became non-active (no bubbles appeared), and the liquid thin film (or menisci) became so thick that the top and bottom liquid films merged into one liquid slug. Then, the liquid invaded the tunnels and filled some of the tunnels. At $q'' = 0$, most tunnels became liquid filled.

Figure 7 shows the boiling on a horizontal tube. The fins on the tube were made with a two lead thread. Therefore, alternate tunnels are physically separated. At low heat flux ($q'' = 5$ kW/m^2), one long vapor region filled 2-to-5 internally connected tunnels, as shown on Fig. 7c. As indicated on the left portion of Fig. 6c, one set of four vapor filled tunnels alternate with another set of liquid filled tunnels. The four vapor filled tunnels were internally connected and did not connect with the liquid filled tunnel between two vapor filled tunnels. A liquid meniscus, which separated the vapor filled and liquid filled regions, oscillated up and down at the end of a long vapor region.

Figure 7 illustrates a fin tunnel, that partly liquid filled and partly vapor filled. In Fig. 7, bubbles emerged from the vapor filled regions. Two oscillating liquid menisci exist at the two ends of the vapor region. When a spherical bubble formed at a pore near the top of the tube, the liquid-vapor interface moved up. As the bubble departed, a liquid-vapor interface moved down immediately, and the length of the liquid region decreased. The length of the vapor filled region is typically more than twice that of the tube perimeter. Therefore, the area of menisci at the two ends of the vapor region were much smaller than the area of the menisci in the corners of the tunnel along the tunnel length.

2.2 BOILING MECHANISM

The visualization experiments of Chien and Webb [8] clearly show that for saturated boiling, an active tunnel is vapor filled with liquid menisci in the corners. The principal boiling mechanism is evaporation from liquid menisci in the corners of the

Fig. 6 Saturated boiling of methanol on a vertical tube at 1 atm: (a) 10 kW/m², (b) 5 kW/m², (c) 3 kW/m².

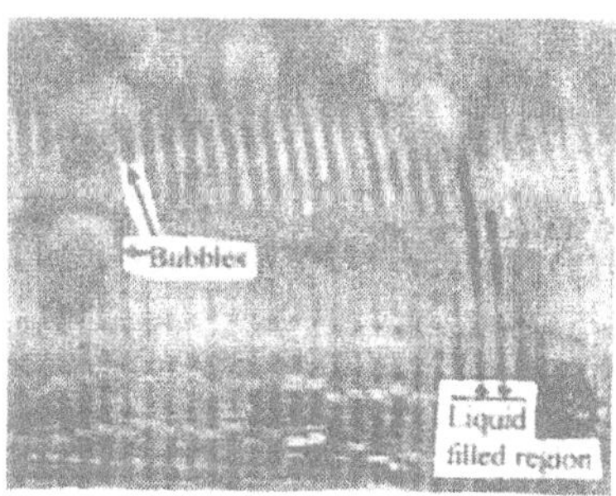

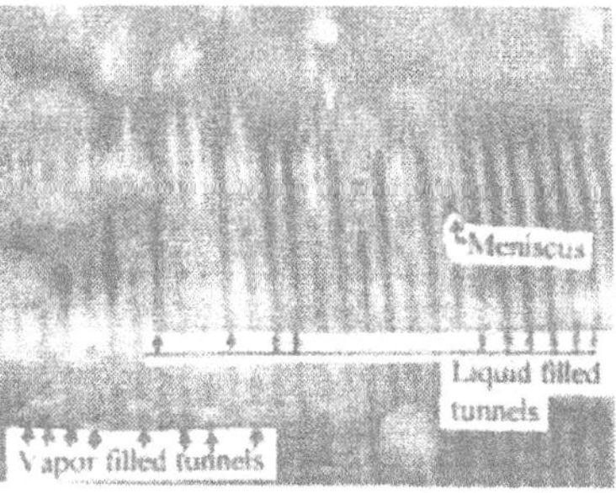

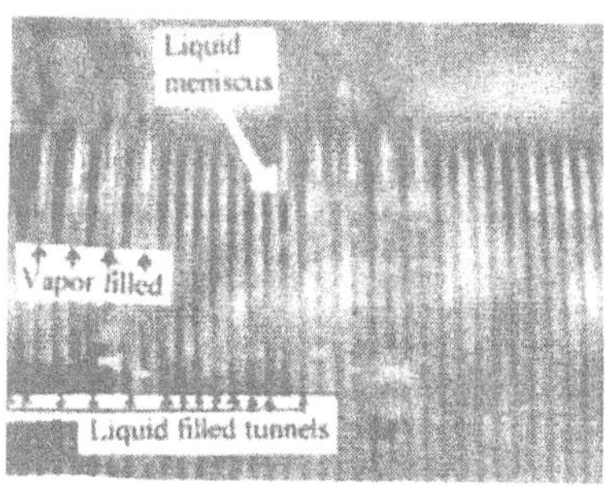

Fig. 7 Saturated boiling of methanol on a horizontal tube at 1 atm: (a) 10 kW/m², (b) 5 kW/m², © 3 kW/m².

tunnel. Liquid is sucked into the tunnel when a bubble departs, and the liquid is then pulled into the corners by surface tension force. The present observations support the existence of the suction-evaporation boiling model as described by Nakayama et al. [11].

3. Chien and Webb Parametric Boiling Studies

The tests were performed by boiling R-11, R-123, R-134a or R-22 on a low, integral-fin tube which was wrapped with a copper foil. The pores were made by piercing the copper foil using a needle. Tunnels were created by the copper foil and the fins. The tunnel pitch and the tunnel height are equal to the fin pitch and height, respectively. Pores of specific diameters (d_p) were pierced in the foil at a specific pore pitch (P_p). The tunnel dimensions were varied by using integral-fin tubes of different fin pitch and fin height.

3.1 ENHANCEMENT GEOMETRIES TESTED

Table 1 summarizes the geometric parameters and test ranges investigated by Chien and Webb [6,7]. The tunnel dimensions $(P_t, H_t,$ and $W_t)$ were defined in Figure 1f, and Figures 8a and 8b illustrate the two tunnel shapes investigated - rectangular and circular. For a circular fin base, the tunnel width is defined as the fin spacing measured at the half fin height as shown on Fig. 7c. Only part of the total investigation of Chien and Webb [6, 7] will be discussed here.

Integral fin tubes were used to provide the tunnel shapes given in Table 1, which includes both the Figure 7a and 7b shapes. The tube was wrapped with the 0.050 mm copper foil and holes were pierced to provide pore pitches of 0.75, 1.5, or 3.0 mm, and pore diameters of 0.12, 0.18, 0.23, or 0.28 mm. R-11 or R-123 were used as the working fluid. Chien et al. [19] provide additional data for R-134a and R-22.

Table 1. Tube Code and Specifications Tested by Chien and Webb [6,7]

Tube code	Fin base shape	Fins/m	P_f (mm)	W_t (mm)	H_f (mm)
1378-0.9	Circ.	1378	0.73	0.40	0.90
1378-0.5	Circ.	1378	0.73	0.40	0.50
1969-0.9	Circ.	1969	0.51	0.25	0.95
1575-0.6	Circ.	1575	0.64	0.33	0.60
1575r-0.6	Rect.	1575	0.64	0.33	0.60

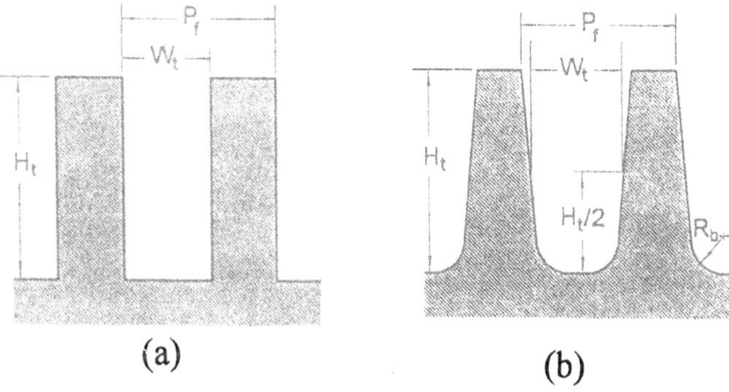

Fig. 8 Tubular surface geometry studied by Chien and Webb [6, 7] (a) Rectangular tunnel shape, (b) Circular fin base.

The first column of Table 1 shows the tube code used to describe the test results. The first number is the fins/m and second is the fin height (mm). The second column shows

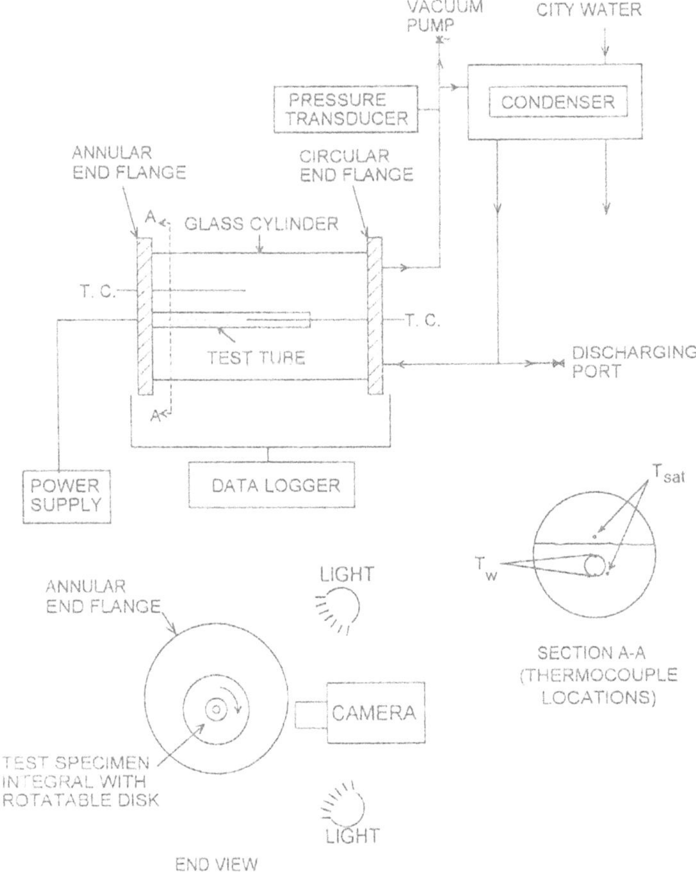

Fig. 9 Schematic of horizontal tube boiling apparatus.

260

the fin base shape. The surface code in the figure legends indicates the tube code followed by the pore diameter and pore pitch. For example, 1969-0.9-0.23-1.5 means a surface, made on a 1969 fins/m and 0.9 mm fin height tube, having 0.23 mm diameter surface pores (d_p) on 1.5 mm pitch (P_p).

A 50 μm (0.002") thick copper sheet was tinned using solder having 165 °C melting temperature. This foil was then wrapped on the finned tube and tied by fiberglass cords. The tube was heated to the solder melting temperature using a 500-W cartridge heater inside the tube. The tube was kept at 190°C and rotated slowly for five minutes to make sure that the solder melted and distributed uniformly. The fiberglass cords were then removed. After completion of the tests and removal of the copper foil, inspection showed that the foil was joined to the fin tips over the entire tube surface area.

3.2 EXPERIMENTAL APPARATUS

The test apparatus is similar to that used by Webb and Pais [15] and is described in detail by Chien and Webb [6]. As shown in Fig. 9, the single-tube pool boiling test cell consists of a 101.6 mm (4.0 in.) outside diameter, 177.8 mm (8.0 in.) long cylindrical glass cell and two brass end flanges. This apparatus allows visualization and photography of the boiling details. One end of the glass cell was clamped with an annular flange. Another end of the glass cell was clamped with a circular end flange. Neoprene gaskets were put between the end flanges and the two ends of the glass cell.

The test tube section was soldered to a 50.8 mm (2 in.) diameter circular brass flange. Two 0.25 mm diameter iron-constantan sheathed thermocouples located below the tube wall at 180 degree opposite positions sensed the tube wall temperature. The brass flange, to which the tube was attached, allowed the test section to be rotated inside the annular end flange. An O-ring seal was used between the two flanges. This

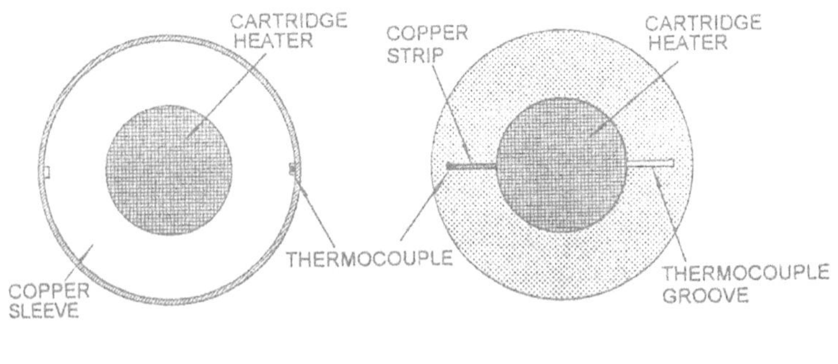

Fig. 10 Thermocouple grooves in tubes.

design allowed measurement of the wall temperature as a function of circumferential angle. The vapor was condensed in an external water-cooled condenser. The test cell pressure was read on a manometer. Two 1.59 mm diameter iron-constantan sheathed thermocouples measured the saturation temperature in the glass boiling vessel. The apparatus also had a charging line, and an evacuation line connected to a positive displacement vacuum pump.

The test tubes were 140 mm long, 9.53 mm inside diameter, and 18-to-19.5 mm outside diameter. A 500-W electric cartridge heater of 9.52 mm diameter and 129 mm length was inserted into the copper tube. Its electric power was controlled by an auto-transformer and measured using a precision voltmeter and ammeter. The heater contains one continuous 101.6 mm long heated section, and two 12.7 mm long unheated sections at each ends. The heater was coated with a silicone-based heat sink compound to enhance thermal contact with the tube.

Two different tube designs (Fig. 10) were used in this work. The different designs were necessary, because different methods were used to make the test tubes. The Method-1 design (Fig. 10a) consisted of a thin wall integral-fin tube and a thick copper sleeve inside the tube. The 15.875 mm inside diameter integral-fin tube had 0.7 mm tube wall thickness. The outside diameter of the copper sleeve is 0.025 mm larger than the inside diameter of the outside tube. The copper sleeve had two 0.51 mm (0.02") wide 0.38 mm (0.15") deep axial thermocouple grooves diametrically opposite in the outer diameter of the copper sleeve. The sleeve was shrink fitted on the tube by cooling it with liquid nitrogen, and then pressing into the outside tube. The Method-1 design was used for the 1378 fins/m tube. Method-2 (Fig. 10b) used a special thick wall tube of 9.52 mm diameter, in which two 0.4 mm wide thermocouple grooves were made by EDM (electric discharging machining). A 0.25 mm sheathed thermocouples were installed in each groove. A 0.3 mm thick copper strip was inserted between the thermocouple and the heater to insure that the thermocouple was located at the base of the groove. The base of the groove is approximately 1.0 mm beneath the root of fins. The measured temperature was corrected for the conduction temperature drop to the tube root diameter. The Method-2 design was used for the 1575 and 1969 fins/m tubes.

The data were taken for 26.7 °C saturation temperature using procedures described by Chien and Webb [6]. The tube wall temperature was determined by extrapolating the thermocouple temperatures to the tube wall using Fourier's law. The transducer and thermocouples were calibrated and checked for repeatability. The uncertainty in the heat transfer coefficient is estimated to be within ±5% at maximum heat flux and within ±10% at low heat flux ($q'' \approx 10$ kW/m²).

3.3 TEST RESULTS (EFFECT OF PORE DIAMETER)

Figure 11 shows boiling curves for different pore diameters using a fixed pore pitch (P_p = 1.5 mm) on a 0.9 mm fin height, 1378 fins/m tube. The surfaces having d_p = 0.18, 0.23 and 0.28 mm give high performance at low heat flux. However, all show a maximum, which indicates the Dry-out-Heat Flux (DHF), and the heat transfer coefficients decrease drastically above their DHF. The DHF is defined as the

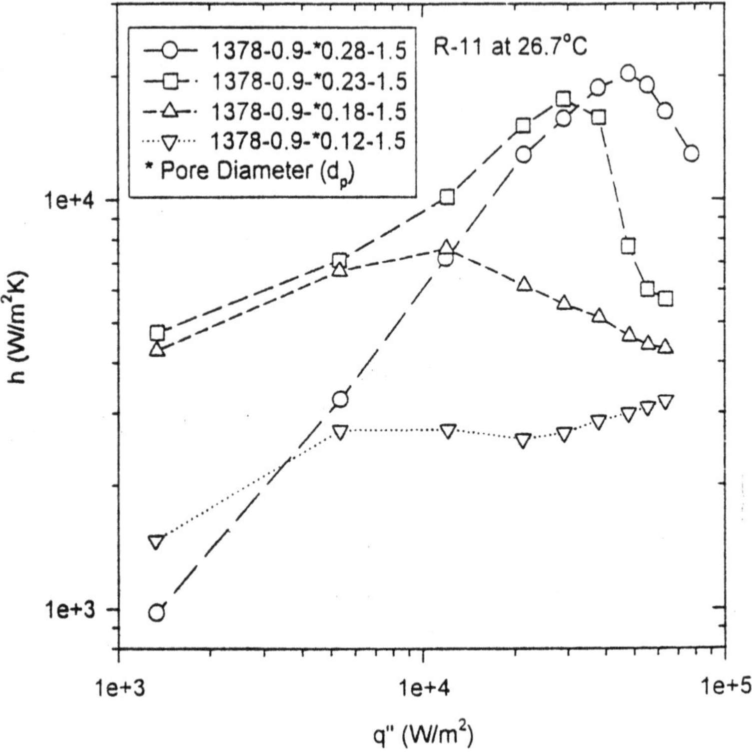

Fig. 11 Effect of pore diameter for fixed pore pitch (1.5 mm) on tube 1378-0.9.

maximum heat transfer coefficient, which occurs when liquid is depleted in the tunnel. Above the DHF, evaporation does not occur in the tunnels, and the vapor in the tunnels causes high thermal resistance. The boiling curves are strongly influenced by the pore diameter. As the pore diameter increases, the boiling curves shift upward and to the right. The DHF increases with increase of the pore diameter. The surface having the largest pores ($d_p = 0.28$ mm) has the greatest heat transfer coefficient at high heat flux and greater DHF. However, the performance of this surface drops rapidly at heat fluxes greater than the DHF. The 0.23 mm pore diameter is the best for $q'' < 35$ kW/m².

3.4 TEST RESULTS (EFFECT OF PORE PITCH)

Figure 12 shows the effect of pore pitch on the 1378-0.9 tube for fixed pore diameter ($d_p = 0.28$ and 0.12 mm) with the pore pitch varied from 0.75-to-3.0 mm. For comparison purposes, test results for the standard 1378-0.9 tube (without copper cover and surface pores) is also shown on Fig. 12. All three curves for $d_p = 0.28$ mm have much higher heat transfer coefficient than the 1378-0.9 tube without the copper cover. For $d_p = 0.28$ mm, the surface having the largest pore pitch ($P_p = 3.0$ mm) has the highest heat transfer coefficient for $q'' < 30$ kW/m².

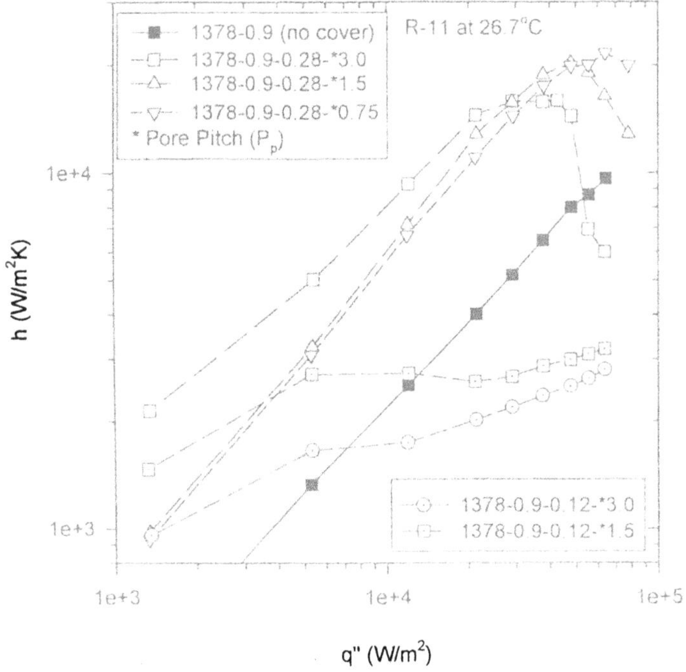

Fig. 12 Effect of pore pitch for fixed pore diameters (0.12 and 0.28 mm) on tube 1378-0.9.

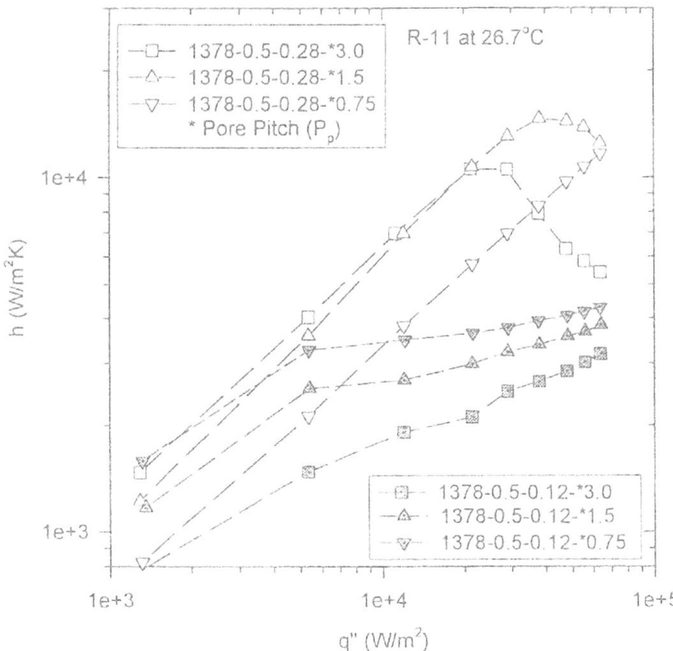

Fig. 13 Effect of pore pitch for fixed pore diameter (0.23 mm) on tube 1378-0.5.

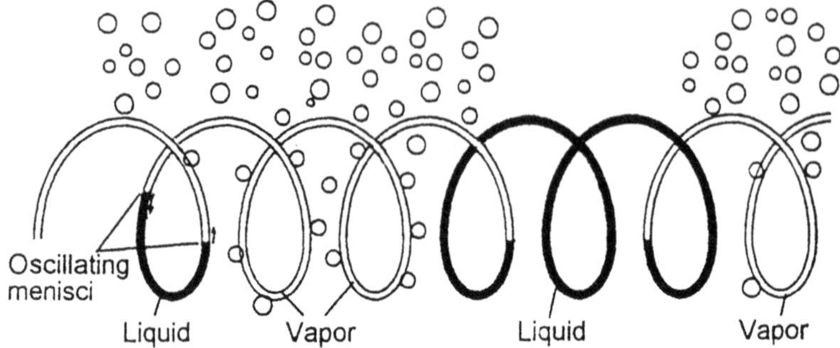

Fig. 14 Boiling mechanism for a horizontal tube at low heat flux.

Figure 13 shows the boiling curves for 0.5 mm fin height (Tube 1378-0.5) as a function of pore pitch for two pore diameters (d_p = 0.12, 0.23 mm). The trends are the same as shown on Fig. 12 for 0.9 mm fin height. As expected, the larger pore pitch yields higher heat transfer coefficient at low heat flux, and a smaller DHF. The smaller pore pitch has higher h at high heat flux, and has lower h at low heat flux. For the smallest pore pitch P_p = 0.75, the DHF was not yet attained at the highest heat flux. However, its heat transfer coefficient is much lower than the other two surfaces at reduced heat fluxes. This is because the total open area for this combination is too large, and part of the tunnel is flooded at any heat flux. The total open area (A_{pt}) increases as the pore pitch is reduced. Hence, the liquid supply rate to the tunnels increases as the total open area increases. For smaller pore diameter (d_p = 0.12 mm), the heat transfer coefficient increases as the pore pitch decreases for all heat fluxes. Because the pore diameter d_p = 0.12 is small, decreasing the pore pitch improves the liquid supply and increases the heat transfer coefficient. At $q'' > 8$ kW/m^2, it is possible that the tunnels are nearly dry. In this case, evaporation does not occur inside the tunnel. The decrease of P_p may increase the number of nucleation sites and increases the external evaporation rate. This may explain the low heat transfer coefficient for the smallest pore diameters.

3.5 DISCUSSION

The previously presented data show that the liquid supply rate to the tunnel is established by the total open area (A_{pt}). This is a function of the pore diameter and pore pitch. Consider for example, use of 1.5 mm pore pitch. Fig. 10 shows that the highest boiling coefficient is obtained by use of the largest pore diameter. However, a smaller pore diameter is preferred for low heat fluxes. A large pore diameter may result in too much liquid supply to the tunnel at low heat flux. This may result in a situation that will decrease the evaporation in the tunnel. Part of the tunnel may be flooded with liquid as shown on Fig. 10 for the 0.28 mm pore diameter.

Figure 14 was prepared based on the visualization observations of Chien and Webb [6] at low heat flux (5 kW/m^2). As shown on Fig. 14, all the tunnels are internally connected. At low heat flux, part of the tunnel is filled with liquid as shown by the darker region on Fig. 14. Oscillating liquid menisci were observed at the two ends of each vapor region. Bubbles emerged from the vapor filled regions and the liquid

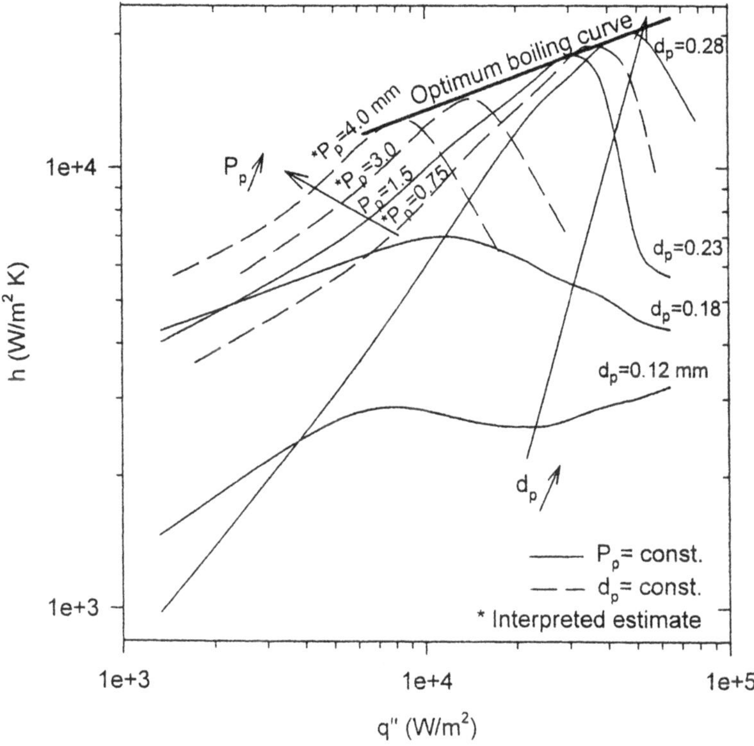

Fig. 15 Map showing combined effect of pore diameter (d_p) and pore pitch (P_p).

filled regions were non-active. The overall heat transfer coefficient decreases because the total evaporation area decreases.

The total open area (A_{pt}) acts to supply liquid to the tunnel. Because $A_{pt} \propto d_p^2/P_p$, the same A_{pt} may exist using closely spaced small pores, or widely spaced large pores. The effect of d_p and P_p are qualitatively explained on Fig. 15 for R-11 boiling on an integral-fin tube having 1378 fins/m and 0.9 mm fin height. This figure illustrates the separate and combined effects of pore pitch (P_p) and pore diameter (d_p). There are two families of curves on Fig. 15:

1. The solid lines are for fixed pore pitch (P_p = 1.5 mm) with increasing pore diameter (d_p). These curves are based on the same data shown on Fig. 11 for P_p = 1.5 mm for the 1378-0.9 tube.

2. The dashed lines show the effect of pore pitch ($0.75 \le P_p \le 4.0$ mm) for fixed pore diameter (d_p = 0.23 mm). Data were obtained only for the d_p = 0.23 mm, P_p = 1.5 and 3.0 mm cases. Based on the data in Chien and Webb [7], the dashed lines are the authors qualitative estimates of the effect of pore pitch.

The salient features shown on Fig. 15 are:

1. As shown by the solid curves for fixed P_p, the DHF increases as the pore diameter increases. A lower DHF occurs for the smaller pore diameters, because they

cannot supply enough liquid to the tunnel at high heat flux. So, the smaller pores become dry and the DHF occurs at a lower heat flux. Conceivably, a higher DHF would be observed for $d_p > 0.28$ mm. However, the maximum possible pore diameter is limited by the fin pitch, 0.5 mm.

2. For fixed P_p and heat fluxes less than the DHF (e.g., 50% of the DHF), the heat transfer coefficient increases as d_p increases, except for the largest d_p (0.28 mm). The lower heat transfer coefficient for the largest d_p occurs because too much liquid is supplied to the tunnels, and floods the intermittent regions. Hence, there is an optimum pore diameter for operation at low-to-moderate heat flux.

3. For fixed d_p (0.23 mm), the DHF decreases as the P_p increases. This is because the tunnel needs greater liquid supply at the higher heat flux. However, the more widely spaced pores are unable to supply the liquid.

4. For low heat flux at fixed d_p (0.23 mm), increasing P_p causes the heat transfer coefficient to increase. At the lower heat flux condition, a smaller liquid supply to the tunnel will prevent tunnel flooding and result in thin liquid films in the tunnel.

At high heat flux, the boiling surface will benefit from increased pore size. However, at reduced heat flux, the same surface may experience liquid flooding in the tunnel. It is possible to optimize the surface performance for operation within a specific heat flux range (below the DHF). This is equivalent to being able to change the slope of the boiling curve.

The "total open area" (A_{pt}) is the parameter that defines the potential liquid supply rate. A larger A_{pt} is needed at higher heat flux than at lower heat flux. If the liquid supply rate exceeds the evaporation rate, the performance will be reduced, because of tunnel flooding. Hence, the preferred pore diameter and pitch depend on the design heat flux range. Although it may be possible to select a d_p and P_p combination that will show a high DHF, this would not be preferred if one desires to operate at a somewhat lower heat flux range. Use of increased pore pitch would yield higher performance at the lower heat flux condition.

The GEWA-TW and bent fin surfaces have continuous open gaps at the fin tips, as opposed to the surface pores of Turbo-B, Turbo-BII, Thermoexcel-E surfaces (described by Webb, [1]) and the present studied geometry. Although comparable data have not been taken on the "continuous fin gap" structures, the authors propose that the understanding developed for the "pored" surfaces can be extended to the "continuous fin gap" structures. The equivalence is found in the total open area, A_{pt}. Assuming the optimum design for the bent fin surface has the same "total open area per unit length," A_{pt}/L, as for the pored surface, the optimum gap width (s_g) of bent fin can be found by

$$s_g = \frac{A_{pt}}{L} = \frac{\pi d_p^2}{4 P_p} \tag{1}$$

The present tests suggest that the optimum combination is $d_p = 0.23$ and $P_p = 1.5$ mm for the pored surface. Substituting these values for d_p and P_p in Eq. (1) yields $s_g = 0.028$

mm, which is within the range recommended by Webb [16], who found that the optimum fin gap for the bent fin surface is between 0.004 and 0.09 mm for R-11.

3.6 OPTIMIZED GEOMETRY

The current technology of making enhanced boiling surface uses uniform pore pitch and pore diameter along the entire tube length. Figure 15 implies that one can improve the boiling heat transfer by selecting the optimum pore diameter and pore pitch for the desired heat flux. A smaller total open area is preferred at a smaller heat flux. For the same pore diameter, the optimum pore pitch is greater at a lower heat flux.

In refrigerant evaporator, heat flux varies from the inlet to the outlet. Near the inlet, the refrigerant is evaporated at high heat flux, large total open area (small pore pitch or large pore diameter) should be selected. However, the heat flux near the refrigerant outlet is much smaller. Therefore, an ideal design involves selecting smaller "total open area" for the tubing at the outlet and larger total open area at the inlet. According to the boiling map of Figure 15, one may obtain the optimum boiling curve by selecting pore diameter and pore pitch for the given heat flux.

Consider a R-123 water chiller tube bundle, for which one seeks to maximize the boiling coefficient (h) over the tube length. In this bundle, the heat flux is 48.0 kW/m^{2}at the water inlet end and linearly drops to 9.0 kW/m^2 at the water exit end. As shown by the thick solid line on Figure 15 labeled "Optimum boiling curve,"the optimum boiling curve would exist, if one used varied pore diameter and/or pore pitch for different heat flux. A practical "optimized" boiling tube may use the following combinations of pore diameter (d_p) and pore pitch (P_p) for the specified heat flux (q"):

1. d_p = 0.28 mm and P_p = 1.50 mm at q" = 48 kW/m^2,
2. d_p = 0.23 mm and P_p = 0.75 mm at q" = 35 kW/m^2,
3. d_p = 0.23 mm and P_p = 1.50 mm at q" = 28 kW/m^2,
4. d_p = 0.23 mm and P_p = 3.00 mm at q" = 15 kW/m^2,
5. d_p = 0.23 mm and P_p = 4.00 mm at q" = 9 kW/m^2.

As implied by Eq. (1), the term A_{pt}/L is defined by d_p^2/P_p. Other combinations of pore diameter and pore pitch may yield similar results if the total pore area (or d_p^2/P_p) remain the same for the given heat flux.

4. Boiling Model

Nakayama et al. [11], Haider [12] and Webb and Haider [13], had proposed mechanistically based analytical models to predict boiling performance of structured enhanced surfaces. The Nakayama et al. model [11] proposed that the "suction-evaporation mode," is active below the DHF, and the "dried-up mode" occurs above the DHF. The Nakayama et al. [4] model assumed evaporation on menisci in the corners of the tunnel for the suction-evaporation mode. The Haider [12] model is based on the "flooded mode," which assumes alternate zones of liquid slugs and vapor plugs

268

in the sub-surface tunnels. The work of Chien and Webb [6, 7, 8] confirm the Nakayama et al. [11] model is correct for saturated boiling.

Nakayama et al. [11] assumed that the total heat flux from structured surfaces is the sum of the following two parts.

1. Tunnel heat flux (q''_{tun}) due to the thin-film evaporation inside the tunnels of the enhanced surface.
2. Sensible heat flux (q''_{ex}) due to the external convection induced by bubble agitation.

$$q'' = q''_{tun} + q''_{ex} \tag{2}$$

Both fluxes are based on the projected surface area. This is expressed by the following equation. Their dynamic model is based on the sequence of events during one complete cycle of bubble growth and departure. According to the tunnel pressure and bubble growth conditions, they divided one bubble cycle into three phases: 1. Pressure build-up phase, 2. Pressure reduction phase, 3. Liquid-intake phase. The temporal variation of the liquid meniscus thickness during a bubble cycle was ignored in their model.

In the Nakayama et al. [4] model, the external convection heat flux does not include the bubble frequency and bubble departure diameter. As discussed by Haider [12], the bubble frequency and departure diameter are important factors for external heat flux. Haider and Webb [14] have formulated an external heat flux model based on transient micro-convection, which accounts for the bubble diameter and frequency. Chien and Webb [10] show applicability of this model to the present structured surfaces.

The present dynamic model accounts for the temporal evaporation rate variation inside tunnels by analyzing meniscus thickness, bubble departure diameter, bubble growth, and the transient convection outside the tunnels.

4.1 OVERALL STRUCTURE OF MODEL

The detailed mathematical model will not be given here, because it is lengthy and is fully described by Chien and Webb [10]. However, we will comment on certain aspects of it, and how it differs from the model previously developed by Nakayama et al. [4]. First, the model has fewer empirical constants, and several key assumptions are different from those used by Nakayama et al. [4].

As given by Eq. (2), the total heat flux is separated into two parts, tunnel heat flux (q''_{tun}) and external heat flux (q''_{ex}). The heat transfer rate in the tunnel is governed by evaporation of liquid menisci. The external heat flux is contributed by transient conduction and convection caused by the departing bubbles.

The key assumptions of the model are:

1. The tunnels are vapor filled, except for liquid menisci in the corners.
2. The distribution of the liquid menisci is uniform along the tunnel length.
3. The latent heat of the vapor bubbles is by evaporation inside the tunnels. Evaporation of the micro layer above the subsurface tunnel is neglected.

4. The evaporation rate of the liquid meniscus is predicted by one-dimensional conduction.

5. For circular fin base tunnels, only two liquid menisci occur in the corners - near the fin tips. For a rectangular tunnel cross-section, liquid menisci exist in each of the four corners.

6. For high thermal conductivity fins (e.g., copper), the wall temperature is constant over the fin height.

4.1.1 *The Bubble Cycle*

A bubble cycle includes three periods: waiting period (Δt_w), bubble growth period (Δt_g) and liquid intake period (Δt_e). Figure 16 shows the process of evaporation in the tunnel during a boiling cycle. A bubble cycle includes the following three periods:

1. Waiting period (Δt_w): In this period (Fig. 16a and b), liquid is evaporated in the tunnel. However, the vapor is constrained inside the tunnel by the surface tension on the pore, so the pressure in the tunnel increases with time. A bubble embryos protrude from a pore when $P_v - P_s \equiv \Delta P_{br} = 4\sigma/d_p$, where ΔP_{br} is called the "break through pressure differential." The radius of the meniscus decreases from $R_{m,i}$ at the beginning of the cycle (Fig. 16a) to $R_{m,g}$ at the end of this period (Fig. 16b). During this period, the tunnel is filled with vapor except for the liquid menisci in the upper corners.

2. Bubble growth period (Δt_g): In this period, vapor passes through surface pores and increases the bubble radius and the bubble grows above the pore. Because the liquid in the tunnel continues to evaporate, the meniscus radius decreases from $R_{m,g}$ to $R_{m,e}$. The effective evaporation temperature increases as the meniscus radius decreases. Meniscus evaporation will stop when the elevation of saturation temperature due to capillary pressure (σ/R) of the meniscus equals the wall superheat. The meniscus radius, at which its saturation temperature attains the wall temperature is R_{ne}, and is

$$R_{ne} = \frac{\sigma}{\Delta T_{ws}} \left(\frac{dT}{dP} \right) \tag{3}$$

Depending on the evaporation rate, $R_{m,e} \geq R_{ne}$. At time t_d (Fig. 16c) bubbles of diameter d_b depart from the surface pores.

3. Liquid intake period (Δt_e): After the bubble departs, the pressure in the tunnel is lower than that of the liquid pool, and liquid flows into the tunnel, and it is retained in the corners. At the end of the liquid intake period, the radius of the meniscus equals $R_{m,i}$ as shown on Figure 16d. The pore diameter and pitch control the amount of the liquid that flows into the tunnel. This period is much shorter than the other two periods and is neglected in calculation of the bubble frequency.

4.1.2 *Wetting and Evaporation on the Tunnel Walls*

At the end of a bubble cycle, liquid is drawn into the tunnel (see Fig. 17a) by inertia force and its movement in the s-direction is impeded by surface tension force. Further, surface tension force acts to spread the liquid in the x-direction. Chien [17] used the surface tension drainage model of Adamek and Webb [18] and showed that the surface tension induced velocity in the x-direction is much larger than s-direction. Liquid inertia will try to spread the liquid along the fin side walls. Once the liquid is on the side wall, surface tension will act to it into the corners at the fin base. However, for a rectangular fin base, substantially more liquid should be retained at the fin tip than at the fin base.

For a circular fin base, the fin base radius is large compared to the meniscus radius at the fin tips. Hence, the ability of surface tension to pull liquid to the fin base is assumed to be negligibly small for a circular fin base. In support of this hypothesis, Chien and Webb [6] show that a surface having a rectangular fin base has higher performance than an identical surface having a circular fin base. The present model treats only the circular fin base geometry and assumes that two liquid menisci exist at the fin tip. No evaporation occurs in the circular fin base.

The model assumes that the evaporation occurs only on liquid menisci in the corners of the tunnel. If a thin film existed on the side walls, it will evaporate more quickly than the thicker liquid menisci. Chien [17] predicted that a thin film on the side wall would evaporate in approximately 10^{-5} s, which is less than 5% of the typical waiting period. Hence, any evaporation that occurs in thin films on the side walls is negligible compared to that evaporated in the relatively thick menisci.

Figure 17b shows a meniscus existing at the top corner of the tunnel. A one-dimensional model is used to calculate the evaporation rate of this meniscus.

4.1.3 *Calculation of Tunnel Heat Flux, q''_{tun}*

The tunnel heat flux in Eq. (2) is the product of the latent heat transferred at the menisci during a bubble cycle and the bubble frequency, as given by,

$$q''_{tun} A_{tun} = (Q''_m) f \tag{4}$$

where Q''_m is the latent heat from the menisci in one bubble cycle. Assuming one-dimensional heat conduction in the liquid, the heat flux is given by

$$q'' = \frac{k_l}{\delta}(T_w - T_s) \tag{5}$$

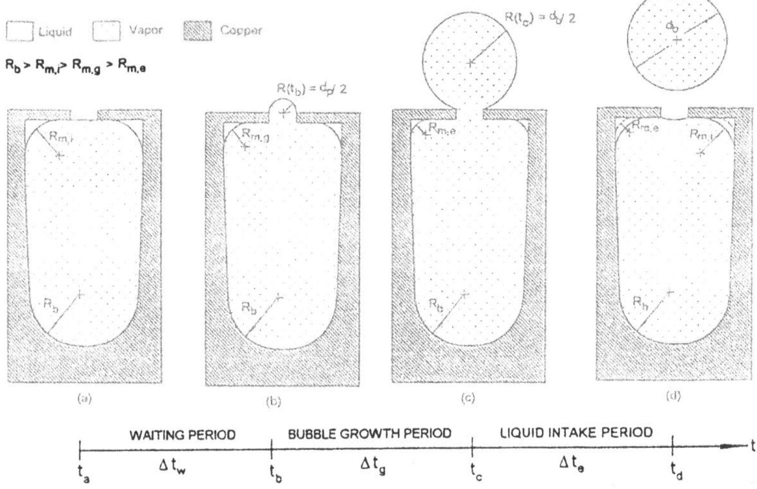

Fig. 16 Process of evaporation in the subsurface tunnel and bubble growth.

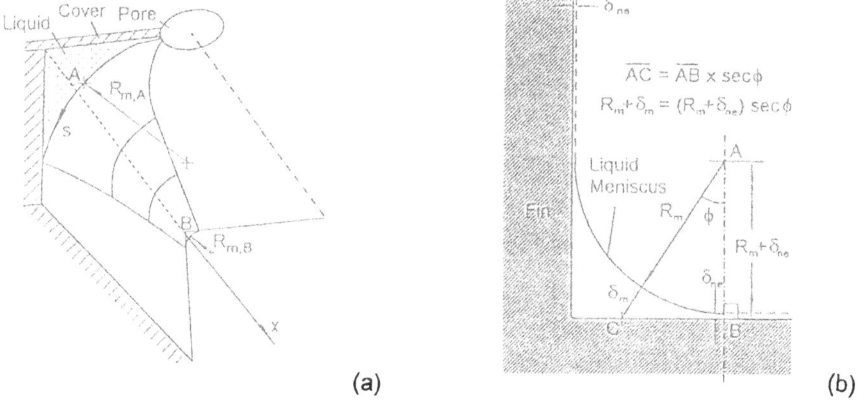

Fig. 17 Geometry of the liquid meniscus: (a) surface tension pulls liquid along the tunnel length, (b) Geometry of evaporating liquid meniscus.

When the liquid meniscus becomes very thin, the saturation temperature elevates because of the disjoining pressure and capillary pressure (σ/R). So, when the elevated saturation temperature equals T_w no further evaporation will occur and the thickness of this non-evaporating film is δ_{ne}. Chien and Webb [10] give methodology to calculate the elevation of T_s as a function of meniscus radius.

The total latent heat in the tunnel during one bubble cycle can be calculated by integrating the latent heat transfer rate at the menisci (dQ_m/dt) during a bubble cycle, and is given by

$$Q_{tun} = \int_0^{1/f} \frac{dQ_m}{dt} dt = \int_0^{1/f} [\Delta T_{ws} \int_0^{A_m} \frac{k_l}{\delta_m(t,\phi)} dA] dt \tag{6}$$

where $\delta_m(t, \phi)$ is the local thickness of the liquid meniscus, and A_m is the meniscus surface per unit tunnel length. For a given meniscus radius (R_m) use of Figure 17b gives $\delta_m(\phi) = (R_m + \delta_{ne})\sec(\phi)$. Because R_m is a function of time, the $\delta_m(t, \phi)$, is given by

$$\delta_m(t,\phi) = [R_m(t) + \delta_{ne}]\sec(\phi) - R_m(t) \tag{7}$$

The integrations in Eqs. (6) were numerically calculated using small increments of time (t) and angle (ϕ). The $R_m(t)$ is updated to a new value, $R_{m,new}$, for each new time step. Substituting Eqs. (6, 7) in Eq. (4) gives the tunnel heat flux. Note that the bubble frequency must be known to evaluate Eq. (6).

The integration of Eq. (6) starts from the initial meniscus radius at the beginning of the bubble cycle ($R_m = R_{m,i}$ on Figure 16a) and stepwise decreases for small time steps until $R_m = R_{m,e}$ (Figure 16c). The initial meniscus radius ($R_{m,i}$) depends on the amount of liquid that flows into the tunnel during a bubble cycle. From geometric analysis, the $R_{m,i}$ is given by,

$$R_{m,i} = \left[R^2_{m,e} + \frac{\Delta A_{l,cyc}}{N_m(1-\pi/4)} \right]^{0.5} \tag{8}$$

where $\Delta A_{l,cyc}$ is the change of the cross section area of the liquid menisci during a bubble cycle and N_m is the number of menisci per tunnel. The N_m requires knowledge of the nucleation site density, which will be addressed later. Chien and Webb [10] developed the following correlation for $\Delta A_{l,cyc}$ based on their experimental data for R-11 and R-123, R-134a and R-22:

$$\Delta A_{l,cyc} = \frac{4-\pi}{2} C_{Rm} \Delta T_{ws}^{n1} d_p^{n2} (H_t + W_t)^{n3} \left(\frac{\sigma^{n5}}{\mu_l^{n4}} \right) \tag{9}$$

where $C_{Rm} = 4.71 \times 10^{-9}$, $n_1 = 0.1882$, $n_2 = 0.609$, $n_3 = 1.49$, $n_4 = 1.771$, and $n_5 = 0.512$. Eq. (11) is applicable to tunnels having a circular fin base, and all fluids and geometric dimensions within the range of the experiments.

4.1.4 External Heat Flux

The sensible heat flux (q''_{ex}) term in Eq. (3) is due to the external convection induced by bubble agitation. The mechanism is similar to pool boiling on a plain surface in that the bubble evaporates into a superheated liquid layer above the base surface. The key difference between the present situation for q''_{ex} and boiling on a plain surface is that the nucleation site density, frequency, and bubble departure diameter are known.

The model used to predict q''_{ex} in Eq. (3) is totally different from the empirical correlation used by Nakayama et al. [4]. The Haider and Webb [14] model is used to predict the external heat flux. This model is based on an extension of Mikic and Rohsenow [20], who developed a model based on transient conduction to the super-heated liquid layer. The Haider and Webb model includes the effects of transient convection to the liquid, caused by convection in the wake of the departing bubbles. An asymptotic correlation is used to patch the transient conduction and the steady-state convection asymptotes to cover the full bubble cycle. Transient micro-convection, rather than transient conduction is the dominant heat transport mechanism.

The model develops the heat transfer rate associated with the area of influence of each bubble site (Q_{bub}). Multiplication of this by the nucleation site density gives q''_{ex}. Its application requires accurate prediction, or measurement, of active nucleation site density (n_s), bubble frequency (f), and bubble departure diameter (d_b). The transient conduction asymptote ($q''_{ex,MR}$) is taken from the Mikic and Rohsenow [20] solution and is given by

$$q''_{ex,MR} = n_s Q_{bub} = f \int_0^{\frac{1}{f}} \left(-k_l \pi d_b^2 \frac{\partial T}{\partial y} \right)_{y=0} dt$$
$$= 2\sqrt{\pi}\sqrt{k_l \rho_l c_p} \sqrt{f}\, d_b^2 n_s (T_w - T_s) \qquad (10)$$

The second asymptote is due to steady state convection due to agitation of the departing bubble. The resulting Haider and Webb [14] formulation for q''_{ex} is given by

$$q''_{ex} = q''_{ex,MR} \left[1 + \left(\frac{0.66\,\pi c}{Pr^{1/6}} \right)^2 \right]^{1/2} \qquad (11)$$

where the empirical constant of $c = 6.42$, was curve fitted from the Chien and Webb [9] and Nakayama et al. [3] bubble formation data. The prediction is within ±25%. One must have d_b, f, and n_s to apply the model.

4.2 PREDICTION OF BUBBLE DYNAMICS PARAMETERS

The equation for prediction of q''_{tun} and q''_{ex} both require knowledge the active nucleation site density (n_s), bubble frequency (f), and bubble departure diameter (d_b). The models used to predict these parameters are now presented. Chien and Webb [9] also performed measurements of these parameters for several refrigerants.

274

4.2.1 *Bubble Departure Diameter*

Using a force balance, Nakayama et al. [4] developed the following empirical equation
for the bubble departure diameter:

$$d_b = C_b \left(\frac{2\sigma}{(\rho_l - \rho_v)g} \right)^{1/2} \tag{12}$$

Their analysis is based on assumed the buoyancy force equals the surface tension force,
and they neglected the inertia force of vapor bubble. The constant C_b was empirical
constant was obtained empirically (0.42 for R-11 and 0.22 for water).

    Chien and Webb [10] formulated the balance between buoyancy and surface
tension forces as a function of d_b, d_p and contact angle (θ), given by

$$\sigma d_p \sin(\theta) = (\rho_l - \rho_v)g \frac{d_b^3}{12} \left(1 + \cos(\theta) + \frac{\cos(\theta)\sin^2(\theta)}{2} \right) \tag{13}$$

Assuming spherical bubbles, Chien and Webb [10] used Figure 18 to show the
following equation, which may be substituted into Eq. (13) to solve the bubble
departure diameter, d_b

$$\theta = Sin^{-1} \left(\frac{d_p}{d_b} \right) \tag{14}$$

Therefore, the bubble diameter d_b can be predicted by Eq. (13) with no empirical
constant.

4.2.2 *Bubble Frequency*

As stated previously, the liquid intake period is much shorter than the waiting period
(Δt_w) and the bubble growth periods (Δt_g). Therefore, $1/f = \Delta t_w + \Delta t_g$. The waiting
period and the bubble growth period are modeled separately.

Waiting Period. Using mass and energy balance equations, Nakayama et al. [4] showed
that

$$\frac{1}{i_{fg}} \frac{dQ_{tun}}{dt} = \frac{dm_v}{dt} = V_v \frac{d\rho_v}{dt} + \rho_v \frac{dV_v}{dt} \tag{15}$$

    The tunnel heat transfer rate (dQ_{tun}/dt) changes because the meniscus radius
changes during the bubble cycle. Integration of Eq. (18) gives

$$\int_0^{\Delta t_w} \frac{dQ_{tun}/dt}{\Delta T_{ws}} dt = V_{vm} i_{fg} \left[\frac{\rho_v (i_{fg} - R_g T_{vo})}{R_g T_{vo}^2} \ln\left(\frac{T_w - T_{vo}}{T_w - T_{vl}} \right) + \frac{\rho_v}{\Delta T_{tl}} \ln\left(\frac{V_{vl}}{V_t} \right) \right] \qquad (16)$$

Eq. (16) is based on a different approach than that used by Nakayama et al. [4]. The right hand side of Eq. (16) can be explicitly calculated for a given wall superheat ΔT_{ws}. The left-hand side of Eq. (16) was calculated using small time steps from the beginning to the end of the waiting period (Δt_w,), until it equals the right hand side of Eq. (16).

Bubble Growth Period. For a plain surface, bubble growth is controlled by inertia force. For a structured surface, the evaporation on the bubble interface is not so important, because the vapor is mainly supplied from the subsurface tunnel. Therefore, inertia controlled bubble growth dominates the entire bubble growth period. For inertia controlled growth, Mikic and Rohsenow [21] showed that the bubble growth on a plain surface is given by

$$\left(\frac{dR}{dt} \right)^2 = \frac{b}{\rho_l} \left(P_v - P_s - \frac{2\sigma}{R} \right) \qquad (17)$$

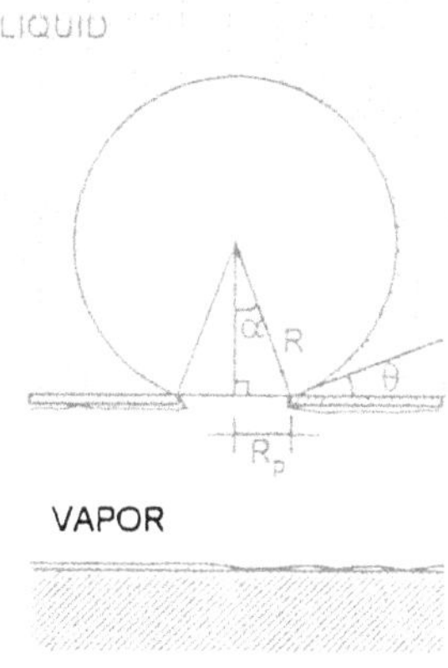

Fig. 18 Bubble growth on a surface pore.

Nakayama et al. [4] concluded that $(P_v - P_s)$ is negligible, based on the assumption $R \gg d_p/2$. However, Chien and Webb [9] found that the bubble growth period is four times greater than the Nakayama et al. [4] prediction and they do not neglect $(P_v - P_s)$ in Eq. (17). It is also reasonable to assume that the $(P_v - P_s)$ increases as the wall superheat $(T_w - T_s)$ increases. Mikic and Rohsenow [21] modified Eq. (17) to get

$$\frac{dR}{dt} = \left(\frac{\pi}{7} \frac{i_{fg}\rho_v \Delta T_{ws}}{\rho_l T_s} \right)^{1/2} \left(\frac{(T_v - T_s) - 2\sigma/(mR)}{T_w - T_s} \right)^{1/2}$$
$$\text{where} \quad T_v - T_s - 2\sigma/(mR) = \left(\frac{T_w - T_s}{T_w^* - T_s} \right) \left(\frac{4\sigma}{m\,d_p} - \frac{2\sigma}{mR} \right) \tag{18}$$

where $m = dP/dT$ at T_s, and T_w^* is a reference wall temperature. At $T_w = T_w^*$, the vapor superheat $\Delta T_{vs} = 4\sigma/(md_p)$. The authors linearize R in the second term on the right side of Eq. (18) and obtain

$$\left(\frac{dR}{dt} \right) = \left(\frac{\pi}{7} \frac{i_{fg}\rho_v \Delta T_{ws}}{\rho_l T_s} \right)^{1/2} C_{tg} \left(\frac{d_b - d_p}{d_b + d_p} \right)^{1/2},$$
$$\text{where} \quad C_{tg} = \left[\frac{4\sigma}{m\,d_p\,(T_w^* - T_s)} \right]^{1/2} \tag{19}$$

The C_{tg} is a dimensionless parameter, which represents the slope of vapor superheat vs. the reference wall superheat $(T_w^* - T_s)$. Since T_w^* is an unknown, the authors obtain C_{tg} by a curve fit of the bubble growth data of Chien and Webb [9]. They found that $C_{tg} = 0.0296$ best fits the data. Setting the boundary conditions for Eq. (19): $R = d_p/2$, at the beginning; and $R = d_b/2$ at the end of the growth period, the growth period is

$$\Delta t_g = 0.0296 \left[\frac{7}{\pi} \frac{\rho_l T_s}{i_{fg}\rho_v \Delta T_{ws}} \frac{(d_b + d_p)}{(d_b - d_p)} \right]^{1/2} \left(\frac{d_b - d_p}{2} \right) \tag{20}$$

4.2.3 *Nucleation Site Density*

The vapor generated in the tunnel is ejected through the surface pores. The bubble diameter and frequency are predicted as discussed above. From an energy balance, $q''_{tun} = i_{fg} V_v \rho_v$. From a mass balance, one obtains the nucleation site density (n_s) by the following equation.

$$n_s = \frac{q''_{tun}}{\rho_v i_{fg} f (\pi d_b^3)/6} \tag{21}$$

4.3 SOLUTION PROCEDURE

The Chien and Webb model predicts the total heat flux (q'') for given wall superheat (ΔT_{ws}) and dimensions (d_b, P_p, P_f, H_t), by the following procedure:

1. Calculate the bubble departure diameter (d_b) for given d_p and fluid properties by Eq. (13).
2. Calculate the bubble growth period (Δt_g) by Eq. (20) for given ΔT_{ws} and fluid properties, using d_b obtained in step 1.
3. Calculate the initial liquid meniscus radius, $R_{m,i}$ by Eq. (8), using a correlation [see Eq. (9)] for prediction of $\Delta A_{l,cyc}$.
4. Calculate the latent heat transferred in the tunnels during the bubble cycle by integrating Eqs. (6) with stepwise decreasing R_m from $R_{m,i}$ to $R_{m,e}$. The Δt during one increment of R_m is then calculated.
5. Calculate the waiting period (Δt_w) by Eq. (16), and find the meniscus radius, ($R_{m,g}$) at the beginning of the bubble growth period.
6. Continue the calculation of the tunnel heat flux as for the bubble growth period, and calculate by decreasing R_m for $\Delta t_w < t < \Delta t_w + \Delta t_g$.
7. Sum the tunnel heat flux (q''_{tun}) during the waiting period (Δt_w) and bubble growth period (Δt_g) for given d_p, P_p, P_f, H_t, ΔT_{ws} and fluid properties, and get the bubble frequency, $f = 1/(\Delta t_w + \Delta t_g)$.
8. Use Eq. (21) to calculate the nucleation site density (n_s) from q''_{tun} and d_b obtained in step 1 and 2, and f obtained in step 7.
9. Calculate the external heat flux (q''_{ex}) by Eq. (11) for a given ΔT_{ws}, and d_b, f, n_s obtained in step 2, 7 and 8.
10. Calculate total heat flux $q'' = q''_{ex} + q''_{tun}$

4.4 PREDICTIVE ABILITY OF THE CHIEN AND WEBB [10] MODEL

The model was used to predict the data of Chien and Webb [6, 7] and Chien et al. [19] for $q'' = 20\%$-to-100% of the DHF. The predictions were limited at the low end, because the model assumes the tunnels are vapor filled, which may not be true at low heat flux. These data include R-11, R-123, R-134a and R-22 pool boiling data on tubular surfaces having a circular fin base for heat fluxes between 10-to-65 kW/m². The range of geometric parameters are shown in Table 2.

The results for the predicted heat flux are shown in Fig. 19. Figure 19 shows that the heat flux is predicted within ±33% absolute error (MSD = 20%), except for the dark symbols, at quite low heat fluxes. The model over predicted these dark points, because some tunnels become partially liquid filled and do not generate bubbles at low heat flux.

Table 2. Range of Parameters in the Data Base

Fluids	d_p (mm)	P_p (mm)	fins/m	W_t (mm)	H_t (mm)
R11 R123	0.12-0.28	0.75-1.5	1378,1575, 1968	0.25-0.4	0.5-1.5
R134a R22	0.18-0.28	0.75-1.5	1575,1968	0.25-0.33	0.6-1.5

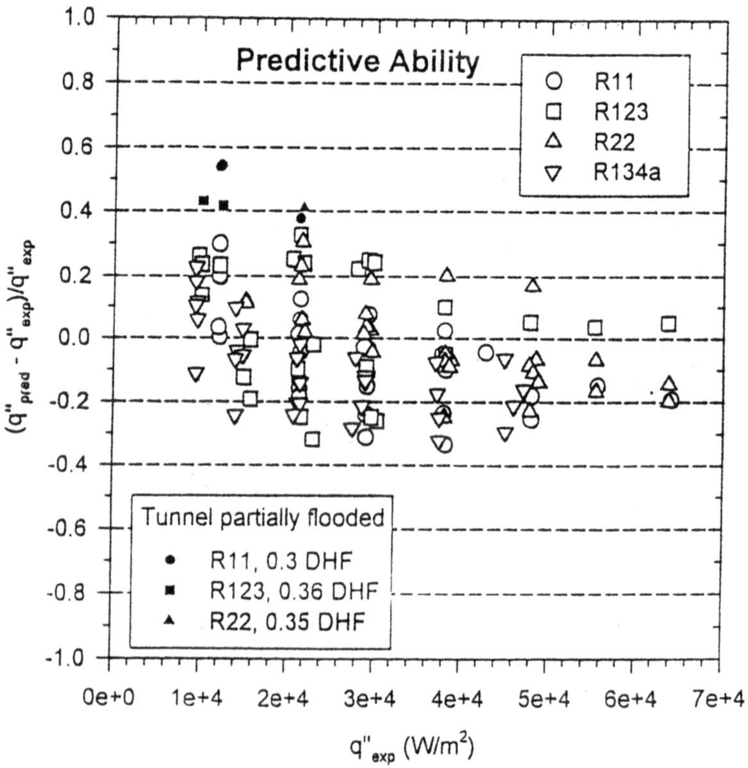

Fig. 19 Predictive error on total heat flux.

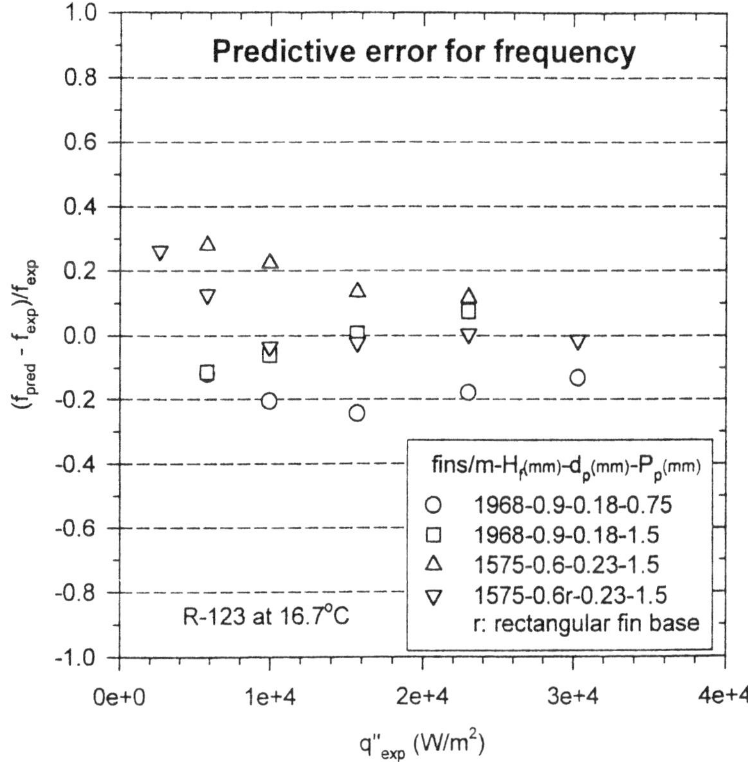

Fig. 20 Predictive error on bubble frequency.

The present model predicts bubble frequency and departure diameter quite well, as shown by Figures 20 and 21. As shown on Figure 20, the model predicts the bubble frequencies of the Chien and Webb [9] R-123 data on four surfaces within ± 30% absolute error (MSD = 15%). Figure 21 shows the error in predicted bubble departure diameter, $(d_{b,pred} - d_{b,exp})/d_{b,exp}$. The Chien and Webb [9] data for R-123 are predicted within ±12% (MSD = 7%). The effect of the heat flux on bubble departure diameter is not considered in the model. However, Figure 21 shows a quite small effect of heat flux on bubble departure diameter.

Table 3 lists the data from Nakayama et al. [4] and the Chien [19] data on water, methanol, R-11, and R-123. As shown by Table 3 the model predicts d_b of the fluids very well.

5. Conclusions

1. Experiments have been performed to define the effect of pore diameter, pore pitch, and tunnel size on performance in structured boiling surfaces.

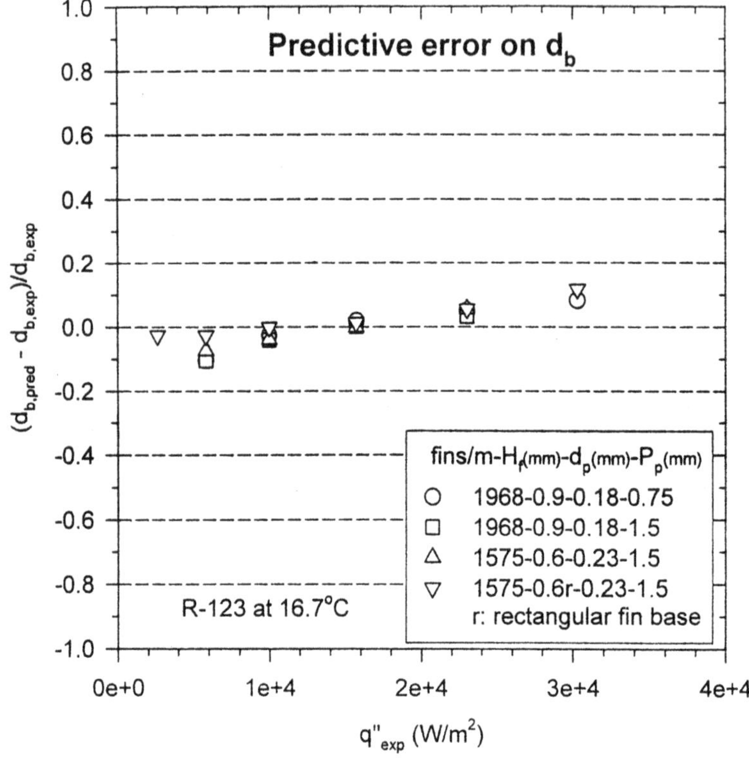

Fig. 21 Predictive error on bubble departure diameter.

2. The Dry-Out Heat Flux increases with the increase of total open area. At a certain reduced heat flux, part of the tunnel will become flooded and the performance will be reduced. Smaller pore size will inhibit flooding at reduced heat flux.

3. For saturated boiling, the tunnel is vapor filled, except for thin liquid films on the tunnel walls and menisci in the corners. Evaporation on liquid menisci in the tunnel corners is the principal boiling mechanism for the structured surfaces.

4. Visualization experiments were performed to determine the bubble departure diameter, bubble frequency, waiting and growth periods, and nucleation site density. These data provided the basis for models to predict these parameters as a function of surface geometry and fluid properties.

Table 3. Comparison of Predictions and Data for Bubble Departure Diameter

Ref.	Fluid	d_p (mm)	$d_{b,exp}$ (mm)	$d_{b,pred}/d_{b,exp}$ (mm)
Chien [6, 7]	Water at 1 atm.	0.1	0.78±0.1	1.00
Nakayama et al.[4]	R11 at 1 atm.	0.1	0.7±0.1	0.76
Chien [6, 7]	R11 at 26.7°C	0.23	0.8±0.1	1.00
Chien [6, 7]	methanol, 1 atm	0.23	1.1±0.1	0.87
Chien [6, 7]	R123 at 26.7°C	0.23	0.77±0.1	1.01
Chien [6, 7]	R123 at 26.7°C	0.18	0.72±0.1	0.93

5. A semi-analytical model was proposed for pool boiling on structured surfaces having a circular fin base. The model assumes evaporation from liquid menisci at the top of the fins and is applicable to surfaces having a circular fin base and an arbitrary combination of d_p, P_p, P_f, H_t. It can predict most boiling heat transfer data for four refrigerants (R-11, R-123, R-134a, and R-22) within ±33% (0.20 MSD).

Nomenclature

A	Surface area, m^2.
A_m	Meniscus surface area, m^2.
A_{tun}	Total surface area inside tunnels.
CHF	Critical heat flux, kW/m^2
c_p	Heat capacity, J/g-K.
d_b	Bubble departure diameter, $d_{b,exp}$ (experimental), $d_{b,pred}$ (predicted),m.
DHF	Dry-out heat flux, kW/m^2
d_p	Pore diameter, m.
dQ_m/dt	Heat transfer rate to a meniscus, J/s.
f	Bubble frequency $f_{b,exp}$ (experimental), $f_{b,pred}$ (predicted), 1/s.
fins/m	Fins per meter.
g	Acceleration due to gravity, m/s^2
h	Heat transfer coefficient, W/m^2-K.
H_f	Fin height, mm
H_t	Tunnel height, m.
i_{fg}	Latent heat, J/kg.
k	Thermal conductivity, k_l (of liquid), W/m-K.

282

L	Tunnel length, m.
m	Slope of P_s vs. T_s ($m = dP/dT$).
MSD	Mean square deviation.
m_v	Vapor mass, kg.
N_m	Number of menisci exist in each tunnel
n_s	Nucleation site density, $1/m^2$.
P	Pressure, P_1 (liquid), P_s (saturation), P_v (of vapor), Pascal.
P_p	Pore pitch, m
Pr	Prandtl number, dimensionless.
P_t	Tunnel pitch, m.
Q	Heat, Q_m (evaporated from meniscus), Q_{bub} (bubble), Q_{tun} (tunnel), J.
q''	Total heat flux based on projected area, kW/m^2
Q''_m	Heat evaporated from liquid menisci in a bubble cycle, J.
q''_{ex}	External heat flux, $q''_{ex,MR}$ (predicted by Eq. 10), W/m^2.
q''_{tun}	Tunnel heat flux, W/m^2.
r	Radius of meniscus interface curvature, m.
R_b	Radius of tunnel base, m.
R_m	Meniscus radius, $R_{m,e}$ (of meniscus at end of bubble cycle), $R_{m,g}$ (meniscus radius at the beginning of the growth period), $R_{m,i}$ (meniscus radius at the beginning of a bubble cycle), R_{ne} (non-evaporation meniscus radius, defined by Eq. (3),
R_g	Gas constant, J/g-K.
R_p	Pore radius, m.
s	Coordinate along the liquid-vapor interface.
S_g	Surface gap width, m.
t	Time, s.
T	Temperature, T_1 (liquid), T_s (saturation), T_v (vapor), T_{vl} (vapor temperature at the end of waiting period), T_{vo} ((vapor temperature at the start of waiting period), T_w (wall), T_w^* (reference wall temperature in Eq. (18), K.
t_a, t_b, t_c, t_d	Periods during bubble growth shown in Figure 16.
u	Velocity of the liquid in the tunnel, m/s.
$u_{s,ave}$	Average liquid velocity in the s direction, m/s.
$u_{x,ave}$	Average liquid velocity in the x direction, m/s.
V_v	Vapor volume, V_b (bubble volume), V_{vl} (at the end of waiting period), V_{vm} (during the waiting period), m^3.
V_t	Tunnel volume, ($V_t = L \cdot W_t \cdot H_t$) m^3.
W_t	Tunnel width, m.
x	Coordinate along the tunnel length.
y	Coordinate normal to the liquid-vapor interface.

Greek Symbols

α	Half angle of the bubble center to pore as defined in Fig. 3, degree.
δ	Liquid film thickness, m.

δ_m Meniscus thickness, m.

δ_{ne} Non-evaporation liquid film thickness, m.

$\Delta A_{l,\,cyc}$ Change of liquid cross section area during a bubble cycle [Eq. (9)], m^3.

ΔP_{br} Break through pressure ($\Delta P_{br} = 4\sigma/d_p$), Pascal.

ΔQ_m Change of evaporation heat during a bubble cycle, J.

Δt Difference of time, Δt_e (liquid intake period),.Δt_g (bubble growth period), t_w (bubble waiting period), s.

ΔT_{tl} Difference of the tunnel temperature during the waiting period, s.

ΔT_{vs} Difference of the saturation and the vapor temperatures ($T_v - T_s$), K.

ΔT_{ws} Difference of the saturation and the tube wall temperatures, K.

θ Contact angle, defined in Fig. 18, degree.

μ Viscosity, μ_l (of the liquid), kg/m^2-s.

ρ Density, ρ_l (liquid), ρ_v (vapor), ρ_{vm} (ave. vapor density during waiting period), kg/m^3.

σ Surface tension, N/m.

φ Angle defined in Fig. 17b.

References

1. Webb, R. L. (1994) *Principles of Enhanced Heat Transfer*, Chapter 11, Wiley Interscience, New York.
2. Thome, J. R., 1990, *Enhanced Boiling Heat Transfer*, Hemisphere Publishing Corp., New York.
3. Nakayama, W., Daikoku, T., Kuwahara, H. and Nakajima, T. (1980) Dynamic Model of Enhanced boiling Heat Transfer on Porous Surfaces Part I: Experimental Investigation, *J. Heat Transfer*, Vol. 102, 445-450.
4. Nakayama, W., Daikoku, T., Kuwahara, H. and Nakajima, T. (1980) Dynamic Model of Enhanced boiling Heat Transfer on Porous Surfaces Part II: Analytical Modeling, *J. Heat Transfer*, Vol. 102, 451-456.
5. Arshad, J. and Thome, J.R. (1983) Enhanced Boiling Surfaces: Heat Transfer Mechanism Mixture Boiling, *Proc. ASME-JSME Therm. Eng. Joint Conf.*, Vol. 1, pp.191-197.
6. Chien, L.-H. and Webb, R. L. (1996) Parametric Studies of Nucleate Pool Boiling on Structured Surfaces, Part I: Effect of Tunnel Dimensions and Comparison of R-11, *ASME Proc. National Heat Transfer Conf.*, Ed., L. Witte et al., Vol. 4, pp. 129-136.
7. Chien, L.-H. and Webb, R. L. (1996) Parametric Studies of Nucleate Pool Boiling on Structured Surfaces, Part II: Effect of Pore Diameter and Pore Pitch, *ASME Proc. National Heat Transfer Conf.*, Ed., L. Witte et al., Vol. 4, pp. 137-144.
8. Chien L.-H. and Webb, R. L. (1998) Visualization of Pool Boiling on Enhanced Surfaces, accepted for publication in *Experimental Thermal and Fluid Science*.
9. Chien L.-H. and Webb, R. L. (1998) Measurement of Bubble Dynamics on an Enhanced Boiling Surface, accepted for publication in *Experimental Thermal and Fluid Science*.
10. Chien, L.-H. and Webb, R. L. (1998) Analytical Model of Nucleate Boiling on Structured, Enhanced Surfaces, to be published in *Int. Journal of Heat and Mass Transfer*.
11. Nakayama, W., Daikoku, T., and Nakajima, T. (1982) Effects of Pore Diameters and System Pressure on Saturated Pool Nucleate Boiling Heat Transfer From Porous Surfaces," *Journal of Heat Transfer*, Vol. 104, pp. 286-291.
12. Haider, I. (1994) A Theoretical and Experimental Study of Nucleate Pool Boiling Enhancement of Structured Surfaces, Ph.D. Thesis, Dept. Mech. Engr. Penn State Univ., University Park, PA.
13. R. L. Webb, R. L., and Haider, S. I. (1992) An Analytical Model for Nucleate Boiling on Enhanced Surfaces," in *Pool and External Flow Boiling*, Eds., V. J. Dhir and A. E. Bergles, ASME, New York, pp. 345-360.
14. Haider, S. I. and Webb, R. L. (1997) A Transient Micro-Convection Model of Nucleate Pool *Int. Jour. Heat and Mass Transfer*, Vol. 40, pp. 3675-3688.

15.   Webb, R. L., and Pais, C. (1992) Nucleate Boiling Data for Five Refrigerants on Plain, Integral-Fin and Enhanced Tube Geometries, *International Journal Heat and Mass Transfer*, Vol. 35, No. 8, pp. 1893-1904.

16.   Webb, R. L. (1972) Heat Transfer Surface Having a High Boiling Heat Transfer Coefficient, U.S. Patent 3,696,861 assigned to The Trane Co.

17.   Chien, L.-H. (1996) Mechanism and Analysis of Nucleate Boiling on Structured Surfaces, Ph.D. Thesis, Dept. Mech. Engr. Penn State Univ., University Park, PA.

18.   Adamek, T. A. and Webb, R. L. (1990) Prediction of Film Condensation on Horizontal Integral-fin Tubes, *Int. J. of Heat Mass Transfer*, Vol. 33, No. 8, 1721-1735.

19.   Chien,L.-H., Zarnescu, V. and Webb, R. L. (1998) Effect of Geometric Parameters on Boiling Performance for R-134a and R-22, submitted to *Journal of Enhanced heat Transfer*.

20.   Mikic, B. B. and Rohsenow, W. M. (1969) A New Correlation of Pool-Boiling Data Including the Effect of Heating Surface Characteristics, *Journal of Heat Transfer*, Vol. 91, 245-250.

21.   Mikic, B. B. and Rohsenow, W. M. (1969) Bubble growth rates in non-uniform temperature field, *Prog. In Heat Mass Transfer*, Vol. 2, 283-292.

HEAT EXCHANGERS FOR THERMOACOUSTIC REFRIGERATORS:

HEAT TRANSFER MEASUREMENTS IN OSCILLATORY FLOW

CILA HERMAN AND MARTIN WETZEL
Department of Mechanical Engineering
The Johns Hopkins University, 3400 North Charles Street
Baltimore, Maryland 21218

Abstract. During the past fifteen years heat transfer in oscillatory flows has become the subject of increasing interest in the engineering community. Applications of oscillatory flows include, for example, the cooling of electronic equipment or alternative, environmentally safe refrigeration technologies, such as thermoacoustic refrigeration, pulse tubes or Stirling refrigerators. Important components of these refrigerators are their heat exchangers. In such devices the working fluid is subjected to oscillatory forcing which is a key part of the process, as opposed to situations where oscillations are generated with the aim to enhance heat transfer. Heat transfer in oscillatory, and often compressible, flows has not yet been completely understood, and the lack of design methodologies for heat exchangers in such flows is one reason that efficiencies of these devices are limited. In this paper, convective heat transfer in steady flow is contrasted to measurements of heat transfer in oscillatory flow obtained by holographic interferometry in operating regimes characteristic of the heat exchangers. The results of the research are expected to lead to guidelines that will allow the design of heat exchangers with improved heat transfer performance.

1. Introduction

In recent years environmental aspects have become an important issue in the design and development of refrigeration systems. Current research efforts focus both on the development of alternative refrigerants and refrigerant mixtures and on alternative technologies, such as Stirling engines, pulse-tube refrigerators, thermoelectric refrigerators, etc. One promising approach in the category of alternative technologies is ThermoAcoustic Refrigeration (TAR), characterized by the naturally occurring oscillatory flow of the working fluid.

TAR was developed during the past two decades as a new, environmentally safe refrigeration technology that employs acoustic power to pump heat and does not rely on refrigerants. Performance calculations predict values of 40% to 50% of the Carnot efficiency for the thermoacoustic core [21]. One reason that these efficiencies have not been achieved in devices built up to date is the poor performance of the heat exchangers. A first step towards improving the performance of thermoacoustic devices was to introduce a systematic design methodology based on the first law analysis of the thermoacoustic refrigerator [21]. This methodology allows independent optimization of

285

S. Kakaç et al. (eds.), Heat Transfer Enhancement of Heat Exchangers, 285–299.

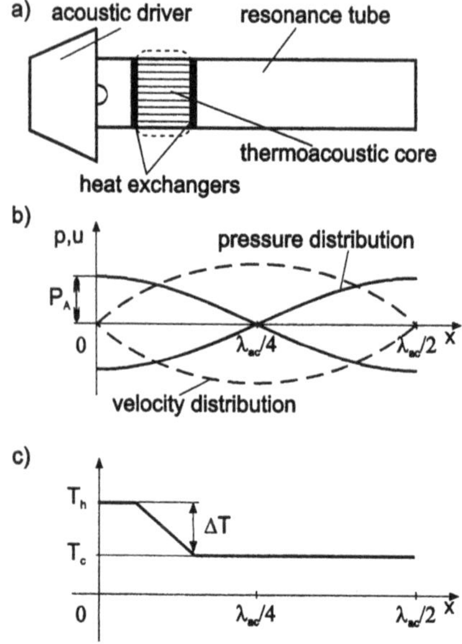

Figure 1. a) Schematic of the thermoacoustic refrigerator. b) Pressure, velocity and c) temperature distributions along the resonance tube.

individual components of the TAR system, such as the heat exchangers. In the next step, heat transfer coefficients necessary to improve the design of the heat exchangers in oscillatory flow in TAR were measured.

2. Thermoacoustic refrigeration

In Figure 1, a schematic of TAR is shown. With an acoustic driver, the working fluid in the resonance tube is excited to generate an acoustic standing wave. The length of the resonance tube corresponds to half the wavelength of the standing wave, $\lambda_{ac}/2$, and the pressure and velocity distributions in it are indicated in Figure 1b. When introducing a densely spaced stack of plates of length Δx at a location specified by the stack center position x_c into the acoustic field, a temperature difference ΔT, shown in Figure 1c, develops along the stack plates in the thermoacoustic core. For a detailed discussion of thermoacoustic theory, we refer to specialized literature [16, 17, 19].

A closer look into the thermoacoustic core, illustrated in the magnified region of Figure 2, reveals the mechanism responsible for thermoacoustic heat pumping, by considering the oscillation of a single gas parcel of the working fluid along a stack plate [17]. The gas parcel starts the cycle at a temperature T. In the first step it is moved to the left, towards the pressure antinode, by the acoustic standing wave. Thus it experiences adiabatic compression, which causes its temperature to rise by two arbitrary units to, T^{++}. In this state the gas parcel is warmer than the stack plate, and irreversible heat transfer towards the stack plate takes place (dQ_h in Figure 2). The resulting temperature of the gas parcel after this step is T^+. On its way back to the initial location, the gas parcel experiences adiabatic expansion and cools down by two arbitrary units, to T^-. At this state the gas parcel is colder than the stack plate and irreversible heat transfer from the stack plate towards the gas parcel takes place (dQ_c in Figure 2). After these steps, the gas parcel has completed one thermodynamic cycle; it has reached its initial location and temperature T, and therefore the cycle can start again. Since there are many gas parcels subjected to this thermodynamic cycle, and heat lost by one gas parcel is transported further by the adjacent parcel (as illustrated in the bottom portion of Figure 2), a temperature gradient ΔT (Figure 1c) develops along the stack plates.

In TAR two characteristic length scales of the thermoacoustic core, the thermal penetration depth $\delta_\kappa \equiv \sqrt{2\alpha/\omega}$ and the viscous penetration depth $\delta_\nu \equiv \sqrt{2\nu/\omega}$

(Figure2), are of importance for design purposes. α and v are thermal and momentum diffusivities, respectively, of the working fluid, and ω is the angular frequency. The thermal penetration depth δ_κ describes the layer around the stack plate where the thermoacoustic effect occurs as the distance heat can diffuse through the fluid during the time interval corresponding to one cycle of oscillations, $1/\omega$. The viscous penetration depth δ_v describes the thickness of the layer of fluid around the stack plates that is restrained in its movement under the influence of viscous forces, thus contributing less to the thermoacoustic effect. The ratio of δ_κ and δ_v, that depends on the Prandtl number, determines the volume of the thermoacoustically active working fluid. Suitable working fluids for TAR are low Prandtl number fluids obtained by mixing noble gases.

3. Heat exchangers for TAR

In order to exploit the thermoacoustic effect for heat pumping, heat exchangers are attached at both ends of the thermoacoustic core (Figure 1a). The cold heat exchanger removes heat from a cold temperature reservoir at T_r and supplies it to the cold side of the thermoacoustic core at T_c. The hot heat exchanger, operating at T_h , rejects the pumped heat and the absorbed acoustic work to the environment at T_e.

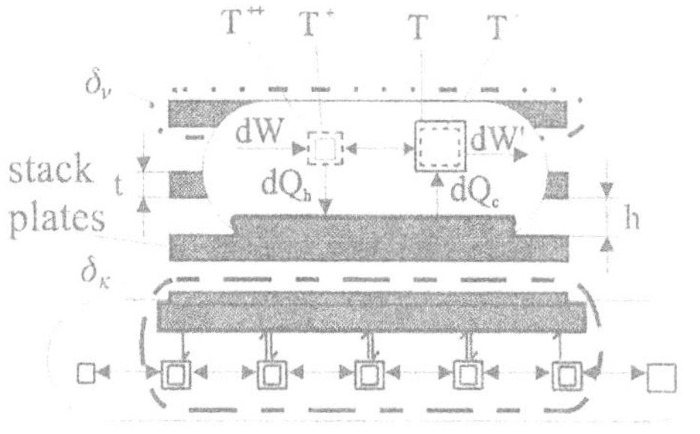

Figure 2. Illustration of the heat pumping cycle along a stack plate by considering the oscillation of one gas parcel.

Up to date, the heat exchangers have been designed without much optimization of their thermal performance, and in some cases they can also cause a substantial flow blockage, with attendant large pressure drop penalty [18]. This pressure drop contributes to the dissipation of acoustic power. Swift [18] measured the acoustic losses in the heat exchangers of a thermoacoustic engine, and found them to be of the order of 25% of the produced acoustic power. The difficulty of modeling thermoacoustic heat transfer in the heat exchangers can be attributed to the fact that we are dealing with oscillatory flow with zero mean velocity. In such situations, standard heat exchanger design methods, such as the effectiveness-NTU (**N**umber of **T**ransfer **U**nits) method or the LMTD (**L**ogarithmic **M**ean **T**emperature **D**ifference) method cannot be applied directly.

In Figure 3 a schematic of the heat exchangers used in Hofler's thermoacoustic refrigerator is shown [9]. Hofler used a long plastic sheet wound spirally around a plastic rod and spaced by fishing wires instead of parallel stack plates. As heat exchangers Hofler used rectangular copper strips. The length scale that determines the width of heat exchangers for TAR is the tidal displacement $x_{td} \equiv (2u_A(x))/\omega$ of the gas parcel at the location of the heat exchanger, where the velocity fluctuation amplitude $u_A(x)$ is obtained from the velocity distribution shown in Figure 1b. The tidal displacement defined in this way describes the peak to peak displacement of the moving gas parcel discussed in section 2. Swift [17] recommends an optimum width for the heat exchangers that matches the tidal displacement, since this allows the gas parcel to be in contact with both the stack plate and the heat exchangers. If the heat exchangers are shorter than the tidal displacement, the gas parcel moves beyond the heat exchanger strip and cannot deposit to or extract heat from the heat exchangers (cold heat exchanger in Figure 3). On the other hand, if the heat exchangers are longer than the tidal displacement, heat picked up from the heat exchanger is dropped back to the same heat exchanger (hot heat exchanger in Figure 3). The design choices in Hofler's design were primarily governed by the amount of heat that had to be rejected to the environment [9]. Since the hot heat exchanger had to reject more heat, it is larger than the cold heat exchanger. The heating and cooling loads of these heat exchangers were of the order of a few watts. Garrett [6] has modified Hofler's design by welding copper tubing onto the copper strips to pump a coolant, and increased the cooling loads to a few hundred watts.

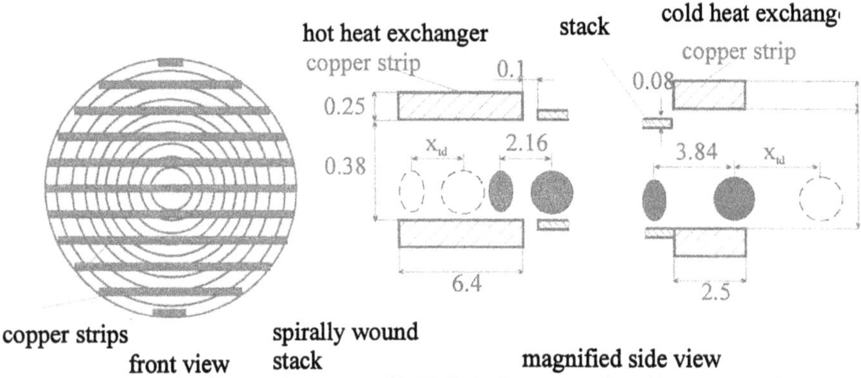

Figure 3: A schematic of heat exchangers used in Hofler's thermoacoustic refrigerator [9]. All dimensions are in millimeters.

From this brief review of thermoacoustic heat exchangers and from Figure 3 it is obvious that the presence of the heat exchangers causes a flow blockage and produces complex flow fields in the neighborhood of the stack. Furthermore, as we will show in this paper, effects at the edges of the stack plates are essential for the heat transfer analysis and design optimization. Thus the design optimization of heat exchangers for TAR is a complex problem that requires heat transfer and pressure drop data as a starting point (not yet available) as well as taking effects specific for thermoacoustic applications into account.

4. First law analysis of TAR

Energy flows in a TAR are displayed in Figure 4. To describe the overall thermodynamic performance of refrigerators, the **Coefficient Of Performance, COP,** defined as

$$\text{COP} \equiv \frac{\text{heat extracted at lower temperature } T_r}{\text{work done on the machine}} \equiv \frac{Q_{load}}{E_{el}}, \tag{1}$$

is typically used. For a TAR, the overall COP is defined as the ratio of the cooling load Q_{load}, introduced into the system through the cold heat exchanger, to the electric power input, E_{el}, into the acoustic driver. The conversions of these two energies within the components of the thermoacoustic refrigerator system are illustrated in the energy flow diagram in Figure 4.

The electric power input E_{el}, introduced into the system, is converted into acoustic power W_{tot} by the acoustic driver. For this energy conversion the electroacoustic efficiency η_{elac} can be defined as

$$\eta_{elac} \equiv \frac{W_{tot}}{E_{el}} = \frac{W_{tc} + W_{dis}}{E_{el}} = \frac{W_{tc} + W_{res} + W_{ex}}{E_{el}} .\tag{2}$$

η_{elac} is of the order of 3% for commercial loudspeakers. For acoustic drivers dedicated for TAR an efficiency of 50% has been demonstrated [5], and values up to 90% should be possible [4].

Figure 4 illustrates that not all of the acoustic power W_{tot} can be exploited to pump heat through the thermoacoustic core, since the thermoacoustic effect occurs not only within the thermoacoustic core but also on the surfaces of the resonance tube and the heat exchangers, where its contribution is dissipative, W_{dis}. The acoustic power losses in the hot portion of the resonance tube and at the hot heat exchanger, W_{hres} and W_{hex}, respectively, act as additional heating loads. The acoustic power losses in the cold portion of the resonance tube and at the cold heat exchanger, W_{cres} and W_{cex}, act as additional cooling loads. The upper limit of the TAR's performance is determined by the performance of the thermoacoustic core, defined as

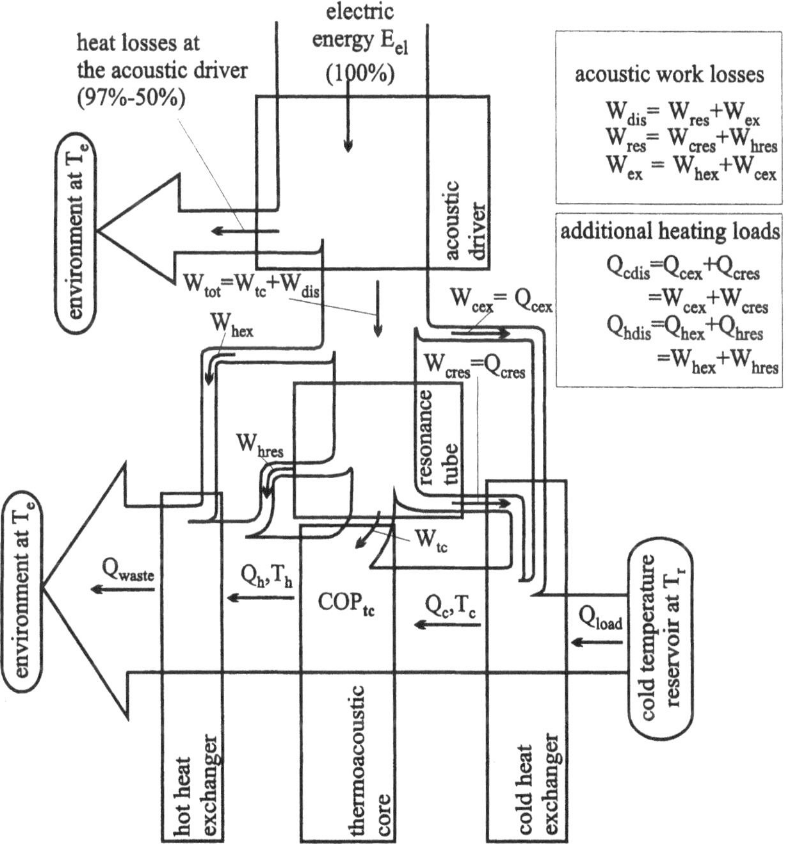

Figure 4: Energy flows in a thermoacoustic refrigerator.

290

$$\text{COP}_{tc} \equiv \frac{\text{heat extracted at lower temperature } T_c}{\text{work done on the thermoacoustic core}} = \frac{Q_c}{W_{tc}}. \tag{3}$$

The COP_{tc} can be estimated by applying a simplified linear model describing thermoacoustic processes, the short stack boundary layer approximation [17].

At this point the effectiveness ε for the cold heat exchanger is introduced as

$$\varepsilon \equiv \frac{\text{heat transferred}}{\text{maximum heat transferable}} \equiv \frac{Q_{load}}{Q_c} = 1 - \frac{Q_{cdis}}{Q_c} = 1 - \frac{Q_{cres} + Q_{cex}}{Q_c}, \tag{4}$$

similar to conventional heat exchanger analysis [10]. Heat transfer through the cold heat exchanger is limited by the amount of heat Q_c the thermoacoustic core is capable of absorbing at T_c. However, because of the dissipation of acoustic power, the transferred heat is limited to the cooling load Q_{load}. After substituting Equations (2), (3) and (4) into Equation (1), the expression for the overall COP of a thermoacoustic refrigerator becomes

$$\text{COP} = \eta_{elac} \cdot \varepsilon \cdot \frac{W_{tc}}{W_{tot}} \cdot \frac{Q_c}{W_{tc}} = \eta_{elac} \cdot \varepsilon \cdot \eta_{ac} \cdot \text{COP}_{tc}. \tag{5}$$

The acoustic power efficiency $\eta_{ac} \equiv W_{tc}/W_{tot}$ in Equation (5) accounts for the fact that not all of the converted acoustic power W_{tot} can be used to pump heat through the thermoacoustic core. η_{ac} achieves its maximum value of one when $W_{dis}=0$. By definition, the maximum possible values for the electroacoustic efficiency η_{elac} and the effectiveness ε of the cold heat exchanger are also one. Thus, the overall COP of the thermoacoustic refrigeration system reaches its maximum value, that corresponds to the COP_{tc} of the thermoacoustic core, when η_{elac}, η_{ac} and ε reach their maximum value of one. This result suggests the option to design and optimize the four main modules of the TAR system, the (i) thermoacoustic core, (ii) resonance tube, (iii) heat exchangers and (iv) acoustic driver, separately at this stage of the research. Improving the design of the cold heat exchanger by decreasing the acoustic power losses W_{cex} improves the effectiveness ε as well as the acoustic power efficiency η_{ac}. By reducing the acoustic power losses W_{cres} in the cold portion of the resonance tube the effectiveness ε of the cold heat exchanger increases as well. The foregoing discussion implies that, at this stage of the analysis, we assume that the four elements of Equation (5), η_{elac}, ε, η_{ac} and COP_{tc} are independent. Since the COP_{tc} for some design solutions may depend on the

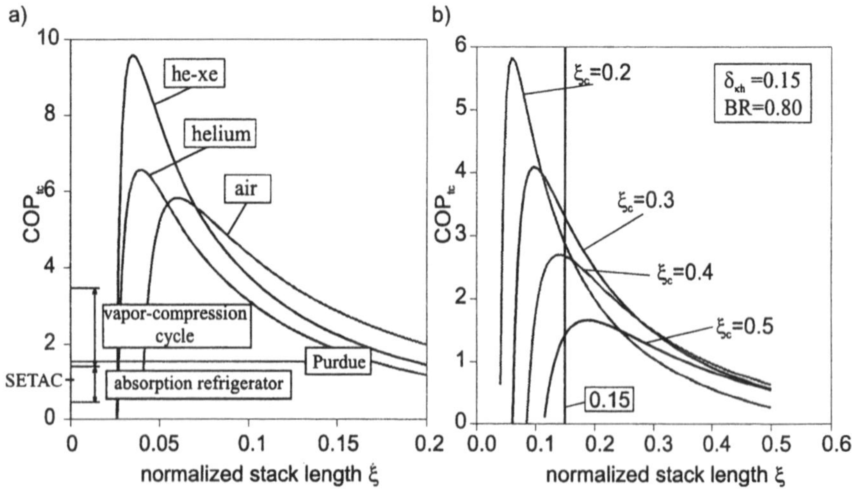

Figure 5. Coefficient of performance of the thermoacoustic core COP_{tc} for a) three working fluids and b) four stack center positions as function of the normalized stack length ξ.

design of the resonance tube and the heat exchangers, interdependence of COP_{tc}, ε and η_{ac} will have to be considered in a more advanced optimization procedure to be developed in the future. Clearly, this would require an accurate model of the heat exchangers, one that is not yet available.

In Figure 5 sample results of performance calculations expressed in terms of the COP_{tc} are shown. Typical values for the vapor compression and the absorption cycles, as well as values achieved by two thermoacoustic refrigerators, the SETAC [13] and the one developed at Purdue University [11] are also indicated in Figure 5a. In Figure 5a, results of performance calculations for three working fluids, a helium-xenon mixture (He-Xe(62%/38%)), pure helium and air, are illustrated. In Figure 5b results that quantify the effect of the stack center position $\xi_c = 2\pi x_c / \lambda_{ac}$ on the COP_{tc} are presented. More details regarding the performance optimization of TARs were reported by Wetzel and Herman [21].

5. Temperature measurements in the stack region

To investigate heat transfer at the edge of the thermoacoustic stack and in the region of the heat exchangers experimentally, it is necessary to measure the temperature distributions in the working fluid, stack plates and heat exchangers. For temperature measurements in the stack plates and heat exchangers, thermocouples were successfully applied in previous experimental investigations [1, 2, 11, 18]. However, because of their intrusive nature, thermocouples cannot be used to measure temperature distributions in the working fluid. In the present study these problems were addressed by applying Holographic Interferometry (HI) combined with high-speed cinematography. For the optical configuration used in the present study the temporal resolution is limited to 0.1ms. The selected experimental approach allowed us to measure local, time averaged heat fluxes and heat transfer coefficients at the edge of the stack plates. For more details on HI and the optical setup used in the present study the reader is referred to the literature [7, 8, 21, 22, 23].

5.1. MEASUREMENT METHOD

HI allows to visualize and measure differences in the refractive index Δn between a reference and an object wave, which can then be related to a single field variable of interest, such as temperature or concentration, to determine its spatial distribution [12, 15, 20]. Until now, most applications of high-speed HI were limited to physical situations in which the difference in the refractive index Δn was caused by a single field variable. In the case of an acoustically driven flow, periodic pressure variations are present in addition to the oscillating temperature field. Consequently, the difference $\Delta n(T, p)$ in the refractive index depends on both field variables, temperature as well as pressure, and the equation of ideal interferometry can be written as

$$S \cdot \lambda = L \cdot \left[\frac{\partial n}{\partial T}\bigg|_p \cdot \Delta T + \frac{\partial n}{\partial p}\bigg|_T \cdot \Delta p \right]. \tag{6}$$

In Equation (6) L denotes the length of the path of the light beam through the phase object, the working fluid in the resonance tube of the TAR, S the fringe order and λ the wavelength of laser light. Equation (6) illustrates that interference fringes cannot be directly interpreted as isotherms or isobars. Wetzel and Herman [22] introduced the concept of "quasi-isotherms", isotherms whose temperature oscillates following the pressure oscillations, to account for this effect.

Wetzel and Herman [22, 23] have introduced new field variable, pressure, into the evaluation procedure. The key results of their approach are two evaluation formulas, one for the time averaged temperature field

$$T_m(x,y) = \frac{T_\infty}{1 + a S_m(x,y)} \, , \tag{7}$$

and one for the temperature fluctuation amplitude

$$\frac{T_A(x,y)}{T_m(x,y)} = \frac{p_A(x,y)}{p_m} - \frac{a S_A(x,y)}{1 + a S_m(x,y)} \, . \tag{8}$$

Equations (7) and (8) indicate that the time averaged temperature fields are independent of the pressure variations, while for the temperature fluctuation amplitude it is necessary to account for the small pressure fluctuations in form of the pressure term $p_A(x,y)/p_m$ in Equation (8). For a more detailed discussion of Equations (7) and (8) the interested reader is referred to the literature [22, 23, 25]. In the present study Equations (7) and (8) were applied to quantify the oscillating temperature fields in a single plate thermoacoustic refrigerator.

5.2. EXPERIMENTAL SETUP

The experimental setup is presented schematically in Figure 6a. The length of the resonance tube ($l = 510\text{mm}$) matches half the wavelength of the acoustic standing wave, $\lambda_{ac}/2$. Since we use air at atmospheric pressure as working fluid, with a speed of sound $c = 343\text{m/s}$ we obtain an operating frequency of 337Hz. The walls in the region of the thermoacoustic core were built using transparent material to allow transillumination by laser light, as indicated in Figure 6a. The cross section of the resonance tube is rectangular in our experiments, and thus the experimental setup represents a two dimensional model of the physical situation in the Cartesian coordinate

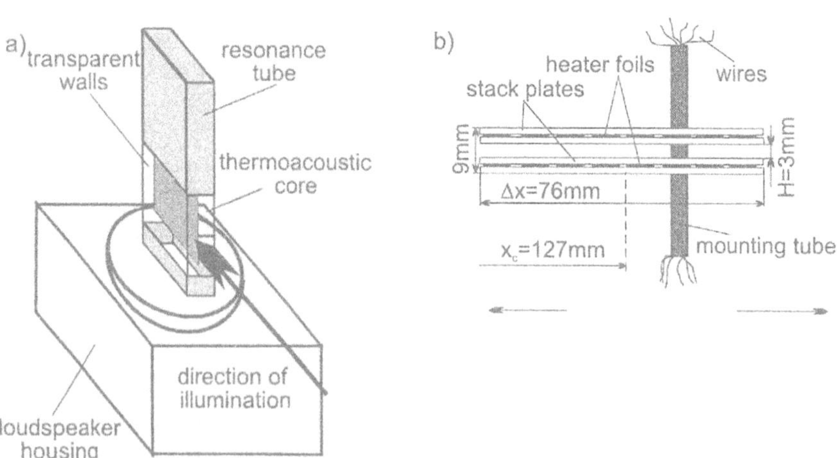

Figure 6. a) Schematic of the experimental setup and b) the thermoacoustic core.

system. The selection of the plate spacing, $H = 3\text{mm}$ in Figure 6b, was a result of optical constraints. In order to keep the physical situation as well as the experimental setup as simple as possible, in the present study we focused our attention on two stack plates only, and generated the linear temperature gradient ΔT that develops along the stack plates during the operation of the TAR (Figure 1c) with heater foils.

The acoustic driver, accommodated in the loudspeaker housing shown in Figure 6a, is a commercially available woofer from ElectroVoice (EV-10). The 337Hz input signal to the loudspeaker is generated by an HP 8116A function generator and then amplified using a Crown DC-300A Series II amplifier. To measure the acoustic pressure generated by the loudspeaker, a dynamic pressure transducer (Sensym SX01) was mounted at the entrance of the resonance tube. The transducer measures the dynamic peak pressure amplitude P_A of the acoustic standing wave. From the peak pressure amplitude, the drive ratio, $DR = P_A / p_m$, is determined as the ratio of peak pressure amplitude P_A to the time averaged pressure p_m in the working fluid. The drive ratio is an important flow parameter that characterizes the operating regime of a thermoacoustic refrigerator. Since our experimental setup represents a thermoacoustic refrigerator, experiments for drive ratios ranging from 0.5% to 3% were conducted. The experimental uncertainties for these experiments were reported in previous publications [22, 23, 25].

5.3. VISUALIZED TEMPERATURE FIELDS

A typical sequence of interferometric images corresponding to one period of acoustic oscillations in the stack region of the TAR is illustrated in Figure 7. The gray bars correspond to the two stack plates and the dark and bright fringes represent quasi-isotherms. With a sampling rate of 5000 frames per second and a frequency of acoustic oscillations of 337Hz, we obtained 15 images representative of 15 time instants t_i of the temperature field for one period of oscillations.

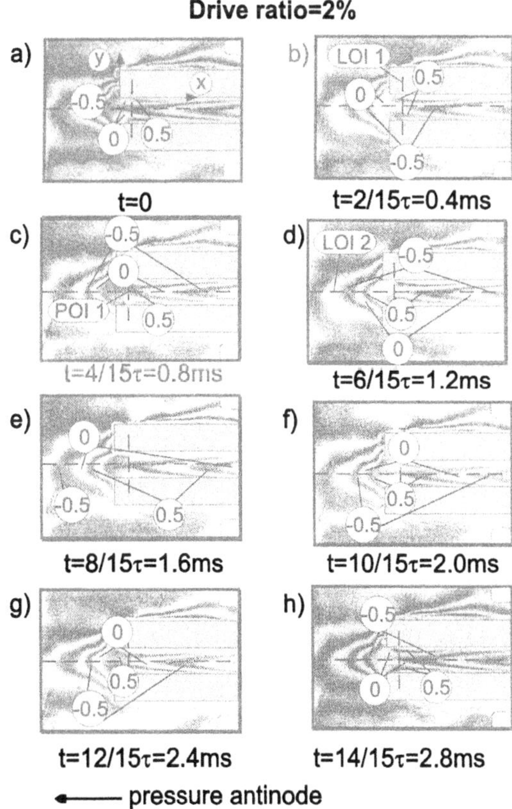

Figure 7. Oscillating temperature fields visualized in the form of quasi-isotherms in the stack region of the thermoacoustic refrigerator model at a drive ratio $DR=2\%$.

Since we have used the infinite fringe field alignment in these measurements, advance knowledge of the temperature distribution is vital in determining the correct interference order S, when applying the evaluation procedure described by Vest [20]. The physical situation yields two essential constraints relevant for the evaluation of interferometric fringes. First, the temperature at the upper stack plate is higher than at the **Point Of Interest 1** (POI 1) in Figure 7c, which is important when evaluating the temperature distribution along the **Line Of Interest 1** (LOI 1) in Figure 7b. The POI 1 denotes the intersection of the LOI 1 and the LOI 2, shown in Figures 7b and d. Because of symmetry, the temperature increases again from POI 1 towards the lower stack plate. This tendency is illustrated in Figure 7a by assigning the interference orders S in decreasing order when the temperature decreases and vice versa. It also follows that, due to symmetry, the two fringes marked with 0 and the two fringes marked with 0.5 (on each side of the LOI 2) in Figure 7a represent the same two temperatures, respectively. The second constraint implies that, when evaluating the temperature distribution along LOI 2, we know that we are analyzing the hot end of the two stack plates. Therefore, the temperature at the right end of LOI 2 is lower than at the POI 1. Somewhere around the POI 1 the temperature reaches its maximum, and then it decreases towards the left end of LOI 2. From this discussion it follows that the fringes marked with 0 in Figure 7c correspond to the same temperature.

We now focus on the displacement of the interferometric fringes numbered with -0.5, 0 and 0.5 in Figure 7, to demonstrate how one period of oscillations was detected. As

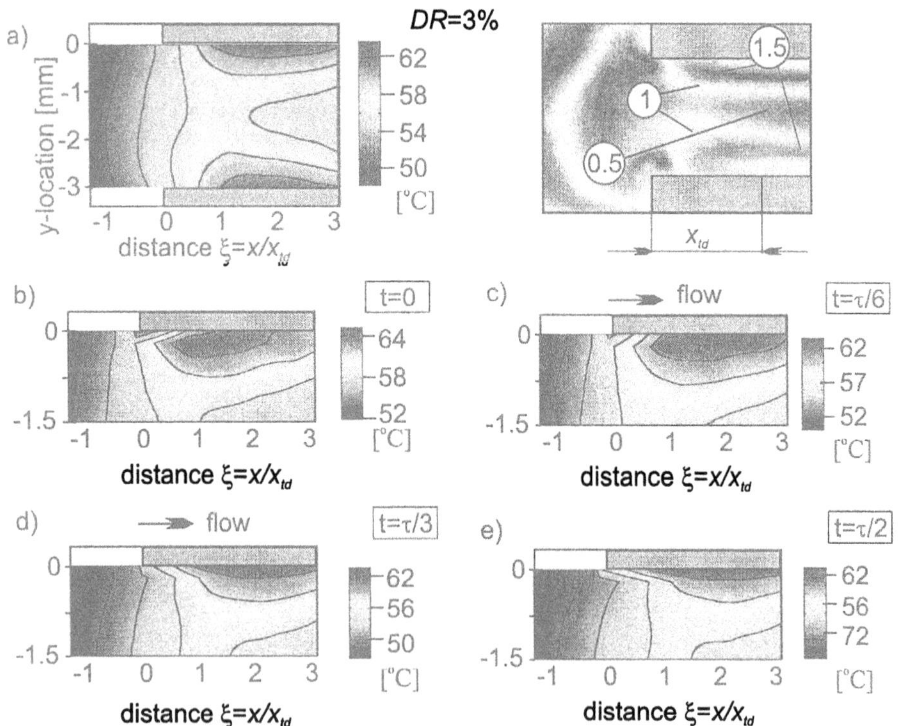

Figure 8. a) Time averaged temperature field and the corresponding interferometric image for the drive ratio $DR = 3\%$. b)-e) Sequence of instantaneous temperature distributions for $DR = 3\%$.

time progresses (Figures 7a-d), the dark fringe numbered with –0.5 in Figure 7a breaks apart to form two fringes that move to the left and right along the LOI 2. During the same time sequence (Figures 7a-d) the pairs of bright and dark fringes numbered with 0 and 0.5 in Figure 7a, respectively, move towards the center of the two stack plates until they merge (Figure 7b for the bright fringe, Figure 7c for the dark fringe) to form a single fringe. Then they break apart too, to form pairs of fringes (Figure 7c for the bright fringes numbered with 0, Figure 7d for the dark fringes numbered 0.5), which now move in the direction perpendicular to the original direction along the LOI 2. Beyond this point (Figures 7e-h) the interferometric fringes numbered with –0.5, 0 and 0.5 start to move in the reverse direction, until they have almost reached their initial location in Figure 7h.

5.4. MEASURED TEMPERATURE FIELDS

The time averaged temperature distribution and the corresponding interferometric image for a drive ratio $DR= 3\%$ as function of axial distance $\xi = x/x_{td}$ are illustrated in Figure 8a. The interferometric image in Figure 8a was recorded at a time instant when the pressure fluctuation $\delta p(x,y,t)$ was approximately zero, and in such a case the interference fringes represent isotherms [22, 23, 25]. In the region $0 \le \xi \le 1$ the temperature of the working fluid is higher than that of the stack plates. In the interferometric image the warmer fringe tagged with 1 rolls over the colder fringe of the order 0.5, which is indicative of heat being transferred into the stack plate.

A key result of the present study is that the heat flow into the stack plates represents a significant departure from conventional boundary layer behavior in the region close to the edge of the plate. Since the stack plates are heated with heater foils, in steady flow one would expect heat transfer from the stack plates to the working fluid. In steady channel flow such a situation would normally be advantageous for heat transfer due to the development of the thermal boundary layer, which is accompanied by high heat transfer rates from a heated surface into the working fluid. Clearly, this is not the case in oscillating flow.

Instantaneous temperature distributions $T(x,y,t)$ in the upper half of the channel for the drive ratio $DR=3\%$ are shown in Figures 8b-e. The bottom portions of the temperature fields mirrors the distributions displayed in Figure 8. The sequence of temperature distributions in Figures 8b-e illustrates one half of the thermoacoustic cycle representing, steps 2 through 4, of the gas parcel model [17] discussed in Section 2. In Figure 8b the working fluid is fully compressed and hotter than the stack plate, which is indicative of heat being transferred into the stack plate (second step in the gas parcel model). As time progresses, as illustrated in Figures 8c-e, the working fluid expands and cools down (third step in the gas parcel model). The fourth step of the gas parcel model is illustrated in Figure 8e. At this time instant the working fluid is fully expanded, it is colder than the stack plate, and heat is being transferred from the stack plate to the working fluid.

296

6. Heat transfer

Recent numerical and theoretical treatments of the thermoacoustic effect [3, 14, 24] have shown that at the edge of the stack plates of a thermoacoustic refrigerator high heat fluxes occur. At the cold side of the refrigerator the direction of these heat fluxes is from the stack plates to the working fluid, and at the hot side their direction is reversed (Figure 9a). Furthermore, the heat fluxes into or out of the stack plates occur roughly within a region of half the tidal displacement $x_{td}/2$ away from the edge of the stack plates [3, 14]. In the middle of the plate the heat fluxes caused by the thermoacoustic effect are parallel to the plates [3]. The direction of the thermoacoustic heat fluxes q''_{th} is indicated schematically in Figure 9a.

Local heat fluxes were determined from the temperature measurements by evaluating the gradients of the time averaged temperature fields in the y direction. The measured local heat fluxes $q''(\xi)$ for

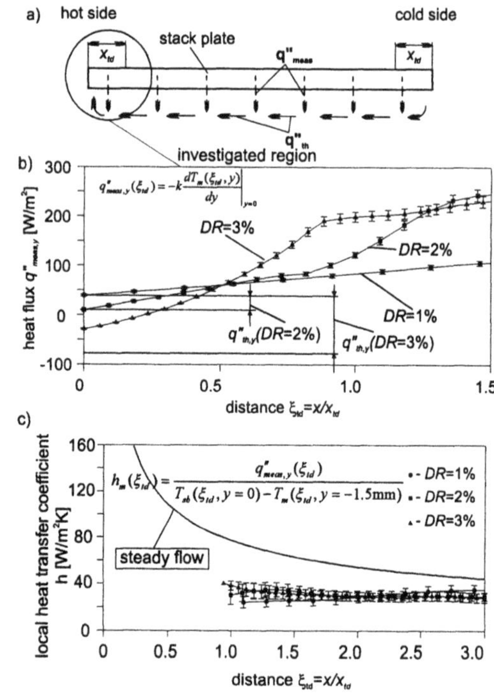

Figure 9. a) Schematic of the heat flux vectors generated by the thermoacoustic effect, q''_{th}, and the heater foils, q''_{hf}. b) Measured local heat fluxes $q''(\xi)$ and c) heat transfer coefficients at the edge of the stack plate.

the drive ratios DR = 1%, 2% and 3% are illustrated in Figure 9b. The results confirm the observations made while analyzing the time averaged temperature fields. For the drive ratios of 2% and 3%, for which the thermoacoustic effect is significant, the heat fluxes are negative at $\xi = 0$, thus indicating that heat is being transferred into the stack.

The reason for this behavior is the following. As long as we consider the heat fluxes $q''(\xi)$ in the middle of the stack plates ($\xi > 0.5$ in Figure 9b) where no edge effects are present, the heat fluxes caused by the thermoacoustic effect q''_{th} are parallel to the stack plate, while the measured heat fluxes are perpendicular to the plates. Thus, they do not affect each other, and the measured heat fluxes $q''(\xi)$ coincide with the heat fluxes q''_{hf} generated by the heater foils. However, as we approach the region $x_{td}/2$ away from the edge of the stack plates ($0 < \xi < 0.5$ in Figure 9b), the heat fluxes caused by the thermoacoustic effect q''_{th} change their direction, as indicated in Figure 9a. Therefore, q''_{th} is perpendicular to the stack plates, and it affects the measured heat flux as

$$q''(\xi) = q''_{hf}(\xi) - q''_{th}. \qquad (9)$$

If the thermoacoustic heat flux q''_{th} in Equation (9) is larger than the heat flux q''_{hf} generated by the heater foils, the measured heat flux $q''(\xi)$ is negative, and we indeed measure heat transfer into the stack plate. This behavior has also been observed in numerical and theoretical treatments of the problem [3, 14, 24]. Our measurements verify, for the first time, the presence of high local heat fluxes into the stack plate at the edge of the stack plate's hot side. The observed behavior does not agree with that expected in a steady developing boundary layer flow.

In Figure 9c the measured local heat transfer coefficient $h_m(\xi_{td})$ is plotted. The heat transfer coefficient remains nearly constant within the region $1 < \xi_{td} < 3$ for all three drive ratios. In the region $0 < \xi_{td} < 1$ the experimental uncertainties have increased (ΔT is very small for the drive ratios 2% and 3%), and it was not possible to reliably measure the heat transfer coefficient. To compare the measured heat transfer coefficient $h_m(\xi_{td})$ to values for steady forced convection, the heat transfer coefficient along a flat plate for the constant heat flux boundary condition [10] is plotted in Figure 9b as well. In order to be able to compare these results to the measurement data, the root mean square value of the velocity oscillation amplitude $u_A(x)/\sqrt{2}$ for the free stream velocity is used as reference. The measured heat transfer coefficients are lower than values for steady boundary layer flow. This result suggests that flow oscillations, that are frequently applied to enhance heat transfer in steady flow, do not lead to enhancement in the physical situation considered in the present study.

7. Conclusions

In the paper the potential of TAR was demonstrated through a first law analysis. The thermodynamic analysis suggests a separate optimization of the four main modules of a TAR; (i) thermoacoustic core, (ii) resonance tube, (iii) heat exchangers and (iv) acoustic driver. The heat pumping capacity of the thermoacoustic core sets an upper limit on the TAR's performance. Calculations of the thermoacoustic core's performance indicate that TAR can achieve COPs competitive to commercially available refrigerators. Of course, when comparing data with other refrigeration systems, we have to keep in mind that the presented calculations do not include energy losses in the other three modules (resonance tube, heat exchangers and acoustic drivers) of a TAR. As emphasized in this paper, a major problem in predicting the overall performance of a TAR is the lack of a suitable model for the heat exchangers.

Next, heat transfer issues essential for developing a systematic design methodology for heat exchangers in oscillatory flows were discussed on the example of TAR. One important step in this direction is to accurately measure the oscillating temperature fields as a function of time. A step towards the development of such a model was achieved by applying HI combined with high-speed cinematography to visualize and quantify the temperature fields at the edge of two stack plates, in the vicinity of the region where the heat exchangers would be inserted. The instantaneous temperature fields revealed that during one acoustic cycle the direction of the heat transfer changes. In other words, depending on the time instant within the acoustic cycle, we can have heat transfer from the working fluid to the plate or vice versa. This behavior could be explained by the gas parcel model of thermoacoustics, and reflects the thermoacoustic effect.

The measured time averaged temperature distribution within a distance of one tidal displacement away from the edge of the stack plates indicates that in this region heat is transferred from the working fluid to the plate, in spite of the heating generated by the heater foils in the plate. Evaluating the local heat fluxes in this region has confirmed this effect. Such a behavior differs in particular from steady state flow situation in forced convection, where high heat transfer rates are achieved at the edge of the plate because of the small thermal resistance of the developing boundary layer. Thus, we conclude that steady state intuitions and correlations cannot be directly applied to oscillatory flows. Measured heat transfer coefficients were lower than in the corresponding steady state situation. They provide basic data necessary to model the heat transfer performance of the heat exchangers.

Finally, we conclude that it is important to carefully investigate the differences between heat transfer in steady state flow and oscillatory flows in the future. The goal of these investigations is to generate databases that provide heat transfer and pressure drop correlations and databases necessary to develop systematic design methodologies for heat exchangers in oscillatory flows.

8. Acknowledgements

This work is supported by the Office of Naval Research.

9. References

1. Atchley, A.A., Hofler, T.J., Muzzerall, M.L., Kite, D., Ao, C. (1990) *Acoustically Generated Temperature Gradients in Short Plates*, J. Acoust. Soc. Am. 88(1), pp.251-263.
2. Brewster, J., Raspet, R., Bass, H. (1997) *Temperature Discontinuities between Elements of Thermoacoustic Devices*, J. Acoust. Soc. Am. 102(6), pp. 3355-3360.
3. Cao, N., Olson, J. R., Swift, G.W., Chen, S. (1996) *Energy Flux Density in a Thermoacoustic Couple*, J. Acoust. Soc. Am. 99 (6), pp. 3456-3464.
4. Corey, J.A., Yarr, G.A. (1992) *HOTS to WATTS: The LPSE Linear Alternator System Re-invented*, Proceedings of the 27th Intersociety Energy Conversion Engineering Conference, p 5.289.
5. Garrett,S.L. (1991) *ThermoAcoustic Life Sciences Refrigerator*, NASA Tech. Report No. LS-10114, Johnson Space Center, Space and Life Sciences Directorate, Houston, TX.
6. Garrett, S.L., Perkins, D.K., Gopinath, A. (1994) *Thermoacoustic Refrigerator Heat Exchangers: Design, Analysis and Fabrication*, 10th International Heat Transfer Conference in Brighton, England, 9-HE-9, August 14.-18, pp 375-380.
7. Hauf, W., Grigull, U. (1970) *Optical Methods in Heat Transfer*, in Adv. in Heat Transfer, Vol. 6, pp. 133-366, Academic Press, New York.

8. Herman, C. V., Mewes, D., Mayinger, F. (1992) *Optical Techniques in Transport Phenomena,* Advances in Transport Processes VIII , Editors A. S. Mujumdar, R. A. Mashelkar, Elsevier Science Publishers, Amsterdam, pp. 1-58.

9. Hofler, T.J. (1986) *Thermoacoustic Refrigerator Design and Performance,* Dissertation, University of California, San Diego.

10. Incropera, F.P., De Witt, D. P. (1990) *Fundamentals of Heat and Mass Transfer,* John Wiley & Sons, New York.

11. Marcotti, G. (1997) *Cool Designs,* Financial Times (London Edition), September, p.24.

12. Mayinger, F., Panknin., W. (1974) *Holography in Heat and Mass Transfer,* Proc. 5 Int. Heat Transfer Conf., Tokyo, Japan, pp. 28-43.

13. McKelvey, D., Ballaster, S., Garrett, S.L. (1995) *Shipboard Electronics Thermoacoustic Cooler,* J. Acoust. Soc. Am., Vol. 98, No.5, p. 2961

14. Mozurkewich, G. (1998) *Time-Average Temperature Distribution in a Thermoacoustic Stack,* J. Acoust. Soc. Am. 103(1), pp. 380-388.

15. Panknin, W. (1977) *Eine holographische Zweiwellenlängen-Interferometrie zur Messung überlagerter Temperatur- und Konzentrationsgrenzschichten,* Dissertation, University of Hannover, Germany.

16. Rott, N. (1980) *Thermoacoustics,* Adv. Appl. Mech., vol.20, pp. 135-175.

17. Swift, G.W. (1988) *Thermoacoustic Engines,* J. Acoust. Soc. Am. 84(4), , pp. 1145-1180.

18. Swift, G.W. (1992) *Analysis and Performance of a Large Thermoacoustic Engine,* J. Acoust. Soc. Am. 92(3), pp. 1551.

19. Swift, G.W. (1995) *Thermoacoustic Engines and Refrigerators,* Physics Today, July, pp.22-28.

20. Vest, C. M. (1979) *Holographic Interferometry,* John Wiley & Sons, New York.

21. Wetzel, M., Herman C. (1997) *Design Optimization of Thermoacoustic Refrigerators,* Int. J. Refrig., Vol. 20, No.1, pp. 3-21.

22. Wetzel, M., Herman C. (1998a) *Limitations of temperature measurements with holographic interferometry in the presence of pressure variations,* Experimental Thermal Fluid Science, Vol. 17, pp. 294-308.

23. Wetzel, M., Herman C. (1998b) *Accurate measurements of high-speed, unsteady temperature fields by holographic interferometry in the presence of periodic pressure variations,* Measurement Science and Technology, Vol. 9, No. 6, pp. 939-951.

24. Worlikar, A. (1997) *Numerical Simulation of Thermoacoustic Refrigerators,* Dissertation, The Johns Hopkins University.

25. Wetzel, M. (1998) *Experimental Investigation of a Single Plate Thermoacoustic Refrigerator,* Dissertation, The Johns Hopkins University.

A STUDY ON THE HIGH PERFORMANCE CERAMIC HEAT EXCHANGER FOR ULTRA HIGH TEMPERATURES

M. KUMADA
Department of Mechanical Engineering
Gifu University
1-1 Yanagido, Gifu City, Gifu Prefecture, 501-1112, Japan

Abstract. A ceramic heat exchanger using a fluidized bed for generating high temperature gas was developed for the coal-fired gas turbine combined cycle. Smooth and externally finned ceramic tubes were used for the heat transfer tubes. The characteristics of heat transfer coefficients and the performance of the heat exchanger were evaluated. It was certified that the heat transfer coefficient on the outside wall of the tube was sharply increased by using both a fluidized bed and the finned heat transfer tubes. But, the overall coefficient of heat transmission was dominated by the heat transfer inside the tube. Therefore, heat transfer enhancement on the inside wall of the tube is needed. This was accomplished by external finning. It was found that relaminarization occurred at high heating rates. Experiments were performed and heat transfer coefficients on the inside wall of the tube at high heating rates were evaluated.

1. Introduction

Recently, power generation has been exposed to the problem of the exhaustion of fossil fuels and global warming. Therefore, the use of renewable energy and the development of atomic energy have become more important. But, under existing conditions, they are not available, because the exploitation of renewable energy is small compared with the total amount of energy consumed, and atomic energy has a problem of safety. Above all, it is important to get rid of the dependence on oil fuel, and to use coal, which is abundant all over the world. In addition, the efficiency of power generation must be improved. High temperature heat exchange technology has become important for improving the performance of power generation. Many in the field have been counting on the development of a heat exchanger for generating high temperature gas. But, it is difficult for the conventional metal heat exchanger to be used at high temperatures or with corrosive gases. Metal heat exchangers have limits in their usage. Ceramic heat exchangers may raise those limits. Under existing conditions, however, the difficulty of manufacturing complicated surfaces with ceramics has prevented heat transfer from improving. By applying a fluidized bed to the ceramic heat exchanger for high

301

S. Kakaç et al. (eds.), Heat Transfer Enhancement of Heat Exchangers, 301–324.
© 1999 *Kluwer Academic Publishers.*

temperatures, improvement in heat transfer is anticipated due to the renewal effect by particles and the radiation from particles. When applying this technology to the multi-tube heat exchanger, however, many problems may arise. These problems include non-uniformity of bed temperature and difficulty of handling.

Many experiments have been published on the use of fluidized beds. These experiments were reviewed by Saxena use standard reference convention[1]. Most of them were performed at lower bed temperatures. It is difficult to extrapolate the results to a heat exchanger generating high temperature gas, because radiative heat transfer hads not been considered. Several experiments[2]-[4] have been published on fluidized beds with relatively high bed temperatures. In these experiments, since water was used as the operating fluid for the cooling tubes, tube wall temperatures were relatively low. In addition, single tubes were used in these experiments. Therefore, the results of these experiments were not sufficient for developing a heat exchanger with a fluidized bed for generating high temperature gas.

The purpose of this study is to develop a heat exchanger for the coal-fired gas turbine combined cycle. In this study, a ceramic heat exchanger using a fluidized bed for generating high temperature gas was developed. Experiments were carried out under conditions of ultra high fluidized bed temperature (i.e. maximum bed temperature was 1373K). The generated gas temperature was more than 1273K. Characteristics of heat transfer coefficients and the performance of the heat exchanger were evaluated. Firstly, a smooth ceramic tube bundle was used. But, in spite of manufacturing difficulties, ceramic tubes with external transverse thin fins, shaped like the usual metal fins, were manufactured for more compactness for the first time apparently. These experiments were made under nearly the same operating conditions. And, a comparison was made to estimate the effect of fins immersed in a fluidized bed.

2. Experimental Apparatus and Method

Experiments were carried out with the system shown in Fig. 1. Air-propane mixed gas was supplied through the distributor plate, and a flame was formed on the plate. Fluidizing gas passed through the free space (240mm height) after the fluidized bed. A porous medium (60mm thick) was set up at the top of the free space to prevent the temperature of the upper bed from dropping and particles from flowing away.

Cooling air supplied to the ceramic tubes was preheated to investigate the effect of cooling air temperature on heat transmission.

The fluidization tower was made of stainless steel. It was insulated with ceramic fibers and had a rectangular cross section (320×120mm). The wall of the fluidizing gas passage was covered with a thin ceramic board. Windows, which were made of quartz glass, were attached in the free space to observe the behavior of fluidization.

The distributor plate was 70mm thick, the 0.9×0.5mm lattice structure block which was made of cordierite, and the opening area was 0.65 of the total area.

Figure 2 shows the finned ceramic heat transfer tube. It was made of silicon nitride. The finned tube consisted of a smooth tube, used first for the experiments, and many

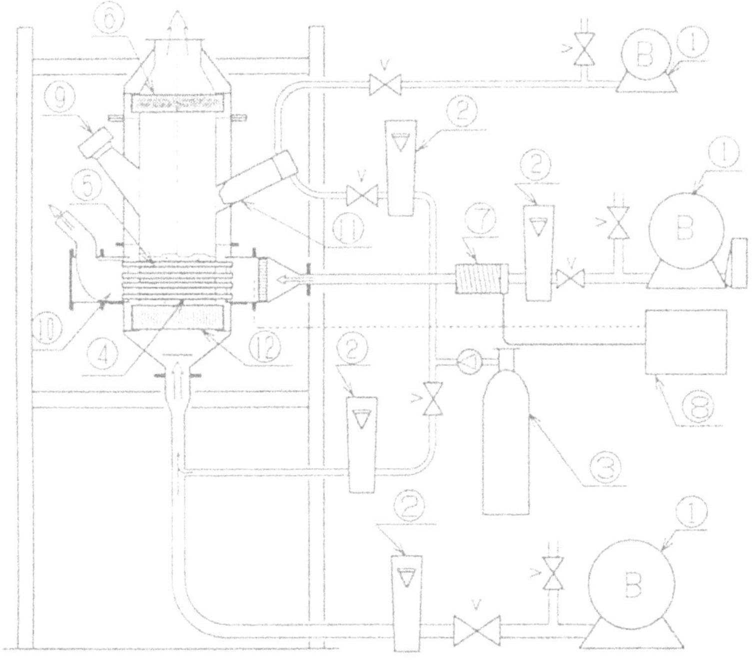

① Blower   ⑤ Ceramic Tubes   ⑨ Observation Window
② Flow Meter   ⑥ Porous Medium   ⑩ Mixing Chamber
③ Propane Gas Cylinder   ⑦ Electric Heater   ⑪ Pilot Burner
④ Fluidized Bed   ⑧ Thermo Controller   ⑫ Distributor Plate

Figure 1: Experimental apparatus

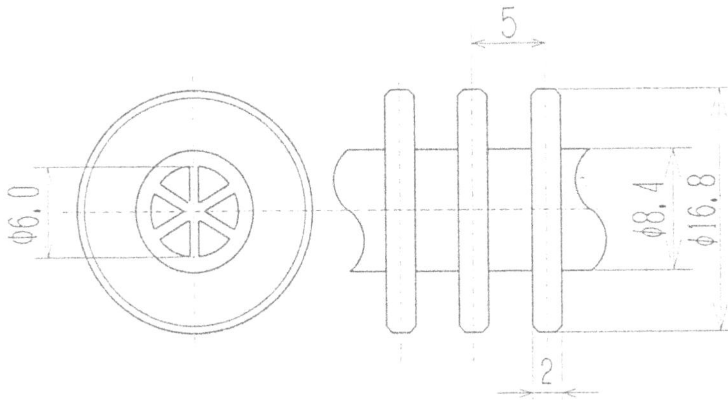

Figure 2: Finned ceramic tube

thin rings. The rings and the smooth tube were fitted together by heating the rings and allowing them to shrink to the smooth tube. The heat transfer tube had six ribs for heat transfer enhancement on the inside wall of the tube. The thickness of each rib was 0.3mm. The smooth tube had 8.4mm outer diameter and 6.0mm inner diameter. The diameter and the thickness of the circular fins were 16.8mm and 2.0mm, respectively. The fin pitch was 5.0mm. Fin shape, surface and interference were selected for their suitability based on preliminary experiments, carried out using 18 kinds of finned tubes. As a result of these experiments, rectangular cross section (but rounded tips), plain surface (without sandblasting or polishing) and 0.18mm interference fins were selected. Smooth and finned tubes were arranged in a staggered arrangement with 16mm and 21mm spacing between centers, respectively. The smooth tube bundle consisted of 33 tubes (four rows), and the finned tube bundle consisted of 22 tubes (five rows).

Particles were made of alumina (Al_2O_3), with a 1.0mm average diameter. The static bed height was maintained at the top of the heat transfer tubes.

The firing of the mixture was performed with a pilot burner attached at the free space. The flame ran across the bed by a flash back, and was stabilized on the distributor plate.

Both the fluidized bed and the tube wall temperatures were measured by R-type thermocouples with a 0.2mm diameter. These thermocouples were located at five positions along the axis of the tube.

In preparatory experiments, bed temperature was measured both by a bare thermocouple and a suction pyrometer, located at the center of the bed with 10mm spacing between each one. The effect of radiation on the reading of a bare thermocouple was negligible, because the difference between them was only about 1K.

Bed temperature was controlled by the method in which fluidizing air flow rate was first fixed, and then, propane gas flow rate was controlled to regulate the equivalence ratio.

Thermocouples for measuring the tube wall temperature were bonded in shallow grooves that were made on the heat transfer tube. Measuring positions were located 20, 90, 160, 230 and 300mm along the axis of the tube from the inner wall of the fluidization tower.

The outlet temperature of the cooling air was measured after the air exited the mixing chamber. This mixing chamber was set at the end of the tubes in order to average the outlet temperatures for the cooling air from all of the tubes. The overall heat transfer coefficients were evaluated with this averaged value.

3. Results and Discussions

3.1. BED AND TUBE WALL TEMPERATURE DISTRIBUTIONS

Typical temperature distributions for both the fluidized bed and the tube wall along the axis of the tube are shown in Fig. 3. Bed temperature was kept uniform in spite of a large change in cooling air temperature. Tube wall temperature was uniform vertically.

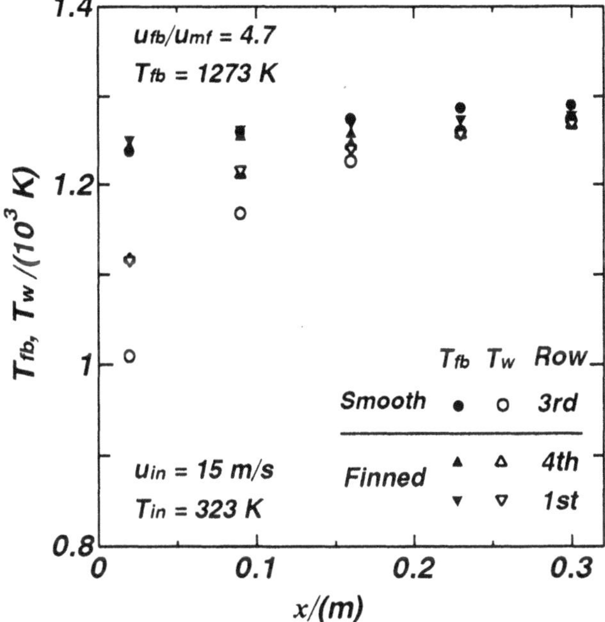

Figure 3: Bed and tube wall temperature distributions

Observations from the window also indicated stable fluidization all over the bed. With the fluidized bed temperature, fluidizing velocity, cooling air temperature and cooling air velocity various values, the temperatures incurred nearly the same trend, except that the tube wall temperature decreased with the cooling air temperature and velocity. The tube wall temperature of the finned tube was higher than that of the smooth tube, because of heat transfer enhancement on the outside wall of the tube.

3.2. HEAT TRANSFER COEFFICIENT ON THE OUTSIDE WALL OF A TUBE

The average heat transfer coefficient was estimated by eq.(1).

$$Q = M_{air} Cp_{air} (T_{out} - T_{in}) = h_{fb} A_o (T_{fb} - T_w) \tag{1}$$

The tube wall temperature and the fluidized bed temperature were the average value at five positions along the axis of the tube.

Figure 4 shows the variation of the Nusselt number on the outside wall of the tube with fluidizing velocity. The fluidizing velocity was non-dimensionalized by the minimum fluidizing velocity. Thermophysical properties were evaluated at the bed temperature.

The Nusselt number of the finned tube was about three times as large as that of the smooth tube.

The heat transfer coefficient of the tube bundle in the case of a pure air flow, calculated by Zukauskas correlation[5], was about 50W/m²K ($Nu_o = 6$) at $u_{fb}/u_{mf} = 4$ and a bed temperature of 1273K. Thus, the heat transfer coefficient of the finned

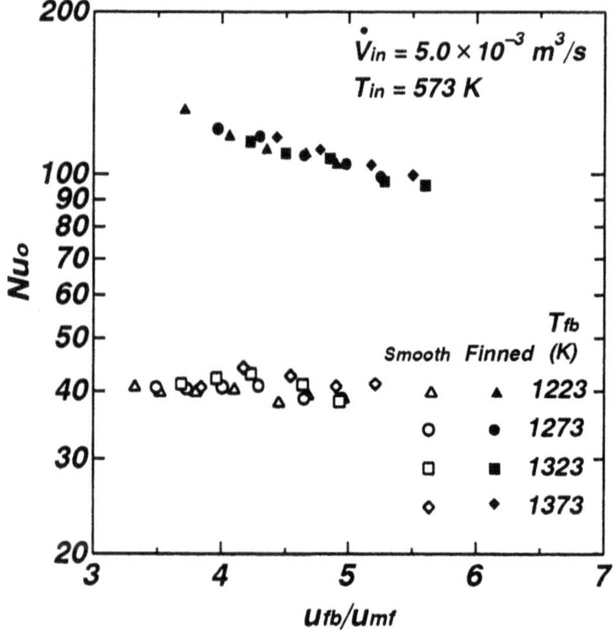

Figure 4: Variation of heat transfer coefficient with
fluidizing velocity

tube in the case of a fluidized bed was about 20 times as large as that in a pure air flow. As the finned tube was four times the surface of the smooth tube and the fin efficiency was about 0.85, the calculated ratio of the heat transfer coefficient of the finned tube to that of a single phase flow is about 3.4. The difference between calculation and experimental ratio was caused by the obstruction to the particle movement by fins.

The Nusselt number slightly decreases with increasing fluidizing velocity, as is the case with low temperature fluidized beds. But the decreasing tendency of the Nusselt number with the fluidizing velocity of the finned tube is larger than that of the smooth tube. The reason was that the density of the mixture decreases with increasing fluidizing velocity.

The Nusselt number for the smooth tube increased with bed temperature, because the effect of radiation increased with bed temperature. The result shows that the Nusselt number on the outside wall of the tube increased by 7% when the bed temperature increased by 100K with this experimental condition. But, the notable influence for the bed temperature on the heat transfer coefficient did not appear in the case of the finned tube.

3.3. OVERALL COEFFICIENT OF HEAT TRANSMISSION

The overall coefficient of heat transmission was evaluated by eq.(2). The logarithmic mean temperature difference, given in eq. (3), was used for the mean temperature

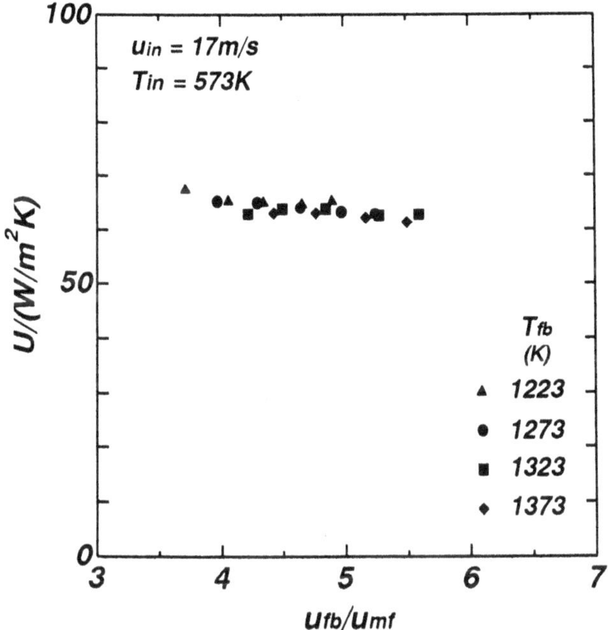

Figure 5: Variation of overall coefficient of heat
transmission with fluidizing velocity and bed temperature

difference.

$$Q = U A_o \Delta T_m \tag{2}$$

$$\Delta T_m = \frac{T_{out} - T_{in}}{ln \dfrac{T_{fb} - T_{in}}{T_{fb} - T_{out}}} \tag{3}$$

Figure 5 shows the variation of the overall coefficient of heat transmission of the finned tube with fluidizing velocity. The fluidizing velocity does not have a large influence on the overall coefficient of heat transmission. Also the influence of bed temperature was not notable. The reason was that the overall coefficients of heat transmission dominated by the heat transfer inside the tube. This was because the heat transfer coefficient on the outside wall of the tube was increased by applying the fluidized bed and fins. In the case of the smooth tubes, there was a similar trend, but we have omitted the results.

Figure 6 shows the variation of the overall coefficient of the heat transmission with the Reynolds number inside the tube. The Reynolds number was defined using the thermophysical properties evaluated at the average value of the inlet and outlet temperatures of the cooling air. The overall coefficient of heat transmission did not depend on the cooling air temperature, and increased with the cooling air velocity, b   of the increase in the heat transfer coefficient on the inside wall of the tube.

         all coefficient of heat transmission of the internally finned tube was a little

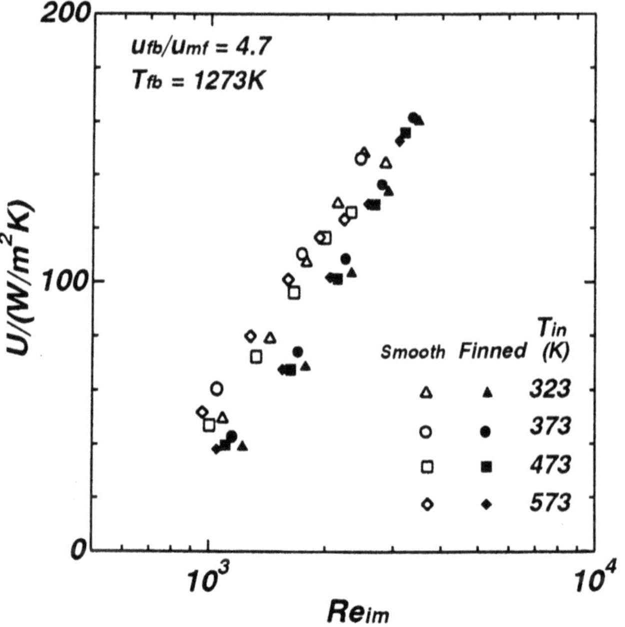

Figure 6: Variation of overall coefficient of heat transmission with cooling air velocity

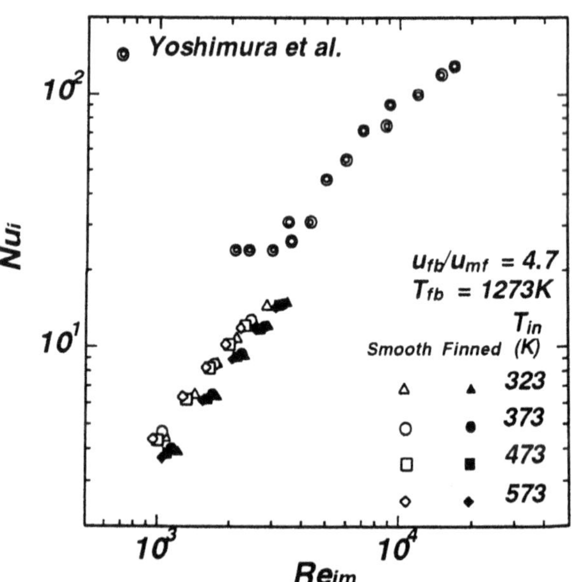

Figure 7: Average heat transfer coefficient on the inside wall of the tube

larger than that of the smooth tube. The reason was that the heat transfer coefficient on the inside wall of the smooth tube was a little larger than that of the finned tube. Furthermore, the effect of heat transfer enhancement on the outside wall of a tube was negligible, because the overall coefficient of heat transmission became dominant by the heat transfer inside the tube.

3.4. HEAT TRANSFER COEFFICIENT ON THE INSIDE WALL OF THE TUBE

Figure 7 shows the variation of the Nusselt number on the inside wall of the tube with the Reynolds number inside the tube. The heat transfer coefficient on the inside wall can be evaluated by eq.(4) using the average tube wall temperature and the average bulk temperature of the cooling air:

$$h_i = \frac{Q}{A_o\left(T_w - T_b\right)} \tag{4}$$

Thermophysical properties were evaluated at the average value of the inlet and the outlet temperature of the cooling air. The result of Yoshimura, et al.[6] using the same shaped smooth ceramic tube ($T_w < 923K$) was also plotted.

The influence of the Reynolds number on the Nusselt number was very large, whether smooth or finned. Nu_i was proportional to about the 1.1 power of Re_{im}. This tendency was not similar to the conventional tendency of a circular tube at low temperatures. This was due to the complicated coexistence of the large change of thermophysical properties of cooling air, the effect of the entrance region and secondary flow caused by the ribs inside the tube. This tendency suggests occurring relaminarization. More detailed investigation is needed to clarify the phenomenon.

The Nusselt number on the inside wall of the finned tube was a little larger than that of the smooth tube. The reason was the difference of the extent of relaminarization, because the heat flux of the finned tube was larger than that of the smooth tube.

The results of this study and Yoshimura et al.[6] showed a similar tendency.

3.5. TEMPERATURE EFFICIENCY

Temperature efficiency, estimated by eq.(5), was used to represent the performance of heat exchangers. T_{fbin} was the presumed temperature of the fluidizing air before heat-exchanging, estimated by eq.(6), using the fluidized bed temperature and heat transfer rate. The adiabatic combustion temperature may represent almost the same temperature as T_{fbin}. T_{fbin}, however, did not include the heat loss from the fluidization tower (the adiabatic combustion temperature is about 20K higher than T_{fbin}).

$$\eta = \frac{T_{out} - T_{in}}{T_{fbin} - T_{in}} \tag{5}$$

$$M_{fb}\, Cp_{fb}\, (T_{fbin} - T_{fb}) = M_{air}\, Cp_{air}\, (T_{out} - T_{in}) \tag{6}$$

Figure 8(a) and 8(b) show the variation of the temperature efficiency with fluidizing velocity

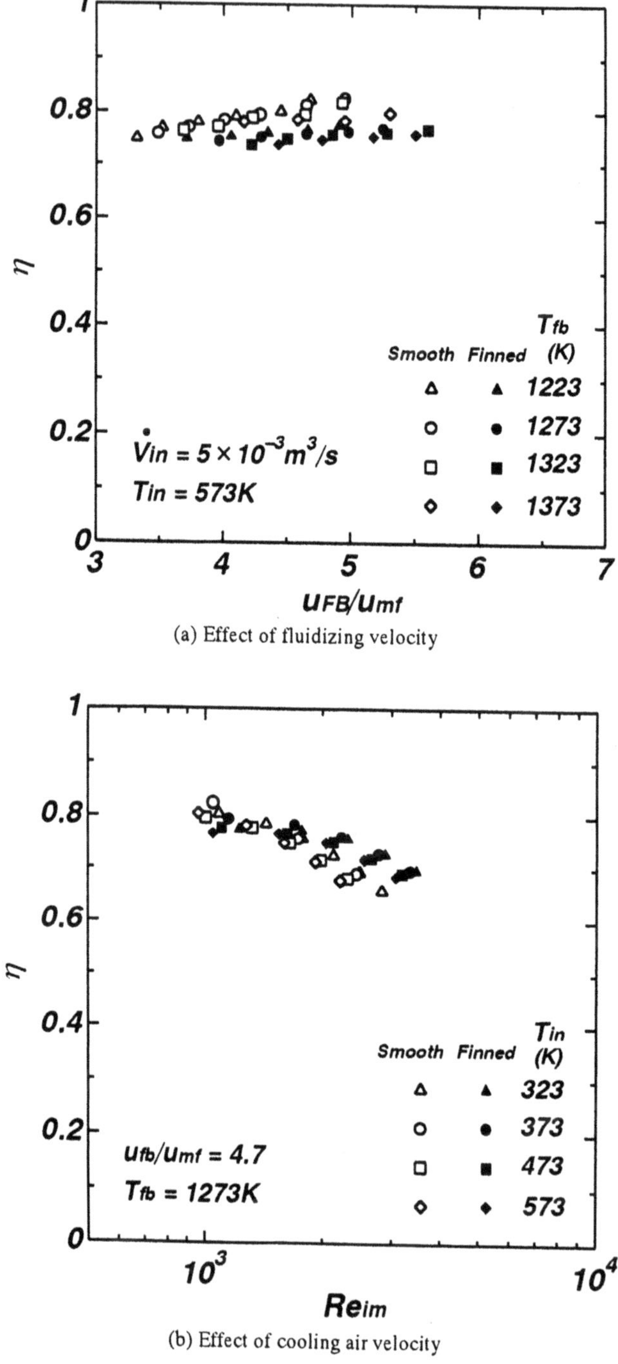

(a) Effect of fluidizing velocity

(b) Effect of cooling air velocity

Figure 8: Temperature efficiency

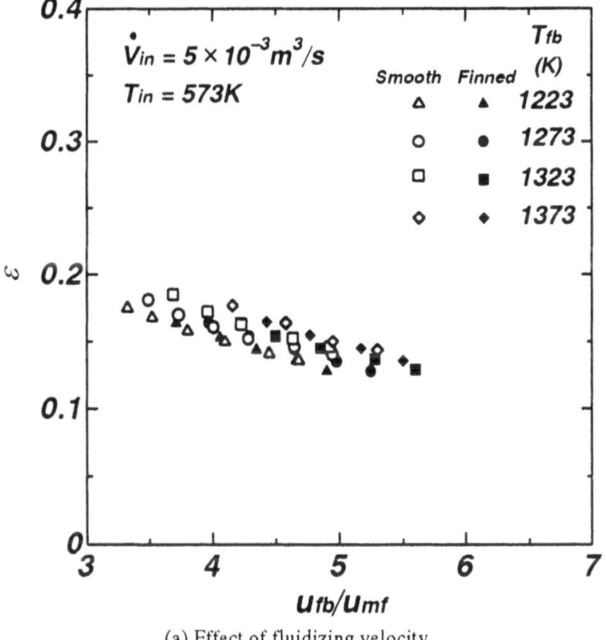

(a) Effect of fluidizing velocity

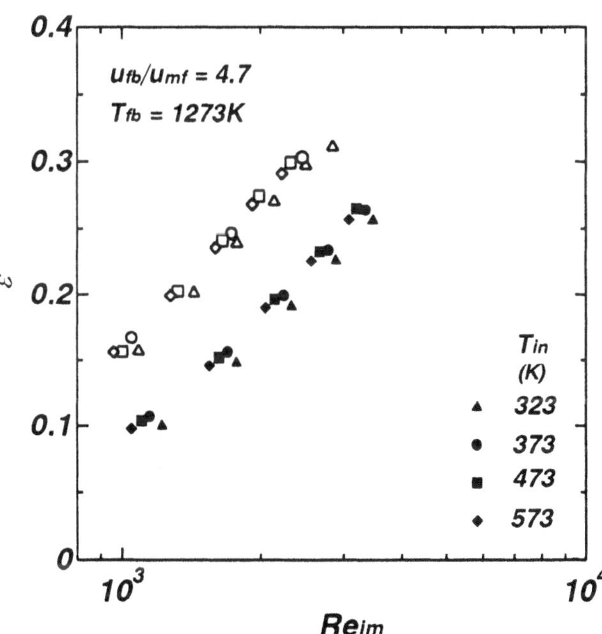

(b) Effect of cooling air velocity

Figure 9: Exergy efficiency

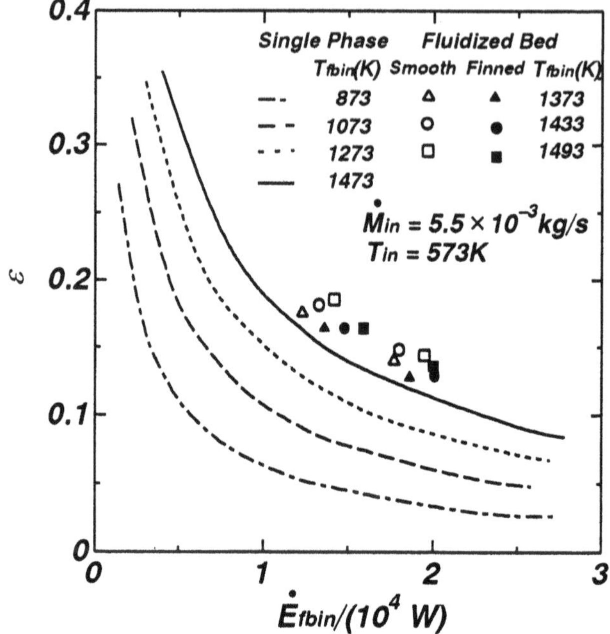

Figure 10: Variation of exergy efficiency with $\dot{E}_{fbin}$

and cooling air velocity, respectively. The temperature efficiency slightly increased with fluidizing velocity, and was almost constant with bed temperature. The temperature efficiency decreased with cooling air velocity.

Temperature efficiencies of the smooth and finned tube displayed nearly the same trend, but could not be compared directly. This is because the total number of finned tubes was not equal to that of the smooth tubes. Bed temperature was affected by the heat transfer rate (i.e., the total number of tubes) if the inlet temperature of the fluidized bed was constant.

3.6. EXERGY EFFICIENCY

Exergy efficiency, estimated by eq.(7), was also used to represent the performance of heat exchangers, because, the purpose of the high temperature heat exchanger in this study was to recover energy from high temperature gas.

$$\varepsilon = \frac{E_{out} - E_{in}}{E_{fbin}} \tag{7}$$

Figure 9(a) and 9(b) show the variation of exergy efficiency with the fluidizing velocity and cooling air velocity, respectively. Exergy efficiency slightly decreases with fluidizing velocity, and slightly increases with bed temperature. Exergy efficiency increases with cooling air velocity, and is almost constant with cooling air temperature.

Exergy efficiencies of smooth and finned tubes have almost the same trend, even though there are some differences in the absolute values. But the temperature efficiency of finned and smooth tubes cannot be compared directly, for the aforementioned reason.

Figure 10 shows the variation of exergy efficiency with exergy supplied to the fluidized bed to compare the heat exchanger in the single-phase flow. Exergy efficiency of the heat exchanger in a single-phase flow was the calculated value using the correlations of the heat transfer coefficient on the outside and inside wall of a low temperature round tube. Exergy efficiency increased with the inlet temperature of the fluidized bed. The exergy efficiency of the heat exchanger using the fluidized bed was larger than that of the single-phase heat exchanger.

But the difference of exergy efficiency between finned and smooth tubes did not appear, because the overall coefficient of heat transmission was dominated by heat transfer inside of the tube.

4. Heat Transfer on the Inside Wall of a Tube at High Heating Rates

It was verified that the heat transfer coefficient on the outside wall of a tube was sharply increased using a fluidized bed and the finned heat transfer tubes. But, the performance of the heat exchanger was not much improved, because the overall coefficient of heat transmission was dominated by the heat transfer inside the tube. Therefore, heat transfer enhancement on the inside wall of the tube is needed.

By the way, as mentioned previously, the influence of the Reynolds number on the Nusselt number was very large and the Nusselt number on the inside wall of the finned tube was only slightly larger than that of the smooth tube. It was supposed that these tendencies were mainly caused by relaminarization. Clarifying the phenomenon is important if the heat transfer coefficient on the inside wall of the tube is to be increased.

Several experimental results about relaminarization have been published[7]-[8]. Relaminarization (the transition from turbulent flow to laminar flow) occurs as a consequence of high heating rates. The authors suggested that relaminarization was evidenced by the departure of the heat transfer results from the turbulent correlations.

The present experiments were performed using round and fan-shaped tubes. They were heated with high heat flux. The local and average heat transfer coefficients and relaminarization limits were evaluated.

4.1. HEAT TRANSFER ON THE INSIDE WALL OF THE ROUND TUBE AT HIGH HEATING RATES

4.1.1. *Experimental Apparatus and Method*
Experiments for measuring the heat transfer coefficient on the inside wall of a tube were carried out with the system shown in Fig. 11. Air was supplied to the horizontal heat transfer tube, which was heated by the nichrome wire. The maximum tube wall temperature in these experiments was about 750K.

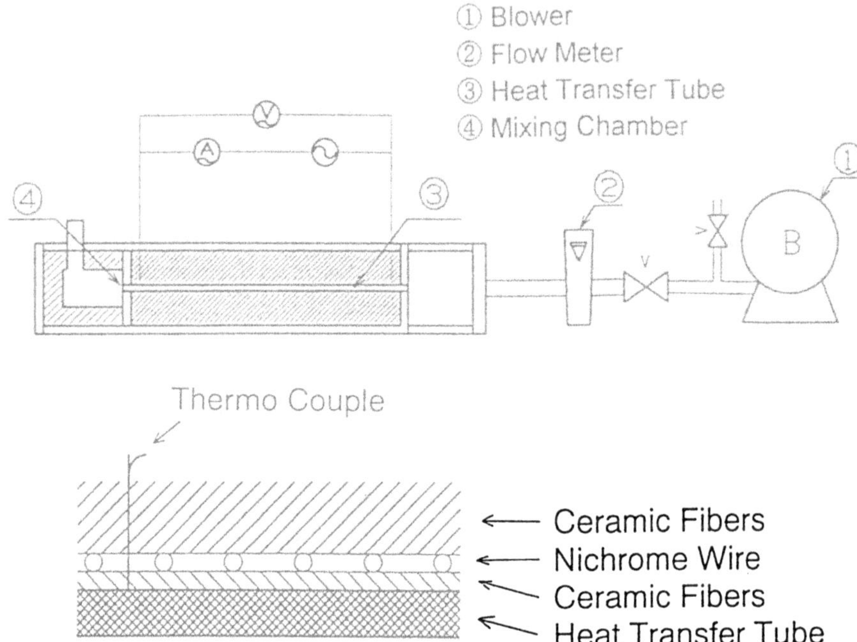

Figure 11: Experimental apparatus for measuring the heat transfer coefficient
on the inside wall of a tube

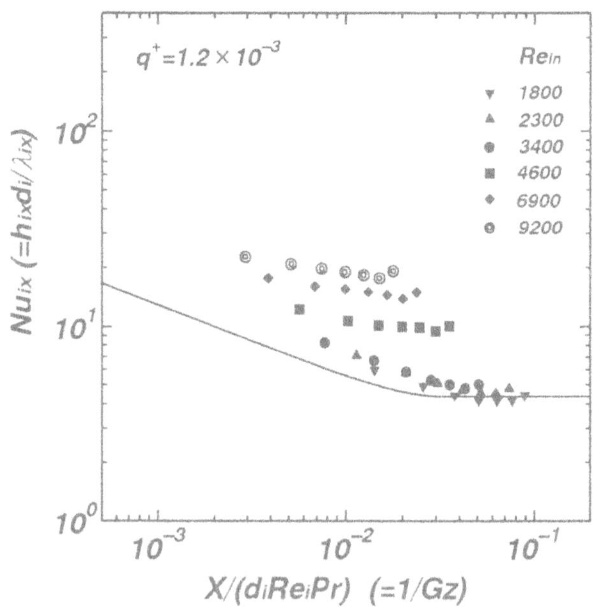

Figure 12: Variation of the local Nusselt number on the inside wall
of the tube with Graetz number

The heat transfer tube was made of stainless steel and had a 8.0mm outer diameter, a 6.0mm inner diameter and a 600mm length. The thermocouples for measuring the tube wall temperatures were welded on the heat transfer tube at 60mm intervals. The details of the heat transfer tube are also shown in Fig. 12. The nichrome wire (0.5mm diameter) was coiled around the heat transfer tube with a 7.5mm pitch. The insulator was made of ceramic fibers.

The inlet temperature of the air was measured by K-type thermocouple. The outlet temperature of the air was measured after the air exited the mixing chamber with an R-type thermocouple.

The local heat transfer coefficient was calculated using the heat transfer rate of the section between adjacent tube wall thermocouples, the tube wall temperature and the local bulk temperature of the air. The local bulk temperature of the air was calculated using the temperature rise evaluated with the local heat transfer rate.

The average heat transfer coefficient was calculated by eq.(8). The non-dimensionalized heat flux parameter defined by eq.(9) was introduced.

$$Q = h_i A_i \frac{(T_{Wout} - T_{out}) - (T_{Win} - T_{in})}{\ln \dfrac{T_{Wout} - T_{out}}{T_{Win} - T_{in}}} \tag{8}$$

$$q^+ = \frac{q}{(GC_pT)_{in}} \tag{9}$$

4.1.2. *Local Heat Transfer Coefficient on the Inside Wall of the Round Tube*

Figure 12 shows the variation of the local Nusselt number on the inside wall of the tube with Graetz number. Figure 12 corresponds to $q^+ = 1.2 \times 10^{-3}$. The line on the figure shows the Nusselt number of laminar flow in a round tube. The Nusselt number on condition of low inlet Reynolds number was almost equal to that of laminar flow. The Nusselt number increased with the inlet Reynolds number. In the case of Fig. 12, the drop of the Nusselt number along the air flow direction was not large on for $Re_{in} \geqq 4600$, because the flow was turbulent. But, the drop of the Nusselt number along the air flow direction was large for $Re_{in} = 3400$. This tendency suggests that relaminarization occurred.

Figure 13 (a) and 13(b) show the variation of the local Nusselt number on the inside wall of the tube with Reynolds number. Figure 13 (a) and (b), respectively, correspond to $q^+ = 1.2 \times 10^{-3}$ and 1.6×10^{-3}. The lines on the figure show the Nusselt number of fully developed laminar flow and turbulent flow in a round tube for constant heat flux.

For large inlet Reynolds numbers, the Nusselt number was in proportion to about the 0.8 power of the Reynolds number. This tendency is similar to the tendency of low heat flux turbulent flow. But, the drop of the Nusselt number along the air flow direction was large for $Re_{in} = 3400$ (Fig. 13 (a)) and $Re_{in} = 4600$ (Fig. 13 (b)). This figure also suggests that relaminarization occurred. In addition, relaminalization easily occurs on condition of large heat flux.

316

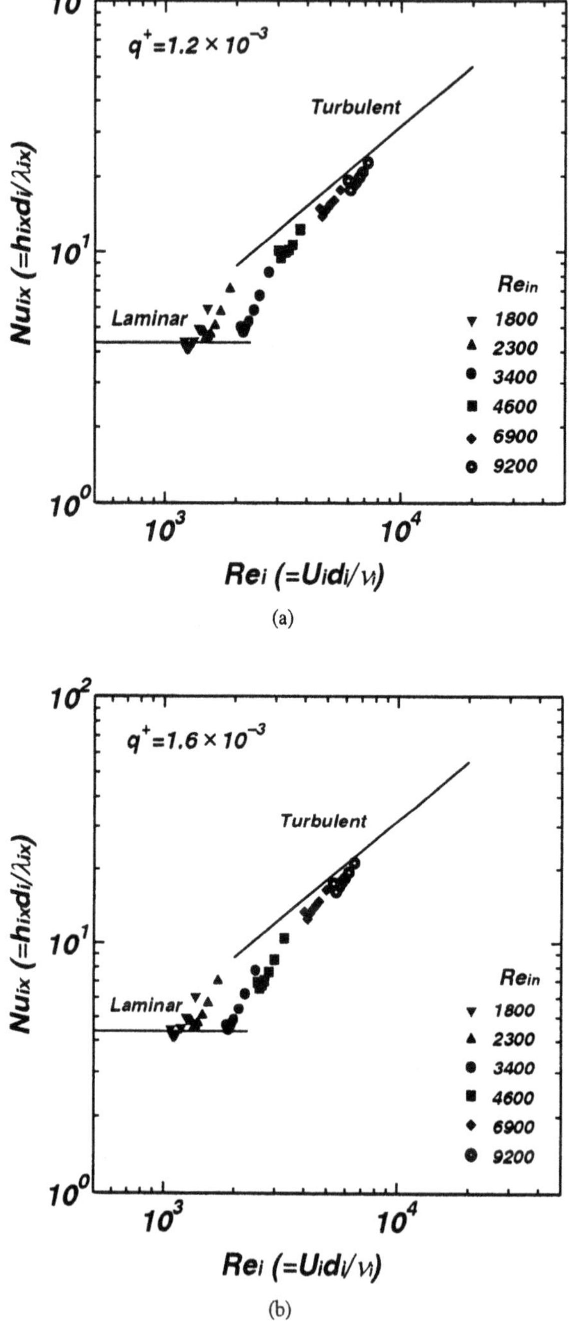

Figure 13: Variation of the local Nusselt number on the inside wall of the tube with Reynolds number

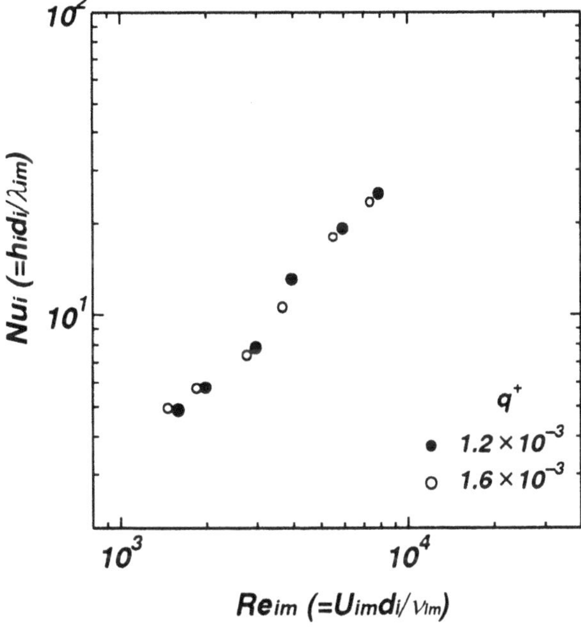

Figure 14: Variation of the average Nusselt number on the inside wall of the tube with Reynolds number

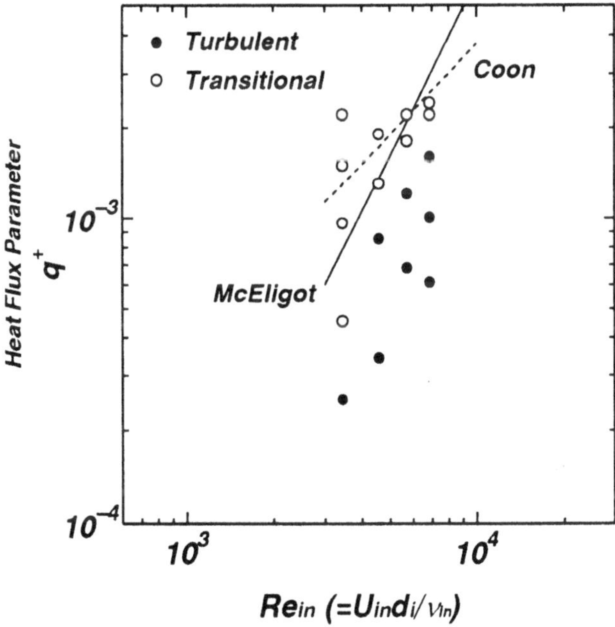

Figure 15: Flow regime classification

318

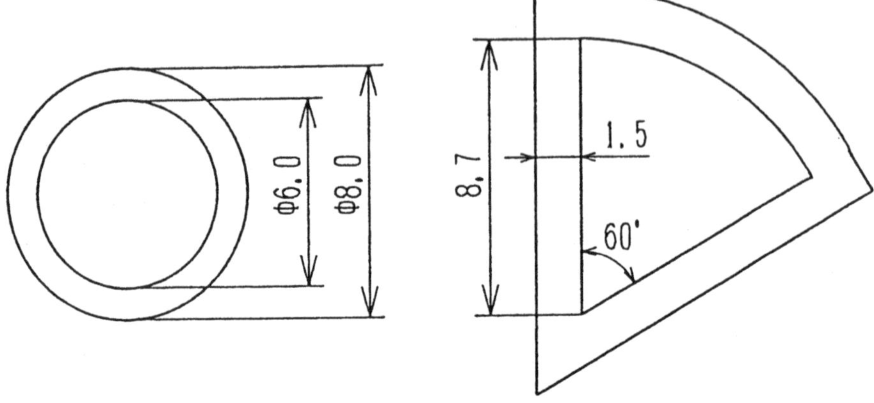

Figure 16: The shape of the heat transfer tube

4.1.3. *Average Heat Transfer Coefficient on the Inside Wall of the Round Tube*

Fig. 14 shows the variation of the average Nusselt number on the inside wall of the tube with the Reynolds number. The dependence of the average Nusselt number on the Reynolds number was larger than that in the case of low heat flux turbulent flow at $Re_{in} \fallingdotseq 4000$ where relaminarization occurred.

4.1.4. *Relaminarization Limits (Flow Regime Classification)*

Figure 15 shows the flow regime classification. The results of McEligot et. al.[8] are also plotted on Fig. 15. The difference of the symbols corresponds to relaminarization or not.

Relaminarization occurred if the heat flux parameter was larger than a certain value. This limit of the heat flux parameter increased with the inlet Reynolds number. The results of this study showed a similar tendency to that of McEligot et.al..

4.2. HEAT TRANSFER ON THE INSIDE WALL OF THE FAN-SHAPED TUBE AT HIGH HEATING RATES

In the sections above, it was confirmed that relaminarization occurred in the round tube at high heating rates. But, in the literature it has been noted that coexistence of laminar and turbulent flows in a tube with triangular cross section occurred (laminar near the apex of the triangle, turbulent at another place)[9]. Therefore, it is expected that the characteristic of the relaminarization in a fan-shaped tube, which is like one section of the ribbed tube, is different from that in a round tube.

In this part of the study, the heat transfer coefficient on the fan-shaped tube at high heating rates was evaluated.

4.2.1. *Experimental Apparatus and Method*

The experimental apparatus for the fan-shaped tube was similar to that for the round tube.

The heat transfer tube was made of stainless steel with a 1.5mm thickness and a 600mm length. Figure 16 shows the shape of the heat transfer tube. The hydraulic diameter of the fan-shaped tube was equal to that of the round tube (6.0mm). The thermocouples for measuring the tube wall temperature were welded at the center of the arc. They were welded at the same intervals as in the round tube. The method for calculating the heat transfer coefficient was also similar to that for the round tube.

4.2.2. *Local Heat Transfer Coefficient on the Inside Wall of the Fan-Shaped Tube*

Figure 17 shows the variation of the local Nusselt number on the inside wall of the tube with the Graetz number. The lines on the figure show the Nusselt number of fully developed laminar flow in both round and triangular tubes for constant heat flux. The Nusselt number increased with the inlet Reynolds number. But, the difference of the increasing tendency with the Graetz number, which appeared in the case of the round tube, did not appear in the case of fan-shaped tube. The local Nusselt number inside the tube sharply decreased as the flow even in the case of large inlet Reynolds numbers.

Figure 18 shows the variation of the local Nusselt number on the inside wall of the tube with the Reynolds number. The lines on the figure show the Nusselt number of fully developed laminar flow in round and triangular tubes and turbulent flow in a round tube for constant heat flux. The local Nusselt number was much influenced by the Reynolds number (in proportion to more than the 0.8 power of Re_{im}) even in the case of large inlet Reynolds numbers. The degree of relaminarization in the case of the finned tube was larger than that in the case of the smooth tube.

Figure 17 and Fig. 18 indicate that relaminarization occurred even in the case of large inlet Reynolds number. This suggests that relaminarization occurs easier in the fan-shaped tube than in the round tube.

4.2.3. *Average Heat Transfer Coefficient on the Inside Wall of the Fan-Shaped Tube*

Figure 19 shows the variation of the average Nusselt number on the inside wall of the tube with the Reynolds number. The average Nusselt number was much influenced by the Reynolds number (in proportion to more than the 0.8 power of Re_{im}). This result suggests that the large dependence of the average Nusselt number on the Reynolds number was due to relaminarization.

4.2.4. *Relaminarization Limits (Flow Regime Classification)*

The relaminarization limits did not appear clearly in the case of the fan-shaped tube. This was due to coexistence of laminar and turbulent flows. It is speculated that relaminarization had an influence on the proportion of laminar to turbulent flow.

Figure 20 shows the variation of the local Nusselt number on the inside wall of the tube with the Graetz number, at constant inlet Reynolds number. The line on the figure shows the Nusselt number of fully developed laminar flow in a round tube. The Nusselt number gradually decreased with increasing heat flux. This tendency suggests that the

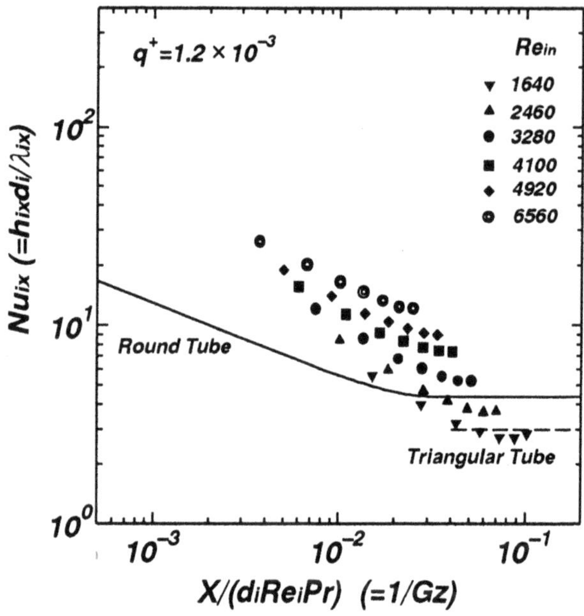

Figure 17: Variation of the local Nusselt number on the inside wall
of the tube with the Graetz number

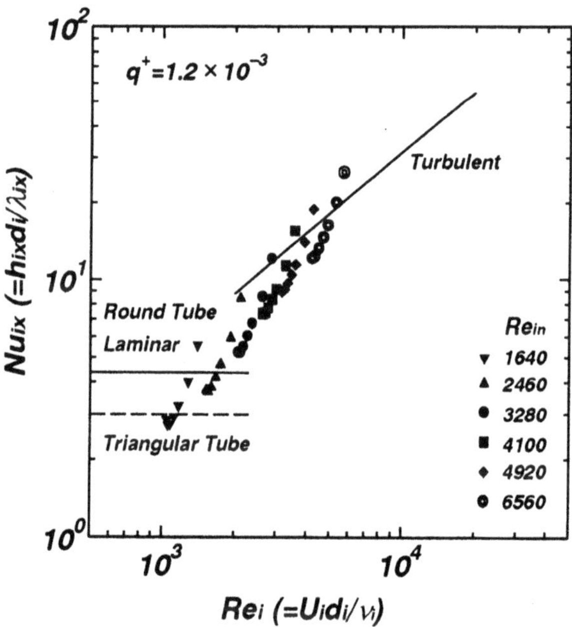

Figure 18: Variation of the local Nusselt number on the inside wall
of the tube with the Reynolds number

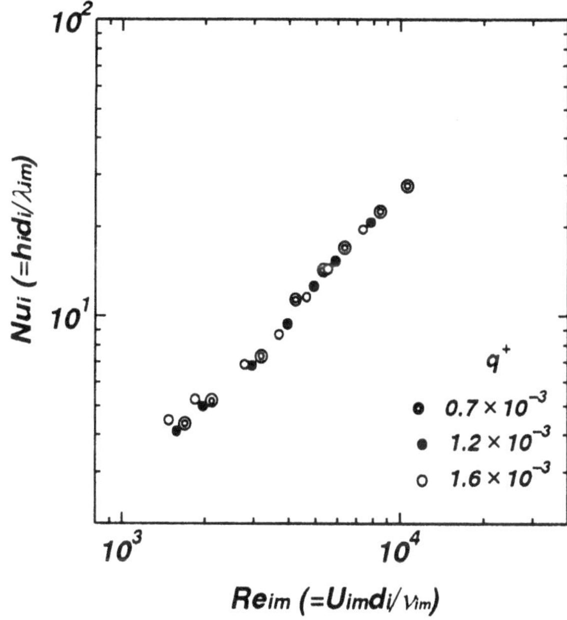

Figure 19: Variation of the average Nusselt number on the inside wall
of the tube with the Reynolds number

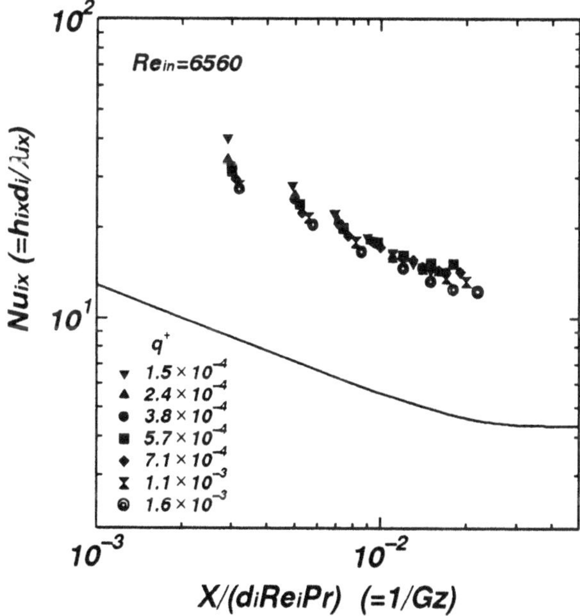

Figure 20: Variation of the local Nusselt number on the inside wall of the tube
with the Graetz number at constant inlet Reynolds number

degree of relaminarization increases with the heat flux parameter. Therefore, the Nusselt number of the finned tube was smaller than that of the smooth tube in the fluidized bed.

5.   Conclusions

A finned ceramic heat exchanger using a fluidized bed (maximum bed temperature was 1373K) was developed, and the characteristics of heat transfer were investigated. The conclusions are as follows:
(1) The heat transfer coefficient on the outside wall of the tube was improved by a factor of three by applying finned tubes and was improved by a factor of more than 20 over a single phase flow by using a externally finned tube.
(2) The heat transfer coefficient on the outside wall of the smooth tubes increased with increasing bed temperature by the effect of radiation. But, that of finned tubes was not influenced very much by bed temperature.
(3) The exergy efficiency increased with inlet temperature of the fluidized bed.
(4) The efficiency of the heat exchanger would be much improved if the heat transfer coefficient on the inside wall of the tube were improved.

   In addition, the influence of the Reynolds number on the Nusselt number on the inside wall of the tube in the high temperature fluidized bed was very large. Accordingly, the characteristics of the heat transfer on the inside wall of the tube at high heating rates were investigated by other experiments. The conclusions are as follows:
(1) It was confirmed that relaminarization occurred at high temperatures.
(2) Relaminarization occurred in the round tube if the heat flux parameter was larger than a certain value.
(3) The relaminarization limits did not appear clearly in the case of the fan-shaped tube. This was due to coexistence of laminar and turbulent flows. Relaminarization had an influence on the proportion of laminar to turbulent flow.
(4) The local and average Nusselt numbers in the fan-shaped tube were much influenced by the Reynolds number (in proportion to more than the 0.8 power of Re_{im}).

6.   Acknowledgements

The authors gratefully acknowledge the financial assistance provided through a Grant-in-Aid for Scientific Research on Priority Areas (No.256) by the Ministry of Education of Japan for fiscal 1994-1996.

Nomenclature

A : Heat transfer area, m^2

Cp : Specific heat at constant pressure, J/kgK
d : tube diameter, m
E : Exergy, W
G : Mass velocity, kg/ m^{2}s
Gz : Graetz number
h : Heat transfer coefficient, W/m^{2}K
M : Mass flow rate, kg/s
Nu : Nusselt number
Pr : Prandtl number
Q : Heat transfer rate, W
q : Heat flux, W/m^2
q^+ : Heat flux parameter
Re : Reynolds number
T : Temperature, K
U : Overall coefficient of heat transmission, W/m^{2}K
u : Velocity, m/s
ΔT_m : Mean temperature difference, K
x : Axial coordinate, mm
ε : Exergy efficiency
η : Temperature efficiency
λ : Thermal conductivity
ν : Kinetic viscosity, m^2/s

Subscripts

air   : Cooling air
fb   : Fluidized bed
i   : Inside of the tube
in   : Inlet of the tube
m   : Mean value
mf   : Minimum fluidization
o   : Outside of the tube
out   : Outlet of the tube
w   : Tube wall
x   : Local value

References

1. Saxena, S.C. (1989) Heat Transfer between Immersed Surfaces and Gas-Fluidized Beds, in *Advances in Heat Transfer* **19**, Academic Press, 97-190.
2. Mathur, A. and Saxena, S.C. (1987) Total and Radiative Heat Transfer to an Immersed Surface in a Gas-Fluidized Bed, *AIChE Journal* **33**, 1124-1135.
3. Chung, T. and Welty, J.R. (1990) Heat Transfer Characteristics for Tubular arrays in a High-Temperature Fluidized Bed : An Experimental Study of Bed Temperature

Effects, *Experimental Thermal and Fluid Science* **3**, 388-394.

4. Goshayeshi, A., Welty, J.R., Adams, R.L. and Alavidadeh, N. (1986) *Trans. of ASME, Journal of Heat Transfer* **108**, 907-912.

5. Zukauskas, A. (1972) Heat Transfer from Tubes in Cross Flow, in *Advances in Heat Transfer* **8**, Academic Press, 93-160.

6. Yoshimura, Y., Itoh, K., Ohori, K. and Hori, M. (1995) The Research and Development of a 30kW-class Ceramic Gas Turbine, *Journal of Gas Turbine Society of Japan* **23**-90, 55-60.

7. McEligot, D.M., Magee, P.M. and Leppert, G. (1965) Effect of Large Temperature Gradients on Convective Heat Transfer: The Downstream Region, *Trans. of ASME, Journal of Heat Transfer* **87**, 67-76.

8. McEligot, D.M., Coon, C.W. and Perkins, H.C. (1970) Relaminarization in Tubes, *Int. J. Heat Mass Transfer* **13**, 431-433.

9. Eckert, E.R.G. and Irvine, Jr., T.F. (1956) Flow in Corners of Passage with Noncircular Cross Sections, *Trans. of ASME* **78**, 709-718.

BOILING AND EVAPORATION OF FALLING FILM ON HORIZONTAL TUBES AND ITS ENHANCEMENT ON GROOVED TUBES

YASUNOBU FUJITA

Department of Energy and Mechanical Engineering, Kyushu University
6-10-1 Hakozaki, Higashi-ku, Fukuoka 812-8581, Japan

Abstract. Thin liquid films are utilized as an important component in various heat transfer processes because of their high heat transfer rate at low feed rates and with small temperature difference. Evaporators employing such characteristics are widely found in refrigeration systems, natural gas and air liquefying facilities, petrochemical and distillation plants, and more recently are proposed for use in OTEC systems. This paper reviews recent developments on evaporation and boiling heat transfer to falling films on horizontal tubes and their enhancement. Main items focused here are; (1) analysis and experiment for evaporative heat transfer to falling films, (2) empirical formulas for predicting heat transfer coefficient, (3) falling film breakdown and resulting deterioration of heat transfer, (4) analysis of enhanced heat transfer in grooved surface, (5) experimental results for heat transfer enhancement on grooved tubes, and (6) enhanced boiling heat transfer in falling films on grooved tubes.

1. Introduction

Thin liquid films have been widely utilized as an effective heat transfer process because of the high heat transfer rates attained at low film flow rates and under small temperature differences. Thus falling film evaporators are employed in refrigeration and air-conditioning systems, natural gas and air liquefying facilities, petrochemical and distillation plants, and recently proposed in OTEC systems.

Film evaporation seems like a very simple phenomenon at a glance but many factors, actually or potentially, influence the evaporation rate of thin liquid films: size and spacing of tubes, surface geometry, tube location in a bundle, feeding method of film, film flow rate, heat flux or wall temperature, operating pressure, and liquid properties. At present the effect of these factors is not fully clarified to establish heat transfer correlations of reasonable accuracy, thus demanding continued investigations.

This article intends to clarify some of the basic heat transfer characteristics in evaporation and boiling of falling films on horizontal plain tubes and to mention

325

S. Kakaç et al. (eds.), Heat Transfer Enhancement of Heat Exchangers, 325–346.

326

the validity of grooved tubes of simple geometry for heat transfer enhancement.

2. Analysis of Film Evaporation on Plain Tubes

2.1 LAMINAR FILM

A thin liquid film at saturation temperature impinges at the top of a horizontal tube of uniform temperature, evaporates while flowing around the tube, and un-evaporated liquid falls from the bottom. The impingement zone is localized to a very small fraction of the total periphery, thus the thermal boundary layer can be assumed to start its growth from the stagnation point ($\theta = 0$) and increases its thickness, δ_t, around the tube periphery until δ_t reaches the film thickness δ. Three regions are successively developed and identified as illustrated in Fig.1. First is the developing region of the thermal boundary layer ($0 \leq \theta \leq \theta_d$) where liquid film is superheated with negligible evaporation until the tip of the thermal boundary layer reaches the film surface. This is followed by the transition region ($\theta_d \leq \theta \leq \theta_f$) that is characterized by superheating of the film and simultaneous evaporation at the liquid surface. The third is the developed region ($\theta_f \leq \theta \leq \pi$) where all heat conducted from the tube wall is consumed in evaporation.

The temperature profile of the liquid film in the developing region is assumed to take a third-order polynomial, while a linear profile is assumed in the developed region. These two profiles are changed in the transition region. Thus temperature profile is written in dimensionless form for the three regions as [1]

$$\frac{T - T_w}{T_w - T_s} = 1 - \frac{3 - \beta}{2}\left(\frac{y}{\delta_t}\right) + \frac{1 - \beta}{2}\left(\frac{y}{\delta_t}\right)^3 \tag{1}$$

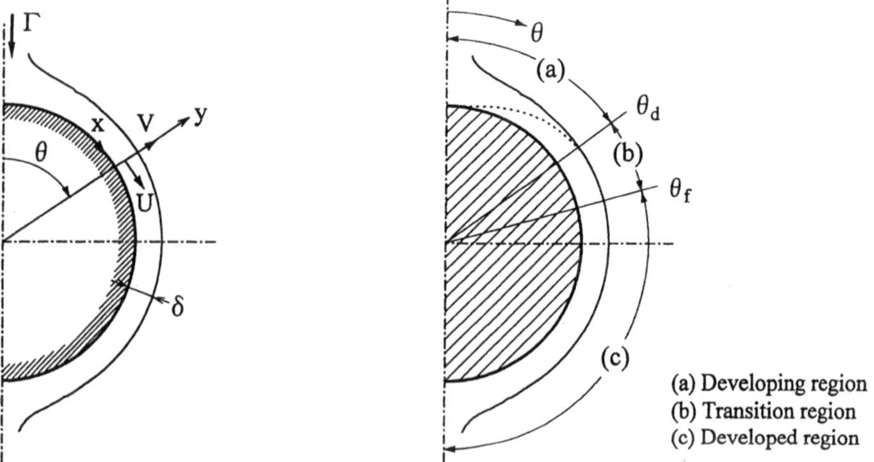

Figure 1. Physical model and three regions around tube

The conditions that should be satisfied in the three regions are as follows.
    (a) the developing region:

$$0 \leq \theta \leq \theta_d, \beta = 0, \delta_t \leq \delta \tag{2}$$

(b) the transition region:

$$\theta_d \leq \theta \leq \theta_f, \beta = \beta(\theta), \delta_t = \delta \tag{3}$$

(c) the developed region:

$$\theta_f \leq \theta \leq \pi, \beta = 1, \delta_t = \delta \tag{4}$$

The integral energy equation of the liquid film is solved with an assumption of a fully developed velocity profile in the liquid film (Couette flow assumption) to yield the local Nusselt number and the average Nusselt number in each region as follows [1].
    (a) the developing region (for $0 \leq \theta \leq \theta_d$):

$$Nu = 0.515 Ar^{-1/3} Pr^{1/3} Re_f^{1/9} \left[\sin\theta / P(\theta) \right]^{1/3} \tag{5}$$

$$Nu_{m,d} = 0.0735 Ar^{-1} Pr Re_f / \theta_d \tag{6}$$

where

$$P(\theta) = \int_0^{\theta} \sin^{1/3}\theta \, d\theta \tag{7}$$

$$\theta_d = P^{-1} \left(0.0294 Ar^{-1} Pr Re_f^{4/3} \right) \tag{8}$$

(b) the transition region (for $\theta_d \leq \theta \leq \theta_f$):

$$Nu_t = 1.10 Re_f^{-1/3} \left(\sin^{1/3}\theta \right) \left\{ 1 + 2.05 \exp\left[-48.0 Ar Pr^{-1} Re_f^{-4/3} P(\theta) \right] \right\} \tag{9}$$

$$Nu_{m,t} = 1.10 Re_f^{-1/3} \frac{P(\theta_f) - P(\theta_d)}{\theta_f - \theta_d}$$

$$+ 2.26 \frac{Re_f^{-1/3}}{\theta_f - \theta_d} \left\{ \int_{\theta_d}^{\theta_f} \left(\sin^{1/3}\theta \right) \exp\left[-48.0 Ar Pr^{-1} Re_f^{-4/3} P(\theta) \right] d\theta \right\} \tag{10}$$

where

$$\theta_f = P^{-1} \left(0.111 Ar^{-1} Pr Re_f^{4/3} \right) \tag{11}$$

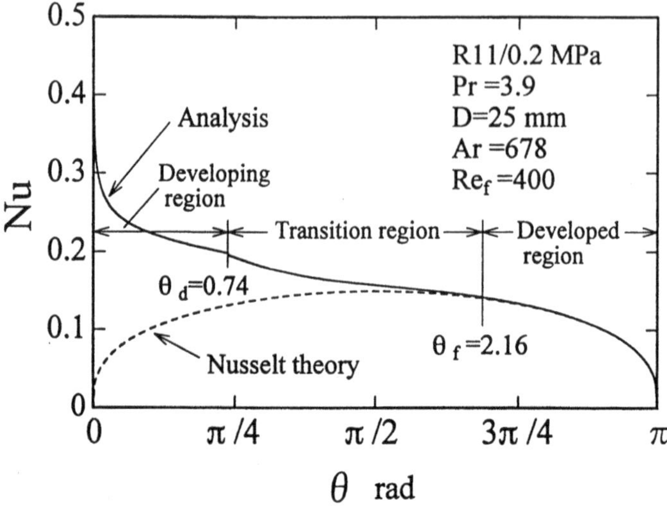

Figure 2. Variation of the heat transfer coefficient around the tube

(c) the developed region (for $\theta_f \leq \theta \leq \pi$):

$$Nu_f = 1.10Re_f^{-1/3}\sin^{1/3}\theta \tag{12}$$

$$Nu_{m,f} = 1.10Re_f^{-1/3}\left(\frac{2.59 - 0.111Ar^{-1}PrRe_f^{4/3}}{\pi - \theta_f}\right) \tag{13}$$

Summation of three contributions in the successive regions gives the average Nusselt number over the whole tube periphery as follows.

$$Nu_m = Nu_{m,d}\left(\frac{\theta_d}{\pi}\right) + Nu_{m,t}\left(\frac{\theta_f - \theta_d}{\pi}\right) + Nu_{m,f}\left(\frac{\pi - \theta_f}{\pi}\right) \tag{14}$$

The local heat transfer coefficient is a function of tube diameter, film flow rate and fluid properties, which are expressed in terms of Ar, Re_f and Pr, and peripheral angle θ. Figure 2 shows a typical variation of the Nusselt number around the tube in dimensionless form. Compared to the classical Nusselt theory[2], indicated by the dotted line, the heat transfer coefficient is much higher in the developing and transition regions, while there is no difference in the fully developed region. An increase in flow rate and/or a decrease in tube diameter extends a relative periphery occupied by the developing region, resulting in a significant deviation from the Nusselt theory. The average Nusselt numbers for laminar flow are predicted from Eq.(14) for three Prandtl numbers and shown against film Reynolds number

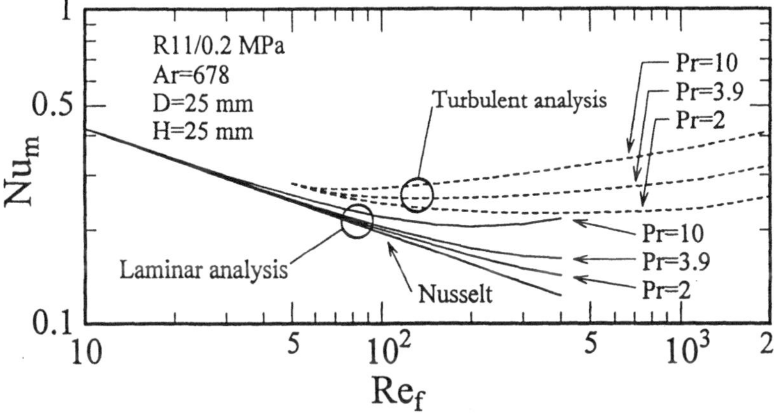

Figure 3. Average Nusselt number as a function of film Reynolds number

TABLE 1. Ratio of theoretically predicted average Nusselt number to Nusselt theory

		Re$_f$			
Ar	Pr	50	100	200	400
	10	1.341	1.806	2.502	3.403
100	3.9	1.133	1.334	1.795	2.486
	2	1.068	1.172	1.431	1.973
	10	1.050	1.127	1.319	1.764
678	3.9	1.020	1.049	1.124	1.314
	2	1.010	1.025	1.064	1.161
	10	1.017	1.043	1.108	1.273
2000	3.9	1.007	1.017	1.042	1.106
	2	1.003	1.008	1.022	1.054

in Fig.3. Table 1 gives the ratio of the average Nusselt number predicted from Eq.(14) to that obtained from the Nusselt theory [2] which is

$$Nu_m = 0.906 Re_f^{-1/3} \qquad (15)$$

Stairs-like line in the Table indicates the boundary for a 10% deviation. The Nusselt theory is clearly limited to low Prandtl number fluids falling on large tubes at low flow rates.

2.2 TURBULENT FILM

If the universal velocity profile for internal turbulent flow in ducts or tubes is applied to film flow without any modification, and the turbulent Prandtl number is assumed unity, then the Nusselt numbers, both local at an arbitrary peripheral angle and averaged over the tube periphery, and the relationship of Reynolds number with dimensionless film thickness are expressed as [1]

$$Nu = \frac{\left(\delta^+\right)^{1/3}\sin^{1/3}\theta}{\int_0^{\delta^+}\dfrac{dy^+}{1+Pr\left[\left(1-y^+/\delta^+\right)/\left(dF/dy^+\right)-1\right]}} \tag{16}$$

$$Nu_m = \frac{0.824\left(\delta^+\right)^{1/3}}{\int_0^{\delta^+}\dfrac{dy^+}{1+Pr\left[\left(1-y^+/\delta^+\right)/\left(dF/dy^+\right)-1\right]}} \tag{17}$$

$$Re_f = 4\Gamma/\mu = \int_0^{\delta^+} F\left(y^+\right) dy^+ \tag{18}$$

Substitution of von Karman's universal velocity profile for $F(y^+)$ into Eqs. (17) and (18) yields,
for $50 \le Re_f \le 1125$, i.e., $5 \le \delta^+ \le 30$:

$$Nu_m = \frac{0.165 Pr\left(\delta^+\right)^{-1/3}}{Pr + \ln\left[1 + Pr\left(\delta^+/5 - 1\right)\right]} \tag{19}$$

$$Re_f = 20\delta^+ \ln\delta^+ - 32.2\delta^+ + 50 \tag{20}$$

and for $1125 \le Re_f$, i.e., $30 \le \delta^+$:

$$Nu_m = \frac{0.165 Pr\left(\delta^+\right)^{-1/3}}{Pr + \ln\left(1 + 5Pr\right) + 0.5\ln\left(\delta^+/30\right)} \tag{21}$$

$$Re_f = 10\delta^+ \ln\delta^+ + 12\delta^+ - 256 \tag{22}$$

The average Nusselt numbers predicted from Eqs.(19) and (21) are shown for different Prandtl numbers by the dotted lines in Fig.3. In the range of high film Reynolds numbers where the falling film is regarded as turbulent, Nu_m is an increasing function of Re_f and Pr.

3. Experiments of Film Evaporation on Plain Tubes

3.1 EFFECT OF FEEDING METHOD AND DUMMY TUBES

In contrast to condensation films, evaporation films are artificially supplied from liquid feeders to the heated tubes so that feeding methods may inevitably affect film flows and, accordingly, heat transfer characteristics.

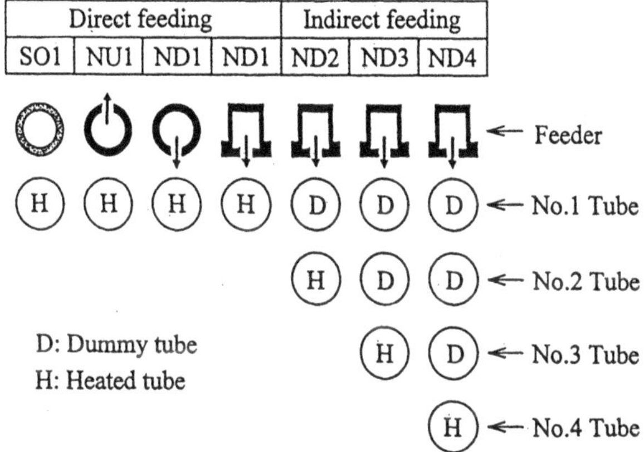

Figure 4. Arrangements of feeder and dummy tubes

Six arrangements of liquid feeders and unheated dummy tubes shown in Fig.4, were tested to clarify their possible effects on heat transfer in film evaporation [3]. The direct feeding tubes were of the same size, 25 mm in diameter, as the heated tube. One was a porous sintered tube, designated to SO1, and another was a tube with small holes in a single row, designated to NU1 or ND1. Holes are open upwards for NU1 and downwards for ND1. The plate type had a row of holes at its center line, designated to ND2 to ND4. For all liquid feeders used in the experiment, stable and uniform liquid feed to the topmost heated tube was

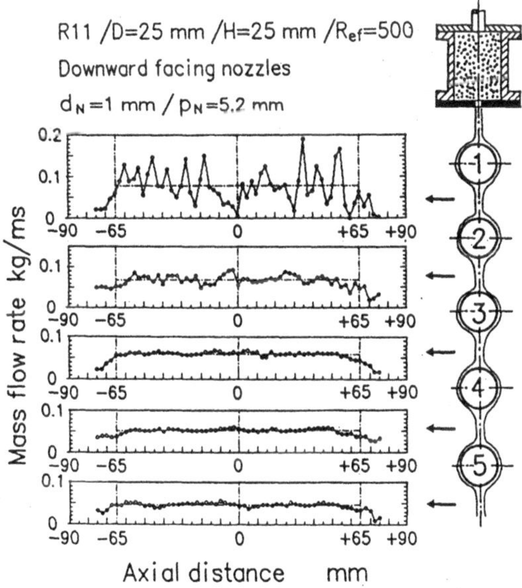

Figure 5. Axial distribution of film flow rate dripping from tube

insured by changing the mesh size of the porous tube or the diameter and pitch of the holes, depending on the supplied flow rate.

It was found in the experiments [3] that the tube just below the feeder shows a little lower heat transfer coefficient compared to other tubes when dummy tubes are arranged above the heated tubes in order to rectify the film flow. A single dummy tube is sufficient to eliminate the effect of the feeding method on heat transfer.

Figure 5 illustrates an axial distribution of flow rate measured under each tube arrayed in a vertical plane [3]. The film flow rate from the bottom of the top tube has remarkable peaks of almost the same pitch as the holes of the feeding plate. For lower tubes, liquid falls more uniformly. Dripping locations from these tubes move freely in the axial direction, keeping a nearly constant pitch between neighboring drips. This random fluctuation of the dripping locations serves to make the time-averaged film flow rate more or less uniform, which is expected in turn to attenuate the effect of feeding method on heat transfer.

3.2 HEAT TRANSFER COEFFICIENT

Heat transfer coefficients in evaporation of falling films were measured on five tubes, numbered 1 to 5 from the top to the bottom, which were successively arrayed in a vertical plane [4]. Different feeding methods were employed to supply liquid to the topmost heated tube. The test liquid was refrigerant R-11 at a saturation temperature under 0.2 MPa. R-11 was supplied from a liquid feeder in the range of 0.001 kg/ms to 0.20 kg/ms for mass flow rates of falling films, which covers film Reynolds numbers ranging from 10 for laminar films to 2000 for turbulent films. Heat flux was varied in the non-boiling range by one order of magnitude from 1.0 to 9.8 kW/m^2 but the measured heat transfer coefficients were almost invariant with heat flux. This suggested no nucleation in the liquid film, so that pure evaporation from the film surface was prevailing in all the experiments.

Figure 6 plots the measured average Nusselt number against film Reynolds number for the topmost (No.1), the middle (No.3), and the bottom (No.5) tubes. Analytical predictions [1] are indicated for reference in the figure. Blackened data appearing in the data plots for No.3 and No.5 tubes are significantly lower than the other. These deviating data indicate deterioration of heat transfer as the result of partial film breakdown as will be mentioned in the next section.

The top tube (No.1 tube) directly below a liquid feeder shows lower heat transfer coefficients than those for the other tubes arrayed below it. Direct supply of liquid from a feeder may be responsible for the lower heat transfer coefficient, because film distribution along the tube axis is not uniform on the topmost tube as seen in Fig.5. The measured data for the top tube is correlated, irrespective of feeding method, as

$$Nu_m = \left(Re_f^{-2/3} + 0.0080 Re_f^{\,0.3} Pr^{0.25} \right)^{1/2} \tag{23}$$

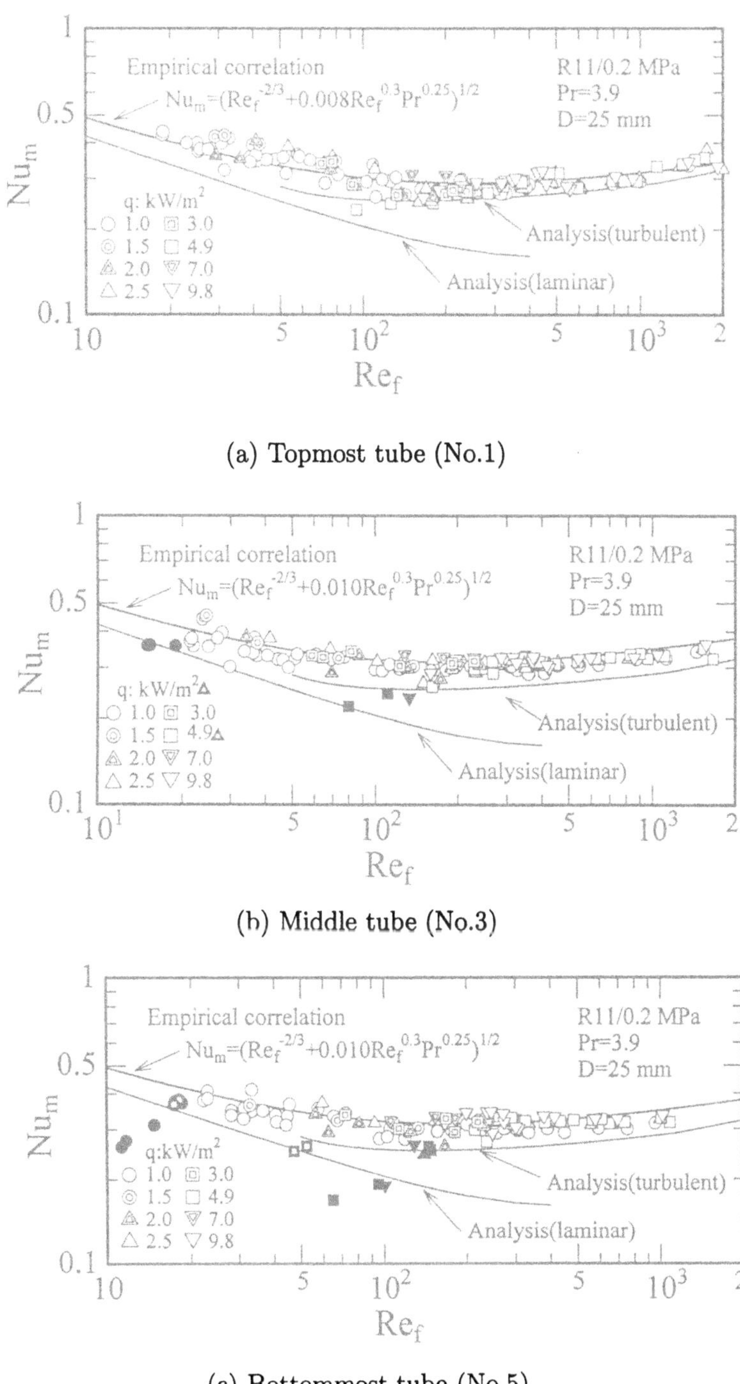

(a) Topmost tube (No.1)

(h) Middle tube (No.3)

(c) Bottommost tube (No.5)

Figure 6. Average Nusselt number as a function of film Reynolds number

334

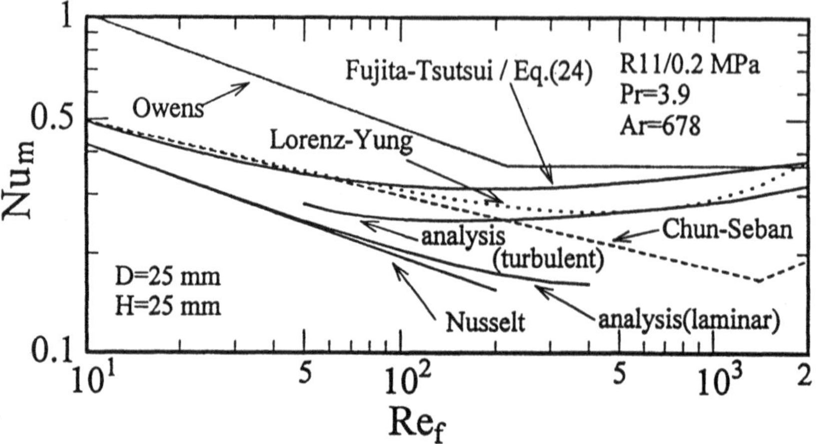

Figure 7. Comparison of various heat transfer correlations for falling film evaporation

Number 2 to No.5 tubes show practically the identical heat transfer coefficient within measurement errors, as long as the tube surface is fully wetted with the liquid film. Such data are correlated to be

$$Nu_m = \left(Re_f^{-2/3} + 0.010 Re_f^{0.3} Pr^{0.25} \right)^{1/2} \tag{24}$$

This formula is considered as a general correlation applicable to tubes for which the effect of feeding method is negligibly small. Figure 7 compares the formula, Eq.(24), with typical existing empirical correlations [5, 6, 7] and theoretical predictions [1, 2] for the same conditions as in the present measurements.

3.3 BREAKDOWN OF FALLING FILM

As a result of evaporation, the falling film becomes thinner and thinner as it flows down the successive heated tubes. Though the film thickness changes with heat flux, an initial flow rate supplied to the topmost tube and the tube location, the liquid film at last breaks down and dry patches appear on the tube surface when the flow rate is reduced below a certain minimum. Thereafter the falling film is unable to cover and wet steadily the whole tube surface. Unwetted dry patches at first appear intermittently while their effect on heat transfer is not so large. They steadily persist with a further reduction of flow rate, and on lower tubes below the tube of dry patch incipience. Thus, the unwetted area is increased with increasing heat flux, decreasing initial flow rate and lower tube location.

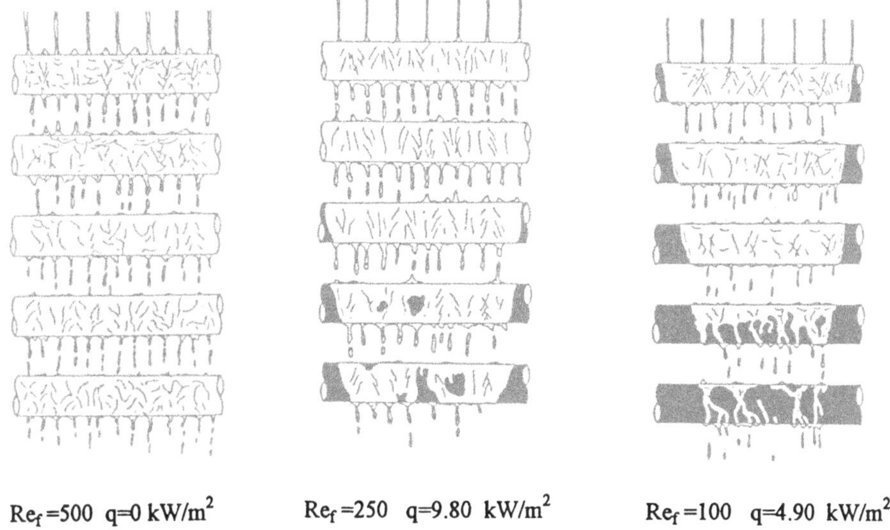

Figure 8. Dry patch pattern as a result of film breakdown

Figure 8 shows sketches of dry patch [4] observed at different heat flux and flow rate. Film Reynolds numbers in the figure denote the value calculated from the supplied flow rate to the topmost tube. Thus the Reynolds number for each tube decreases with lower tubes. The wetted area fraction of the total tube surface, determined from photographs [4], is illustrated in Fig. 9. As seen in the figure, the wetted area fraction is a decreasing function of heat flux. The area fraction also decreases with decreasing flow rate and lower tube location.

3.4 DETERIORATION OF HEAT TRANSFER DUE TO FILM BREAKDOWN

As the wetted area fraction is reduced appreciably, it influences the heat transfer coefficient because the heat transfer coefficient is defined by assuming the whole tube surface is covered by liquid. Figure 10 shows the measured Nusselt number for the lowest tube, No.5, as a ratio of the measured to the predicted value from the empirical formula, Eq.(24). Data clustered around a horizontal line of unity correspond to the case of the tube surface being fully wetted. Blackened data deviating downwards from the horizontal line indicate heat transfer deterioration caused by the persistent dry patch formation as the result of film breakdown.

These lower points vary with heat flux and film flow rate. Figure 11 gives heat flux for the inception of film breakdown as a function of film Reynolds number. An empirical relation for film breakdown is

$$q_{\mathrm{dry}} = 0.048 Re_f \tag{25}$$

336

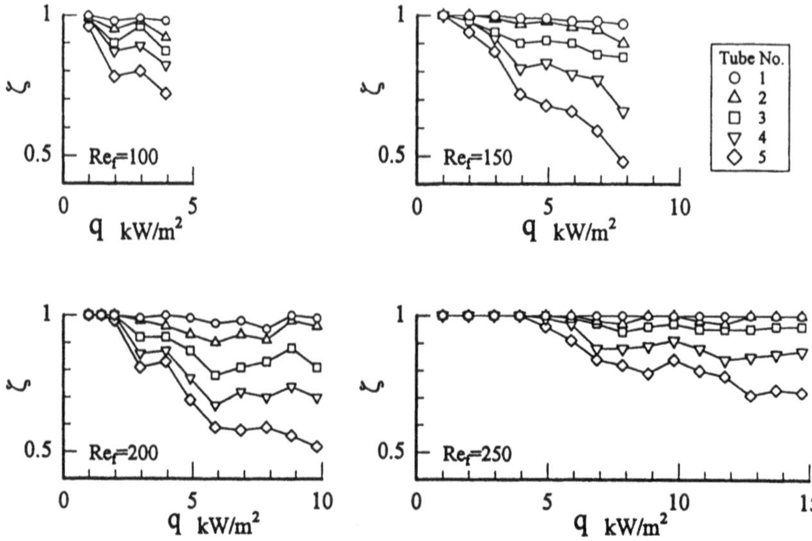

Figure 9. Wetted area fraction as a function of flow rate, heat flux, and tube location

Przulj and Ganic [8] obtained an empirical formula similar to Eq.(25) for the breakdown of subcooled falling liquid film. Their formula has the exponent of Reynolds number as 1.05, but it predicts very high heat flux, as much as five to seven times the data in Fig. 11. Film breakdown in the experiment of Przulj and Ganic seems to be caused by boiling bubble generation in subcooled liquid films, resulting in very high heat flux for the film breakdown in their experiment.

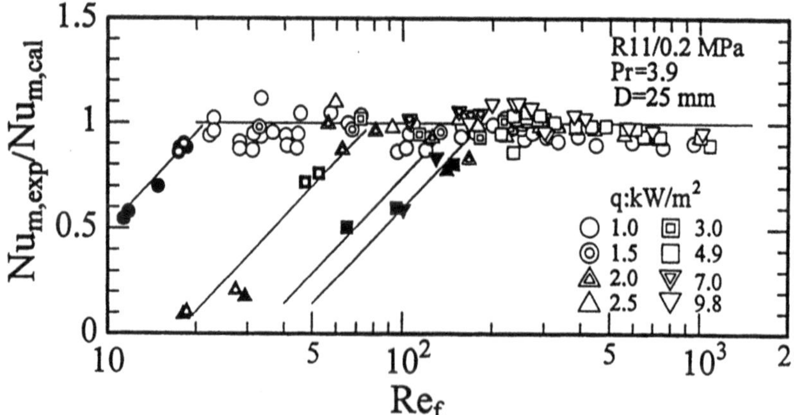

Figure 10. Heat transfer deterioration caused by dry patches

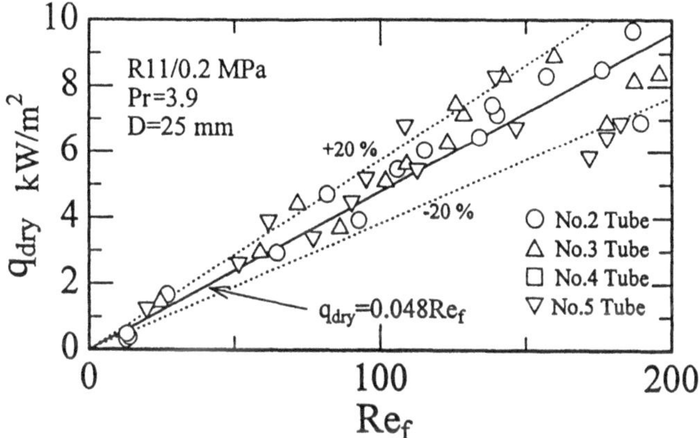

Figure 11. Heat flux for film breakdown as a function of film Reynolds number

A thinner falling film is favorable because of its high heat transfer characteristics. However it is vulnerable to heat transfer deterioration when the film breakdown occurs and dry patches are steadily formed. Thus one of the key issues in designing the falling film evaporators is to maintain all the heated tube surfaces fully wetted for any given operation conditions.

4. Heat Transfer Enhancement on Grooved Tubes

4.1 ANALYSIS OF LAMINAR FILM IN GROOVE

Conduction resistance mainly dominates evaporation of falling films. Grooved surfaces seem to be useful in reducing the conduction resistance, because capillary force acting on the film surface extends the film-covered area and makes the film thickness nonuniform, i.e., thicker in the groove valley and thinner in the outlying portion. This nonuniformity of film thickness is expected to reduce effectively the overall conduction resistance less than that for a uniform film on plain tubes.

The Nusselt theory [2] predicts the angular variation of film thickness and heat transfer coefficient for laminar falling films on a uniform wall temperature horizontal tube as

$$\delta = \left(3\mu\Gamma/\rho^2 g \sin\theta\right)^{1/3} \tag{26}$$

$$\alpha = \frac{\lambda}{\delta} = \lambda\left(\rho^2 g \sin\theta/3\mu\Gamma\right)^{1/3} \tag{27}$$

Substitution of $\theta = \pi/2$ in the above equations gives the same results for the film thickness and heat transfer coefficient as those for a falling film on a vertical

338

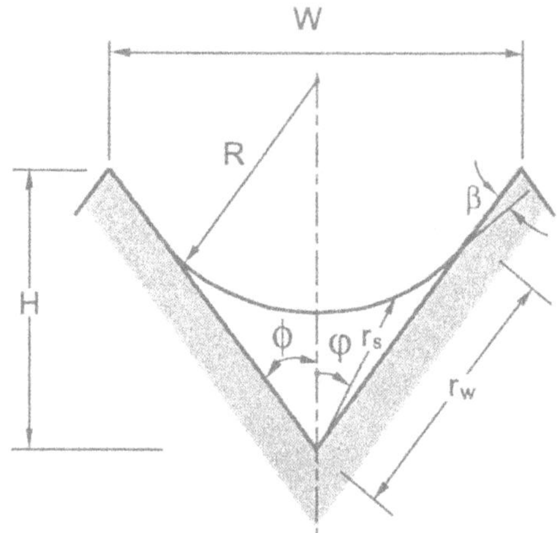

Figure 12. Geometry of triangular groove and nonuniform liquid film in the groove

flat surface. To validate the enhancement of a grooved surface for heat transfer enhancement, a laminar film falling in a vertical triangular groove depicted in Fig.12, is analyzed and compared with a plain vertical surface.

In a small groove, the meniscus of falling film at relatively low velocity has a nearly constant curvature radius of R, and contacts the groove wall at a contact angle of β. The distance from the groove vertex to the meniscus is given as

$$r_s = F\left(\varphi\right) r_w \tag{28}$$

Here for $\beta + \phi \neq \pi/2$:

$$F\left(\varphi\right) = \frac{\cos\beta\cos\varphi - \left(\sin^2\phi - \cos^2\beta\sin^2\varphi\right)^{1/2}}{\cos\left(\beta + \phi\right)} \tag{29}$$

and for $\beta + \phi = \pi/2$:

$$F\left(\varphi\right) = \frac{\cos\phi}{\cos\varphi} \tag{30}$$

If the inertia and convective terms are neglected, as done in the Nusselt's analysis, the momentum and energy equations for laminar film in the groove can be written in the $r - \varphi$ coordinates as

$$\frac{\partial^2 u}{\partial r^2} + \frac{\partial u}{r\partial r} + \frac{\partial^2 u}{r^2\partial\varphi^2} = \frac{\rho g}{\mu} \tag{31}$$

$$\frac{\partial^2 T}{\partial r^2} + \frac{\partial T}{r\partial r} + \frac{\partial^2 T}{r^2\partial\varphi^2} = 0 \tag{32}$$

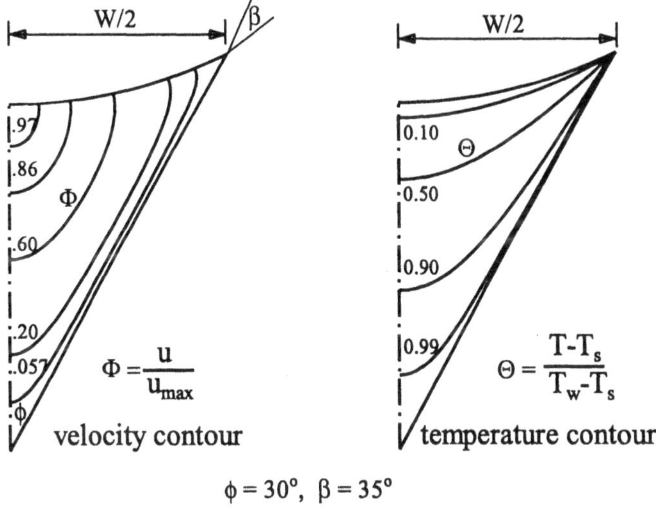

Figure 13. Dimensionless velocity and temperature contours

The boundary conditions are non-slip and uniform temperature on the groove surface, and no shear force and saturation temperature on the film surface. Figure 13 indicates dimensionless velocity and temperature contours obtained from numerical solution of the above equations [9].

If the groove pitch is used as the reference length to define the film flow rate and heat flux, the flow rate per unit tube length and heat transfer coefficient are determined from integration of the numerical results as

$$\Gamma_W = \frac{\rho u_m A_c}{W} = \frac{2\rho}{W} \int_0^\phi \int_0^{r_s} u(r,\varphi) r \, dr \, d\varphi \tag{33}$$

$$\alpha_W = \frac{Q}{W(T_w - T_s)} = -\frac{2\lambda}{W(T_w - T_s)} \int_0^{r_w} \left(\frac{\partial T}{\partial n}\right)_{\varphi=\phi} dr \tag{34}$$

The heat transfer coefficient for the grooved surface determined in this way is shown in Fig. 14 as a function of the film flow rate. The right end of each curve corresponds to the flooding point at which the film tip, i.e., the triple interface, reaches the top of the groove slope, i.e., $r_w = H/\cos\phi$. The heat transfer coefficient for a plain surface, calculated from the Nusselt theory of Eq.(15), is also given for reference. As is clear from these figures the grooved surface certainly enhances the heat transfer coefficient, except for very small film flow rate where liquid film covers only the bottom portion of the groove and other portion is dry and inactive for heat transfer. As seen in Fig.14, enhancement is dependent upon the groove vertex angle 2ϕ, width W, depth H, the liquid-surface contact angle β

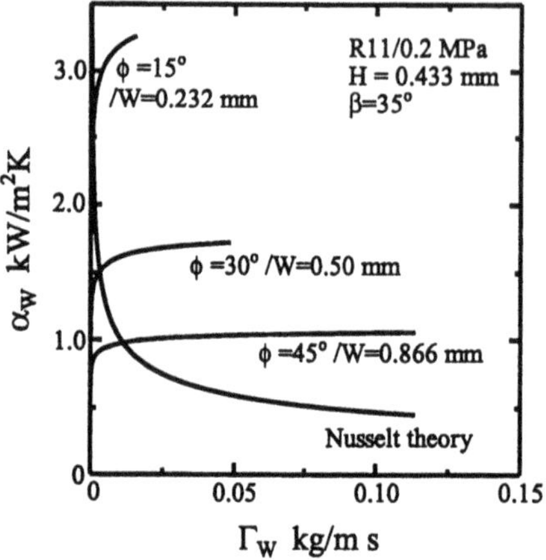

(a) Effect of groove width, W, on heat transfer

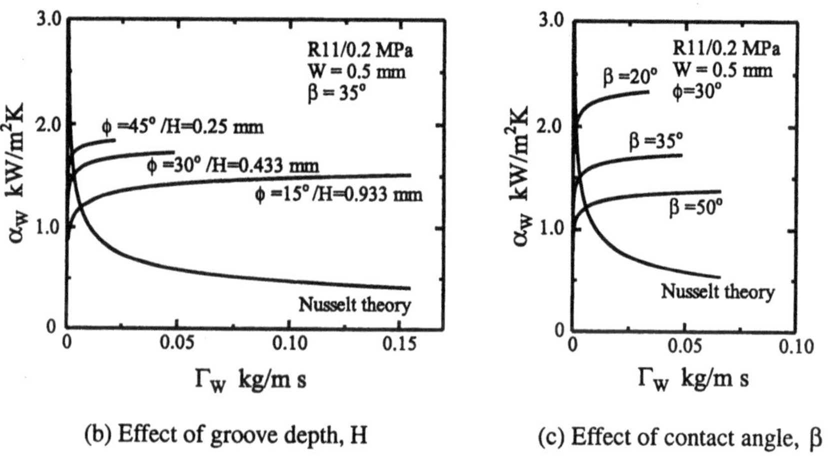

(b) Effect of groove depth, H

(c) Effect of contact angle, β

Figure 14. Predicted heat transfer coefficient for laminar falling film in a groove. W: groove width, H: groove depth, β: contact angle, ϕ: half vertex angle of groove

and surface tension σ. The flooding flow rate also varies with these parameters as is clear in Fig. 14.

4.2 EXPERIMENTAL VERIFICATION OF ENHANCEMENT ON GROOVED TUBES

The effectiveness of the grooved surface for heat transfer enhancement was experimentally confirmed. Test [7] was done for three grooved tubes referred to as

TABLE 2. Geometry of grooved tubes and their dimensions

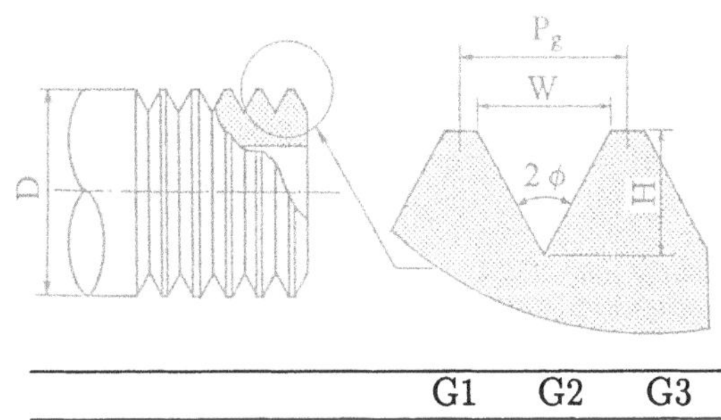

		G1	G2	G3
P_g	mm	0.5	1.0	2.0
W	mm	0.5	0.7	1.50
H	mm	0.43	0.6	1.30
2ϕ	deg	60	60	60
P_g/W		1.00	1.67	1.54
area increase ratio		2.03	1.74	1.87

G1, G2 and G3. The geometry and dimensions of the grooved tubes are specified in Table 2. A cylinder of diameter D-2H was used as the reference surface to determine heat flux and tube wall temperature.

The heat transfer coefficient for each grooved tubes is shown in Fig.15 together with the foregoing theoretical results and the predicted coefficient from the empirical correlation, Eq.(24), for plain tubes. All the grooved tubes show substantial increase in heat transfer coefficient for a wide range of flow rate, as much as about four times for G3 to six times for G1 compared to a plain tube.

The measured enhancement exceeds not only the increase in surface area of the grooved tubes ranging 1.74 to 2.03 but also the theoretically predicted heat transfer coefficient for laminar film flowing in the triangular grooves.

The observed enhancement is partly attributable to rather uniformly distributed liquid in each groove. The main enhancement mechanism seems an intermittent wetting of the upper part of the grooves by disturbance waves and the subsequent evaporation of the very thin film left behind after the film tip retreats on the groove slope.

Among the three tubes, the G1 tube, with the smallest grooves, demonstrates the greatest enhancement over the G2 and G3 tubes. This difference may arise because of the distribution of the film in the grooves. In smaller grooves, bridging of the film across the groove crests becomes easier, leading to an increase in the wetted surface area fraction.

The heat transfer coefficient for grooved tubes depends weakly on the heat flux and tends to decrease as the heat flux increases. Partial dry patches occurring at the crests of grooves at higher heat flux is one possible reason for the heat flux

342

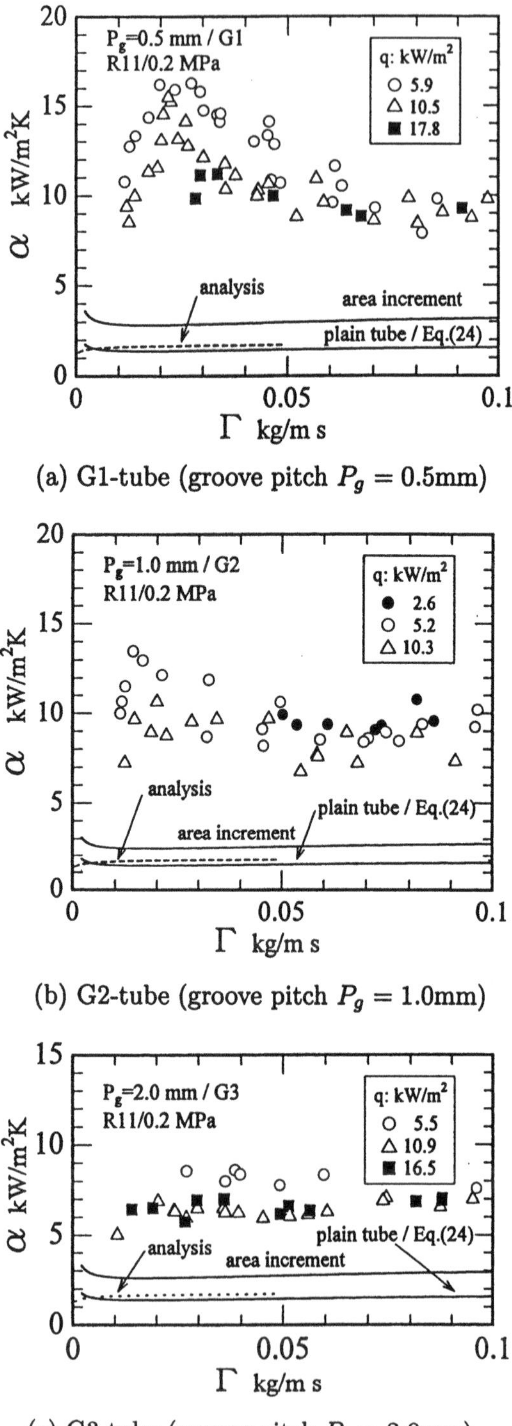

(a) G1-tube (groove pitch $P_g = 0.5$mm)

(b) G2-tube (groove pitch $P_g = 1.0$mm)

(c) G3-tube (groove pitch $P_g = 2.0$mm)

Figure 15. Measured enhanced heat transfer coefficients on three grooved tubes

dependency of the heat transfer coefficient. As film flow rate is decreased below a certain limiting value, the heat transfer coefficient abruptly decreases. This is accompanied by a sharp increase of dry patch area on the grooved tube surface, as observed similarly on plain tubes and shown in Fig. 8. The grooved tubes are vulnerable to dry patch formation at low flow rate because liquid flow is more concentrated in the grooves and mutual liquid supply across the groove crests is more difficult compared to plain tubes.

4.3 ENHANCEMENT OF NUCLEATE BOILING

When the tube surface temperature exceeds the nucleation temperature, boiling bubbles are generated on the tube surface so that the heat transfer mode changes from evaporation to nucleate boiling. The heat transfer performance in nucleate boiling in the falling film on plain tubes is basically not so different from that for pool boiling on a plain smooth surface. For enhancement of nucleate pool boiling, several kinds of micro-structured surfaces are proposed and some of them are practically used in flooded evaporators. Compared to such micro-structured surfaces, a grooved tube is very simple in the structure and effective in enhancing nucleate boiling in a falling film. Figure 16 shows the measured heat transfer coefficient for the grooved tube G1 specified in Table 2 and for a plain tube. Compared to the plain tube, the heat transfer coefficient in nucleate boiling on

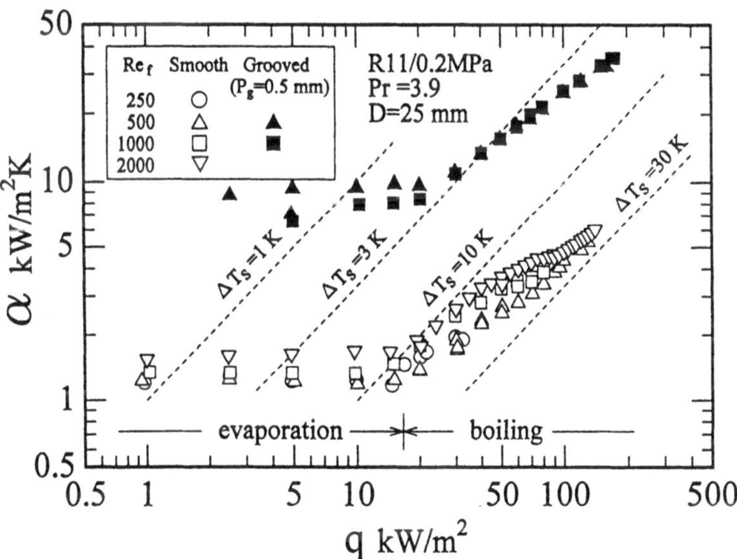

Figure 16. Enhancement of nucleate boiling in falling film on a grooved tube (G1)

344

the grooved tube is enhanced more than five times. This amount is the same order enhancement obtained for several micro-structured surfaces.

5. Concluding Remarks

The heat transfer characteristics in evaporation of falling films on horizontal tubes were presented mainly on the basis of the author's experimental and analytical investigations. The main feature of falling film evaporators is that the film is artificially supplied from a liquid feeder to the topmost tube in a tube array, so that the feeding method influences heat transfer. But its influence exerts only on the tube located just below a liquid feeder. For other tubes located at further lower position, the heat transfer coefficient is basically a function of film flow rate and is predictable from the proposed empirical correlation, Eq.(24), as long as the tube surface of interest is fully covered with a liquid film.

Dry patches are formed at smaller flow rate, at higher heat flux and on tubes at lower positions. The heat flux for dry patch formation is related to film flow rate with a nearly linear relationship expressed by Eq.(25). Incomplete wetting of the tube surface as the result of dry patch formation substantially reduces heat transfer. Hence avoidance of dry patch formation is needed to maintain the advantageous performance of evaporators of the falling film type.

Grooved tubes of simple geometry are effective in enhancement of the evaporation heat transfer. The experimentally observed enhancement is well beyond the area increase of the grooved tube. Theoretical predictions are based on a meniscus extension model in a groove. The enhancement is dependent on the size of the groove, with smaller grooves showing higher heat transfer coefficients. A grooved tube is vulnerable to dry patch formation and therefore to heat transfer deterioration. Boiling heat transfer in a falling film is also enhanced on grooved tubes to the same extent as in evaporation.

The thermal design and performance evaluation of falling film evaporators are expected to be further improved through continuous investigations on the evaporation and boiling in film flow.

Nomenclatures

Ar dimensionless tube radius, $R/(\nu^2/g)^{1/3}$
d_N nozzle diameter of feeder, m
D tube diameter, m
$F(y^+)$ dimensionless velocity, $u/(\tau_w/\rho)^{1/2}$
g acceleration of gravity, m/s^2
H triangular groove depth or vertical pitch of tubes, m
Nu local Nusselt number, $\alpha(\nu^2/g)^{1/3}/\lambda$
Nu_m average Nusselt number on the whole tube surface
p_N nozzle pitch of feeder, m
p_T vertical pitch of tube array, m
P pressure, Pa

P_g groove pitch, m
Pr Prandtl number
q heat flux, W/m^2
q_{dry} heat flux for inception of dry patches, W/m^2
r radial distance, m
r_w wetted length on the groove, m
R tube radius, m
Re_f film Reynolds number, $4\Gamma/\mu$
T temperature, K
T_s saturation temperature, K
T_w wall temperature, K
u velocity of falling film, m/s
W groove top width, m
y^+ dimensionless distance, $y(\tau_w/\rho)^{1/2}/\nu$

Greek symbols
α heat transfer coefficient, $q/(T_w - T_s)$, W/m^2·K
β contact angle, deg
δ film thickness, m
δ_t thermal boundary layer thickness, m
ϕ half vertex angle of the groove, deg
φ angle, deg
Γ mass flow rate of liquid film per unit length of the tube, kg/m·s
λ thermal conductivity, W/m·K
μ dynamic viscosity, Pa· s
ν kinematic viscosity, m^2/s
θ peripheral angle, deg
ρ density, kg/m^3
ζ wetted area fraction

References

1. Fujita, Y. and Tsutsui, M. (1994) Evaporation Heat Transfer of Falling Film on Horizontal Tube (1st Report, Analytical Study), *Trans. Japan Society of Mech. Engrs.* **60**, 3469-3474.

2. Nusselt,W. (1916) Die Oberflächenkondensation des Wasserdampfes, *VDI Zeitschrift* **60**, 569-546 and 569-575.

3. Fujita, Y., Tsutsui, M. and Zhou, Z. Z. (1994) Evaporation Heat Transfer of Falling Films on Horizontal Tube (2nd Report, Experimental Study), *Trans. Japan Society of Mech. Engrs.* **60**, 3475-3480.

4. Fujita, Y. and Tsutsui, M. (1994) Evaporation Heat Transfer of Falling Films on Horizontal Tubes (3rd Report, Heat Transfer of Tube Array), *Trans. Japan Society of Mech. Engrs.* **63**, 1701-1706.

5. Chun, K.R. and Seban, R.A. (1971) Heat Transfer to Evaporating Liquid Films, *Trans. ASME, J. Heat Transfer*, **93**, 391-396.

6. Owens, W. L. (1978) Correlation of Thin Film Evaporation Heat Transfer Coefficients for Horizontal Tubes, *Proc. 5th OTEC Conf.* **3**, 71-89.

7. Lorenz, J.J. and Yung, D. (1978) Combined Boiling and Evaporation of Liquid Films on Horizontal Tubes, *Proc. 5th OTEC Conf.* **3**, 46-70.

8. Przulj, V. and Ganic, E. N. (1991) Breakdown of Subcooled Falling Liquid Film on Horizontal Tube,*Proc. 2nd World Conf. on Experimental Heat Transfer, Fluid Mechanics, and Thermodynamics* **1**, 792-799.

9. Fujita, Y., Bai, Q. and Tsutsui, M. (1997) Falling Film Evaporation on Horizontal Smooth Tubes and Its Enhancement on Grooved Tubes, *Proc. 11th Int. Symp. on Transport Phenomena in Thermal Science and Process Engineering* **2**, 379-384.

NUMERICAL AND EXPERIMENTAL INVESTIGATION OF ENHANCEMENT OF TURBULENT FLOW HEAT TRANSFER IN TUBES BY MEANS OF TRUNCATED HOLLOW CONE INSERTS

TEOMAN AYHAN, YUSUF AZAK and CEVDET DEMIRTAS
Karadeniz Technical University Mechanical Engineering Department
61080 Trabzon/TURKEY

BETUL AYHAN
Karadeniz Technical University Ordu Technical Center
52000 Ordu/TURKEY

Abstract. In this study, the incompressible turbulent flow in a circular channel with a series of conical hollow inserts, placed axisymmetrically, is investigated numerically and experimentally. In computations, FLUENT, a commercially available CFD code, is used. The inserts are installed in a diverging structure towards the incoming flow. The effect of the inserts geometry on the heat transfer at the wall is evaluated for a range of Reynolds number between 3000 and 20000. It is demonstrated that a significant enhancement in heat transfer from the tube wall occurs in comparison with the conventional plain tube, and the level of the enhancement depends on the flow rate, the type of the inserts and the installation positions of the inserts. In order to see the flow fields, the velocity profiles are shown. Using the entropy minimization technique the optimum geometries and the installation distances were estimated. The results obtained through the numerical simulation are compared with the experimental ones and, a satisfactory agreement is found.

1. Introduction

The economic benefits of energy and material savings have prompted and received great attention in order to increase convective heat transfer rates in the process equipment. Enhancement of convective heat transfer process is usually gained at the expense of surface and inserts, and requires external power. The use of truncated hollow cone inserts is one of most promising techniques for augmenting convective heat transfer in terms of energy cost.

Recorded attempts to increase normal heat transfer coefficients have been analyzed and classified by Bergles [1]. This special area of heat transfer is of great current interest. Recent works in this category include the papers by Yilmaz and Ayhan [2] and, Ayhan and Karabay [3].

In this study, an optimum type of the conical hollow cone turbulators, which enhances the heat transfer is determined experimentally and numerically. Despite the heat transfer enhancement in comparison with the plain tube, pressure drops are increased approximately 8-9 times. The enhancement in the heat transfer results from the secondary flows the region of the conical hollow cone turbulators near the wall. In the buffer region between the two turbulator flows perpendicular to the main flow occur. This result is also supported by the flow visualization experiments [4].

S. Kakaç et al. (eds.), Heat Transfer Enhancement of Heat Exchangers, 347–356.
© 1999 *Kluwer Academic Publishers.*

348

2. Experimental Set-up

A schematic diagram of the experimental set-up with the basic components and instrumentation is shown in Figure 1. It consist of a number of components, including an air filter, inlet section, test section, mixing room, thermocouples, orifice plate and a fan. Experiments are conducted in the suction mode.

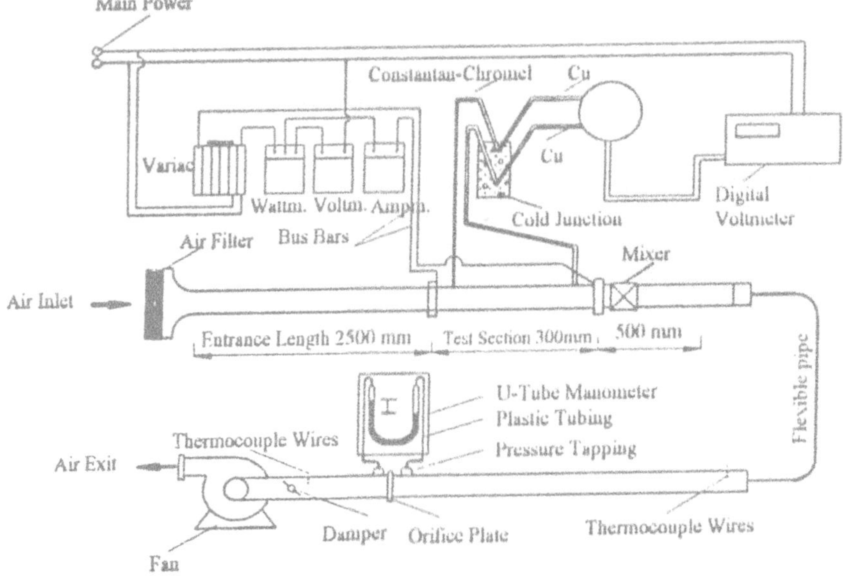

Figure 1. Schematic diagram of the experimental set-up.

The air entering the apparatus firstly encounters an air filter which is used to intercept any particles from the atmospheric air in the laboratory. The air then flows into a air filter which is designed to provide smooth air flow into the inlet section. The air traverses the inlet and test section then leads to flow metering section (orifice plate), a damper and then to the fan.

The experiments are conducted on both the thermal and combined entrance regions. Air flows from the air filter on the inlet section of the duct (2770 mm) where the velocity field is fully developed. Then the flowing air travels into the test section of the thermal entrance region. At the end of the test section, the flowing air enters a flexible pipe. The pipe (2.04 m in length) is connected to the end of the test section through a flexible pipe. A set of companion flanges are used to house the orifice plate.

At the downstream of the companion flanges, another section of the pipe (0.84 m long) is used as a feeder to the fan. This section contains a damper which could be adjusted to produce the desired air flow rates.

The test section is based on a 0.058 m o.d, stainless steel tube with i.d. of 0.055 m to provide a sung fit with the truncated hollow cone tested. The tube is 0.30 m long and is supported by using asbestos flanges at the ends of the tube; to minimize axial heat loss, also to facilitate installation and removal of inserts. The constructional details of the truncated hollow cone inserts are shown in Figure 2. A numerical values and legends of devices experimentally tested are given in Table 1.

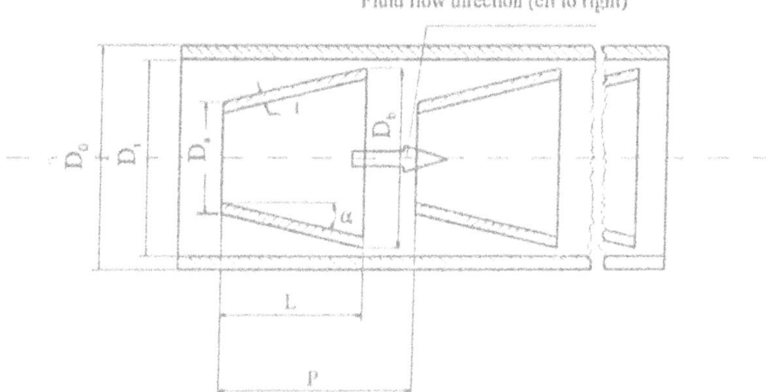

Figure 2. Characteristic parameters of the truncated hollow cone inserted tubes.

Symbol arrow (→) on Table 1 denotes the direction of the flow. The pressure drop across the test section was measured with conventional inclined U-tube manometer. Heating of the active length of the test section was achieved electrically using a Nichrome resistance wire spirally wound over the outer periphery of the tubes. The outer surface of the test section was covered with a layer of glass-wool to prevent external (radial) heat loss and the entire section was covered with a protective shell of stainless steel.

Eleven equally-spaced chromel-alumel thermocouples are mounted along the effective length of the test section to permit an accurate measurement of wall temperature distribution, and other two thermocouples also permitted the measurement of the air temperature at entry to and axis from the test section. At the exit end a series of baffles in the tube enable the bulk temperature of the fluid to be more deadly assessed. Each end of the tube is fitted with three pressure taps spaced equally around the circumference.

The mass flow rate was calculated from the pressure drop across the orifice plate obstruction meter which was constructed from brass and machined to standard specifications given by ASME standard [6].

Experiments were conducted with one smooth tube and seven conical ring inserted tubes for a wide range of Reynolds number. For each Reynolds number a range of heater dissipation rates were examined.

For a specified run the power dissipation by the heater is not totally transferred to the coolant owing to inevitable external heat losses. These losses may be broadly attributed to conduction at the each end of the test section. Also, together with an external convective loss was estimated from a series of calibration tests undertaken prior to the main experimental program.

From steady state measurements of the flow rate, power dissipation, temperature levels, etc. it is possible to determine the nondimensional groups such as Nusselt, Prandtl and Reynolds number.

3. Mathematical Formulation and Numerical Procedure

A commercially available CFD code, FLUENT is used for all the computations. The thermophysical properties of the fluid and solid regions are assumed to be constant. The flow under consideration is governed by two dimensional form of the continuity, the time averaged Navier-Stokes, energy, kinetic energy of turbulence, and its dissipation rate equations. In a cylindrical coordinate system for steady-state, incompressible and axisymmetric flow neglecting buoyancy, these equations can be written in the following general form:

Table 1. Characteristic dimensions of truncated hollow cone inserts

Tube type	Legend	D_0 (mm)	D_i (mm)	D_b (mm)	D_a (mm)	L (mm)	P (mm)	α^0	t (mm)	Flow direction
TYPE AB0	0	58	55	-	-	-	-	-	-	→
TYPE A01	◊	58	55	52	42	20	31	14	1	→
TYPE B01	□	58	55	52	32	40	52	14	1	→
TYPE B02	●	58	55	50	32	36	52	14	1	→
TYPE B03	×	58	55	48	32	32	52	14	1	→
TYPE B05	Δ	58	55	40	15	48	50	14	1	→
TYPE B06	+	58	55	34	15	36	50	14	1	→

$$\frac{1}{r}\frac{\partial}{\partial r}(\rho r v \varnothing) + \frac{\partial(\rho u \varnothing)}{\partial x} = S_\varnothing + \frac{1}{r}\frac{\partial}{\partial r}\left(\Gamma_\varnothing r \frac{\partial \varnothing}{\partial r}\right) + \frac{\partial}{\partial x}\left(\Gamma_\varnothing \frac{\partial \varnothing}{\partial x}\right) \tag{1}$$

Where $\varnothing$ stands for the dependent variables u, v, k, ε and T. u and v are the local time averaged velocity in the x and r directions respectively; $\Gamma_\varnothing$ and $S_\varnothing$ are the corresponding turbulent diffusion coefficient and the source term, respectively. The equations are summarized in Table 2. The turbulent viscosity μ_T and μ_e can be written as:

$$\mu_T = \rho v_T = \rho C_\mu \frac{k^2}{\varepsilon} \tag{2}$$

$$\mu_e = \mu_l + \mu_T = \mu_l + \rho c_p \frac{k^2}{\varepsilon} \tag{3}$$

The constants of the k-ε turbulence model suggested by Launder and Spalding [7] are shown in Table 3.

Table 2. Conservation Equations

Equation	$\varnothing$	$\Gamma_\varnothing$	$S_\varnothing$
Mass	1	0	-
x-momentum	u	μ_e	$-\dfrac{\partial P}{\partial x}+\dfrac{1}{r}\dfrac{\partial}{\partial r}\left(r\,\mu_e\dfrac{\partial v}{\partial x}\right)+\dfrac{\partial}{\partial x}\left(\mu_e\dfrac{\partial u}{\partial x}\right)$
r-momentum	v	μ_e	$-\dfrac{1}{r}\dfrac{\partial P}{\partial r}+\dfrac{1}{r}\dfrac{\partial}{\partial r}\left(r\,\mu_e\dfrac{\partial v}{\partial r}\right)-\dfrac{2\mu_e v}{r^2}+\dfrac{\partial}{\partial x}\left(\mu_e\dfrac{\partial u}{\partial r}\right)$
Energy	T	$\dfrac{\mu_e}{\sigma_e}$	-
Turbulent kinetic energy k		$\dfrac{\mu_e}{\sigma_k}$	$G-\rho\varepsilon$
Turbulent energy dissipation rate	ε	$\dfrac{\mu_e}{\sigma_\varepsilon}$	$\dfrac{\varepsilon}{k}\left(C_1 G-C_2\rho\varepsilon\right)$

Where

$$G=\frac{\mu}{k}\left\{2\left[\left(\frac{\partial u}{\partial x}\right)^2+\left(\frac{\partial v}{\partial r}\right)^2+\left(\frac{v}{r}\right)^2\right]+\left(\frac{\partial u}{\partial r}+\frac{\partial v}{\partial x}\right)^2\right\} \tag{4}$$

Table 3. Constants of the k-ε Model

C_μ	C_1	C_2	σ_k	σ_e
0.09	1.44	1.99	1.0	1.3

At the inlet boundary, uniform flow conditions are imposed. The near-wall region was treated by two zone models, i.e. viscous suplayer and fully turbulent zone, and the wall function method was used to bridge the viscous sublayer.

The flow field can be regarded as fully developed when the outlet is located far away from recirculation region. Therefor, zero normal gradients at the outlet plane are used in the following forms:

The numerical method used in the present study is based on the SIMPLEX procedure for handling the velocity-pressure linkages. The conservation equations are discretized by a control volume based finite difference method with a power-law scheme.

$$\frac{\partial u}{\partial x}=0, \qquad \frac{\partial v}{\partial x}=0, \qquad \frac{\partial T}{\partial x}=0 \tag{5}$$

The set of difference equations are solved iteratively using a line by line solution method in conjunction with a tridiagonal matrix algorithm. The solution is considered to be converged when the normalized residual of the algebraic equation is less then a prescribed value of 0.001. Details of the method for discritezation of conservation and energy equations for velocity components, pressure and temperature are given [4] and will not be repeated here.

4. Results and Discussion

A series of experimental tests and computations were done with the truncated hollow cone inserts fitted inside the tube. The experiments and the computations are carried out for the range of Reynolds number between 3000 and 20000 and the Prandtl number of 0.72. The air flow rates were selected so that they were nominally the same as those used for the plain cylindrical tubes.

The enhancement in heat transfer which results from fitting the truncated hollow cone inserts is demonstrated in Figure 3, which shows the variation of mean Nusselt number with nominal Reynolds number for tubes given in Table 1. For comparison, the corresponding results for the plain tube are shown. Figure 4 shows the details of turbulent flow measurements with truncated conical hollow cone inserts in Table 1. The data are presented in the form of a friction factor. As expected, a significant increase in the flow resistance occurs owing to the inclusion of the truncated hollow cone inserts.

The computer code was thoroughly tested for the flow through a smooth tube before computing results for the truncated hollow cone inserts. Results for different fluid flow domain in truncated hollow cone inserted tubes, also fluid flow domain which is defined in Table 1 are produced. Velocity vector distributions through the truncated hollow cone inserted tubes were shown in Figure 5.

Figure 6 and 7 depict the variations of mean Nusselt numbers and friction factors with regard to Reynolds numbers for the optimal design of the truncated hollow cone inserted tubes.

The results of theoretical optimization presents the most convenient type of truncated hollow cone insert for turbulent flow heat transfer. Experimentally supported data give a good agreement for the theoretically optimized one. It's geometric dimensions are L=0.032 m , D_b=0.052 m, D_s=0.030 m and P=0.045 m.

For the results obtained correlation equations are obtained, which are given in the following;

$$Nu = 0.0215\,Re^{0.73}\,Pr^{0.4}\left(\frac{D_b}{D_s}\right)^{2.13}\left(\frac{P}{D_0 - D_b}\right)^{0.15} \tag{6}$$

$$f = 11.28\,Re^{-0.316} \tag{7}$$

These equations are available for the tube length 0.35 m and tube internal diameter 0.055 m, using theoretically produced data. Using entropy generation minimization method by Bejan [5], the optimal type of truncated holow cone insert is also approved.

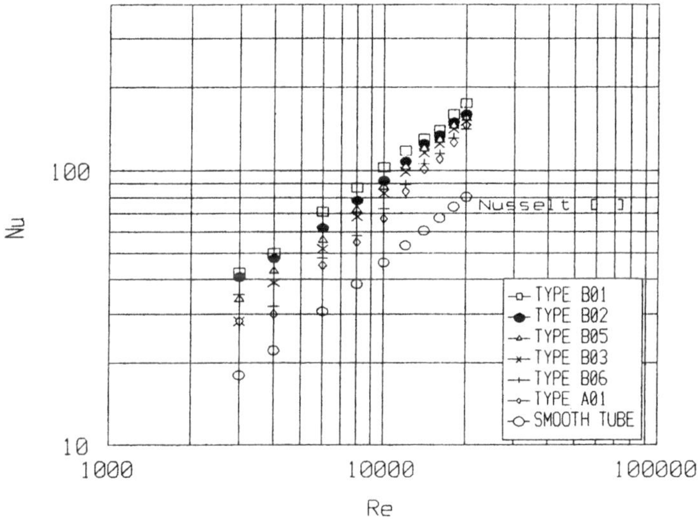

Figure 3. Typical enhancement of heat transfer in truncated hollow cone inserted tube compared with the smooth cylindrical tube.

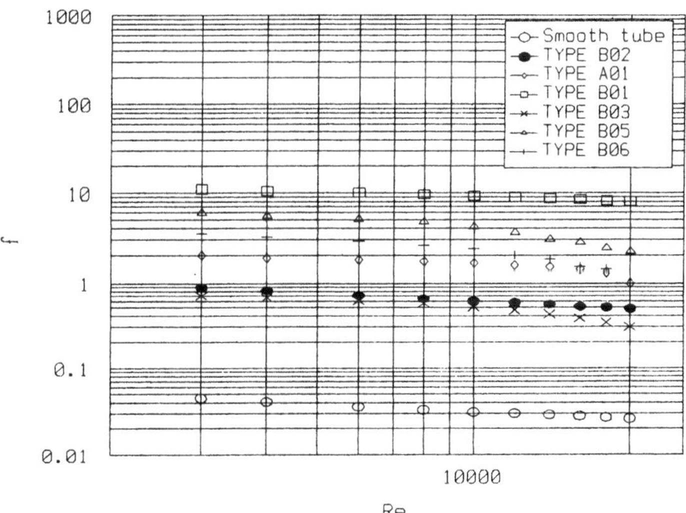

Figure 4. Comparison of friction factor versus Reynolds number for the truncated hollow cone inserted tubes and the smooth cylindrical tube.

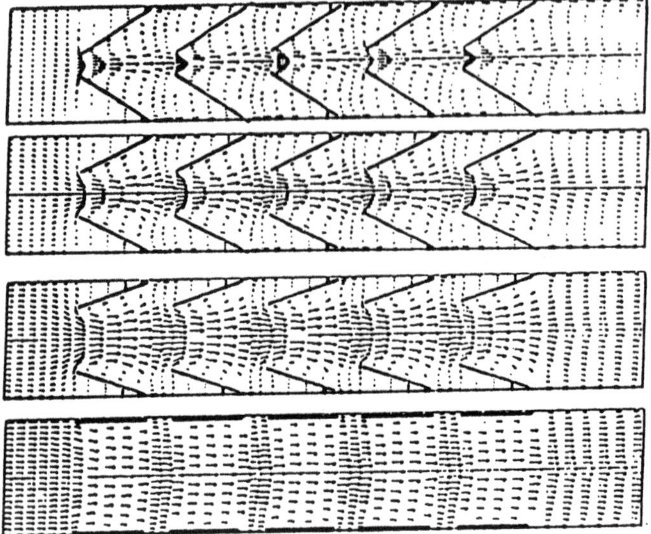

Figure 5. Velocity profiles generated by the numerical solution.

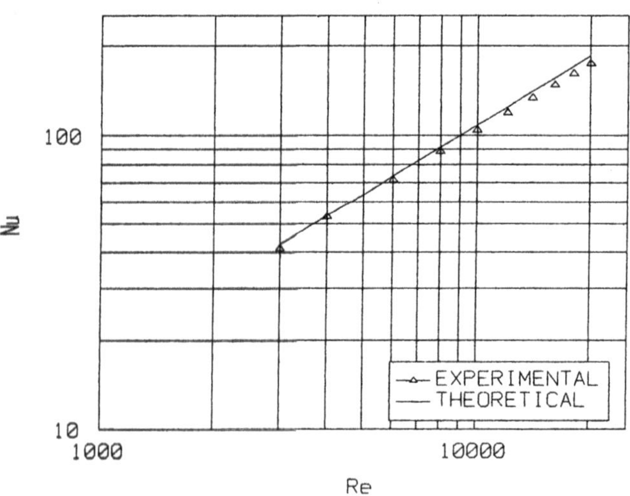

Figure 6. Mean Nusselt number versus Reynolds number for the optimal type.

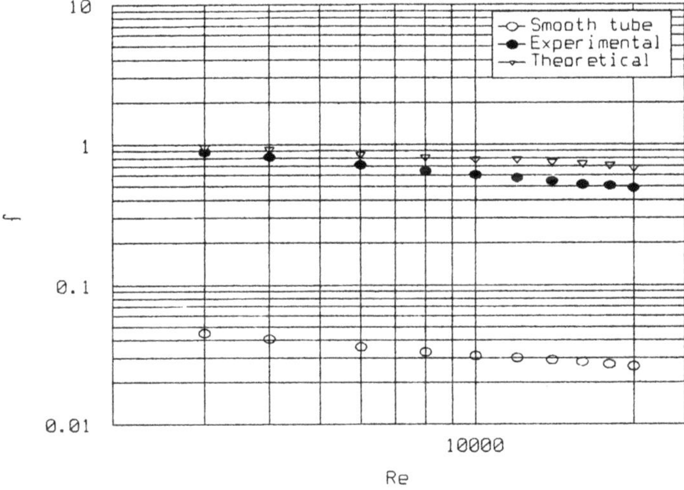

Figure 7. Comparison of friction factor versus Reynolds number for the optimal type.

5. Conclusions

The present results show that the truncated conical hollow cone inserts tested provide considerable enhancement of convective heat transfer coefficients. Design parameters are presented for the optimal type of truncated hollow cones and the existing correlation for the truncated hollow cone is confirmed in pipe flow. It is apparent that resistance in the flow is always greater than the associated enhancement in the heat transfer. Also, it is seemed that the experimental and numerical results correspond very well for the optimum type.

Nomenclature

C_1, C_2, C_μ	turbulence constants
c_p	specific heat at constant pressure
D_b	diameter of large cross section of truncated hollow cone insert
D_s	diameter of small cross section of truncated hollow cone insert
D_0	outer diameter of tube
f	friction factor
G	production of turbulent kinetic energy
k	turbulent kinetic energy
Nu	Nusselt number
P	pitch of truncated hollow cone insert
p	pressure
Pr	Prandtl number
Re	Reynolds number

356

S_ϕ	source term
T	temperature
u	axial velocity
v	radial velocity
α	inclination angle of truncated hollow cone insert
ρ	density of the fluid
ϕ	dependent variable
r	radial coordinate
x	axial coordinate
Γ_ϕ	exchange coefficient
μ_e	eddy viscosity
μ_l	laminar viscosity
μ_T	turbulent viscosity
ε	dissipation of turbulent kinetic energy
$\sigma_k, \sigma_\varepsilon$	effective Prandtl numbers
v	kinematic viscosity

References

1. Bergles, A.E., Nirmalan, V., Junkhan, G.H. and Webb, R.L. (1983) Bibliography on augmentation of convective heat and mass transfer II, *Heat Transfer Laboratory Report HTL-31, ISU-ERI-Ames 84222, DE-84018484*, Iowa State University, Ames.
2. Yilmaz, T. and Ayhan, T. (1983) Heat transfer and pressure drop in adjacent channels with periodic enlargements, *4th National Congress of Thermal Sciences an Technology*, pp. 133-149, Gaziantep, Turkey.
3. Ayhan, T. and Karabay, H. (1990) Observation on the influence of contracting and enlarging conical surfaces on heat transfer in the pipe flows, *Journal of Thermal Sciences and Technology*, 11, 4, pp. 39-43, Turkey.
4. Azak, Y., Demirtas, C. and Ayhan B., (1996), A second law analysis of the optimum design and operation of conical ring inserted tubes for turbulent flow heat transfer, *Proceedings of the TIEES-96*, vol. 2, pp. 651-656, Turkey.
5. Measurements of fluid flow in pipes using orifice, nozzle and venturi, *ASME Standard*, MFC-3M, (1984).
6. Spalding, D.B. and Launder, B.E., (1974), The numerical computations of turbulent flows, *Computer Methods in Applied Mechanics and Eng.*, vol. 3, pp. 269-289.
7. Bejan, A., (1996) Entropy Generation Minimization, CRC Press Inc., New York.

MODERN ADVANCES IN OPTICAL MEASURING TECHNIQUES TOOLS TO SUPPORT ENERGY CONSERVATION

F. MAYINGER
Lehrstuhl A für Thermodynamik
Technische Universität München
Boltzmannstraße 15
D-85748 Garching

Abstract. Transport processes – heat and mass – play an important role in energy conservation. To optimise these processes, optical measuring techniques can be of great help. An overview is given on various optical measuring techniques, which give insight into transport processes, especially with heat transfer and with combustion. A short introduction into holography, holographic interferometry as well as Rayleigh Scattering and Laser Induced Fluorescence show, together with examples from several fields of thermo-fluiddynamic research work, the capabilities of optical measuring techniques to improve transport processes.

1. Introduction

As long as our energy demand is mostly provided via heat production, the optimisation of heat transfer plays an important role in energy conservation. Heat is generated by combustion processes to a large extent, and low combustion efficiency, with non-negligible content of partially unburned components in the exhaust gas, has a strong negative effect on energy conservation.

In heat recovery, the optimum design of heat exchanging components with respect to augmenting heat transport and minimising pressure drop plays an important role. Holographic interferometry provides detailed insight into the local conditions of the boundary layer in which heat transfer and friction are controlled. So these optical measuring techniques allow a well aimed strategy for improving heat exchanger performance.

The turbulent transport of mass and heat in and in front of the flame has a strong influence on the combustion process, and if both are too weak or if one prevails over the other, incomplete combustion is the consequence. Optical techniques, like Raman Scattering, Laser Induced Fluorescence or Auto Fluorescence allow us to follow the combustion process in detail and give precise and instantaneous information about the flame structure. These optical techniques do not only monitor the combustion process itself; they also give very valuable hints to how and where to influence the process in a posi-

357

S. Kakaç et al. (eds.), Heat Transfer Enhancement of Heat Exchangers, 357–379.

tive way to achieve high combustion efficiency and low content of toxic components in the exhaust gas.

If liquid fuel is used for combustion processes, adequate spray-formation is the key for efficient burning of the flame. Droplet distribution and droplet movement in the spray can be conveniently monitored and measured by holography especially by double-pulse holography.

During the last 5 or 10 years, the development of optical methods has been supported by the availability of new equipment like high energy light sources, intensified electronic camera systems and electronic devices as well as new software for image evaluation. This allows us to reduce the time of data processing, which has been very time consuming in the past.

measuring technique	physical effect	application	dimensions	real-time application
schlieren and shadow	light refraction	heat, mass transfer	2D (integ)	yes
holography	holography	particle size, velocity	3D	no
interferometry	change of light velocity	heat, mass transfer	2D (integ)	yes
laser Doppler velocimetry	Mie scattering	flow velocity	point	yes
Raman scattering	Rayleigh scattering	density, temperature	point-2D	yes
dynamic light scattering	Raman scattering	mol. concentration, temperature	point-1D	no
laser induced fluorescence	fluorescence	concentration, temperature	point-2D	no
absorption	absorption	concentration, temperature	point-2D (integ.)	yes
pyrometry	thermal radiation	temperature	1D	yes
thermography	thermal radiation	temperature	2D (integ)	yes
self fluorescence	therm. fluorescence, chemoluminiscence	concentration, temperature	2D (integ)	yes

TABLE 1. Over-view of optical measuring techniques

It is well known that optical methods work in a non-invasive and inertialess way, and therefore do not influence the thermo-fluiddynamic process that has to be investigated.

In principle, one can distinguish between imaging and non-imaging techniques. Imaging techniques provide simultaneous information on a large area, and use any kind of conventional or electronic means to store this information. Non-imaging optical methods are of local nature, i.e., they work with focussed light beams (mostly laser) and usually register data of a very small volume, usually smaller than 1 cubic-millimeter. Table 1 presents a list of examples for modern optical measuring techniques used in fluiddynamic processes and in combustion. This table also gives information about the physical effect used in the various techniques, and presents examples for application. As it is not possible to discuss all the optical techniques, known from the literature, in detail here, emphasis is given to

- Holography,

- Holographic Interferometry,

- Raman Scattering,

- Laser Induced Fluorescence (LIF, LIPF) and

- Auto Fluorescence.

A detailed overview over several optical measuring techniques, their working principle and their application is given in [1].

The evaluation of optical data, for example, from a hologram, from an interferogram, from Raman Spectroscopy, or from Laser Induced Fluorescence, has become much faster in the last years than it could be done in the past. A few years ago it took hours to evaluate an interferogram. Today the same work is done by a computer within a few seconds. But also the huge storage capacity of modern computers – even of the PC type – was an important requirement for preparing the way for the revival of optical methods.

2. Holography

In the middle of this century Gabor [2] invented a new method for recording and storing optical information, which was called holography. Unlike photography, which can only record the two-dimensional distribution of the radiation emitted by an object, holography can store and reconstruct three-dimensional pictures. The name holography comes from the ability of the method to record the totality (holos) of the light information, respectively, of the wave front, namely the amplitude (as brightness), the wave-lengths (as colour) and the phase position of the light. By using these possibilities, completely new interference methods can be developed. Holography, however, demands a source emitting coherent light. Therefore, holography, as invented by Gabor in 1949, could only be widely used when the laser was developed, which was 10 years later.

The general theory of holography is very comprehensive and the interested reader is referred to the literature [3, 4, 5]. Here only a set-up for monitoring spray systems will be briefly explained. To create a hologram of any object, this object has to be illuminated by a monochromatic light source, usually a laser. The reflected or scattered light (object wave) has a very complicated wave form. According to the principles of Huygens, one can regard it as the superposition of many elementary spherical waves. Fig. 1 shows the optical arrangement for so-called "through-light" holograms, consisting of illuminating lasers, various lenses and mirrors. The object wave in this arrangement is produced by a pulsed ruby laser, and this object wave travels via a diverging lens, a beam splitter and a ground glass through the spray (measured object) onto a photographic plate, called in this case hologram. In the beam splitter, part of the laser beam is deflected, bypasses the measured object and falls also onto the hologram. This wave is called reference wave. The superposition of the object wave and the reference wave

360

produces an extremely fine (microscopic) fringe pattern on the photographic plate, the hologram.

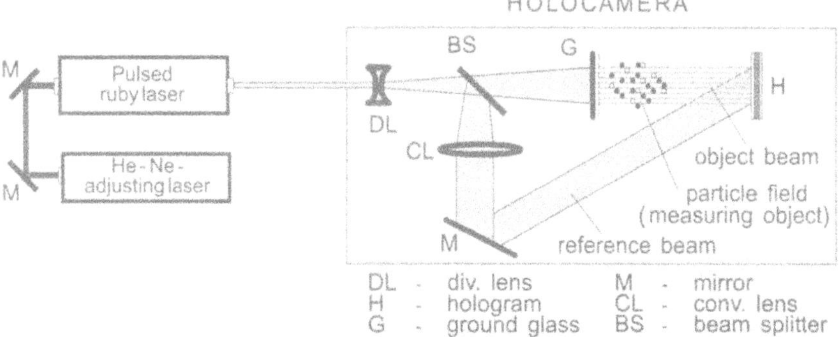

Figure 1. Holographic set-up for through-light, ultra-short time exposures with a pulsed laser

The pulsed ruby laser produces an ultra-short monochromatic light flash of 30 ns duration and a wave-length of 693 nm. The helium-neon laser, also shown in Fig. 1, is only used for adjusting the optical arrangement. The helium-neon laser produces continuous red light of similar wave-length as the ruby laser. The ruby-rod is transparent for the light of the helium-neon laser, so the adjusting light is exactly in line with the recording light.

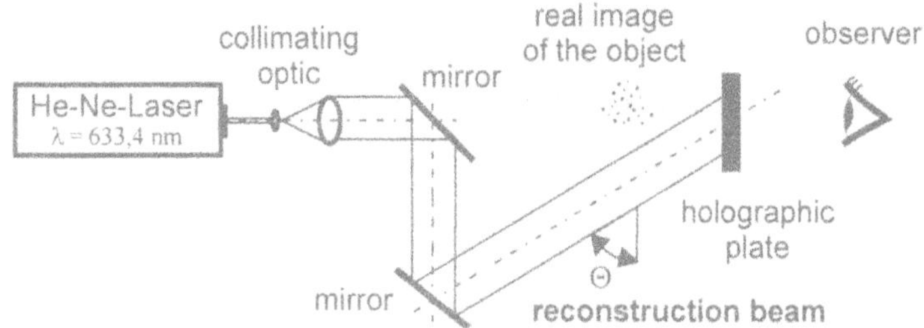

Figure 2. Optical arrangement for the reconstruction of pulsed laser holograms

After the illumination, the photographic plate is chemically processed - developed, fixed and dried - and it is then repositioned into the light-beam of the helium-neon laser, as shown in Fig. 2. However, it is illuminated only with the reference wave, which is called reconstruction wave now. This illumination by the reconstruction wave produces two images of the former measured object - the spray - a virtual image and a real one. These images appear on different sides of the hologram. The virtual image can be observed with the naked eye, and it provides, to a certain extent, a three-dimensional information of the droplet distribution in the spray. The real image - see Fig. 3 - can be looked at by using a microscope or also any camera, for example a video camera. This real image has a three-dimensional extension, and by using a very long focus lens for the camera, only one plane in the real image of the spray will be recorded sharply. If we now place our video camera on a traversing mechanism (Fig. 3), we can take two-dimensional pictures of several planes of the measured object by moving the video camera forward and backward, and so we can store the three-dimensional information in a computer. One can argue that this recording, plane by plane, could also be done by changing the focal length of the lens in the camera. In this case, however, we would get different reproduction scales for the planes.

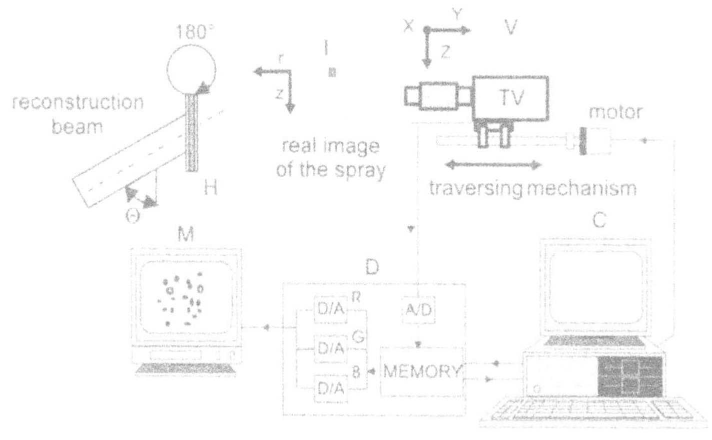

C   Personal Computer AT   H   Hologram
D   Digitizer PCVISIONplus   M   RGB-Graphic Monitor
I   Reconstructed Image   V   Video Camera

Figure 3. Digital Image-Processing System for the evaluation of pulsed laser holograms

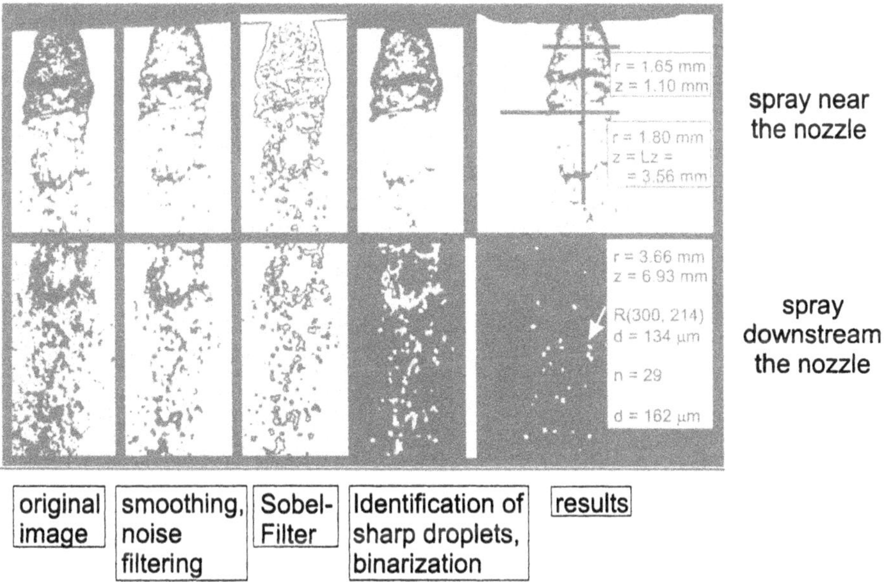

Figure 4. Steps of Image-Processing of a hologram
Upper row: spray near the nozzle
Lower row: spray downstream of the nozzle

The digital image processing system, together with the camera monitoring the real image, coming out from the hologram via the reference wave, is also shown in Fig. 3. The data from the video camera - a CCD camera - are given to a digitiser and are reprocessed in a personal computer. In the computer a complicated image evaluating process is performed. Fig. 4 gives a visual impression of how the numerical procedure, going on in the computer, changes the original photographic picture into a computerised one.

A spray veil and a swarm of particles with different grey values in the range from black to white (via grey) representing droplets of this spray and a pattern of fine grain, which formed the optical noise (speckle noise), is seen in the first scene (left) in Fig. 4. This disturbing noise is produced by the diffusive monochromatic illumination necessary for recording the hologram. This noise at first is removed in the computer by a so-called noise filter (second scene from left). After the separation of the droplet images from the background, the identification of sharply focussed droplets, the measurement of their projected areas, and the evaluation of their equivalent diameters and centre-points can be performed (third and fourth scene in Fig. 4). An important step within this procedure is to eliminate all optical reproductions of droplets which are out of the focus plane. This is done by determining the gradient of greyness at the boundary of the reproduction of the droplet. Those droplets which are in the focus plane have a large gradient of greyness. For a detailed description of this image processing, reference is made to [1 (chapter 7)].

Finally, the computer reproduces the image of one single plane out of the spray, as shown in the right scene of Fig. 4. In this example the focussed plane represents a slice out of the spray, thinner than 0,5 mm.

From this figure one can now derive the length of the liquid veil, until it disrupts in droplets, and the thickness of its liquid layer. It can be clearly seen, that the veil has a wavy form and its liquid flow is characterised by a week pulsation with an oscillatory production of more or less dense droplet clouds. Far downstream of the veil one can identify the droplets with respect to their diameter and their concentration in the cloud.

To get information about the velocity of the droplets we produce the so-called double-pulse holograms. To do this, one illuminates the photographic emulsion of the holographic plate twice before processing it. This can be done for example with a ruby laser, which allows the emission of more than one laser pulse within a short period of time (usually 1 - 800 μs). For slow-moving particles one can use an argon-laser emitting continuous light which, however, is periodically interrupted via a mechanical device or an optic chopper. The evaluation of such a double-pulse hologram is - as it is easily understandable - even much more complicated than that of a single-pulse hologram. At first, the procedure is the same as sketched out above. Then the computer produces connecting lines between each pair of optically reproduced droplets, without regarding whether it is the reproduction of the first or of the second illumination. The distance and the orientation (angle) of each connecting line are then calculated and stored in the computer. With the help of a Fourier-analysis, the normalised frequency of the angle and of the distance is then plotted versus the angle and the distance of the reproductions. If the flow in this spray is not too chaotic, two maxima are found, one for the angle and the other one for the distance. Both maxima can be used as average values for the spray angle and for the droplet velocity, because this average velocity can be easily calculated by dividing the mean distance by the delay time between the first and the second illumination.

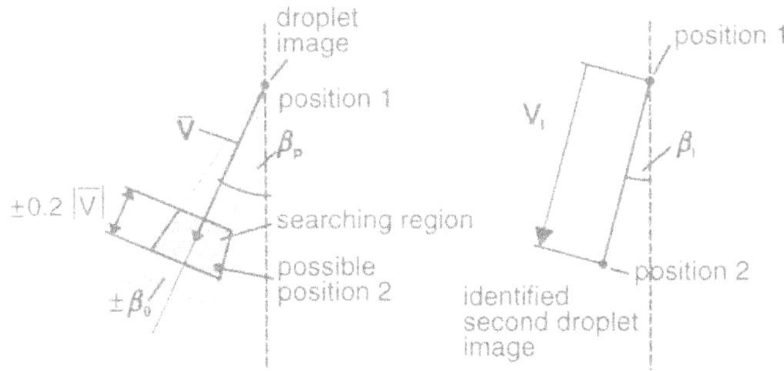

Figure 5. Method for evaluating single particle velocity from double-pulse holograms

One can now proceed one step further to calculate the real velocity of each single droplet. For this purpose one defines a searching area, as briefly lined out in Fig. 5. The centre of this searching area is given by the average values for the velocity and the angle of the droplets, and the extension of this area depends on the droplet concentration on the spray, because there should be only one optical reproduction of a droplet in this area. Therefore, for very dense sprays or particle clouds this procedure has some limitations.

In total, this holographic velocimetry gives reasonable results, as Fig. 6 proves. In this figure the average droplet velocity of a spray is plotted versus the mass flow rate with the pressure of the spray atmosphere as the parameter. The measuring technique clearly records the influence of the spray atmosphere and presents a reduction of the droplet velocity with increasing the air density, due to a higher pressure.

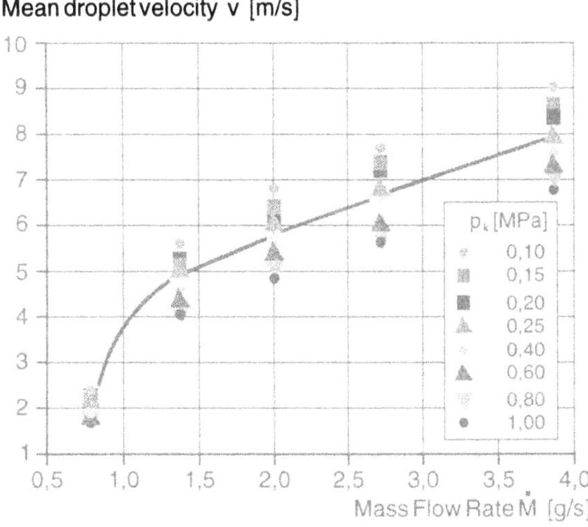

Figure 6. Mean droplet velocity in a spray as a function of the flow rate at different ambient pressures

Of course similar results can also be gained by Particle-Image Velocimetry (PIV) in which a thin light sheet, produced by cylindrical lenses, penetrates the droplet cloud, illuminating the particles there. The particles which are momentarily present in this light sheet can be recorded with a normal high-speed camera. In this case, however, only the velocity parallel to the light sheet can be determined. Particles moving orthogonal to the light sheet cannot be followed up. With the double-pulse holography these particles can be found by evaluating adjacent planes successively after the treatment of the first plane. The procedure is mathematically complicated. A brief description can be found in [6, 7].

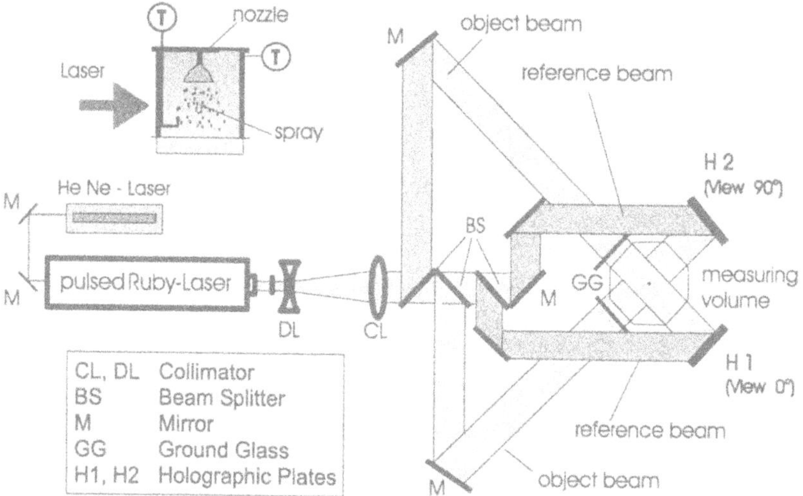

Figure 7. Holographic set-up for stereo-matching of strongly three-dimensional measurement problems
The two holographic plates are exposed at the same time

For a real stereo matching of strongly three-dimensional measurement problems a holographic set-up can be used, as shown in Fig. 7. Two holograms are recorded simultaneously perpendicular to each other. Both holograms have to be scanned and digitised by a camera, which is again focussed stepwise along the depth coordinate in order to record the entire three-dimensional information contained in the holographic images. The procedures for evaluating these images and for reprocessing the data are briefly described in [8]. This optical set-up and its data processing can be used for investigating three-dimensional bubble flow, for example in stirrers.

3. Holographic Interferometry

By using the recording capabilities of holography, different waves - even those shifted in time - can be stored in the same holographic plate, as we learned when discussing double-pulse holography in chapter 2. If the developed holographic plate is illuminated with the reference wave, all object waves are reconstructed simultaneously. Where they differ only slightly from each other, interference patterns are observed. These are the fundamentals of holographic interferometry.

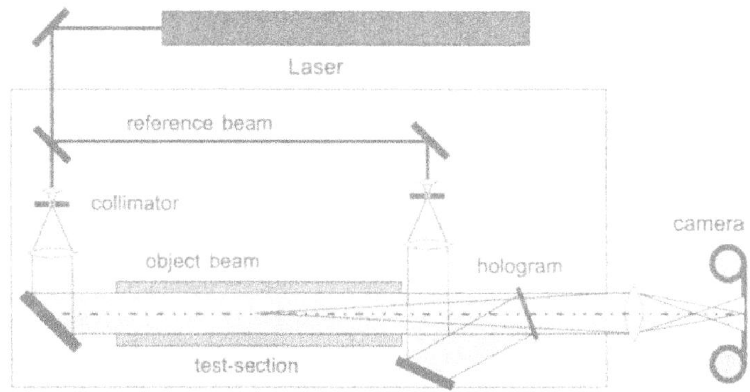

Figure 8. Optical set-up for holographic interferometry

In heat and mass transfer the temperature and the concentration distribution in a fluid are of special interest. To investigate processes with heat or mass transfer a so-called through-light method is used, where the object wave is irradiating through a volume in which the transport processes take place. A simple exemplary arrangement for holographic investigations is shown in Fig. 8. A beam splitter divides the laser-beam into an object wave and into a reference wave, which is also called comparison wave. Both waves are expanded to parallel wave bundles behind the beam splitter via lenses, usually consisting of an arrangement of a microscopic lens and a collecting lens. The expanded and in a parallel way organised object wave travels through the space of the research object of interest - the test section - in which the distribution of temperature or concentration is studied. The reference wave bypasses the test section and falls directly onto the photographic plate.

The benefit of holographic interferometry compared to other interferometric methods - like Mach-Zehnder Interferometry - is that there is no need for a high optical quality of the optical components, because only relative changes of the object wave are recorded, and optical errors are automatically compensated in this interferometric method. On the other hand, the monochromatic light producing the wave front has to be very stable, therefore a laser of good coherency is needed as a light source. For more details on holographic interferometry reference is made to the literature [1, 9, 10].

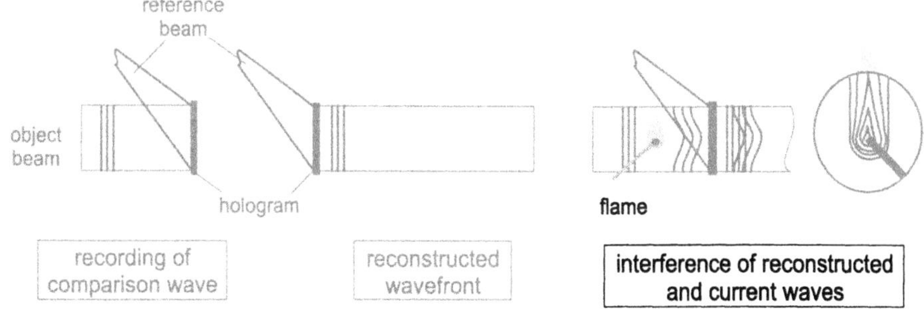

Figure 9. Real-time method for holographic interferometry

Several procedures exist to produce interferograms. Here, only the so-called "real-time method" will be explained, which can also be used in connection with high-speed cinematography. This method observes the process to be investigated in real-time and continuously. The method is illustrated in Fig. 9. After the first exposure by which the comparison wave is recorded and during which no heat transfer is going on in the test section, the hologram is developed and fixed. Afterwards, accurately repositioned, the comparison wave is reconstructed continuously by illuminating the hologram with the reference wave. This reconstructed wave showing the situation without heat transfer in the test section can now be superimposed onto the momentary object wave. If the object wave is not changed, compared to the situation before the chemical developing process, and if the hologram is precisely repositioned, no interference fringes will be seen on the hologram. This feature is a valuable help to replace the hologram accurately.

Now the heat transfer process can be started. Due to the heat transport, a temperature field is formed in the fluid and the object wave receives an additional phase shift when passing through this temperature field. Behind the hologram, both waves interfere with each other and the changes of the interference pattern can be continuously observed or photographed.

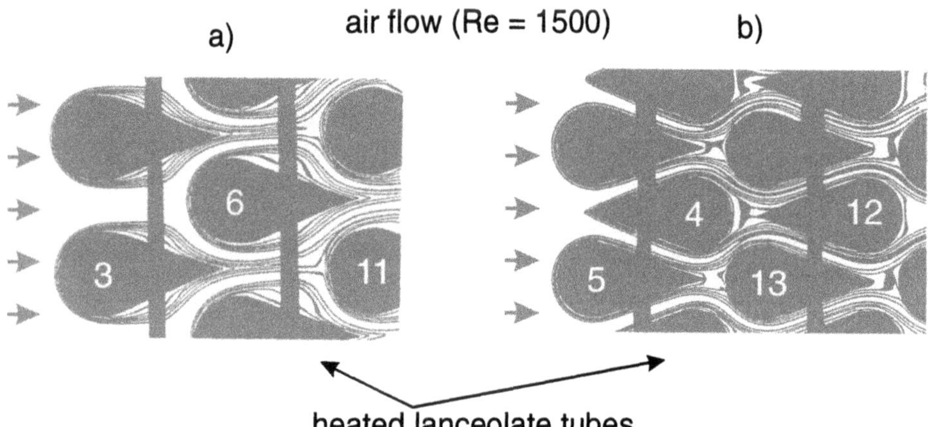

Figure 10. Interferograms of cross-flow through bundles of lanceolate tubes
a) uniformly orientated; b) mixed orientation (even rows reversed)

It has to be mentioned that interference fringes are created by density gradients and these can originate from temperature-, pressure- or concentration-differences. In a single component fluid with flow velocities far away from the sound velocity, interference fringes are only the product of temperature differences, and each line - black or white - represents - in a first approximation - an isotherm in the fluid. When the isotherms are densely packed, there are steep temperature gradients and therefore a high local heat transport. For the evaluation of such interferograms, reference is made to [1 and 10]. Examples of holographic interferograms, taken with the real-time method, are illustrated in Fig. 10.

The task of the research project, in which the interferograms in Fig. 10 were produced, was to develop a compact heat exchanger operating at very low air pressure (cross-flow) with a good overall performance, i.e., maximum total heat flux and low friction. One has to be reminded that the total heat flux is the product of the average heat transfer coefficient and the heat transferring surface to the ambient air. The low ambient pressure resulted in low Reynolds numbers.

Tubes of lanceolate cross-section can be much more densely packed than tubes of circular cross-section. They also have a lower pressure drop at the same air flow rate. The most dense package can be achieved if every second row of the bundle has reversed orientation, as shown in Fig. 10b. This arrangement has a high density of heat transferring surface per volume of the heat exchanger. The question is what happens to the heat transfer coefficient or to the Nusselt number, if we reverse the orientation of the tubes in the even rows with respect to the odd ones.

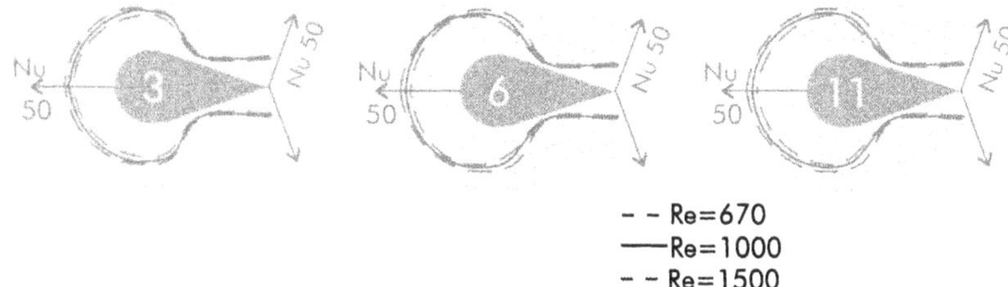

Figure 11. Evaluation of interferograms in Figure 10a, local Nusselt numbers around lanceolate tubes in bundle (positions see Figure 10a)

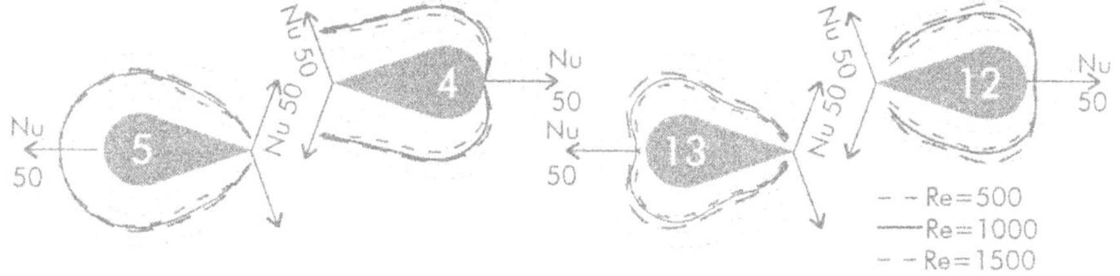

Figure 12. Evaluation of interferograms in Figure 10b, local Nusselt numbers around lanceolate tubes in bundle (even rows reversed, positions see Figure 10b)

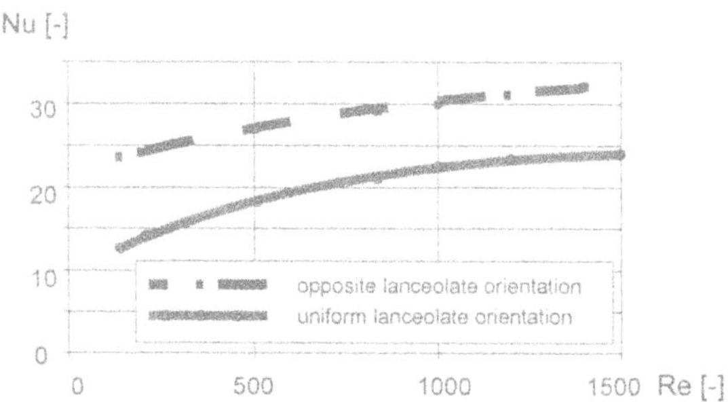

Figure 13. Average Nusselt numbers in cross-flow bundles with lanceolate tubes

From the interferograms in Fig. 10, reliable and accurate local Nusselt numbers can be evaluated around the lanceolate tubes, as shown in Fig. 11 and Fig 12. If all the tubes have the same orientation in the bundle, the position does not have too much influence on the Nusselt number, as Fig. 11 demonstrates. There the local Nusselt number versus

368

the circumference of the tubes in the rows 1, 2 and 3 (positions 3, 6, 11 in Fig. 10a) are plotted. One can clearly see that only two third of the surface of each tube participates in the heat transport.

If we turn the orientation of the tubes in every second row (even rows, positions 4, 5, 12, 13 in Fig. 10b), then the situation changes remarkably. As Fig. 12 demonstrates, the lanceolate tubes now have a much better heat transfer performance. This is clearly expressed, if we plot the average Nusselt numbers (averaged over the whole heat transfer area of the bundle) versus the Reynolds number, as done in Fig. 13. The improvement due to the reversed orientation of the tubes holds for a wide range of Reynolds numbers. As it is easily understandable, it is a little larger at very low Reynolds numbers than at higher ones. Therefore the reversed orientation of the tubes does not only allow a more compact design, it also improves the heat transfer performance.

Figure 14. Holographic interferograms of the detachment and re-condensation of bubbles from a heated wall in sub-cooled water

With the simple holographic arrangement, shown in Fig. 8, not only can the heat transfer coefficients in a single-phase flow be measured; it can also be used for studying phenomena with phase change, for example with boiling. Fig. 14 shows a sequence of interferograms which were taken within a period of less than 5 μs in a horizontal water flow. The bulk temperature of the water flowing from right to left was slightly sub-cooled and the lower wall of the channel was heated to such an extent, that in spite of the fact that the bulk temperature was below the saturation temperature, sub-cooled boiling occurred at the wall.

If we look at the wall-near boundary layer we can detect that there is one area where a plume of hot liquid is a little more extended into the bulk region. In the period between 0,0 and 0,3 ms, a bubble is created and grows from the wall into this hot liquid plume. When the bubble penetrates the plume boundary, it comes into contact with the sub-cooled bulk and starts to condense (sequences from 1,9 to 4,7 ms) again. The whole span of life-time of this bubble is approximately 4,5 ms.

We can observe a second bubble which starts growing at a position of the wall where the boundary layer has a very steep temperature gradient (see right side of scene 0,3 ms). Apparently, the liquid adjacent to the wall at this position is much hotter than at the position of the first bubble, and therefore this second bubble grows much faster. However, it reaches the sub-cooled bulk earlier, which results in a very rapid condensation, and the life-time of this bubble is only in the order of 2 ms.

So holographic interferometry, together with high-speed cinematography, gives a very good insight into the phenomena of boiling and demonstrates that bubble growth and bubble condensation are not only governed by determining parameters, such as the heat flux or the sub-cooling, but has a stochastic nature, too. This stochastic nature originates from fluctuations in the temperature gradients of the boundary layer and its thickness. These fluctuations are influenced by the turbulence and also by the "history" of the bubble detachment.

From the interferograms in Fig. 14 we can also learn a limiting phenomena of this method. The laser light travelling through the fluid is not only shifted in phase, but it is also deflected by the density gradients. This light deflection is the reason why in Fig. 14 we do not get any information in the immediate neighbourhood of the heated wall.

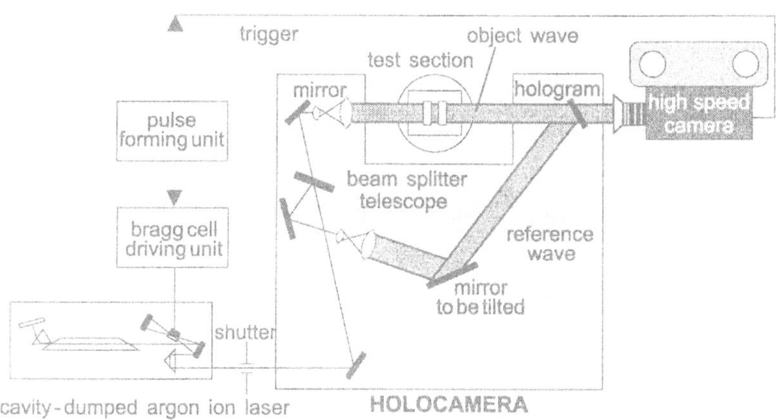

Figure 15. Arrangement of holographic interferometry with finite-fringe method

With very high heat transfer coefficients, the boundary layer at a heat transferring surface becomes very thin down to a few hundredths of a millimetre. In this case, it is difficult to evaluate the interference pattern, if it is registered with the procedure described up to now. A slightly altered method, the so-called "finite-fringe method" offers some benefits. In this method, after the reference hologram has been produced, a pattern of parallel interference fringes is created by tilting the mirror in the reference wave of Fig. 15 or by moving the holographic plate there within a few wave-lengths. The direction of the pattern can be selected as one likes, and it only depends on the direction of the movement of the mirror or of the holographic plate. This pattern of parallel interference fringes is then distorted by the temperature field due to the heat transport process. The distortion or deflection of each fringe from its original parallel direction is a measure for the temperature gradient at this spot and allows one to deduce the heat flux and therefore the heat transfer coefficient [11].

370

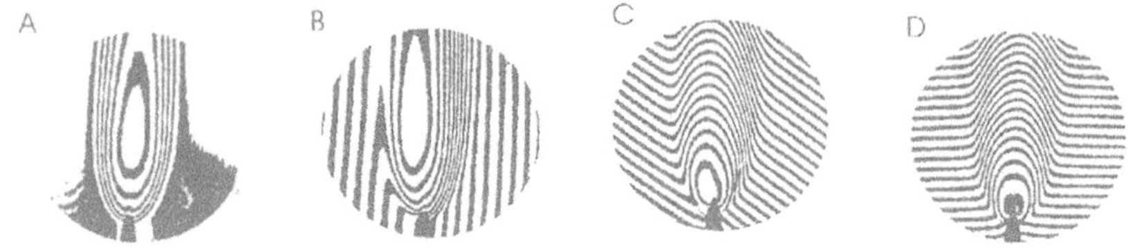

Figure 16. Comparison of an Infinite-Fringe-and a Finite-Fringe-Interferogram
(various inclination angles of the mirror to be titled)

An example of how these "finite-fringe interferograms" differ from the method described before – infinite-fringe interferograms - is shown in Fig. 16. On the left upper side of Fig. 16 an infinite-fringe interferogram of the temperature distribution in a small gas flame is shown. All fringes represent lines of constant temperature, and as to be expected, the highest temperature is recorded in the centre line of the flame. The other 3 interferograms of this figure show finite-fringe interferograms of the same flame. These fringes do not represent lines of constant temperature, but the deviation of the fringes from their original direction (vertical, inclined or horizontal) gives the local temperature gradient in the flame. From this local temperature gradient it is easy to calculate the heat transport.

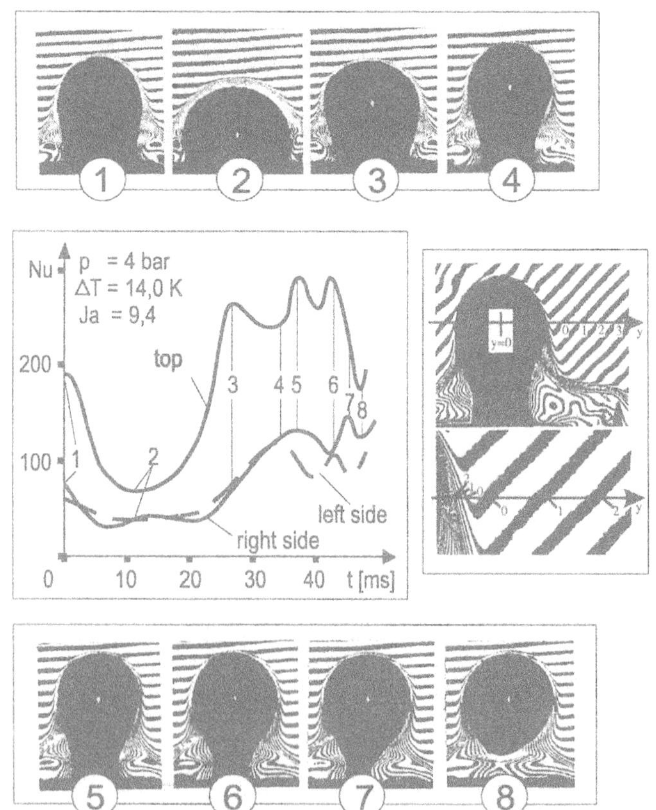

Figure 17. Heat transfer at the phase-interface of a steam bubble, condensing in sub-cooled water, deduced from a sequence of interferograms

Fig. 17 demonstrates an example for using these techniques in a flow with a bubble condensing in a liquid. The combination of this method with high-speed cinematography allows an inertialess and precise calculation of the heat transfer coefficient at the phase-interface of the condensing bubble. Steam is flowing out of a thin capillary against a slow downward water flow with a temperature, which is 14 K below the saturation temperature of the steam. With the penetration of the steam into the water the condensation starts, and after a few ms the reduction of the bubble volume by condensation is larger than the volumetric flow rate of the steam fed through the capillary. The bubble stops growing and a thin layer of saturated liquid is formed around its surface. This results in a sudden reduction of the heat transfer through the phase interface until the steam flow through the capillary can overcome the condensation rate again. The heat transfer increases and continues for a while with an oscillating character [11].

So holographic high-speed interferometry allows insights into thermo- and fluiddynamic phenomena, which could never be gained by conventional measuring techniques.

4. Light Scattering Methods

Light as a sensor can provide several information, and not only can the refractive index and the phase-shift be used to get information about the distribution of concentration or temperature in a substance. Besides the phase-shift, the effect of scattering is most commonly used to get information about the chemical and physical conditions in a gas or a liquid. Raman Scattering is a method which allows to measure the variety of substances present in a mixture, the concentration of each species and, under certain circumstances, also the temperature. Rayleigh Scattering can be used as a non-intrusive method for measuring the temperature. The fluorescence indicates the kind and the concentration of atoms and molecules and also allows me to deduce the temperature. There are also other scattering methods like Mie-Scattering used in Phase-Doppler-Velocimetry, Bragg-Scattering for detecting density fluctuations and for investigating the structure of crystals and finally also sound absorption. Here only the fluorescence used together with a light sheet method and Raman Scattering will be discussed.

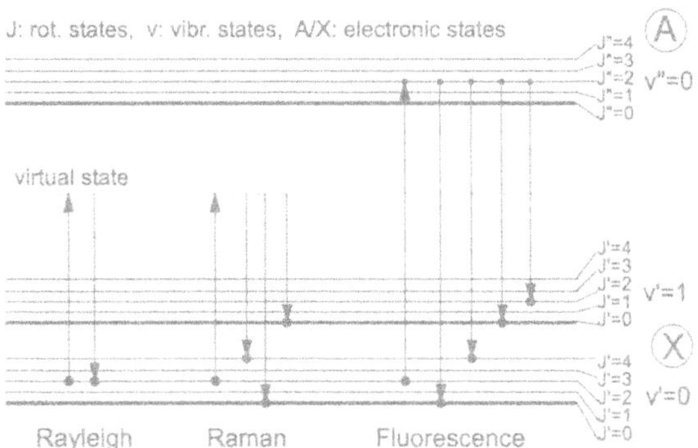

Figure 18. Energy diagrams for Rayleigh, vibrational Raman Scattering and Laser Induced Fluorescence

372

It should be briefly mentioned, that Mie-Scattering has been recently used for flow visualisation in Particle-Image Velocimetry (PIV). A laser-beam is formed into a very thin light sheet, illuminating a plane within the volume of interest. For one measurement two consecutive laser pulses are fired within a very short time interval, and the radiation scattered by the particles in the illuminated area is recorded by means of a two-dimensional camera. The diameter of scattering particles is usually between 3 μ and 300 μm.

If the particles, scattering the light, are smaller than the wave-length of the light, Rayleigh Scattering occurs. Typical scattering particles in this case are molecules. There are two models to describe the interaction between the incoming light and the molecule, namely the model of oscillating dipoles and the simplified quantum mechanical model. For details, reference is made to Bohren and Huffman [12].

In Raman Spectroscopy a molecule initially absorbs a photon of the light hitting it. For visible light the energy of this photon is higher than what can be stored by rotation or vibration in the molecule, but often lower than the difference between the ground level and the first electronic state of the photon. Therefore the molecule is lifted up to a highly unstable virtual level by absorbing the photon for a moment. The molecule afterwards immediately drops back to a stable energy level by emitting a photon. If this new level is identical with the original level, Rayleigh Scattering will be observed. The principle of this change in energy is shown in Fig. 18. If the new level is lower than the original level the scattered light has a different frequency than the incidenting light. The shift in frequency is referred to as the Raman Shift and it is proportional to the energy difference between the two molecular levels involved. One has to distinguish between rotational levels and vibrational levels of the molecule, and the difference between two rotational levels is much smaller than that of two vibrational levels. In Fig. 18 the change in energy is shown for Rayleigh Scattering, for Raman Scattering and also for Laser Induced Fluorescence within 2 vibrational levels.

Since every species has its own energetic structure, the observed frequency shift can be related to certain molecules. Raman Scattering therefore provides an excellent possibility for detecting the concentration of several species in a gas mixture of interest.

With Laser Induced Fluorescence the molecule under consideration also absorbs one photon of the incoming laser light. In this case, however, the energy of the photon must be equal to the energetic difference of two energy levels, the original level in the ground electronic state and the corresponding level in the first electronic state (see Fig. 18).

4.1 RAMAN SPECTROSCOPY

The theory of Raman Spectroscopy is well described in the literature [13, 14, 15]. Raman Spectroscopy is applicable to any transparent medium regardless of its physical state. In heat and mass transfer the applications are usually concentrated to liquid or gaseous states while chemical and biological applications of it deal with solid samples.

Based on the Raman effect various methods for the measurement of temperature and concentration have been developed, and some of them rely on special molecular resonance effects. The most widely used methods are Spontaneous Raman Scattering (SRS) or the Coherent Anti-Stokes Raman Scattering (CARS). The latter one provides light signals of higher intensity, which is especially useful for gaseous applications. On the other hand the set-up and the evaluation of the results are much more complicated for CARS than for SRS.

An example for a set-up to measure species concentration and temperature with Raman Spectroscopy is shown in Fig. 19. Raman Spectroscopy is a spot-wise measuring technique and the laser beam is therefore focussed on a small spot, of which the diameter forms the measuring volume. The signal collection is usually arranged at a 90° angle to the direction of the laser beam. At this direction the signal intensity reaches a maximum, and the size of the volume observed can be very well determined. If the vessel or channel enclosing the medium to be investigated allows the installation of a convex mirror opposite to the signal collecting lens, the obtained signal intensity can be remarkably increased. In order to analyse the scattered intensities, the collected light has to be resolved spectrally. This is accomplished either by diffraction units (for example polychromators) or selective filters. The intensity of the light at the selected wavelength is then converted into electronic signals, digitised and then numerically processed.

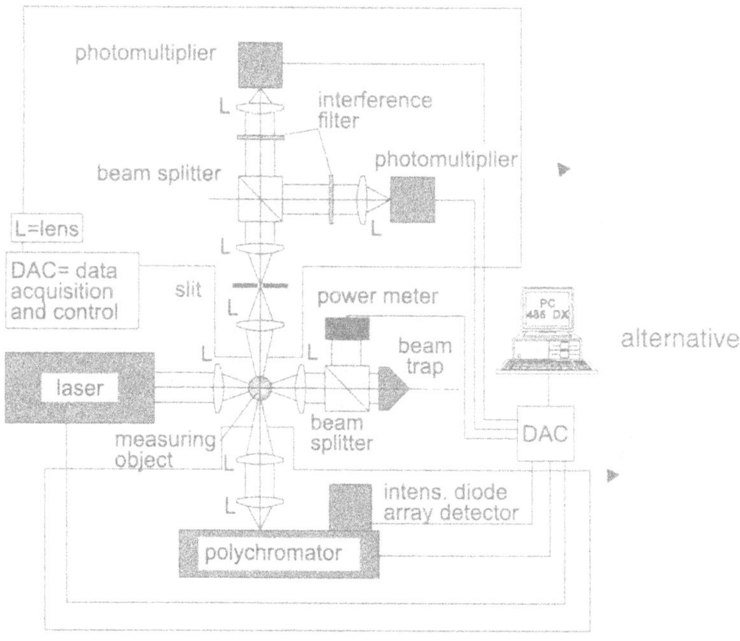

Figure 19. Typical Raman Set-up for point measurements

An example of an investigation with Raman Spectroscopy is shown in Fig. 20. The chemical reaction of a hydrogen flame was analysed there. The experimental set-up was operated under steady-state conditions with a closed tube type burner having a rectangular cross-section. A metal grid was used to stabilise the flame. The pre-mixed unburned gas with a hydrogen concentration of 12 volumetric % approaches the grid upward with a velocity of 17 m/s. There is a separate turbulent flame stabilised behind each opening in the grid, and the burnt gases leave the flame area upward.

374

The concentration distributions are depicted for each species in the form of levels with various distances from the grid. The first spectrum is taken before the grid; no combustion has taken place up to now, and so there is no steam present. The second spectrum is taken from the main reaction zone. Some of the hydrogen has reacted with parts of the oxygen in the air to form steam as a product. All four species (H_2, O_2, OH, H_2O) and the nitrogen content of the air can be seen in the spectrum. The last spectrum is from a location above the reaction zone, where the under-stoichiometric combustion is complete. Only steam, nitrogen and the surplus oxygen are present there. Assuming the nitrogen behaves like an inert gas - or at least its oxidation is negligible - the temperature can be derived from the change of the integrated area under the nitrogen peak. Of course the temperature of the mixture before the combustion starts has to be known.

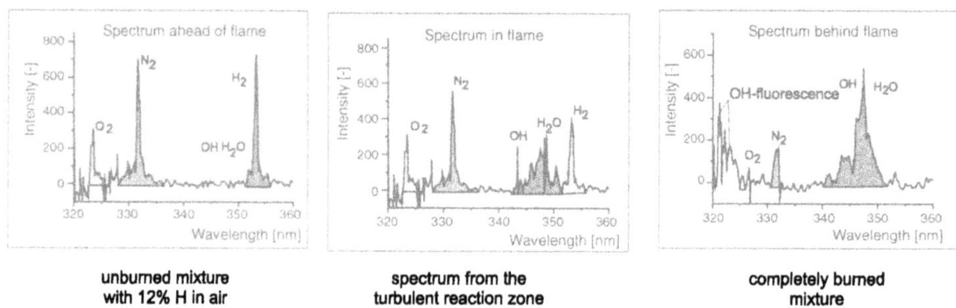

Figure 20.   Typical spectra from representative points in a burner: Unburned mixture with 12% H_2 in air
Middle: spectrum from the turbulent reaction zone
Bottom: completely burned mixture

Finally a more practical application of Raman Spectroscopy is presented in Fig. 21, which shows the Raman spectrum of the gas flow in the exhaust pipe of a small motorbike engine. A lot of unburned hydro-carbons can be detected in this gas flow, which proves that this small piston engine has a low burning efficiency and that it emits toxic gases [16].

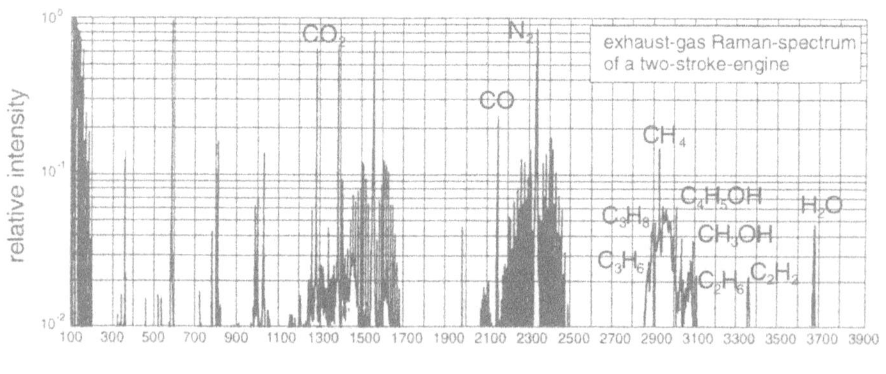

Figure 21.   Exhaust-gas Raman Spectrum of a small two-stroke engine (according to Algermissen [16])

4.2 LASER INDUCED FLUORESCENCE

Laser Induced Fluorescence (LIF) and Laser-Induced-Pre-dissociated Fluorescence (LIPF) can be used to measure local concentrations in mass transfer processes or in chemical reactions. A typical arrangement of optical components for performing LIF or LIPF measurements is shown in Fig. 22. A laser beam must be used for the electronic excitation of the molecules of interest. The wave-length of the laser must be adjusted to the substance that is to be investigated. For studying concentration fields in combustion processes, the most active radical of interest in this connection is the OH-radical in hydrogen combustion. The excitation of the OH-radical has to be done with a wave-length of 308 nm and the wave-length of the fluorescing light is of the same value, therefore one has the problem with LIF to distinguish carefully between the primary light of the exciting laser and the secondary light of the OH-fluorescence.

For LIPF-measurements the two wave-lengths are different, for example 248 nm for the exciting light and 290-304 nm for the fluorescing light. Here it is easier to distinguish between primary and secondary light. However, the energy for producing LIPF is much higher than for LIF. To perform LIF for example, an EXCIMER-Laser filled with XeCl can be used, which has a bandwidth of 0,27 nm around the wave-length of 308 nm. To illuminate an area of approximately 50 cm^2 a pulse energy of 150-200 mJ is needed. The pulse duration of EXCIMER-Lasers is around 20 ns.

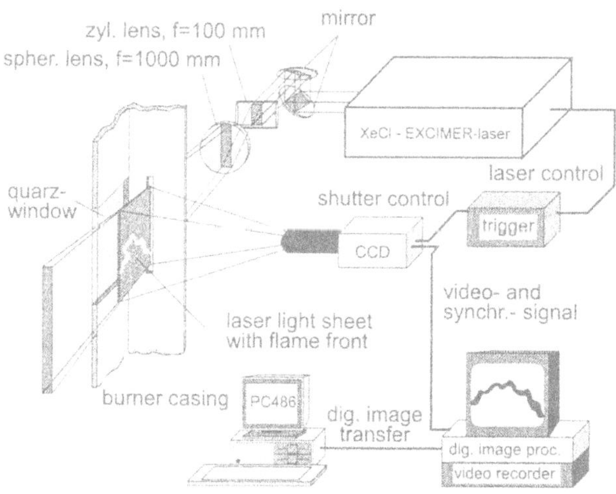

Figure 22. Optical set-up for Laser Induced Fluorescence

Fig. 22 demonstrates how the laser-beam originating from the EXCIMER-Laser is expanded and deformed into a thin light sheet with a height of approximately 5 cm and a thickness less than 0,7 mm. This light sheet travels through a quartz-window into the reaction zone where the fluorescence is produced in this thin layer. The fluorescence is observed and recorded in the perpendicular direction to this light sheet with the aid of a CCD-camera, which is, in this case, intensified in the ultraviolet range. To get a good view of the fluorescing area, large quartz-windows have to be used and the parts, where no optical observations are intended, have to be painted carefully with black colour to avoid or damp the light scattering in the reaction zone. The video signal of the camera is processed in an image-evaluating unit and transformed into pseudo-colour pictures. These pictures are recorded by a SVHS-video recorder, working in an analogue way. The time-wise coordination of the laser and the camera is done via a triggering unit, which is synchronised by the video-camera.

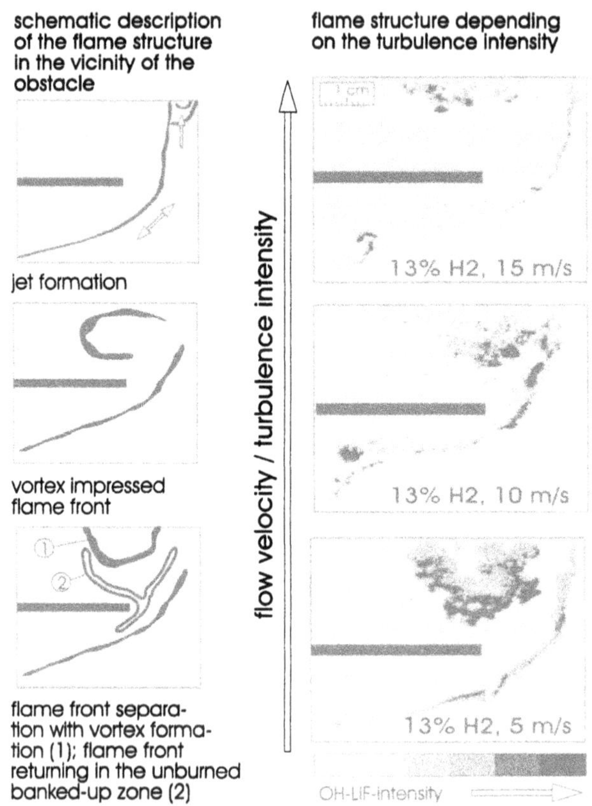

Figure 23. Influence of turbulence on flame formation (H_2-flame)

For LIPF-measurements the same arrangement can be used. However, a laser must be available, emitting light of shorter wave-lengths and higher energy density. This can be done for example by using a Nd-YAG Pump-Dye-Laser or a KrF-EXCIMER-Laser.

This method is very capable, and it is possible to resolve even very thin flame fronts with it. In Fig. 23, the influence of turbulence on the stability of an hydrogen flame is demonstrated [17]. The flame front is travelling in a flowing H_2-air mixture around an obstacle. The obstacle - a rectangular plate - blocked approximately 2/3 of the cross-section of the channel and therefore produced strong turbulence. The H_2-concentration in the air was kept constant at 13% and the flow velocity before the obstacle was varied between 5 and 15 m/s. The laminar flame velocity is the same in all 3 cases of Fig. 23. However, the turbulence changes the true flame velocity remarkably, not only in a positive direction, but it has also the effect of quenching. At low flow velocities we can observe a continuous flamelet passing the area of the obstacle and behind the obstacle a burning zone, such as in a well stirred reactor. With increasing the flow velocity the flamelet shows more and more gaps of unburned or quenched zones. This is a simple but convincing demonstration that turbulence does not only enhance the combustion; it can also deteriorate it.

4.3 AUTO FLUORESCENCE

Auto Fluorescence can be observed during exothermal chemical reactions. It is based on the fact that some of the radicals or molecules appearing during the chemical reaction already arise in an electronically excited state. A part of these excited radicals radiates energy by spontaneous emission of photons in connection with a change in the energetic state of the molecules which is referred to as chemo-luminescence. The emitted energy is always equal to the difference of the two energetic states involved.

Compared to Laser Induced Fluorescence the Auto Fluorescence has the disadvantage of low spectral signal intensities with the consequence of limited time and special resolution. On the other side, however, it is easy to handle, and it only needs a simple experimental set-up for obtaining insight into the overall structure and behaviour of fast-reacting systems.

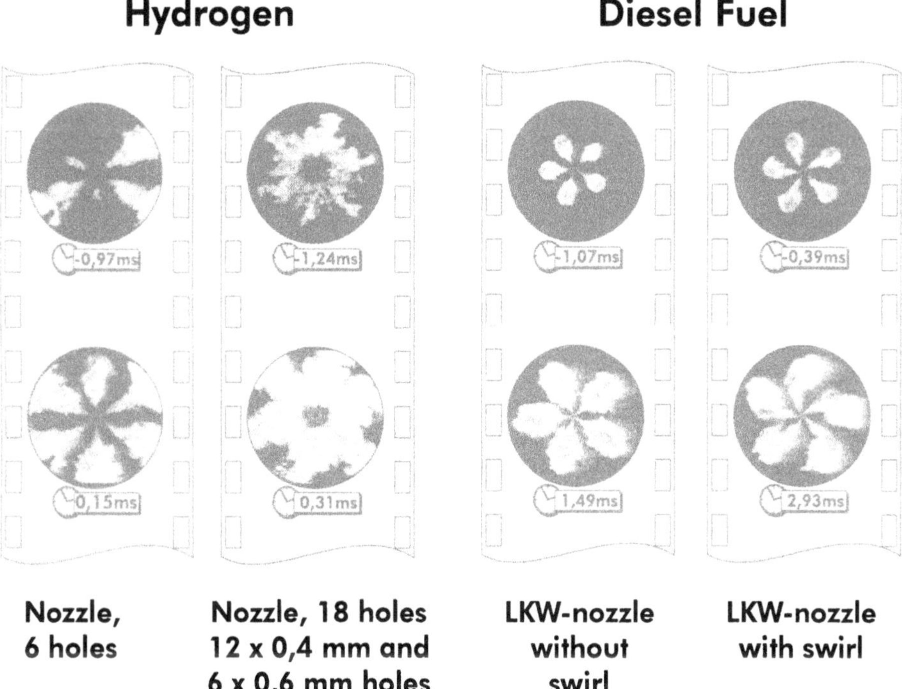

Figure 24. Comparison between the combustion processes of hydrogen and Diesel fuel in a Rapid Compression Machine (The data for the time count from the moment on, when the piston is in the inner dead centre)

An example for a fast-reacting process is the ignition and the combustion in a Diesel-engine in which fuel is injected into a highly compressed air. Due to the compression the temperature of the air is above the ignition temperature of the fuel. This ignition process and the flame propagation is strongly dependent on the injection of the fuel, but also on the properties of the fuel itself. In Fig. 24 examples of such ignition processes and of the flame propagation are presented for 2 different fuels, namely for the usual Diesel-oil and for pure hydrogen. Diesel-oil starts to ignite immediately at the moment when the injection starts and burns smoothly along the injection sprays. If there is a rotating flow in the cylinder, it cannot act at the beginning of the injection onto the flame propagation, because the momentum of the injected spray is too large. After the spray-cloud has expanded, the rotating flow can overcome the radial jet forces and the flame starts to rotate as well.

Hydrogen is much more sensible to the injection mode in spite of the fact, that it is injected in its gaseous phase. There is a big difference in the flame propagation if the same hydrogen amount is injected through 6 holes for example or through 18 holes. In the 6-hole nozzle the hydrogen jets are rather massive and only a few of them start to ignite spontaneously. Only after all 6 jets have been reflected at the cylinder, resulting in an improved mixing, the combustion is well distributed over the circumference. Injecting through 18 holes results in a much better mixing between compressed air and hydrogen from the early beginning, and this affects the combustion process positively. The Auto Fluorescence pictures, shown in Fig. 24, were all photographed through the piston in a specially designed engine, and the time indicated in each picture counts from the moment when the piston reaches the inner dead centre. A negative time indicates a situation before that moment [18].

5. Concluding Remarks

"Computational engineering" is the modern and promising trend not only in research but also in the daily industrial praxis. Without any doubt, modern computers offer very powerful possibilities to predict even complicated thermo-and fluiddynamic processes. We can trust on these predictions only if the physical phenomena are well known so that we can describe them with our "mathematical language". For unknown physical phenomena we have to rely on experiments and here optical measuring techniques provide the very best input for mathematical modelling.

6. References

1. Mayinger, F. (Ed.) (1954) Optical Measurements, Techniques and Applications, *Springer Verlag*, Berlin, Heidelberg.
2. Gabor, D. (1948) A New Microscopic Principle, *Nature* 161; (1949) Microscopy by Reconstructed Wavefronts, *Proc. Roy. Soc.* A 197; (1951) Microscopy by Reconstructed Wavefronts II, *Proc. Roy. Soc.* A 197.
3. Caulfield, H.J.; Sun Li (1970) The Applications of Holography, *Wiley*, New York.
4. Collier, C.B.; Burckhardt; Sun Li (1971) Optical Holography, *Academic Press*, New York.
5. Smith, H.M. (1977) Holographic Recording Materials, *Springer Verlag*, Berlin, Heidelberg, New York.
6. Gebhard, P.; Mayinger, F. (1996) Evaluation of Pulsed Laser Holograms of Flashing Sprays by Digital Image Processing, Flow Visualisation and Image Processing of Multiphase Systems, *ASME*, FED-Vol. 209.
7. Gebhard, P. (1996) Zerfall und Verdampfung von Einspritzstrahlen aus lamellenbildenden Düsen, *Dissertation*, Technische Universität München.
8. Feldmann, O.; et al. (1997) Evaluation of Pulsed Laser Holograms of Flashing Sprays by Digital Image Processing and Holographic Particle Image Velocimetry, *Proc., CSNI Specialist Meeting on Advanced Instrumentation*, Santa Barbara, USA.
9. Panknin, W.; Mayinger, F. (1974) Holography in Heat and Mass Transfer, 5th *Int. Heat Transfer Conference*, Tokio, VI 28.
10. Panknin, W. (1977) Eine holographische Zweiwellenlängen Interferometrie zur Messung überlagerter Temperatur- und Konzentrationsgrenzschichten, *Dissertation*, Universität Hannover.
11. Nordmann, D.; Mayinger, F. (1981) Temperatur, Druck und Wärmetransport in der Umgebung kondensierender Blasen, *VDI-Forschungsheft* 605.
12. Bohren, C.F.; Huffmann D.R. (1983) Absorption and scattering by small particles, *John Wiley and Sons*, New York.
13. Long, D.A. (1977) Raman spectroscopy, *Mc Graw-Hill*, London.
14. Alonso, M.; Finn, E.J. (1988) Quantum Physics, *Addison-Wesley*.
15. Hanson, R.K.; Seitzmann, J.M.; Paul, P.H. (1990) Planar fluorescence imaging of combustion gases, *Applied Physics*, Vol. B 50.
16. Algermissen, Personal communication, University Stuttgart.
17. Ardey, N; Mayinger, F. (1995) Influence of transport phenomena on the structure of lean premixed hydrogen air flames, 11th *Proc. Of Nuclear Thermal Hydraulics*, San Francisco, California, Oct. 29 – Nov. 2 and *Amer. Nuclear Soc.*, S. 33-41 (ANS Order No. 700227).
18. Dorer, F.; Prechtl, P.; Mayinger, F. (1997) Investigation of mixture formation on combustion processes in a hydrogen fuelled Diesel engine, *HYPOTHESIS II, Intern. Symposium*, Aug. 18-22, Grimstadt, Norway (to be published).

ENHANCEMENT OF COMBINED HEAT AND MASS TRANSFER IN ROTARY EXCHANGERS

U. DINGLREITER, F. MAYINGER
Lehrstuhl A für Thermodynamik
Technische Universität München
Boltzmannstraße 15
D-85748 Garching, Germany

Abstract. The principal objective of this study is the analysis and improvement of the combined heat and mass transfer of various newly designed rotary solid storage elements.

Rotary exchangers usually consist of a combination of a carrying material and an adsorbing storage element. The heat and mass transfer surface is of cellular structure usually referred to as matrix. Using the holographic interferometry experiments were conducted in order to determine the most effective duct geometry for advanced heat and mass transfer. Therefore various duct geometries like sinusoidal, rectangular, triangular and semicircular ducts were investigated at different temperatures and flow velocities. According to the configurations that were examined experimentally a numerical calculation has been performed.

In order to improve the heat and mass transfer in rotary exchangers several new combinations of carrying materials and adsorbents were developed and compared to a reference material. Ten combinations which proved to be most effective were scrutinized. For this reason the water adsorption and regeneration behavior as well as the pressure drop was measured. Cycle experiments were conducted in order to determine the suitability of a material combination as a rotary humidity exchanger. The results demonstrate a good applicability of most of the new materials.

1. Introduction

Rotary solid storage elements are used in a great multitude of technical processes such as air dehumidification or solvent recovery. The design showing the best prospects is the rotary counter flow conception which combines high performance with compactness. The regenerator generally consists of a matrix of flow channels which have, depending on the manufacturing process, a rectangular, triangular or sinusoidal shape. The disk exchanger is separated by flow channels into two sections, the regeneration section and the adsorption section, in which the process air and the regeneration air are

381

S. Kakaç et al. (eds.), Heat Transfer Enhancement of Heat Exchangers, 381–394.
© 1999 *Kluwer Academic Publishers.*

in counter flow with the matrix. In the adsorption section the adsorbate (humidity or solvent) from the air is stored to the matrix walls, whereas the preheated air passing through the regeneration section of the rotor heats up the material and regenerates it. After several revolutions a steady state with a constant mass transfer from adsorption air to regeneration air is achieved. The volume flow through the regeneration area is about a quarter of the volume flow through the adsorption area. For this reason the humidity or the solvents are considerably more concentrated in the regeneration air. This results in a better performance of a condensation procedure or a catalytic oxidation.

For the development of further appropriate adsorbens-carrying material combinations, firstly the influence of the geometry of the flow channels and of the flow velocity on the heat transfer was investigated at various temperatures. Considering the opposite direction of the heat and the mass transfer in the adsorption process, the analogy between heat and mass transfer can be used to determine the mass transfer coefficient (Krischer [1]).

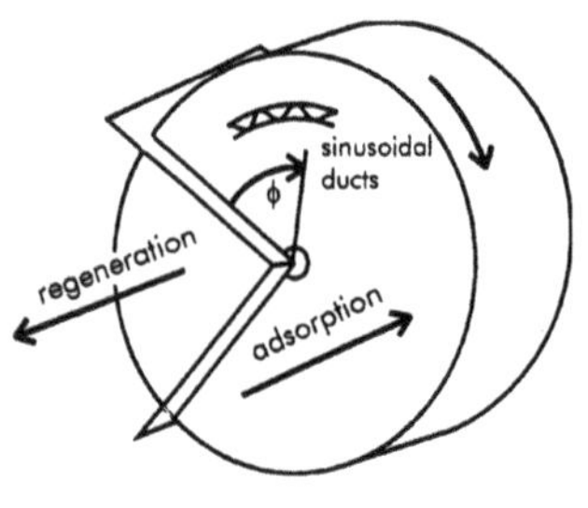

Figure 1. Rotary mass exchanger

Both the heat transfer coefficient and the mass transfer coefficient are required as input data to perform a numerical simulation of the combined heat and mass transfer in rotary exchangers. Subsequently samples of various new material combinations and geometries were manufactured and investigated with respect to their suitability as rotary humidity exchangers. The adsorption and regeneration behavior as well as the pressure drop were investigated. By means of cycle experiments the applicability of the new materials could be determined.

2.   Rotary Exchanger Duct Geomtries

Three main aspects have to be considered when designing single ducts in the rotary adsorbers:

- heat and mass transfer
- pressure drop
- producibility

As a first step the literature (e.g., Shah and London, [10] [11]) provides data on mean heat transfer and pressure drop in ducts with different cross sectional geometries for fully developed laminar flow and allows a pre-selection of duct cross-sectional geometries.

In order to get detailed information about the local heat transfer in the duct, also for higher flow rates (especially in the transition region from laminar to turbulent flow) and the combined (thermal and velocity) entry length problem, additional experiments and numerical calculations have been performed.

duct geometry	2b/2a (height/bottom)	Nu_T	Nu_{H1}	L_{hy}^+	ζRe
rectangular	1,00	2,98	3,61	0,0340	56,91
circular	1,00	3,66	4,36	0,0500	64,00
semicircular	0,50	---	4,09	---	63,07
sinusoidal	0,50	2,12	2,62	0,0464	44,83
sinusoidal	0,56	2,17	2,69	0,0453	45,81
sinusoidal	0,64	2,24	2,78	0,0439	47,13
sinusoidal	0,68	2,27	2,83	0,0432	47,79
sinusoidal	1,00	2,45	3,10	0,0400	52,09
sinusoidal	1,50	2,60	3,27	0,0394	56,09
triangular (60°)	0,86	2,47	3,11	0,0398	53,33
triangular (90°)	0,50	2,34	2,98	0,0421	52,61

TABLE 1. Heat Transfer (Nu_T: T_W=const. Nu_H: q = const.) and pressure drop coefficient ζRe for fully developed, laminar flow (Shah und London, [10]). Hydrodynamical entry length L_{hy}^+ for various channel geometries

2.1. EXPERIMENTAL SET-UP FOR HOLOGRAPHIC INTERFEROMETRY

The influences of the duct geometry and of the flow velocity on heat transfer were investigated at various temperatures by holographic interferometry. Ambient air from the climate controlled laboratory ($T_\infty = 20°C$) is drawn directly into the test section by means of a compressor which is equipped with a flow-regulating throttle. Considering the original rotary adsorber arrangement no inlet sections or any flow homogenizers are applied to the duct. The test section consists of a heated aluminum duct with uniform wall temperature. The duct outlet is connected to a funnel with glass walls that allows optical access in the axial direction of the duct, which is also the main flow direction (figure 2b).

a)

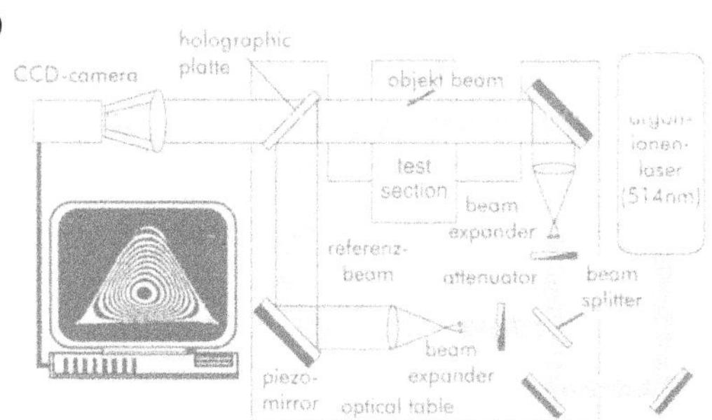

b)

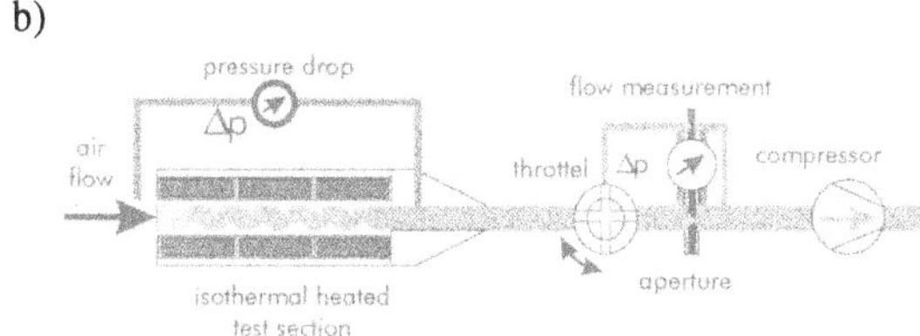

Figure 2. a) Setup for holographic interferometry (real-time)
b) Setup for the investigation of single ducts

Two-dimensional temperature fields, and thus the local heat transfer in the duct, were obtained by holographic interferometry measurements. The experimental set-up for the holographic interferometry (real-time method) is illustrated in figure 2a. This optical measurement technique allows the continuous visualization of the two-dimensional temperature field in the flow, avoiding any influence on the effect to be measured. This technique shows the thermohydraulic flow pattern within the duct in real time. Detailed information about this measurement technique and its applications are given by Mayinger et al. [4], [5],[6],[7] and Tauscher and Mayinger [12]. For the basic understanding of the interferograms presented in this paper (figures 3 and 4) it is only important to mention, that the interference lines (black and white fringes in the images) of the interferograms are approximately equal to the isotherms of the investigated flow (temperature step: $\Delta T = 2.3$ K). Due to the experimental set-up these isotherms represent the axially integrated temperature in the duct. From the distance between these isotherms and the known wall temperature, the temperature gradient and, therefore, the local heat flux, the local heat transfer coefficient and the local Nusselt number can be calculated. For the evaluation of the interferograms, such as the measurements of the fringe distances along lines which are perpendicular to the wall, a home made digital image processing system was used.

In table 2 the experimentally investigated arrangements are shown. Due to space restrictions only the most important results concerning local heat transfer will be presented in the following.

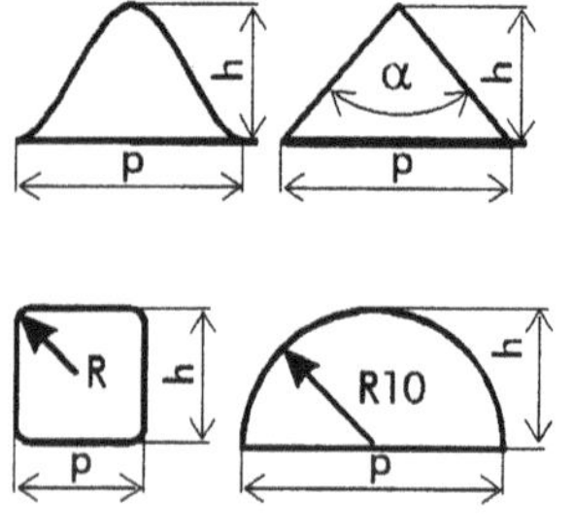

duct geometry:		Variation
sinusoidal:	h/p:	0,54 / 0,56 / 0,68 / 1,00 / 1,50
rectangular:	R:	1 / 3 / 5 mm
triangular:	h/p:	1,12; α=60° / 0,5; α=90°
semicircular:	h/p:	0,5
length:	L:	100 / 200 / 300 mm
wall temperature:		30 / 50 / 70 / 90 °C
Reynolds number:		50 – 50 000

TABLE 2. Experimentally investigated arrangements

2.2. EXPERIMENTAL RESULTS

Figures 3 and 4 show interferograms of the air flow in different sinusoidal, semicircular, triangular and rectangular ducts at different Reynolds numbers.

It is apparent that the flow temperature gradient reaches the highest values in linear perimeter sections, while minima can be observed in the corners. With increasing the flow rates the air flow is able to reach also corners and to use also these duct areas for heat transfer. As mentioned above the local Nusselt number over the duct perimeter is obtained by evaluating the interferograms. An exemplary evaluation is shown in figure 5. By integrating the local heat transfer the mean heat transfer for each duct is gained. This can be used to verify the measurements by comparing them to global calorimetric measurements.

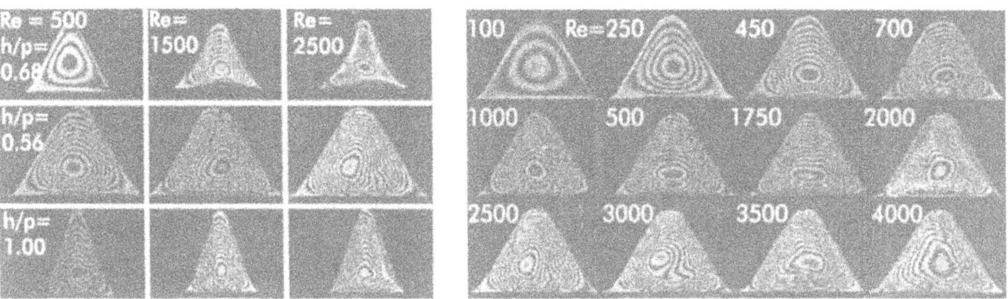

Figure 3. Temperature field in sinusoidal ducts at different Reynolds numbers

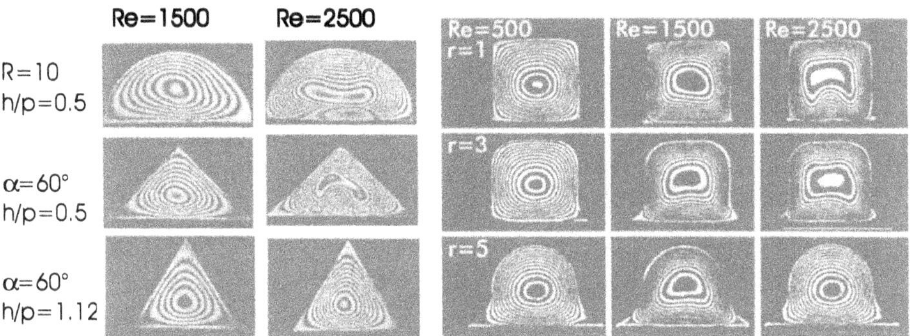

Figure 4. Temperature field in semicircular, triangular and rectangular ducts at different Reynolds numbers

Due to the fact that the optical experiments are very time consuming, also numerical calculations have been performed (figure 6). Once the numerical results have been verified by the experimental data, further parameter variations can be supplied by computer only. The numerical simulation of the fluid flow in the ducts was performed by means of the commercially available CFD-code CFX. This code works with a finite volume method based on the finite elements. The boundary conditions and the geometry were modeled exactly as in the experimental set-up.

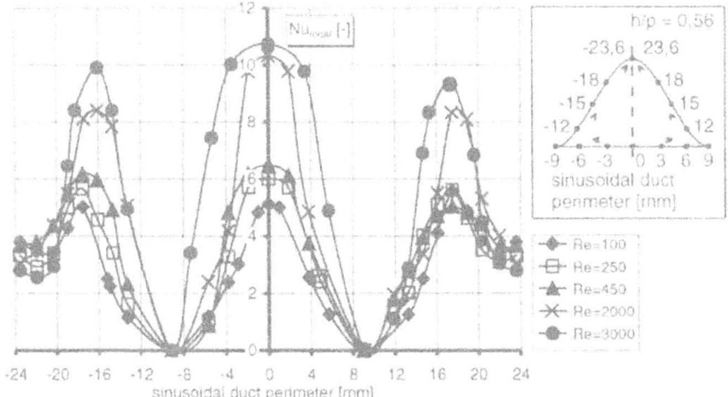

Figure 5. Local Nusselt number over sinusoidal duct perimeter

The symmetry of the ducts was used to model only one half of the duct in order to increase the calculation speed. The difference between the numerical calculation and the experimental results is less than 12% for each Reynolds number. The experimental mean Nusselt numbers have always a lower value than the ones which are calculated. This behavior becomes evident when regarding the trends of the local Nusselt numbers. Although the maxima in the experimental Nusselt numbers are slightly higher than the calculated ones, they are not that great. This results possibly from the fact, that, due to limitations of the digital image processing system, only few temperature gradients have been evaluated along the duct perimeter.

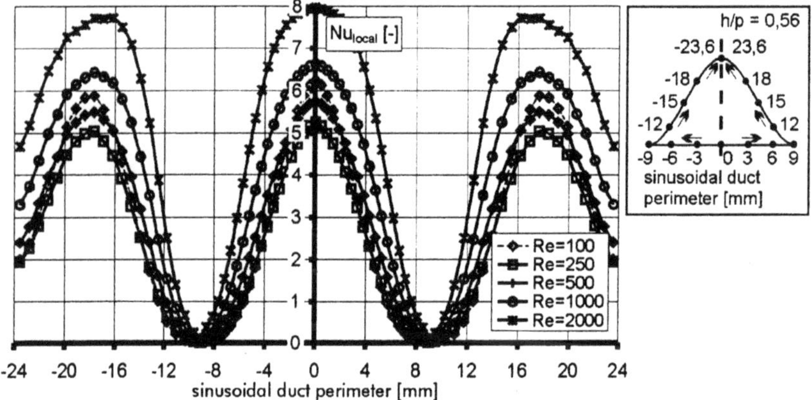

Figure 6. Local Nusselt number for a sinusoidal duct (h/p = 0.56) obtained from simulation

An additional fact is that the axial progression of the heat transfer coefficient, which is required for the numerical calculation of the combined heat and mass transfer of rotary exchangers, is a result of the numerical simulation. In figure 7 the perimeter averaged Nusselt number is depicted.

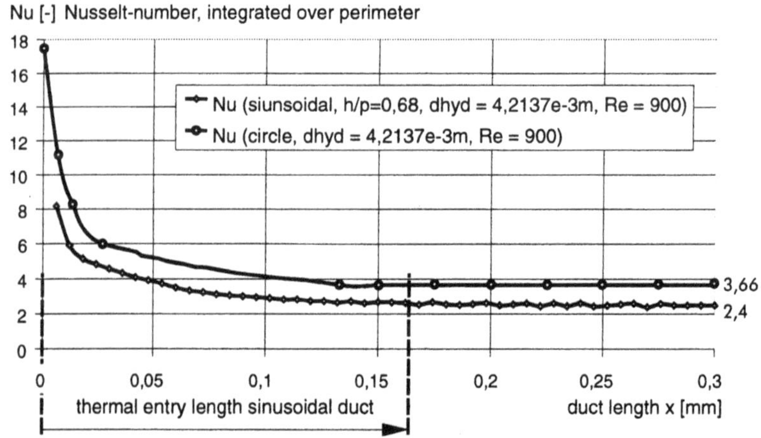

Figure 7. Nusselt number integrated over perimeter for circular and sinusoidal duct

As a result of these experiments and calculations for each duct geometry, best practice parameters were obtained.

For sinusoidal ducts a height to bottom length ratio of 0,6 up to 0,7 is most effective.

3.   Investigation of Rotary Exchanger Samples

3.1. ROTARY EXCHANGER SAMPLES

sample	description	duct geometry	manufacturer	cross section [dm^2]	number of ducts
1P	silicagel-coated. adsorption paper	sinusoidal small	Lehrstuhl A für Thermodynamik	0,32	1160
2P	silicagel-coated. adsorption paper	sinusoidal medium	Lehrstuhl A für Thermodynamik	0,31	730
3P	adsorption paper not coated.	sinusoidal medium	Lehrstuhl A für Thermodynamik	0,30	705
4P	silicagel-coated. adsorption paper	sinusoidal large	Lehrstuhl A für Thermodynamik	0,29	333
K1	synthet. silicagel-on ceramic	sinusoidal medium	reference material	0,32	1057
K2	ceramic, LiCL-coated	rectangular	R. Scheuchl GmbH	0,32	1884
Z1	extruded zeolite (monolith)	rectangular	IKT (Stuttgart University)	0,32	1952
Z2	zeolite-coated adsorption paper	sinusoidal medium	Lehrstuhl A für Thermodynamik	0,31	712
Z3	zeolite-coated. adsorption paper	sinusoidal medium	Lehrstuhl A für Thermodynamik	0,30	733
Li I	adsorption paper LiCl-coated	sinusoidal medium	Lehrstuhl A für Thermodynamik	0,29	626

TABLE 3. Characterization of the rotary exchanger samples

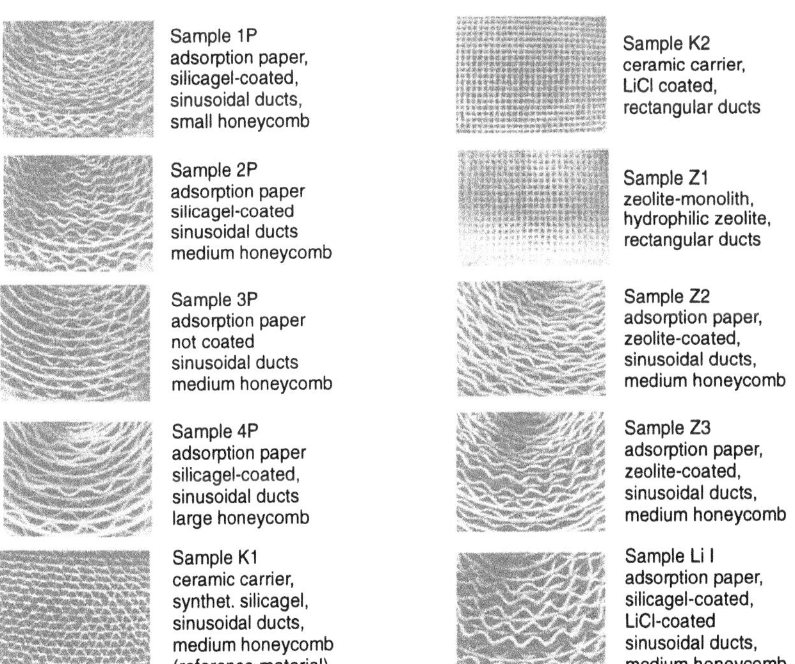

Figure 8. Investigated rotary exchanger samples

388

In order to gain information about the pressure drop, the loading capabilities and re-generability of several different material combinations, rotary exchanger samples with a diameter of 70mm and a duct length of 200mm were designed. In the following experiments ten different combinations of carrying structures and adsorbents as shown in figure 8 were investigated. Table 3 characterizes the different samples referring to duct geometry, manufacturer, cross section and number of ducts.

Most of the samples consist of a new non-flammable adsorption paper which was developed at Lehrstuhl A für Thermodynamik. In order to improve the adsorption properties the production process of conventionally used non flammable paper was modified. This modification leads to a silicagel portion of almost 50% in the paper itself. For some applications the portion of adsorbents is increased by an additional silicagel, zeolite or LiCl-coating. In figure 9 SEM-pictures of a conventional paper, of the modified adsorption paper, and of the adsorption paper after an additional silicagel coating are shown.

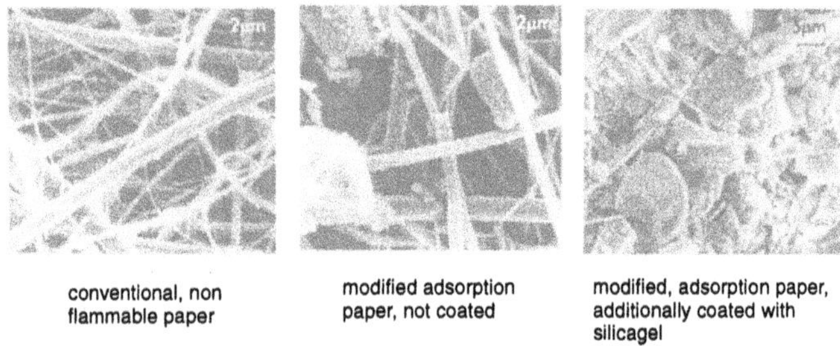

Figure 9. SEM-Pictures of conventional non flammable paper (left), adsorption paper (middle) and additionally coated adsorption paper (right)

The samples 1P, 2P, 4P differ only in the dimensions of the sinusoidal duct. There is no difference in the coating. Sample 3P consists of uncoated adsorption paper, whereas sample LI I was first coated with silicagel and afterwards soaked in a 13% LiCl-solution. Sample K1 was bought and serves as a reference material to the newly developed combinations. Sample K2 consists of a extruded ceramic carrier that was soaked in the LiCl-solution. Together with sample K2 the ducts of sample Z1 are rectangular. Sample Z1 consists of a extrudated hydrophilic zeolite. It was developed at the Institut für Kunststofftechnologie, Universität Stuttgart and placed at our disposal for further measurements. The samples Z2 and Z3 are adsorption papers coated with different, hydrophilic molecular sieves.

3.2. EXPERIMENTAL SET-UP

Figure 10 shows the following experimental rig which was used for the investigation of the combined heat and mass transfer in rotary exchanger samples.

The holding device for the sample is composed of a tube jointing sleeve with a diameter of 70 mm and a length of 220 mm in which the sample is inserted. The air temperature and the humidity are adjusted in a conditioning section. Thermocouples and humidity sensors for the determination of the air temperature and humidity are positioned at a short distance in front of and after the rotary exchanger sample.

Along a surface line in a flow channel of the sample there are seven thermocouples arranged at an equal distance from one another. They measure the temperature distribution in the sample during the adsorption or desorption process. By means of a high-accuracy weighing machine the loading progression of the sample can be measured.

The experimental rig enables the investigation of sample diameters up to 70 mm and of flow velocities up to 10 m/s. Inlet-temperatures between 0°C and 80°C and air humidity up to 98 % at a temperature of 45°C can be adjusted.

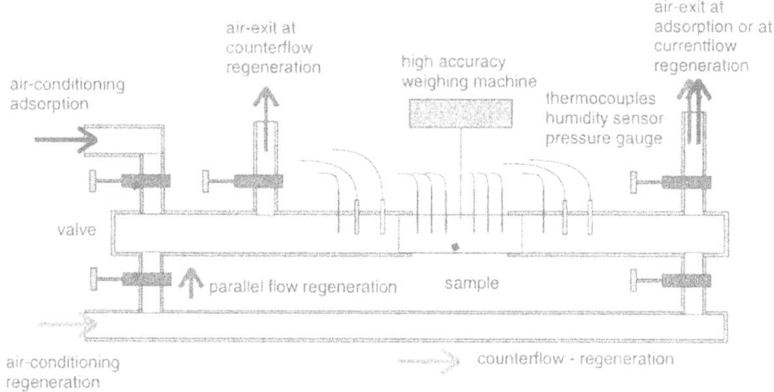

Figure 10. Experimental rig for the investigation of rotary exchanger samples

Before an experiment is conducted the samples are dried in a vacuum oven (pressure < 100 mbar) at a temperature of 130°C for at least 24 hours.
By measuring the pressure drop and the adsorption isotherms a first evaluation of the loading behavior of the samples can be realized. The regeneration behavior of the material combinations is then investigated. In order to prove the suitability of a material as a rotary exchanger, several cycle-experiments were conducted.

3.3. PRESSURE DROP OF THE SAMPLES

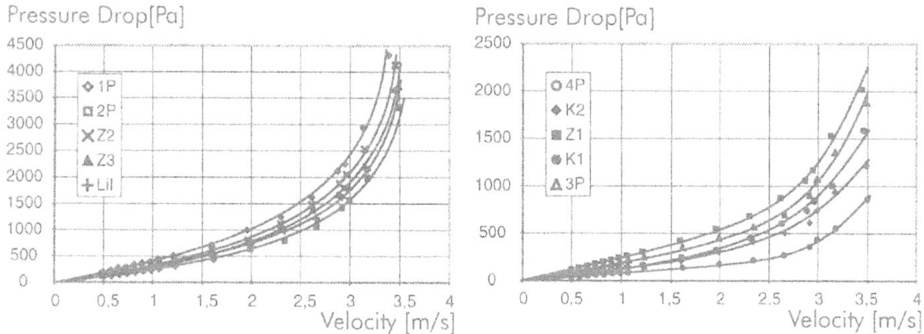

Figure 11. Pressure loss of all samples versus air velocity

In order to avoid high pumping power, duct geometries with a maximum in mass transfer combined with minimum in pressure drop have to be developed. In figure 11

390

the pressure drops of all samples are shown. In the important velocity range up to 2m/s the pressure drop of the samples depicted in the right graph are comparable to that pressure drop of the reference sample K1 which has the lowest of all samples. The pressure drop of the samples 1P, 2P, Z2, Z3 and Li I is significantly higher. This is due to smaller honeycomb ducts together with a thick coating with silicagel or molecular sieve.

3.4. ADSORPTION EQUILIBRIUM

In order to get detailed information about the adsorption behavor the adsorption isotherms for air temperatures of 25°C, 35°C, 45°C and 55°C at relative humidities from 0% to 90% were measured.

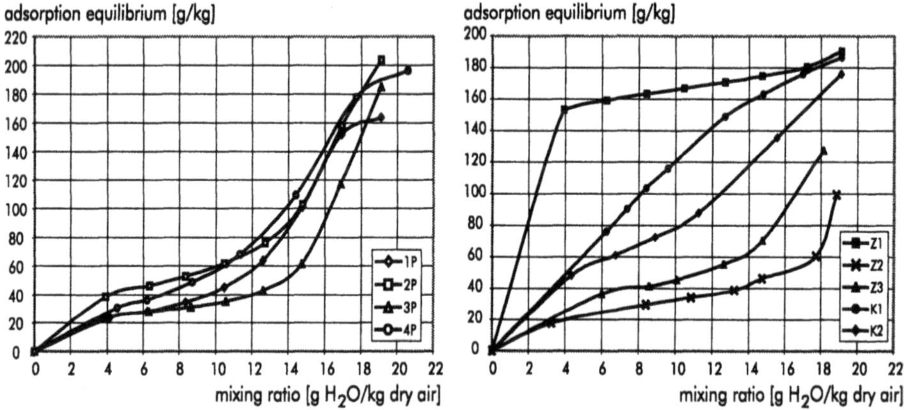

Figure 12. Adsorption isotherms at a temperature of 25°C

In figure 12 the 25°C-isotherms are presented for all samples except for the sample Li I. The isotherm of this material exceeds the scale in a linear way up to a equilibrium of about 1600g water/kg dry material at a mixing ratio of 16g water/kg dry air.

The adsorption isotherm of the sample Z1 rises steeply up to a proportion of ingredients of 5kg water per kg dry air and then rises steadily until it reaches the maximum loading (19.05%). The progression of the reference sample is characterized by a steady rise up to the maximum loading of approximately 19.0%. The isotherm of sample K2 shows almost the same behavior, with a slight decrease at the mixing ratio between those of 5 and 12g/kg. The adsorption paper based samples (1P, 2P, 3P, 4P, Z2 and Z3) show a similar progression with a flattening at a mixing ratio of 4 to 5 g/kg and a steep increase at a mixing ratio of approximately 12 to 14 g/kg.

All material combinations show good adsorption behavior.

3.5. REGENERATION BEHAVIOR

In addition to the adsorption capabilities the regeneration behavior is to be considered.

Therefore, in the experimental setup depicted in figure 9 the samples underwent the following measurement cycle:

- regeneration   until the samples are dry (24 hours in a vacuum oven 130°C)
- adsorption   air flow 30°C, 60% relative humidity, 1m/s, 20 minutes
- regeneration   air flow 130°C, 0% relative humidity, 1m/s, 20 minutes
- adsorption   air flow 30°C, 60% relative humidity, 1m/s, 10 minutes
- regeneration   air flow 130°C, 0% relative humidity, 1m/s, 10 minutes

The results of these investigation are presented in figure 13. The samples which consist of the new adsorption paper (left graph: all samples, right graph: sample Z3) except the sample Li I reach their dry mass at least within the first half of the regeneration time. Within a quarter of the regeneration time the loading of sample 4P decreases below 15%, the loading of the other adsorption paper samples decreases below 5% of their maximum loading. Compared to the reference (sample K1) these results indicate good suitability. By means of a high heat capacity the regeneration behavior of sample Z1 is inadequate for usage in a cycle process with high frequencies. Within the full time range it is not possible to regenerate this sample to its dry mass.

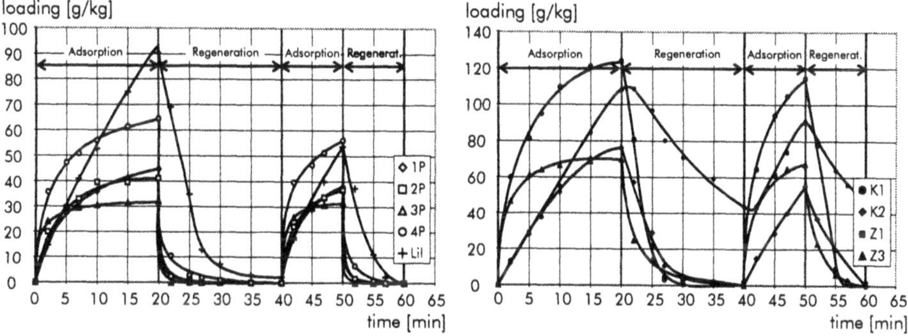

Figure 13. Regeneration behavior after 20min and after 10min water adsorption

3.6. CYCLE EXPERIMENTS

In the cycle process the matrix is frequently loaded and regenerated. In order to concentrate humidity or solvent in the regeneration air, the regeneration period is usually reduced to a quarter of the adsorption period. According to literature the optimal number of revolutions of rotary mass exchangers is between 18 and 25. For the reference material (sample K1) an efficient number of revolutions of 24 was investigated. Therefore, for each duct a retention time in the adsorption area of 112,5 seconds is obtained, whereas in the regeneration area the retention time is 37,5 seconds. In order to determine the applicability in comparison to the reference sample under these conditions, all newly designed material combinations were investigated in a cycle experiment. In the adsorption period air with a velocity of 1m/s, a temperature of 30°C and a relative humidity of 60% flows through the sample within 112,5 seconds. After the sample has been weighed, the regenerating air, with a velocity of 1m/s, a temperature of 130°C and a relative humidity of 0%, flows in the opposite direction through the sample for another 37,5 seconds. Afterwards the sample is weighed again and the cycle starts at the beginning. In advance the samples were dried in the vacuum oven at a temperature of 130°C.

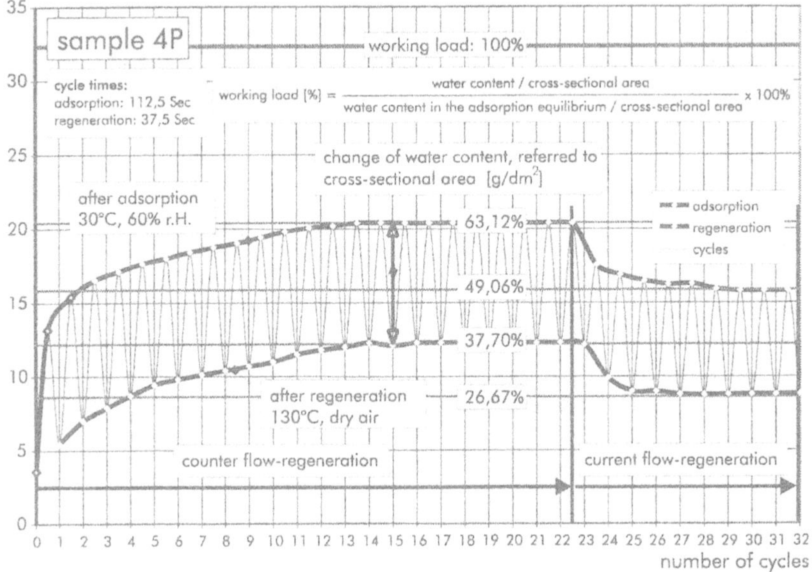

Figure 14. Cycle experiment of sample 4 – Explanation of table 4 and table 5

| Cycle-Experiments | adsorption: | 112,5 sec | 30°C | 60% r.H. | 1m/s | | | | |
| | regeneration: | 37,5 sec | 130°C | 0% r.H. | 1m/s | | | | |

		sample	1P	2P	3P	4P	K1	K2	Z1	Z3	Li I
counter flow-regen-eration		number of cycles	9	9	5	14	15	16	20	13	130
		absolute water content[g]	1,66	1,75	1,69	2,41	2,48	2,10	1,55	1,92	1,45
		change in loading [g/kg]	13,05	14,43	20,90	22,99	15,29	6,17	4,89	15,77	8,17
		water change/cross section [g/dm²]	5,16	5,60	5,56	8,22	7,78	6,54	4,81	6,31	4,93
		working load after adsorption [%]	49,35	47,17	85,70	63,12	95,21	78,56	97,16	76,72	50,14
		working load after regeneration [%]	28,78	27,84	36,35	37,70	84,61	72,71	94,23	48,10	49,40
parallel flow-regen-eration		sample	1P	2P	3P	4P	K1	K2	Z1	Z3	Li I
		number of cycles	7	4	1	6	9	7	9	5	---
		absolute water content[g]	0,93	1,28	1,59	2,13	2,15	1,40	1,15	1,20	---
		change in loading [g/kg]	7,32	10,58	19,61	20,25	13,25	4,11	3,63	9,91	---
		water change/cross section [g/dm²]	2,89	4,10	5,22	7,24	6,74	4,36	3,57	3,96	---
		working load after adsorption [%]	33,39	35,99	83,90	49,06	62,18	45,13	90,64	59,16	---
		working load after regeneration [%]	21,86	21,82	37,61	26,67	52,99	41,23	88,47	41,18	---

TABLE 4. Cycle experiment: all samples, counter flow and parallel flow-regeneration; adsorption at 30°C, 60% r.H.

The cycle was repeated until a steady state was obtained. Then the cycle experiments were continued with a current flow regeneration until a new steady state was reached.

The most important criterion for the application as a rotary exchanger is the change in the water content related to the flow surface of the material between adsorption and regeneration.

In addition to that change of the water content, the number of cycles to steady state, the absolute water content, the change of loading and the working load are listed in table 4 and table 5 for counter flow and parallel flow regeneration and for different regeneration air humidities. The listed sizes are explained in figure 14 with sample 4P as an example.

Cycle-Experiments	adsorption: 112,5 sec 30°C 80%/30% r.H. 1m/s regeneration: 37,5 sec 130°C 0% r.H. 1m/s		
	sample	4P	K1
	number of cycles (steady state)	14	15
adsorption at a relative humidity of 80%	absolute water content [g]	2,41	2,48
	change in loading [g/kg]	22,99	15,29
counter flow-regeneration	water change/cross section [g/dm^2]	8,22	7,78
	working load after adsorption [%]	63,12	95,21
	working load after regeneration [%]	37,70	84,61
	sample	4P	K1
	number of cycles (steady state)	6	9
adsorption at a relative humidity of 30%	absolute water content [g]	2,13	2,15
	change in loading [g/kg]	20,25	13,25
parallel flow-regeneration	water change/cross section [g/dm^2]	7,24	6,74
	working load after adsorption [%]	49,06	62,18
	working load after regeneration [%]	26,67	52,99

TABLE 5. Cycle experiment: samples 4P, K1, counter flow and parallel flow-regeneration; adsorption at 30°C, 80% r.H and 30% r.H.

Table 4 shows a maximum of the water change over all samples of 8.22g/dm^2 for sample 4P. According to the water change of the reference sample K1 (7,78) an improvement of approximately 5,66% was achieved. Considering all relative humidities of the adsorption air (table 5), sample 4P shows better values than the reference sample. In industrial applications a water change related to a cross section of about 5g/dm^2 at the adsorption air temperature of 30°C and a humidity of 60% is used to calculate the device. Only the samples Z1 and Li I do not achieve this value. For these two samples the cycle has to be modified to lower cycle frequencies and maybe to a higher adsorption to regeneration time relation.

4.   Concluding Remarks

Several different duct geometries were investigated by means of holografic interferometry. Due to manufacturing requirements all adsorption-paper-based materials

394

should have a matrix with sinusoidal ducts. The value of the height-to-bottom length ratio should be between 0,6 and 0,7. In addition to the optical experiments the numerical simulation of the heat transfer provides values for the mean axial heat transfer coefficient. With respect to the analogy between heat and mass transfer, the mass transfer coefficient can be obtained additionally. Both values are required for a numerical simulation of the combined transport processes concerning rotating exchangers. This simulation serves as an important tool for the design of industrial devices.

Most of the new material combinations show good performance in the cycle experiments, which represents a relevant test for the application of rotary exchangers. Additionally, the mechanical properties of all samples are very satisfactory. As a consequence of the presented results a serial production of rotary exchangers has started. The exchangers are used as dehumidifiers particularly in desiccant cooling systems or other industrial applications.

The modified production process of the adsorption paper and the following coating process allow the production of hydrophobic exchanger materials as well. The portion of hydrophilic and hydrophobic adsorbents is variable in a wide range. This fact enables the design of rotary exchangers for both dehumidification and solvent recovery in a single device.

There is a need for further research in order to optimize the material combination, as well as the geometry according to each application.

5. References

1. O. Krischer, W. Kast (1978) Die wissenschaftlichen Grundlagen der Trocknungstechnik, Dritte Auflage, *Springer Verlag*, Berlin

2. Hauff, W., Grigull, U., Mayinger, F. (1991) Optische Meßverfahren in der Wärme- und Stoffübertragung, *Springer-Verlag*, Berlin

3. Klas, J. (1993) Wärmeübergang in Strömungskanälen ohne und mit Turbulenzpromotoren *Diss.* TU-München

4. Mayinger, F. (Ed..) (1994) Optical Measurements, Techniques and Applications, *Springer-Verlag*, Berlin

5. Mayinger, F., Chen, Y.M. (1985) Holographic Interferometry Studies of the Temperature Field Near a Condensing Bubble, Optical Methods in Dynamics of Fluids and Solids, *Proc. of an Int. Symposium IUTAM*, Berlin

6. Mayinger, F., Klas, J. (1993) Investigation of Local Heat Transfer in Compact Heat Exchangers by Holographic Interferometry, *Proc. of the 1st Int. Conf. on Aerospace Heat Exchanger Technology*, pp. 449-465, Palo Alto

7. Mayinger, F., Panknin, W. (1994) Holography in Heat and Mass Transfer , *Proc. of the 5th Int. Heat Transf. Conf.*, Tokyo

8. Panknin, W., Mayinger, F. (1978) Anwendung der holographischen Zweiwellenlängeninterferometrie zur Messungüberlagerter Temperatur- und Konzentrationsgrenzschichten, *Verfahrenstechnik*, Bd. 12,9, pp. 582-589

9. C.P. Howard (1965) Heat Transfer and Flow-Friction Characeristics of Skewed-Passage and Glass-Ceramic Heat-Transfer Surfaces,
   Int. Journal Heat Mass Transfer - Oxford 1. S 72-86

10. Shah, R.K., London, A.L. (1978) Laminar Flow Forced Convection in Ducts
    Academic Press, New York

11. R.K. Shah (1981) Thermal Design Theory for Regenerators,
    Heat Exchangers, Thermal Hydraulic Fundamentals and Design, edited by S. Kakac, A.E. Bergles and F. Mayinger, pp. 721 - 763, *Hemisphere*, New York

12. Tauscher, R., Mayinger, F. (1997) Advances in Heat Transfer by Optical Techniques, *Proc. of the 2nd Int. Symposium on Heat Transfer and Energy Conservation*, Guangzhou, pp.14-29

ADVANCES IN UNDERSTANDING OF FLAME ACCELERATION FOR THE IMPROVING OF COMBUSTION EFFICIENCY

C. GERLACH, A. EDER, M. JORDAN, N. ARDEY, F. MAYINGER
Lehrstuhl A für Thermodynamik, Technische Universität München
85747 Garching, Germany

Abstract. The propagation of gaseous explosions is governed by the interaction of chemical kinetics with the molecular and turbulent heat and mass transport. Combustion processes like deflagration and detonation depend on the different valence of physical effects under certain conditions. Geometry and the expansion flow of the flame itself affect the turbulence and therefore the transport of fuel into the reaction zone. The present paper discusses the different hydrogen combustion processes and reports on the experimental investigations of transport phenomena during flame propagation with highly blocking obstacles. Several facilities have been operated with sophisticated optical measurement techniques like high speed schlieren videographie, laser induced predissociation fluorescence and laser doppler velocimetry to obtain detailed information about the combustion process. It will be shown that the turbulent quenching of flames leads to an amount of free radicals resulting in sensitive clouds of those radicals with corresponding high chemical reaction rates, which has a strong influence on the efficiency of the combustion processes.

1. Introduction

Combustion processes can be devided into two different kinds, deflagration and detonation. The deflagration is due to the velocity (subsonic speed) of the flame front more controllable compared to a fast deflagration or detonation (supersonic speed). A detonation leads to highest pressure peaks, which is desired in many technical applications. The transition from the deflagration to the detonation (DDT) is difficult to predict due to the dependency on several parameters like geometry, mixture and the thermody-

S. Kakaç et al. (eds.), Heat Transfer Enhancement of Heat Exchangers, 395–406.

namical quantities. Therefore DDT processes are of great interest for the improving of the combustion efficiency as well as for the safety design of industrial systems and buildings.

Several projects for the observation of hydrogen combustion effects have been carried out at the Lehrstuhl A für Thermodynamik ([1], [2], [3]). At the moment several facilities are in operation for the observation of turbulent deflagrative combustion as well as DDT processes like shock focusing, critical flame acceleration and hot-jet ignition.

2. Deflagration, Detonation and the Transition from Deflagration to Detonation

2.1. DEFLAGRATION

A laminar flame front propagates transverse to the flame surface with a characteristic laminar flame velocity s_l depending on the reaction rate, mass and thermal diffusivity, temperature and pressure of the unburned gas.

A relative small Lewis number of $Le = 0.3 - 0.5$ (ratio of thermal diffusivity a to mass diffusivity D) expresses the more important mass diffusion term for a H_2/Air mixture. The unburned gas is ignited by the heat release of the combustion process. Due to the viscosity and the heat loss by the surrounding walls, the surface of the flame front tends to curve. So the effective flame velocity increases due to a ratio of the flame surface A_l to the cross-section of a geometry A_0.

Pressure waves are originated from the flame itself with ongoing propagation. The flame front starts to wrinkle and to create a cellular structure. A higher pressure gradient across the flame front leads to a more cellular structure [7]. Figure 1 shows this effect for several H_2 concentrations in an explosion tube.

At higher velocities of the expansion flow, the influence of turbulence increases. Macro scale vortices (approximately 10% of the tube dimension) extend the surface of the flame front, which leads to a higher effective flame velocity. Micro vortices in the dimension of the reaction zone (Kolmogorov micro scale λ_k) increase the mixture of burned and unburned gas and also accelerate the combustion process. Nevertheless the flow consists of a continous spectrum of vortices from micro to macro scale, a classification in length scales eases the characterization of the combustion process.

If the chemical reaction time exceeds the lifetime of the micro vortices, which is expressed in the Karlovitz number $Ka > 1$ (ratio of chemical reaction time τ_c to the lifetime τ_k of the Kolmogorov micro vortices), the influence of quenching effects extends. The high turbulence leads to a quenching of the flame due to the mixing of cold unburned gas with the

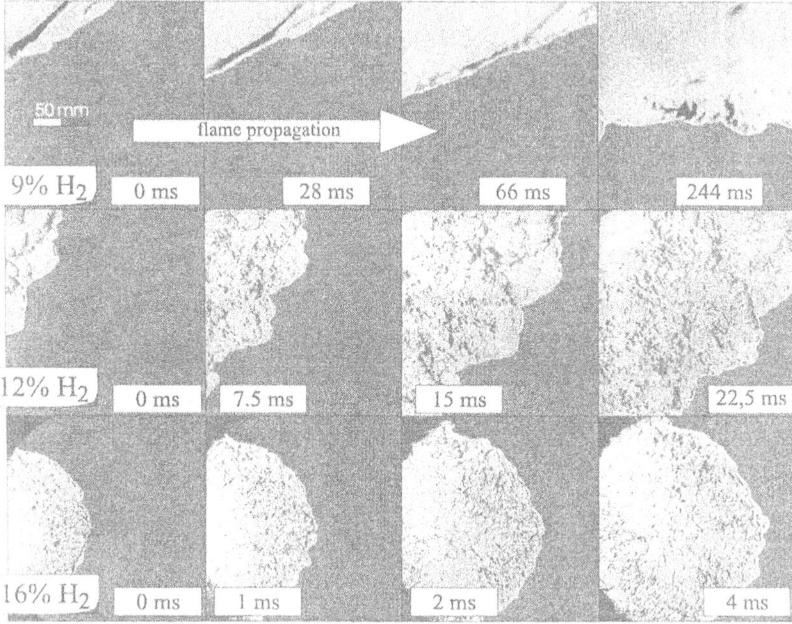

Figure 1. High speed schlieren photographs of H_2/Air flames at several H_2 concentrations

reaction zone. Therefore the temperature decreases and the ignition temperature cannot be reached. This leaves free radicals (OH, O, H etc.) that can build highly sensitive areas with the surrounding unburned gas.

2.2. DETONATION

The detonation differs essentially from the deflagration process. The flame front is strongly coupled to the leading shock front. Due to the increasing temperature of the gas mixture in the shock system, it ignites after a characteristic induction time (depending on the mixture) behind the shock front. The reaction zone has to release as much energy as needed for the shock system to maintain the ignition conditions for a stable detonation.

The detonation also occurs due to transverse components of the shock front which are reflected at the surrounding walls. They superimpose with other reflected shock components. The resulting regular pattern of the superimposing shock system can be visualized with smoked foils, and gives a good impression of the course of the detonation. The cell size λ_d of this pattern only depends on the gas mixture. The minimum tube diameter to keep up a detonation were determined in different scales of explosion tubes ([13], [14], [16]).

398

Figure 2 shows the typical propagation of a detonation in the *Pre*h*eatable*
*D*etonation Tube (PhD) at the Lehrstuhl A für Thermodynamik. It has a
length of 6 m and a inner diameter of 66 mm. The photo is taken by means
of schlieren techniques through a window section of 60 × 50 mm.

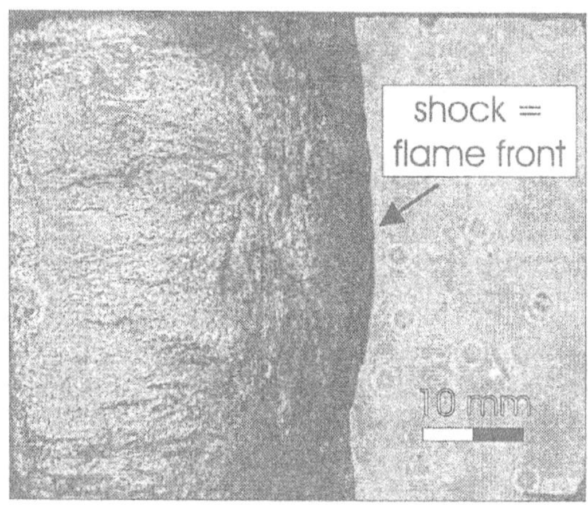

Figure 2.   Schlieren image of a detonation propagating in the PhD Tube (20 H_2 vol.-%
in air)

2.3. DEFLAGRATION TO DETONATION TRANSITION (DDT)

Under some circumstances a deflagration with its precursor pressure wave
system can turn into a detonation. The influence of turbulence makes it
very difficult to predict a possible DDT process precisely. Three different
DDT mechanisms are to be explained:

- shock focusing
- exceeding critical flame speed
- local explosions resulting from reignition of partially quenched volumes
  (hot-jet ignition)

As mentioned above, the superposition of shock fronts allows to reach
the conditions for the initiation of a detonation process. Special geometries
like baffles or cups provoke a focusing of a shock system, and lead to an
ignition of the burnable gas.

A deflagration produces a strong precursor pressure wave system. With
the ongoing propagation in combination with the energy release of the
flame front, the pressure wave can increase to a shock front. A supersonic
flame with an uncoupled shock system is referred to as fast deflagration.
If the flame behind the shock system is strongly accelerated by turbulence

it can couple with the shock system and develop into a detonation. The transition from deflagration to detonation is visualized in figure 3. The flame accelerates over a run-up length of 3 m with annular obstacles ($BR = 60\%$, obstacle path length 2 m, distance between the obstacles $d = 185\ mm$).

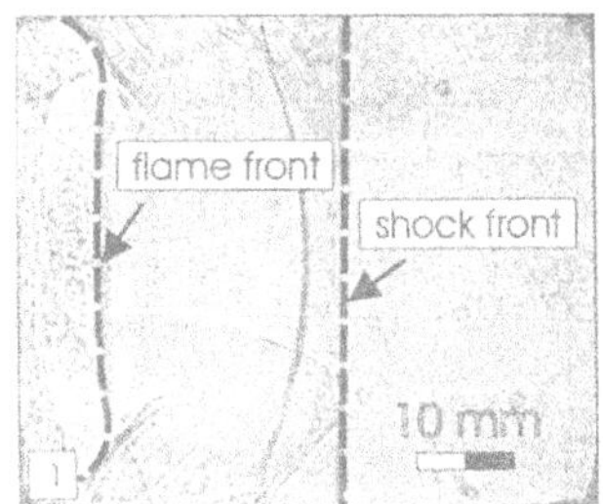

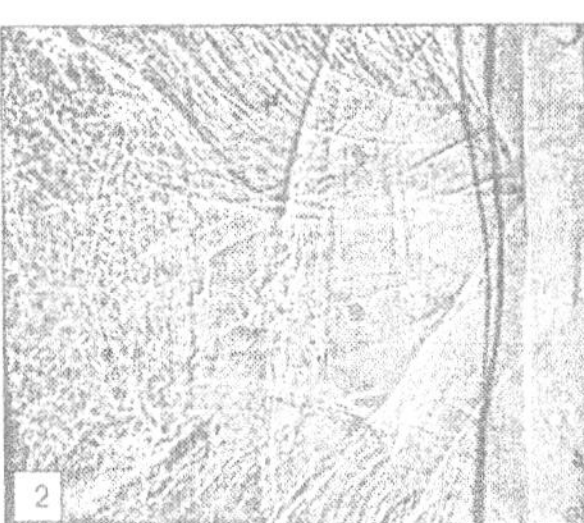
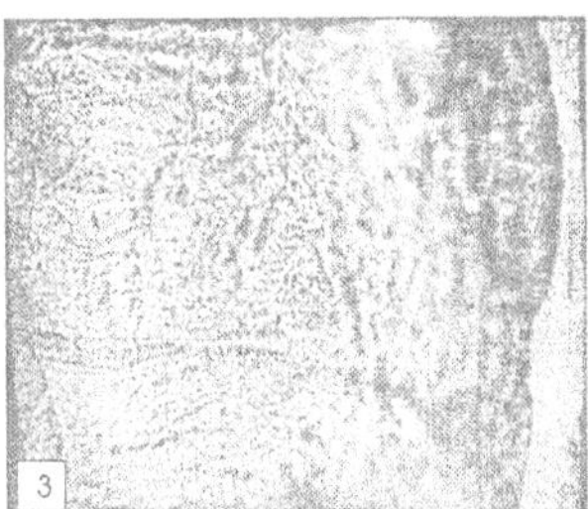

Figure 3. Selected frames of the transition from deflagration to detonation by flame acceleration in the PhD tube (15.9, 16.0 and 16.1 vol.-% H_2)

For highly blocking obstacles (BR$\geq$90%) the observed combustion processes cannot be described only by increasing of the turbulence. The effective turbulent burning velocity behind a jet-producing orifice differs from the turbulent velocity induced by the expansion flow. Several experiments have been carried out to obtain more information about the hot-jet ignition mechanism, which will be described in the following.

3. Experiments at the L.View Facility

The L.View facility in Pisa consists of a rectangular test section (677 × 677 × 3200 mm), which is divided into two chambers. The first chamber has a length of 1050 mm, and is seperated from the second chamber by a wall with a central round orifice with a diameter of 100 mm, resulting in a blockage ratio of $BR = 98.3\%$.

The second chamber has a weak rupture disk to the ambient with the dimensions of 300×300 mm. Two axial fans inside the dividing wall ensure a homogenous H_2/air mixture of the same equivalence ratio in both chambers before ignition. To visualize the flame propagation, a video camera with a frame rate of 25 Hz is used. The combustion is recorded simultaneously through the front windows and, reflected by a 45° mirror, through the top windows. To enhance the contrast, aerosols of a NaCl solution are added, which are stimulated to emit light at the high combustion temperature. The flow velocity is measured inertialess and non-intrusively for the horizontal component parallel to the main flow and the vertical component at three different positions ($\sharp1 - \sharp3$ in figure 4) with a two component LDV system.

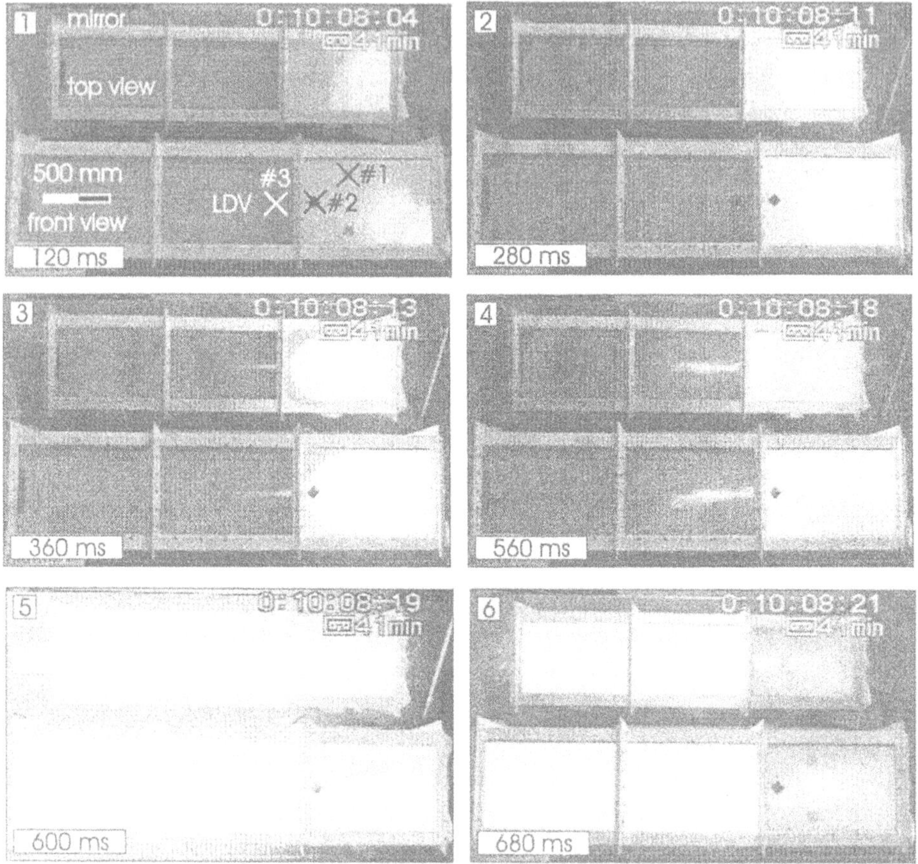

Figure 4. Selected frames of a 9 vol.-% H_2 in air flame at the L.View facility with an orifice ($BR = 98.3\%$)

Within figure 4 selected frames of a H_2/air flame (9 vol.-%) are shown. The mixture was ignited at the lower right corner.

About 300 ms after ignition, the flame reaches the orifice. Due to the high penetration velocity of the gas through the orifice, a sudden ignition in the second chamber would be expected. However, during 240 ms between picture 3 to 5, a flame jet is observed in the second chamber, but no extending flame propagation. Then, 600 ms after the ignition of the first chamber, the gas in the second chamber ignites with a very high reaction rate.

The relatively low repetition rate of the video camera allows inadequate resolution of the process. With a methane/air mixture (lower laminar burning velocity s_l) it can be observed that the ignition process in the second chamber happens at distinct points behind the orifice. High-speed video records will be taken at the L.View facility to resolve these effects on hydrogen, too.

4. Experiments at the PuFlaG Facility

Corresponding to the investigations at the L.View facility, measurements have been made at the PuFlaG (*Pulsed Flame Generator*) facility. Due to the smaller scale of the experimental setup it is possible to apply highly sophisticated optical measurement techniques. The tube with an inner diameter of 80 mm is equipped with four quartz windows (190×60 mm). Different obstacles with a central orifice of varying diameters cause a blockage ratio between 95% and 99.7%.

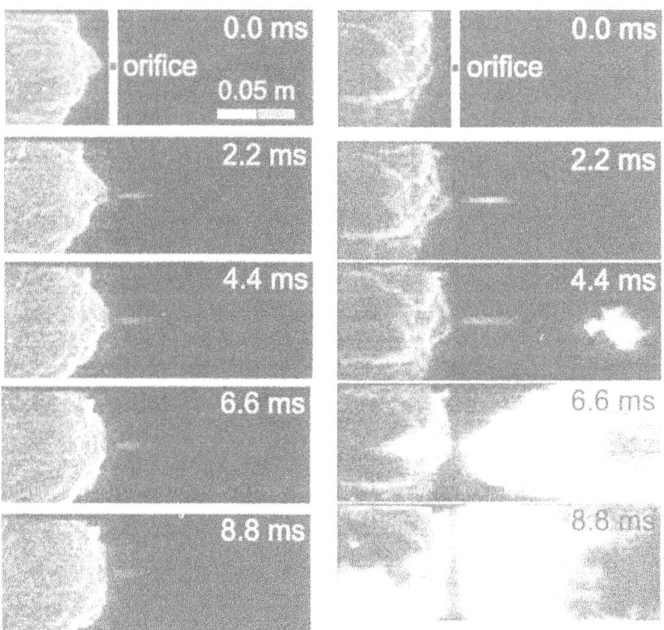

Figure 5. Selected frames of a flame (self fluorescence); left: 12 vol.-% H_2 in air; right: 12.2 vol.-% H_2 in air flame at the PuFlaG facility ($BR = 99.7\%$)

The total gas volume is about 25 liters in front of the obstacle and about 36 liters behind the obstacle. Four high-speed piezo-capacitive pressure transducers are used to record the pressure history during the combustion process.

Due to the character of schlieren records, visualizing only the density gradient, it is impossible to distinguish between exhaust gas and the flame itself. Therefore self fluorescence of the OH-radical, which is a characteristic species in the chemical reaction of the hydrogen/air combustion processes, has been used to obtain information about the flame itself. The very weak emission of the radical had to be amplified by a two-stage intensifier in front of the video camera.

In figure 5 selected frames of a 12 vol.-% H_2 flame compared to a 12.2 vol.-% H_2 flame are shown. Both flames have approximately the same velocity in front of the orifice (approximately 3.3 m/s) but as the richer flame reignites after a certain time (2 ms after the flame reaches the orifice), the leaner mixture is quenched completely. The ignition of the second chamber occurs approximately 20 orifice diameters behind the obstacle. After an ignition in the second chamber, a very high combustion rate can be observed.

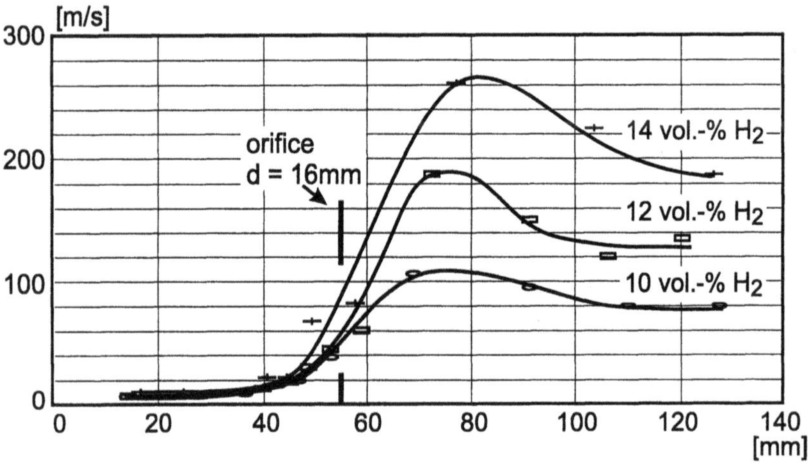

Figure 6. Propagation velocities during the combustion of various H_2/air mixtures through an orifice ($BR = 97\%$)

The resulting high velocities shown in figure 6 have been obtained by measuring the propagation of the flame from frame to frame in the schlieren records. The velocity strongly increases to a maximum behind the orifice. With the expansion of the flame to the wall the axial propagation decreases. DDT processes are not object of the investigations at this facility and have not been observed.

5. Experiments at the MuSCET Facility

Experiments in the MuSCET (*Mu*nich *S*quare *C*ross-section *E*xplosion *T*ube) facility, a horizontal square cross section explosion tube (268 × 268 mm) with a length of 3.5 m, were performed to improve the understanding of the influence of flow obstacles on flame propagation.

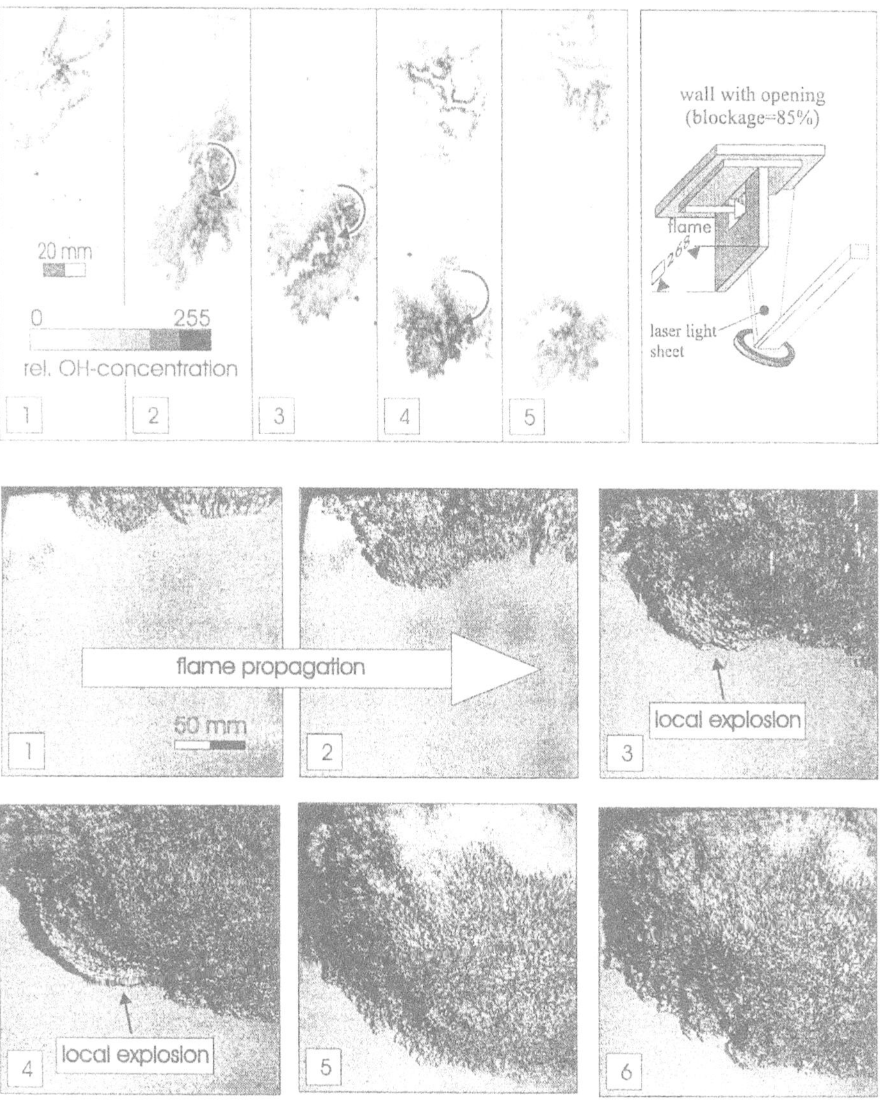

Figure 7. LIPF and high speed schlieren records behind the orifice ($BR = 85\%$, 16 vol.-% H_2 in Air)

For obstacles with a blockage ratio above 50% a strong contraction of the expansion flow in front of the obstacle could be observed. The obstacle leads to an acceleration of the flow in combination with an increasing turbulence in the shear layer of the jet after the orifice. In this case the turbulence is sufficient for the quenching of the flame.

Figure 7 shows the accumulation of highly reactive OH-radicals in a zone below the orifice. For the visualization of the OH-radicals *Laser Induced Predissociation Fluorescence* (LIPF) is used. The OH-radicals are excited within a thin lightsheet (thickness ≤ 1 mm) by a wavelength of 248 nm with a tunable KrF excimer laser. The excitiation is realized by the transition $A^2\Sigma^+, V = 3 \longleftarrow X^2\Pi, V = 0$ ($P_1(8)$). The fluorescence of the OH-radical ($A^2\Sigma^+, V = 3 \longrightarrow X^2\Pi, V = 2$) at 295 - 304 nm is recorded by an UV-intensified CCD camera applied with a special reflectance filter to avoid additional fluorescence and rayleigh scattering signals. The pulse of the laser has a duration of about $1.7 \cdot 10^{-8}$ s, the lifetime of the excited state is in the range of $10^{-10} - 10^{-5}$ s [19]. Therefore the flow situation seems to be 'frozen' at the time of record.

The mixture of the OH-radicals with the surrounding unburned gas builds a high sensitive cloud that simultaneously ignites, forming a local explosion as seen in the Schlieren records. The LIPF pictures were taken as a series of single shots triggered at different times to give an impression of the propagating combustion process.

Several local explosions can amplify themself, transitioning to a global detonation in the tube if the sensitive cloud volume is sufficiently large. Initial conditions for the spherical detonation from a local ignition were mentioned by Desbordes [17]. The process to the fully developed detonation is referred to as 'Hot-Jet Ignition'.

6. Conclusion

The main goal of this experimental work is to deliver sufficient data for a numerical simulation of all kinds of combustion processes. At this time moment, different codes for deflagration and detonation processes must be used due to the differing nature of deflagration and detonation. So far qualitative information as well as conservative limits for the DDT had been collected, but more quantitative data is needed to correctly predict a DDT. It has been shown that the DDT process consists of several physical phenomena, which are difficult to describe. These effects have a very strong influence on the combustion process and the transition point between deflagration and detonation, so that the effects have to be taken into consideration for the further development of integrated deflagration/detonation codes. For the validation of quenching effects at larger scales, a new facility called Hot-Jet Explosion Tube (transition from a tube with a diameter of 200 mm and a length of 4 m via an orifice to a tube with a diameter of 500 mm and a length of 3 m) has been set into operation at the Lehrstuhl A für Thermodynamik.

Acknowledgements

It is gratefully acknowledged, that the work presented in this paper has been supported by the German Ministry of Education, Science, Research and Technology BMBF and the European Commission.

Nomenclature

Le	Lewis number, $Le = a/D$
a	thermal diffusivity
D	mass diffusivity
s_l	laminar burning velocity
A_l	Surface of the flame
A_0	Cross-section of the tube
ρ	density
p	pressure
λ_k	Kolmogorov micro scale
Ka	Karlovitz number, $Ka = \tau_c/\tau_k$
τ_c	Chemical reaction time
τ_k	Lifetime of the Kolmogorov micro vortices
λ_d	Detonation cell width
BR	blockage ratio, ratio between blocked and unblocked area

References

1. Brehm N. and Mayinger F., Limits for the transition from Deflagration to Detonation in Hydrogen - Air -Steam Mixtures, final report BMFT RS 1500712, FIZ4, Karlsruhe, 1988

2. Strube G., Struktur und Brenngeschwindigkeit turbulenter, vorgemischter Wasserstoff-Flammen, Dissertation TU München, 1993

3. Beauvais R., Brennverhalten vorgemischter, turbulenter Wasserstoff-Luft-Flammen in einem Explosionsrohr, Dissertation TU München, 1994

4. Strube G., Beauvais R., Mayinger F., Derzeitiger Wissensstand über den Verlauf der Grenze für den Übergang einer Deflagration in eine Detonation (DDT) im Dreistoff-Diagramm Wasserstoff/Luft/Wasserdampf nach Shapiro/Moffette, Abschlußbericht, TU München, 1988

5. Ardey N., Durst B. and Mayinger F., Influence of Flame-Obstacle-Interaction on the Structure of Turbulent Deflagrations, Proceedings of the Int. Cooperative Exchange Meeting on Hydrogen in Reactor Safety, Toronto, 1997

6. McIntosh A. C., Influence of Pressure Waves on the Initial Development of an Explosion Kernel, AIAA Journal, Vol. 33, No. 9, September 1995

406

7. Ardey N. and Mayinger F., Flame Acceleration by Turbulent Promotion and Jet Ignition, submitted to Progress in Energy and Combustion Science, Pergamon Press, Oxford, 1998

8. Jordan M. et al., Influence of Turbulence on the Deflagrative Flame Propagation in Lean Premixed Hydrogen Air Mixtures, FISA-97Symposium on EU Research on Severe Accidents, Luxembourg, 1997

9. Jordan M., Ardey N., Mayinger F., Effect of Turbulent Transport and Mixing on Flame Acceleration through Highly Blocking Obstacles, 11th International Heat Transfer Conference, Korea, 1998

10. Kuo K. K., Principles of Combustion, John Wiley & Sons, New York, 1986

11. Klein R., Breitung W., Rehm W., Olivier H., He L., Armand P., Ang M., Models and Criteria for Prediction of Deflagration-to-Detonation Transition (DDT) in Hydrogen-Air-Steam Systems under Severe Accident Conditions, FISA-97 Symposium, Luxembourg, 1997

12. Chan C. K., Lau D., Radford D., Transition to Detonation Resulting from Burning in a Confined Vortex, 23rd International Symposium on Combustion, The Combustion Institute, pp. 1797-1804, 1990

13. Dupré G., Unstable Detonations in the Near-limit Regime in Tubes, 23rd International Symposium on Combustion, The Combustion Institute, pp. 1813-1820, 1990

14. Vendel J., Armand P., Large Scale Hydrogen-Air-Steam DDT Experiments in the RUT Facility, Hydrogen Combustion Work Group Meeting, Nuclear Safety and Protection Institute, Tokyo, 1978

15. Peters N., Laminar Flamelet Concepts in Turbulent Combustion, 21st. International Symposium on Combustion, The Combustion Institute, Pittsburgh, 1986

16. Ciccarelli, Ginsberg T., Boccio J., Economos C., Sato K., Kinoshita M., Detonation Cell Size Measurements and Predictions in Hydrogen-Air-Steam Mixtures at Elevated Temperatures, Combustion and Flame Vol. 99, pp. 212-220, 1994

17. Desbordes D., Critical Initiation Conditions for Gaseous Diverging Spherical Detonations, Journal de Physique IV, Colloque C4, suplément au Journal de Physique III, Vol. 5, 1995

18. Abdel-Gayed R. G., Bradley D., Combustion Regimes and the Straining of Turbulent Premixed Flames, Combustion and Flame, Vol. 76, pp. 213-218, 1989

19. Eckbreth A. C., Laser Diagnostics for Combustion Temperature and Species, Abacus Press, Tunbridge Wells, UK, 1988

HEAT AND MASS TRANSFER
WITH DRYING OF WATER-BASED VARNISHES

J. MINTZLAFF AND F. MAYINGER
Lehrstuhl A für Thermodynamik
Technische Universität München
85747 Garching, Germany

Abstract. This study deals with the drying of a dispersed varnish which is applied as a thin film on paper. Global measurements are performed in order to record the drying curves, whereas single phenomena are examined by local measurements. Their results are shown, and the basic equations for the calculation of the drying are presented.

1. Introduction

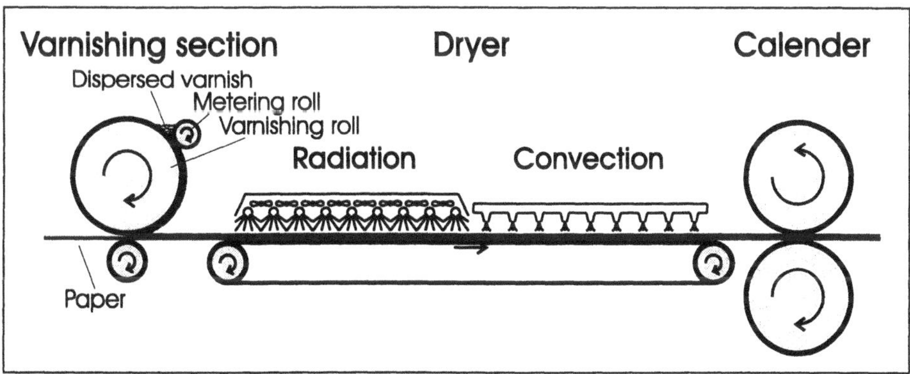

Figure 1. Varnishing machine

The objective of print finishing is the improvement of the optical, mechanical and physical properties of paper by applying a coating. Examples of such process products are high quality packings, money bills, or playing

S. Kakaç et al. (eds.), Heat Transfer Enhancement of Heat Exchangers, 407–421.
© *1999 Kluwer Academic Publishers.*

408

cards. New laws concerning environmental protection have led to a change from solvent- to water-based varnishes within the last years. These lacquers consist of water and thermoplasts, one half of which is dispersed and the other is solved in the water. Their drying behaviour is different from that of the formerly used varnishes: the solvent volatility is lower. The drying process requires much energy, and thus contains a high potential for energy conservation.

Varnishing consists of three steps: Applying of the coating, drying and, if required, calendering as depicted in figure 1. This investigation focuses on the phenomena in the drying section of a varnishing machine, which incorporates convective and radiative modules. These two drying methods, convection and radiation, are investigated in order to optimize the drying section and to provide an instrument for the design of a combined convective and radiative dryer. As a result the size, the energy consumption and, as a consequence, the costs of the varnishing machines can be reduced.

2. Method of Investigation

In order to compare the different drying methods and to get an idea of their function, global measurements have been performed. The drying curves have been taken with different combinations of the following parameters: the temperature, the velocity and the humidity of the drying air, the spatial arrangement of the nozzles, the radiative power and wavelength and the thickness of the applied varnish.

The knowledge of the thermal and hydrodynamical boundary layer of the drying air flow over the varnished paper as well as of the processes within the drying coating is of major importance for the effective enhancement of the heat and mass transfer in the drying section. Therefore local investigations of the thermal boundary layer have been performed using the optical measurement technique of holographic interferometry in combination with balancing measurements. In order to obtain the required parameters for the numerical simulation of the drying process, further local measurements have been carried out: micrographs of the coated paper, gravimetrical investigations of the desorption isotherms, examinations of the radiation field, and an analysis of the absorption spectrum of the drying material.

Finally the drying process is simulated numerically using the results of the local measurements. These calculations are verified by comparing their results to the global and industrial investigations. The simulation program is then implemented for the design of varnishing machines.

3. Global Investigations

3.1. EXPERIMENTAL SET-UP

For these measurements an experimental facility has been designed and set up as it is shown in figure 2. Process air passes a conditioning unit where it adopts a defined humidity and temperature. It is then either blown into the radiative segment or through a heating element into the convective drying segment. Having passed the coated paper, it is drawn off.

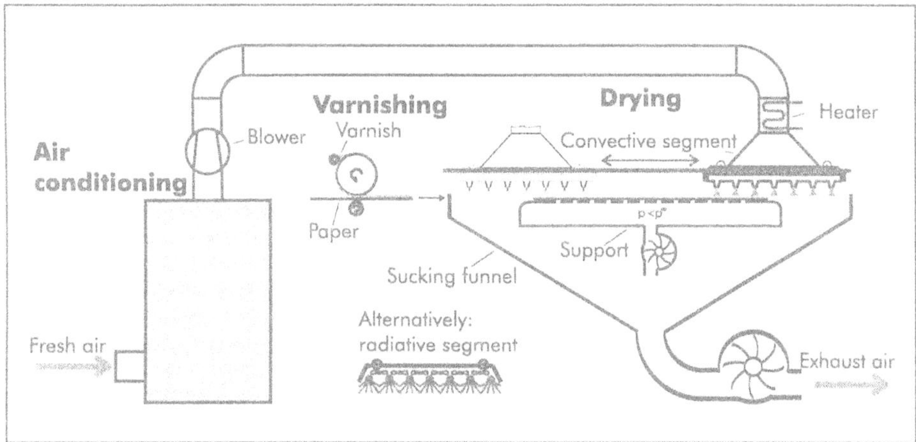

Figure 2.   Experimental installation

The paper which has a size of 900 mm vs. 195 mm and a basis weight of 180 g/m^2 is coated in the varnishing unit, while the drying segment is on the opposite side of the unit. As soon as the coated paper is placed on the support, the drying segment starts moving back and forth above the paper in order to simulate a continuous movement of the paper in the industrial varnishing machine.

The moisture of the coated paper has been determined by means of a photometer by IR-reflection. The surface temperature has been measured with a pyrometer.

3.2. CONVECTIVE DRYING

The examinations have been performed by varying the temperature, the velocity, and the humidity of the drying air, and the spatial arrangement of the nozzles as illustrated in figure 3. The distance between the nozzles and the paper was 31 mm.

410

In terms of the drying behaviour, it has been proved that the drying of the homogenious varnish, as will be seen later, shows the same behaviour as porous bodies (figure 4). A starting period is followed by a constant rate period and a falling rate period afterwards [3, 5].

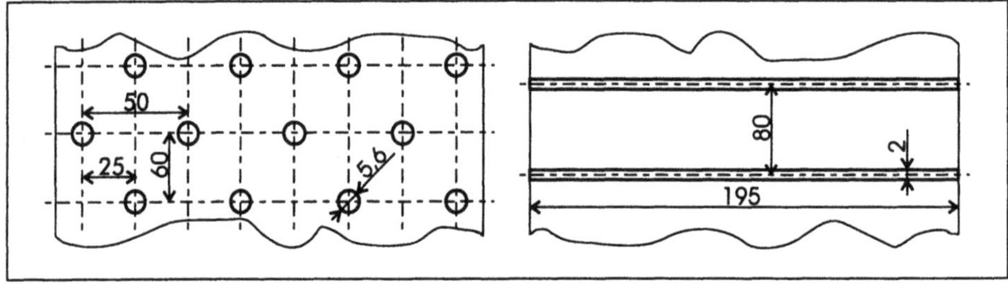

Figure 3.   Nozzle arrangment of the convective drying segment

When using high temperatures or high airflows, a higher drying rate is achieved. This fact leads to a prolongation of the starting period. This might cause the onset of the falling rate period before the steady state conditions of the constant rate period are established, which can be seen in figure 5. By drying with a significant drying rate, only the equilibrium moisture which the pure paper had before varnishing can be achieved. This is the level 12.5 g/m^2 in the diagram.

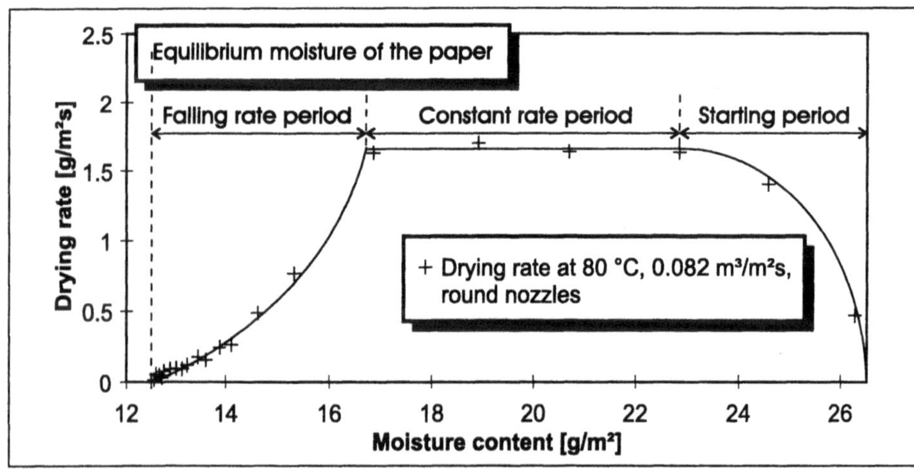

Figure 4.   Drying behaviour of the water borne coating

Comparing the different spatial arrangements at equal temperatures and equal air flows, the round nozzles show better results as depicted in figure

5. As expected, this effect can only be observed in the constant rate period, with the mass transfer resistance lying in the hydrodynamical boundary layer [3, 5]. In the falling rate period the drying rates of both types of nozzles are identical.

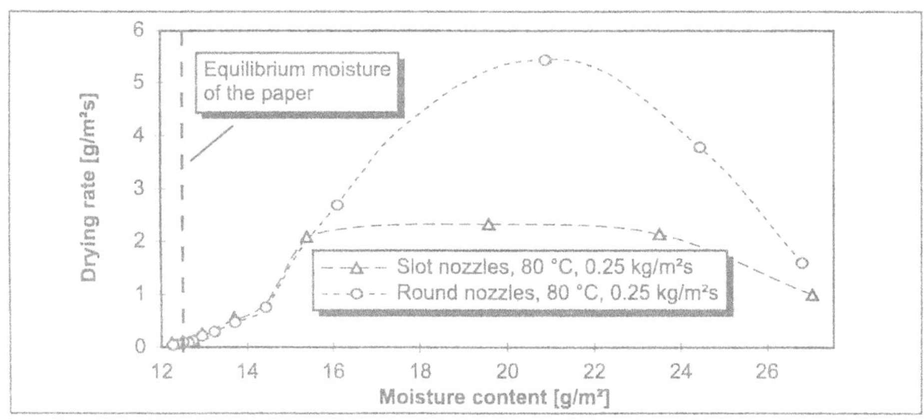

Figure 5.   Drying curves of round and slot nozzles at 80 °C and 0.25 $m^3/m^2 s$

3.3. RADIATIVE DRYING

The radiative segment is based on five quartz glass IR tubes with a distance of 100 mm to one   another and with an angle of 60 ^{o} in the direction of

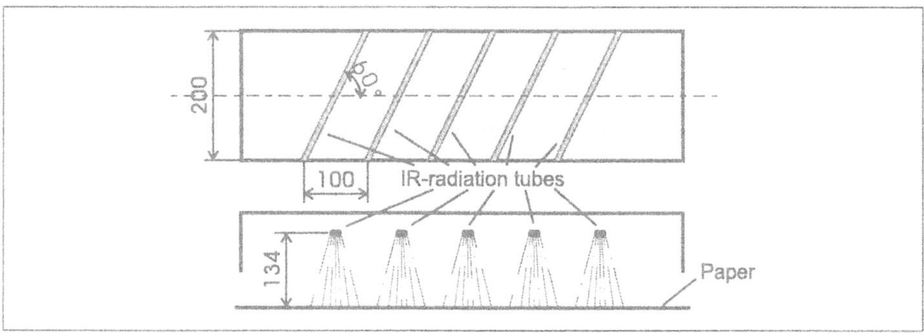

Figure 6.   Drying segment for IR-radiation

the longitudinal axis. They are located at a distance of 134 mm above the paper, as illustrated in figure 6.

412

The drying curves for short- and medium-wave IR-radiation, with an electrical power of 25 kW/m^2, can be seen in figure 7. Wet varnish of an amount of 26 g/m^2 has been applied. The striking difference from the convective drying is the fact that the coated paper can be dried to a moisture content underneath the equilibrium level of the pure paper. It can be over-dried. The maximum drying rate is lower than that with convection, but at low moisture contents the drying rate is higher with radiation.

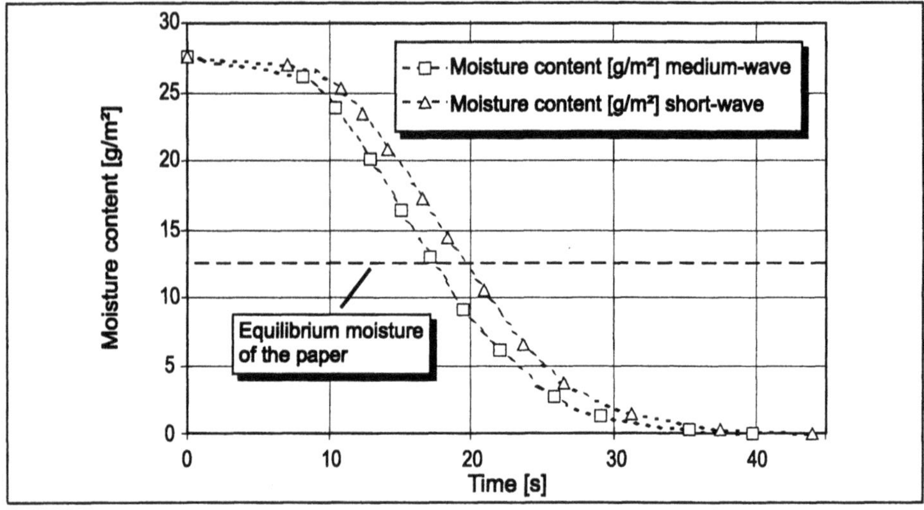

Figure 7. Drying curves for medium- and short-wave IR-radiation at 25 kW/m^2

A comparison with the two radiation curves shows that 12.5 % less time for the drying down to equilibrium moisture with medium-wave IR-radiation is required. This fact is caused by two phenomena [8]. The absorption bands of water are stronger in the medium-wave spectrum. The short-wave radiation penetrates more deeply into the coated paper. The heat then has to be conducted to the surface of the lacquer while a portion of the heat gets lost from the bottom surface of the paper. The surface temperature is 80 ^oC when the equilibrium moisture is reached.

3.4. COMBINED CONVECTIVE AND RADIATIVE DRYING

In order to incorporate the advantages of both types of drying segments, a combined segment has been set up as shown in figure 8. Drying air reaches a common space above the IR- tubes and is spread among the round nozzles, which are inbetween the radiative tubes. The air becomes hot in the common space by the waste heat of the IR-tubes and in the nozzles which are affected by the radiation field. Therefore no additional heating element is required.

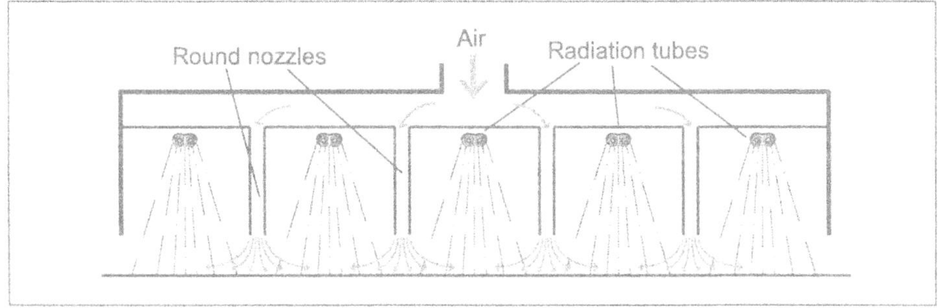

Figure 8. Combined drying segment for convection and IR-radiation

Figure 9 shows the drying curves of the combined drying segment with medium-wave tubes for two different air flows. The amount of 26 g/m^2 has again been applied. The bigger air flow causes a shorter starting phase and a higher drying rate at a high moisture content. At lower moistures

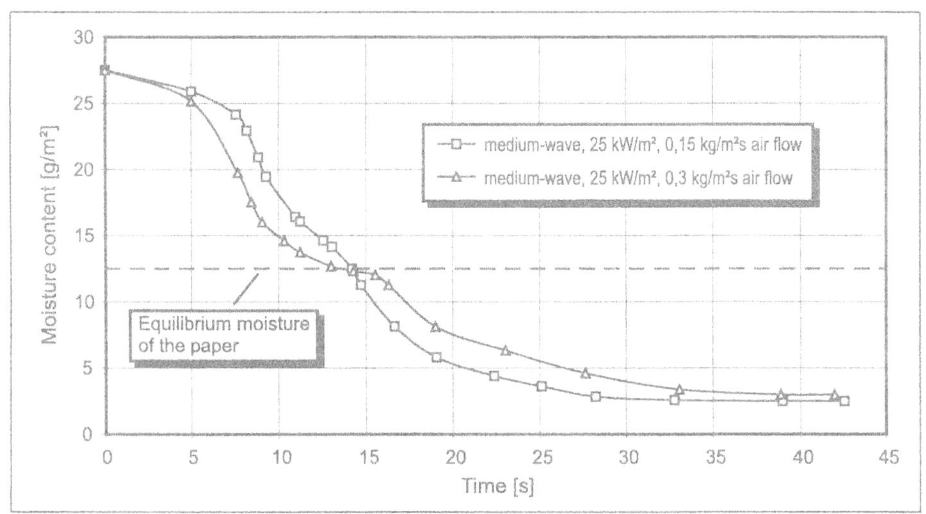

Figure 9. Drying curve of the combined drying segment

the drying rate increases with the lower air flow. If the moisture content is below the equilibrium moisture of the paper, drying with a lower air flow is even faster. Compared to pure radiative drying t,he time that is needed to achieve the equilibrium moisture is achieved is shorter by 20 %. The surface temperature at equilibrium is 60 ^oC, 20 K less than with pure radiation. This means that fewer cooling segments are required behind the drying section in the industrial process.

3.5. COMPARISON BETWEEN THE DIFFERENT DRYING SEGMENTS

The drying rates vs. the moisture content of the three drying segments are shown in figure 10. For all three drying methods the starting moisture was 15.0 g/m^2 above the equilibrium moisture of the pure paper, which is 27.5 g/m^2. The comparison shows that the maximum of the drying rate of the combined segment, which is caused by the convectional part, is not as high as the maximum of the pure convectional drying. In the region of the lower moisture contents, however, the drying rate is nearly as high as that with pure radiation but with a surface temperature at the end of the drying that is 20 K lower.

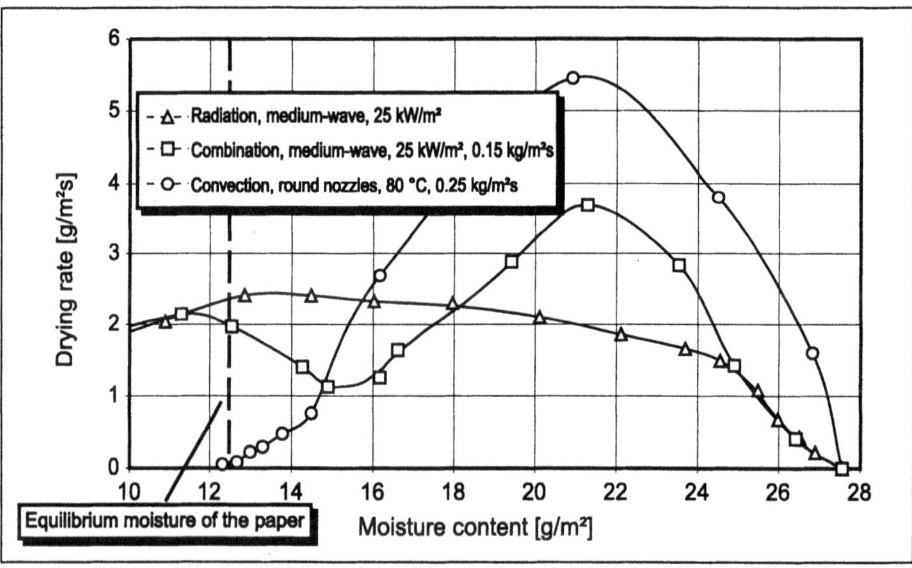

Figure 10.   Drying curves for convective, radiative and combined drying

An optimized line-up for the dryer of an industrial varnishing machine starts with convective drying to utilize its high drying rates in the first drying section. Combined segments follow, in which the air flow is adapted to the moisture content.

4. Local Investigations

For a better understanding, local investigations have been carried out to examine the single phenomena involved in the drying process.

4.1. COATING

For the later calculation of the drying it was of major importance to know the microscopical structure and the penetration behaviour of the varnish in the paper. Therefore scanning electron micrographs have been made of the cross section of the coated paper (figure 11) which have proved the varnish to be homogenious without any pores. The mass transfer of the water in the coating has to be calculated according to the laws of liquid diffusion [2, 6].

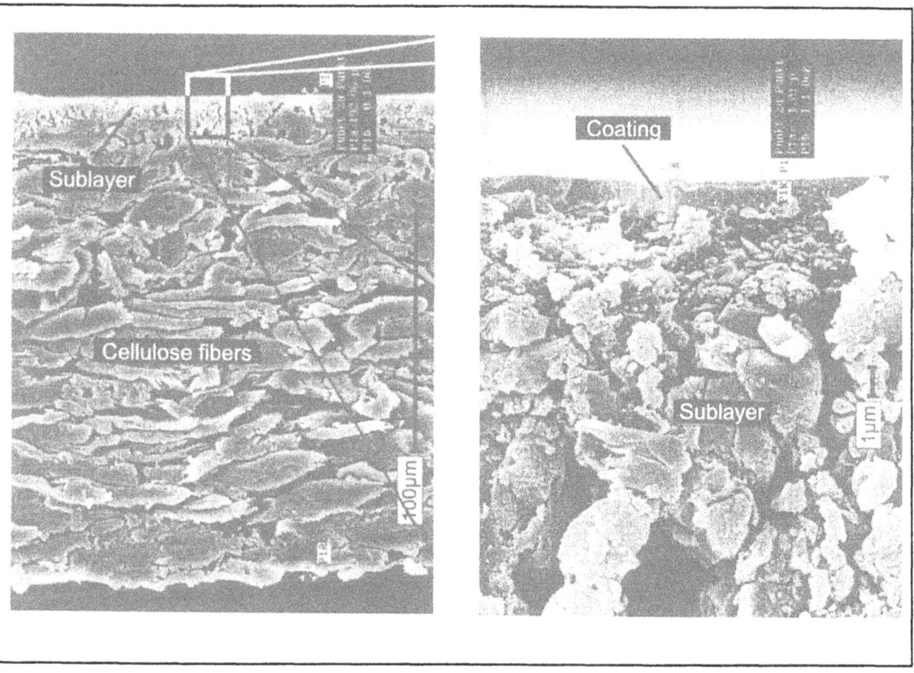

Figure 11.   Scanning electron micrograph of the cross section of a coated paper

From several micrographs with different thicknesses of the film, a connection with the penetrated ratio could be obtained (figure 12). It is proved that within the examined range, it can be assumed with a good precision that the ratio is constant.

416

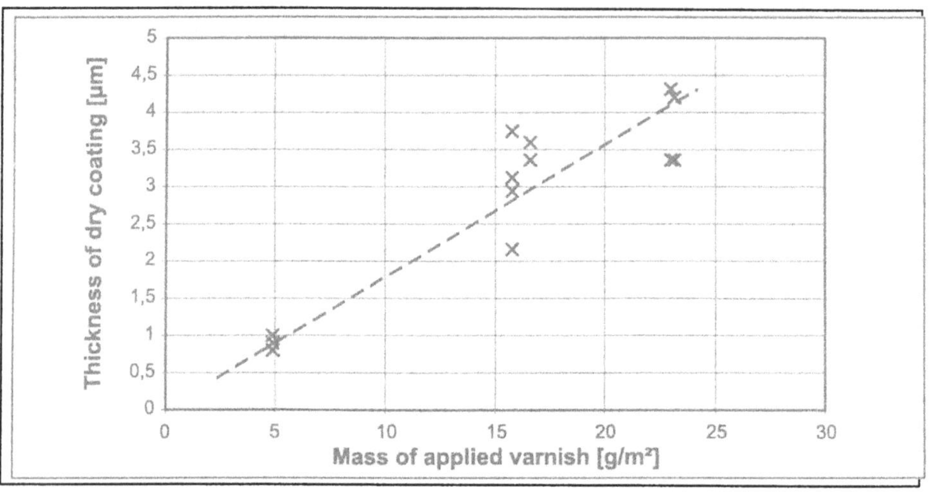

Figure 12.   Thickness of dry film vs. applied wet mass

4.2. BOUNDARY LAYER OF THE DRYING AIR

Knowledge of the boundary layer of the drying air above the varnish is essential for the determination of the heat transfer coefficient. On the other hand the Lewis number Le, which is the ratio a/D of the thermal diffusivity a and the mass diffusion coefficient D, equals 0.937 which is close to 1 [1]. This means that the analogy of heat and mass transfer can be used to calculate the mass transfer coefficient from the heat transfer coefficient [5].

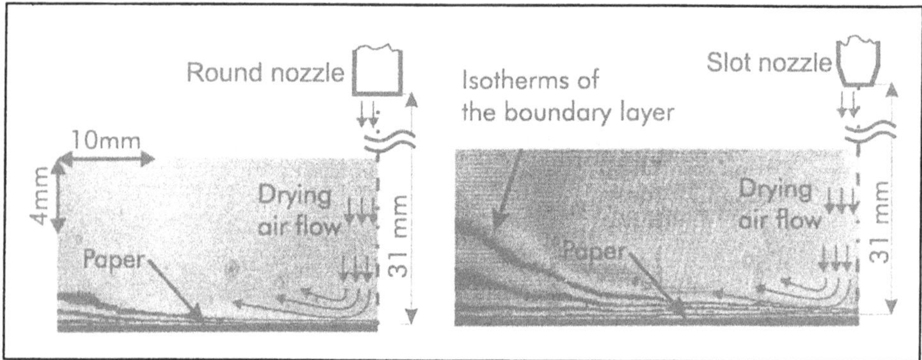

Figure 13.   Interferograms for convection with round and slot nozzles

By the means of holographic interferometry it is possible to record a two dimensional temperature field [4]. Figure 13 shows the interferograms of the boundary layer of half of the space between two slot nozzles and two rows of round nozzles respectively at the same air flow. The dark lines can be regarded as isotherms.

It can easily be seen that the considerably thinner boundary layer of the round nozzles causes a better heat and mass transfer. In balancing examinations the heat transfer coefficients have been determined, as shown in figure 14. These results correspond with the drying curves in figure 5.

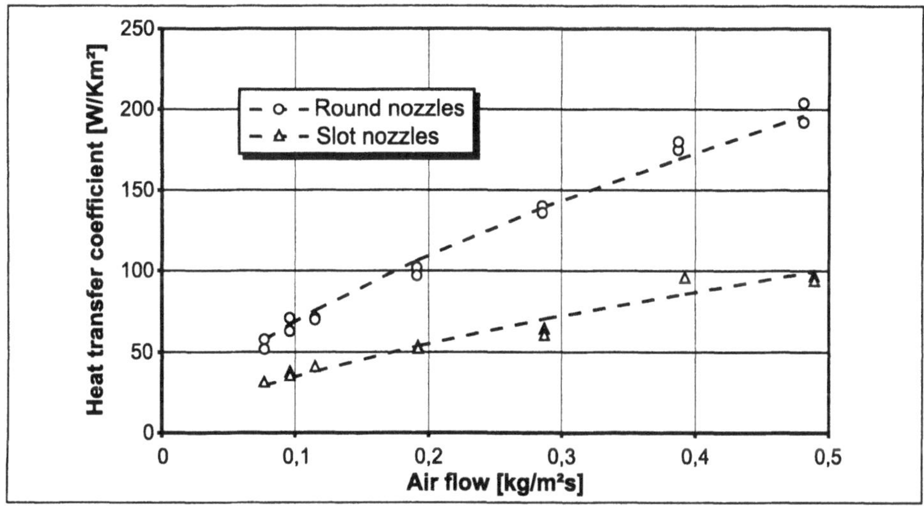

Figure 14. *Value of the heat transfer coefficient vs. air flow for round and slot nozzles*

The thinner boundary layer of the round nozzles is a result of the higher vertical speed of the air at the surface underneath the nozzles. In order to obtain equal velocities of the air jet at equal air flow rates the exiting nozzle area has to be of the same size. A nozzle width of 0.6 mm would have to be realized then. Considering the costs of fabrication of the drying section of an industrial varnishing machine, a width of 2 mm was chosen. As a consequence, the starting velocity of the jet at the exit of the slot was about one third of the velocity of the round jet. An additional fact is that the boundary area relative to the cross-sectional area of the jet is bigger at the slot nozzle. Therefore the momentum exchange of this jet with the surrounding is greater as well as its deceleration.

4.3. FURTHER LOCAL INVESTIGATIONS

Besides the examples shown, further local investigations have been carried out, such as measurements of the activity of the water in the varnish a_W, examinations of the radiation underneath the drying segment or an analysis

418

of the absorbing behaviour of IR-radiation of the coated paper. All these results are used for the numerical simulation, which enables the calculation of the drying process.

5. Numerical Simulation - Basic Equations

After the experimental results the governing equations for the drying process are introduced, especially to show what the results of the local measurements are required for.

5.1. ASSUMPTIONS

In order to enable the calculation of the drying some assumptions have been made to simplify the problem:

- Only transport processes rectangular to the surface are considered.
- The coated paper is divided into three zones: pure varnish, paper penetrated with varnish and pure paper.
- The mass transfer within the coating is merely caused by liquid diffusion because the lacquer does not have any pores.
- The bottom of the paper is adiabatic.
- The varnish is shrinking while drying.
- A system of coordinates based on the dry body is used, which requires a transformation of coordinates. Water is then the only component which is moving within the grid .
- The convective energy term caused by diffusing water is negligible compared to the conducted energy.

5.2. MASS CONSERVATION

Since water is the only moving component in the system of coordinates based on the dry body the mass equation only has be set up for water [7].

$$\frac{\partial X}{\partial t} = \frac{\partial}{\partial \zeta}\left(\frac{D_{WL}}{(1+X\epsilon)^2}\frac{\partial X}{\partial \zeta}\right) \tag{1}$$

Here ϵ is the relation of the specific partial volumes.

$$\epsilon = \frac{\hat{V}_W}{\hat{V}_{L,dr}} \tag{2}$$

The transformed spatial coordinate is calculated by

$$\partial \zeta = \frac{\partial z}{1+X\epsilon} \tag{3}$$

The boundary conditions for $t > 0$ are at the bottom

$$\frac{\partial X}{\partial \zeta} = 0 \tag{4}$$

and at the top

$$\frac{\partial X}{\partial \zeta} = \frac{(1 + X\epsilon)^2}{D_{WL}\rho_{L,dr}} \cdot \dot{m}_{W,ev} \tag{5}$$

The evaporated flux of water $\dot{m}_{W,ev}$ can be calculated as follows [5]:

$$\dot{m}_{W,ev} = \frac{\beta_h}{R_V T_m}(a_W p_{Vp} - p_{V,No}) \tag{6}$$

The activity a_W is the ratio of the vapour pressure of the water above the surface of the lacquer $p_{V,L}$ and the vapour pressure of pure water p_V:

$$a_W = \frac{p_{V,L}}{p_V} \tag{7}$$

a_W is taken from the local measurements. The mass transfer coefficient for the semipermeable boundary layer β_h can be calculated from the heat transfer coefficient α [5] which also has been measured:

$$\beta_h = \frac{\alpha}{\bar{c}_{a,m}\bar{\rho}_{a,m}Le\left(1 - \frac{p_{Vp}}{p}\right)} \tag{8}$$

According to the analogy of heat and mass transfer this calculation is possible since the air flow is turbulent and the Lewis number $Le = 0.937$ is close to 1 [1].

5.3. ENERGY CONSERVATION

The energy equation is

$$\frac{\partial(\rho_{i,W}c_{i,W}) + \rho_{i,L}c_{i,L}) + \rho_{i,P}c_{i,P})}{\partial t} = \frac{\partial}{\partial z}\left(\lambda\frac{\partial T}{\partial z}\right) \tag{9}$$

The boundary conditions for $t > 0$ are at the bottom

$$\frac{\partial T}{\partial z} = 0 \tag{10}$$

and at the top

$$\frac{\partial T}{\partial z} = \alpha(T_{No} - T_{surface}) - \dot{m}_{W,ev}[r_o + c_{p,V}(T_\infty - T_{surface})] \tag{11}$$

This equation considers the convective heat transfer to the coated paper, the evaporation enthalpy of the evaporated water and the energy flux caused by the vapour flux from the surface to the ambient air. The vapour reaches the temperature of the ambient air, which leads to a reduction of the convective heat transfer [3].

These equations are sufficient for a numerical simulation of the drying process. They provide an instrument for the future design of the optimized drying section of a varnishing machine.

6. Conclusions

Drying in general is quite a complex combined heat and mass transfer process. The difficulties arise from the complexity of the process and from the changing properties which are strongly affected by temperature and humidity. In case of the drying of a water-borne coating applied on paper, an attempt has been made here to show the aspects for an entire treatment of this subject.

Global measurements provide the base of the investigation. They enabled the finding of the best nozzle arrangment for the convective segment and the optimum wave length for the radiative segment. With this knowledge a combined segment was created and tested. Regarding the requirements of each drying phase, an optimized line up of the segments in the drying section of an industrial varnishing machine was found.

For the calculation of the drying process in local investigations the individual phenomena involved in the drying process were examined. Furthermore the results proved to be in good correspondance with the global measurements.

Finally the basic equations for a calculation of the drying process have been introduced. They enable the numerical simulation and provide an instrument for the future design of the optimized drying section of a varnishing machine.

Nomenclature

a	thermal diffusivity, $\mathrm{m^2/s}$
a_W	activity of the water in the lacquer, -
c	specific heat, $\mathrm{J/kg}$
$\bar{c}_{a,m}$	mean specific heat of the air in the boundary layer, $\mathrm{J/kg}$
D_{WL}	Fickian diffusion coefficient of the water in the lacquer, $\mathrm{m^2/s}$
Le	Lewis number, -
$\dot{m}_{W,ev}$	evaporated flux of water, $\mathrm{kg/m^2 s}$
p_V	saturated state pressure of the vapour, Pa
$p_{V,L}$	partial pressure of vapour above the laquer, Pa
$p_{V,No}$	partial pressure of vapour in the drying air, Pa
R_V	specific gas constant of vapour, $R_V = 462 J/kgK$
t	time, s
$\hat{V}_n$	specific partial volume of component n, $\mathrm{m^3/kg}$
X	humidity referred to dry body, $\mathrm{kg/kg}$
α	heat transfer coefficient, $\mathrm{W/m^2K}$
β_h	mass transfer coefficient for the semipermeable boundary layer, $\mathrm{m/s}$
ϵ	ratio of the specific partial volumes, -
λ	thermal conductivity, $\mathrm{W/Km}$
$\rho_{L,dr}$	density of the dried lacquer, $\mathrm{kg/m^3}$
$\rho_{i,n}$	partial density of component n, $\mathrm{kg/m^3}$
$\bar{\rho}_{a,m}$	mean density of the air in the boundary layer, $\mathrm{kg/m^3}$
ζ	transformed spatial coordinate, m

References

1. Baehr, H. D., Stephan, K. (1994) *Wärme und Stoffübertragung*. Springer-Verlag, Heidelberg
2. Crank, J., (1979) *The Mathematics of Diffusion*. Oxford University Press, London
3. Kast, W., Krischer, O., Kröll, K. (1992) *Trocknungstechnik, Bd. 1 Die wissenschaftlichen Grundlagen der Trocknungstechnik*. Springer-Verlag, Heidelberg
4. Mayinger, F. (Ed.) (1994) *Optical Measurements*. Springer-Verlag, Heidelberg
5. Mersmann, A. (1980) *Thermische Verfahrenstechnik*. Springer-Verlag, Heidelberg
6. Reid, R. C., Prausnitz, J. M., Poling, B. E. (1987) *The Properties of Gases and Liquids*. McGraw-Hill Book Company, New York
7. Saure, R. (1995) *Zur Trocknung von lösemittelfeuchten Polymerfilmen*. VDI-Verlag, Düsseldorf
8. Siegel, R., Howell, R. H., Lohrengel, J. (1988) *Wärmeübertragung durch Strahlung, Teil 1 Grundlagen und Materialeigenschaften*. Springer-Verlag, Heidelberg

ENERGY CONVERSION IN A HYDROGEN FUELED DIESEL ENGINE:
OPTIMIZATION OF THE MIXTURE FORMATION AND COMBUSTION

PETER PRECHTL; FRANK DORER; FRANZ MAYINGER
Lehrstuhl A für Thermodynamik, Technische Universität München,
D-85747 Garching, Germany
C. VOGEL; V. SCHNURBEIN
MAN B&W DIESEL AG, Stadtbachstr. 1, D-86153 Augsburg, Germany

Abstract. A large-bore, four-stroke engine, with high pressure injection and compression ignition, running with hydrogen, is a modern concept for clean energy conversion without CO_2 emission. The key processes for reliable engine operation are a good mixture formation, a reliable ignition and efficient combustion. The investigations of these processes are carried out in a rapid compression machine, with modern optical measurement techniques.

1. Introduction

The simple reaction of hydrogen and oxygen into water as a clean method for energy conversion, the high energy density, the wide ignition ranges and the high burning velocities have been the reasons for a long time for people to investigate the usability of hydrogen as a fuel for internal combustion engines. Short overviews are given in [1,2]. In recent years, extensive research has been done with different concepts on the application of hydrogen for engine use. A summary of German activities is given in [3]. The method of internal mixing and compression ignition (C.I.) combines the advantages of the high efficiency of a diesel engine and the clean combustion of hydrogen and oxygen. For this combustion method, however, one main problem has to be solved which is the auto-ignition of hydrogen. A variety of studies on this topic have shown that a reliable running C.I. engine cannot be realized yet [4,5,6,7]. In a cooperation of MAN B&W Diesel AG and two departments of the Technische Universtät München, a large-bore, four-stroke engine running with hydrogen is being developed. The research project is supported by the Bayerische Forschungsstiftung (Bavarian Research Foundation). The Lehrstuhl für Verbrennungskraftmaschinen und Kraftfahrzeuge (LVK, department of internal combustion engines and motor vehicles) is investigating

423

S. Kakaç et al. (eds.), Heat Transfer Enhancement of Heat Exchangers, 423–432.

the behavior of a single cylinder test engine. The Lehrstuhl A für Thermodynamik (LAT, department of thermodynamics) is investigating the injection and combustion process with modern optical measurement techniques in a rapid compression machine (RCM) under realistic conditions. As mentioned above, the ignition of hydrogen under internal mixing conditions is the key process for a reliable engine operation. Due to this fact it is main subject of the current research activities at the LAT.

2. Experimental Investigations

In a first phase the experiments are carried out in the RCM by varying of the compression ratio, the load pressure and the flow condition (variation of the swirl and the turbulence). The compression ratio reaches a value up to 22, and the compression end pressure is between 5 and 11 MPa. The temperature reaches values over 950 K. The hydrogen is injected with a pressure between 25 and 33 MPa, and a maximum injection duration of 15 ms. In the next step the influence of the injector geometry is analyzed. For these experiments nozzles with a hole number of 6 up to 24 and a slot nozzle with nearly equivalent cross-sections were manufactured. These measurements were carried out under the same pressure conditions.

2.1.   EXPERIMENTAL SETUP

The experimental setup used for these investigations is based on an fully optically accessible rapid compression machine (RCM). The machine simulates a single compression stroke under realistic conditions. This experimental setup allows a variety of investigations under different conditions. It is simple to replace the injection device or the nozzle geometry. The compression ratio, the load, the intake temperature as well as the injection timing can be varied. In addition to that fact a swirl or a turbulence can be engendered in the combustion chamber. On the top of the combustion chamber the electro-hydraulic-controlled injection device for pressurized hydrogen is mounted. The hydrogen is supplied from single bottles and from bundles. The hydrogen is compressed with a 30 MPa piston compressor, and it is stored in a small high pressure tank before the injection. The optical access to the combustion chamber of the RCM is realized through a large quartz window in the bottom of the piston and additional windows in the cylinder walls. The RCM allows a maximum combustion pressure over 20 MPa. This setup allows the use of optical measurement techniques, such as laser-induced fluorescence (LIF) [8] for the detection of several species like OH, and fast imaging techniques, such as the high speed digital video and Schlieren techniques to visualize the mixing and combustion process. A typical setup using the high speed video camera is shown in figure 1. The dimensions of the machine are in a 1:1 scale to the single cylinder test engine, which has a piston displacement of about 14 liters and a power output of approximately 180 kW. The velocity of the piston of the rapid compression machine corresponds to the piston speed of the engine with 800 – 900 RPM near the top dead center (TDC). The emitted light (self-fluorescence) from the ignition and the combustion process is observed using a digital high speed camera at a frame rate of 13,500 Hz. In addition to the conventional data, such as the pressure, the piston displacement, the needle lift and other required signals, the images are stored digitally. In the next step the data is analyzed, combined with the images and recorded to a digital mpeg-film. These films give a good impression of the location and the spatial distribution of the ignition and combustion process. This method contributes to a better understanding and an efficient improvement of the combustion concept.

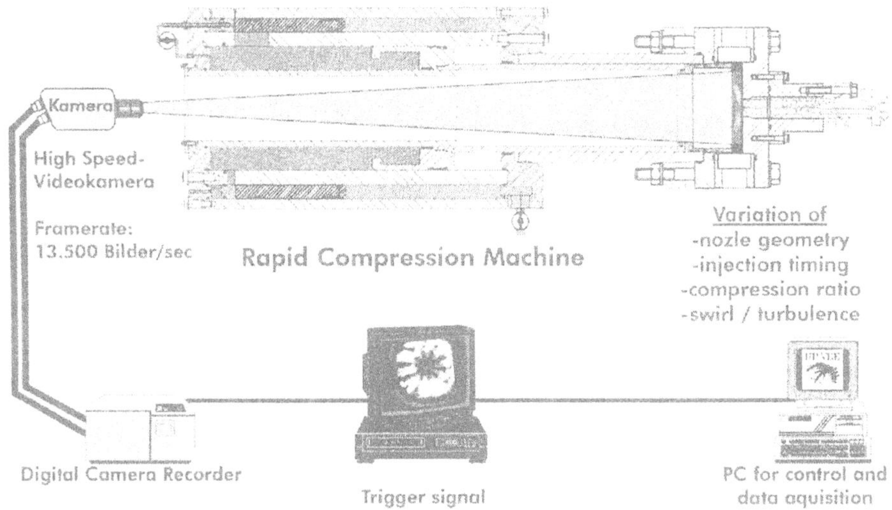

Figure 1: Experimental setup using the rapid compression machine (RCM) and a high speed video camera to observe the ignition and combustion process

2.2.   RESULTS AND DISCUSSION

In the first series of experiments it has been shown that the compression ratio has a significant influence on the ignition delay. Higher temperatures lead to shorter ignition delays. However, large variations of the ignition delays have been observed in all the experiments. The ignition occurs statistically distributed in the combustion chamber. The analysis of the films proves the assumption that an ignition of a single jet does not lead to the ignition of other hydrogen jets, as is shown in figure 2. It has also been observed that several jets have not ignited. It can be inferred that different areas of combustible hydrogen mixtures of the injected jets in a single compression cycle can have different ignition delays. As a result, cycle-to-cycle variations in the pressure recordings are expected, and the possibility of miss-fire can not be excluded. Possible reasons can be statistical effects on the ignition due to the compression end temperatures being near the auto-ignition temperature of hydrogen and the varying mixing conditions. It is likely that there is an influence of the purity of the pressurized hydrogen, the intake-air or of the leakage in the injection system on the ignition. All three possibilities were analyzed. The results show that there is no contamination of the hydrogen compression and of the high pressure storage system. Furthermore, the air supply system also has no contamination. The injection device has an oil sealing system which was suspected to leak impurities of oil into the injected hydrogen.

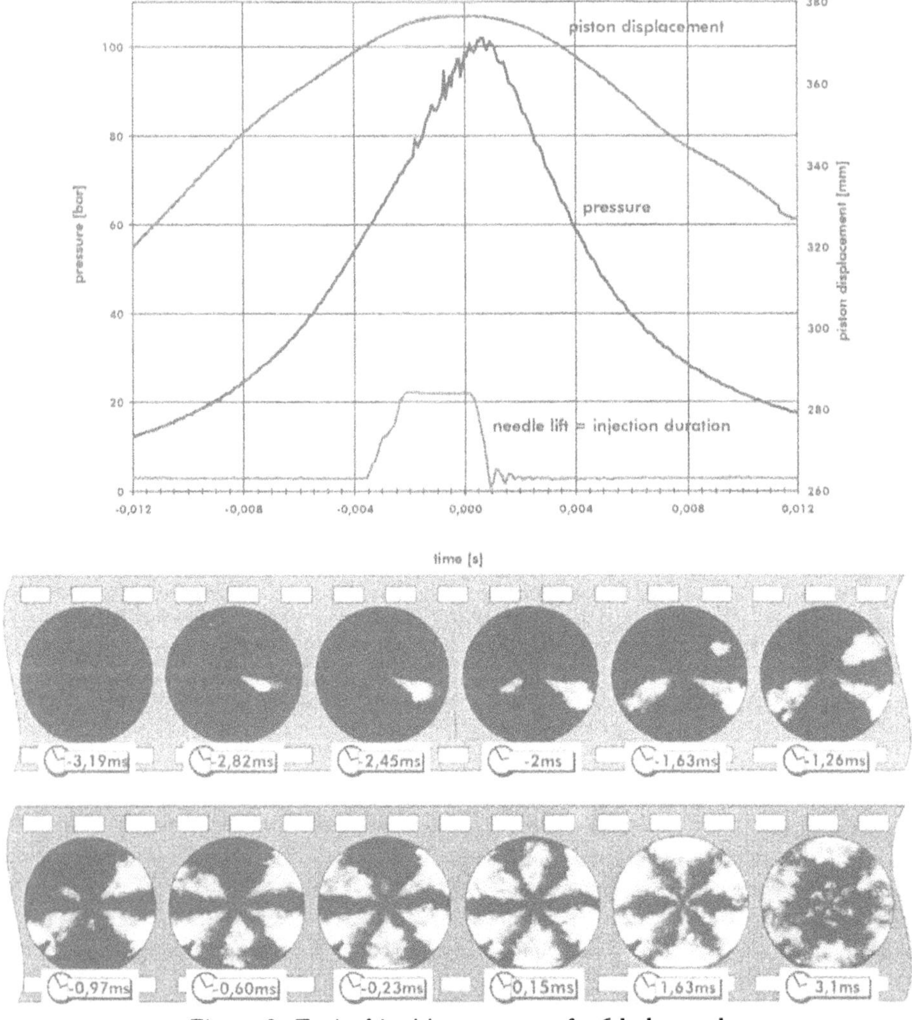

Figure 2: Typical ignition process of a 6 hole nozzle

The installation of another injector with dry sealings has no significant impact on the statistical ignition behavior. But it has to be taken into account that the hydrogen is used as it is delivered from a commercial gas supplier. That means it is delivered as a compressed gas, and it is not as chemically clean as liquid hydrogen (the boiling temperature of liquid hydrogen is 20 K). The phenomenon of auto-ignition, which is difficult to realize, and the variations of the ignition delays were also observed and discussed by other researchers working on the topic [6, 7, 4, 2]. The ignition delays estimated from calculations of hydrogen air mixtures with zero dimensional calculations [9-11] are in a range of 2 ms at compression end temperatures of 1050 K. Siebers [12] reports ignition delays in his experiments of 0.5 ms at temperatures of 1200 K.

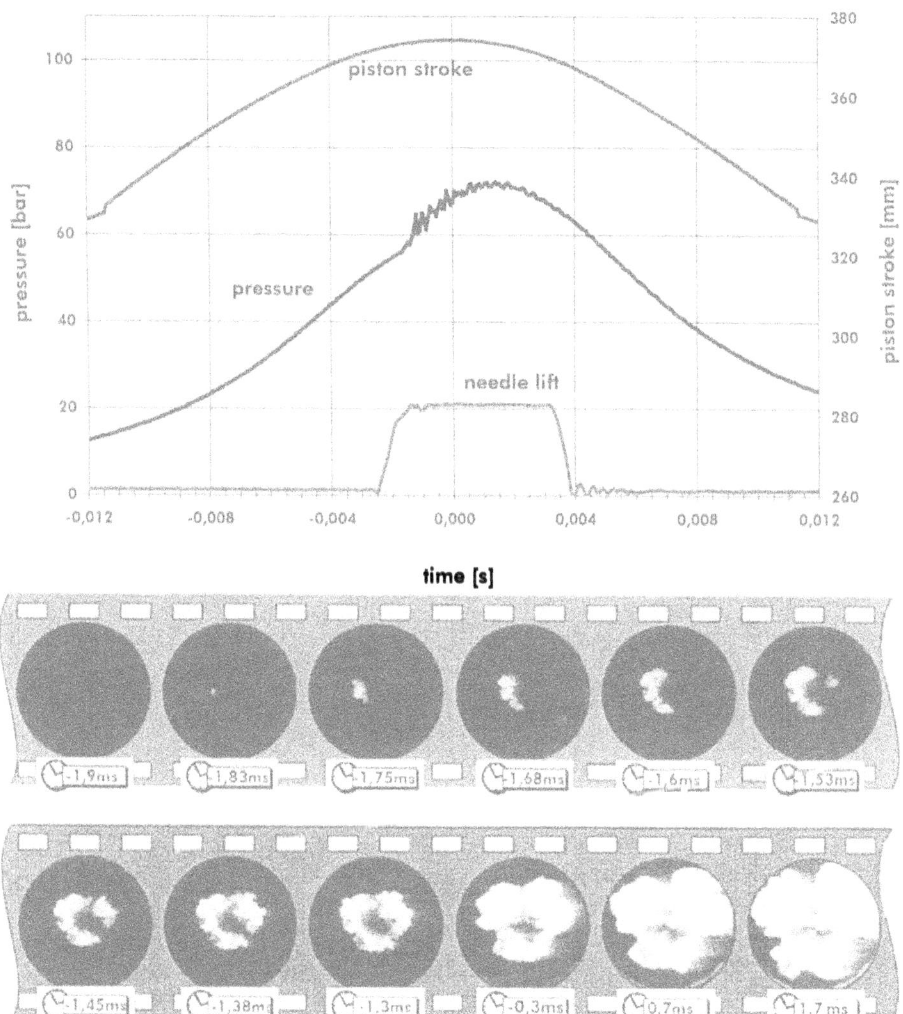

Figure 3:Typical ignition process of a slot nozzle with a slot height of 100 um; flame propagation around the nozzle within 0.4 ms

An enhanced flow field in the combustion chamber, which can be realized through a swirl or a turbulence, improves the reliability of the ignition and reduces the variations of the ignition delays. Due to the critical-flow injection of the hydrogen [13] into the combustion chamber and its high speed in the nozzle outlets of about 1400 m/s (which is the speed of sound), the injected jets are not deflected in the vicinity of the outlets. Owing to the strong slow-down of the injected hydrogen, a significant influence on the penetration direction can be observed at a distance of 50 mm from the nozzle.

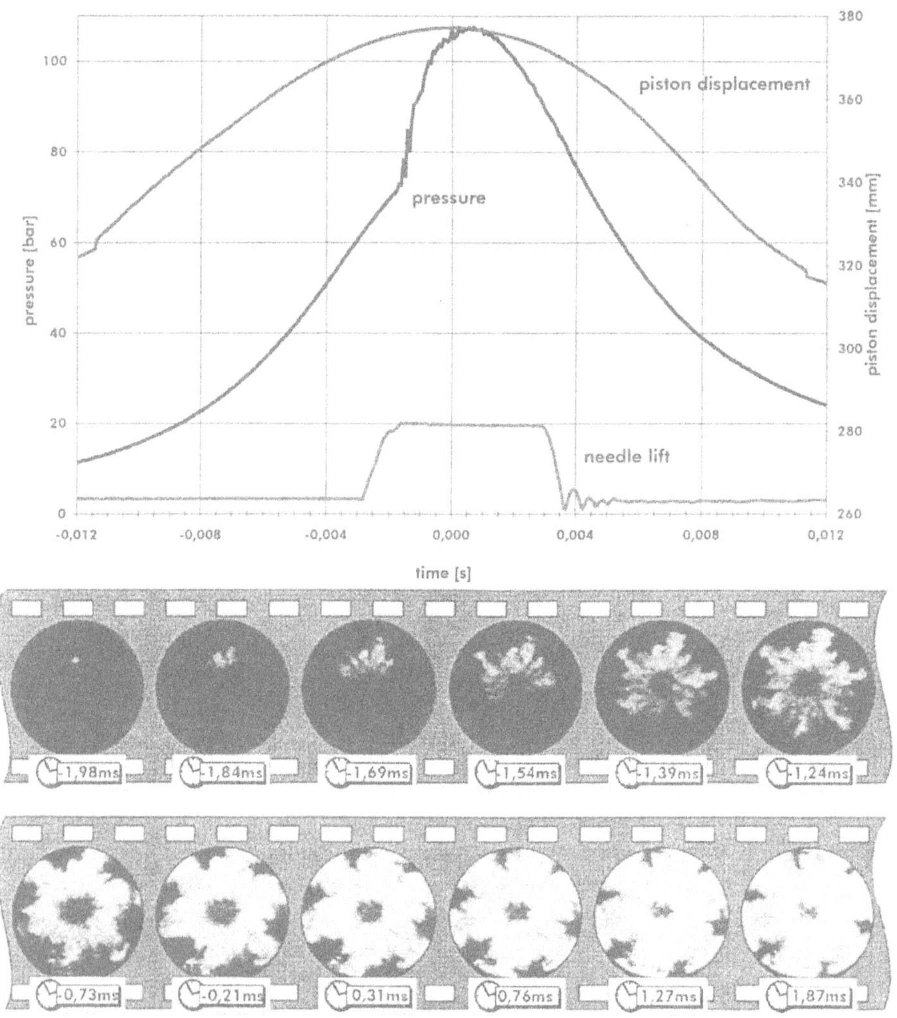

Figure 4: Typical ignition process of a combined nozzle with 6 x 0.6 and 12 x 0.4 mm; fast flame propagation, fast combustion and good spatial use of the entire combustion chamber

In order to run the engine under similar conditions like a diesel engine with a high compression ratio and late gaseous injection, a nozzle geometry with 6 or more holes is set up. All the experiments with hole numbers up to 10 show that a burning jet hardly ignites the neighbor jet. As discussed above, this behavior leads to large variations in the pressure recordings. As a result of these observations a nozzle system with a spatially connected combustible mixture distribution was set up. Experiments with nozzles with 18 holes or a slot show a much better combustion behavior than those with 6 to 10 holes. Once auto-ignition occurs, the flame propagates very fast over all injected jets. Due to the equivalent cross-section of the nozzles the diameter of the holes decreases with the increase of the number of holes. A smaller bore diameter reduces the momentum of the injected hydrogen significantly ($\sim d^2$). Owing to a smaller momentum the jet is slowed down faster and a reduced penetration speed of the jets can be observed. Due to the smaller speed the auto-ignition begins in the vicinity of the nozzle. The best ignition behavior is observed with a slot nozzle. Figure 3 illustrates the ignition process of a nozzle with a slot height of 100 um. Under these conditions an earlier and a better reproducible ignition can be achieved. On the other hand a longer combustion duration is observed. The flow conditions have significant influence on the combustion process. Nozzle geometries with low jet velocities and large ignitable areas have good auto-ignition properties, but induce insufficient conditions for a fast and effective overall combustion. The flow energy for the mixing process is lost in highly throttling nozzles. The nozzles with 6 holes show, in contrast to nozzles with more bore holes, better combustion properties as soon as the hydrogen is ignited. The combination of both advantages leads to the application of nozzles with a series of large and small bores. The existence of the small bores ensure an early and reliable ignition in the vicinity of the nozzle and a fast flame propagation, which ignites the surrounding jets. The large bores, however, cause a high momentum of the injected hydrogen, which leads to good mixing. Figure 4 illustrates the ignition and the combustion process of a combined 18-hole nozzle with 6 x 0.6 mm and 12 x 0.4 mm bores. A fast flame propagation around the nozzle is achieved within 0.6 milliseconds. In addition to that fact a faster penetration of the jets from the large bore holes to the wall of the cylinder is observed, which results in a more efficient use of the whole combustion chamber. Nozzles of these types are, at present, the best combination of a reliable ignition and of an efficient combustion. However, the variations of the ignition processes and the pressure rise rates are higher than those observed in real diesel engines. In addition to that fact an improvement of the ignition of the hydrogen near the nozzle can be expected using slow-opening valves. These types of valves are used in modern combustion concepts of diesel engines with Common-Rail injection systems, with which the pressure rise rates can be controlled and the emissions can be reduced.

Although an operation of a C.I. engine using hydrogen seems to be possible under the above discussed conditions, alternative ignition devices should still be considered. The low ignition energy of hydrogen suggests the use of a spark plug ignition under late internal mixing conditions. The varying auto-ignition behavior of the hydrogen, because of possible impurities, can lead to pre-ignitions under early-injection conditions. Due to this fact S.I. (spark ignition) hydrogen engines have a reduced compression ratio of about 8-10 [14 15, 16]. The possibility of setting up a configuration with late internal mixing and spark ignition, as that used in a new generation of GDI (gasoline direct injection) engines, is also of great interest. For this reason experiments have been set up with a small modification to the cylinder head of the RCM, the insertion of a spark plug. With the same nozzle geometry and a reduced compression ratio, the auto-ignition can be avoided, which enables a reliable and a fast

430

ignition of the hydrogen. Due to the optimized arrangement of the holes in the nozzle, a fast flame propagation around the nozzle is achieved within 1 ms. Figure 5 shows a spark ignited combustion. The spark-plug timing is set 1 ms after the beginning of the hydrogen injection. With a late injection near TDC and an immediate ignition with a spark plug, high compression ratios can be realized, as opposed to external or early internal mixing.

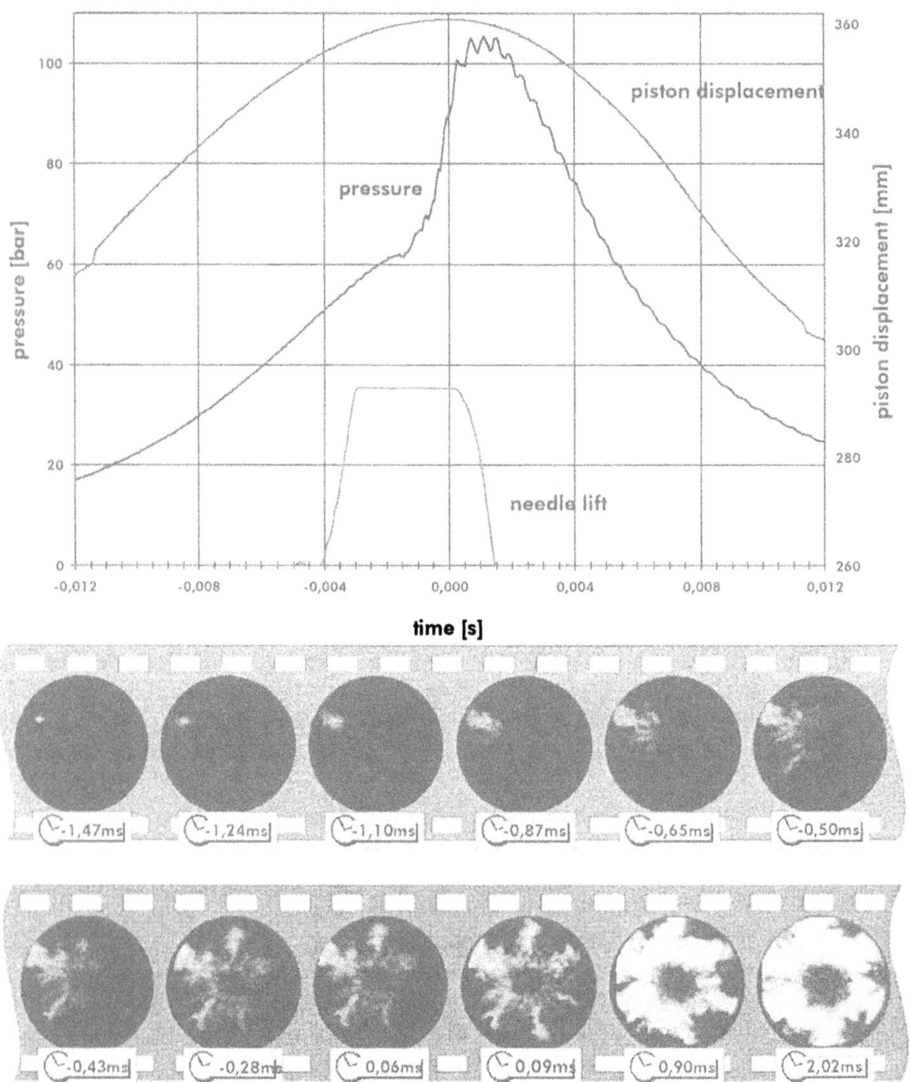

Figure 5: Spark ignition of a late injected hydrogen jet

3. Conclusions

The developed setup for the investigations of a hydrogen-fueled large-bore C.I. engine with a high pressure injection system allows a detailed analysis of all relevant processes. The modern optical techniques enable the analysis of the mixing, the ignition and the combustion processes. The results obtained with this setup contribute to a better understanding of the very fast processes. Furthermore this setup allows an efficient testing and development of engine components. The auto-ignition of hydrogen has been observed. High pressures and temperatures have a positive influence on a short ignition delay. Turbulence and swirl support the propagation of the flame over the whole combustion chamber and regulate the combustion for smoother pressure rates. At present, combined nozzles containing 18 bores with different diameters seem to be the best injection geometry to assure small variations of the ignition delay and of the pressure. However, the current ignition behavior of hydrogen in a C.I. engines without ignition sources is not applicable yet. Therefore further improvement for a smooth and safe engine operation is still required. Modern injection concepts with their possibilities of pilot injection and slow-opening rates offer additional options for improvement. However, further ignition sources for hydrogen under late internal mixing conditions have to be considered, and should be investigated.

References

1. Das L. M., Hydrogen Engines: A view of the Past and a look into the future, Int. J. Hydrogen Energy, Vol. 15, 425-443, 1990
2. Wong J. K. S. Compression ignition of hydrogen in a direct injection diesel engine modified as a low heat-rejection-engine Int. J. Hydrogen Energy, Vol. 15, No. 7, 507-514, 1990
3. TÜV Bayern, Energieträger Wasserstoff 1996, TÜV Bayern Sachsen Forschungsprojekte und Anwendungen, Westendstr. 199, 80686 München, Germany
4. Furuhama S. and Kabayashi Y. Development of a hot surface ignition hydrogen injection two stroke engine Proc. 4 th World Hydrogen Energy Conference Vol. 3, California. 1982
5. Karim G. A., Klat S. R., Experimental and analytical studies of hydrogen as fuel in Compression Ignition Engines. ASME Paper No. 75-DGP-19, 1975
6. Homan H. S., Reynolds R. K., De Boer P. C. T. and Mclean W. J. Int. J. Hydrogen Energy, Vol. 4, 315-325, 1979
7. Ikegami M., Miwa K. Shioji M. A study of hydrogen fueled compression ignition engines Int. J. Hydrogen Energy, Vol. 4, 341-353, 1982
8. Eckbreth A.C.   Laser Diagnostics for Combustion Temperature and Species, Abacus Press, 1988
9. Warnatz J., Maas U. Technische Verbrennung, Springer-Lehrbuch, 1993
10. Maas U., Warnatz J. Ignition Processes in Hydrogen-Oxygen Mixtures, Combustion and Flame 74, 53-69, 1988

432

11. Dorer F., Prechtl P., Mayinger F.   Investigation of Mixture Formation and Combustion Processes in a Hydrogen Fueled Diesel Engine, Hypothesis II, August 1997
12. Siebers D., Naber J. Hydrogen Combustion under Diesel Engine Conditions, XI World Hydrogen Conference, 1503-1512, 1996
13. Stephan K., Mayinger F.   Thermodynamik, 1/2. edition, Springer, Berlin, 1986
14. Digeser, Jorach, Enderle, Daimler Benz AG The intercooled hydrogen truck engine with early internal fuel injection – a means of achieving low emissions and high specific output Hydrogen Combustion under Diesel Engine Conditions, XI World Hydrogen Conference, 1537-1546, 1996
15. Knorr, Held, Prümm, MAN Nürnberg, The MAN hydrogen system for city buses, XI World Hydrogen Conference, 1611-1620, 1996
16. Peschka W, Hydrogen the future cryofuel in internal combustion engines, (DLR, BMW) XI World Hydrogen Conference, 1611-1620, 1996

ENHANCEMENT OF HEAT TRANSFER WITH HORIZONTAL PROMOTERS

S.U. ONBAŞIOĞLU and A. N. EĞRİCAN
Mechanical Engineering Department
Istanbul Technical University
Gumussuyu 80191 Istanbul TURKEY

Abstract. The installation of adiabatic horizontal partition plates between fins at constant intervals has been investigated. Both the partition plate height and the pitch values, the parameters used for the computation, affect the magnitude of the local heat transfer coefficient. Exceeding certain values for the height of the partition plates and the pitches, transition to turbulence has been observed. Since the plates are adiabatic they did not work as extended heat transfer surfaces but as promoters. Thus, the promoters redirecting the flow can be used to enhance the heat transfer in both laminar and turbulent cases. In this study, the flow and temperature fields around the plates have been computed numerically and a correlation between the partition plate height, the pitch and the heat transfer performance has been obtained.

1. Introduction

The vertically installed fins working as extended heat transfer surfaces are not applicable to the tall vertical plates due to the thick boundary layer developed over the fins. The height of the fin attached to the base plate should be larger than the boundary layer, and in the cases of the taller base plates the gap between the fins should be larger to prevent the choking of the boundary layer. The taller the base plate, the larger the fin height and the horizontal pitches are. As the limit, the fins are no longer practical from the viewpoints of both compactness and heat transfer performance.
As effective cooling of electronic devices and transferring heat in other applications become increasingly important, the problem has been realized and much emphasis has been directed to alternative enhancement techniques [1,2].

433

S. Kakaç et al. (eds.), Heat Transfer Enhancement of Heat Exchangers, 433–446.

434

If the partition plates are installed horizontally on the vertical plates, the flow along the vertical plate will be stagnated in front of the partition plates and separated in the rear of these plates. Experimental techniques are used and reported by several researchers [3-6] to examine the effect of the similar plates. The problem is generally called natural convection with large-scale roughness elements. Hung and Shiau [3], who performed their experiments inside a channel, have obtained reasonable differences between the heat transfer rates in the cases with and without the rectangular ribs. However, their results are not applicable to the present problem because the natural convection problem along a vertical plate differs from that within a channel and inside an enclosure. Burak *et al.*[4] have studied experimentally the natural convective heat transfer on a vertical plate with constant heat flux in the presence of one or more steps. They have concluded that the heat transfer process on the step surfaces has been weakened. Bhavnani and Bergles [5] have performed their experiments by using an interferometric technique, and reported that the heat transfer performance decreased as the rib pitch-to-height ratio decreased, and the surfaces with low thermal conductivity ribs had a heat transfer performance improved relative to the surfaces with high thermal conductivity ribs. Their comment on the topic is that the literature seems contradictory with some studies reporting increases of 100% and others reporting no increases or even decreases. Misumi *et al.* [6,7] investigated the problem experimentally and their results showed that it is possible to increase the overall and local heat transfer coefficients by 20 and 40 percent. The computational analysis of their case for a single partition plate has been performed and extended for various plate heights and locations [8]. In the present study, on the other hand, the objective is to use the partition plates as promoters, not as extended surfaces. Since it seems that the main problem resulting in the contradiction is due to the conductivity of the ribs, the use of adiabatic partition plates, called as promoters hereafter, will resolve these issues. Special care should be given to parameters such as the height and the pitch of the promoters.

2. Theoretical Model

The problem under consideration is a vertical plate with repeated horizontal, rectangular promoters (*Figure 1*). The thickness of the promoters (s) is very small compared to the height. Since the transition to turbulence has been predicted, the flow is assumed to be governed by equations with Reynolds stress terms.

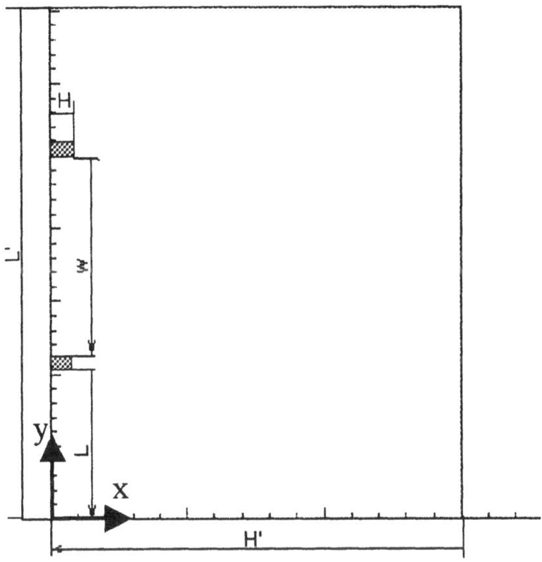

Figure 1. Computational domain.

2.1 GOVERNING EQUATIONS

The governing equations of the buoyant turbulent flow are as following:
Continuity:

$$\frac{\partial U_i}{\partial x_i} = 0 \tag{1}$$

Momentum:

$$\frac{DU_i}{Dt} = \frac{\partial}{\partial x_j}\left[\nu_1 \frac{\partial U_i}{\partial x_j} - \overline{u_i u_j} \right] - \frac{1}{\rho}\frac{\partial P}{\partial x_i} + g_i \tag{2}$$

Energy:

$$\frac{DT}{Dt} = \frac{\partial}{\partial x_i}\left[\alpha \frac{\partial T}{\partial x_i} - \overline{u_i T'} \right] \tag{3}$$

The Boussinessq approximation has been assumed for the equation of state, $\rho = \rho(T)$.

2.2 BOUNDARY CONDITIONS

The computational domain shown in *Figure 1* has three types of boundaries, namely, wall, blockage and freestream boundaries. Boundary conditions can be summarized as:

Walls:

$0 \leq y < L$ and $x = 0$	$u = v = 0$; $-k(\partial T/\partial x)=q''$
$L+s < y \leq L+s+w$ and $x=0$	$u = v = 0$; $\partial T/\partial x=q''$
$L+2s+w < y \leq L'$ and $x=0$	$u = v = 0$; $-k(\partial T/\partial x)=q''$
$0 \leq x \leq H'$ and $y = 0$	$u = v = 0$; $\partial T/\partial y=0$

Blockage:

$L \leq y \leq L+s$ and $0 \leq x \leq H$	$u=0$; $v=0$; $\partial T/\partial x=0$; $\partial T/\partial y=0$
$L+s+w \leq y \leq L+2s+w$ and $0 \leq x \leq H$	$u=0$; $v=0$; $\partial T/\partial x=0$; $\partial T/\partial y=0$

Freestream:

$0 < x \leq H'$ and $y = L'$	$u = v = 0$; T=const.
$0 < y < L'$ and $x = H'$	$u = v = 0$; T=const. $= T_\infty$

Since numerical computations in natural convection gravity flows do not set all initial velocity values to zero, the initial velocities are taken as very small and constant. The initial and ambient temperatures are assumed to be 300 K.

The reason for assuming the bottom wall as adiabatic is for the convenience of comparison with the experimental results of Misumi and Kitamura [6].

2.3 TURBULENCE MODEL

The flow is simulated by means of the standard k-ε model with buoyancy terms included. The general structure of the model is as follows:

$$\frac{Dk}{Dt} = \frac{\partial}{\partial x_j}\left[(\frac{\nu_t}{\sigma_k} + \nu_1)\frac{\partial k}{\partial x_j}\right] + \nu_t(P_k + G_k) - \varepsilon \tag{4}$$

$$\frac{D\varepsilon}{Dt} = \frac{\partial}{\partial x_j}\left[(\frac{\nu_t}{\sigma_\varepsilon} + \nu_1)\frac{\partial \varepsilon}{\partial x_j}\right] + c_{\varepsilon 1}\frac{\varepsilon}{k}(P_k + c_{\varepsilon 3}G_k) - c_{\varepsilon 2}\frac{\varepsilon^2}{k} \tag{5}$$

where P_k and G_k are shear production and buoyancy production/destruction terms, respectively:

$$P_k = -\overline{u_i u_j}(\partial U_i / \partial x_j) = \frac{\partial U_i}{\partial x_j}\nu_t(\frac{\partial U_i}{\partial x_j} + \frac{\partial U_j}{\partial x_i}) \tag{6}$$

$$G_k = -\beta g_i \overline{u_i T'} = \beta g_i \frac{\nu_t}{\sigma_\theta}\frac{\partial T}{\partial x_i} \tag{7}$$

Turbulent viscosity v_t is defined as:

$$v_t = c_\mu k^2 / \varepsilon \tag{8}$$

The boundary conditions for the turbulent variables are assumed to be $\partial k/\partial n = \partial \varepsilon_e/\partial n = 0$ at the wall and blockage-type boundaries, and $k = \varepsilon = 0$ on the freestream type boundaries. The model constants are: $c_\mu = 0.09$; $c_{\varepsilon 1} = 1.44$; $c_{\varepsilon 2} = 1.92$, $c_{\varepsilon 3} = 0.6$. $\sigma_k = 1.0$; . $\sigma_\varepsilon = 1.0$; . $\sigma_\theta = 0.8$

3. Computational Technique

Computations are performed by a commercial code, CFD2000. The code has been customized for the present study by discretization of the governing equations on a collocated grid arrangement and using the finite-volume technique suggested by Peric *et al.* [9].
Convection is formulated by hybrid and diffusion terms, using an arithmetic mean scheme. The solution algorithm is the PISO (Pressure Implicit with Splitting of Operations), which is basically a time-marching procedure. This algorithm, which is developed by Issa [10], is similar to the SIMPLER algorithm of Patankar [11].   However, unlike SIMPLER, which is semi-implicit in the pressure-velocity coupling, PISO is fully implicit, and by this method the fields obtained at the end of each step are close approximations of the exact solutions. Incomplete factorization for the pressure and Alternate Directions Implicit (ADI) method for the other equations are used. Incomplete factorization follows the LU decomposition of the direct solver and performs better than ADI for the solution of the pressure equation. The ADI method is a semi-iterative one, in which the equations are solved at each step by maintaining the full implicitness in one direction.

4. Discussion

4.1 VALIDATION OF THE CODE

Comparison with the measurements reported by Bhavnani and Bergles [5] has validated the code. For this comparison, the flow and temperature fields along a ribbed vertical plate have been computed. For the computational case the ribs with low thermal conductivity and with a rib height-to-pitch value of 8 have been considered. *Figure 2* illustrates that the computed profiles compare well with the available experimental data.

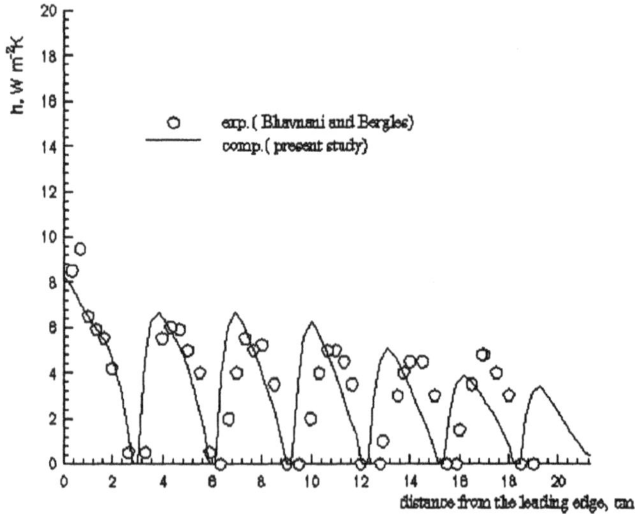

Figure 2 Comparison of the code predictions with the experimental data reported in the literature [5].

4.2 RESULTS

In *Figure 3* the local heat transfer coefficients for various promoter height and pitch values are illustrated. For the sake of simplicity, the location of the lower promoter (L) has been kept constant for each case. The effect of the promoter height has been observed at constant pitch value in *Figure 4*. Compared to the case without promoters, it is seen that the heat transfer performance is always improved above the upper promoter. If *Figure 3* is taken into consideration with the streamline diagrams and temperature maps (*Figures 5 and 6*) of the corresponding cases, it is concluded that each promoter causes main reverse flow zones. These zones which occur adjacent to the intervals of the promoters, thereby increase the rate of the convection. However as the reverse flow moves downwards, the reattachment of high temperature fluid flow to the plate instead of the low temperature one will decrease the heat transfer coefficient. The circulation zone occurring above the upper promoter will fall towards the promoter, and the similar enhancement effect will be observed in this region.

If the space between the promoters is not large enough these circulation zones are not observed. These zones are due to the separation and the falling down of the high temperature fluid from the reattaching region. Thickening the boundary layer, this separation at first decreases the local heat transfer coefficient and then since the low temperature flow entering the separation zone gets accelerated, the heat transfer rate increases. A larger pitch causes the regions of high heat transfer to be larger than the decreasing heat transfer regions. Thus replacement of the low temperature fluid into the space between the promoters increases the heat transfer performance and maintains the heat transfer coefficient greater than the one for the plate without promoters.

In the upstream region and the region between the two promoters, the increase in heat transfer coefficient depends also on the promoter height. One of the reasons of this dependence is that as the promoter height increases, the pitch value gets larger and the flow will attach to the upper promoter, causing a secondary flow on the lower one.

The second effect of the promoter height is related to the turbulent behavior of the motion. As the height of the promoter increases, the separation zones which redirect the high temperature fluid outwards get larger and the flow becomes turbulent. This effect is further clarified in *Figure 4*, illustrating the cases for constant pitch values instead of the constant height-to-pitch values.

Computations were made for the cases of only two promoters on a flat plate of 0.35 m., which is the experimental test case, for a single partition plate, described by Misumi and Kitamura [6]. Since the height of the flat plate is not enough to show the effect of higher pitch values such as the ones for the promoter height-to-pitch values of 70 mm/ 350 mm, 50 mm/350 mm and 70 mm/490 mm, these cases are presented separately in *Figure 7*. The characteristics of the curves in this figure imply that when certain pitch and promoter height are exceeded, no improvement in the heat transfer performance can be observed.

The correlation between the geometric parameters and the heat transfer performance, computed in this study, have been obtained for laminar and turbulent flow cases as follows:

$$Nu = 55 \frac{1}{Ra_*^{1/10}} \left(\frac{w}{H} \right) \qquad Ra_* < 10^{10} \qquad (9)$$

$$Nu = 0.2 \, Ra_*^{1/5} \left(\frac{w}{H} \right)^{1/2} \qquad Ra_* \geq 10^{10} \qquad (10)$$

where Ra_* is the modified Rayleigh number based on the pitch:

$$Ra_* = \frac{g\beta q'' w^4}{k\alpha v} \qquad (11)$$

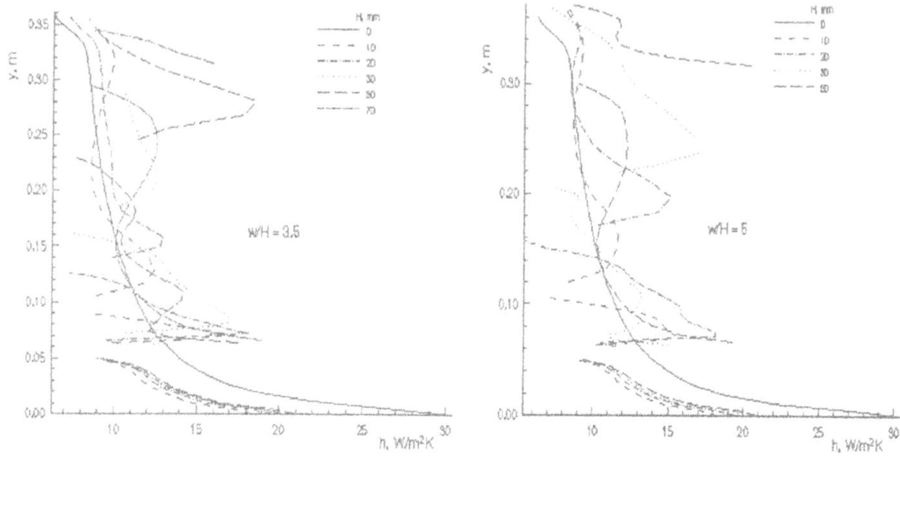

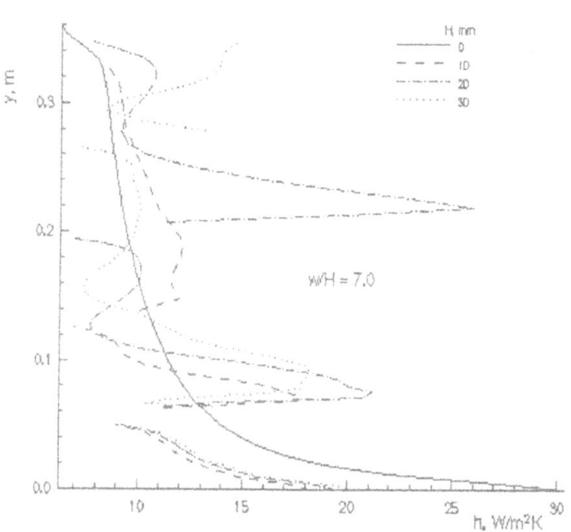

Figure 3 Effect of promoter height and the pitch value on the local heat transfer coefficient.

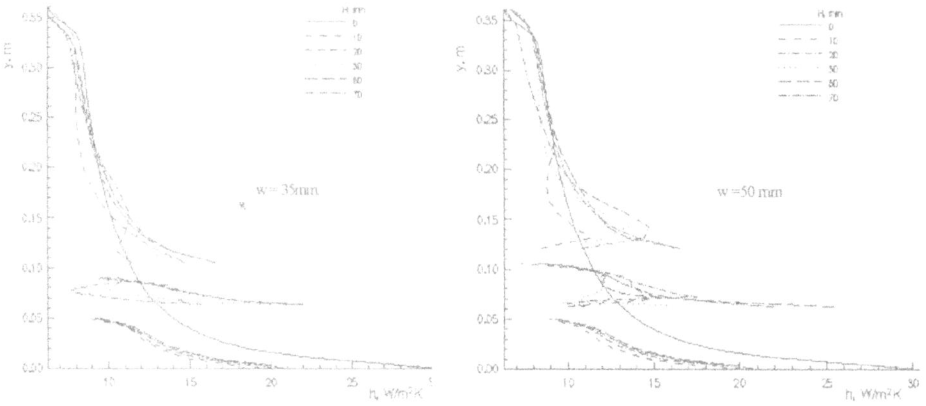

Figure 4 Effect of promoter height on the local heat transfer coefficient at constant pitch.

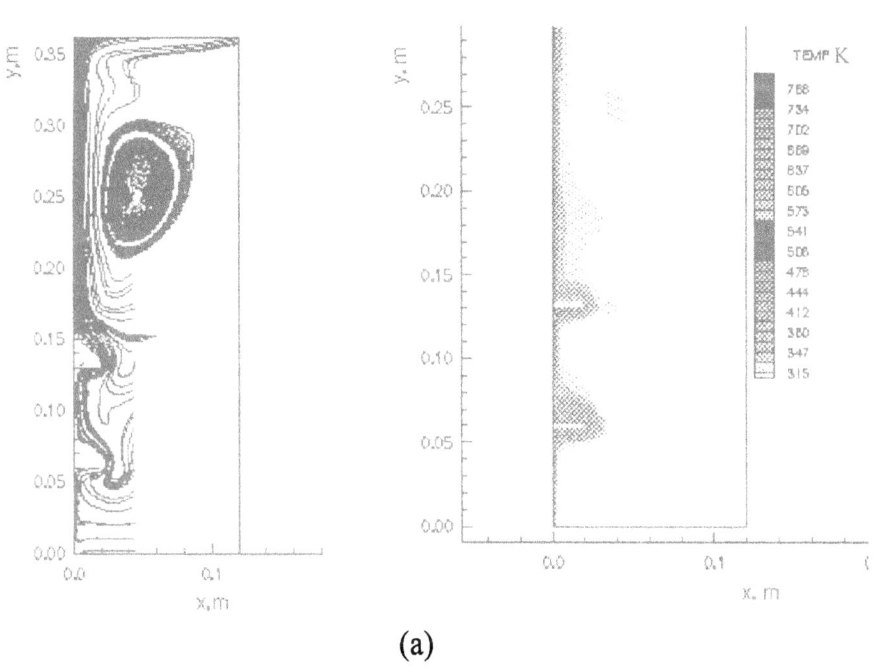

(a)

Figure 5 Effect of pitch on the flow and temperature fields for H = 20 mm. a) w/H = 3.5

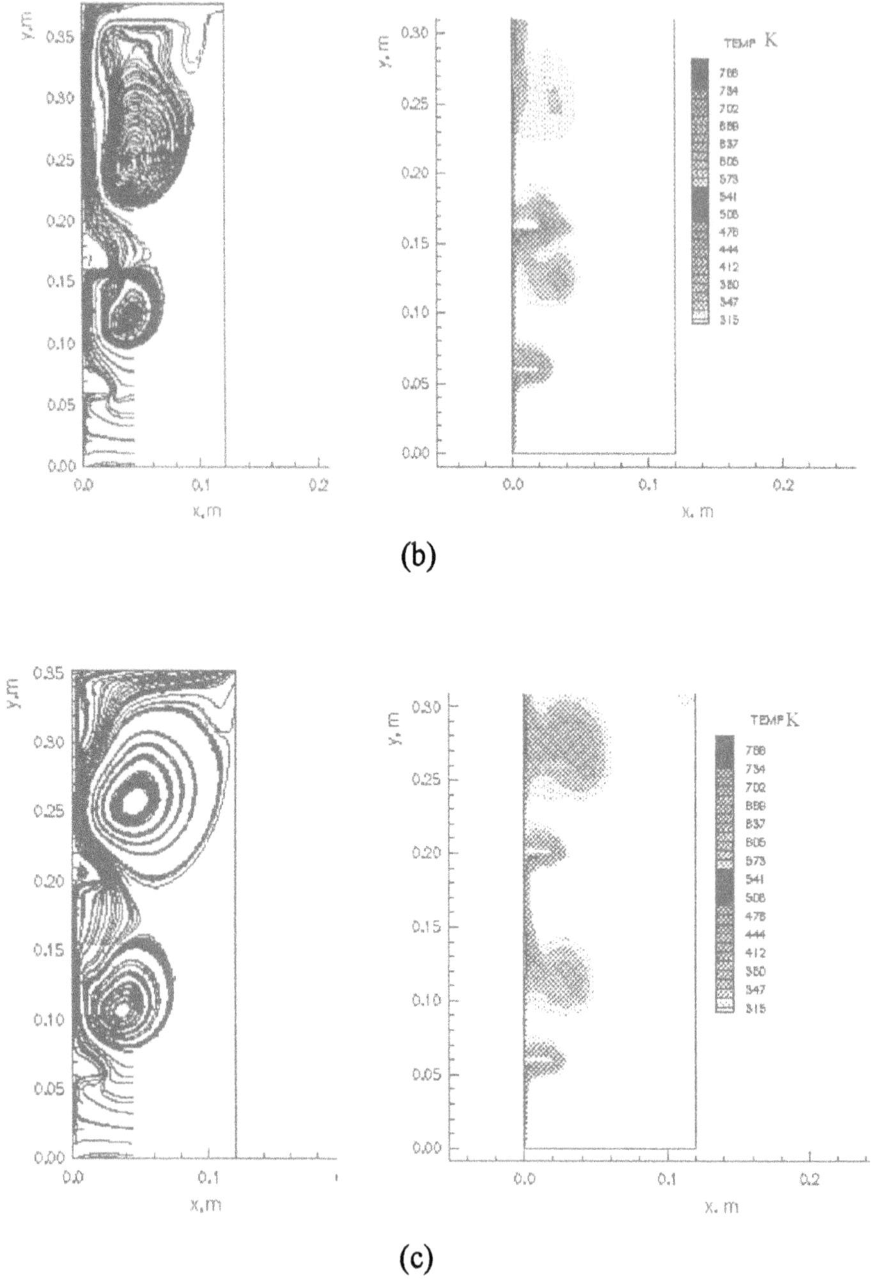

Figure 5 continued. b) w/H = 5. c) w/H=7. (Streamlines on the left, temperature contours on the right.)

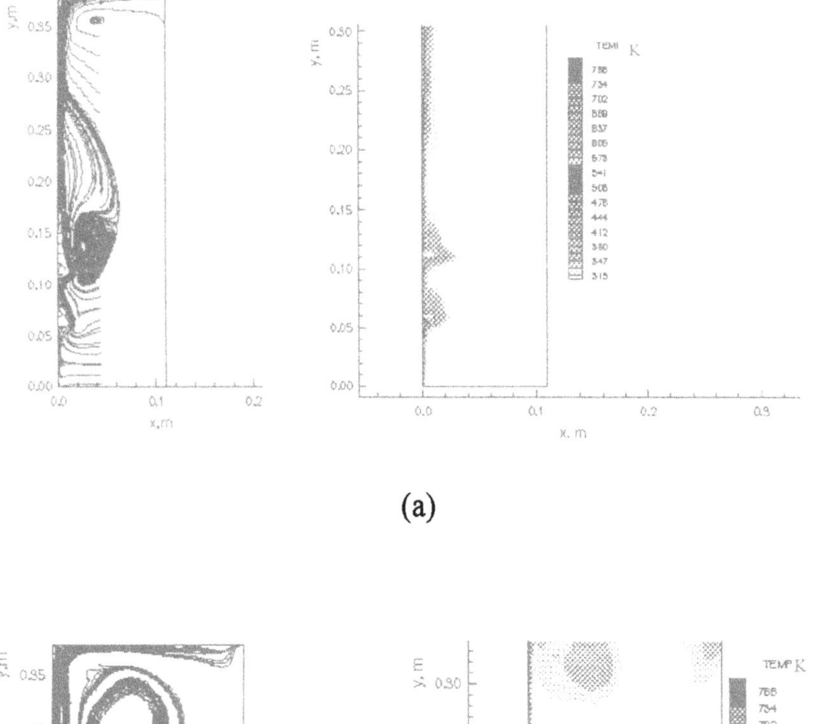

(a)

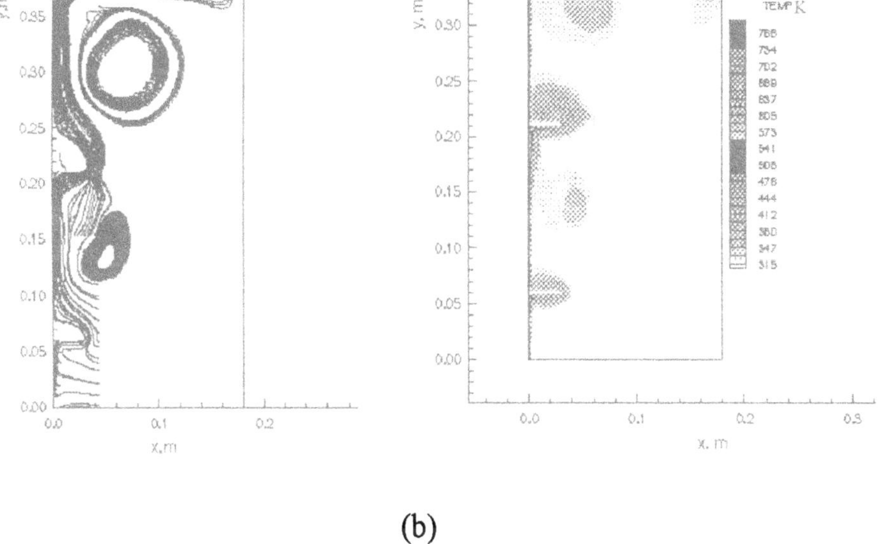

(b)

Figure 6 Effect of the promoter height on the flow and temperature fields for w/s = 5
a) *H =10 mm. b) H = 30 mm. (Streamlines on the left, temperature contours on the right.)*

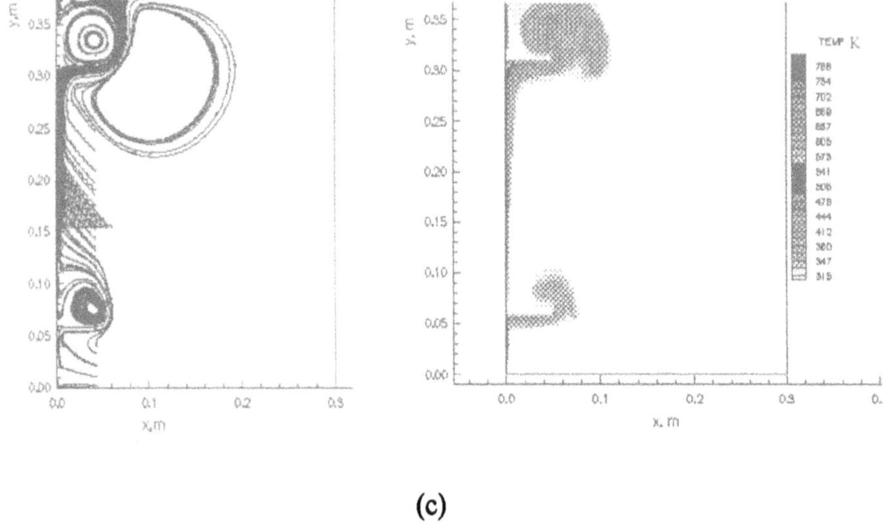

(c)

Figure 6 continued. c) H = 50 mm. (Streamlines on the left, temperature contours on the right.)

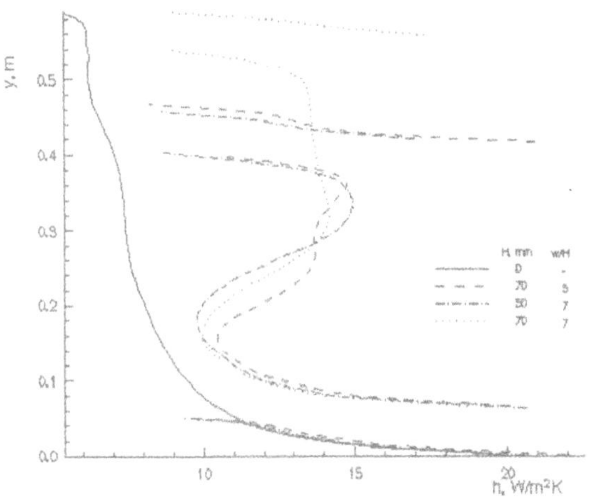

Figure 7 Change of local heat transfer coefficients for larger promoters and pitches.

4.3 CONCLUDING REMARKS

The present paper reports the numerical predictions of natural convective heat transfer from a flat plate on which the horizontal promoters are installed at various intervals and heights. The promotion is realized by redirecting the high temperature boundary layer flows adjacent to the heated vertical plates and by introducing the low-temperature fluid flow into the near wall region. The separation and reverse flow zones achieve the redirection.

The main points related to the heat transfer improvement are summarized as follows:

i-   The increased heat transfer rate along the vertical fins attached to the vertical plates is sensitive to the partition plate orientation and material. Since the high conductivity of these plates decreases the heat transfer coefficients by lowering the fluid velocity, adiabatic promoters should be installed instead of enhanced surfaces.

ii-   The space between these promoters should be held large enough to allow the high temperature fluid to reverse. Thus the promoters should be thin, which implies that their orientation should be horizontal.

iii-   Since the height of the promoters is effective on increasing the number of separation zones and enlarging the space occupied by these zones, this variable plays an important role on the local heat transfer coefficients.

REFERENCES
[1] Kitamura, K., Nagae, N., and Kimura, F. (1994) Enhancement of Natural Convection Heat Transfer from a Horizontal Heated Plate Using Grid Fins, *Heat Transfer- Japanese Research*, **23**, No. 8, 756-768.

[2] Hung, Y.H., and Shiau, W.M. (1988) Local Steady-State Natural Convection Heat Transfer in Vertical parallel Plates with a two Dimensional Rectangular Rib, *Int. J. Heat Mass Transfer*, **31**, 1279-1288.

 [3] Hung, Y.H., and Shiau, W.M. (1989) An Effective Model for Measuring Transient Natural Convective Heat Flux in Vertical Parallel Plates with a Rectangular Rib, *Int. J. Heat Mass Transfer*, **32**, 863-871.

[4] Burak, V.S., Volkov, S.V., Martynenko O.G., Khramtsov, P.P., and Shikh I.A. (1995) Experimental Study of Free-Convective Flow on a Vertical Plate with a Constant Heat Flux in the Presence of one or More Plates, *Int. J. Heat Mass Transfer*, **38**, No. 1, 147-154.

[5] Bhavnani, S. H and Bergles, A. E. (1990) Effect of Surface Geometry and Orientation on Laminar Natural Convection Heat Transfer from a Vertical Flat Plate with Transverse Roughness Elements, *Int. J. Heat Mass Transfer*, **33**, No. 5, 965-981.

[6] Misumi, T., and Kitamura, K. (1990) Natural Convection Heat Transfer from a Vertical Heated Plate with a Horizontal Partition Plate, *Heat Transfer- Japanese Research*, **19**, No. 1, 57-72.

[7] Misumi, T., and Kitamura, K.(1994) Enhancement Technique for Natural Convection Heat Transfer from Vertical Finned Plates, *Heat Transfer- Japanese Research*, **23**, No. 6, 513-524.

446

[8] Onbasioglu, S.U. (1998) Enhancement of Heat Transfer from Vertical Fins with Horizontal Promoters, submitted to *Journal of Heat Transfer*.

[9] Peric, M., Kessler, R., and Scheuerer, G. (1988) Comparison of Finite Volume Numerical Methods with Staggered and Collocated Grids, *Computers and Fluids*, **16**, 389-403.

[10] Issa, R.I.(1985) Solution of the Implicit discretized Fluid Flow Equations by Operator Splitting, *Journal of Computational Physics*, **62**, 40-65.

[11] Patankar, S.V.(1980) *Numerical Heat Transfer and Fluid Flow*, Hemisphere Publishing Corporation, NY.

NOMENCLATURE

g_I	gravitational accelaration, m/s^2
H	heat transfer coefficient, W/m^{2}K
H	partition plate height, m
H'	fin height,m
K	thermal conductivity, W/mK
L	location of the partition plate, m
L'	width of the computational domain
N	normal direction
Nu	Nusselt number, hL'./k
q''	heat flux through the fin (W/m^2)
Ra$_*$	modified Rayleigh number based on w $(g\beta q''w^4 Pr/kv^2)$
S	thickness of the promoter
t	time, s
T	temperature, K
u_I	fluctuating velocity, m/s
U_i	mean velocity, m/s
w	pitch, m.
x	horizontal coordinate
y	vertical coordinate
β	thermal expansion coefficient, 1/K
v	kinematic viscosity, m^2/s
ρ	density, kg/m^3

Subscripts

l	laminar
t	turbulent

THE EFFECT OF AUGMENTED SURFACES ON TWO-PHASE
FLOW INSTABILITIES

SADIK KAKAÇ
Department of Mechanical Engineering
University of Miami
Coral Gables, FL 33124

Abstract. The effect of different heater surface configurations on two-phase flow instabilities has been investigated in single channel, forced convection , up-flow and horizontal systems. Freon-11 is used as the test fluid, and different heater tubes with various surface configurations have been tested at different heat inputs. The effect of augmentation on flow instabilities is studied with three different tubes made of different inner surfaces, at various inlet temperatures. All experiments are carried out at constant system pressure and exit restriction. Steady-state characteristics of the system composed of different tubes are found and presented in pressure-drop versus mass flow rate diagrams. Dynamic instabilities, such as pressure-drop-type, density-wave-type instabilities and thermal oscillations are found to occur in every system, and boundaries for the appearance of these oscillations are found. The dependence of the characteristics of these oscillations on the augmented surfaces is emphasized, and comparisons between the bare tube and the augmented tubes are made.

1. Introduction

Two-phase flow instabilities may cause operational and safety problems in heat exchange equipment involving two-phase flow, such as boiling water nuclear reactors, steam generators, boilers, evaporators, and condensers.

Research has been carried out in the field of boiling two-phase flow instabilities with either vertical or horizontal in-tube flow boiling systems. More information can be found in [1-14].

Lin et al. [4] did experimental and analytical studies on pressure-drop type instabilities in a horizontal hairpin tube. A homogenous phase equilibrium model for pressure-drop type instabilities was adopted, and computer programs for steady and unsteady states were developed. Mentes et al. [7] carried out experimental research to investigate the effects of heat transfer augmentation and inlet subcooling on two-phase flow instabilities in a vertical boiling channel.

Lin et al. [13] studied the effect of surface configurations on heat transfer and proposed a correlation. The details of the effect of augmented surfaces on two-phase flow instabilities in horizontal and vertical systems can be found in [7] and [14].

Wang et al. presented experimental results of pressure-drop-type

447

S. Kakaç et al. (eds.), Heat Transfer Enhancement of Heat Exchangers, 447–465.
© 1999 *Kluwer Academic Publishers.*

oscillations in a medium-high pressure system. They studied the effect of different ratios of channel-to-exit restriction diameters. The result was that when this ratio was increased, the magnitude of the negative slope also increased and made the system more unstable. Kakaç et al. [3] performed an experimental study on thermal instabilities in forced convection boiling. Experiments with various heat inputs and inlet temperatures were conducted, and a theoretical model was developed to predict the two-phase flow instabilities in a single-channel up-flow system. Liu and Kakaç[3] investigated two-phase flow instabilities in a single channel, forced convection up-flow system, and compared the data for different augmented surfaces. Different heater tubes were tested at four different inlet temperatures with constant heat input. In this paper they defined a thermal oscillation as one that accompanies the pressure-drop type oscillations. Padki et al.[9] developed a numerical model to predict the steady-state characteristics of the forced convective two-phase flow and the pressure-drop-type and thermal oscillations in a single channel. Xiao [12] carried out an experimental study on a steam-water high pressure system and determined the steady-state characteristics and two-phase flow instabilities. Ding [1] performed an experimental work on two-phase flow instabilities in a horizontal in-tube boiling system. He studied the effect of system pressure, heat input, inlet temperature and exit restriction on steady-state characteristics. Dynamic instabilities, oscillation boundaries and their dependence on various system parameters were also investigated. Some other relevant work and recent research on dynamic instabilities were done by Parros et al. [10], Liu [6], and Ozawa et al. [8].

In practical applications, however, it is not always possible to work in the stable regions. It is the purpose of this paper to show the effect of tube surface augmentation and inlet subcooling on the magnitude of amplitudes and periods of system parameters, such as inlet pressures and wall temperatures, in vertical and horizontal in-tube flow boiling systems. Knowing the characteristics of the oscillations and the boundaries, important conclusions can be drawn for the safe operation of various types of two-phase flow heat exchanger systems.

2. Experimental System and Procedure

2.1 EXPERIMENTAL SYSTEMS

Schematic diagrams of the experimental set-ups are shown in Figs. 1 and 2. The set-ups were specially designed and built to generate the main types of two-phase flow oscillations, and allow the changing of the heater tubes without disturbing other parts of the system.

Basically the experimental set-ups are forced convection boiling systems operating between constant inlet and exit pressures. The loops can be divided into three parts; **fluid supply system, test section** and the **recovery section**. Fluid supply system includes a main tank, a filter, an inlet control valve and a subcooling controller, installed in that order. A pressurized nitrogen supply connected to the main tank provides the necessary pressure to pump the liquid through the system. Two series of experiments were conducted at different main tank pressures of 5.15 bars and 6.53 bars.

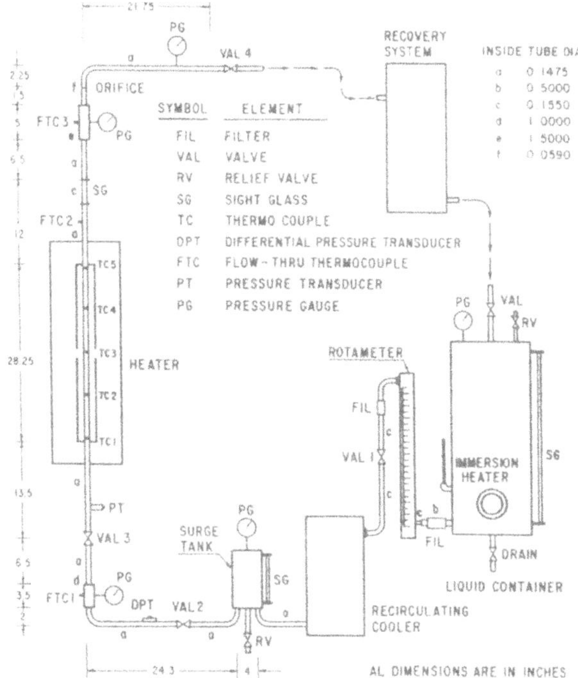

Figure 1. Schematic drawing of the vertical experimental system.

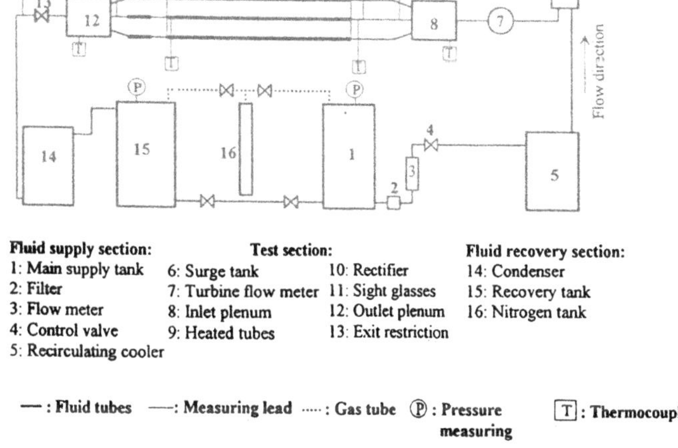

Fluid supply section:
1: Main supply tank
2: Filter
3: Flow meter
4: Control valve
5: Recirculating cooler

Test section:
6: Surge tank
7: Turbine flow meter
8: Inlet plenum
9: Heated tubes
10: Rectifier
11: Sight glasses
12: Outlet plenum
13: Exit restriction

Fluid recovery section:
14: Condenser
15: Recovery tank
16: Nitrogen tank

— : Fluid tubes   — : Measuring lead   ⋯⋯ : Gas tube   (P) : Pressure measuring   [T] : Thermocouple

Figure 2. Schematic diagram of the horizontal experimental set-up.

The test section is the part of the loop where two-phase flow oscillations occur under controlled conditions. It includes a surge tank, an inlet throttling valve, the heater tube, and an exit restriction. The main body of instrumentation is also clustered in this section.

The recovery section consists of a condenser and a recovery tank. The recovery tank is kept open to atmosphere to ensure a constant level of exit pressure.

Flow rate and heat input, along with pressures and temperatures at appropriate locations, are measured during each run. Temperature measurements are made by standard Copper-Constantan thermocouples. Two flow-through thermocouples, one before the heater and one after the heater, are used to measure the fluid bulk temperatures at these locations. Pressures are measured at four different points by Bourdon type pressure gauges. In addition to those, a strain gauge type pressure transducer is used to record pressure variations at the heater inlet. Flow rate measurements are made using a calibrated rotameter for mean values and a differential pressure transducer for instantaneous variations in mass flow rate at the heater inlet.

A four-channel, amplifier-recording unit was used to record the mass flow rate, inlet pressure, exit fluid temperature, and any one of the heater wall temperatures continuously. The heat input to the system is obtained by measuring the current and the voltage drop across the heater tube.

Six different heater tubes, which are shown in Fig.3 were prepared and used during the experiments with vertical system. Each heater was made of 7.493 mm inside and 9.525 mm outside diameter, 605 mm long Nichrome tubes. Tubes are classified according to their effective diameters, which is defined as:

$$d_e = \sqrt{\frac{4V}{\pi L}} \tag{1}$$

where, V is the net inside volume and L is the length of the heater tube.

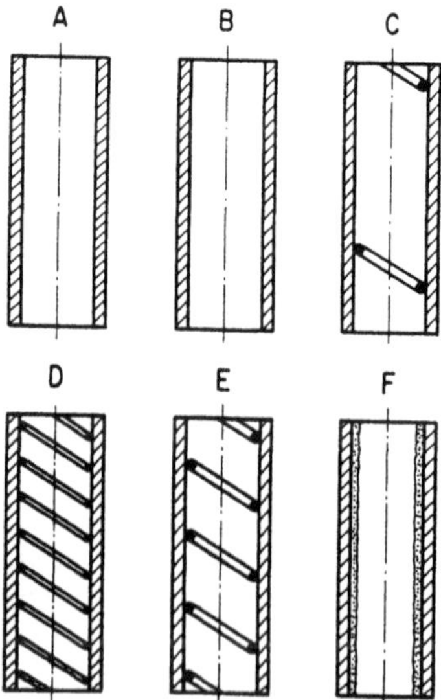

Figure 3. Heater Surface Configurations for vertical systems.

Tube A is the bare tube, which is used for comparison purposes. Descriptions of the tubes are given in Table 1.

Table 1 Description of the Heater Tubes for the vertical system

Tube	Description of the tube	d, mm
A	Bare	7.493
B	Threaded, 7.938 mm - 16 threads per 25.4 mm	7.619
C	With internal spring of 0.794 mm wire diameter and 19.05 mm pitch	7.446
D	With internal spring of 0.432 mm wire diameter and 3.175 mm pitch	7.401
E	With internal spring of 1.191 mm wire diameter and 6.350 mm pitch	7.192
F	Coated with: Union Carbide Linde High Flux	7.073

The transparent part of the systems was used to observe and to take pictures of the two-phase flow system behavior during operation.

More detailed information on the set-up and the instrumentation can be found in Ref. [1-7].

3. Experiments

For each heater tube, six different experiments, one without heat input, and five with different heat inputs, were carried out. Each experiment was composed of a sufficient number of tests to cover a wide range of boiling regimes from single-phase liquid up to very high qualities.

Oscillations were identified by the cyclic variation of the test-section pressure, the mass flow rate and the temperatures. Instability thresholds were located by disregarding short life transients and considering only sustained oscillations [7].

3.1 EXPERIMENTAL PROCEDURE

The experimental procedure for vertical and horizontal systems is as follows: for each heater tube, an initial experiment was conducted without heat input to find the single-phase characteristics. Experiments with heat input were started with high mass flow rate and continued to lower mass flow rates to cover the whole boiling region.

The test procedure can be outlined as follows:
1. With enough liquid in the main tank, the tank was pressurized using nitrogen gas.
2. The surge tank was half filled with liquid Freon-11 and pressurized to a predetermined value by nitrogen gas.
3. Flow rate and heat input were increased gradually to the desired starting point, and the system was allowed to become steady, as indicated by recordings of system pressure, temperature and flow rate.
4. Measurements of temperature, pressure, flow rate and heat input were taken and critical observations were noted.
5. The mass flow rate was reduced by a small amount using the inlet control valve and the system was allowed to become steady, and the readings were taken. This procedure was repeated starting

from step 4 until sustained oscillations were observed. After reaching the unstable region, mass flow rate was first increased and then decreased very slowly to locate the boundaries.

6. While operating in the unstable region, first recordings of the flow rate, the heater inlet pressure, the heater exit fluid temperature and a selected heater wall temperature were made. Then, the system was stabilized by closing the inlet throttling valve slowly and readings were taken. After taking the readings, the inlet throttling valve was brought into full open position and the mass flow rate was reduced by a small amount.

The above procedure was repeated for each tube and heat input. Experiments were stopped after reaching dry-out. A continuous film was made during some of the tests, while the recorder simultaneously kept a continuous history of the flow variables during the filming.

3.2 EXPERIMENTAL RESULTS

Two basic types of oscillations, namely the pressure drop type and the density-wave type, and various forms of superimposition of these oscillations were observed during the experiments. Experimental results are presented in graphical forms. For Steady-state characteristics of the various tubes were obtained with the system pressure drop versus mass flow rate as coordinates. Oscillation boundaries are also indicated on the figures. These are the typical sets of curves showing the total system pressure drop versus mass flow, which is parabolic with zero heat input and S-shaped for two-phase flow for various values of the heat input [7].

Boundaries for the pure pressure drop and superimposed oscillations for different tubes are collected in Figs. 4 and 5. An examination of these figures shows that the curves for different heater tubes exhibit a similar pattern, with internally springed tubes having narrower unstable regions. The order of tubes is the same for both figures.

An examination of Figs. 4 and 5 and Table 1 shows that that the heater tubes with internal springs, namely tubes C, D, and E, have narrower unstable regions in this order and thus, more stable than the others which are bare, threaded, and coated ones. The order of the curves in Figs. 4 and 5 suggests that for the tubes with internal springs, the stability decreases with decreasing effective diameter, but there is not a consistent pattern for the other tubes.

From the investigation of Figs. 4 and 5 one can deduce that among the tested tubes, the tube with the Linde High Flux Coating is the most unstable one. For this tube, for every heat input, oscillation boundaries were located at higher mass flow rates than the other tubes, but on the other hand, the amplitudes of the pressure-drop and the density-wave oscillations were smaller than those for the other tubes tested.

Heat transfer coefficients change during the oscillation and among the tubes. Comparing the bare tube with the other tubes, it has been found out [8] that the time average local heat transfer coefficients during oscillations increased 8% for different spring configurations, about 20% for internally threaded tube, and up to 94% for the coated tube (Fig. 6).

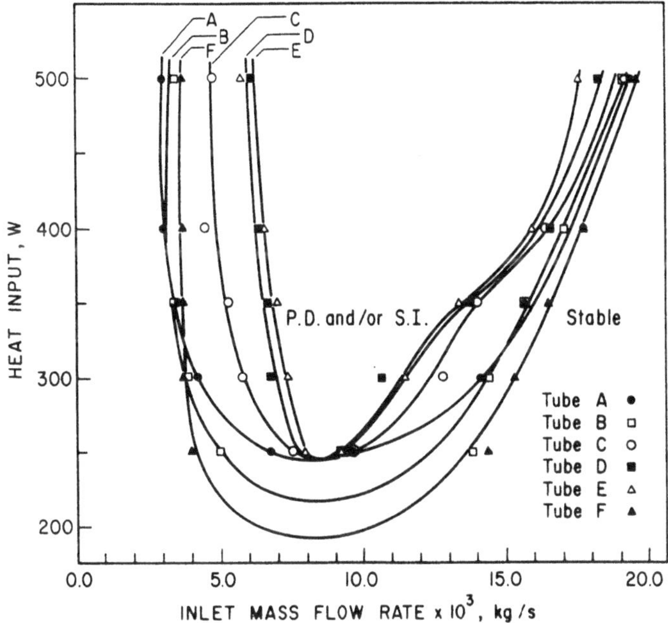

Figure 4. Boundaries of the pressure-drop type oscillations in the vertical system.

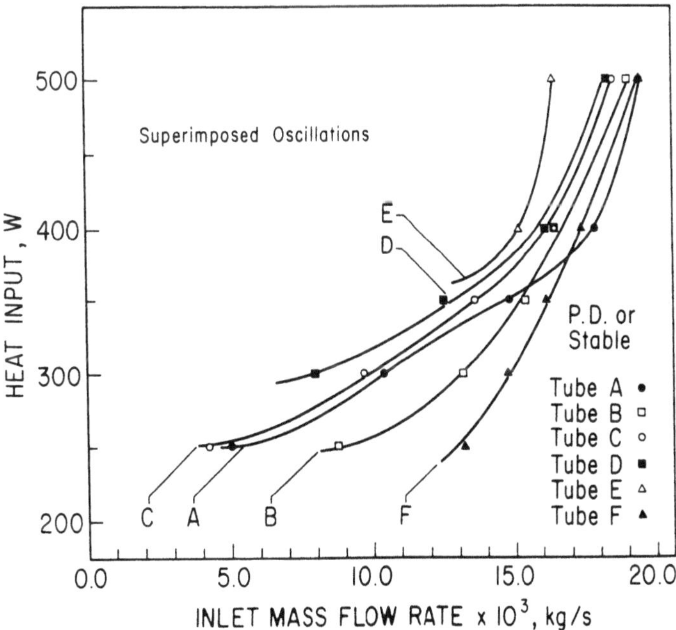

Figure 5. Boundaries of the pressure-drop type oscillations with superimposed
density-wave type oscillations in the vertical system.

454

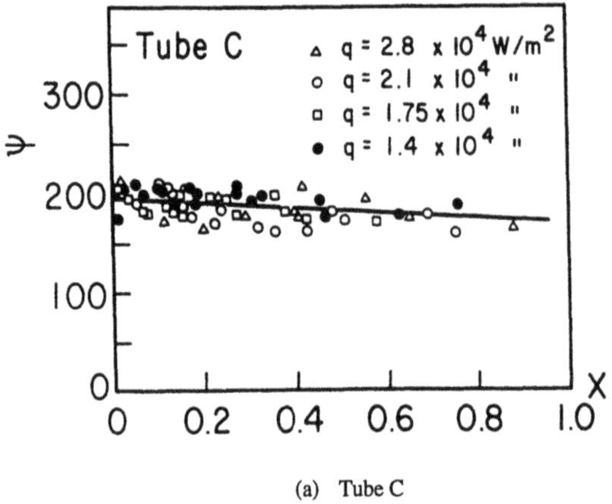

(a) Tube C

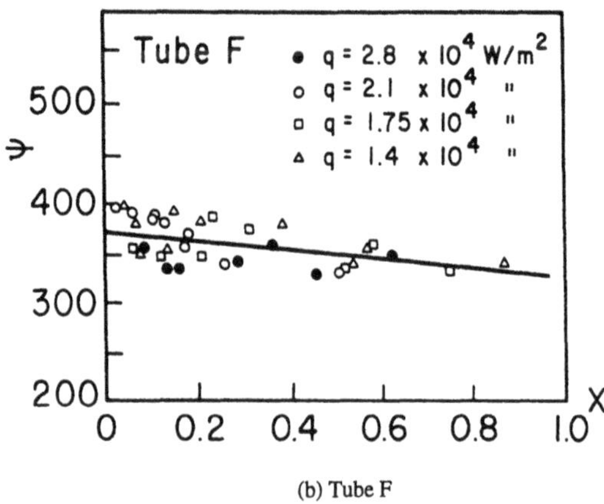

(b) Tube F

Figure 6. Variation of $\psi = Nu / [(qD_e / \Delta h_v \mu_l)^{0.7}(P / P_c)^{0.25}]$ with quality.

Another interesting observation is the variation in oscillation periods. For the density-wave oscillations, every tube had similar periods, but for the superimposed oscillations, there was a great difference among the tubes although the amount of air in the surge tank was kept constant for each tube. The period of the oscillations in Tube E was more than four times the period of the oscillations in Tube F.

These interesting behaviors may be due to the wave propagation lags and feedback effects in the system as a result of the combined interaction between mass flow, the vapor generation pressure drop and the variation of compressible volume within and upstream of the two-phase flow system. Further research is being conducted along these lines and this is one of a series of papers planned on the subject of two-phase flow instabilities with internally augmented surfaces.

4. Horizontal System

Figure 2 shows the set-up for the experimental investigation of two-phase flow boiling instabilities in a horizontal test section. This set-up is especially designed to generate the main types of two-phase flow oscillations, while allowing a change of heater tubes without disturbing other parts of the system. The set-up consists of the fluid supply, test and fluid recovery sections. It is basically a loop, forced convection boiling system operating between two constant pressures, established in the main tank and recovery tank as described in vertical system.

The heater tube ($L = 1060$ mm, $d_o = 12.5$ mm, $d_i = 10.9$ mm, thickness = 0.8 mm) was uniformly electrically heated and insulated with fiberglass wool tape to ensure a constant heat input. The heat input was determined by measuring the current and the voltage drop across the heated section. For this study, experiments with only a single-tube system were performed. To measure the oscillations of the wall temperature, two thermocouples were fixed at the end of the tube, one at the bottom and one at the top of the outer surface. To avoid floating voltage effects, the thermocouple bead was insulated from the electrically heated tube wall surface by a dab of electrically non-conductive paste. Perfect insulation was needed between the thermocouple beads and the wall surface. To accomplish this, the first layer of thermal insulation with fiberglass strip (2.5 mm thickness) was made by wrapping the strip around the tube very tightly. At the location where temperature measurement was required, holes of 3 mm diameter were cut through strip all the way to the tube surface. Next, the holes were filled with electrical insulation paste, which also exhibited high thermal conductivity. The thermocouple beads were then pressed into the center of each hole. A second layer of thermal insulation was applied over the first one to secure the thermocouple installations. The heater tube was followed by a sight glass for visual inspection of the flow. The exit restriction is defined as the ratio between the inner diameter of the orifice plate and the inner tube diameter and simulates pressure drops for a long tube. After the exit restriction, the exit pressure was measured with a pressure gauge. Downstream of the test section the working fluid, mostly a mixture of liquid and vapor, was sent to a condenser and the condensed fluid was collected in a recovery tank. The condenser and the recovery tank were parts of the fluid recovery section.

4.1 DETAILS OF EXPERIMENTS

Three different heater tubes, described in Fig. 7, were tested separately under the same heat flux, exit restriction, inlet subcooling and system pressure, for five different inlet temperatures. By inserting springs with different pitches in the tube, augmented surfaces were created. Springs with pitches of 5 mm for tube 2 and pitches of 15 mm for tube 3 were used. Figure 6 shows a description of the tube used, where they are classified according to their effective diameter, which is defined by Eq. (1) (Table 2).

The steady-sate characteristics of the system, composed of different tubes were first determined, and then dynamic instabilities were investigated.

Table 2 Description of heater tube for horizontal system.

Tube	Description of the tube	d_e (mm)
1	smooth surface	10.9
2	with internal spring of 1.0 mm diameter and pitch of 5 mm	10.50
3	with internal spring of 1.0 mm diameter and pitch of 15 mm	10.77

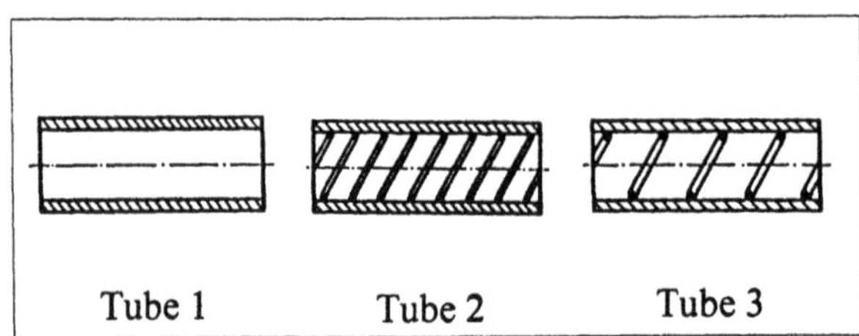

Figure 7. Longitudinal cross section of the heater tube used in horizontal system.

Before stating the experiments, the main tank was pressured to 6.9 bar using high pressure industrial grade nitrogen gas. The heat input was set at the desired value. For experiments at low temperatures, the re-circulating cooler was used to cool the working fluid. For the steady state characteristics, the surge tank was by-passed, and the control valve was opened. Then the mass flux was decreased gradually in steps such that the system underwent changes from single-phase to two phase regions. The readings of the rotameter which showed the mass flux, and the pressure drop, which was indicated by the difference between the surge tank and exit pressures, was plotted in a pressure drop versus mass flow rate diagram. For the experiments to investigate the oscillations, the surge tank was filled with the working fluid and pressurized to a predetermined value by nitrogen gas. The control valve was opened, so that the system worked in the single-phase liquid region. In this region no oscillations was expected. Then the mass flux was reduced by a small amount, using the inlet control valve and the system behavior was observed. When reaching the two-phase flow region, oscillations started and the data, which includes the inlet pressure, wall temperatures and mass flux were supplied to a recorder. A variable-speed drive for the recorder paper movement, and a time marker, combined with separate gain controls for each channel, permitted detailed recordings of signal histories. All data were also observed on a digital voltmeter. Several different mass fluxes was utilized to cover the whole unstable region. When entering the single-phase vapor region, attention must be paid to the wall temperatures to avoid tube burnout, which is likely in this region.

4.2 MEASUREMENTS AND UNCERTAINTIES

The temperature measurements were made by E-type thermocouples with 0.25 mm diameter. The bias uncertainty associated with temperature measurement is calculated as ±0.5 °C. The fluid inlet temperature were controlled within ±0.5 °C, over the experimental range.

The heat input is calculated as the product of the current supplied by the rectifier and the voltage drop across the heater tube. The overall uncertainty for heat input is about ± 2%.

Pressure measurements were made by both pressure gauges and transducers. The pressure gauges were the typical primary pressure measurement instruments serving two purposes. First, they were installed along the tube at several locations to give visual observation of the pressure variation, second, they were used as the calibration instrumentation for the secondary pressure sensors. The pressure gauges had a range from 0 to 10 bar with subdivisions of 0.01 bar, and an accuracy of ±0.1% full scale. The pressure transducers employed as secondary sensors were of the resistive strain-gage type. The response time constants of the pressure transducers were 0.1 ms and their full scale accuracy was ±0.25%. The differential pressure transducers, also secondary sensors, were based on the principle of variable inductance caused by the movement of a diaphragm. Their response time was 0.4 ms, and their full scale accuracy was ±0.5%. The shortest period of the oscillations studied was in the order of 0.5 second. Therefore, the instruments used were capable of dynamic measurements.

4.3 EXPERIMENTAL RESULTS OF HORIZONTAL SYSTEM

To study the effects of tube augmentation, three different inner tube surfaces each of them with five different inlet temperatures, were used. The pressure drop, defined as the difference between the pressure in the surge tank and exit of the heated tube, is plotted versus the mass flow rate for different tube configurations. The curves show an inverted "S" shape for all investigated inlet temperatures and tube surfaces (Fig. 8).The flow is stable along the right hand side portion of the steady-state characteristic curve, where the slope is positive. Near the minimum point of the steady-state characteristic curve, the first bubbles were observed and the flow enters the two-phase region. The pressure drop starts to increase with decreasing mass flow rate. As the mass flow rate is reduced further, most of the liquid evaporates when reaching the exit of the heater tube, and the curves start to round up from its local maximum. Then the system operates in the single-phase vapor region. While operating in the two-phase region, where the slope is negative which becomes steeper with decreasing inlet temperature. With decreasing inlet temperature, boiling starts at a lower mass flow rate, which results in the minimum values of the steady-state characteristic curves corresponding to a lower mass flow rate.

4.3.1 *Effect of Surface Augmentation on the Steady-state Characteristics*

For tubes with augmented inner surfaces, the pressure drop is higher as can be seen in Figure 9. The additional pressure drop, which is due to friction caused by the augmented surface is small. This is to be expected since the frictional pressure drop over the test section of the system is small compared to the total pressure drop The negative slope of the curves is steeper for tubes 2 and 3, and, as a consequence, the system becomes less stable. When measuring Δp, which is the pressure difference between the maximum and the minimum of the steady-state curve, it can be noticed that there are differences among the three tubes. The values of Δp for the bare tube are not increasing as much as for the tubes with augmented surfaces when decreasing the inlet temperatures. The tubes with the augmented surface, however, show a different Δp with different inlet temperatures. In these tubes, Δp becomes larger when decreasing the inlet temperature. Comparing these results with earlier studies by Liu (1993) for a vertical tube and by

458

Ding (1994) for a horizontal tube, steady-state characteristic curves for horizontal tubes with augmented surfaces are similar to vertical tubes.

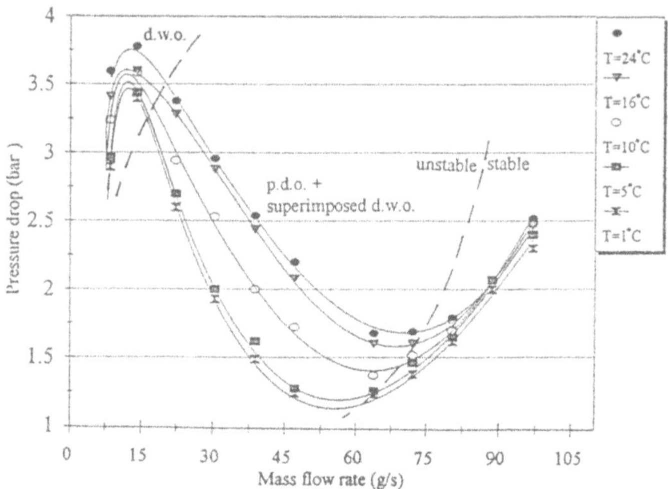

Figure 8. Steady-state characteristic curve-tube 2, Q = 2500 W, β=0.24.

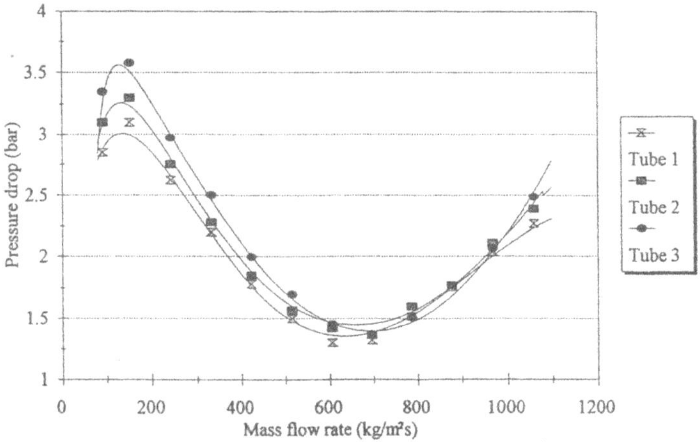

Figure 9. Steady-state characteristic curves for different tubes, T_{in}=10 °C, m =39 g/s, Q =2500 W, β =0.24.

4.3.2 *Stability Boundaries for the Oscillations*

To avoid or control flow oscillations, it is important to know the flow conditions, such as mass flow rate, fluid inlet temperature, exit restriction and system pressure, which create the basis for the origin of oscillations.

During these experiments, three different modes of oscillations were observed. They are identified as pressure-drop type, density-wave type instabilities and thermal oscillations. In Fig. 8, the boundaries for the oscillations are shown by dashed lines. Note that 'p.d.o' refers to pressure-drop-type oscillations, and 'd.w.o' refers to density-wave-type oscillations. When decreasing the mass flow rate, oscillations were observed right after the lowest point of the steady-state characteristic curve. They are pressure-drop-type oscillations with superimposed density-wave-type oscillations. For tubes with augmented surfaces, pressure-drop-type oscillations occurred almost in the entire negative slope region. The same occurrence of pressure-drop-type oscillations was found by Liu (1993) for vertically arranged tubes. For the bare tube, pressure-drop type oscillations ended somewhere in the middle of the negative slope region. At lower flow rates, density-wave-type oscillations occurred for all three heater tubes. Density-wave-type oscillations occurred even with small mass flow rates that corresponded to the single-phase vapor region. Further reduction of the mass flow rate might have resulted in a tube burnout. Figure 10 shows the pressure-drop-type oscillation boundaries for all tested tubes at different inlet temperatures as a function of the mass flow rate. For tubes with augmented surfaces, pressure-drop-type oscillations start at lower mass flow rate and cover a wider range of mass flow rates than for the bare tube. This is a result of unevenness of the tube surface, creating less stratification of the flow than in a smooth tube.

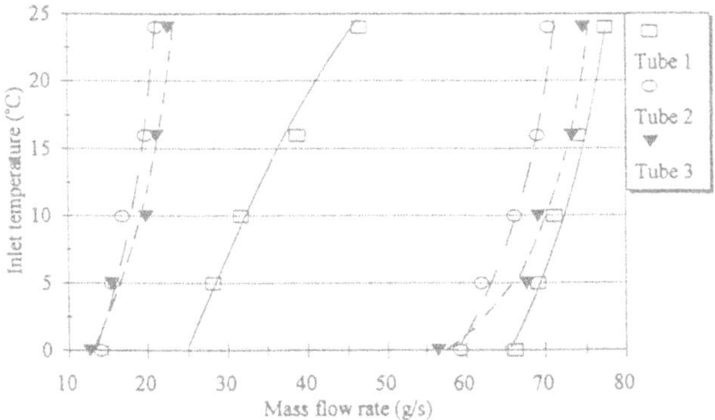

Figure 10. Boundaries of pressure-drop type oscillations for different tubes.

Pressure-Drop Type and Thermal Oscillations. The first oscillations were observed right after the minimum point of the steady-state characteristic curves and are of the pressure-drop-type with superimposed density-wave type oscillations. Oscillations of wall temperatures, which are named thermal oscillations, are in phase with the oscillations of the inlet pressure, while the phase relationship between the inlet pressure and the mass flux

460

variation is about 180°. Figure 11 shows a typical recording of pressure-drop type oscillations for tube 1. Figure 12 shows recordings for inlet pressure and bottom wall temperature for tube 2 which has augmented surface under the same flow conditions. Periods and amplitudes of the inlet pressure and wall temperatures are higher for tubes with augmented surfaces.

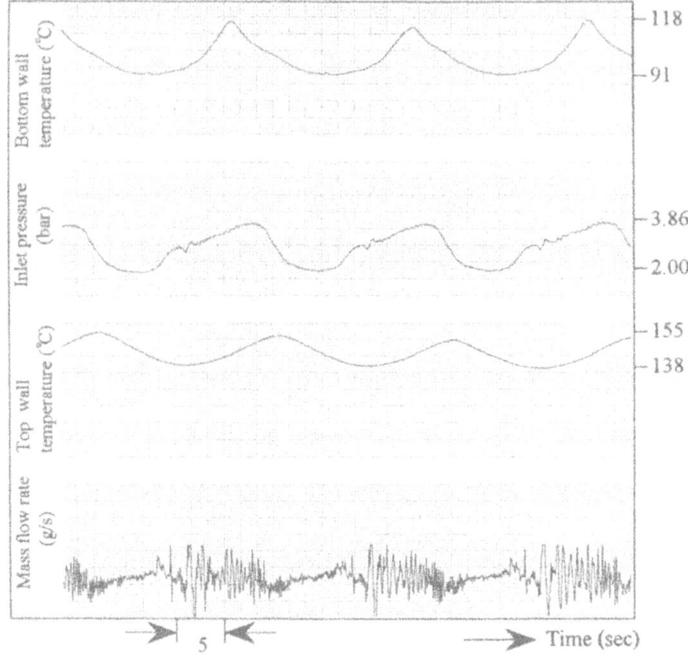

Figure 11. Recordings of pressure-drop-type oscillations-tube 1, $T_{in} = 10\ ^{\circ}C$, $m = 39$ g/s.

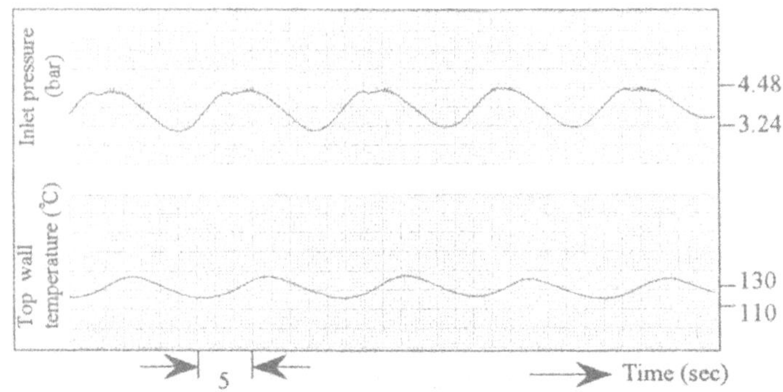

Figure 12. Recordings of pressure-drop-type oscillations-tube 2, $T_{in} = 10\ ^{\circ}C$, $m = 39$ g/s.

Results for the periods of pressure-drop type oscillations are compared in Fig. 13. With decreasing mass flow rate, the periods of the oscillations increase for tubes with augmented surfaces while for the bare tube the periods decrease slightly. The periods for tubes with augmented surfaces are about twice those of the bare tube.

Figure 14 shows a diagram for the amplitude of the bottom wall temperature versus mass flow rate. The shapes of the curves are similar to those for the periods of the oscillations. During the rise of the surge tank pressure the mass flow rate drops to very low values, and a flow regime of dry-out is established. The heat transfer coefficient drops steadily and the wall temperatures increase to sustain the constant heat flux. The longer the period of the oscillations the longer the tube has the dry-out condition and the wall temperatures increase to high values. When the surge tank pressure decreases again, the mass flow rate increases and the wall temperatures decrease due to the higher heat transfer coefficient in two-phase flow and nucleate boiling.

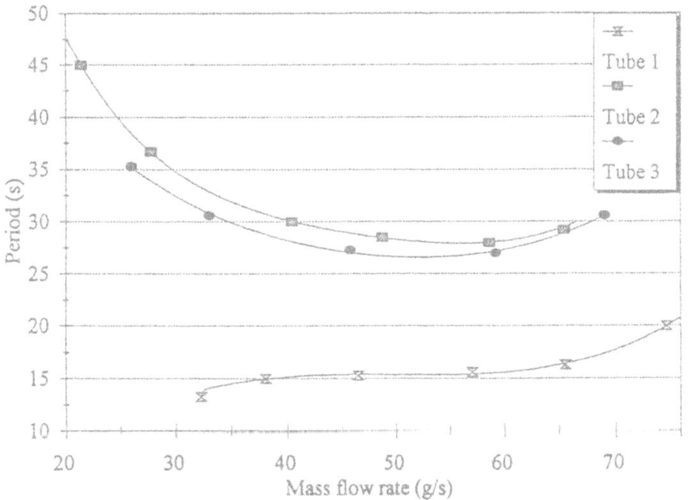

Figure 13. Periods of pressure-drop-type oscillations for three tubes,
$T_{in} = 10\ ^{\circ}C$, $Q = 2500$ W, $\beta = 0.24$.

Figure 15 shows the amplitude of inlet pressure versus mass flow rate for three different tubes. There are drops in the amplitude of inlet pressure at low and high mass flow rates for each tube during pressure-drop type oscillations. This can be explained with the steady-state characteristic curves in Fig. 5. The drops in the amplitude of the pressure occur when operating near the maximum and the minimum point of the steady-state characteristic curve. In these regions, the limit cycle is not fully achieved and the amplitudes of the inlet pressure oscillations are smaller than those occurring in the middle of the negative slope region.

462

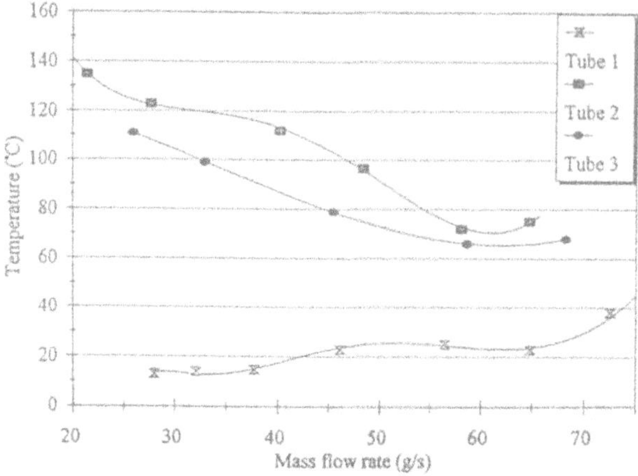

Figure 14. Amplitude of bottom wall temperature of pressure-drop-type oscillations
for three tubes, $T_{in} = 10\ ^{\circ}C$, $Q = 2500$ W, $\beta = 0.24$.

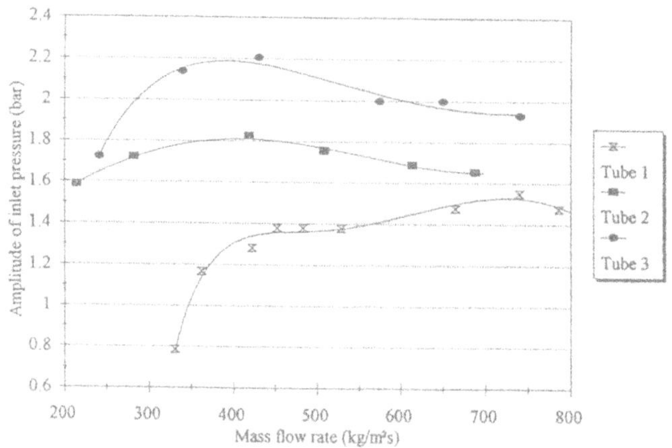

Figure 15. Amplitude of inlet pressure oscillations for three tubes,
$T_{in} = 10\ ^{\circ}C$, $Q = 2500$ W, $\beta = 0.24$.

The differences in the range of inlet pressure amplitudes among the different tubes can be explained with Fig. 9. The steady-state characteristic curves for $T_{in} = 10^{\circ}C$ of three different tubes are plotted on this figure. When measuring the pressure difference between the minimum and maximum of these curves, ΔP is of the same amount as the amplitude of inlet pressure for each tube (Fig. 15). The difference in the pressure drop on the steady-state characteristic curves is a consequence of the different extent of the flow stratification. Tubes with augmented surfaces have less flow stratification than the tube with a smooth surface.

Density-wave-Type Oscillations. Density-wave-type oscillations occur after the pressure-drop-type oscillations. Figure 16 shows recordings of density-wave-type oscillations in the tube with smooth surface. Oscillations of mass flow rate and inlet pressure are in phase. Both top and bottom wall temperatures are not oscillating. However, the wall temperature at the top of the tube is higher than at the bottom of the tube, and the top of the tube is in a wall dry-out condition. Density-wave type oscillations have smaller periods and smaller amplitudes than pressure-drop-type oscillations. When comparing the steady-state characteristic curves it can be seen that density-wave-type oscillations occur almost always on the positive slope, which represents the single-phase vapor region, for tubes with augmented inner surfaces. The result that density-wave- type oscillations are located on the positive slope region was found by Liu [6] who investigated instabilities in a vertical tube. This shows that horizontal tubes with augmented surfaces have characteristics similar to vertical ones. In the smooth tube, however, these oscillations start in the negative region.

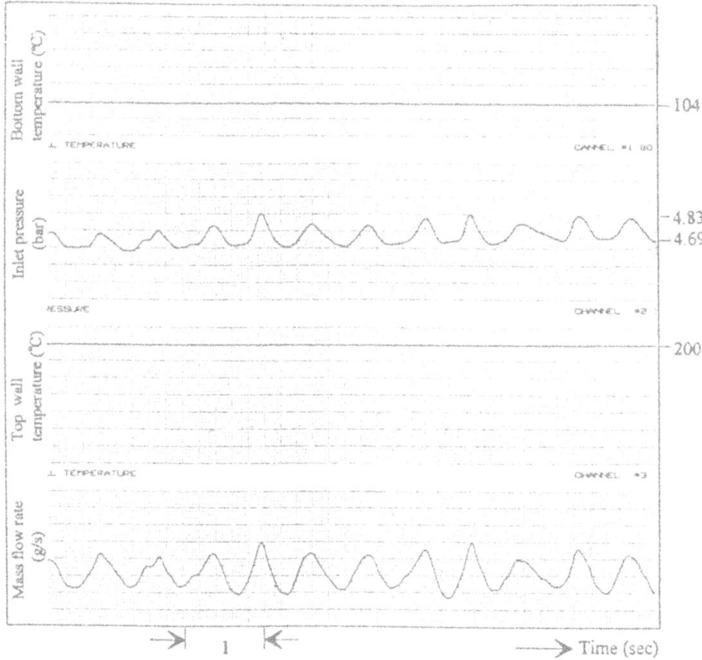

Figure 16. Recordings of density-wave-type oscillations-tube 1, T_{in} = 16 °C, m = 17 g/s, Q = 2500 W, β = 0.24.

464

5. Practical Significance

As in this study, under certain operating conditions, two-phase flow systems may be unstable and cause various problems, such as mechanical vibration, problems with system control and tube failure. At present, two-phase boiling flow instabilities are, in most cases, avoided by using high inlet pressure drops, which waste a considerable amount of energy. Under some circumstances, unstable operation may not cause any problems if the oscillation amplitudes and periods are within certain limits. The information presented in this paper could be used for the verification of the predictions obtained from the modeling of two-phase flow instabilities, and it may lead to more systematic and elaborated correlations for the amplitudes and periods of thermal oscillations, which could be used as guidelines in the design and operation of boiling two-phase flow systems.

The modeling of the horizontal in-tube boiling systems is being studied further. Table 1 shows the results of experimental and theoretical studies of pressure-drop type oscillations in a horizontal single channel flow by the use of the homogeneous model. Stratification was not present under the experimental conditions studied, Kakaç et al. [2].

6. Conclusions

- Three modes of oscillations are found, pressure-drop type, density-wave type and thermal oscillations for all investigated inner tube surfaces and inlet temperatures.
- Periods and amplitudes of wall temperatures during pressure-drop type oscillations are larger for tubes with augmented surfaces.
- Periods and amplitudes of pressure-drop type oscillations increase with decrease mass flux in tubes with augmented surfaces, whereas in the smooth tube periods and amplitudes decreasing with decreasing the mass flux.
- Wall temperatures during pressure-drop-type oscillations are higher for tubes with augmented surfaces.
- The unstable region for pressure-drop-type oscillations is larger for tubes with augmented surfaces than for the smooth tube. The stability boundaries of pressure-drop type oscillations move to lower mass fluxes for tubes with augmented surfaces.
- The horizontal tube with an augmented inner surface behaves like a vertical tube, because of the lesser degree of stratification of the flow than for a smooth horizontal tube.
- No end of density-wave type oscillations with decreasing mass velocity was found for all tube surface configurations.

Nomenclatures

d diameter of the orifice plate used as an exit restriction, m
d_i inner tube diameter m
d_o outer tube diameter, m
d_e effective inner diameter, m
D_e effective diameter, m
L tube length, m

m mass flow rate, kg/s
Δp difference between the inset pressure and the outlet pressure.
p fluid pressure, bar
P_c critical pressure, bar
$P.D.$ pressure-drop-type
q heat flux, W/m^2
Q heat transfer rate, W
SI superimposed instabilities
T temperature, °C, K
T_{in} fluid inlet temperature, °C, K
U velocity, m/s
V net inside volume of the tube, m^3/s
Z distance between two thermocouples, m
$\bar{z}$ ratio Z/L

Greek Symbol

β ratio of the diameter of the orifice plate (used as an exit restriction) to the inside diameter of the tube, d/d_i
ρ fluid density, kg/m^3

References

1. Ding, Y. (1993) *Experimental investigation of two-phase flow phenomena in horizontal convective in-tube boiling system*, Ph.D. dissertation, University of Miami. Coral Gables. FL.
2. Kakaç, S., Mullur, V. R., and Chen, X. J. (1994) Finite-difference modeling of two-phase flow instabilities m a horizontal boiling system, *Proceedings of Workshop/Symposium on Boiling-Condensation and Two-Phase Flow*, Andhra University, Vishakapatnam, India January 10-11.
3. Kakac, S., Veziroglu. T. N., and Padki, M. M. (1990) Investigation of thermal instabilities in a forced convection upward boiling system, *Experimental Thermal and Fluid Science*, Vol. 3, pp. 365-376.
4. Lin, Z. H., Zhang, X., Chen, X. J., Veziroglu, T. N. and Kakaç, S. (1988) An analytical study of the pressure-drop type instabilities in a horizontal hairpin tube, *Int. J. Engineering Fluid Mechanics*, Vol.14, pp. 427 444.
5. Liu, H. T., and Kakaç, S. (1991) An experimental investigation of thermally induced flow instabilities in a convective boiling up-flow system, *Wäirrne- und Stoff übertragung*, Vol. 26, pp. 365-376.
6. Liu, H. (1993) *Pressure-drop type and thermal oscillations in convective boiling systems*, Ph. D. dissertation, University of Miami Coral Gables, FL.
7. Mentes, A., Kakaç, S., Veziroglu, T. N., and Zang, H.Y. (1989) Effect of inlet subcooling on two-phase flow oscillations in a vertical boiling channel, *Warme-and Stof Jubertragung*. Vol. 24, pp. 25-36.
8. Ozawa, M., Akagawa, K. and Sakaguchi, T. (1989). Flow instabilities in parallel-channel flow systems of gas-liquid two-phase mixtures, *Int. J. of Multiphase Flow*, Vol. 15, No.4. pp. 639-657.
9. Padki, M.M., Liu, H.T., and Kakaç, S. (1991) Two-phase flow pressure-drop type and thermal oscillations, *Int. J. Heat and Fluid Flow*, Vol. 12. No 3. pp. 2448.
10. Parros, J.C. and Gentile. D. (1991) An experimental investigation of transition boiling in subcooled freon-113, *J. Heat Transfer*, Vol. 113, pp. 459-462.
11. Wang, Q., Veziroglu, T.N., Kakaç, S. and Chen. X J. (1990) An investigation on density-wave oscillations, *Proceedings of the 2nd International Symposium on Multiphase Flow and Heat Transfer*, 18-21 Sept., 1989, Xi'an, China. Hemisphere, New York.
12. Xiao, M. 1991. *Investigation of high pressure steam-water two phase flow instabilities in parallel boiling channels*, Ph. D. dissertation. Xi'an Jiaotong University, Xi'an. China.
13. Lin, Z. H., Veziroglu T. N., Kakaç. S., Gurgenci, H., and Mentor, A. (1982) Heat transfer in oscillating two-phase flows and effect of heater surface conditions, *Proceedings of the 7th International Heat Transfer Conference*, München, Hemisphere, New York, N.Y.
14. Widmann, F., Comakli, Ö, C. Gavrilescu, Ding, Y. and Kakaç, S. (1995) The effect of augmented surface on two-phase flow instabilities in a horizontal system, *J. Enhanced Heat Transfer*, Vol. 2, No. 4, pp. 263-271.

FLOW BOILING INSIDE MICROFIN TUBES: RECENT RESULTS AND DESIGN METHODS

JOHN R. THOME
International Consulting Engineer
Via S. Angela Merici 30, Rome 00162, ITALY
Present Address: Professor, LTCM-DGM, EPFL, CH-1015 Lausanne

Abstract: Microfin tubes have become the most important type of heat transfer augmentation for intube evaporation of refrigerants in direct-expansion evaporators, climate-control systems and heat pumps. These tubes are characterized by a large number of small fins inside small diameter tubes, providing heat transfer augmentations in the range from 1.5 to 4 times with only modest two-phase pressure drop penalties. The most recent experimental research on evaporation of pure fluids in microfin tubes is reviewed and the latest prediction methods are critically assessed.

1. Introduction

The topic of this review is intube evaporation of pure refrigerants (CFC, HFC and HCFC refrigerants and ammonia) inside microfin tubes. Previous reviews on this subject have been presented by Thome [30, 32, 34] and consequently the present review will only cover the more recent work on this topic. A companion lecture at this NATO meeting covers intube evaporation of refrigerant-oil mixtures in microfin tubes.

Intube evaporation inside microfin tubes has become an important research topic because of their wide-spread application in direct-expansion evaporators in recent years. These are all horizontal units and hence all the published test data for microfin tubes presented here will be for that orientation. The only vertical microfin tube data apparently available in the literature are those of Kattan, Thome and Favrat [13] for R-134a and R-123 for those so interested.

Microfin tubes are characterized by numerous small internal fins of 0.1-0.4 mm height that can be either longitudinal or helical and either two-dimensional (i.e. plain microfins) or three-dimensional (i.e. crosscut or notched microfins). The fins have a height-to-width at base ratio of about 1.0 and their fin profiles tend to be trapezoidal, triangular or rectangular depending on the particular tooling of the manufacturer. Most of the tubes are seemless (manufactured by drawing a plain tube over an externally grooved mandrel) but several manufacturers now produce these tubes from

S. Kakaç et al. (eds.), Heat Transfer Enhancement of Heat Exchangers, 467–486.

468

strip by first rolling the enhancement geometry onto one side of the strip and then forming a longitudinally seamed tube from the strip in a continous process. Microfin tubes are most commonly available in copper in diameters from about 5 to 16 mm outside diameter. Several versions are available in high alloys with 14-22 mm internal diameters (stainless steels and titanium) and also in carbon steel and aluminum (for ammonia systems). Nearly all microfin tubes tested in university labs have been used in their original production form; in some industrial applications of microfin tubes, such as air-conditioning coils with external aluminum fins, they are mechanically expanded into the coils and hence their fin heights are reduced by a small margin and their internal diameters increased.

2. Microfin Heat Transfer Enhancement Mechanisms

The significant enhancement in evaporation performance of microfin tubes relative to plain tubes, with only a small adverse augmentation of their two-phase pressure drops, has been attributed to the following heat transfer enhancement mechansims by Thome [30] and Thome, Kattan and Favrat [36]:

- *Extended surface area effect*: Internal wetted surface area ratios for microfin tubes relative to their equivalent plain bore tubes range from about 1.3 to 1.9, depending on the number of fins, their shape, height and helix angle (typically the area ratio represents the *lower* bound on the heat transfer enhancement level);
- *Enhanced convective heat transfer*: Microfins augment the convective heat transfer process across the annular liquid film, similar to internal ribs used for single-phase flow inside tubes, and this increases the two-phase convection contribution to annular flow boiling;
- *Flow pattern effect*: The helical microfins tend to convert what would otherwise be a stratified-wavy flow in plain tubes into the more thermally effective annular flow regime, meaning that all the internal tube wall perimeter is wetted and active rather than only the lower wetted fraction in plain tubes; this represents the principal reason for their very large augmentation ratios at low mass velocities;
- *Nucleate boiling heat transfer*: The microfins may favor the activation of nucleation sites by partially shielding the potential cavities from the flow; nucleate boiling will also occur on the total wetted extended surface area;
- *Swirl effects*: Swirl imparted to the annular liquid film by helical microfins may retard the onset of dryout to higher vapor qualities; swirl of the continuous vapor-phase in mist flow and annular flow with partial dryout will augment vapor-phase heat transfer to the dry wall similar to a single-phase turbulent flow inside an internally ribbed tube; also swirl will drive entrained droplets to the tube wall;
- *Grigorig film thinning effects*: The convex contours at the tips of microfins will tend to thin the evaporating liquid film on the fins, similar to that which occurs for film condensation on external low finned tubes, increasing the evaporating heat transfer coefficient on the microfins significantly (schematic diagram in Figure 1).

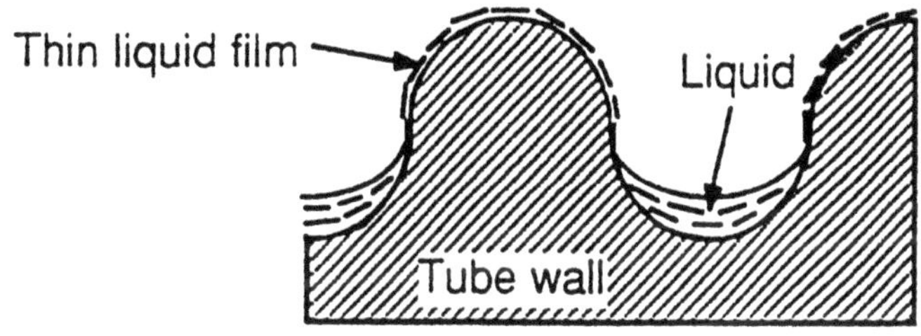

Figure 1. Thinning of annular liquid film on microfins
from Thome, Kattan and Favrat [36].

3. Recent Microfin Experimental Tests for Pure Fluids and Azeotropes

Experimental research on intube evaporation of pure refrigerants is surveyed in this section. The experimental studies published since the previous review by Thome [34] was prepared for publication (circa mid-1995), together with some earlier works missed in that review, are covered here.

Eckels and Pate [6] have provided mean heat transfer coefficient data for a microfin tube [8.72 mm ID with 60 fins of 17° helix angle of 0.20 mm height and area ratio of 1.5]. In their tests they compared R-134a to R-12 at 5°C, 10°C and 15°C saturation temperatures, which were a continuation of their earlier study on a 8.00 mm ID plain tube in Eckels and Pate [5]. They obtained mean heat transfer coefficients using counter-current flow of hot water in an annulus, utilizing a modified-Wilson plot to obtain these data. They reported slightly lower microfin enhancement factors for R-134a than for R-12 at the same mass velocity but essentially identical factors when compared at equal heat duties. The augmentation ratios at low mass velocities were from 2.3 to 2.6 while about 1.7 to 1.9 at higher mass velocities.

Torikoshi, Kawabata and Ebisu [38] presented mean evaporation data for R-134a at three saturation temperatures (5, 15 and 30°C) for a plain tube [8.7 mm ID] and a microfin tube [8.8 mm ID with 60 fins of 18° helix angle and 0.20 mm height, no area ratio cited]. Counter-current hot water heating in an annulus was used but in these tests neither measured wall temperatures were measured nor were modified-Wilson plot type liquid-liquid tests run to obtain the water-side heat transfer coefficients; instead, they calculated their values from the Dittus-Boelter tubular correlation using the hydraulic diameter of the annulus without using the VDI-Heatatlas annulus correction factor.

Eckels, Pate and Bemisderfer [7] ran a comparative study on five 9.525 mm OD microfin tube geometries and also on three smaller diameter 7.94 mm OD microfin tubes. All the tubes were mechanically expanded before testing with R-22 at two saturation temperatures (2°C and 7°C). Counter-current, hot water heating and a modified-Wilson plot method were used to measure mean heat transfer coefficients. The tube internal diameters were however not cited nor was it stated if the fin dimensions provided were before or after mechanical expansion. The helix angle was said to be about 18° for all the tubes but the number of internal fins was not cited. Fin profiles studied included triangular fins, equilateral trapezoidal fins and a rectangular trapezoidal type of fin, which was tested in both flow orientations. Over their range of test conditions, the performances of the five larger tubes varied less than 25% while those of the three smaller tubes by 17% or less, thus showing that fin geometry does have an effect on performance but is it is a secondary influence.

Eckels, Doerr and Pate [4] reported new mean heat transfer coefficients for R-134a at 1°C evaporating inside a smooth tube [8.0 mm ID] and a microfin tube [8.92 mm ID with 60 fins of 17° helix of 0.20 mm height with an area ratio of 1.5] using counter-current, hot water heating. Their R-134a plain tube data compared reasonably well with existing correlations, i.e. within about ±20%. The heat transfer augmentation ratio versus the smaller ID plain tube ranged from 1.9 at low mass velocities (85 kg/m^{2}s) down to 1.45 at high mass velocities (375 kg/m^{2}s), which is similar to the area ratio of 1.5. The augmentation factor of 1.9 at low G is lower than the typical range found in our studies of 2.5 to 4.

Torikoshi and Ebisu [39] reported mean flow boiling coefficients for R-22 evaporating inside a small diameter microfin tube [6.40 mm ID with 50 fins of 18° helix angle of 0.18 mm height, no area ratio cited] at 5°C with hot water as the heating source. They again directly calculated the water-side coefficient with the Dittus-Boelter tubular correlation using the hydraulic diameter of the annulus in order to back out their boiling coefficients from the overall heat transfer coefficient. In any case, their heat transfer data were plotted on ill-sized log-log graphs which makes reading values of their data points all but impossible.

Chamra and Webb [1] measured local coefficients for R-22 at a relatively high saturation temperature of 24.4°C for a crossgrooved microfin tube using hot water heating. Their 14.66 mm ID tube had 15° helix angle microfins of 0.35 mm height that were crosscut at an opposing angle to make three-dimensional fins. Figure 2 shows their flow boiling data compared to analogous condensation data. Essentially these data illustrate that in annular flow the heat transfer process for evaporation and condensation are similar, except for the effect of nucleate boiling. They also presented some mean heat transfer data for the same crossgrooved tube and a plain tube (no diameter given) at 2.2°C, where augmentation levels ranging from 3.5 at low mass velocities of about 50 kg/m^{2}s down to 1.7 at 200 kg/m^{2}s.

Kuo et al. [22] measured local evaporation coefficients for R-22 at three saturation temperatures (2°C, 6°C and 10°C) inside a microfin test section comprised of four 1.2 m long subsections connected in series with hot fluid heating [6.50 mm ID with 60 fins of 18° helix angle and 0.15 mm high with an area ratio of 1.49]. Comparing their microfin data to the Kandlikar [10] plain tube correlation's predictions, they observed essentially no augmentation at low vapor qualities but large microfin augmentation at intermediate and high vapor qualities. This is quite different than what is observed in completely experimental comparisons which typical display significant augmentation at low vapor qualities, and their observations are probably only the result of an erroneous slope in h vs. x obtained from the plain tube correlation.

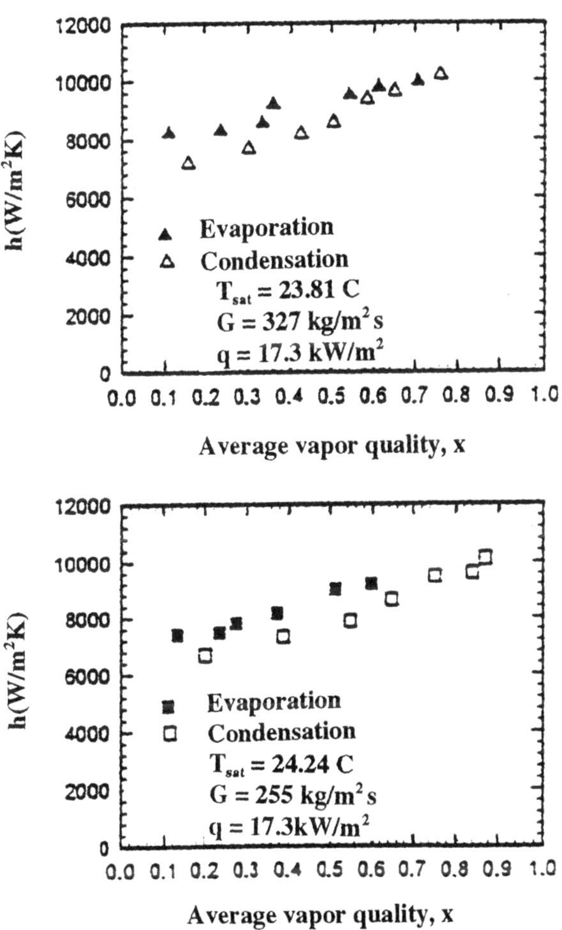

Figure 2. Chamra and Webb [1] evaporation and condensation
data for R-22 at two mass velocities.

Kido et al. [18] experimentally determined the relative performances of seven microfin tubes for R-22 evaporating at a saturation pressure of 4.9 bar. They tested microfins

472

with 60 to 100 fins with helix angles from 3° to 18°, fin heights from 0.15 to 0.21 mm and area ratios from 1.63 to 2.49. They condensed R-114 on the outside of their 7.0 mm OD tubes as their heating source and soldered thermocouples around the tube perimeter at three axial locations. Their heat transfer data, heat fluxes and mass fluxes were reported based on the mean internal diameter, i.e. that at half the height of the internal fins, rather than the usual maximum internal diameter at the roots of the fins. The heat flux may not be uniform around the circumference of the tube with external film condensation but they did not address this issue. Figure 3 depicts their data at one mass velocity at a heat flux of 9.30 kW/m^2 for the following microfin tubes: #1 (60 fins/18°helix/0.15 mm height/1.63 area ratio/6.32 mm ID), #2 (70/11°/0.21/2.21/6.29), #5 (85/17°/0.16/2.13/6.38), #6 (85/7°/0.21/2.49/6.27) and a plain tube (6.40 mm ID). These coefficients are defined based on the nominal diameter at half the fin height, not at the root of the fins as in most other tests. Peaks in local heat transfer coefficient vs. vapor quality were found for all the tubes at about x = 0.8 and also the level of augmentation increased with increasing vapor quality up to these peaks. Microfin #6 outperformed the other tubes. As general trends in this and their other graphs, they found that (i) for a nearly fixed helix angle (7° and 9°) and fixed number of fins (85) that an increase in fin height increased performance and (ii) for tubes with an equal number of fins and fin height, enlarging the helix angle increased performance.

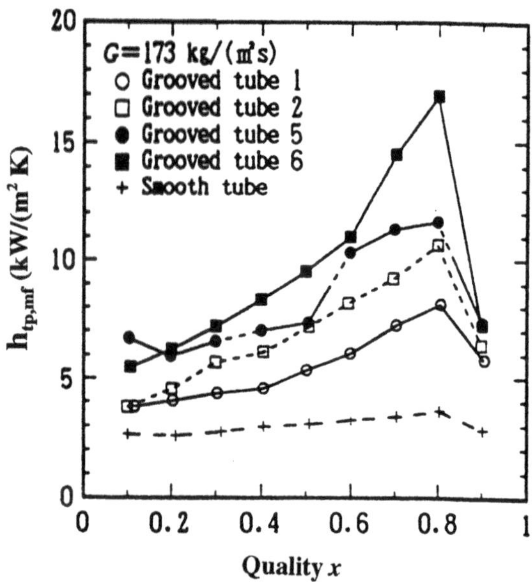

Figure 3. Microfin data of Kido et al. [18] for R-22 at 4.9 bar.

Koyama et al. [19] ran R-22, R-134a and R-123 boiling tests inside a microfin tube [8.475 mm maximum ID with 60 fins of 18° helix angle of 0.17 mm height and area ratio of 1.52]. They used hot water heating and tube surface thermocouples to obtain local test data. They also defined their internal heat transfer coefficients, heat fluxes and mass fluxes with an equivalent internal diameter rather than using the maximum internal diameter as is the standard in other studies. Only one set of local R-134a data at 6.55 bar at one mass velocity was shown in a h vs. x format while all the other data were depicted in a correlated format.

Uchida et al. [40] measured both local and mean boiling coefficients for R-22 evaporating inside two small microfin tubes [6.40 mm ID with 60 fins of 18° helix angle of 0.163 mm height but no area ratio cited; 6.10 mm ID with 43 crossgrooved fins of 19° helix angle of 0.29 mm height without an area ratio cited]. Their short 0.6 m (1.97 ft) test sections were heated with an electrical heater coiled around the outside and they soldered thermocouples onto the outside of the tube for wall temperature measurements. No saturation temperature nor test pressure was cited in the paper making these data impossible to use.

Kuo and Wang [20] and Kuo and Wang [21] reported local boiling coefficients for R-22 at 6 bar for a microfin tube [8.92 mm ID with 60 fins of 18° helix angle and 0.20 mm height with an area ratio of 1.57] using a test section with an effective heat transfer length of 1.3 m. They used hot water heating and modified Wilson plots in these tests. The Shah [27], Gungor-Winterton [9] and Kandlikar [10] horizontal tube correlations did not predict their R-22 data very well, with the majority of the data off by more than 25%. Comparing here these data to the microfin data of Eckels, Pate and Bemisderfer for R-22 for similar diameter microfin tubes at approximately similar test conditions, the Kuo and Wang mean data were about 20-30% higher.

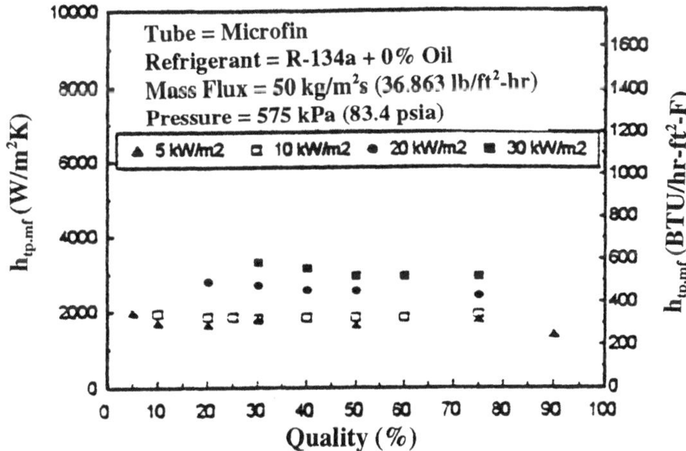

Figure 4. Singh, Ohadi and Dessiatoun [28] test data for R-134a in a microfin tube.

Sundaresan et al. [29] presented some more mean test data for R-22 at 7°C for a smooth tube [8.0 mm ID] and a microfin tube [8.72 mm ID with 60 fins of 17° helix angle and 0.20 mm high with an area ratio of 1.5] as part of a larger test program on zeotropic refrigerant mixtures. They used countercurrent hot water heating in a 3.67 m (12 ft) long test section.

Singh, Ohadi and Dessiatoun [28] measured local and mean boiling coefficients for R-134a at 5.75 bar for a microfin tube [11.78 mm ID with 60 fins of 18° helix angle and 0.30 mm height with no area ratio cited] using an electrical heating tape wrapped around their 495 mm long test section. Figure 4 depicts their data at four heat fluxes at a small mass velocity. There appears to be an inflection point in the heat flux effect on heat transfer between 10-20 kW/m², where heat transfer coefficients above this level increased with heat flux while below they did not. It could be that the threshold for the onset of boiling nucleation was surpassed in this heat flux range or that for this flow regime the nucleate boiling in the stratified liquid became progressively 'explosive' enough to wet the otherwise dry top perimeter of the tube.

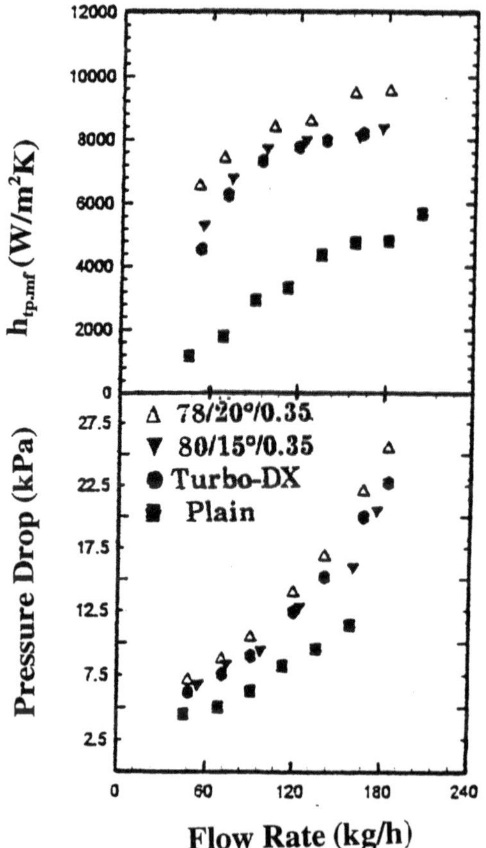

Figure 5. Chamra, Webb and Randlett [2] R-22 data for microfin tubes compared to Thors and Bogart [37] data.

Chamra, Webb and Randlett [2] have reported some mean R-22 evaporation data for four two-dimensional microfin tubes and four similar crossgrooved microfin tubes, all at 2.2°C, comparing the performance of the plain two-dimensional microfin tubes to identical tubes with crosscut fins. They used hot water heating and made accurate measurements. Their corresponding plain tube data compared well with those of Thors and Bogart [37]. Figure 5 shows a comparison of their 78 fin/20° helix angle/0.35 mm high crossgrooved tube (designated as MCG-20® in their paper) and their 80/15°/0.35 (MX-15®) tube to some Turbo-DX tube data of Thors and Bogart [37]. The MCG-20 tube thermally outperformed the Turbo-DX by 23% at low mass velocities but by only 6% at high mass velocities and had a higher two-phase pressure drop. Consequently, crossgrooving has some enhancement potential but it appears to be less significant than might be otherwise envisioned.

Kaul, Kedzierski and Didion [17] undertook a comprehensive experimental study on R-22, R-134a, R-32 and R-125 at a saturation temperature of 4.4°C. A 8.925 mm ID microfin tube was tested but no dimensional information was cited for the microfins themselves. Also, the definitions of their heat fluxes and heat transfer coefficients are unclear since in their paper it was written "The heat flux was calculated as:" and then the ensuing information was missing in the manuscript. The tests covered intermediate vapor qualities from 0.2 to 0.6 and a wide range of heat fluxes. Interestingly, they utilized an enthalpy profile approach for the counter-current hot water in the annulus and thus local heat transfer coefficients were measured without resorting to electrical heating. R-32 significantly outperformed the other pure refrigerants R-125, R-22 and R-134a, in that order. The data were empirically curvefit for future use.

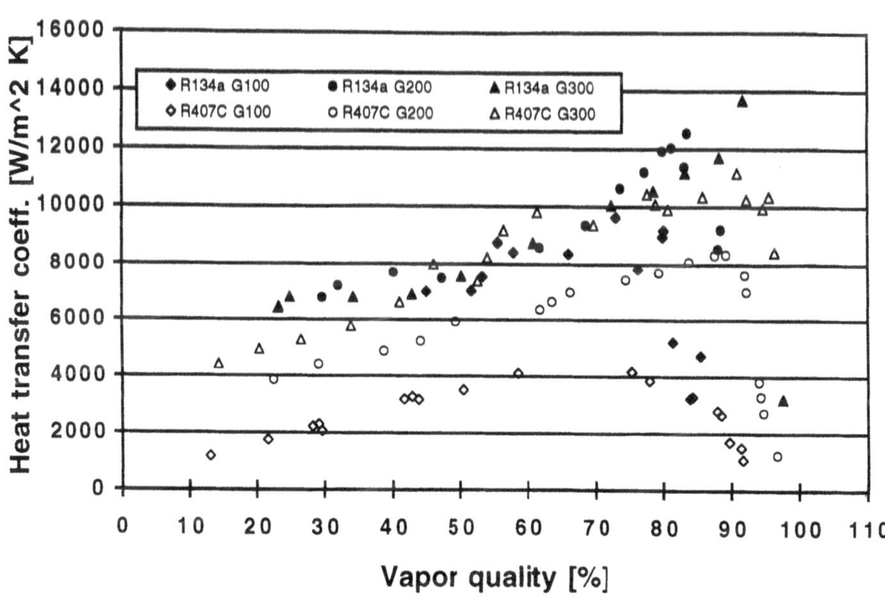

Figure 6. Zürcher, Thome and Favrat data comparison of R-134a to R-407C for a microfin tube from Thome [35].

Nidegger, Thome and Favrat [24] and Zürcher, Thome and Favrat [42] measured local flow boiling coefficients in a horizontal microfin tube [11.90 mm ID with 18° helix

476

angle and 0.25 mm height with area ratio of 1.739] and a similar plain tube [10.92 mm ID] for R-134a at 4.4°C, using hot water heating and a modified-Wilson plot approach. Their local plain tube data were comparable to the local R-134a data of Wattelet et al. [41] for a similar tube diameter and test conditions and, determining mean coefficients from their local data, they would be very similar in value to the mean coefficients measured by Eckels and Pate [5]. Comparing the microfin data to the plain tube data in their study, the level of heat transfer augmentation increased with increasing vapor quality, reaching a maximum just before the plain tube reached its point of onset of dryout. The R-134a data of Nidegger, Thome and Favrat [24] for the microfin tube demonstated a peak in h vs. x similar to that found in plain tubes, and hence the microfin tube also reached a threshold for onset of dryout at high vapor qualities. Figure 6 shows a comparison of their pure R-134a results to similar tests for the same tube with R-407C at the same nominal test conditions obtained by Zürcher, Thome and Favrat [44]. At 300 kg/m²s the two fluids gave similar performances and at 200 kg/m²s R-407C performed moderately below R-134a but at 100 kg/m²s R-407C performed much worse than R-134a. The reasons for this probably lie in the adverse mass transfer effects in R-407C, a topic beyond the scope of the present review.

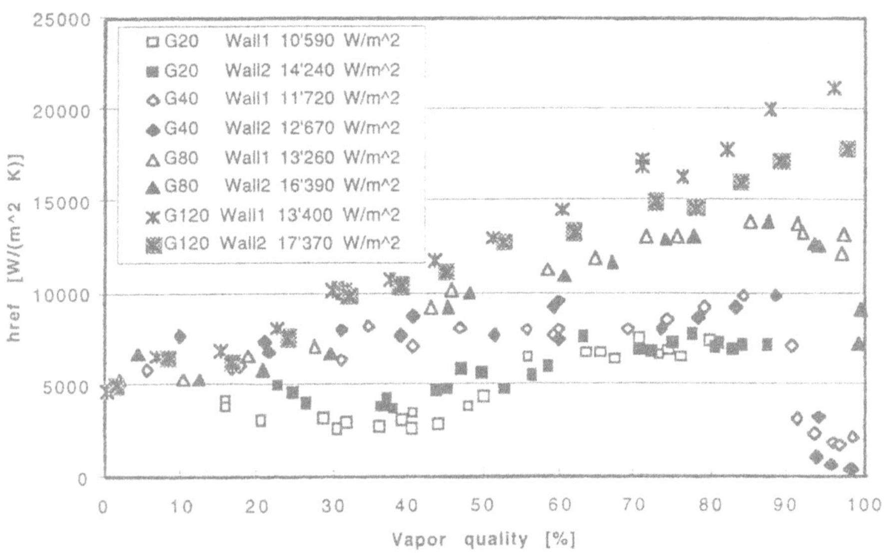

Figure 7. Zürcher, Thome and Favrat [43] ammonia flow boiling data
for microfin tube at 4°C.

Zürcher, Thome and Favrat [43] recently presented local flow boiling data for ammonia evaporating at 4°C for mass velocities from 20-120 kg/m²s inside a plain tube [14.0 mm ID]and a *Microrib* tube [13.46 mm ID with 34 fins of 18° helix angle and 0.33 mm high with an area ratio of 1.33; (note: these are hard alloy microfin tubes developed by High Performance Tube Inc. and J.R. Thome, patent pending)]. The test sections were made of 439 grade stainless steel with hot water heating, both 3 m long.

Four local thermocouples were installed in the tube wall around the circumference of each tube at two different axial locations remote from the tube entrance for tube wall temperature measurements. Using four sets of thermocouples in the hot water flow at four axial locations, the water temperature profile was obtained and curvefit with a spline. The local enthalpy gradient in the water-side could then determined from the local slope of the spline, allowing local heat fluxes to be calculated at the two wall thermocouple locations, such that true local heat transfer coefficients were obtainable with hot water heating rather than "quasi-local" values when applying the modified Wilson plot method. Figures 7 and 8 show some of their enhanced tube test data and the corresponding enhancement ratios with respect to their plain tube tests, which augmented local heat transfer coefficients with respect to the plain tube by a factor of about 2.2 at low vapor qualities while up to 7.7 times at high vapor qualities, and yielded mean augmentation ratios from 4-5 at $G < 80$ kg/m^{2}s, flow rates typical of ammonia direct-expansion evaporator applications, which are much larger than the corresponding area ratio of 1.33. Curiously, at $G \geq 80$ kg/m^{2}s, no augmentation was observed, perhaps because of ammonia's already very high plain tube performance.

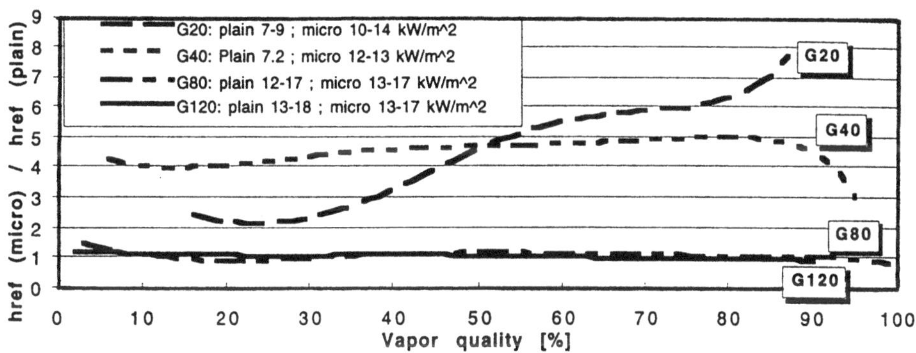

Figure 8. Zürcher, Thome and Favrat [43] heat transfer augmentation
ratios for ammonia in a microfin tube at 4°C.

Muzzio, Niro and Arosio [23] recently reported R-22 evaporation data for a plain tube, a new type of microfin tube with alternating fins of 0.23 mm and 0.16 mm (0.009 and 0.006 in.) height, a second microfin with a conventional fin profile, and a third microfin with a *screw* profile. Heat transfer coefficients were measured in test sections with a vapor quality change of 30% from inlet to outlet. The twin height microfin tube gave significantly higher heat transfer performance than the other two tubes.

4. Microfin Flow Boiling Models and Methods

The first flow boiling model specifically for microfin tubes was apparently that developed by Thome in 1990 that became commercially available in Thome [31],

478

which was a completely general method applicable to any fluid for the complete range of microfin dimensions used industrially, which can be interactively input by the user into the accompaning software program, covering a wide range of helix angles ($0°$-$30°$), number of microfins, fin heights and internal surface area ratios. It was based on test data for R-11, R-12, R-113 and R-22 available at that time for numerous microfin tube geometries and tube diameters. It has since been improved to include new refrigerants such as R-134a, R-123, etc. in its database (and zeotropic refrigerant mixtures such as R-407C). The details of the method are not in the open literature but a parametric study of its predictions were shown in Thome [33] for R-134a.

Kandlikar [11] next presented a curvefitting approach for microfin tubes, using fluid specific correction factors, i.e. 1.5 for R-12, 2.2 for R-22, 1.3 for R-113, 1.9 for R-123 and 1.63 for R-134a, similar to the approach in his plain tube flow boiling method. This was not really a design method but an empirical method for curvefitting a specific data set.

Fujii et al. [8] proposes the following a microfin correlation:

$$Nu_{microfin} = Nu_L \, (4.6/X_{tt}) \tag{1}$$

$$Nu_L = 0.045 \, Re_L^{0.8} \, Pr_L^{0.4} \tag{2}$$

where the liquid Reynolds number Re_L was defined as

$$Re_L = G(1-x) \, D_{mean}/\mu_L \tag{3}$$

In these equations: D_{mean} refers to the mean microfin diameter apparently at half the fin height, Nu_L is the liquid fraction Nusselt number, i.e. based on $G(1-x)$, that was correlated with the Dittus-Boelter form of turbulent correlation to specific test data for one microfin tube to obtain the leading factor of 0.045 [hence the method is *not* a general type of method but is only applicable to that particular microfin tube] and X_{tt} is the standard form of the Lockhart-Martinelli parameter. The other parameters are those standard to the literature.

Kido et al. [18] experimentally measured the relative performances of seven microfin tubes for R-22 at one saturation pressure (4.9 bar), one heat flux (9.3 kW/m^2) and three mass velocities (86, 173 and 345 kg/m^{2}s) and showed that the microfin evaporation correlation of Fujii et al. [8] {they gave the Fujii et al. equations in their paper} did not agree with their data at all. Kido and coworkers then proposed a new microfin correlation that fit 80% of their data to within ±20%; it incorporates the microfin dimensions and hence has a general format. Yet, it is completely empirical with no attempt to physically model any thermal enhancement mechanisms and must still be compared to other heat fluxes, pressures, fluids, tube diameters, microfin geometries and *longer* test sections [their test section was only 30 cm long and was heated by condensing R-114 on the outside surface].

Koyama et al. [19] ran R-22, R-134a and R-123 evaporation tests inside a microfin tube. They proposed a new microfin flow boiling correlation based on these data, but apparently it can only be used if corresponding liquid-liquid tests are run on the particular microfin tube to determine the leading empirical coefficient to input into the Dittus-Boelter form of correlation. Besides the mean internal diameter, the only other dimensional related factor that can be input into their correlation is the surface area ratio and hence fin height and helix angle effects on heat transfer are ignored. Evenso, they demonstrated that this method worked reasonably well for some sets of independent data too, but it is not clear how they obtained the leading empirical constant for each of these different tubes or if they used the same one for their own one microfin tube geometry.

Kandlikar and Raykoff [12] have presented a revised version of the earlier Kandlikar [11] curvefitting approach for microfin tubes, retaining the fluid specific correction factors from that study, i.e. 1.5 for R-12, 2.2 for R-22, 1.3 for R-113, 1.9 for R-123 and 1.63 for R-134a, similar to the approach in Kandlikar's plain tube flow boiling method. For microfin tubes, each particular microfin geometry (this means *each* individual microfin tube, *each* tube diameter and *each* fluid) required an additional set of three new empirical constants to fit the equations to each data set. In addition, experimental data for single-phase turbulent flow in the particular microfin tube are required as an input into their method [hence this is not a general design method]. Once all this is done, their simple set of equations that can be fit to a particular microfin tube for one fluid at one saturation temperature, but the method still cannot predict the experimental peak in h vs. x observed experimentally nor high vapor quality data with partial tube dryout.

Since these authors have persisted with this unusual approach to calculating microfin heat transfer coefficients, it is worth spending a little time to analyze the negative aspects of their method. For example, the empirical convective boiling correction factors required to fit this method to individual microfin tube experimental data ranged in value from 0.0241 to 11.0, i.e. a ratio of 456-to-1! Similarly, the nucleate boiling correction factors for various tubes ran from 0.00524 to 5.8, i.e. a ratio of 1107-to-1! These tremendous variations in 'empirical' constants are proof in themselves that this method is not capable of capturing the trends in the data but has to be forcefit to each set of test data for each fluid and each tube. Similarly, while on this topic, there seems to be no benefit to use the corresponding plain tube Kandlikar [10] correlation with its fluid specific constants either since: (i) other correlations without resorting to such factors provide equal or better accuracy, (ii) the fluid specific factors change substantially from one fluid to another even though the method uses the physical properties of the fluid, (iii) the method's choice of the larger of two calculated heat transfer coefficients originally introduce by Shah [27] is now out-of-date since test data show that a smooth transition occurs from the nucleate boiling to the convective evaporation dominated regime, not a step-change in heat transfer coefficient, (iv) the method is not amenable to extension of zeotropic blends, refrigerant-oil mixtures, new untested fluids, tube diameters, etc.

Thome, Kattan and Favrat [36] have presented a new microfin flow boiling model with local predictions of microfin coefficients, valid for vapor qualities from 0.15 to 0.81, heat fluxes from 2-47 kW/m^2 and mass velocities from 100-501 kg/m^{2}s. Their method is based on data for R-134a (T_{sat} from -1.35°C to +10.6°C) and R-123 (T_{sat} of 29.6°C) and pressures from 1.07 to 4.22 bar (it is now being compared to their new R-407C and ammonia microfin data). In its present form it is used without a flow pattern map since flow regime observations showed that the microfins converted stratified-wavy flows to annular flows, i.e. one of the physical mechanisms that makes microfins so thermally effective. The method is an outgrowth of the first flow boiling model developed for evaporation in microfin tubes available in the commercial software EHT developed by Thome [31] and described in Thome [33]. The microfin tube in the tests had the following geometric characteristics: 70 internal microfins, 18° helix angle, microfin height of 0.25 mm, internal diameter at root of microfins of 11.9 mm, and an internal surface area of 0.065 m^2/m (i.e. I.D. area ratio of 1.739).

The new Thome, Kattan and Favrat [36] microfin flow boiling model is a heat transfer augmentation approach incorporated into an asymptotic flow boiling model, where the local microfin flow boiling heat transfer coefficient $h_{tp,mf}$ is determined from the following equation

$$h_{tp,mf} = E_{mf} \left[(h_{nb})^3 + (E_{RB} \, h_{cb})^3 \right]^{1/3} \tag{4}$$

where h_{nb} is obtained with the Cooper [3] nucleate pool boiling correlation for pure fluids given as

$$h_{nb} = 55 \, P_r^{0.12} \, (-\log_{10} P_r)^{-0.55} \, M^{-0.5} \, q^{0.67} \tag{5}$$

where h_{nb} is in W/m^{2}K (note: his correlation is not dimensionless), P_r is the reduced pressure, M is the molecular weight and q is the total local heat flux in W/m^2 determined from the total internal surface area (not the nominal I.D. area). Since the microfins are small in height and made from copper, their fin efficiencies are very close to 100% and this is what is assumed. Consequently, nucleate boiling is augmented in the microfin tube by the increase in internal wetted surface area relative to that of a plain bore tube.

The convective flow boiling contribution to horizontal flow boiling in a microfin tube ($E_{RB} \, h_{cb}$) is obtained from the turbulent film flow model for annular flow developed by Kattan, Thome and Favrat [14, 15, 16] for plain bore tubes. Their plain tube convective flow boiling coefficient is h_{cb} and it is multiplied by the microfin convection factor E_{RB}. This is the enhancement factor in the Ravigururajan and Bergles [25] single-phase turbulent tube flow correlation for internally ribbed tubes and is given as:

$$E_{RB} = \left\{ 1 + \left[2.64 \, (Re_D)^{0.036} \, (e_f/D_f)^{0.212} \, (p_f/D_f)^{-0.21} \, (\alpha_f/90°)^{0.29} \, (Pr_L)^{-0.024} \right]^7 \right\}^{1/7} \tag{6}$$

where e_f is the microfin height (in m), p_f in the axial pitch from fin to fin (in m), D_f is the internal tube diameter at the root of the fins (in m), α_f is the helix angle of the microfins (in °). Pr_L is the liquid Prandtl number and Re_D in Eq. (6) is the liquid-phase tubular Reynolds number defined as

$$Re_D = G\ (1\text{-x})\ D_f/\mu_L \tag{7}$$

where G is the mass velocity based on the total flow of liquid and vapor inside a tube of maximum internal diameter D_f (in kg/m^{2}s), x is the local vapor quality and μ_L is the liquid dynamic viscosity (in Ns/m^2).

The turbulent film flow model for annular flow presented in Kattan, Thome and Favrat [16] for plain bore tubes was developed from an experimental database for refrigerants R-123, R-134a, R-502, R-402A and R-404A (since shown to predict R-407C and ammonia flow boiling data without modification in two new papers currently under review for publication). Their convective heat transfer coefficient for the annular film, h_{cb}, is obtained from the turbulent film flow equation:

$$h_{cb} = 0.0133\{(Re_L)_{film}\}^{0.69}\ (Pr_L)^{0.4}\ (\lambda_L/\delta) \tag{8}$$

where λ_L is the liquid thermal conductivity (note: the constants 0.0133 and 0.69 are those they developed from their original plain tube database, not the present microfin database). In this expression, the liquid film Reynolds number is determined from the mean velocity of the liquid in the annular film using the local void fraction as:

$$(Re_L)_{film} = 4\ G\ (1\text{-x})\ \delta/[(1\text{-}\alpha)\ \mu_L] \tag{9}$$

where G is the mass velocity as defined above, α is the local void fraction, and δ is the local thickness of the annular liquid film (ignoring any effect of the microfins on α and δ in the present application). The local void fraction is determined using the Rouhani and Axelsson [26] void fraction correlation for plain tubes since no microfin method is available

$$\alpha = (x/\rho_v)[\{1{+}0.12(1\text{-x})\}[(x/\rho_v){+}((1\text{-x})/\rho_L)]$$
$$+\{1.18(1\text{-x})[g\sigma(\rho_L\text{-}\rho_v)]^{0.25}/(G\rho_L^{0.5})\}]^{-1} \tag{10}$$

while the local annular liquid film thickness is calculated from the cross-sectional area occupied by the liquid phase assuming uniform thickness of the film around the tube perimeter as:

$$\delta = D_f\ (1\text{-}\alpha)/4 \tag{11}$$

where ρ_L is the liquid density, ρ_v is vapor density, g is gravitational acceleration (9.81 m/s^2), and σ is the surface tension of the fluid (all in SI units).

482

The Grigorig effect draws liquid from the microfin tips towards their root area, which enhances film evaporation, similar to film condensation on low finned tubes. In addition, the convection enhancement factor E_{RB} is for *tubular* flow, not *film* flow. Thus, an addition enhancement factor for microfin tubes, E_{mf}, has been required to account for these two effects (which is the only factor added to the present model specifically based on the present microfin test data). E_{mf} is given as

$$E_{mf} = 1.89 \, (G/G_{ref})^2 - 3.7 \, (G/G_{ref}) + 3.02 \tag{12}$$

G_{ref} is a reference value introduced to non-dimensionalize the expression with G_{ref} set to the maximum value of the present test values ($G_{ref} = 500$ kg/m^{2}s). Since normal operating conditions for microfin tubes in direct-expansion evaporators vary from 50-500 kg/m^{2}s, this expression is sufficient for now until a mechanistic model can be developed. The value of E_{mf} ranges from 2.36 at 100 kg/m^{2}s to 1.21 at 500 kg/m^{2}s.

A statistical comparison to a total of 362 local heat transfer coefficients a standard deviation, a mean deviation and an average deviation for the R-134a data of 18.5%, 12.8% and 2.0%, respectively, while for R-123 the corresponding values were 12.9%, 11.8% and 6.4%. Note that this new prediction method is a generalized form and only microfin physical dimensions, fluid physical properties and operating conditions (G, x and q) are entered, not fluid specific nor microfin specific constants.

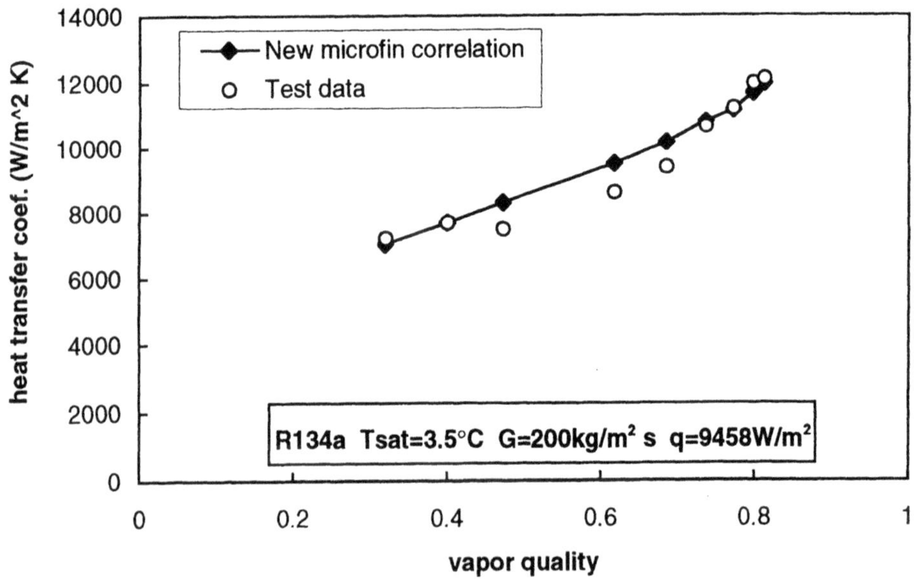

Figure 9. Thome, Kattan and Favrat [36] microfin model compared to R-134a data.

Figure 9 depicts a comparison of the microfin tube model to R-134a test data at G = 200 kg/m^{2}s. Not only are the predictions accurate, but the slope of the change in h vs.

x is also modelled well. Figure 10 shows the predicted local augmentation ratios averaged over a vapor quality from 0.15-0.85 and plotted as a function of mass velocity for R-134a, showing that the model predicts large augmentations at low mass velocities and tends towards the microfin's internal area ratio at high mass velocities, which is the typical trend in nearly all published experimental results. Future work is required to compare it to more microfin geometries and test fluids and to extend it to high vapor qualities by including an onset of dryout criterion, like that in the new annular film flow model of Kattan, Thome and Favrat [16] for plain tubes.

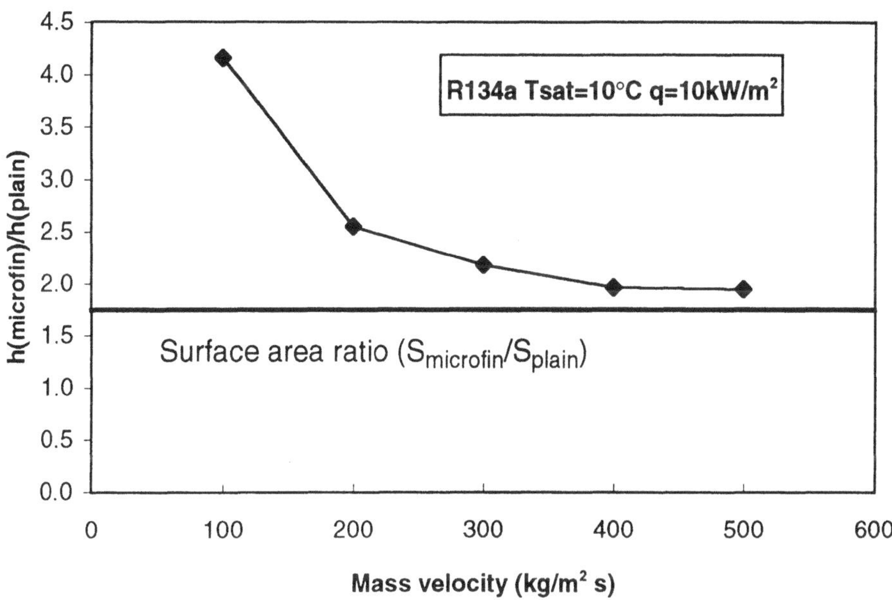

Figure 10. Augmentation ratios predicted with Thome,
Kattan and Favrat [36] microfin model.

5. Conclusions

The augmentation of microfin heat transfer coefficients relative to plain tube values was shown to result from:

- Increase in the convective contribution by the single-phase effect of the microfins;
- A rise in the nucleate boiling contribution by the additional internal surface area;
- Enhancement of annular film evaporation by the liquid film thinning effect of the microfins, i.e. Gregorig effect;
- The thermally effective conversion of stratified-wavy flow into annular flow at low mass velocities.

Recent experimental studies on evaporation inside microfin tubes were surveyed and trends in their results discussed. Future work should concentrate more on measurement of local values using hot water heating rather than measurement of mean data or use of electrical heating. Also tests should be run up to vapor qualities approaching 1.0. Single-phase tests for subcooled liquid and superheated vapor would also be useful for model building.

Heat transfer models, empirical methods and curvefitting approaches available for evaporation inside microfin tubes were reviewed. Some promising developments and advances have been made, but much more work is required in this area.

6. Nomenclature

D_f	maximum internal diameter at root of the fins [m]
D_{mean}	internal diameter at half the fin height [m]
E_{mf}	additional enhancement factor [-]
E_{RB}	single-phase tubular flow enhancement factor [-]
e_f	fin height [m]
G	mass velocity [kg/m^{2}s]
G_{ref}	= 500 kg/m^{2}s
h_{cb}	convective boiling heat transfer coefficient [W/m^{2}K]
h_{nb}	nucleate boiling heat transfer coefficient [W/m^{2}K]
$h_{tp,mf}$	local two-phase, microfin heat transfer coefficient [W/m^{2}K]
M	molecular weight of liquid [-]
Nu_L	liquid Nusselt number [-]
$Nu_{microfin}$	microfin Nusselt number [-]
p_f	axial pitch of fins [m]
P_r	reduced pressure [-]
Pr_L	liquid Prandtl number [-]
q	heat flux [W/m^2]
Re_D	tubular liquid Reynolds number in Eq. (7) [-]
Re_L	liquid Reynolds number in Eq. (3) [-]
$(Re_L)_{film}$	liquid Reynolds number of film in Eq. (9) [-]
X_{tt}	Lockhart-Martinelli parameter [-]
x	vapor quality [-]
α	void fraction in Eqs. (9-11) [-]
α_f	helix angle of microfins [°]
δ_L	thickness of liquid film [m]
μ_L	dynamic viscosity of the liquid [Ns/m^2]
ρ_L	density of the liquid [kg/m^3]
ρ_v	density of the vapor [kg/m^3]
σ	surface tension [N/m]

7. References

1. Chamra, L.M. and Webb, R.L. (1995). Condensation and Evaporation in Micro-Fin Tubes at Equal Saturation Temperatures, *J. Enhanced Heat Transfer* **2** (**3**), 219-229.
2. Chamra, L.M, Webb, R.L. and Randlett, M.R. (1996). Advanced Micro-Fin Tubes for Evaporation, *Int. J. Heat Mass Transfer* **39**(**9**), 1827-1838.
3. Cooper, M.G. (1984). Saturated Nucleate Pool Boiling-A Simple Correlation, Int. Chem. Engng. Symp. Ser. **86**, 785-792 and also in *Advances in Heat Transfer*, Academic Press, Orlando **16**, 157-139.
4. Eckels, S.J., Doerr, T.M. and Pate, M.B. (1994). In-Tube Heat Transfer and Pressure Drop of R-134a and Ester Lubricant Mixtures in a Smooth Tube and a Micro-Fin Tube: Part I-Evaporation, *ASHRAE Trans.* **100**(**2**), 265-282.
5. Eckels, S.J. and Pate, M.B. (1991). An Experimental Comparison of Evaporation and Condensation Heat Transfer Coefficients for HFC-134a and CFC-12, *Int. J. Refrig.* **14**(**3**), 70-77.
6. Eckels, S.J. and Pate, M.B. (1991). Evaporation and Condensation of HFC-134a and CFC-12 in a Smooth Tube and a Micro-Fin tube, *ASHRAE Trans.* **97**(**2**), 71-81.
7. Eckels, S.J., Pate, M.B. and Bemisderfer, C.H. (1992). Evaporation Heat Transfer Coefficients for R-22 in Micro-Fin Tubes of Different Configurations, *J. Enhanced Heat Transfer*, ASME HTD **202**, 117-125.
8. Fujii, T., Koyama, S., Inoue, N., Kuwahara, K. and Hirakumi, S. (1993). An Experimental Study of Evaporation Heat Transfer of Refrigerant HCFC22 inside an Internally Grooved Horizontal Tube, *Trans. of JSME* **59**(**562**), 2032-2042 (in Japanese).
9. Gungor, K.E. and Winterton, R.H.S. (1986). A General Correlation for Flow Boiling in Tubes and Annuli, *Int. J. Heat Mass Transfer* **29**(**3**), 351-358.
10. Kandlikar, S.G. (1990). A General Correlation for Saturated Two-Phase Flow Boiling Heat Transfer inside Horizontal and Vertical Tubes, *J. Heat Transfer* **112**, 219-228.
11. Kandlikar, S.G. (1991). A Model for Predicting the Two-Phase Flow Boiling Heat Transfer Boiling Heat Transfer Coefficient in Augmented Tubes and Compact Evaporator Geometries, *J. Heat Transfer* **113**, 966-972.
12. Kandlikar, S.G. and Raykoff, T. (1997). Predicting Flow Boiling Heat Transfer of Refrigerants in Microfin Tubes, *J. Enhanced Heat Transfer* **4**(**4**), 257-268.
13. Kattan, N., Thome, J.R. and Favrat, D. (1995). Boiling of R-134a and R-123 in a Microfin Tube, Proc. 19th Int. Congress of Refrigeration, The Hague, IVa, 337-344.
14. Kattan, N., Thome, J.R. and Favrat, D. (1998). Flow Boiling in Horizontal Tubes. Part 1: Development of a Diabatic Two-Phase Flow Pattern Map, *J. Heat Transfer* **120**(**1**), 140-147.
15. Kattan, N., Thome, J.R. and Favrat, D. (1998). Flow Boiling in Horizontal Tubes. Part 2: New Heat Transfer Data for Five Refrigerants, *J. Heat Transfer* **120**(**1**), 148-155.
16. Kattan, N., Thome, J.R. and Favrat, D. (1998). "Flow Boiling in Horizontal Tubes. Part 3: Development of a New Heat Transfer Model Based on Flow Patterns," *J. Heat Transfer* **120**(**1**), 156-165.
17. Kaul, M.P., Kedzierski, M.A. and Didion, D.A. (1996). Horizontal Flow Boiling of Alternative Refrigerants within a Fluid Heated Micro-Fin Tube, in *Process, Enhanced, and Multiphase Heat Transfer, A Festschirft for A.E. Bergles*, Begell House, New York, 167-173.
18. Kido, O., Taniguchi, M., Taira, T. and Uehara, H. (1995). Evaporation Heat Transfer of HCFC22 inside an Internally Grooved Horizontal Tube, *ASME/JSME Thermal Engineering Joint Conference*, Maui, March 19-24, **2**, 323-330.
19. Koyama, Sh., Yu, J., Momoki, S., Fujii, T. and Honda, H. (1996). Forced Convective Flow Boiling Heat Transfer of Pure Refrigerants inside a Horizontal Microfin Tube, *Convective Flow Boiling*, Eds. J.C. Chen et al., Taylor & Francis, 137-142. (Proc. of Convective Flow Boiling Conf. in Banff, Canada, April 30-May 5, 1995).
20. Kuo, C.S. and Wang, C.C. (1996). Horizontal Flow Boiling of R22 and R407C in a 9.52 mm Micro-Fin Tube, *Applied Thermal Engineering* **16**(**8/9**), 719-731.
21. Kuo, C.S. and Wang, C.C. (1996). In-tube Evaporation HFC-22 in a 9.52 mm Micro-Fin/Smooth Tube, *Int. J. Heat Mass Transfer* **39**(**12**), 2559-2569.
22. Kuo, C.S., Wang, C.C., Cheng, W.Y. and Lu, D.C. (1995). Evaporation of R-22 in a 7-mm Microfin Tube, *ASHRAE Trans.* **101**(**2**), 1055-1061.
23. Muzzio, A., Niro, A. and Arosio, S. (1998). Heat Transfer and Pressure Drop during Evaporation and Condensation of R22 inside 9.52-mm O.D. Microfin Tubes of Different Geometries, *J. Enhanced Heat Transfer* **5**(**1**), 39-52.
24. Nidegger, E., Thome, J.R. and Favrat, D. (1997). Flow Boiling and Pressure Drop Measurements for R-134a/Oil Mixtures Part 1: Evaporation in a Microfin Tube, *HVAC&R Research*, ASHRAE, **3**(**1**), 38-53.

25. Ravigururajan, T.S. and Bergles, A.E. (1985). General Correlations for Pressure Drop and Heat Transfer for Single-Phase Turbulent Flow in Internally Ribbed Tubes, *Augmentation of Heat Transfer in Energy Systems*, ASME HTD **52**, 9-20.

26. Rouhani, Z. and Axelsson, E. (1970). Calculation of Volume Void Fraction in the Subcooled and Quality Region, *Int. J. Heat and Mass Transfer* **13**, 383-393.

27. Shah, M.M. (1982). Chart Correlation for Saturated Boiling Heat Transfer: Equations and Further Study, *ASHRAE Trans.* **88(1)**, 185-196.

28. Singh, A., Ohadi, M.M. and Dessiatoun, S. (1996). Flow Boiling Heat Transfer Coefficients of R-134a in a Microfin Tube, *J. Heat Transfer* **118**, 497-499.

29. Sundaresan, S.G., Pate, M.B., Doerr, T.M. and Ray, D.T. (1996). A Comparison of the Effects of POE and Mineral Oil Lubricants on the In-Tube Evaporation of R-22, R-407C and R-410A, *Proc. 1996 International Refrigeration Conference at Purdue*, West Lafayette, IN, July 23-26, 187-192.

30. Thome, J.R. (1990). *Enhanced Boiling Heat Transfer*, Hemisphere, New York.

31. Thome, J.R. (1991). *Enhanced Heat Transfer Software*, John Thome Inc., 256 Lafayette Street, Ionia, MI 48846.

32. Thome, J.R. (1994). Two-Phase Heat Transfer to New Refrigerants, Special Keynote Lecture, *Proc. 10th International Heat Transfer Conf.*, Brighton, **1**, 19-41.

33. Thome, J.R. (1994). High Performance Augmentations for Refrigeration System Evaporators and Condensers, *Enhanced Heat Transfer* **1(3)**, 275-286.

34. Thome, J.R. (1996). Boiling of New Refrigerants: A State-of-the-Art Review, *Int. J. Refrig.* **19(7)**, 435-457.

35. Thome, J.R. (1997). Heat Transfer and Pressure Drop in the Dryout Region of Intube Evaporation with Refrigerant/Lubricant Mixtures, ASHRAE Final Report of Project 800-RP, February.

36. Thome, J.R., Kattan, N. and Favrat, D. (1997). Evaporation in Microfin Tubes: A Generalized Prediction Method, Convective Flow and Pool Boiling Conference, Engineering Foundation (NY), Kloster Irsee, May 18-23, Paper VII-4.

37. Thors, P. and Bogart, J.E. (1994). In-Tube Evaporation of HCFC-22 with Enhanced Tubes, *J. Enhanced Heat Transfer* **1(4)**, 365-377.

38. Torikoshi, K., Kawabata, K. and Ebisu, T. (1992). Heat Transfer and Pressure Drop Characteristics of HFC-134a in a Horizontal Heat Transfer Tube, *Proc. 1992 International Refrigeration Conf. at Purdue*, West Lafayette, IN, **1**, 167-176.

39. Torikoshi, K. and Ebitsu, T. (1994). In-Tube Heat Transfer Characteristics of Refrigerant Mixtures of HFC-32/134a and HFC-32/125/134a, *Proc. 1994 International Refrigeration Conf. at Purdue*, July 19-22, West Lafayette, IN, 293-298.

40. Uchida, M., Itoh, M., Shikazono, N. and Kudoh, M. (1996). Experimental Study on the Heat Transfer Performance of a Zeotropic Refrigerant Mixture in Horizontal Tubes, *Proc. 1996 International Refrigeration Conf. at Purdue*, July 23-26, West Lafayette, IN, 133-138.

41. Wattelet, J.P. et al. (1994). Heat Transfer Flow Regimes of Refrigerants in a Horizontal-Tube Evaporator, ACRC Report TR-55, University of Illinois at Urbana-Champaign, Urbana, IL.

42. Zürcher, O., Thome, J.R. and Favrat, D. (1997). Flow Boiling and Pressure Drop Measurements for R-134a/Oil Mixtures Part 2: Evaporation in a Plain Tube, *HVAC&R Research*, ASHRAE, **3(1)**, 54-64.

43. Zürcher, O., Thome, J.R. and Favrat, D. (1997). Flow Boiling of Ammonia in Smooth and Enhanced Horizontal Tubes, *Compression Systems with Natural Working Fluids*, IEA Annex 22 Workshop, Gatlinburg, TN, USA, October 2-3.

44. Zürcher, O., Thome, J.R. and Favrat, D. (1998). Intube Flow Boiling of R-407C and R-407C/Oil Mixtures Part I: Microfin Tube, *HVAC&R Research*, ASHRAE, **4(4)**,{see Thome [35]}.

FLOW BOILING OF REFRIGERANT-OIL MIXTURES IN PLAIN AND ENHANCED TUBES

JOHN R. THOME
International Consulting Engineer
Via S. Angela Merici 30, Rome 00162, ITALY
Present Address: Professor, LTCM-DGM, EPFL, CH-1015 Lausanne

Abstract: Research on heat transfer to refrigerants has gained importance in the 1990's in response to the massive conversion of the refrigeration and air-conditioning industry from refrigerants 'retired' by the Montreal protocol (R-11, R-12, R-502, etc.) to more environmentally-safe refrigerants (R-134a, R-123, R-407C, R-410A, ammonia, hydrocarbons, etc.). Lubricating oil from these systems' compressors, absorbed by the refrigerant charge circulating in these systems, can have a dramatic effect on boiling performance and represents one of the old, unresolved problems of refrigeration heat transfer. The present survey presents a summary of the recent research on evaporation of refrigerant-oil mixtures inside plain and microfin tubes, including experimental studies and new prediction models.

1. Introduction

Intube evaporation inside plain and enhanced tubes is an important design problem for the thermal design of direct-expansion evaporators widely used for refrigeration, air-conditioning and heat pump systems. The tubes are nearly always horizontal and may or may not have internal augmentation. The most commonly used internal augmentation is now the microfin geometry, which refers to numerous small internal fins of about 0.1-0.4 mm height that can be longitudinal or helical and either two-dimensional (i.e. plain microfins) or three-dimensional (i.e. crosscut or notched microfins). These types of tubes are commonly available in copper but some versions are available in high alloys (for example stainless steels and titanium) and in carbon steel and aluminum (for ammonia systems). Plain tubes are also used in these applications, especially for air-heated coils where the limiting thermal resistance is on the outside of the tube; however, even for these units, microfins are often used to achieve a more economical and compact design.

One major longstanding problem with thermal design and operation of these units is the effect of lubricating oil on the evaporation performance and two-phase pressure drop. Lubricating oil from the compressor's bearings can enter the refrigerant as a

S. Kakaç et al. (eds.), Heat Transfer Enhancement of Heat Exchangers, 487–513.

mist in a vapor compression cycle, dissolve into the liquid refrigerant in the condenser and then end up circulating all the way around the flow circuit to the evaporator. The amount of oil may seem to be negligible, often ranging from only 0.5-1.0 wt.% but up to 3-5 wt.% in severe cases, yet it can have a very detrimental effect to evaporator performance.

The normal boiling point of a lubricating oil is about 300°C (572°F) or higher; hence it is very high relative to that of a refrigeration system evaporator, and thus the oil is essentially involatile at these operating conditions. Hence, this means that the oil remains in its liquid-phase, either dissolved in the liquid refrigerant (if miscible) or in the form of entrained droplets at the exit of the evaporator. Therefore, for a certain mass of oil in the refrigerant charge, the oil concentration in the liquid-phase will tend to increase along the evaporation flow path through a direct-expansion evaporator as more and more liquid refrigerant is evaporated. Consequently, the highest oil concentration in the liquid-phase will be at the outlet and its lowest concentration at the inlet.

As a further illustration of the variation of oil concentration within an evaporator, assume for example that there is 1.0 wt.% oil in the total mass flow of subcooled liquid leaving the condenser and heading towards the expansion device, i.e. 1 g of oil for every 100 g of liquid refrigerant and oil. Leaving the expansion device (valve or capillary tube) at say a local vapor quality of 0.25, the local oil concentration in the remaining liquid will be 1.33 wt.%, i.e. 1 g of oil per 75 g of the remaining liquid (74 g of liquid refrigerant + 1 g of oil) at that location. In the evaporator at a local vapor quality of 0.95, the local oil concentration in the remaining liquid will be 20 wt.%, i.e. 1 g of oil in every 5 g of liquid. At the outlet of the evaporator, the local vapor quality when oil is present cannot reach 1.0 since the oil cannot be evaporated (its vapor pressure is only on the order of 1 millionth that of the refrigerant). In fact, for a nominal 1.0 wt.% oil circulating in the unit, the highest local vapor quality that can be attained is somwhat less than 0.99. Since the local oil concentration increases with local vapor quality, it can thus be expected to have its dominiant effect on thermal performance (heat transfer, two-phase pressure drop and local boiling point temperature) at the high quality end of a direct-expansion evaporator.

The evaporation of refrigerant/lubricating oil mixtures is thus a thermodynamic problem as well as a heat transfer problem, analogous to evaporation of zeotropic refrigerant mixtures except that a refrigerant-oil mixture has a temperature glide on the order of 300°C since the oil increases the local saturation temperature with increasing oil concentration. Furthermore, to be thermodynamically correct and to respect the definition of the boiling heat transfer coefficient, this local bubble point temperature of a refrigerant-oil mixture should be used in calculating the boiling wall superheat between the tube wall and the refrigerant-oil mixture, not the saturation temperature of the pure refrigerant. In addition, the physical properties of the liquid can be significantly affected by the oil, especially the liquid dynamic viscosity where a lubricating oil has a viscosity on the order of 1000-1200 times that of a liquid refrigerant at normal evaporation temperatures. Therefore, a nominal 1.0 wt.% oil mixture will substantially increase the local liquid viscosity at the point in the evaporator where the local vapor quality reaches 0.95 and thus the oil's local concentration in the liquid is 20 wt.%, such that the local refrigerant-oil mixture can attain liquid viscosities as high as 35 cp compared to 0.2-0.35 cp for pure refrigerants such as R-134a or R-22, i.e. an increase of 100-175 times! Other key physical properties are also affected, but to a lesser extent.

The topic of this review is the influence of oil on intube evaporation of refrigerant-oil mixtures. Previous reviews on this subject have been presented by Schlager, Pate and Bergles [16] and by Thome [19, 20, 21, 23]. Consequently, the present review will only cover the more recent work on this topic.

2. Summary of Oil Effects on Intube Evaporation

Figure 1 from Zürcher, Thome and Favrat [27] depicts local flow boiling data for R-134a/oil mixtures evaporating inside a plain horizontal tube of 10.92 mm ID at 3.43 bar, where the local measured heat transfer coefficients with and without oil (h oil) have been normalized using calculated values for pure R-134a (h no oil) obtained with the new intube flow boiling model of Kattan, Thome and Favrat [10, 11, 12]. The test conditions were at two fixed mass velocities of 100 and 300 kg/m^{2}s and nominal inlet oil concentrations of 0, 0.5, 1.0, 3.0 and 5.0 wt.% oil. This figure summarizes many of the trends typically observed in local flow boiling data for refrigerant-oil mixtures inside plain tubes:

- At low mass velocity and intermediate vapor qualities, the oil enhances heat transfer by up to about 60% depending on inlet oil concentration;
- At higher mass velocity and intermediate vapor qualities, the oil still enhances heat transfer but by a much smaller amount (0-20%);
- At high vapor qualities for all mass velocities, the oil has a very detrimental influence on local heat transfer, resulting in performance degradations of up to 80% or more.

Hence, the effect of oil on evaporation heat transfer can either be very positive, insignificant or very negative, depending on the local operating conditions at that point in an evaporator. Also, notice in Figure 1 that the new flow boiling model predicts these new pure R-134a data (0% curves) all to within -8% and +21%, where the flow pattern is stratified-wavy at 100 kg/m^{2}s and annular at G=300 kg/m^{2}s over the wide vapor quality range from 0.18-0.93. Hence, for accurate thermal design of direct-expansion evaporators with inevitably some oil present in the refrigerant charge, whose amount depends on the type of compressor and effectiveness on the oil separator (if present), the influence of oil on the local heat transfer coefficient must be taken into account, especially at high vapor qualities where the local coefficient becomes very low and is hence the controlling thermal resistance in the overall heat transfer coefficient.

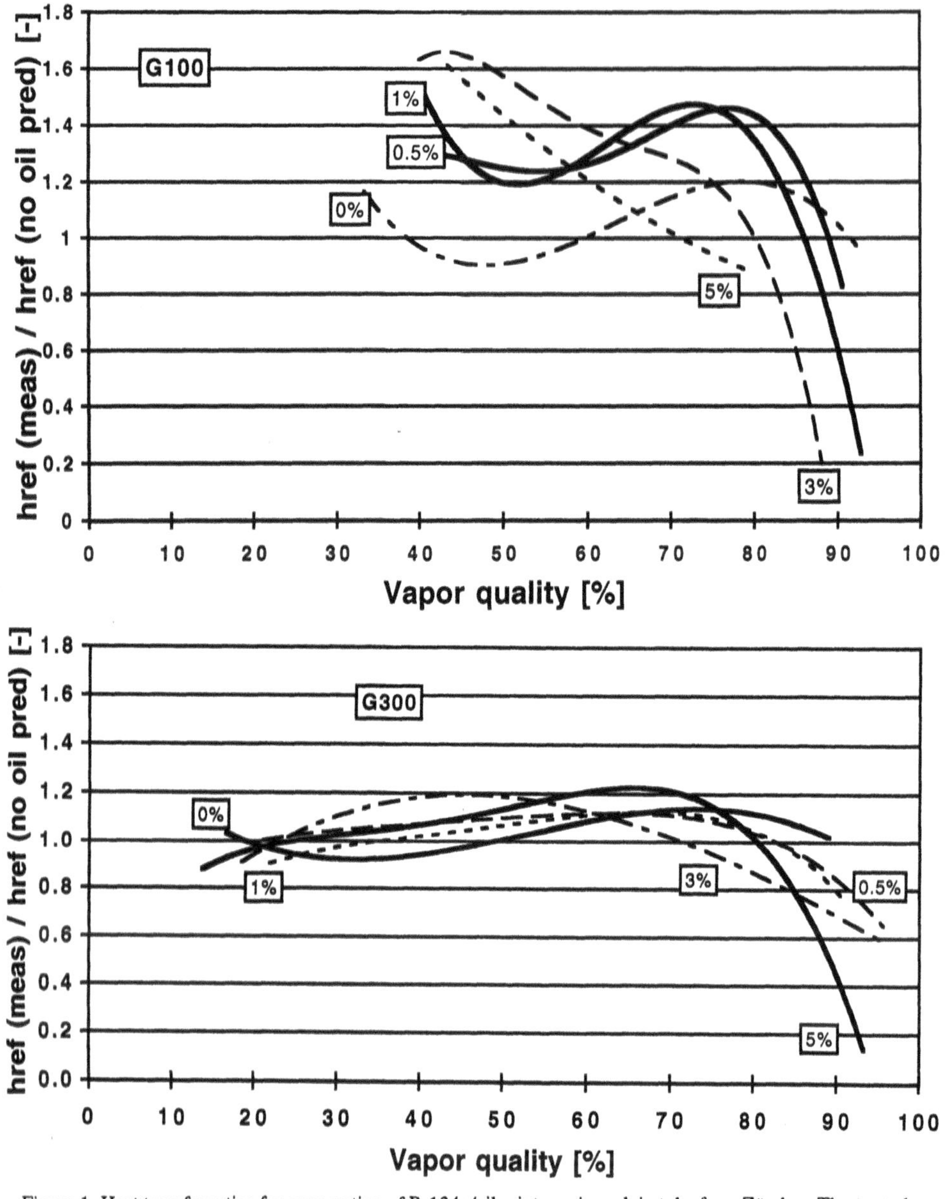

Figure 1. Heat transfer ratios for evaporation of R-134a/oil mixtures in a plain tube from Zürcher, Thome and Favrat [27] at two mass velocities at 3.43 bar.

For evaporation inside microfin tubes, Figure 2 from Zürcher, Thome and Favrat [28] {or see Thome [24]} illustrates the typical effects of oil in the same format as in Figure 1 but for R-407C/oil mixtures. The microfin tube tested had an 11.90 mm internal diameter (at root on internal fins) with 70 helical fins of 0.254 mm height, 18° helix angle, and internal area ratio of 1.74. The local heat transfer coefficients with oil were normalized using the experimental data for pure R-407C. The test conditions were at two fixed mass velocities of 100 and 300 kg/m²s and nominal inlet oil concentrations of 0, 0.5, 1.0, 3.0 and 5.0 wt.% oil. These graphs illustrate the trends typically observed in local microfin flow boiling data for refrigerant-oil mixtures:

- At low mass velocities and intermediate vapor qualities, the oil can either have a very detrimental effect or nearly no effect on heat transfer such that the influence varies from about -30% to -1%;

- At higher mass velocity and intermediate vapor qualities, the oil still degrades heat transfer but by a smaller amount (about -20% to +10%);

- At high vapor qualities for all mass velocities, the oil has a very detrimental influence on local heat transfer, resulting in performance degradations of up to 90%.

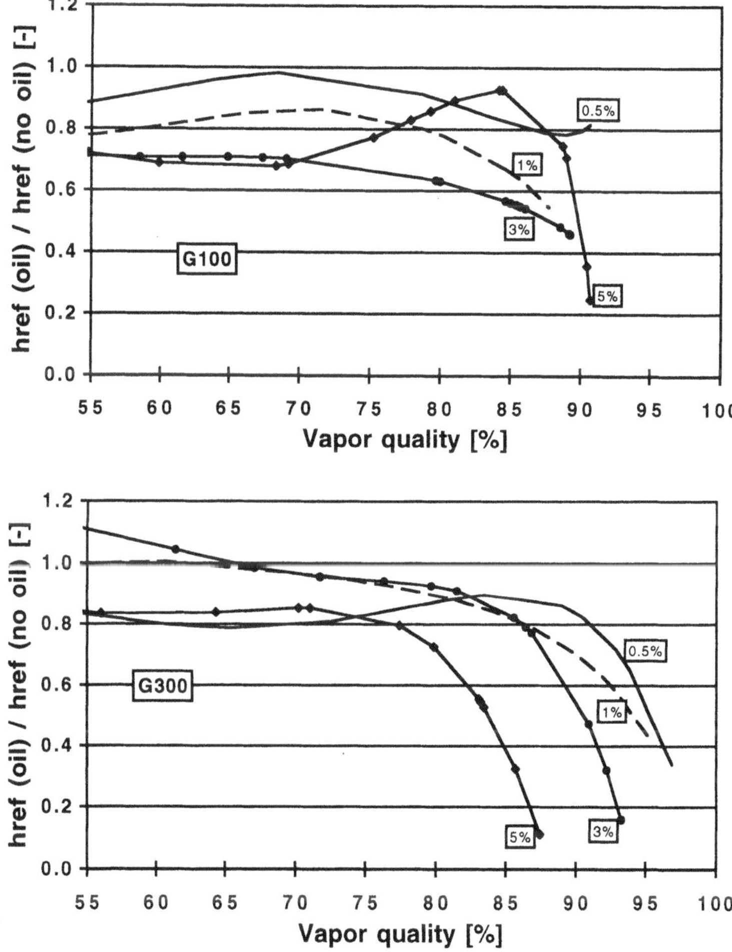

Figure 2. Heat transfer ratios for evaporation of R-407C/oil mixtures in a microfin tube from Zürcher, Thome and Favrat [28] {or see Thome [24]} at two mass velocities at 6.45 bar.

Hence, the effect of oil on evaporation heat transfer can either be slightly positive, insignificant or very negative, depending on the local operating conditions at that point in an evaporator.

Twelve effects of lubricating oil on intube flow boiling heat transfer are identified and summarized below, which are thought to be important to understanding and explaining the above trends in $h_{oil}/h_{no\ oil}$ and similar ones in other published test data for plain and enhanced tubes:

Bubble point temperatures of refrigerant-oil mixtures. Oil has a much higher boiling point than common refrigerants and hence adding oil to a refrigerant increases its bubble point temperature relative to the saturation temperature of the pure refrigerant. This effect becomes increasingly important at local oil mass fractions greater than 10-20 wt.% oil.

Oil effects on thermodynamic and transport properties. Lubricating oils have much larger molecular weights than common refrigerants and much lower critical pressures. This has a significant affect on liquid density (often the density difference is as much as 300-400 kg/m^3), liquid viscosity, liquid specific heat, critical pressure and surface tension. Since the oil is essentially involatile at refrigeration operating conditions, the oil only effects liquid-phase properties but not vapor-phase properties since only very trace amounts of oil enter the vapor-phase as a vapor.

Oil effects on enthalpy. Adding oil to a refrigerant has primarily two effects: it increases the bubble point temperature (as noted above) and it changes the liquid specific heat as a function of oil concentration. The combined influence of these two effects is that the change in enthalpy during the evaporation of a refrigerant-oil mixture includes both sensible and latent heat while isobaric evaporation of a pure refrigerant only involves latent heat transport.

Oil viscosity effects on convection. The rapid rise in local liquid viscosity with local oil mass fraction as local vapor quality approaches 1.0 has a detrimental effect on the liquid only heat transfer coefficient, h_L, utilized in modelling flow boiling heat transfer (h_L in turbulent flow is inversely proportional to μ_L);

Oil holdup effects. The actual local oil mass fraction can be much higher than expected at some flow conditions because the vapor shear is not strong enough to effectively drag the viscous refrigerant-oil liquid film along the tube, especially in microfin tubes, and thus higher local oil mass fraction increases the local liquid viscosity and bubble point temperature with respect to the values otherwise expected;

Foaming effects. Some oils promote foaming with particular refrigerants, which may influence the onset of dryout in evaporating annular flows and shift its location to lower than expected vapor qualities and/or it may convert a stratified flow into an unstratified flow by virtue of the height of the foam above the normal liquid level.

Significant increases in local evaporating heat transfer coefficients by the oil at intermediate vapor qualities are often ascribed to foaming effects (however, since these are also observed for systems without foaming, foaming alone cannot explain this enhancing effect). Foaming probably has an effect on local void fractions and the transition from one two-phase flow pattern to another;

Wetting effects of oil. The higher surface tension of oil and its surfactant effect will tend to increase the wetted fraction of the tube perimeter in stratified flows in horizontal tubes and hence could be a key mechanism responsible for an oil's heat transfer augmentation effect at low to medium vapor qualities in plain tubes mentioned above;

Flow pattern changes due to the oil. The physical property changes with oil concentration have the potential to modify the threshold boundary curves between different flow patterns. For example Manwell and Bergles [14] have reported two-phase flow patterns for R-12/oil mixtures and discussed what changes they observed [I personally have viewed hundreds of such video sequences for refrigerant-oil mixtures, which are waiting to be systematically analyzed];

Mass transfer effect of oil on heat transfer. Oil builds up at evaporating interfaces just like the heavier component(s) in zeotropic refrigerant blends, creating a concentration gradient and thus increasing the bubble point temperature at the vapor-liquid interface, which in turn reduces the superheat driving the evaporation process and consequently adversely affects heat transfer;

Viscous transition to laminar film flow. Since an oil is essentially involatile, it becomes increasingly concentrated in an evaporating annular liquid film, i.e. annular flow, and prevents the film from drying out at high vapor quality as would be usual for a pure refrigerant. Consequently, for refrigerant-oil mixtures the less profound peaks in h vs. x observed experimentally may represent a transition in the liquid annular film from a turbulent regime to the less thermally effective laminar regime.

Nucleate boiling augmentation. An oil may increase the nucleate boiling contribution to flow boiling over the low and intermediate vapor quality range {as discussed for nucleate pool boiling of refrigerant-oil mixtures in Thome [23]}.

Solubility effects. Some refrigerant-oil combinations are soluble (and miscible) over the entire concentration range while others are only soluble for oil mass fractions less than 5-10 wt.% depending on the temperature. Consequently, at high vapor qualities where the liquid can reach 30-60 wt.% oil, the local mixture may no longer be in the soluble range {for instance, refer to Hewitt and McMullan [9]}.

The complex trends observed experimentally for evaporation inside plain and microfin tubes with oil dissolved in the refrigerant are a composite result of these effects, whose individual impacts on heat transfer depend on local flow conditions (oil concentration, mass velocity, vapor quality, heat flux, flow pattern, plain or enhanced surface, foaming if any, etc.).

3. Thermodynamics of Refrigerant-Oil Mixtures

3.1. DEFINITION OF HEAT TRANSFER COEFFICIENT

Thome [22] has presented a detailed discussion of the thermodynamic effects on boiling points and enthalpy caused by oil. To include oil effects correctly in measuring and modelling flow boiling coefficients, the local bubble point temperatures and the Temperature-enthalpy-vapor quality (T-h-x) relationship respecting the variation in the refrigerant-oil concentration in the liquid-phase should be calculated and be used for determining the local temperature differences and energy balances to determine local vapor quality, analogous to how these concepts are applied to zeotropic refrigerant mixtures. Since the boiling heat transfer coefficient h is defined as h = $q/(T_{wall}-T_{bub})$, using the higher value of T_{bub} as appropriate for refrigerant-oil mixtures rather than T_{sat} for pure refrigerant results in an increase in the heat transfer coefficient. On the other hand, use of T_{bub} rather than T_{sat} for the pure refrigerant in an incremental LMTD calculation for an evaporator will decrease the value of the LMTD, which can partially compensate for the larger boiling coefficient when using T_{bub} in the definition of h. In any case, it should be remembered that the difference between using T_{bub} or T_{sat} typically only becomes significant when reducing test data at vapor qualities greater than about 80%.

3.2. BUBBLE POINT TEMPERATURES

Takaishi and Oguchi [18] developed an empirical method for predicting bubble point temperatures for a specific lubricating oil mixed with R-22. The vapor pressure equation for predicting the bubble point temperature from their measurements is as follows:

$$\ln (P_{sat}) = [A(w_{oil})/T_{bub}] + B(w_{oil}) \tag{1}$$

where P_{sat} is the saturation pressure in MPa and T_{bub} is the bubble point temperature in K. Equation (1) can be rearranged in order to calculate the bubble point temperature for a given saturation pressure and oil concentration as:

$$T_{bub} = A (w_{oil})/[\ln (P_{sat}) - B(w_{oil})] \tag{2}$$

where the oil in the liquid-phase is given by w_{oil}, i.e. in mass *fraction*, not weight *percent*. $A(w_{oil})$ and $B(w_{oil})$ are given by the following empirical expressions:

$$A(w_{oil}) = a_o + a_1 w_{oil} + a_2 w_{oil}^3 + a_3 w_{oil}^5 + a_4 w_{oil}^7 \tag{3}$$

$$B(w_{oil}) = b_o + b_1 w_{oil} + b_2 w_{oil}^3 + b_3 w_{oil}^5 + b_4 w_{oil}^7 \tag{4}$$

where the values of the empirical constants are:

$$a_o = -2394.5 \quad b_o = 8.0736$$
$$a_1 = 182.52 \quad b_1 = -0.72212$$
$$a_2 = -724.21 \quad b_2 = 2.3914$$
$$a_3 = 3868.0 \quad b_3 = -13.779$$
$$a_4 = -5268.9 \quad b_4 = 17.066$$

Evaluating Equation (2) at a saturation pressure of 5.5 bar, Table 1 depicts the effect of oil concentration on the bubble point temperature. The right column shows the difference between the bubble point temperatures of the mixtures and the saturation temperature of pure R-22 at this pressure.

Table 1. Bubble point temperatures of
R22/oil mixtures at 5.5 bar.

w_{oil} (mass)	T_{bub} (°C)	$T_{bub}-T_{R-22}$ (°C)
0.00	2.99	0.00
0.01	3.01	0.02
0.02	3.03	0.04
0.03	3.04	0.05
0.04	3.06	0.07
0.05	3.09	0.10
0.06	3.11	0.12
0.07	3.13	0.14
0.08	3.15	0.16
0.09	3.17	0.18
0.10	3.19	0.20
0.20	3.44	0.45
0.30	3.79	0.80
0.40	4.31	1.32
0.50	5.25	2.26
0.60	7.22	4.23
0.70	11.53	8.54
0.80	19.95*	16.96*

• Extropolated value.

The above method has been generalized by Thome [22] for use with refrigerants other than R-22 and also temperatures outside the original experimental range. This was accomplished by replacing their values of a_o and b_o for R-22 by determining new ones for the desired refrigerant, such as R-134a. Since the vapor pressure of oil is very small compared to that of the refrigerant and he also demonstrated that the effect of the specific type of oil on the empirical constants a_1 to a_4 and b_1 to b_4 was negligible for oil fractions below 0.50 (50 wt.% oil). Rather than using fixed values of a_o and b_o for a particular pure refrigerant, another important improvement was the utilization of a more accurate equation of state for the pure refrigerant vapor pressure curve to determine the values of a_o and b_o accurately at the specified saturation pressure instead of the simple Antoine expression, Eq. (1). Using these new values of a_o and b_o, Equations (1) and (2) are now accurate for predicting saturation temperatures and bubble point temperatures over narrow temperature ranges for essentially any miscible mixture of refrigerant and oil. Thome [24] has also applied this method to zeotropic refrigerant blends mixed with miscible oils (in particular R-407C/oil mixtures) and showed that it accurately matched experimental measurements over a wide range of oil concentrations.

The procedure to use Equation (2) for any refrigerant is thus:

1. Determine the pure refrigerant saturation temperature and pressure just above and just below the design pressure with an accurate equation-of-state of the user's choice;
2. Use these two sets of values for T_{bub} and P_{sat} to solve for a_o and b_o in Eq. (2) with w_{oil} set to zero. The solution is straightforward with two equations and two unknowns.
3. These new values of a_o and b_o for the refrigerant in question replace those of $a_o = -2394.5$ and $b_o = 8.0736$ for R-22, respectively. All the other values of a_1 to a_4 and b_1 to b_4 remain the same as in the original correlation since they only refer to the effect of the oil on T_{bub}.
4. Equations (2), (3) and (4) can then be evaluated for the desired oil wt. Fraction, w_{oil}, to obtain T_{bub} for the desired value of P_{sat} and w_{oil}; instead, P_{sat} can be obtained for specified values of T_{bub} and w_{oil} using Eqs. (1), (3) and (4).

3.3 LOCAL OIL CONCENTRATIONS

Preparation of a T-h-x curve and determination of local thermodynamic and transport properties of refrigerant-oil mixtures require knowledge of the local oil concentration in the liquid at locations along the evaporator tube. From a thermodynamic viewpoint, the ideal evaporation process is infinitesimally slow such that thermodynamic equilibrium exists and consequently the liquid phase has a uniform oil concentration and temperature at any cross-sectional location along the tube. Enthalpy curves are prepared assuming that thermodynamic equilibrium exists throughout the process, exactly the same as assumed for evaporating a pure fluid, i.e. where thermal boundary layers are ignored.

The liquid-phase concentration of the oil circulating in a refrigeration system is a function of its location in the system. Thus, to unequivocally define this concentration, a point is chosen where all the circulating fluid is in the liquid-phase, which occurs in the refrigerant line between the exit of the condenser and the entrance into the expansion valve or expansion device. The oil concentration at this location will be defined as w_{inlet}. This is the value that is obtained when measuring the concentration of a fluid sample of a system's refrigerant charge at this location. After the expansion device, the local vapor quality will typically range from 0.20 to 0.30 depending on operating conditions and the oil concentration in the liquid-phase will have increased. Along the evaporator tube the oil concentration continues to increase as the refrigerant evaporates into the vapor-phase.

When the local vapor quality x is zero, the local oil concentration w_{local} is equal to w_{inlet}. If it were possible to evaporate all of the refrigerant leaving brhind only the oil in the liquid-phase, then the local vapor quality x would be equal to $(1-w_{inlet})$ at that point since all of the refrigerant would be in the vapor-phase and all the oil in the liquid-phase; w_{local} would be equal to 1.0 at this point since only oil remains in the liquid-phase. Hence the maximum exit vapor quality is equal to $(1-w_{inlet})$. This value in fact <u>cannot</u> be reached since the oil will start evaporating into the vapor-phase too as the bubble point temperature approaches the boiling point of the pure oil, which is far above temperatures occurring in evaporators, condensers and compressors of refrigeration, air-conditioning and heat pump systems. This is an important limiting point to remember because in designing a system with say 3 wt.% oil, the exit vapor quality from the evaporator has to be less than 0.97. Also, for refrigerant-oil flow boiling experiments the maximum outlet vapor quality must be less than $(1-w_{inlet})$.

From a design viewpoint, the local oil concentration as a function of vapor quality is required. The expression that relates the local oil concentration, w_{local}, to the local vapor quality, x, and the inlet oil concentration w_{inlet} is obtained from a conservation of mass and is a simple expression:

$$w_{local} = w_{inlet}/(1-x) \qquad (5)$$

As an illustration, assume that 100 g of refrigerant-oil mixture enters the expansion valve with an oil concentration, w_{inlet}, equal to 0.05 (thus, there are 5 g of oil and 95 g of refrigerant):

1. Before the expansion valve the local vapor quality is 0.0 and $w_{local} = w_{inlet} = 0.05$ as indicated by Eq. (5);
2. For a local vapor quality equal to 0.20 immediately after the expansion valve, w_{local} is 0.0625 [i.e. 0.05/(1 0.20)];
3. For a local vapor quality of 0.90, w_{local} is 0.50 or 50 wt. % oil, i.e. [0.05/(1-0.90)];
4. If it were possible to evaporate all the refrigerant from the liquid mixture, the local vapor quality would be $(1-w_{inlet})$ or 0.95 and Eq. (5) appropriately predicts that w_{local} for this situation is 1.0 [i.e. 0.05/(1-0.95)].

For a pure refrigerant without any oil, w_{local} is always equal to zero. From a practical standpoint, when programming Eq. (5) into a computer code the right side becomes undefined for a pure refrigerant ($w_{inlet} = 0.0$) at x = 1.0 because the expression results in dividing 0.0 by 0.0. Thus whenever $w_{inlet} = 0.0$ the computer code should set w_{local} to zero and not evaluate Eq. (5) to avoid a "machine crash".

3.4. ENTHALPY CURVES

The change in enthalpy of a mixture during evaporation or condensation is referred to in the literature by various names: condensation curve, heat release curve, evaporation curve and enthalpy curve. The term "evaporation curve" is easily confused by non-specialists with "boiling curve" so this name is better avoided. The term *enthalpy curve* will be used here.

The change in enthalpy, dh, of a mixture during evaporation of a mixture with a temperature glide with respect to its original enthalpy for 100% saturated liquid, i.e. with respect to h at x=0, is comprised of three contributions:

1. The latent heat to the fraction of liquid vaporized, x;
2. The sensible heat to the fraction of fluid in the liquid phase (1-x);
3. The sensible heat to the fraction of fluid in the vapor phase x.

In mathematical terms for a refrigerant-oil mixture this is:

$$dh = h_{LV} x + (1-x) dT_{bub} (c_p)_L + x dT_{bub} (c_p)_V \tag{6}$$

where x is the local vapor quality, h_{LV} is the latent heat of vaporization of the pure refrigerant since on oil enters the vapor-phase, $(c_p)_L$ is the specific heat of the liquid-phase refrigerant-oil mixture and $(c_p)_V$ is the specific heat of the pure refrigerant vapor. The values of h_{LV} and $(c_p)_V$ are obtained from equations for the pure refrigerant at the local saturation temperature while $(c_p)_L$ is a function of the local oil composition and bubble point temperature. Equation (6) reduces to only the latent heat for a pure fluid or an azeotropic mixture.

A heat release curve is not actually determined as a curve, but instead as a series of points at a set interval of temperature or vapor quality that gives the amount of heat absorbed by the fluid per unit mass (i.e. dh is in J/kg or Btu/lb) relative to its inlet state together with the bubble point temperature and vapor qualities that correspond to these points. At the inlet no heat has yet been added so the heat absorbed is zero.

As an illustration at a standard evaporation temperature of 4.44°C (40°F) in a refrigeration system, for pure R-134a this corresponds to a saturation pressure of 3.43 bar absolute. For a mixture with 3 wt.% oil in R-134a at this saturation pressure, Table 2 lists the enthalpy curve in tabular form at this pressure covering the vapor quality range from 0.15 to 0.95 utilizing the method of Thome [22]. The oil is Mobil Arctic EAL 68 whose density is 971 kg/m^3 at 15.56°C {Note: the oil's density is utilized to predict its liquid specific heat, cf. Thome [22] and thus that of the refrigerant-oil mixture}. The first column in Table 2 shows the local vapor quality intervals along the evaporator tube, the next two columns list T_{bub} and w_{local}, and the total heat absorbed relative to the inlet condition is shown in the fourth column (in kJ/kg). The contributions of the heat absorbed as latent heat and as sensible heating of the liquid and vapor are given in the last two columns (in kJ/kg), respectively. As can be seen from these tabular values, the rise in the bubble point temperature and oil concentration in the liquid is sharpest at high vapor qualities. The contribution of sensible heat is mostly at high vapor qualities since its effect is directly dependent on the rise in T_{bub}.

A tabular heat release curve such as that in Table 2 can be generated by a computer program for essentially any desired condition: inlet oil concentration, inlet and outlet vapor quality and pressure. Thus it can be made to match actual experimental test conditions in order to reduce raw test data to boiling heat transfer coefficients or those specific to the design of an evaporator.

Table 2. An enthalpy curve for R-134a/3 wt.% oil mixture at 3.43 bar.

x	T_{bub}	w_{local}	Q_{total}	Q_{latent}	$Q_{sensible}$
0.150	4.509	3.53	0.00	0.00	0.00
0.230	4.516	3.90	15.64	15.63	0.01
0.310	4.525	4.35	31.28	31.26	0.02
0.390	4.536	4.92	46.92	46.89	0.03
0.470	4.550	5.66	62.56	62.51	0.05
0.550	4.570	6.67	78.20	78.13	0.07
0.630	4.598	8.11	93.85	93.75	0.10
0.710	4.643	10.34	109.52	109.37	0.15
0.790	4.729	14.29	125.23	124.99	0.24
0.870	4.954	23.08	141.05	140.59	0.46
0.950	8.289	60.00	159.76	156.07	3.69

Table 2 also illustrates the typical influence of oil on T_{bub} with respect to the pure refrigerant (R-134a) saturation temperature T_{sat} of 4.444°C. For x < 0.50, the increase in boiling point tends to be 0.1°C or less, for 0.50 < x < 0.80, the rise increases up to about 0.3°C, and for x > 0.80 the elevation of the boiling point can become very significant (3.845°C at x = 0.95 in Table 2!). The temperature difference between T_{bub} and T_{sat} thus depends on local vapor quality and the inlet oil concentration.

4. Flow Boiling Experimental Studies

Only new experimental test results on intube evaporation of refrigerant-oil mixtures published since the review of Thome [23] will be discussed here, although some earlier publications missed in that survey will be included. Some studies still investigate the effect of oil on *mean* evaporating coefficients rather than on *local* coefficients, the latter which are more useful for developing design methods. As illustrated earlier, the influence of oil is very dependent on the local vapor quality since it in turn controls the local oil concentration through Equation (5). Thus, it does not make much sense to continue to measure mean heat transfer coefficients over an evaporating quality range from 0.1-0.15 at the inlet of a test section to 0.8-0.85 at the outlet since these may or may not include the sharp falloff in the ratio of $h_{oil}/h_{no\ oil}$ at high vapor quality and may mask enhancement effects at intermediate vapor qualities.

In addition, it should be pointed out that only the studies by Thome and coworkers have correctly reduced their raw test measurements to refrigerant/oil boiling heat transfer coefficients utilizing bubble point temperatures and enthalpy curves as described in the preceeding section. The heat transfer coefficients reported in other studies are based on using the pure refrigerant saturation temperature T_{sat} in place of T_{bub}, and since $T_{bub} > T_{sat}$, the coefficients reported are smaller than the thermodynamically correct values, especially for vapor qualities greater than 0.8 (for x < 0.8, the influence of oil on T_{bub} is relatively small, especially for mean data). The difference becomes particularly significant for high oil concentrations and also for microfin tubes which have small boiling wall superheats, and will affect quantitative trends in the test data.

Yoshida, Matsunaga, Hong and Miyazaki [26] presented both mean and local heat transfer coefficients for two plain tubes of 10.6 and 15.4 mm internal diameter for R-22/Suniso 3GS oil mixtures in 3 and 4 m long test sections heated by alternating electrical current. They tested oil concentrations of 0, 1, 3 and 6 wt.% over the heat flux range from 5-30 kW/m^2. At the low vapor quality of x = 0.2 over their range of mass velocities from 100 to 300 kg/m^{2}s, they found that the oil always increased the local heat transfer coefficient, except for a few cases of 1 wt.% oil; the oil increased performance with increasing heat flux too. On the other hand, at the higher vapor quality of x = 0.7, the oil drammatically decreased heat transfer performance by as much as 70-80%, except at the lowest mass velocity tested, i.e. 100 kg/m^{2}s. Based on mean coefficients, the oil increased performance at low mass velocities but degraded it at higher ones, which may suggest that in stratified-wavy flow oil is beneficial while in annular flow it is not. Circumferential heat transfer coefficients were also measured and two-phase flow pattern photographs taken and foaming was observed to occur, which they indicated as being a possible factor in improving tube wetting in the stratified flows and hence enhancing heat transfer. They also calculated refrigerant-oil properties (no methods cited) and, substituting these values into a pure refrigerant flow boiling correlation, qualitatively predicted the adverse effect of oil on heat transfer at high vapor qualities.

Eckels, Zoz and Pate [6] measured bubble point temperatures of R-134a mixtures with 169 SUS and 369 SUS ester lubricating oils as a function of local vapor quality during evaporation and plotted their data as $(T_{bub}-T_{sat})$, where the bubble point temperature was that measured for the refrigerant-oil mixture while the saturation temperature was that calculated from the local vapor pressure. The effect of oil was to create a temperature glide whose magnitude depended on the inlet oil concentration and local vapor quality, similar to Table 2.

Eckels, Doerr and Pate [4] measured mean evaporating coefficients for R-134a/oil mixtures at a nominal temperature of 1°C for a smooth tube and a microfin tube. They tested two penta erythitol ester mixed-acid type oils with viscosities of 169 SUS and 369 SUS for 0-5 wt.% oil mixtures. They reduced data to heat transfer coefficients using a modified-Wilson plot approach, but curiously used the *mean* of the inlet and outlet temperatures of the refrigerant-oil mixtures rather than the inlet and outlet temperatures (i.e. this does not give the same result). They stated that the effect of oil on bubble point temperatures in these tests were accounted for (but this fact is not clear since some contradictory statements were made in the article) while the effect of enthalpy change for sensible heating of the oil was only included in the subcooled liquid-phase in the preheater energy balance but was ignored in determining local vapor qualities. However, this would have little effect on their data since they were mean values with inlet vapor qualities from 5-11% (±3%) and outlet vapor qualities from 82-85% (±8%).

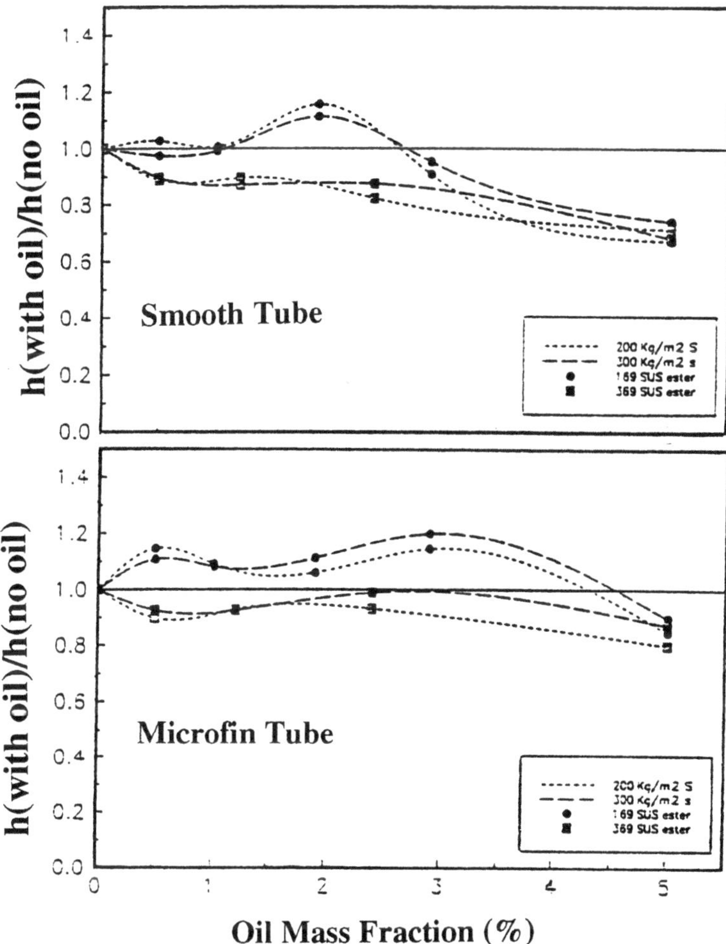

Figure 3. Eckels, Doerr and Pate [4] R-134a/oil mean boiling coefficients with two oils.
Top: Plain tube; Bottom: Microfin tube.

Figure 3 shows the ratios of $h_{with\ oil}/h_{no\ oil}$ for their plain and microfin tubes for the two oils. For the plain tube the less viscous oil provided a little peak in performance while the heavier oil resulted in a monotonic decrease in heat transfer with increasing oil concentration. The trends for the microfin tube were similar to their plain tube. The plain tube results without oil also compared well to several correlations, atestign to the

502

reliability and accuracy of their mean data. Finally, they prepared curvefits to these data points together with those at other mass velocities, but their empirical equations do not include either the refrigerant-oil nor the oil viscosity (nor any other physical properties) and hence cannot be used for other refrigerant/oil combinations.

Sundaresan et al. [17] carried on the above study with additional tests with R-22, R-407C and R-410A but using a POE and a mineral oil with each refrigerants. The data reduction method was similar to that above and no special consideration was apparently taken in reducing the R-407C data with its temperature glide. Any amount of oil reduced performance for all three refrigerants with the exception of 0.2% POE in R-407C. R-410A had the highest performance of the refrigerants without oil, and it was more affected by the oil than was R-407C. For these mean coefficients the maximum degradation for R-410A was -54% while for R-407C it was -27%. For R-410A and R-407C, the polyolester type of oil was superior to the mineral oil, i.e. less degradation in heat transfer performance, while the opposite was true for R-22. The oil viscosities were not cited unfortunately and no explanation was offered as to why one type of oil was better than another. In addition, as is typical in many articles, it was not specifically stated whether or not the oils were miscible in the respective refrigerants (Was the mineral oil miscible in R-410A and R-407C?), so this may have been a factor here too.

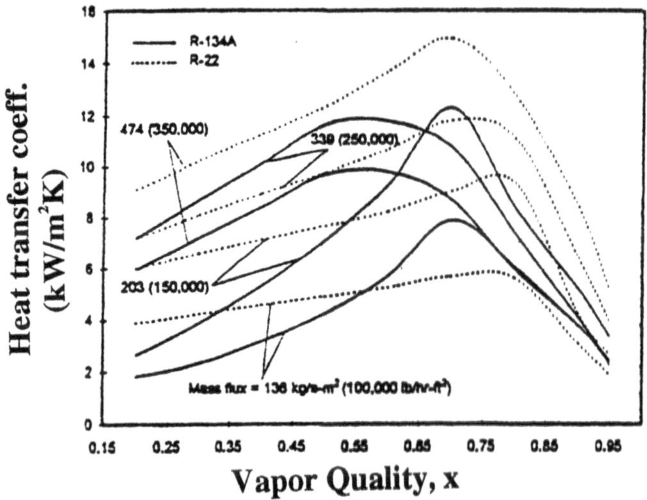

Figure 4. Liu [13] local evaporating heat transfer coefficients at four mass velocities.

Liu [13] has reported flow boiling coefficients for R-22 and R-134a with a nominal 1% oil in the refrigerant in a microfin tube of 8.82 mm internal diameter with 72 longitudinal fins of 0.185 mm height. His tests utilized six 2.2 m (7.2 ft) logn microfin tube subsections connected in sequence with 7.6 cm (3 in.) radius U-bends with a 1.8 m (5.9 ft) heated length in each. The oil was an AB/cs68 oil for R-22 while a

POE/cs68 oil was used for R-134a. A modified-Wilson plot approach was used to determine the evaporation coefficients in each subsection, using pure refrigerant saturation temperatures and latent heats without taking into account oil effects, which would only be significant here at very high vapor qualities since only 1% oil was tested. Figure 4 depicts his data curvefits for the R-22 and R-134a mixed with an inlet oil concentration of 1% oil. All his data display the characteristic peak in h vs. x at high vapor quality found in pure refrigerant tests, where the amplitude of the peak increases with mass velocity as predicted by the Kattan, Thome and Favrat [12] flow boiling model.

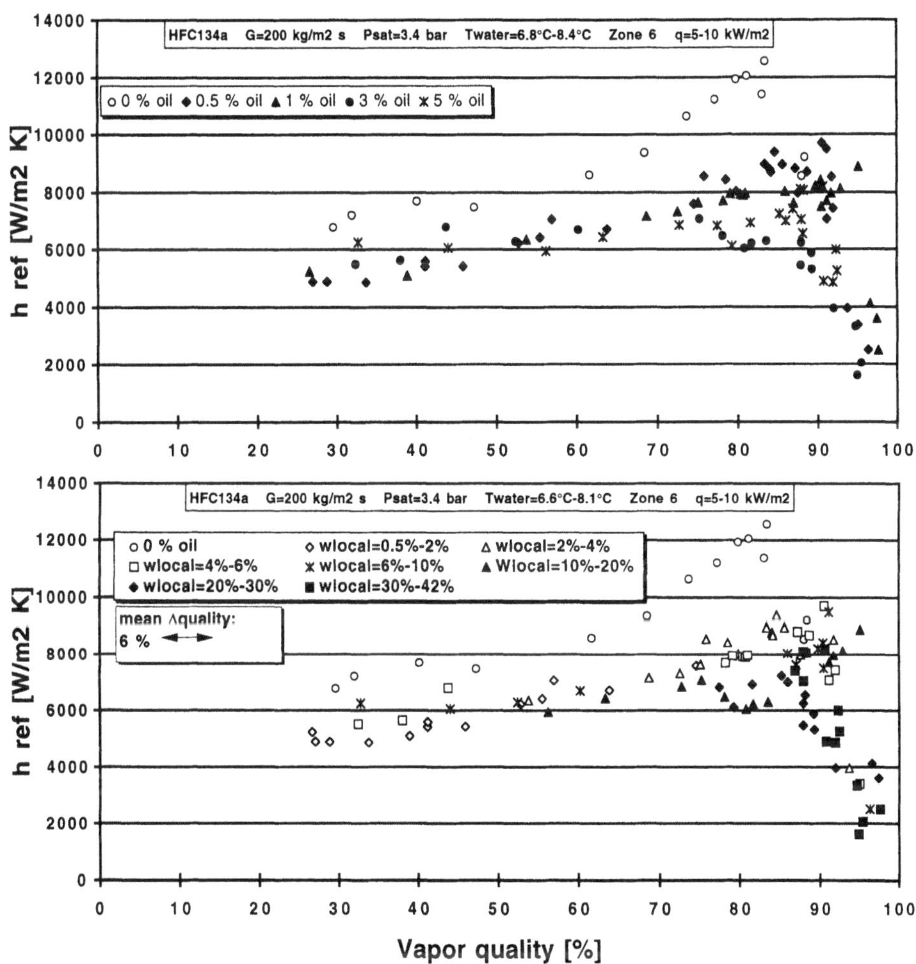

Figure 5. Nidegger, Thome and Favrat [15] flow boiling data for R-134a/oil mixtures in a microfin tube. Top: Plotted as inlet oil concentration; Bottom: Plotted as local oil concentration[reprinted from *HVAC&R Research*].

Nidegger, Thome and Favrat [15] ran very accurate local R-134a/oil evaporation tests for a microfin tube (internal diameter of 11.90 mm with seventy 18° helical fins of

0.254 mm and internal area ratio of 1.74) with nominal oil concentrations of 0, 0.5, 1.0, 3.0 and 5.0 wt.%, measuring the actual oil concentrations with an accurate online density flowmeter {cf. Bayini, Thome and Favrat [1]} rather than assuming its value from the concentration charged to the loop as is typical of other tests, since the nominal concentration can change with test conditions if oil holds up anywhere in the loop. They reduced their *quasi-local* data using the Thome [22] thermodynamic approach for refrigerant/oil mixtures with the heat transfer coefficient determined from bubble point temperatures of the local refrigerant-oil mixture in the inlet and outlet of their test zones. They tested Mobil Arctic Grade EAL 68 ester oil with a kinematic viscosity of 62.5 mm^2/s at 40°C (100°F) that was completely miscible over all test conditions. Their data were quasi-local data obtained with hot water heating for vapor quality changes from 4.5-12% in their test section zones, measured with modified-Wilson plot tests. Figure 5 shows their data at a mass velocity of 200 kg/m^2s plotted in two different ways, first in the tradional way as a function of the nominal inlet oil concentration in the subcooled liquid and then as a function of the local oil concentration corresponding to the local vapor quality. Adding oil to R-134a always reduced heat transfer at vapor qualities below the peak in h vs. x, but above the peak a few data were higher than the pure R-134a values, apparently because a small amount of oil retarded the onset of dryout (or eliminated it since the refrigerant-oil mixture cannot dry out) while a high local oil concentration is very detrimental to performance. They reported that no foaming occurred in any of their R-134a/oil tests.

Zürcher, Thome and Favrat [27] ran the associated plain tube tests with R-134a/oil at the same test conditions, i.e. constant inlet saturation pressure of 3.40 bar corresponding to T_{sat} = 4.4°C for pure R-134a. Figure 6 depicts their test data for a plain tube of 10.92 mm internal diameter. The oil concentrations of 0.5 and 1.0% tended to increase boiling performance at low to intermediate vapor qualities while 3.0% and 5.0% oil tended to decrease it. At high vapor qualities the 0.5% oil mixture had a performance similar to pure R-134a until after the peak in h vs. x. The other tests suffered very severe performance degradations from the oil at high vapor quality, falling by as much as -90% and shifting in the peak in h vs. x towards lower vapor qualities and to lower amplitude peaks. Their pure R-134a data agreed reasonable well with previously published R-134a data of Eckels and Pate [5] and Wattelet et al. [25] for similar size tubes. Their annular flow data for pure R-134a was also accurately predicted by five different existing flow boiling correlations, atesting to its reliability.

Zürcher, Thome and Favrat [28] {or see Thome [24] reported local R-407C/oil evaporation tests in a microfin tube test section for nominal inlet oil concentrations of 0, 0.5, 1.0, 3.0 and 5.0 wt.% at 6.45 bar.} The microfin tube was the same tested previously with R-134a/oil mixtures. They reduced their data using the Thome [22] thermodynamic approach for refrigerant/oil mixtures together with inlet oil concentrations measured online and with the heat transfer coefficients determined from bubble point temperatures of the local refrigerant-oil mixture at the inlet and outlet of their test zones. In fact they presented temperature measurements that verified their calculated values of bubble point temperatures of the temperature glides for both R-

407C and R-407C/oil mixtures as a function of vapor quality. [They presented a complete set of thermodynamic equations for describing R-407C mixtures valid from -5°C to +50°C (23-122°F) for reducing test data and for thermal design of R-407C evaporators and condensers]. The same Mobil Arctic Grade EAL 68 ester oil with a kinematic viscosity of 62.5 mm^2/s at 40°C (100°F) as in the prior R-134a tests was used (always completely miscible according to the oil manufacturer). Their data were quasi-local data obtained with modified-Wilson plot tests. Figure 2 presented earlier depicts their data as normalized values relative to pure R-407C. Adding oil to R-407C always reduced heat transfer at all vapor qualities except for a minor increase of 6-8% over part of the range for 3.0 wt.% oil that probably is not significant. This combination of refrigerant and oil caused a lot of foaming to occur, which seemed to intensify with increases in oil concentration, mass velocity and local vapor quality.

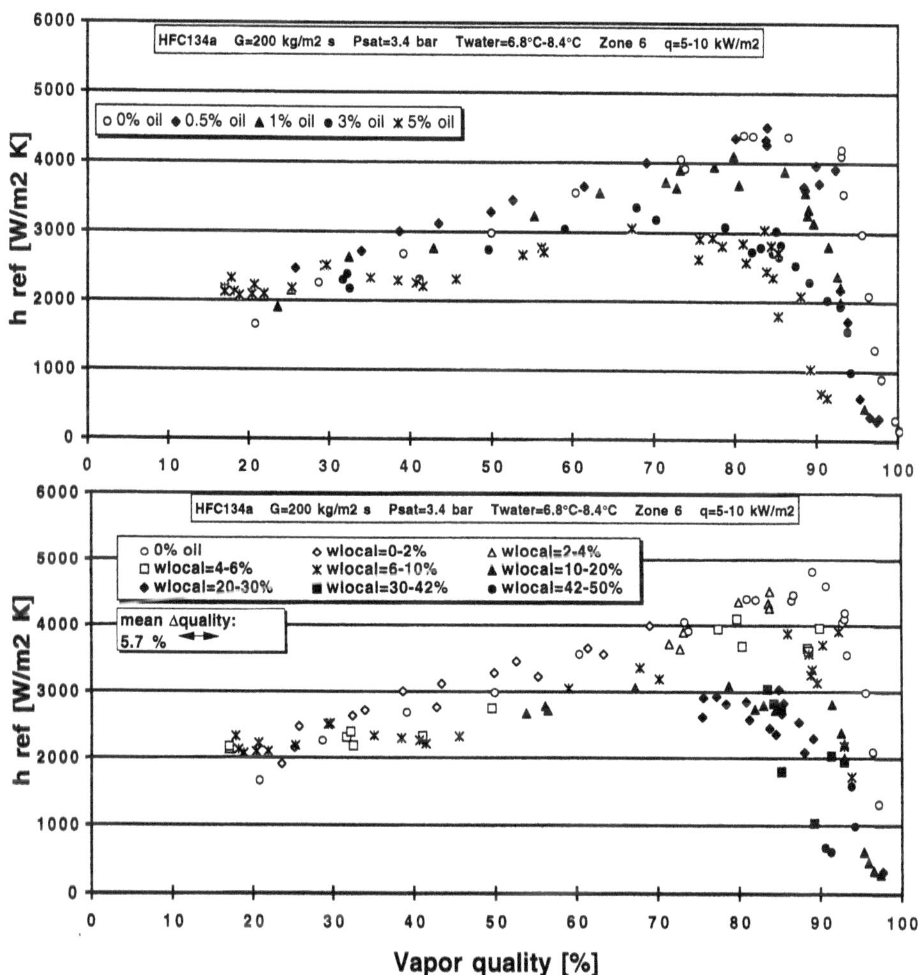

Figure 6. Zürcher, Thome and Favrat [27] flow boiling data for R-134a/oil mixtures in a plain tube. Top: Plotted as inlet oil concentration; Bottom: Plotted as local oil concentration [reprinted from *HVCA&R Research*].

506

Zürcher, Thome and Favrat [29] also ran plain tube tests with R-407C/oil at similar test conditions, i.e. constant inlet saturation pressure of 6.45 bar corresponding to a saturation temperature of 4.4°C for pure R-407C at a vapor quality of x equal to 0.0. Figure 7 depicts their 10.92 mm bore, plain tube test data at 300 kg/m^{2}s that have been normalized by the Kattan, Thome and Favrat [12] flow boiling model. At high vapor qualities the refrigerant-oil mixtures suffered very severe degradations from the oil with increasing oil concentration, falling by as much as -90%. Note that the flow boiling model predicts their pure R-407C data to within a maximum error of ±20% and a mean error of less than 10%.

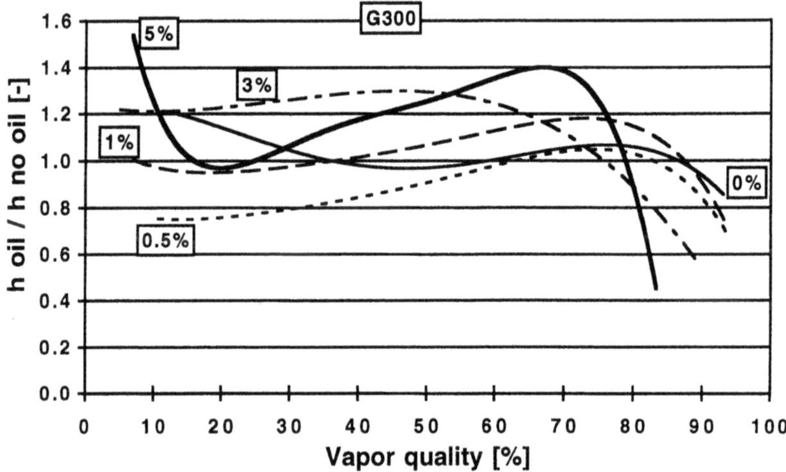

Figure 7. Zürcher, Thome and Favrat [28] flow boiling data for R-407C/oil mixtures in a plain tube plotted as normalized coefficients at 6.45 bar.

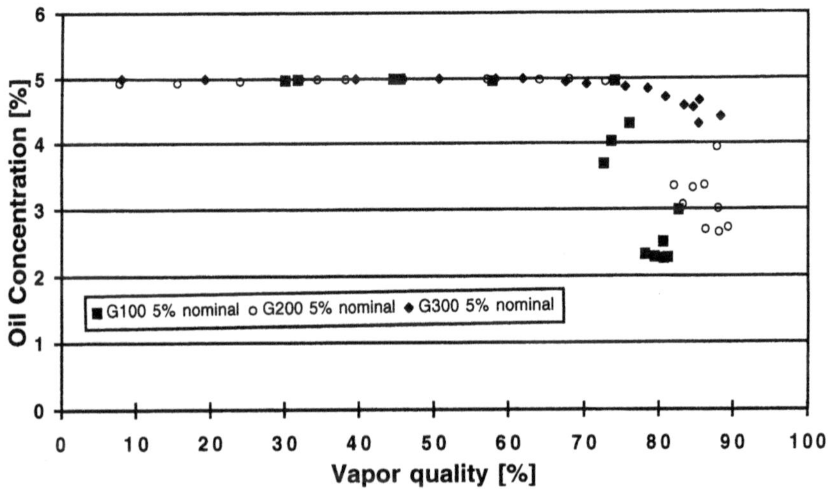

Figure 8. Thome [24] online oil concentration measurements illustrating the effect of mass flow rate on oil holdup in their microfin tube test section with R-407C for a nominal concentration of 5 wt.% oil.

The phenomenon of oil holdup in microfin tubes during evaporation was a topic addressed in Thome [24] and Zürcher, Thome and Favrat [28] for the first time. Figure 8 depicts their oil concentrations measured online for subcooled refrigerant-oil liquid flowing into their preheater compared to the 5 wt.% oil concentration charged into the flow boiling loop. At local vapor qualities below about 65%, no oil held up in the two microfin tube test sections connected in series, as indicated by the online value which matched that charged into the loop. In sharp contrast, at outlet high vapor qualities, oil did hold up in the microfin test section, becoming more severe with decreasing mass velocity. Hence, it appears that very viscous refrigerant-oil mixtures at high vapor qualities flowing as a liquid film on the microfins tend to get retained by their higher surface tension forces, increasingly so as the vapor shear on the vapor-liquid interface decreases with decreasing mass velocity. As a consequence, the local oil concentration at high vapor qualities in microfin tubes are higher than those calculated assuming no holdup, i.e. with Equation (5), which means that the actual local bubble point temperature is higher than that estimated assuming no holdup.

5. Modelling Effects of Oil on Intube Flow Boiling Heat Transfer

5.1. PLAIN TUBES

For intube evaporation of refrigerant-oil mixtures in plain tubes, in Thome [24] and Zürcher, Thome and Favrat [29], it was demonstrated that the sharp falloff in heat transfer at high vapor quality with oil can be accurately predicted if the pure refrigerant's liquid viscosity is replaced with that of the local refrigerant-oil viscosity in the Kattan, Thome and Favrat [10, 12] flow boiling model and flow pattern map.

For the mass velocities of 200 and 300 kg/m^{2}s for R-134a/oil and R-407C/oil at x > 0.70, i.e. where there is the most severe falloff, they predicted these data to a standard deviation of 22.3%, which is only slightly higher than for the same refrigerants without oil, i.e. about 17%. The high vapor quality, local data are difficult to predict more accurate that this in any case because of the very steep slope in h vs. x after the peak. The local refrigerant-oil viscosity was calculated using the Arrhenius mixing law, which was rearranged to the simpler to use form of:

$$\mu_{\text{ref-oil}} = (\mu_{\text{ref}})^{(1-w)}(\mu_{\text{oil}})^{w} \tag{7}$$

where w is the local oil concentration in weight fraction determined from the local vapor quality with Equation (5), $\mu_{\text{ref-oil}}$ is the liquid dynamic viscosity of the refrigerant-oil mixture, μ_{ref} is the liquid viscosity of the pure refrigerant, and μ_{oil} is the liquid viscosity of the pure oil, all evaluated at the local bubble point temperature of the mixture. Their data cover local liquid viscosities up to about 35 cp, i.e. up to 150 times the values of pure R-134a and R-407C.

Figure 9 shows his simulation applying this method to R-134a for the same EAL 68 oil as used in their tests, which predicts the trends in the experimental data at high vapor

508

qualities very well, i.e. the shift of the peak in h vs. x to lower vapor qualities with smaller amplitude peaks.

As an alternative, Thome [24] and Zürcher, Thome and Favrat [29] also showed that adding the multiplying factor

$$F_{oil} = [\mu_{ref}/\mu_{oil}]^{0.26w} \tag{8}$$

to the Kattan-Thome-Favrat annular liquid film flow convective boiling coefficient h_{cb} in their asymptotic model

$$h_{wet} = [(h_{nb})^3 + (h_{cb})^3]^{1/3} \tag{9}$$

yielded very good predictions too, using the same parameter definitions as above. While this F_{oil} factor could be added to other plain tube correlations as a temporary 'fix' to model oil effects, such as that of Gungor-Winterton [8], these other correlations do not in fact predict *pure* refrigerant test data at high vapor qualities very accurately and thus this course of action is not recommended.

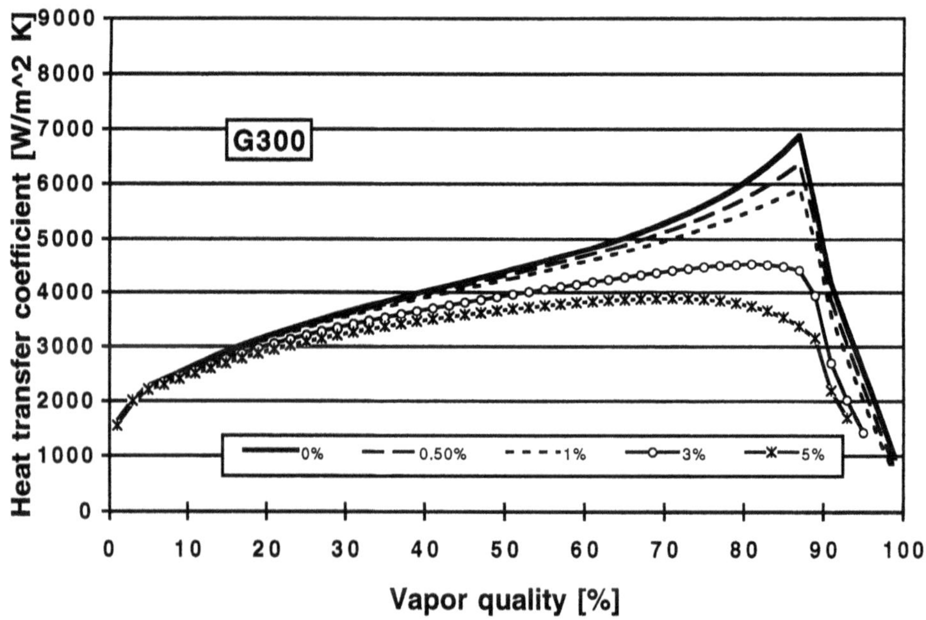

Figure 9. Predicted variation in R-134a/oil flow boiling coefficients in a plain tube at
$q = 10 \ kW/m^2$ at 3.43 bar.

5.2. MICROFIN TUBES

No prediction method for refrigerant-oil effects on microfin tubes has yet been proposed, although the multiplier $[\mu_{ref}/\mu_{oil}]^{0.26w}$ developed in Thome [24] for plain tubes could be tentatively applied as a correction to local pure refrigerant microfin experimental data as an interim approach until then, where w corresponds to w_{local} in Equation (5).

6. Modelling Effects of Oil on Intube Two-Phase Flow Pressure Drops

6.1. PLAIN TUBES

Numerous test measurements of two-phase pressure drops for evaporation of refrigerant-oil mixtures are available for plain and microfin tubes, both for quasi-local pressure drops over a narrow change in vapor quality and mean pressure drops over a wide change in vapor quality. Space limitations do not allow all this experimental work to be reviewed here. Similar to evaporation, quasi-local two-phase pressure drop data are more valuable than mean data since the higher concentration of oil in the liquid-phase at high vapor qualities results in a significant increase in two-phase frictional pressure drop while only a minor effect of oil occurs at lower vapor qualities.

As an example, Figure 10 depicts the plain tube pressure drop data for R-134a and R-407C mixed with Mobil Arctic EAL 68 oil obtained by Nidegger, Thome and Favrat [15] and Zürcher, Thome and Favrat [29] {or see Thome [24]}, respectively, during their boiling tests described earlier. Their quasi-local refrigerant-oil measurements were normalized by the corresponding pure refrigerant values, and polynomial curvefits are shown in the graphs at a mass velocity of 300 kg/m²s for vapor qualities greater than 55%. The influence of oil increases with its local oil concentration in the liquid-phase, where the local refrigerant-oil liquid viscosity can be calculated by Equation (7).

Friedel [7] is just one of many two-phase frictional pressure drop correlations available in the literature for plain tubes and has been recommended for general use in Collier and Thome [2,3]. The frictional two-phase pressure drop gradient $(dP/dz)_F$ is defined as

$$(dP/dz)_F = [(2\ f_{fo}\ G^2)\ (\rho_L\ D]\ \phi_{fo}^2 \tag{10}$$

where ρ_L is the density of the liquid-phase, f_{fo} is the Fanning friction factor for turbulent flow for all flow as liquid, G is the total mass velocity, D is the tube internal diameter, and ϕ_{fo}^2 is the two-phase friction multiplier.

To include the influence of the oil on the two-phase pressure drop, its effect on liquid viscosity will be the dominant factor. Adding an empirical correction parameter n to the Arrhenius mixing law for liquid viscosities, i.e. Equation (7), and rearranging, the following factor was obtained by Zürcher, Thome and Favrat [29] {or see Thome [24]}:

$$f_{oil} = [\mu_{ref\text{-}oil}/\mu_{ref}]^n = [\mu_{oil}/\mu_{ref}]^{nw} \tag{11}$$

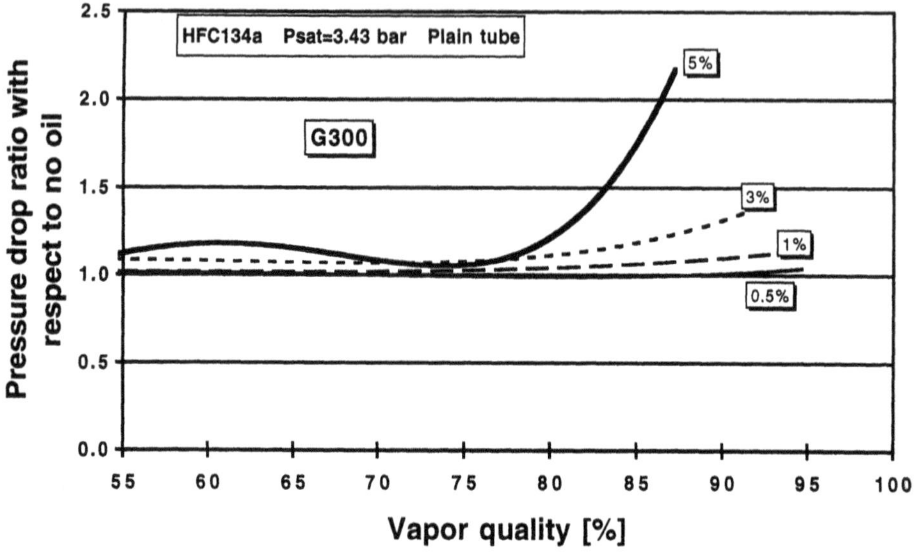

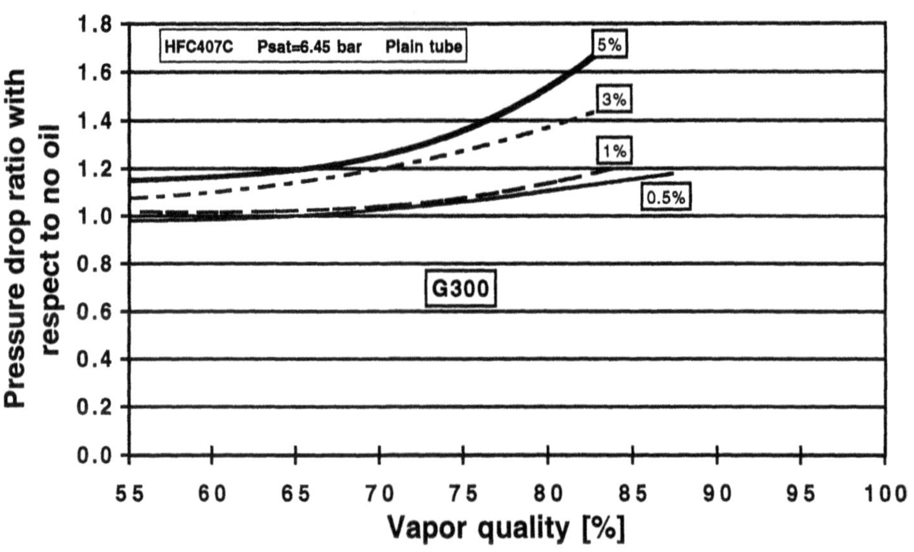

Figure 10. Two-phase pressure drop ratios for evaporation of refrigerant-oil mixtures in a 10.92 mm diameter tube at 300 kg/m^{2}s. Top: R-134a; Bottom: R-407C.

where w is the local oil fraction in the liquid-phase determined with Equation (5). Thus the two-phase friction multiplier for refrigerant-oil mixtures becomes:

$$(\phi_{fo}^2)_{ref\text{-}oil} = (\phi_{fo}^2)_{ref}\, f_{oil} = (\phi_{fo}^2)_{ref}\, [\mu_{oil}/\mu_{ref}]^{nw} \tag{12}$$

where

$$n = 0.18355 \tag{13}$$

without foaming (R-134a/oil mixtures) and

$$n = w_{local}\,(3.583\, w_{inlet} + 0.0616) \tag{14}$$

with foaming (R-407C/oil mixtures). Overall standard deviations at mass velocities of 200 and 300 kg/m^{2}s with inlet oil fractions from 0.005 to 0.05 and w_{local} from 0.01-0.4 were 9.2% and 7.3% for R-134a/oil and R-407C/oil mixtures, respectively. Thus, the effect of oil can easily be included in local two-phase pressure drop calculations.

6.2. MICROFIN TUBES

For microfin tubes, the corresponding test data of Thome and coworkers, and numerous studies by others, have still not been analyzed. In the meantime, the above approach can be utilized to estimate the influence of oil on microfin tubes.

7. Conclusions

The general trends for evaporation of refrigerant-oil mixtures in plain tubes are that (i) oil increases the local boiling heat transfer coefficient at low to intermediate vapor qualities on the order of 10-30% on average and (ii) then a sharp falloff at high vapor qualities. The general trend for microfin tubes is (i) little (5-10%) or no enhancing effect of the oil on heat transfer at these low and medium vapor qualities but sometimes a detrimental influence up to about -30% and (ii) then a substantial falloff at high vapor qualities. So far, no model can reliably explain the enhancing effect of oil at the low and medium vapor qualities while the rapid rise in local liquid viscosity at high vapor qualities has been used to accurately predict the sharp falloff in heat transfer coefficients at high vapor qualities. Thus, the new flow boiling model and flow pattern map of Kattan, Thome and Favrat [10, 12] utilizing the refrigerant-oil viscosity, $\mu_{ref\text{-}oil}$, in place of the pure refrigerant value, μ_{ref}, shows promise for describing the falloff in heat transfer coefficients at high vapor qualities for plain tubes. Oil also increases two-phase pressure drops sharply at high local vapor qualities for plain and microfin tubes. Using a locally-applied, oil-to-refrigerant viscosity correction factor proposed by Thome [24, 29], two-phase pressure drops are able to be adequately predicted for foaming and non-foaming refrigerant-oil mixtures evaporating inside plain tubes.

8. Nomenclature

A	empirical constant [-]
$a_0\text{-}a_4$	empirical constants [-]
B	empirical constant [-]
$b_0\text{-}b_4$	empirical constants [-]
$(c_p)_L$	liquid specific heat [J/kgK]
$(c_p)_V$	vapor specific heat [J/kgK]
D	inside tube diameter [m]
F_{oil}	mixture factor on heat transfer [-]
f_{oil}	mixture factor on pressure drop [-]
f_{fo}	Fanning friction factor [-]
G	mass velocity [kg/m^{2}s]
h	heat transfer coefficient [W/m^{2}K]
dh	change in enthalpy [J/kg]
h_{cb}	convective boiling heat transfer coefficient [W/m^{2}K]
h_{lv}	latent heat of vaporization [J/kg]
h_{nb}	nucleate boiling heat transfer coefficient [W/m^{2}K]
h_{wet}	heat transfer coefficient on wetted perimeter of tube [W/m^{2}K]
n	empirical exponent [-]
P_{sat}	saturation pressure [MPa]
$(dP/dz)_F$	frictional pressure gradient [Pa/m]
q	heat flux [W/m^2]
T_{bub}	bubble point temperature of mixture [K]
T_{sat}	saturation temperature of pure refrigerant [K]
w	local oil mass fraction [kg/kg]
w_{inlet}	inlet oil mass fraction [kg/kg]
w_{local}	local oil mass fraction [kg/kg]
w_{oil}	oil mass fraction [kg/kg]
x	vapor quality (including oil in liquid-phase) [-]
μ_{oil}	dynamic viscosity of pure oil [Ns/m^2]
μ_{ref}	dynamic viscosity of pure refrigerant [Ns/m^2]
$\mu_{ref\text{-}oil}$	dynamic viscosity of local refrigerant-oil mixture [Ns/m^2]
ρ_L	liquid density [kg/m^3]
$\phi_{fo}^{\,2}$	two-phase friction multiplier [-]

9. References

1. Bayini, A., Thome, J.R. and Favrat, D. (1995). Online Measurement of Oil Concentrations of R-134a/Oil Mixtures with a Density Flowmeter, *HVAC&R Research*, ASHRAE, **1(3)**, 232-241.
2. Collier, J. G. and Thome, J. R. (1994). *Convective Boiling and Condensation*, 3rd Edition, Oxford University Press, Oxford.
3. Collier, J. G. and Thome, J. R. (1996). *Convective Boiling and Condensation*, 3rd Edition, Oxford University Press, Oxford (paperback edition).

4. Eckels, S.J., Doerr, T.M. and Pate, M.B. (1994). In-Tube Heat Transfer and Pressure Drop of R-134a and Ester Lubricant Mixtures in a Smooth Tube and a Micro-Fin Tube: Part I-Evaporation, *ASHRAE Trans.* **100(2)**, 265-282.

5. Eckels, S.J. and Pate, M.B. (1991). Evaporation and Condensation of HFC-134a and CFC-12 in a Smooth Tube and a Micro-Fin tube, *ASHRAE Trans.* **97(2)**, 71-81.

6. Eckels, S.J., Zoz, S.C. and Pate, M.B. (1993). Using Solubility Data for HFC-134a and Ester Lubricant Mixtures to Model an In-Tube Evaporator or Condenser, *ASHRAE Trans.* **99(2)**, 383-391.

7. Friedel, L. (1979). Improved Friction Pressure Drop Correlations for Horizontal and Vertical Two-Phase Pipe Flow, European Two-Phase Flow Group Meeting, Ispra, Italy, June, Paper E2.

8. Gungor, K.E. and Winterton, R.H.S. (1987). Simplified General Correlation for Saturated Flow Boiling and Comparisons of Correlations with Data, *Chem. Eng. Res. Des.* **65**, 148-156.

9. Hewitt, N.J. and McMullan, J.T. (1995). Refrigerant-Oil Solubility and Its Effect on System Performance, *Proc. 19th International Congress of Refrigeration*, The Hague, **IVa**, 290-296.

10. Kattan, N., Thome, J.R. and Favrat, D. (1998). Flow Boiling in Horizontal Tubes. Part 1: Development of a Diabatic Two-Phase Flow Pattern Map, *J. Heat Transfer* **120(1)**, 140-147.

11. Kattan, N., Thome, J.R. and Favrat, D. (1998). Flow Boiling in Horizontal Tubes. Part 2: New Heat Transfer Data for Five Refrigerants, *J. Heat Transfer* **120(1)**, 148-155.

12. Kattan, N., Thome, J.R. and Favrat, D. (1998). "Flow Boiling in Horizontal Tubes. Part 3: Development of a New Heat Transfer Model Based on Flow Patterns," *J. Heat Transfer* **120(1)**, 156-165.

13. Liu, X. (1997). Condensing and Evaporating Heat Transfer and Pressure Drop Characteristics of HFC-134a and HCFC-22, *J. Heat Transfer* **119**, 158-163.

14. Manwell, S.P. and Bergles, A.E. (1990). Gas-Liquid Flow Patterns in Refrigerant-Oil Mixtures, *ASHRAE Trans.* **96(2)**, Paper SL-90-1-4.

15. Nidegger, E., Thome, J.R. and Favrat, D. (1997). Flow Boiling and Pressure Drop Measurements for R-134a/Oil Mixtures Part 1: Evaporation in a Microfin Tube, *HVAC&R Research*, ASHRAE, **3(1)**, 38-53.

16. Schlager, L.M., Bergles, A.E. and Pate, M.B. (1987). A Survey of Refrigerant Heat Transfer and Pressure Drop Emphasizing Oil Effects and In-Tube Augmentation, *ASHRAE Trans.* **93(1)**, 392-416.

17. Sundaresan, S.G., Pate, M.B., Doerr, T.M. and Ray, D.T. (1996). A Comparison of the Effects of POE and Mineral Oil Lubricants on the In-Tube Evaporation of R-22, R-407C and R-410A, *Proc. 1996 International Refrigeration Conference at Purdue*, West Lafayette, IN, July 23-26, 187-192.

18. Takaishi, Y. and Oguchi, K. (1987). Measurements of Vapor Pressures of R22/Oil Solutions, *18th International Congress of Refrigeration*, Vienna, **B**, 217-222.

19. Thome, J.R. (1990). *Enhanced Boiling Heat Transfer*, Hemisphere, New York.

20. Thome, J.R. (1994). Two-Phase Heat Transfer to New Refrigerants, Special Keynote Lecture, *Proc. 10th International Heat Transfer Conf.*, Brighton, **1**, 19-41.

21. Thome, J.R. (1995). Flow Boiling of Refrigerant-Oil Mixtures: a Current Review, Convective Flow Boiling Conf., Banff, Canada, April 30-May 5: proceedings published in *Convective Flow Boiling*, Eds. J.C. Chen et al., Taylor & Francis, Washington (1996), 285-290.

22. Thome, J.R. (1995). Comprehensive Thermodynamic Approach to Modelling Refrigerant-Oil Mixtures, *HVAC&R Research*, ASHRAE, **1(2)**, 110-126.

23. Thome, J.R. (1996). Boiling of New Refrigerants: A State-of-the-Art Review, *Int. J. Refrig.* **19(7)**, 435-457.

24. Thome, J.R. (1997). Heat Transfer and Pressure Drop in the Dryout Region of Intube Evaporation with Refrigerant/Lubricant Mixtures, ASHRAE Final Report of Project 800-RP, February.

25. Wattelet, J.P. et al. (1994). Heat Transfer Flow Regimes of Refrigerants in a Horizontal-Tube Evaporator, ACRC Report TR-55, University of Illinois at Urbana-Champaign.

26. Yoshida, S., Matsunaga, T., Hong, H.P. and Miyazaki, M. (1991). An Experimental Investigation of Oil Influence on Heat Transfer to a Refrigerant inside Horizontal Evaporator Tubes, *Heat Transfer-Jap. Res.* **20(2)**, 113-129.

27. Zürcher, O., Thome, J.R. and Favrat, D. (1997). Flow Boiling and Pressure Drop Measurements for R-134a/Oil Mixtures Part 2: Evaporation in a Plain Tube, *HVAC&R Research*, ASHRAE, **3(1)**, 54-64.

28. Zürcher, O., Thome, J.R. and Favrat, D. (1998). Intube Flow Boiling of R-407C and R-407C/Oil Mixtures Part I: Microfin Tube, *HVAC&R Research*, ASHRAE, **4(4)**, {see Thome [24]}.

29. Zürcher, O., Thome, J.R. and Favrat, D. (1998). Intube Flow Boiling of R-407C and R-407C/Oil Mixtures Part II: Plain Tube Results and Predictions, *HVAC&R Research*, ASHRAE, **4(4)**, {see Thome [24]}.

INFLUENCE OF CONFINEMENT ON FC-72 POOL BOILING FROM A FINNED SURFACE

M.MISALE, G.GUGLIELMINI, and M. FROGHERI
Dipertimento di Termoenergetica w Condizionamento Ambientale
University of Genoa
Via All'Opera 15-a (I) 1645 Genova - Italy

A.E . BERGLES
Department of Mechanical Engineering, Aeronautical Engineering and Mechanics
Rensselaer Polytechnic Institute
Troy (NY), 12180-3590, USA

Abstract This paper presents an experimental study dealing with basic nucleate boiling from a finned surface placed in a narrow channel. The influence of the channel width, the orientation of the base surface (horizontal or vertical), and the pressure are discussed. The experiments were performed in a saturated pool of FC-72, while the channel widths were varied from 40 mm up to 0.5 mm. The experimental data are compared with those obtained in the case of the unconfined extended surface. Channel width reduction does not affect the heat transferred to the liquid in the case of vertical orientation of the base surface, but it causes a drastic reduction in the heat transfer behavior in the case of a horizontal base surface. A critical distance of the channel wall from the top of the finned surface (horizontal orientation of the base surface), which causes the beginning of the heat transfer reduction, is experimentally determined, for pressures ranging from 0.5-2.0 bar.

1. Introduction

In numerous technological applications the thermal problem consists of removing very high heat fluxes with acceptable difference temperatures between the surface and the fluid. This problem is very important in the cooling of electronic devices where the reliability of the system can be improved by the choice of the cooling technique.

S. Kakaç et al. (eds.), Heat Transfer Enhancement of Heat Exchangers, 515–528.
© *1999 Kluwer Academic Publishers.*

Improvements in cooling technologies for electronic devices are continually required by the increasing miniaturization of integrated circuit components. It has been predicted that in the coming years, heat fluxes in excess of 100 W/cm^2 will have to be dissipated while keeping the device temperature at 85 °C [1].

Direct immersion cooling in inert dielectric liquids with phase change has been recognized as one of the most effective methods of thermal control, owing to its efficient heat transfer mechanism [2]. Nevertheless, in order to accommodate the high heat flux and keep the surface temperature low, the immersion cooling may involve a finned surface attached to the electronic component. Moreover, it is possible to transfer up to 130 W/cm^2 when the dielectric liquid is highly subcooled [3]. But, in many cases, the space where the electronic-component finned surface assembly is placed is not very large, so it is not clear what is the thermal behavior of the assembly when it is placed in a narrow channel. Many studies are available in the literature on the nucleate boiling in narrow channels involving flat surfaces or tubes [4-11], but only few papers are focused on the thermal performance of enhanced surfaces in a narrow channel. Ishibashi et al. [12] studied the influence of the gap on the water nucleate boiling of two kinds of horizontal tubes (smooth and roll-worked surfaces) under reduced pressure conditions. The boiling data obtained by the roll-worked surface show that the reduction of gap size (up to 1.6 mm) causes a slightly better performance than that measured in the case of the unrestricted situation. Nowell et al. [13] analyzed the thermal performance of a vertical enhanced surface realized by a microfigured heat sink immersed in a FC-72 pool. By means of an opposite unheated wall, channels of different width (6.0, 4.0, 2.0, and 1.0 mm) were formed. The lowest heat transfer coefficient was measured for a channel width of 6.0 mm; as the gap size was changed from 4.0 mm to 1.0 mm, the heat transfer increased, reaching nearly the unconfined boiling value.

Misale et al. [14] performed a preliminary study on the boiling of two finned surfaces placed in narrow channels of 2 mm and 0.5 mm. The base surface was displaced horizontally and vertically. They observed a reduction of heat transfer when the wall channel is close to the finned surface in the case of horizontal orientation of the surface, whereas the influence of the channel width is negligible for the vertical orientation.

The scope of this paper is to study experimentally the thermal behavior of a finned surface. placed in a narrow channel, for different values of the saturation pressure and different orientation of the tests section (horizontal or vertical). These data are then compared with those obtained in the case of an unrestricted finned surface.

2. Experimental set-up and operating procedures

The experimental apparatus consists of a hermetic stainless steel vessel containing the test-section assembly immersed in a pool of FC-72, a typical "electronic liquid". In Fig.1 the schematics of the experimental apparatus and test section are depicted. An auxiliary heater is wrapped around the vessel to pre-heat the liquid bulk and to compensate for heat losses. At high heat fluxes, saturation conditions are maintained by a water-cooled condenser placed in the vapor space.

The heat source module consists of a copper block, used as a heat flux meter, thermally insulated. The boiling area is the circular upper surface of the heat flux meter, which has a diameter of 30 mm. The heat flux through the copper test section and the surface temperature were calculated on the basis of the temperatures measured by nine K-type thermocouples lodged inside the copper flux meter at different depths, and located at three planes spaced 18 mm apart, starting 3 mm below the boiling surface. The position of the test section inside the vessel could be varied, by fitting hangers, enabling the base surface orientation to be changed from horizontal to vertical.

Liquid and vapor temperatures were measured by four K-type thermocouples, three of them immersed into the liquid at different depths. During the tests, the pressure inside the vessel was maintained at atmospheric pressure.

All temperatures were recorded by means of a high-precision data acquisition system. The chain of instruments used for the detection of the signal given by the thermocouples produces an accuracy of $\pm0.1°C$ on absolute temperature values and $\pm0.02°C$ on differential values. Considering the heat conduction through the test section, in the region of nucleate boiling, the maximum error associated with the heat flux was calculated to be less than 5%, while the maximum uncertainty associated with the surface temperature was estimated to be less than ±0.2 K. Further details about the apparatus are reported in [15].

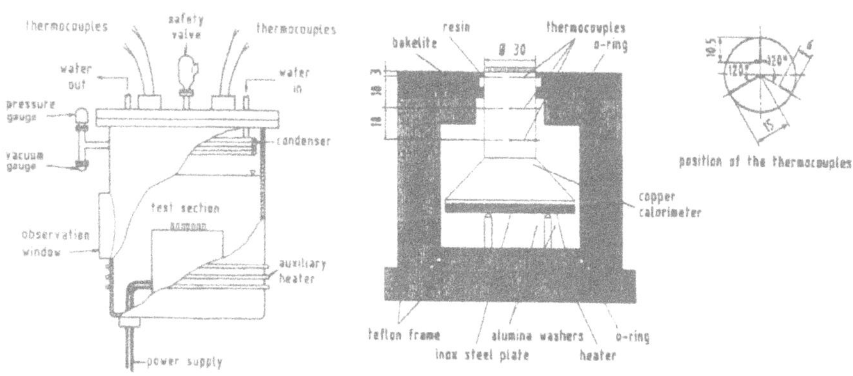

Figure 1 - Experimental apparatus and test section

518

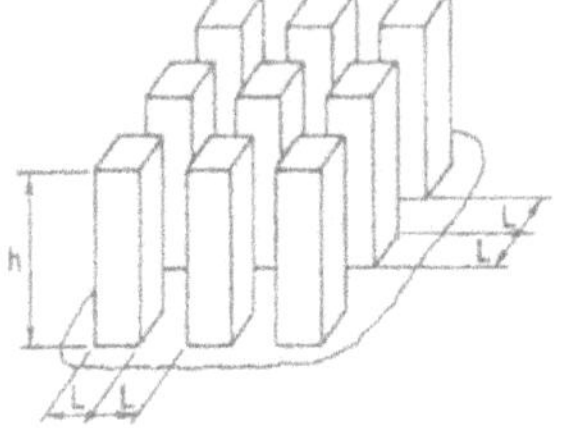

surface	h (mm)	L (mm)	A_r
0.8 spines	3	0.8	4.75

Figure 2 - Geometrical dimensions of the extended surface

The extended surface used during the experiments was made up of straight spines with square cross-section. Figure 2 shows both the sketch and geometrical dimensions of the extended surface. The same figure also gives the ratio A_r between the total heat transfer area of the extended surface and the base area.

A plexiglas wall was used to make the narrow channel with the following geometrical dimensions: 5 mm thickness and exactly the same diameter as the bakelite disk (90 mm). The gap space s between the top of the finned surface and the plexiglas wall was set by interposing a space adjuster and fixing the gap space by screws. The accuracy in the channel width is ±0.02 mm. A sketch of the channel is shown in Fig. 3.

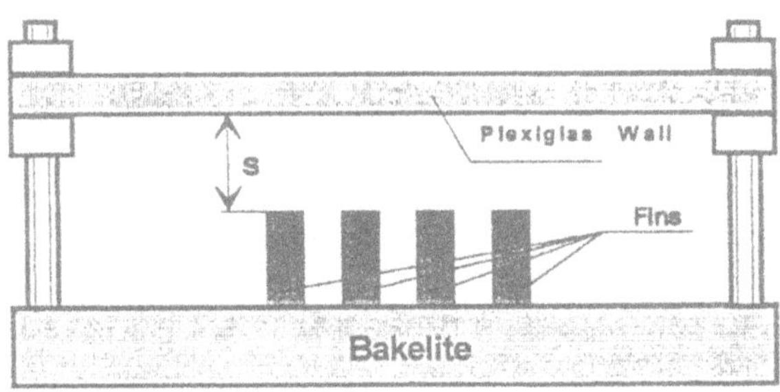

Figure 3 - Particular of the test section

For both orientations of the test section (horizontal or vertical), the tests were also conducted with open periphery conditions of the channel. Nevertheless, in the case of vertical orientation, the spines were positioned so that the lateral fin surfaces were parallel and perpendicular to the gravity vector.

The operating procedures followed before each test include a degassing operation that was carried out until correspondence between pressure and saturation temperature was obtained.

Two series of tests were performed following different operating procedures:

1st series of tests. During these experiments the following parameters were investigated:

- orientation of the base surface: horizontal or vertical
- saturation pressure: 1.0 bar
- channel width: unconfined, s=2 mm, and s=0.5 mm
- Operating procedure **[A]** - Before the run, the heating surface was left immersed in the saturated liquid overnight, with zero heat flux; the run was carried out by supplying the electric heater with gradually increasing power. Experiments were run increasing the heat flux by small steps. As soon as conditions of heat flux close to the critical value had been achieved, the voltage applied to the heater was reduced by steps of the same order. In all the tests, the liquid level was maintained at approximately 80 mm above the boiling surface or above its centerline when in the vertical orientation.

2nd series of tests. During the experiments the following parameters were investigated:

- orientation of the base surface: horizontal
- saturation pressure: 0.5 bar (T_{sat}=37 °C), 1.0 bar (T_{sat}=56 °C), and 2.0 bar (T_{sat}=78 °C)
- channel width: unconfined, s=40 mm, s=30 mm, s=20 mm, s=10 mm, s=8 mm, s=7 mm, s=6 mm, s=4 mm, s=2 mm, and s=0.5 mm
- Operating procedure **[B]**. Before the run the finned surface was boiled at 7.7 W/cm^2 for 30 min, then the experiment was started from 1.0 W/cm^2 and the heat flux increased in small steps.

As can be noted, the second series of tests was performed only for the horizontal orientation of the base surface, because during the first series, experiments conducted with the vertical orientation showed negligible effects of the channel width.

3. Results and discussion

The experiments, conducted at saturation conditions, were performed starting from lower heat fluxes and gradually increasing the voltage applied to the heaters. When conditions of heat flux close to the critical value had been achieved, the voltage to the heaters was reduced by steps of the same order. The typical boiling curve of highly-

520

wetting liquids was thus obtained. The surface temperature overshoot at the incipience of boiling and the hysteresis phenomenon, i.e., distinctly different boiling curves in the cases of increasing and decreasing heat flux, were readily apparent.

The results obtained during the different tests are presented as boiling curves. The abscissa gives the difference between the average surface temperature and the saturation temperature of the liquid (T_w-T_{sat}), while the ordinate gives the heat flux (q"). The latter was based on the plain surface of the copper block (7.07 cm^2).

First series of experiments

The results of the tests performed with the 0.8 mm finned surface are reported in Fig. 4 as a function of both the channel width and the orientation of the base surface.

As can be noted, in the case of the vertical orientation of the base surface, the influence of the channel width is negligible. In fact there is no significant difference between the boiling data measured in the unrestricted situation and those obtained for the two values of the gap size, s=2.0 mm and s=0.5 mm. This thermal behavior may be

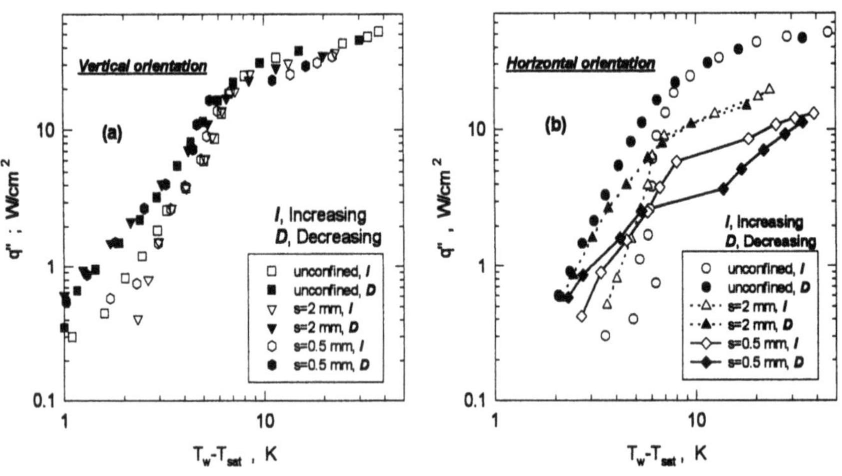

Figure 4 - Boiling data in the case of vertical (a) or horizontal (b) orientation of the base surface for three values of channel width: unconfined, s=2.0 mm, and s=0.5 mm

due to the fact that only a small amount of the boiling surface really looks at the channel wall; in fact only the top of the spine is positioned at a fixed value of the channel width, while the lateral surface of the spine and the base surface all around the spine are in the same condition as in the unrestricted situation (the ratio between the total boiling surface and the top surface of the fin is 34).

A slight influence of the channel width on the hysteresis phenomenon was observed at low heat flux. This result, however, has to be confirmed by repeating the experiments because, as is well known, the hysteresis phenomenon can be expressed in terms of probability [16].

The influence of the channel width became more evident when the experiments were conducted with the horizontal orientation of the base surface. Comparison of the data obtained in the case of s=2.0 mm or s=0.5 mm shows a drastic reduction in the heat transferred to the liquid, passing from the unrestricted situation to the lowest gap size. The maximum heat flux reached during the tests performed at s= 0.5 mm was 14.2 W/cm^2; this value is lower than the flux that can be exchanged with nucleate boiling from a flat surface in the case of the unrestricted situation at the same saturation pressure [17].

A particular behavior was detected when the gap size was reduced. In fact, after the maximum heat flux is reached, the boiling data are partly located to the right of those measured during the increase of heat flux. This unexpected behavior is probably caused both by the stagnation of the vapor in the gap after the maximum heat flux has been reached and of the thermal inertia of the copper flux meter.

During these experiments a drastic increment in the time necessary to perform the tests was observed. In fact, the time necessary to reach the steady-state condition was several hours for the step back in the heat flux, after the maximum heat flux was reached, while it was about 30 minutes for the other boiling data.

Second series of experiments

During the second part of the study, it was decided to conduct a systematic investigation on the influence of the channel width on nucleate pool boiling from the finned surface. Observing the data obtained during the first series of tests, especially those with vertical orientation of the base surface, it was decided to concentrate the study only on the horizontal orientation of the base surface (the vertical orientation showed negligible effects) and to adopt the operating procedure to avoid unexpected behavior. Hence the tests were performed according to procedure **[B]**.

The boiling data, obtained for all the investigated values of channel width and for a saturation pressure of 0.5 bar, 1.0 bar and 2.0 bar, are shown in Figs. 5, 6, and 7, respectively.

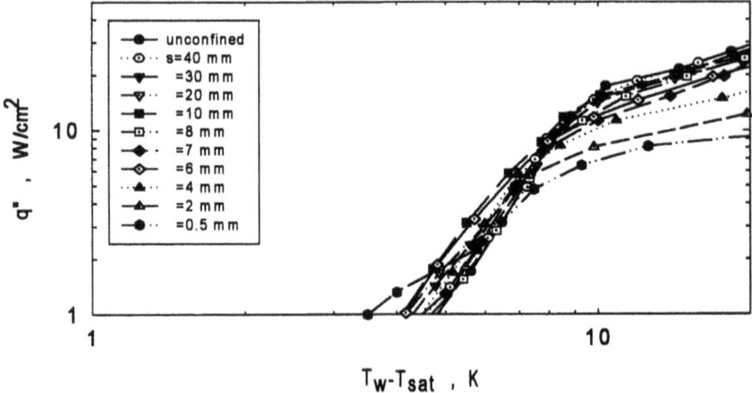

Figure 5 - Comparison of the boiling data for different values of channel width: saturation pressure 0.5 bar.

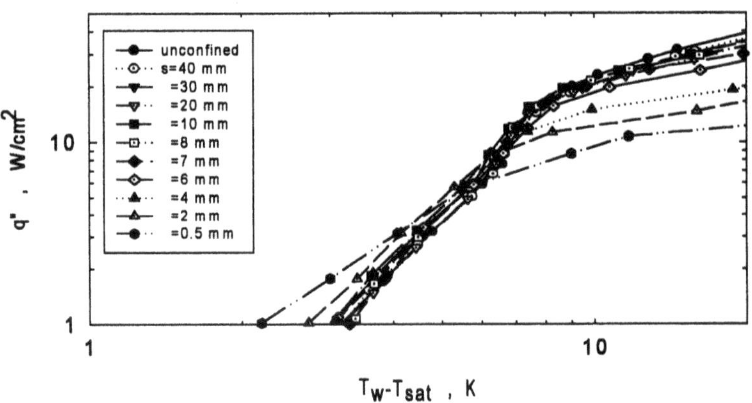

Figure 6 - Comparison of the boiling data for different values of channel width: saturation pressure 1.0 bar.

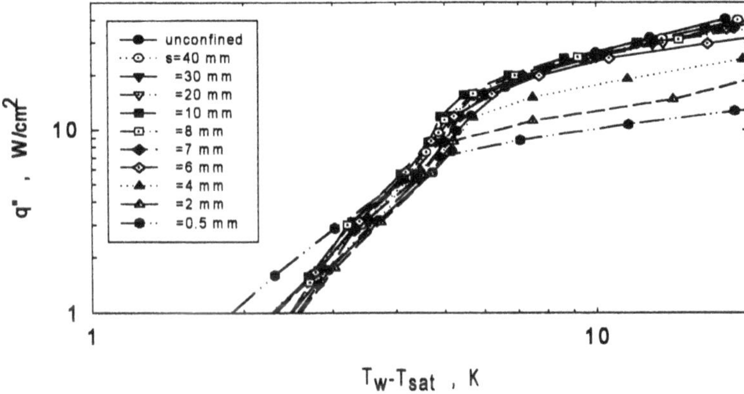

Figure 7 - Comparison of the boiling data for different values of channel width:
saturation pressure 2.0 bar.

As can be observed, there is a reduced influence on the heat transfer behavior for the channel width s=40 mm, s=30 mm, s=20 mm, s=10 mm, s=8 mm, whereas a constant reduction of the heat transfer coefficient was detected in the case of s=7 mm, s=6 mm, s=4 mm, s=2 mm, s=0.5 mm.

To quantify this reduction, the boiling data were processed, i.e., the ratios between the heat flux measured in the confined situation and that measured in the unconfined situation, for the same surface superheat, were calculated.

In Fig. 8 (a, b, c) these ratios are drawn for the three investigated pressures. As can be seen, for all pressures, a critical value of the gap can be specified below which it is possible to observe the transition between the influence and the no influence of the channel width. This transition value is about 7-8 mm.

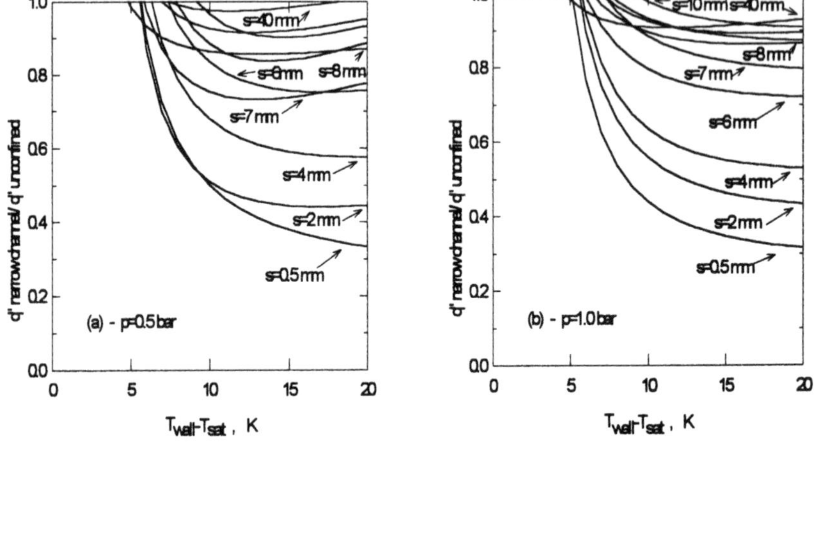

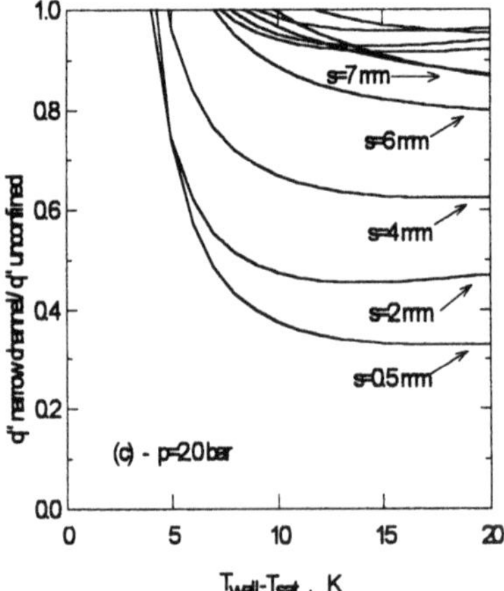

Figure 8 - Ratio between the heat flux (narrow channel) and heat flux (unconfined) for different values of channel width: (a) p=0.5 bar, (b) p=1.0 bar, and (c) p=2.0 bar.

The last analysis of the experimental data was performed to verify if the well known pressure effect on the pool boiling is influenced by the smaller channel widths. In Fig. 9 the boiling data for all the investigated pressures and three values of channel width (s=40 mm, s=4 mm, and s=0.5 mm) are reported. As can be seen, the influence of the pressure is not dependent on the value of the narrow channel in case of gap size higher than 4 mm for the s=2 mm and s=0.5 mm experiments this behavior was observed only for low wall superheats, whereas at high values of wall superheats it seems that for p=1.0 bar and p=2.0 bar the influence of the pressure is negligible only for p=0.5 bar the heat flux is always reduced, probably in this situation the bubble departure diameter is larger than the gap size.

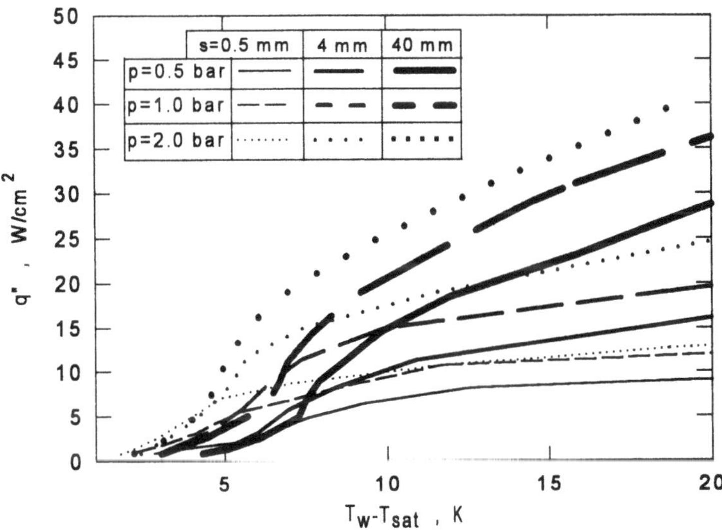

Figure 9 - Effect of the pressure and channel width on the heat transfer behavior.

4. Conclusions

An experimental investigation of the influence of channel width was carried out for nucleate boiling of a dielectric fluid (FC-72). The experiments were performed using a finned surface, and the gap sizes investigated were as follows: unconfined, s=40 mm , s=30 mm s=20 mm s=10 mm s=8 mm s=7 mm s=6 mm s=4 mm s=2 mm and s=0.5 mm. During the tests the pressure assumed the values 0.5 bar, 1.0 bar, and 2.0 bar.

Two series of experiments were performed. During the first one, a limited number of runs were carried out. In particular, the two gap channels s=2 mm and s=0.5 mm were investigated. Channel width reduction does not affect heat transfer in the case of vertical orientation of the base surface, while it causes a drastic reduction in the thermal behavior in the case of a horizontal orientation of the base surface. Furthermore, for the latter test conditions, an unexpected behavior was observed in the case of the lowest gap size s=0.5 mm. In fact, the wall superheats measured after the reaching of the maximum heat flux are shifted to the right of those measured at the same heat flux but during the heat flux increase. This behavior is probably caused by vapor stagnation in the gap and thermal inertia of the copper flux meter.

During the second series of tests a systematic investigation of the influence of channel width was performed. For all the runs the base orientation of the finned surface was maintained horizontal. A critical value of the channel width was found. Above this value (7-8 mm) the influence of the gap size is negligible, whereas a drastic reduction of the heat transfer behavior was detected below this value. This critical value seems not influenced by the pressure, in fact it was always the same (7-8 mm) even though the pressure changes from 0.5 bar to 2.0 bar.

Nomenclature

A_r ratio between the total heat transfer area of the extended surface and the base area, dimensionless

h   fin height, mm

L   fin thickness, mm

p   pressure, bar

q"   heat flux, W/cm^2

s   channel width, mm

T_{sat} saturation temperature, K

T_w surface temperature, K

Acknowledgments

This work was supported by grant "Cofinanziamento Murst 1997 - Progetto Termofluidinamica mono e bifase". The authors thank Dr. R. Van San from 3M Corporation for supplying the fluid.

REFERENCES

1. Bar-Cohen, A., (1992), State-of the art and trends in the thermal packaging of the electronic equipment" *J. of Electronic Packaging*, **114**, 257-270.
2. Bergles, A.E,. and Bar-Cohen., A., (1994), *Cooling of electronic systems*, Kluwer Academic Publishers, Dordrecht, 539-621.
3. Mudawar, I., and Anderson, T.M., (1993), Optimization of enhanced surfaces for high flux Chip cooling by pool boiling, *J. of Electronic Packaging*, **115**, 89-100.
4. Katto, Y., and Yokoya, S., (1966), Experimental study of nucleate pool boiling in case of making interference-plate approach to the heating surface", *Proc. Third International Heat Transfer Conference*, 219-227.
5. Hung, Y.-H., and Yao, S.-C., (1985), Pool boiling heat transfer in narrow horizontal annular crevices", *J. of Heat Transfer*, **107**, 656-662.
6. Bar-Cohen A. and Schweitzer H., (1985), Thermosyphon boiling in vertical channels, *J. Heat Trans.*, **107**, 772-778.
7. Fujita, Y., Ohta, H., Uchida, S., and Nishikawa, K., (1988), Nucleate boiling heat transfer and critical heat flux in narrow space between rectangular surfaces, *Int. J. Heat Mass Trans.*, **3 1**, 229-238.
8. Katsuta M. and Nagata, K., (1992), Boiling induced heat transfer enhancement using a narrow space, in *Proc. Eng. Foundation Conf. Pool and External Flow Boiling,*, 381-386.
9. Xia, C., Guo, Z., and Hu, W. (1992), Mechanism of boiling beat transfer in narrow channels, in *Proc. 28th Nat. Heat Trans. Conf*, 111-119.
10. Rampisela, P.F., Berthoud, G., Marvillet, C., and Bandelier, P., (1993), Enhanced boiling heat transfer in very confined channels, *Aerospace Heat Exchanger Technology* , R.K. Shah and A.Hashemi, Eds., Elsevier Science Publishers, New York, NY, 227-236.
11. Misale, M., and Bergles, A.E., (1997), The influence of channel width on natural convection and boiling heat transfer from simulated microelectronics components, *Experimental Thermal and Fluid Science*, **14**, 187-193.
12. Ishibashi, E., Liu, Z.-A. , Chifiru, T., and Murakami, N., (1993), "Pool-boiling heat-transfer characteristics of the surface-worked heat-transfer tubes in narrow spaces under reduced pressure conditions, *Heat Transfer-Japanese Research*, **22**, 703-715

13. Nowell, R. M. Jr., Bhavnani, S.H., and Jaeger, R.C., (1995), Effect of channel width on pool boiling from a microconfigured heat sink", *IEEE Transactions on Components, Packaging, and Manufacturing Technology*, Part A, **18**, N. 3, 534-539.

14. Misale, M., Gugleumini, G., and Bergles, A.E., (1996), FC-72 Pool boiling from finned surfaces placed in a narrow channel: preliminary results, VLSI and Microsystem Packaging Techniques and Manufacturing Technologies, Baveno, Italy.

15. Misale, M., Ferrando, F., Guglielmini, G., and Schenone, C., (1991), An experimental apparatus for studying dielectric fluids boiling, *Energy Engineering Department*, EGR/16, University of Genoa, Italy.

16. Bar Cohen, A., You, S.M., Simon, T., (1990), Boiling incipience and nucleate boiling heat transfer of highly wetting dielectric fluids from electronic materials, *IEEE Transaction on Components, Hybrids, and Manufacturing Technology*, **13**, 1032-1039.

17. Guglielmini, G., Misale, M., and Schenone, C., (1993), Pool boiling heat transfer of dielectric fluids for immersion electronic cooling: effect of pressure", *Proc. of Eurotherm Seminar n. 29*, Thermal Management of Electronic Systems, 243-253.

PREDICTION OF CONDENSATION AND EVAPORATION IN MICRO-FIN AND MICRO-CHANNEL TUBES

Ralph L. Webb
Pennsylvania State University
University Park, PA, 16802 U.S.

Abstract. This paper surveys methods to predict condensation and evaporation in plain, and in micro-fin and micro-channel tubes. Hydraulic diameters as small as 1.0 mm are of interest. To date, the Shah correlation has been well accepted for prediction of condensation in plain tubes. This work shows apparent deficiencies of the Shah equation for $p/p_{cr} > 0.44$. An improved predictive model called the "equivalent Reynolds number model" is introduced Use of the new model requires prediction of the single-phase heat transfer coefficient and the two-phase pressure gradient. For condensation in micro-fin tubes, both vapor shear and surface tension forces contribute to the condensing coefficient. The equivalent Reynolds number model predicts the vapor shear component. An existing theory of Adamek and Webb is applicable for the surface tension contribution. The vapor shear model is also applicable to evaporation inside tubes; however, it is not applicable after dryout. A nucleate boiling component will modestly add to the evaporation coefficient. Data are shown that suggests the nucleate boiling contribution is less than 15%, and that this contribution exists only at vapor qualities less than 50%.

1. Introduction

One consideration in the search for alternate refrigerants includes how the use of a particular fluid will affect the heat transfer coefficient and friction factor for condensation, evaporation, and single-phase flow. Unless this can be predicted, expensive tests are required. A complicating factor is that enhanced surfaces are typically used in refrigeration systems. Hence, one must be able to predict the performance for complex surface geometries. This paper will address the ability to predict heat transfer and friction as a function of 1) Fluid properties, and 2) Surface geometry in enhanced tubes used in refrigeration evaporators and condensers.

Natural refrigerants are also of interest. These are typically pure fluids, which include isobutane, propane, carbon dioxide, water, and ammonia. However, the new HCFC refrigerants are typically refrigerant mixtures, called "blends." For boiling and

S. Kakaç et al. (eds.), Heat Transfer Enhancement of Heat Exchangers, 529–550.
© 1999 *Kluwer Academic Publishers.*

530

condensing processes, existing correlations that apply to pure fluids are not applicable to mixtures. This is because a mass transfer "diffusion" resistance exists in a vaporizing or condensing mixture. For single-phase heat transfer or friction, mixtures introduce no problems in predicting performance, if the mixture fluid properties are evaluated at the weighted molar fraction of the mixture.

The predictive methods of interest here are for vaporization and condensation in round micro-fin tubes, and in small hydraulic diameter extruded aluminum tubes. We will survey predictive methods for single-phase flow, vaporization, and condensation for the above tube geometries. Single-phase heat transfer and friction correlations are included, primarily because they are the primary "building block" for convective condensation and evaporation predictive models. The reader may wish to consult Webb [1] who reviewed predictive methods for a wide range of enhanced surface geometries, including air-side surfaces.

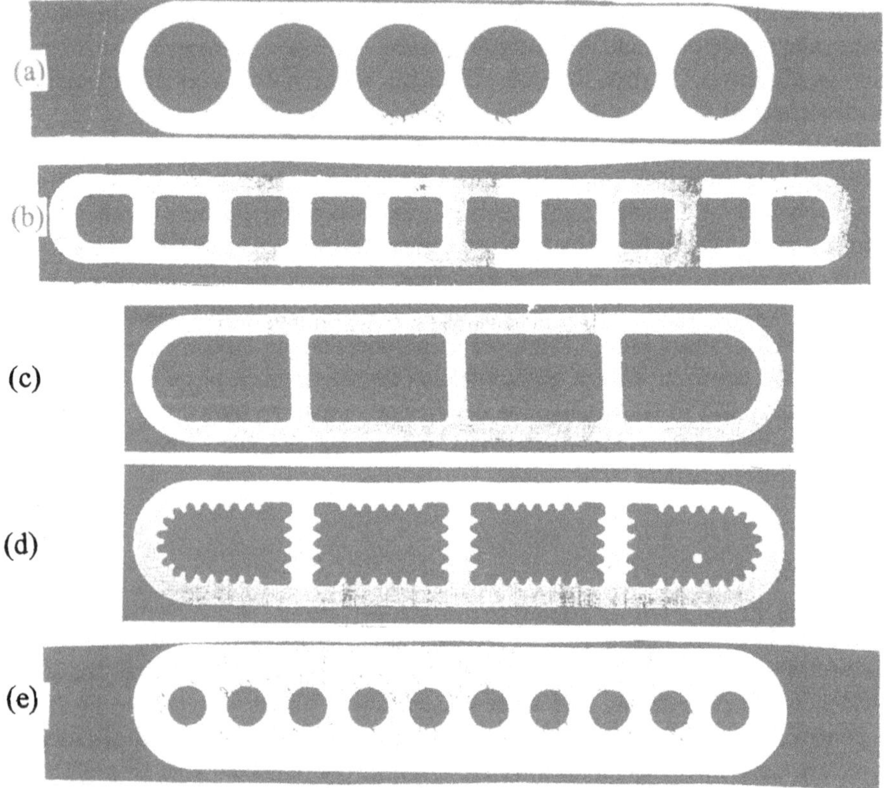

Fig. 1Cross-section photos of multi-port extruded aluminum tubes. a) 3-16-6p (D_h = 2.13 mm), b) 2-20-10p (D_h = 1.33 mm), c) 3-16-4p (D_h = 2.64 mm), d) 3-16-4m, (D_h = 1.56 mm), e) 3-16-10p (D_h = 0.96 mm)

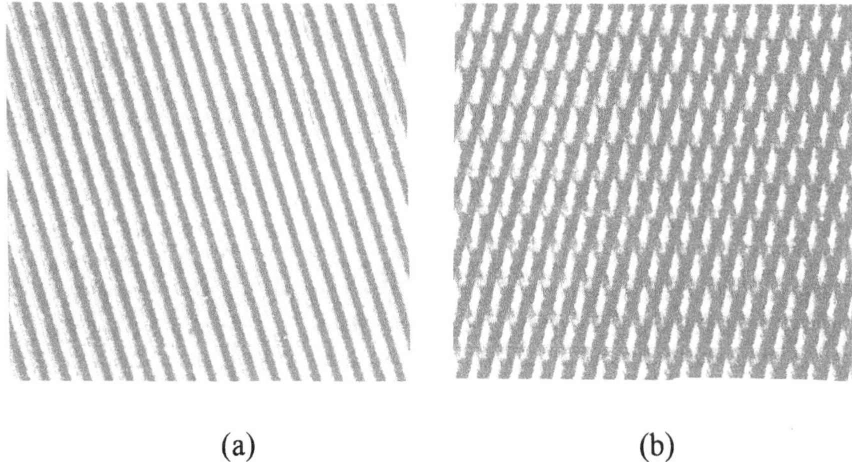

(a)   (b)

Fig. 2Copper micro-fin tubes. a) Single grooved, and b) Cross-grooved.

2. Single-phase Flow

2.1 INSIDE PLAIN TUBES

In this work, the term "plain tubes" is taken to mean plain channels of circular or non-circular shape. One important application for non-circular channels is extruded aluminum tubes used in brazed aluminum refrigerant condensers. Figure 1 illustrates such extruded tubes, which typically have hydraulic diameters less than 3.0 mm. Figure 1 also shows extruded tubes having internal micro-grooves, which will be discussed in the next section.

Turbulent flow is of primary interest. For turbulent flow in circular tubes, the recommended equations are Petukhov [2] for heat transfer, and the associated Filonenko for friction. These equation are described by Webb [1]. The simple empirical Blasius friction equation ($f = 0.079 \, \mathrm{Re}^{-0.25}$) is also suitable. The same equations are applicable for non-circular tubes if one uses the hydraulic diameter concept.

2.2 MICRO-GROOVED TUBES

The tube geometries of interest is shown in Figure 2 and Figure 1d. Figure 2a shows a round, single helical micro-fin tube having 0.3 mm groove depth. This tube is used for tube-side evaporation and condensation of refrigerants in virtually all evaporators and condensers. Used in air cooled evaporators or condensers, the tube outside surface is smooth, and the tube is expanded to provide good contact with aluminum fins on the outer surface. The tube provides a heat transfer coefficient 2-to-3 times that of a plain

532

tube. The Figure 2a geometry is made in diameters between 4 and 16 mm. The cross-grooved tube shown in Figure 2b is also commercially available. Their performance has been reported by Chamra et al.[3,4].

For a plain I.D. round tube, the characteristic dimension is the diameter (D_i). However, for a micro-grooved tube, such as Figure 1d or Figure 2a, the detailed groove dimensions are also involved. Many workers define the measured condensation coefficient in terms of the "nominal" internal area ($A_n/L = \pi D_i$). This essentially ignores the internal micro-fin dimensions. In the present predictive work, we use the total internal surface area, which is defined in terms of the detailed micro-fin dimensions. These detailed dimensions (see Figure 3) are the micro-fin height (e), the fin pitch (p), the fin base thickness (t_b), the apex angle (β) of the micro-fin, the fin tip radius (r_t), and the helix angle (α) for a round tube. One desires to predict the performance as a function of the detailed internal geometry.

Brognaux et al. [5] reported single-phase data on a 14.9 mm inside diameter round micro-fin tube (0.35 mm fin height and 0.6 mm fin pitch) with the helix angle varied between 17 and 27 degrees. The effect of Prandtl number was investigated for turbulent flow with $0.70 \le Pr \le 7.85$. The present authors have updated the data to include helix angles between 0 and 34 degrees. Using the Petukhov equation with hydraulic diameter to predict the data shows that the heat transfer coefficient is over predicted nearly 30%. This is because the large shear stress in the interfin region slows the velocity causing reduced heat transfer coefficient. Brognaux et al. developed an empirical correlation to account for Prandtl number and helix angle. These data were correlated by the empirical equation

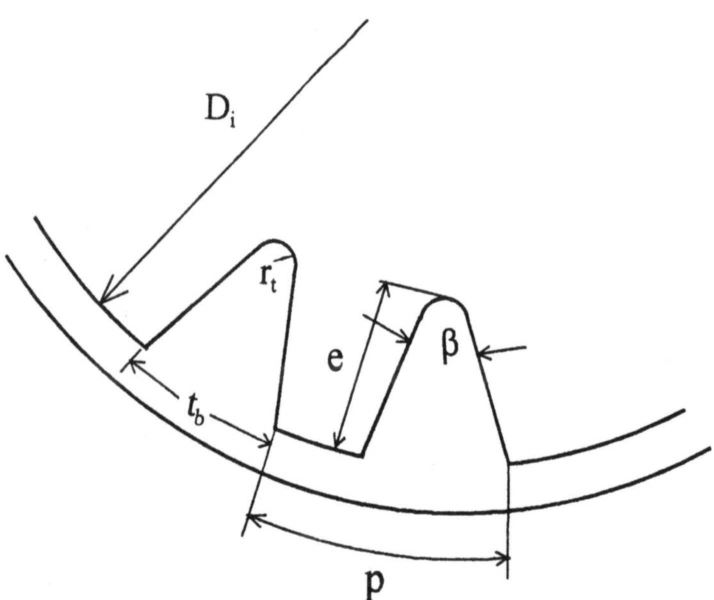

Fig. 3Micro-fin dimensions

$$Nu = C\,Re^{0.81}\,Pr_l^{0.55} \tag{1}$$

where $C = 0.02271 + 3.72E\text{-}05\,\alpha - 9.337E\text{-}7\,\alpha^3$, α is the helix angle, the characteristic diameter is the "melt-down" diameter, D_e (the inside diameter if the fins were melted and returned to the tube wall). Work is in progress to extend the correlation to account for the effect of the micro-fin dimensions (fin height, fin pitch, and apex angle).

Yang & Webb [6] report single-phase [R-134a and R-12] data on the Figure 1c and 1d extruded aluminum tubes. They find that the Petukhov and Blasius equations reasonably predict the single-phase heat transfer and friction, respectively. Note that the Figure 1d tube has axial grooves ($\alpha = 0$). The Petukhov equation works better for the small hydraulic diameter Figure 1d tube than for the 14.9 mm diameter micro-fin tube having $\alpha = 0$. This is apparently because the e/D_i of the extruded aluminum tube is much higher than for the round 14.9 mm diameter tube. The greater fin height prevents significant bypassing of the flow in the interfin region.

3. Condensation in Plain Tubes

3.1 SHAH EQUATION

Popular correlations for use with plain round tubes are those of Shah [7] and Cavallini [8]. Moser et al. [9] have shown that the Shah correlation is valid for condensation in plain round copper tubes with diameters down to 3.14 mm for refrigerants condensing at saturation temperatures below 40°C. This evaluation included refrigerants R-11, R-12, R-113, R-22, R-125, R-134a, and R-410). However Yang et al. [6] found that the Shah correlation over-predicted their R-134a condensation data in the Figure 1c multi-port aluminum tubes ($D_h = 2.64$ mm) by 80% for at $T_{sat} = 65$°C.

Zhang [10] has further evaluated the Shah correlation using R-134a at 40 and 65 °C in plain, round copper tubes of 6.2 and 3.2 mm, and the Figure 1a 2.13 mm multi-port aluminum tubes. Figures 4 and 5 shows that the ability of the Shah correlation to predict condensation for the 3.2 mm I.D. single copper tube, and the Figure 1a extruded aluminum tube ($D_i = 2.13$ mm) . These figures show that the Shah correlation increasingly fails as the saturation temperature increases above 40°C. The Shah correlation is given by

$$Nu = Nu_l\left[1 + a\left(\frac{p}{p_{cr}}\right)^{-b}\left(\frac{1}{1-x}\right)^{0.76}\right] \tag{2}$$

where Nu_L is for all liquid flow, with $a = 3.8$ and $b = 0.38$. The Shah database includes data for $0.001 \le p/p_{cr} \le 0.44$. All of the data were in the range $0.10 \le p/p_{cr} \le 0.44$, except for the R-11 data, which had $p/p_{cr} \le 0.001$. Using the original Shah database

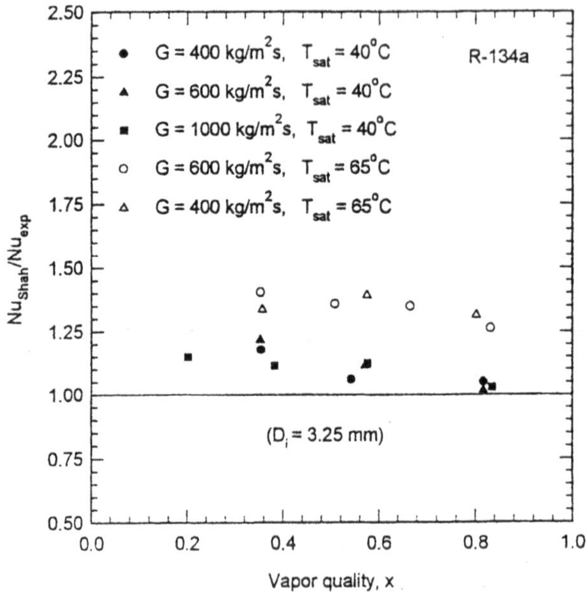

Fig. 4 Ability of the Shah [1979] correlation to predict the R-134a condensation coefficient in a 3.2 mm internal diameter plain copper tube.

with the R-11 data excluded, Zhang [10] found that changing the exponent on p/p_{cr} from 0.38 to 0.8 and using a = 2.35 substantially improved the correlation of the data in the Shah database. Further, the data shown on Figures 4 and 5 were well predicted. This modified Shah correlation works very well for $0.10 \le p/p_{cr} \le 0.80$. For R-134a at 65°C, $p/p_{cr} = 0.47$. Hence, it appears that the Shah correlation increasingly fails for $p/p_{cr} > 0.44$.

3.2 EQUIVALENT REYNOLDS NUMBER MODEL

Moser et al. [11] have developed a theoretically based predictive model for condensation in plain tubes, which they describe as the "equivalent Reynolds number concept." This model is based on the analogy between heat and momentum transfer. To be able to predict condensation in micro-channels or micro-fin tubes, one must be able to predict condensation in plain round tubes of any diameter. Hence, this predictive model for plain round tubes forms the "foundation" of our analytical understanding. This model assumes annular flow and is based on the following concepts (or assumptions):

1. That the condensation coefficient is directly related to the heat transfer coefficient for single-phase flow in the same geometry. The linkage is the "Equivalent single-phase Reynolds number (Re_{eq})." At the Re_{eq}, the single-phase flow will have the same wall shear stress as for the condensing flow.
2. Assuming the equivalent single-phase flow is turbulent, one uses the Petukhov equation to calculate the single-phase heat transfer coefficient (h_{sp}) .

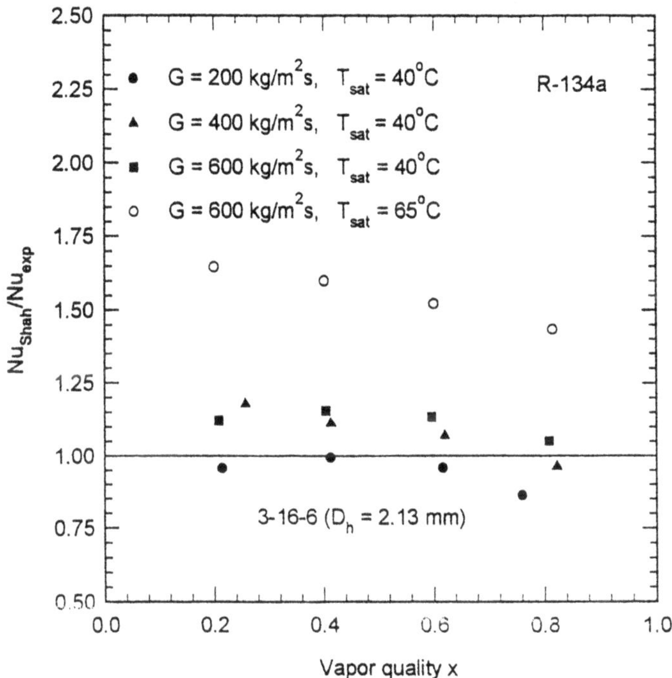

Fig. 5 Ability of the Shah [1979] correlation to predict the R-134a condensation coefficient in the Figure 1a extruded aluminum tube having $D_i = 2.13$ mm.

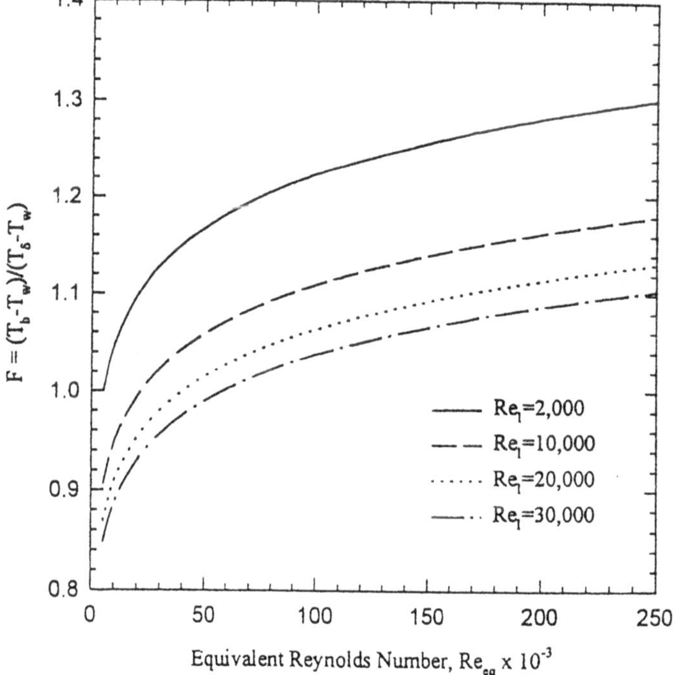

Fig. 6 Plot of the F-factor (Eq. 4) vs. Re_{eq} vs. Re_l for $Pr_l = 3.0$.

536

3. The condensing coefficient (h_{cond}) is related to the calculated single-phase heat transfer coefficient by the following analytically based equation given by Moser et al. [11]:

$$\frac{h_{cond}}{h_{sp}} = \frac{T_b - T_w}{T_s - T_w} = F \tag{3}$$

Figure 6 shows a graph of the F-factor vs. Re_{eq} for $Pr_l = 3.0$. For $30{,}000 \leq Re_{eq} \leq 200{,}000$, and $Re_l \geq 10{,}000$, the figure shows that $0.95 \leq F \leq 1.15$. By curve fitting the calculated F-factors, the correction factor is given by the following expression:

$$F = 1.31\,Pr_l^{-0.185}(R^+)^A\,Re_l^{\,B} \tag{4}$$

where $R^+ = 0.0994\,Re_{eq}^{\;7/8}$, $A = 0.126\,Pr_l^{-0.448}$, and $B = -0.113\,Pr_l^{-0.563}$. As shown by Figure 6, the Equation 4 F-factor is a "second order" term. The Re_{eq} is defined by the following equation:

$$Re_{eq} = \left[\frac{1}{2}\frac{D^3\rho_l}{f_l\mu_l^{\,2}}\left(\frac{dp}{dz}\right)_f\right]^{\frac{1}{2}} \tag{5}$$

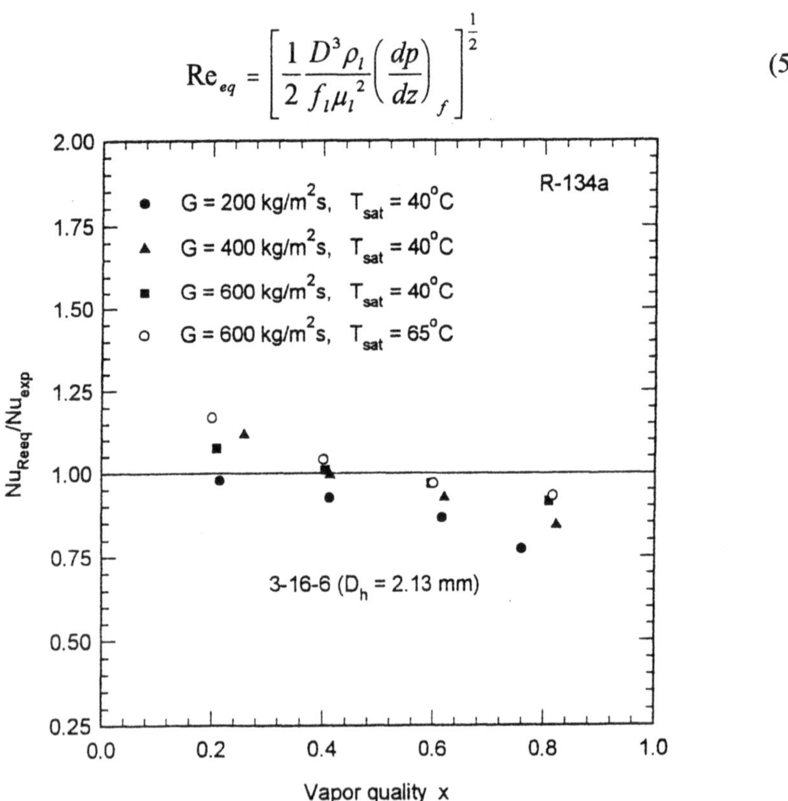

Fig. 7 Ability of the equivalent Reynolds number correlation to predict the R-134a condensation coefficient in the Figure 1a extruded aluminum tube having D_i = 2.13 mm.

here $(dp/dz)_f$ is the frictional pressure gradient for the two-phase (condensing) flow. There are two possible methods to get $(dp/dz)_f$:

1. Predict it. For plain, round tubes one may use a correlation such as that proposed by Friedel [12].
2. Use the experimental value. This gives a precise test of the validity of the model. However, it is preferred to use a predictive correlation, if an accurate one exists.

Moser et al. [11] used the Friedel correlation to predict the two-phase friction pressure gradient for single, round tubes having of 3.14-to-20 mm inside diameter. Local and average condensation data were predicted within 14%.

Zhang and Webb[13] used the "equivalent single-phase Reynolds number model" to predict the condensing coefficient for a variety of tube geometries and refrigerants. The conditions predicted are shown in Table 1. The Re_{eq} was predicted using measured pressure drop. The single-phase heat transfer coefficient was calculated using the Petukhov equation. Figure 7 shows the ability of the model to predict the R-134a condensing coefficient in the Figure 1a multi-port tubes having 2.13 mm diameter round holes for R-134a (T_{sat} = 40 and 65 °C). For all of the data listed in Table 1, the condensation data were generally predicted within ±20.

Zhang [10] also used the "equivalent single-phase Reynolds number model" with the pressure gradient predicted by the Friedel [12] equation to predict the condensation data. This resulted in an over prediction of the condensing coefficient by 20-to-50%. This suggests that the Friedel equation does not accurately predict the R-134a pressure gradient at high saturation temperature (65°C) for round tube diameters below 2.13 mm. Pressure drop prediction will be discussed in the next section.

3.3 FRIEDEL PRESSURE DROP CORRELATION

The Friedel [12] correlation is a "two-phase multiplier" type of correlation based on a data base of 25,000 points. The correlation is given by the following equation:

$$\frac{\Delta p}{\Delta p_{lo}} = \phi_{lo}^2 = C_{F1} + \frac{3.24 C_{F2}}{Fr^{0.045} We^{0.035}} \tag{6}$$

where Fr $(= G^2/gD_i\rho_{tp})$ and We $(= G^2 D_i/\sigma\rho_{tp})$ are the Froude and Weber numbers, respectively. The terms C_{F1} and C_{F2} are defined as

$$C_{F1} = (1-x)^2 + x^2 \left(\frac{\rho_l}{\rho_v}\right)\left(\frac{f_{vo}}{f_{lo}}\right) \tag{7}$$

$$C_{F2} = x^{0.78}(1-x)^{0.24}\left(\frac{\rho_l}{\rho_v}\right)^{0.91}\left(\frac{\mu_v}{\mu_l}\right)^{0.19}\left(1-\frac{\mu_v}{\mu_l}\right)^{0.7} \tag{8}$$

538

The ρ_{tp} is the homogeneous two phase density. This equation is given on page 429 of Carey [14].

Table 1
Data predicted by Zhang & Webb [13] Using the
Equivalent Reynolds Number Model

Tube Geometry	Inside Diameter (or D_h), (mm)	Refrigerants	T_{sat} (°C)
Round tube	6.2	134a	40, 65
Round tube	3.25	134a, 22, 404a	25-65
Fig. 1a type multi-port tube	0.96, 1.45, 2.13	134a	40, 65
Fig. 1b multi-port tube	1.33	134a	65

3.4 MODIFIED FRIEDEL CORRELATION FOR SMALL DIAMETER TUBES

When developing his correlation, Friedel [12] used the following two theoretical limits: 1) At $x = 0$ (all liquid flow), $\phi_{lo}^2 = 1.0$, 2) At $x = 1$ (all vapor flow), $\phi_{lo}^2 = (dp/dz)_{vo}/(dp/dz)_{lo} = (\rho_l/\rho_v)(f_{vo}/f_{lo})$. Except for the two theoretical limits, all of the coefficients and exponents were based on best fit analysis. The reduced pressure (p/p_c) can be used to replace the property ratio $(\rho_v/\rho_l)^{0.5}(\mu_l/\mu_v)^{0.1}$. Also in the Friedel correlation, the exponents on Weber number (We), and Froude number (Fr) are very small (0.045 and 0.035). Zhang [10] used the above observations to develop a modified form of the Friedel correlation valid for the tube geometries listed in Table 1, plus multi-port tubes: 3-16-8p, 3-16-10p. His result, which correlated 120 data points for the Table 1 geometries with 11.5% absolute deviation, is given by:

$$\phi_{lo}^2 = (1-x)^2 + 2.20x^2\left(\frac{p}{p_c}\right)^{-0.94} + 2.60x^{0.8}(1-x)^{0.25}\left(\frac{p}{p_c}\right)^{-1.44} \tag{9}$$

4. Condensation in Small Diameter Micro-fin Tubes

Yang and Webb [15] developed a predictive model for multi-port extruded aluminum tubes having micro-grooves, such as shown in Figure 2d. Surface tension force contributes to condensate removal from the convex fin shape. The predictive model calculates the vapor shear (h_{sh}) and the surface tension (h_{st}) contributions by the following equation:

$$h^2 = h_{sh}^2 + h_{st}^2 \qquad (10)$$

The vapor shear component was calculated using the "equivalent Reynolds number model" with the Re_{eq} as defined by Akers et al. [16]. As discussed by Moser et al. [11] the Akers equivalent single-phase Reynolds definition is flawed. Use of the Moser et al. [11] equivalent Reynolds number model is recommended. The surface tension component (h_{st}) is calculated using a linear surface tension drainage model described by Adamek and Webb[17]. This model predicted the R-134a 95% of the condensation data of Yang and Webb [6] within ±16%. An empirical correlation for the pressure gradient developed by Yang and Webb [6] was used to obtain Re_{eq}.

Figure 8 shows the R-12 data at T_{sat} = 65 °C of Yang and Webb [18], where the ordinate is the Nusselt number (hD_h/k) based on hydraulic diameter. At low vapor quality and all mass velocities, the Nusselt numbers for both tubes are nearly equal. The most dramatic difference occurs at low mass velocity, and vapor qualities greater than 0.5. The Nusselt number sharply increases with increasing vapor quality. The Nu_{Dh} enhancement (relative to the plain tube) is attributed to the effect of surface tension drainage force.

Using the Zivi [19] void fraction equation, they found that fin tips would be flooded with condensate, if $x < 0.5$. For $x > 0.5$, a surface tension induced pressure gradient acts to drain condensate from the small radius fin tip into the concave drainage channel at the base of, and between the fins. As shown in Chapter 12 of Webb [1], this pressure gradient may be approximated as

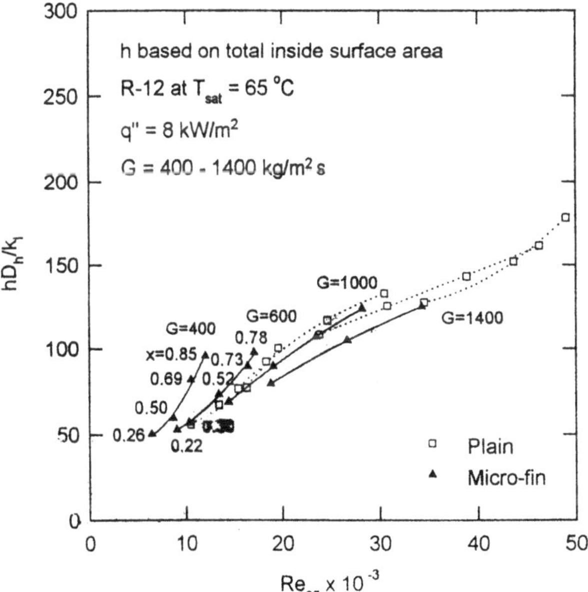

Fig. 8 Comparison of condensation Nusselt number in the Fig. 1c plain (D_h = 2.64) and 1d micro-fin (D_h = 1.564) tubes., where Re = $D_h G/\mu$.

540

$$\frac{dp}{ds} = -\sigma \left| \frac{d(1/r)}{ds} \right| \tag{11}$$

where r is the local radius of the fin surface. The term s is the unflooded length of the fin side. The condensation coefficient is given by $h = k_l/\delta$ for laminar condensate film, where δ is the condensate film thickness. Surface tension force acts to maintain a smaller film thickness on the micro-fins than exists on the surface of the plain tube (Figure 1a). Hence, the condensation coefficient is increased. This effect is significant at low mass velocity (400 kg / m^2-s) for vapor qualities greater than 0.5. At high mass velocity (1,000 kg / m^2-s), this effect is not as strong, because the vapor shear force is much higher at the high vapor qualities.

For x < 0.5, the micro-fins are flooded by the condensate, so surface tension drainage can act. Only vapor shear forces are important. Then, the heat transfer is "shear controlled."

5. Condensation in Round Micro-fin Tubes

5.1 EQUIVALENT REYNOLDS NUMBER MODEL

We have applied the "equivalent single-phase Reynolds number model" to predict the condensing coefficient in 14.8 mm inside diameter round microfin tubes. R-22

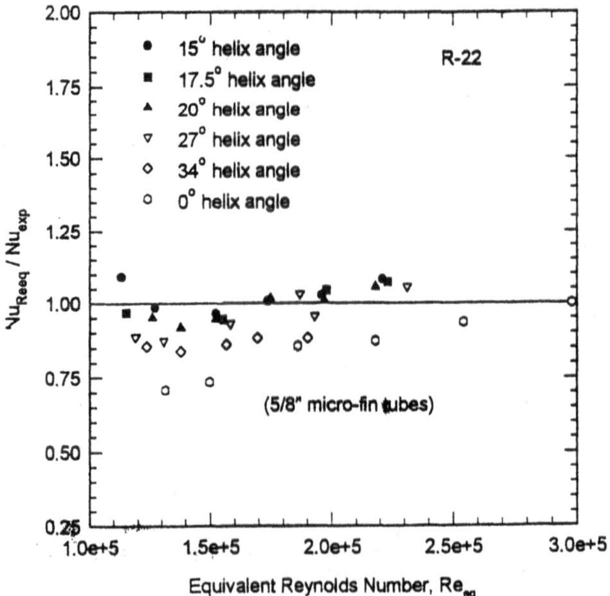

Fig. 9 Predictions for R-22 condensation at 24° C in 14.8 mm inside diameter micro-fin tubes .

condensation at 27 °C, and single-phase data were taken on five microfin tubes, whose helix angle varied from 0.0-to-34 degrees. The single-phase data for most of these tubes are reported by Brougnaux et al. [5]. The Re_{eq} was predicted using measured pressure drop. The single-phase heat transfer coefficient was calculated using an empirical correlation based on tests using water and air. Figure 9 shows the ability of the model to predict the R-22 condensing data. The figure shows that more than 90% of the data are predicted within ±15%. Note that this model does not account for the possible effect of surface tension induced condensate drainage from the fin tips.

Based on the data shown in Figure 8 for the 1.56 mm hydraulic diameter multiport tube, one may think that neglect of the surface tension component in the Figure 9 predictions will result in underprediction of the results. The cross-section area of the interfin region is only 6.25% of the total flow area for the 14.88 mm diameter micro-fin tube. Using the Zivi [19] equation, one can calculate the vapor quality (x_{fill}) that will result in 0.935 (1.0 - 0.0625) void fraction. This calculation shows that $x = 0.70$ at $\alpha = 0.935$ for R-22 at $T_{sat} = 25°C$. That means the surface tension effect should be negligible for $x < 0.70$. Because the Figure 9 data were taken for inlet and exit vapor qualities of 80% and 20%, it is not expected that surface tension will be of major influence. Liquid entrainment in the vapor core can also result in underprediction or the equivalent Reynolds number model. If entrainment exists, the underprediction should be more obvious for high mass velocities with high vapor qualities.

A similar evaluation of the vapor quality, at which the interfin space will be condensate filled has been done for other micro-fin geometries and tube diameters. Figure 10 shows the predicted vapor quality at which the interfin region of the microfin tube will be filled (x_{fill}). The figure shows the calculated results for R-22 condensing in three different microfin tube geoemtries and spans 25-to-75 °C. The void fraction was calculated using the Zivi [1964] void fraction. As saturation temperature increases, x_{fill} increases. Note the much smaller values of x_{fill} for the Figure 1d small hydraulic diameter, extruded aluminum tube.

5.2 KEDZIERSKI AND GONCALVES MODEL

Kedzierski and Goncalves [20] obtained condensation data at 40 °C for five refrigerants in a single 9.52 mm O.D. micro-fin tube having 60 fins 0.2 mm high at 18 degree helix angle. The data were obtained for R-22, R-32, R-125, R-134a, and R-410a. They developed multiple regression equations to predict the condensation coefficient and pressure drop as a function of fluid properties in the one tube tested ($D_i = 8.91$ mm, $N_f = 60$, $e = 0.2$ mm). Because the equations do not contain geometry data, they are not applicable to predict the condensation coefficient or pressure drop for micro-fin tubes of different internal geometries.

5.3 CAVALLINI ET AL. MODEL

Recently, Cavallini et al. [21] have developed an empirical correlation to account for fluid properties and internal geometry in a micro-fin tube. This was based on

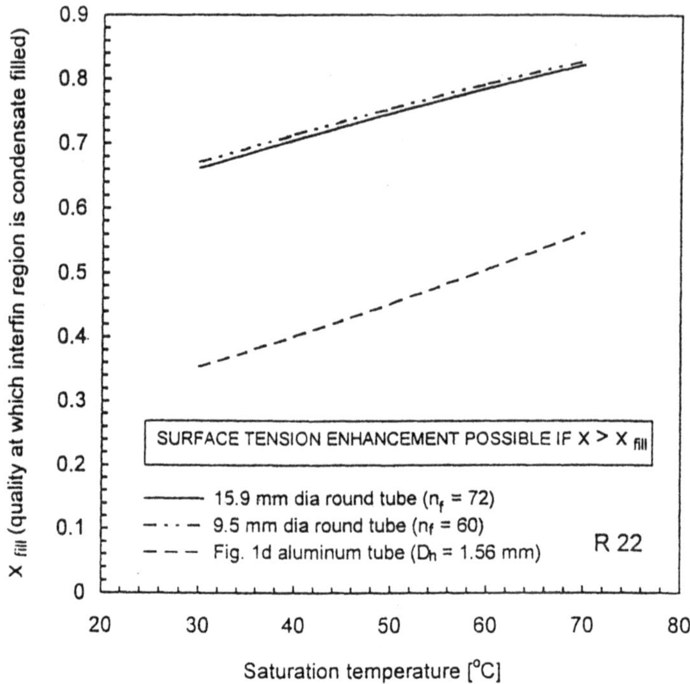

Fig. 10   Predicted R-22 vapor quality at which the interfin region of the microfin tube
will be filled (x_{fill}).

approximately 300 data points for seven refrigerants condensing in micro-fin, low-fin
(e/D > 0.04), and cross-grooved micro-fin tubes. Nearly all of the data points were for
the average condensing coefficient with high entering, and low leaving vapor quality
(e.g., 80-20%), and the average vapor quality is typically 0.50-0.60. The vast majority
of the data points are for the single-grooved micro-fin tube. Most of the tubes tested were
commercially produced single-groove micro-fin tubes, which do not span a wide range
of dimensional variables. The refrigerants and dimensional range of the tubes are shown
in Table 2. Table 2 shows that only two "low fin" tubes ($e/D_i > 0.04$) are in the database.

The correlation is an extension of the Cavallini and Zecchin [8] correlation for plain
tubes, which is

$$Nu = 0.05 \, Re_{eq}^{0.8} \, Pr_l^{1/3} \tag{12}$$

where Re_{eq} is calculated as specified by Akers et al. (16). The Cavallini et al. [8] microfin correlation adds an empirical "geometry factor" (total/nominal surface area at fin
tips), and the (Bo Fr) group to account for surface tension effects (Bond number) and
stratification (Fr). Their correlation for singe and cross-grooved tubes is:

$$Nu = 0.05 \, Re_{eq}^{0.8} \, Pr_l^{1/3} (Bo \cdot Fr)^t Rx^s \tag{13}$$

Table 2. Summary of the tube geometries of the experimental data in Cavallini et al. [20] model (+ Low-fin tube, * Cross grooved geometry)

Source	Ref.	D_t (mm)	N_t	e (mm)	β (deg)	α (deg)
Tatsumi et al. [27]	22	8.32 - 8.52	65	0.15 - 0.25	45	7 - 25
Shinohara & Tobe [28]	22	8.52 - 8.68	60 - 65	0.12 - 0.20	25 - 80	7 - 25
Yasuda et al. [29	22	6.94 - 8.52	50 - 60	0.20 - 0.25	40	18 - 30
Hori & Shinohara [30	22	5.95 - 6.21	30 - 60	0.15 - 0.20	33 - 60	8 - 18
Schlager et al. [31]	22	7.75+, 8.32	21, 60	0.38, 0.20	0, 50	30, 18
Schlager et al. [32	22	8.52 - 8.62	60	0.15 - 0.20	50	15 - 25
Eckels & Pate [33]	134a, 12	8.32	60	0.20	50	17
Kaushik & Azer [34]	113	13.3+	24	0.635	0	30
Torikoshi et l. [35, 36]	32/125/ 134a, 32/134a	6.14	50	0.18	40	18
Eckels et al. [37]	134a	8.52	60	0.20	50	17
Arosio et al. [38]	22	8.52	60 - 65	0.15 - 0.20	53 - 90	18 - 25
EPRI [39]	22, 502	8.52	60	0.20	53	18
Chamra et al. [4]	22	14.18	74 - 80	0.35	30	15 - 27

544

where the "diameter" in Nu and Re is the diameter between fin tips (D_t), which is somewhat unconventional. Others typically use the diameter to the base of the fins. The Re_{eq} is the Akers [16] equivalent Re given by $D_h G_{eq}/\mu$, where

$$G_{eq} = G\left[(1-x) + x(\rho_l / \rho_v)^{1/2}\right] \tag{14}$$

The Rx parameter accounts for the micro-fin geometry and is defined as $Rx = A_{tot}/A_{Dt}$, where A_{tot} is the total internal surface area, $A_{Dt}/L = \pi D_t$. The exponent (s) on Rx depends on the basic micro-groove geometry, and are given in Table 3.

Table 3 Exponents of Equation 13

Exponent	Low-fins (e/D > 0.04)	Micro-fins (e/D < 0.04)	Cross-grooved
s	1.40	2.00	2.10
t	-0.08	-0.26	-0.26

Note that the exponents on the micro-fin and cross-grooved tubes are nearly equal. The Bond number is the ratio of gravity and surface tension forces, and is defined as $g\rho_l \pi e D_t/8\sigma N_f$. Equation 13 suggests that the Nu will increase with increasing surface tension. Because the range of surface tension values for the test fluids is very small, one should be careful in drawing this conclusion. Equation 13 correlated the Table 2 data generally within ±20%.

5.4 CAVALLINI ET AL. PRESSURE DROP MODEL

This 1998 model by Cavallini et al.[22] is a modified plain tube correlation based on the plain tube correlations of Friedel [12] and Sardesi et al. [23]. In his model, the two-phase multiplier ϕ_l^2 is calculated using Eq. 6, and the single-phase friction factors (f_{vo} and f_{lo}) are determined using an equivalent friction factor based on the Moody friction factor vs. Reynolds number chart for flow in commercially rough pipes (c.g. p. 425 of Incropera and DeWitt [24]). They developed an empirical correlation to define the "equivalent Moody chart roughness" (e_m/D_t) of the micro-fin tubes in their database. The e_m/D_t was obtained from regression analysis and is given by

$$\frac{e_m}{D_t} = A \frac{e/D_t}{0.1 + \cos\alpha} \tag{15}$$

Where A is an empirical constant, e is micro-fin height, D_t is the fin tip diameter, and

α is the helix angle. Two different values were found for the constant A: 0.18 for condensation and 0.30 for evaporation. The following equation developed by Colebrook [25] may be used to calculate the Moody friction factor

$$\frac{1}{2f^{1/2}} = 1.74 - 2\ln\left(\frac{2e_m}{D} + \frac{18.7}{2\,\text{Re}\,f^{1/2}}\right) \tag{16}$$

Note that the Equation 16 does not account for apex angle, or the fin pitch. Cavallini et al. [23] show that Equation 16 predicts all of the data within ±20%.

6. Vaporization Inside Tubes

It is generally assumed that convective vaporization involves both nucleate boiling and thin film evaporation modes. As described by Webb and Gupte [26] the combined modes may be written in terms of an "asymptotic" model given by

$$h = (h_{nb}^{2} + h_{sh}^{2})^{1/2} \tag{17}$$

where h_{nb} and h_{sh} are the nucleate boiling and convection (shear) terms, respectively. It is expected that the nucleate boiling contribution will be small for micro-fin tubes. This is because the surface geometry of the micro-fin tube should not provide high nucleate boiling performance.

If $h_{sh} \gg h_{nb}$, the shear term for vaporization and convective condensation in a tube would be governed by the same equation. The heat transfer mechanism for vapor shear controlled condensation and evaporation of a thin liquid film is the same, so the h_{sh} in Equation 17 can be predicted using the same rules as for the h_{sh} term in Equation 10. The equation for h_{sh} would be applicable up to the dryout point in the evaporator tube.

To examine the validity of the hypothesis that the mechanism of condensation and evaporation coefficients in shear controlled flow are the same, Chamra and Webb [27] took R-22 data for condensation and evaporation in a micro-fin tube at the same saturation temperature. The condensation and evaporation data were taken in the same tube at 24.4 °C saturation temperature, for a range of mass velocities (150-327 kg/s-m^2), and vapor qualities (0.1-0.9). Figure 11 shows the test results for condensation and evaporation at $T_{sat} = 24$ C and $q = 17.3$ kW/m^2 heat flux for two mass velocities. In both figures, the condensation and evaporation coefficients are very nearly equal at vapor qualities greater than 50%. However, at $x < 0.50$, the evaporation coefficient exceeds the condensation coefficient by 10-15%. This greater value of the evaporation coefficient is apparently because of the nucleate boiling contribution. The figures support that the mechanism of condensation and evaporation are the same. That is, evaporation/condensation across a thin annular liquid film.

The thin liquid film is acted upon by both vapor shear and surface tension forces. As shown by Yang and Webb [15] for condensation, the surface tension contribution is significant only at low mass velocities. At the mass velocities shown on Figure 11, it is expected that the heat transfer mechanism will be vapor shear dominated. However, Figure 11 does not provide proof that the effect of surface-tension forces is similar for both condensation and evaporation.

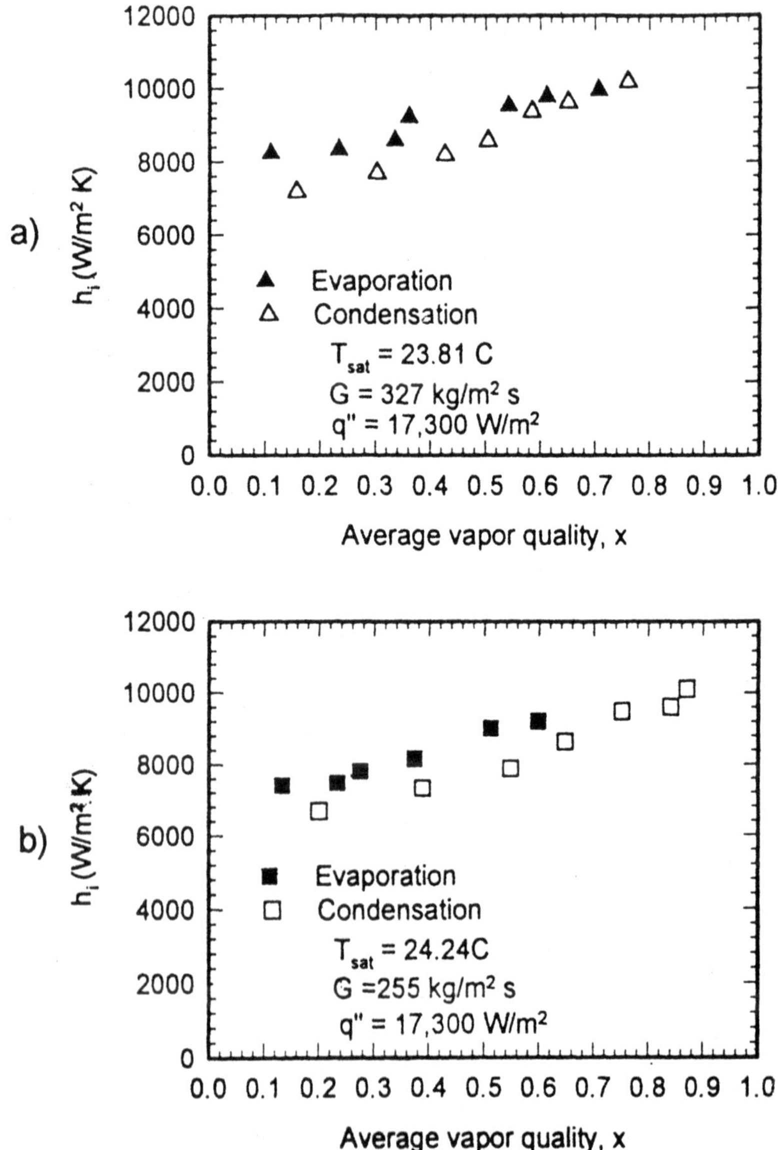

Fig. 11 Heat transfer coefficients for evaporation and condensation at Tsat =24°C in the Figure 2b micro-fin tube. (a) G=327 kg/m²-s, (b) G=255 kg/m²-s

7. Conclusions

This manuscript discusses rationally based methods to predict condensation and evaporation in micro-fin tubes. The details of the predictive model are described, and illustrations of its successful use are provided for round micro-fin tubes, and for extruded aluminum micro-channel tubes. The shear-controlled component is predicted using a two-phase heat-momentum transfer analogy model, which is called the "equivalent Reynolds number" model. This model has been validated by Moser et al. [11] for plain, round tubes. In this paper, we show that the model is also applicable to small hydraulic diameter extruded aluminum tubes having either circular or non-circular passages.

The authors show how one may account for the effect of surface-tension-drainage force on the micro-fins. Further argument is provided to show that the shear controlled model should also apply to predict the shear component in convective vaporization. In the event that a nucleate boiling contribution is significant, the authors show how this contribution may be added. However, data for a cross-grooved suggest that the nucleate boiling contribution is limited to vapor qualities less than 50%, and that its additive effect is less than 15%.

Nomenclature

A	Surface area, A_n (nominal surface area), A_{tot} (total surface area), $[m^2]$
Bo	Bond number
D	Diameter, D_o (outside diameter), D_h (hydraulic diameter), D_i (inside diameter), D_t (fin tip diameter)[m], D_e (inside diameter if the metal in the fins were returned to the tube wall), [m]
e	Fin height, e_m (equivalent roughness height defined by the Moody chart, [m]
(dp/dz)	Pressure gradient, $[Pa\ m^{-1}]$
F	Correction factor defined by Eq. 3, dimensionless
f	Fanning friction factor, dimensionless
Fr	Froude number $(G^2/gD_i\rho_{tp})$, dimensionless
G	Mass velocity in tube, $[kg/m^2\text{-s}[$
g	Acceleration due to gravity, $[m/s^2[$
h	Heat transfer coefficient, $[W\ m^{-2}\ K^{-1}]$
k	Thermal conductivity, [W/m-K[
L	Flow length, m
Nu	Nusselt number, hD/k_l, dimensionless
Pr	Prandtl number, dimensionless
p	Groove pitch normal to fins, [m]
p	Pressure, p_{cr} (critical pressure), [Pa]
r	Local radius of the fin surface, [m]
Re	Reynolds number, defined where used, dimensionless
Re_{eq}	Equivalent all liquid Reynolds number, dimensionless

548

r_t Fin tip radius, [m]

Rx $(A_{tot}/A_{Dt})/\cos \alpha$ as used in Eq. 13, dimensionless

s   Coordinate distance along curved condensing profile, [m]

T   temperature [°C]

t_b Fin base thickness, [m]

We   Weber number $(G^2 D_i/\sigma\rho_{tp})$, dimensionless

x   Vapor quality, x_{fill} (vapor quality at which the interfin region is liquid filled), dimensionless

GREEK SYMBOLS

φ_l^2 two-phase multiplier, φ_{lo}^2 (two-phase multiplier based on all-liquid flow), dimensionless

α helix angle, [degrees]

β fin apex angle, [degrees]

δ condensate film thickness, [m]

ε void fraction, dimensionless

μ dynamic viscosity, [kg m^{-1} s^{-1}]

ρ density, ρ_{tp} (homogeneous two-phase density), [kg/m^{-3}]

σ surface tension, [N/m]

π Pi, 3.1416

SUBSCRIPTS

b   bulk fluid

cond   condensation

exp   experimental data

f   friction

i   inside

l   liquid

lo   all liquid flow

n   nominal

nb   nucleate boiling

sat   saturation condition

sh   shear controlled term

sp   single phase

st   surface tension controlled term

tp   two-phase

v   vapor

vo   all vapor flow

w   wall

References

1.   Webb, R. L. (1994) "Advances in Modeling Enhanced Heat Transfer Surfaces," *Heat Transfer 1994, Proc. Tenth Int'l. Heat Transfer Conf.*, Vol. 1, pp. 445-460.

2.   Petukhov, B. S. (1970) "Heat Transfer and Friction in Turbulent Pipe Flow with Variable Physical

Properties," *Advances in Heat Transfer*, Vol. 6, Academic Press, New York.

3. Chamra, L. M., Webb, R. L. and Randlett, M. R, (1996a) "Advanced Micro-Fin Tubes for Evaporation," *Int. Jour. Heat and Mass Transfer*, Vol. 39, No. 9, pp. 1827-1838.

4. Chamra, L. M., Webb, R. L. and Randlett, M. R (1996b). "Advanced Micro-Fin Tubes for Condensation," *Int. Jour. Heat and Mass Transfer*, Vol. 39, No. 9, 1839-1846.

5. Brognaux, L., Webb, R. L. and Chamra, L. M. (1997) "Single-phase Heat Transfer In Micro-fin Tubes," accepted for publication in *Int. Journal of Heat and Mass Transfer.*

6. Yang, C. Y. and Webb, R. L. (1996a) "Condensation of R-12 in Small Hydraulic Diameter Extruded Aluminum Tubes with and without Micro-fins," *Int. Jour. Heat and Mass Transfer*, Vol. 39, No 4, pp. 791-800.

7. Shah, M. M. (1979) "A General Correlation for Heat Transfer During Film Condensation Inside Pipes," *Int. Jour. Heat and Mass Transfer*, Vol , pp. 547-556.

8. Cavallini, A. and Zecchin, R. (1974) "A Dimensionless Correlation For Heat Transfer in Forced Convection Condensation," *Proc. of 15th Unione Italiana di Termofluodinamica National Heat Transfer Conf.*, Torino, Italy, 19-20n June 1994, vol. I, pp. 521-531.

9. Moser, K., Webb, R. L., and Na, B. (1998) "A New Equivalent Reynolds Number Model for Condensation in Smooth Tubes," *Journal of Heat Transfer*, Vol. 120.

10. Zhang, M. H. (1997) "A New Equivalent Reynolds Number Model for Vapor Shear-Controlled Condensation Inside Smooth and Micro-Fin Tubes," Ph.D. Thesis, Penn State University, University Park, PA.

11. Moser, K., Webb, R. L., and Na, B., (1998) "A New Equivalent Reynolds Number Model for Condensation in Smooth Tubes," *Journal of Heat Transfer*, Vol. 120.

12. Friedel, L. (1979) "Improved Friction Pressure Drop Correlations for Horizontal and Vertical Two Phase Pipe Flow," Paper E2, European Two Phase Flow Group Meeting, Ispra, Italy.

13. Zhang, M. and Webb, R. L. (1998) "Condensation Heat Transfer in Small Diameter Tubes," to be published in *Proc. 1998 International Heat Transfer Conference*, Korea.

14. Carey, V. P. (1992) *Liquid-Vapor Phase-Change Phenomena*. Hemisphere Publication Corp.

15. Yang, C. Y. and Webb, R. L. (1997) "A Predictive Model for Condensation in Small Hydraulic Diameter Tubes having Axial Micro-Fins,"*Journal of Heat Transfer*, Vol. 119, pp. 776-782.

16. Akers, W. W., Deans, H. A., and Crosser, O. K. (1959) "Condensing Heat Transfer within Horizontal Tubes," *Chemical Engineering Progress Symposium Series*, Vol. 55, No. 29, pp. 171-176.

17. Adamek, T. A. and Webb, R. L., 1990. "Prediction of Film Condensation on Vertical Finned Plates and Tubes - A Model for the Drainage Channel," *Int'l. Journal of Heat and Mass Transfer*, Vol. 33, No. 8, pp. 1737-1749.

18. Yang, C. Y. and Webb, R. L. (1996b) "Friction Pressure Drop of R-12 in Small Hydraulic Diameter Extruded Aluminum Tubes with and without Micro-fins," *Int. Jour. Heat and Mass Transfer*, Vol. 39, No 4, pp. 801-809.

19. Zivi, S. M. (1964) "Estimation of Steady State Steam Void-Fraction by Means of Principle of Minimum Entropy Production," *Jour. Heat Transfe*, Vol. 86, pp. 247-252.

20. Kedzierski, M. A., and Goncalves, J. M. (1997) "Horizontal Convective Condensation of Alternative Refrigerants Within a Micro-fin Tube," NISTIR 6095, U.S. Dept. of Commerce, Washington, D.C.

21. Cavallini, A., Col, D.Del, Doretti, L., Longo, G. A. and Rossetto, L. (1998) "A New Computational Procedure for Heat Transfer and Pressure Drop During Refrigerant Condensation Inside Enhanced Tubes," to be published in *J. Enhanced Heat Transfer.*

22. Sardesi, R. G., Owen, R. G. and Pulling, D. J. (1982) "Pressure Drop for Condensation of a Pure Vapor in Downflow in a Vertical Tube," *Proc. 8th Int. Heat Transfer Conf.*, Munich, pp. 139-145.

23. Incropera, F. P., and DeWitt, D. P. (1996) *Fundamentals of Heat and Mass Transfer*, 4th ed., John Wiley, New York.

24. Colebrook, F. (1939) "Turbulent Flow in Pipes with Particular Reference to the Transition Region between Smooth and Rough Pipe Laws," J. Inst. Civ. Eng., Vol. 4, 14-25.

25. Webb, R. L., and Gupte, N. S. (1992) "A Critical Review of Correlations for Convective Vaporization in Tubes and Tube Banks," *Heat Transfer Engineering*, Vol. 13, No. 3, pp. 58-81.

26. Chamra, L. M., Webb, R. L., and Randlett, M. R. (1996) "Advanced Micro-Fin Tubes for Condensation," *Int. Jour. Heat and Mass Transfer*, Vol. 39, No. 9, 1839-1846.

27. Tatsumi, A., Oizumi, K., Hayashi, M. and Ito, M. (1982) "Application of Inner Groove Tubes to Air Conditioners." *Hitachi Rev.*, Vol. 32, No. 1, pp. 55-60.

28. Shinohara, Y. and Tobe, M. (1985) "Development of an Improved Thermofin Tube." *Hitachi Cable Review*, No. 4, pp. 47-50.

29. Yasuda, K., Ohizumi, K., Hori, M., and Kawamata, O. (1990) "Development of Condensing Thermofin-HEX-C Tube," *Hitachi Cable Review*, No. 9, pp. 27-30.

30. Hori, H. Shinohara, (1991) "Internal Heat Transfer Characteristics of Small-Diameter Thermofin Tubes. Hitachi Cable Rew," Vol. 10, pp. 85-90

31. Schlager, L.M., M. B. Pate, Bergles, A.E. (1989) "Effect of oil on Heat Transfer and Pressure Drop Inside Augmented Tubes during Condensation and Evaporation of Refrigerants (RP-469)." Final Report, College of Eng. Iowa State Univer.

32. Schlager, L. M., Pate, M. B. and Bergles, A. E. (1989) "Heat Transfer and Pressure Drop During Evaporation and Condensation of R22 in Horizontal Micro-Fin Tubes," *Int. J. Refrig., Vol. 12, pp. 6-14.*

33. Eckels, S. J., and Pate, M. B. (1991) "An Experimental Comparison of Evaporation and Condensation Heat Transfer Coefficients for HFC-134 and CFC-12." *Int. Journal of Refr.*, Vol.14, pp. 70-77.

34. Kaushik, N., and Azer, N. Z. (1988) "A General Heat Transfer Correlation for Condensation Inside Internally Finned Tubes," *ASHRAE Transactions*, Vol. 94, Part 2, pp. 261-279.

35. Torikoshi, K., Ebisu, T. (1993) "Heat Transfer and Pressure Drop Charachteristics of R134a, R32, and a Mixture of R32/R134a inside a Horizontal Tube," *ASHRAE Trans.*, Vol. 99, pp. 90-96.

36. Torikoshi, K., Ebisu, T. (1994) "In-Tube Heat Transfer Charachteristics of Refrigerant Mixtures of R32/R134a and FHC32/125/124a. Int. Refr. Conf. Purdue, pp. 293-298.

37. Eckels, S. J., Doerr, T.M., Pate, M.B. (1994) "In-tube Heat Transfer and Pressure Drop of R-134a and Ester Lubricant Mixtures in a Smooth Tube and a Micro-Fin Tube: Part II - Condensation," *ASHRAE Trans.*, Vol. 100, No, 2, pp. 2833-293.

38. Arosio, S., Muzzio, A., Niro, A. (1996) "Evaporation and Condensation Heat Transfer and Pressure Drop of R22 Inside Micro-Fin Tubes." 14th Nat. Heat Transfer Conf., Rome, pp. 227-233.

39. EPRI, 1996. "Heat Transfer Characteristics of Alternative Refrigerants. Project 3412-6424," Final Report, Vol. 1.

PERFORMANCE ENHANCEMENT OF HEAT EXCHANGERS FOR SEMICONDUCTOR - CHIP MANUFACTURING

WEN-JEI YANG
> *Department of Mechanical Engineering and Applied Mechanics,*
> *University of Michigan*
> *Ann Arbor, Michigan, 48109, U.S.A.*

SHUICHI TORII
> *Department of Mechanical Engineering,*
> *Kagoshima University*
> *Korimoto, Kagoshima, 890, JAPAN*

Abstract. The semiconductor and its uses in transistors are briefly introduced, followed by the introduction of heat exchangers (often referred to as process chambers/reactors for IC manufacturing) for manufacturing semiconductor chips These heat exchangers can be classified, according to their functions, into six major categories: (i) heat exchangers, for the growth of thin silicon dioxide layers on a silicon wafer (also called high-temperature film furnaces), (ii) heat exchangers for doping of the masked substrate with appropriate ions (referred to as ion-implantation chambers), (iii) heat exchangers for wet/dry etching (etch chambers), (iv) heat exchangers for chemical vapor deposition (CVD reactors), (v) heat exchangers for the metallization process, and (vi) heat exchangers for heating chemical fluids. All these types of heat exchangers have one thing in common: they are all "single, solid (or tube), single-fluid" in nature, performing heat transfer by conduction, convection, or radiation. This is in sharp contrast to conventional heat exchangers in the power and refrigeration industry that consist of "double-tube (including shell-and-tube type), two-fluid" types.   Therefore, heat exchangers for high-tech applications are radiative-heat-transport-controlled and are, in general easier to design and control, and environmentally clean.   This lecture emphasizes how heat transfer performance can be enhanced in heat exchangers for manufacturing semiconductor chips.

1.   Introduction

The semiconductor and its uses in electronic computing machines are briefly presented, followed by the introduction of heat exchangers used in manufacturing semiconductors and their chips. The word "transistor" is composed of "transfer" and "resistor" from its transferring an electrical signal across a resistor. It is an electronic device that is similar to the electron tube in use for samplification and rectification, and consists of a small block of a semiconductor (such as germanium) with at least three electrodes. Today,

551

S. Kakaç et al. (eds.), Heat Transfer Enhancement of Heat Exchangers, 551–577.
© 1999 *Kluwer Academic Publishers.*

transistors are packed by the millions onto microprocessors to run car engines, cell phones, missiles, satellites, gas pumps, ATM machines, microwave ovens, computers, CD players, and many electronic toys and tools. Indeed, the transistor has changed the way we drive, bank, cook, communicate, listen to music and watch television.

Semiconductors are materials whose properties lie midway between those of an electrical insulator and a conductor. In 1947, W. Brattian, and J. Bardeen at Bell Labs invented the "point-contact" transistor which boosted power 450 percent. The key realization was that "holes" (quantum-mechanical entities that are the absence of electrons) carried current in silicon. In 1948, their colleagues at Bell Labs invented a "junction" transistor, that imitated a sandwich. The bread would be semiconductor material with an excess of electrons; it was dubbed "n-type". The meat would be "p-type", with an excess of positively charged holes. When wires were attached and voltage was applied, holes streamed across the n-material into the p-area. Thus, current was amplified like it was by the "point-contact" transistor.

J. Kilby and R. Noyce, working concurrently, developed a microchip in 1958. On the drawing board, it was called the monolithic integrated circuit (IC): an entire miniaturized electrical circuit on a single, fingernail-sized wafer of crystalline silicon. It has been getting smaller, faster, and more complex ever since. "How a chip is made" is briefly described in the followings:

Step 1:   Pure silicon, a semiconductor, is sliced into wafers and coated with a layer of electrical insulator.

Step 2:   Circuit patterns are stenciled onto the wafer using a "mask", or outline, and light-sensitive plastic.

Step 3:   Solvents strip away any plastic not exposed to light, revealing a pattern in the silicon dioxide.

Step 4:   Layers of pattern are created by adding, masking and stripping new layers of silicon.

Step 5:   Exposed silicon is made more conductive by adding "impurities". Metal strips connect the layers.

Step 6:   Individual chips are split off from the finished wafer; connecting wires are bonded onto each.

These microchips were used to construct a computing machine with the scheme of: memory, stored programs and a central processor for number crunching. Finally, it led to the development of the personal computer (PC), a development which was made possible by the invention of the microprocessor by T. Hoff in 1971. A microprocessor is the computer's central processing unit (CPU). It is a computer on a chip.

It is imperative to recognize an important distinction between two categories of industrial heat exchangers: For applications in power, refrigeration, manufacturing, and the petroleum and chemical industries, the performance enhancement of heat exchangers is based on convective mechanisms. For applications in the high-tech industries, however, it is based on the radiative-mechanism, since heat transfer in the heat exchangers for chemical liquid heating as well as IC manufacturing is radiative-component controlled. In the former case, both the active and passive enhancement methods are related solely to convective transport, and thus the fouling of heat exchange surfaces is of great concern. In contrast, the major concern with heat exchanges for semiconductor manufacturing is radiative heat transport between the heating and heated surfaces, which is subjected to interruption (such as absorption and scattering) by the medium between them.

Based on their structure, heat exchangers can be classified into the single-tube, single-fluid type and the double-tube, two-fluid type, as illustrated in Figs. 1-a and -b, respectively. In the former case, heat from a heat source may be transferred to a fluid by means of conduction, convection or radiation. In the latter case, two fluids of different temperatures are separated by a solid wall; conventionally one fluid flows in a tube or a multiple of tubes called a tube bank while the other fluid flows in a space enclosed by another tube of a different size. Thus, they form either a double (or concentric) pipe heat exchanger or a shell-and-tube-type heat exchanger.

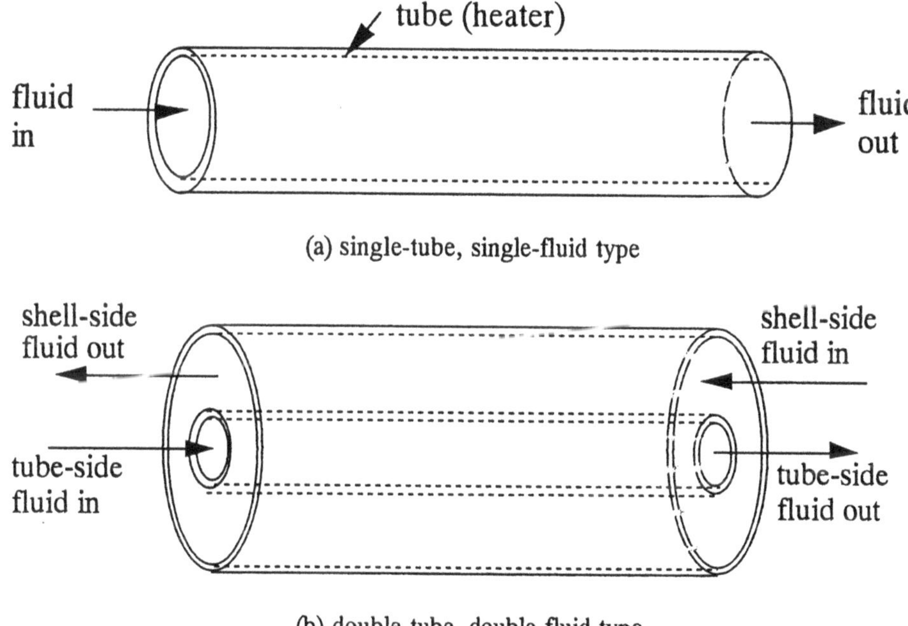

Figure 1. Classification of heat exchangers (a) single-tube, single-fluid type and (b) double-tube, double-fluid type.

Unlike the process and power industries, which utilize the double-tube, two-fluid type heat exchangers, the semi-conductor manufacturing industry employs single-tube, single-fluid type heat transfer equipment. That equipment: (i) heat exchangers for the growth of a thin silicon dioxide layer on a silicon wafer (the so called high temperature film furnaces); (ii) heat exchangers (referred to as ion-implantation chambers) for doping of the masked substrate with appropriate ions. This process is called ion-implantation, where ionized atoms are accelerated towards the wafer's surface at an energy level high enough to penetrate beyond the surface region and become implanted. (iii) Heat exchangers (called etch chambers) used to selectively remove layers of unmasked dielectric or metal films from the substrate surface by either wet or dry etching; (iv) heat exchangers for chemical vapor deposition (CVD), also referred to as CVD reactors. Selective epitaxy growth and/or second and subsequent layer (s) of oxide growth can be achieved through CDV. (v) Heat exchangers for the metallization process, which deposits a thin layer of metal as a liner and/or a barrier by CVD or by physical vapor deposition (PVD, also known as sputtering), and (vi) heat exchangers for heating chemical liquids.

The design, dynamic analysis and control of the single-tube, single-fluid type are easier than for the double-tube, two-fluid type.

2.   Single-Tube/Single-Fluid Heat Exchangers

The simplest heating device consists of a single fluid being heated (or cooled) in a single tube. Heat from one heat source can be transferred to the fluid either at rest or flowing inside the tube by means of (i) conduction, (ii) convection or (iii) radiation or some combination of them. The construction of each heating method may be classified into types depending on relative positions of the heater and the fluid: (a) internal and (b) external heating types. Figure 2 illustrates both internal and external heating types for the three distinct heating methods.

Chemical liquids such as wafer-rinsing liquid, etching liquid, resist acid stripper, film remover, plating solutions and cleaning chemicals, must be heated to a range of about 50°C to 150°C depending on their function. In contrast to conventional liquid heaters, chemical liquid heaters for semiconductor manufacturing must be in a highly purified state. Because of this requirement, the compatible surface materials are severely restricted, as listed in Table 1. It shows that quartz and teflon are the only choices.

Quartz glass is employed in Fig. 2 (shown as blank space), while the liquid space is dotted and the darkened area represents the heat source. The conductive-heating heat exchanger is disadvantaged by the low thermal conductivity of the quartz. Nevertheless, it can achieve a moderate liquid temperature level. Figure 2-b shows a schematic of an external-heating type convective-heating method. The liquid flows through a quartz glass, which is in direct contact with a nichrome heater that is surrounded by an insulating material. Such a device has the advantage of a better heat transfer performance resulting from a higher heat conductance. In Fig. 2-a, the internal heating type radiative heater has a light source (i.e., a lamp) placed inside an inner quartz tube around which the liquid flows. Radiative heat from the light source can penetrate the quartz glass to directly heat the liquid. It results in the best performance. Important features in radiative heating include the need for a very small heat transfer area, but at the expense of a very high source temperature.

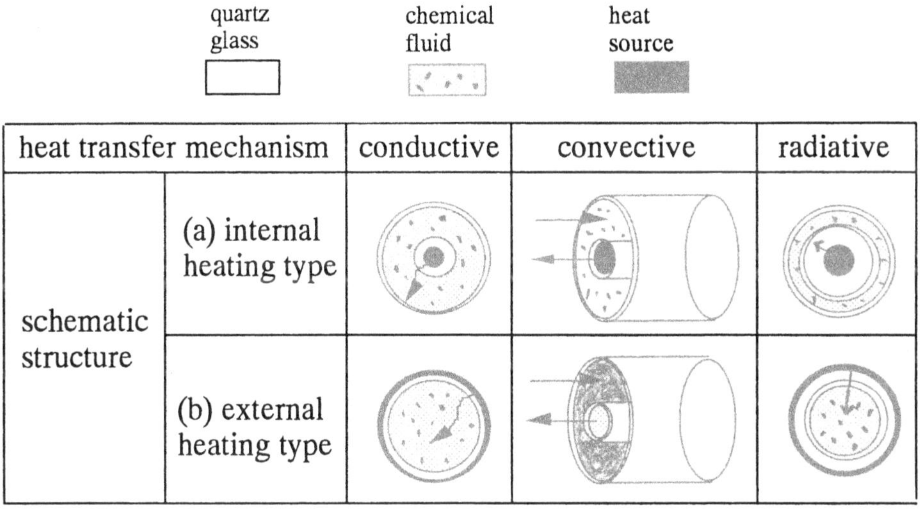

heat transfer mechanism	conductive	convective	radiative
schematic structure	(a) internal heating type		
	(b) external heating type		

Figure 2. Heat transfer mechanisms for heating chemical liquids and a comparison of schematic structures.

Table 1. Compatibility between chemical liquids and surface materials

Chemical Liquid	Temperature	Quartz for Semiconductors	Teflon	Plexiglass
Ammonia peroxide	80°C	good	good	poor
Sulfuric acid peroxide	120-150°C	good	good	poor
Hydrochloric acid peroxide	80°C	good	good	poor
Hot sulfuric acid	150-160°C	fair*	fair*	poor
Sulfuric acid	120°C	good	good	poor

*good under certain conditions

2.1. STATUS OF CHEMICAL LIQUID HEATING

One special feature in chemical liquid heaters for semiconductor manufacturing that is distinct from conventional liquid heaters is the very strict requirement to reduce pollution to an extremely low level. Hence, materials that can meet this stringent restriction are very limited (again, see Table 1). The construction of these heat exchangers is thus substantially limited unlike the conventional ones, which can be freely installed.

So far, two kinds of direct heating apparatus have been in use with chemical liquids as illustrated in Fig. 3: One has a nichrome-wire heater covered with quartz-glass or Teflon, and is installed in a treatment bath filled with a chemical liquid. This type, shown in Fig. 3-a, is referred to as the "injection" type. The other has the treatment bath made of quartz-glass or Teflon with a nichrome-wire heater placed along its outside surface. This is called "acu-bath" type shown in Fig. 3-b. While the "injection" type suffers from temperature non-uniformity and higher liquid pollution, the "aqu-bath" type is disadvantageous in its temperature non-uniformity and poor dynamic response. A recent trend is to apply an in-line arrangement, indirect heating method. Figure 4 shows an example of such a device for heating chemical liquids. It consists of a treatment bath, a pump, a heater, a filter, a controller, and a sensor. Its merits include a higher flexibility in heating power, higher efficiency, faster response, and cleaner products. Figures 5-a and -b show the heaters in use: an external-heating, convective type and an internal-heating, radiative type, respectively. Figure 6 compares the features of the radiative and convective type devices for heating chemical liquids, obtained from a one-dimensional analysis for a heat load of 10 kW. It was discovered that the radiative type Fig. 6-b, requires a heat transfer area of about 1/10 that for the convective type, Fig. 6-a, resulting in a compact size. Another difference is the average glass temperature which is 80°C in the radiative type in contrast to about 290°C for the convective type.

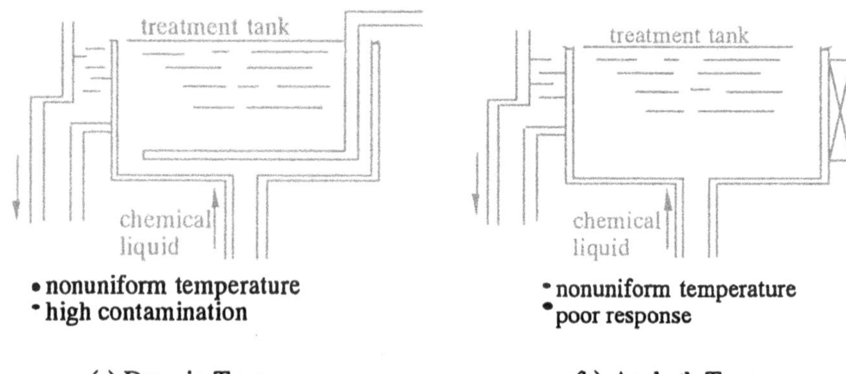

(a) Drop-in Type   (b) Aqubath Type

Figure 3. Method of direct heating of chemical liquids and their special features, (a) direct injection type and (b) aqubath type.

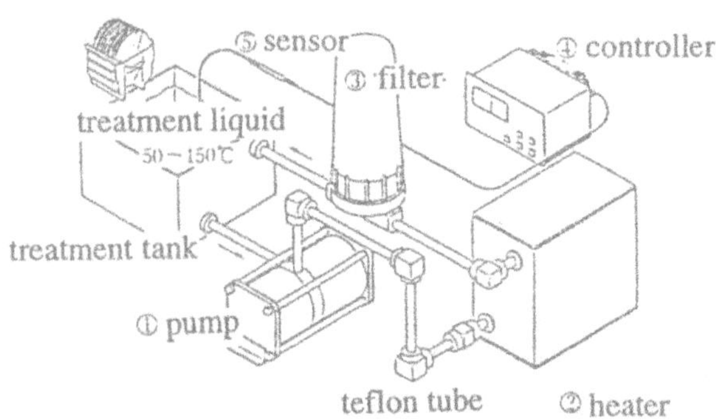

Figure 4. Indirect heating method of chemical liquid and its special features.

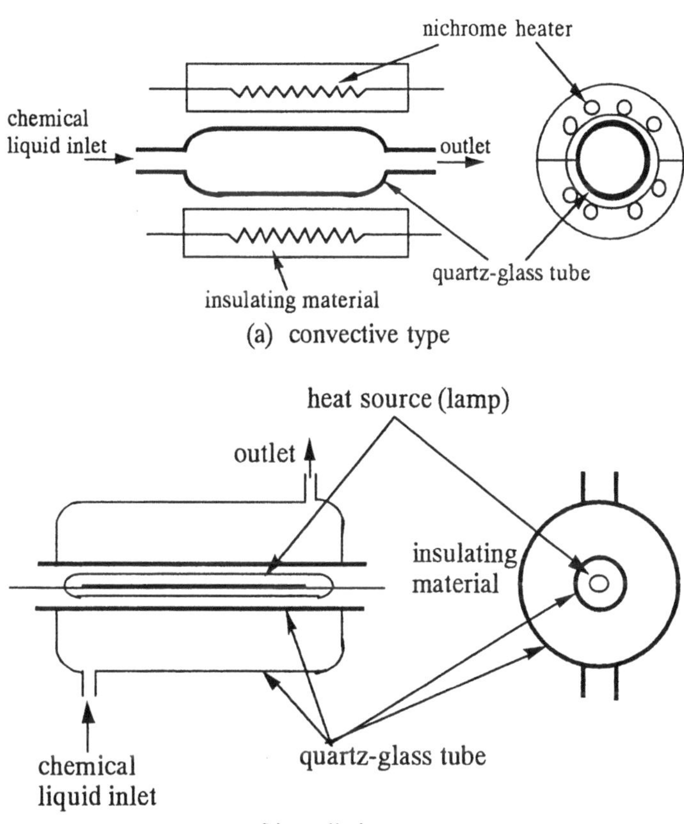

Figure 5.   Heat transfer mechanisms of in-line type chemical liquid heaters and a comparison of their schematic structures, (a) convetive type and (b) radiative type.

In view of the setting up and controlling the heating of the liquid, the internal heating arrangement is preferred over the external one, which is, however, better for selecting light source geometry and easier for the installation of insulating material.

There are a number of steps in the manufacturing of semiconductor integrated circuits (IC). The first is wafer production. Figure 7 illustrates a typical cross-section of the constituents of the IC's basic level. It also shows the various processes required to achieve each step in this configuration. Usually, a silicon wafer is first oxidized to create a thin dielectric silicon dioxide (SiO_2) layer. Subsequently, the layer is selectively etched to remove the oxide from the surface and create holes which are called "contacts". The contacts are ion-implanted with the appropriate doping element to create either p- or n-junction. They are filled with a metal to create interconnects. The oxide layer may also be masked to selectively deposit an epitaxy (single crystal Si) or a polysilicon layer on it. The wafer is typically held by a susceptor inside a vacuum chamber equipped with hardware to produce plasma flow reactive gases, and/or heated to desired temperatures. Most of these steps involve intermediate masking steps achieved through photolithography (ion-beam or optical) to facilitate the deposition or etch.

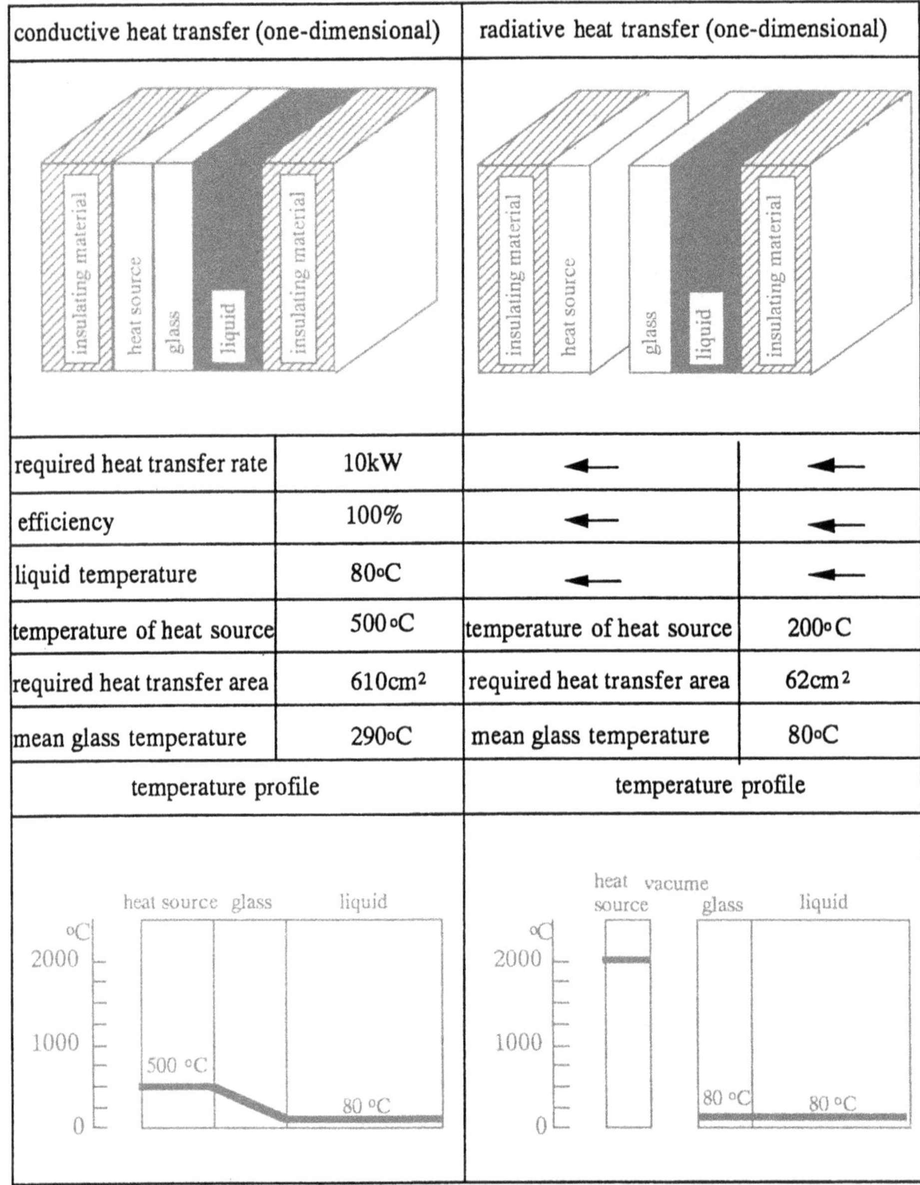

* quartz-glass thickness of 3 mm

Figure 6. A comparison of suitabilities of heat transfer mechanisms for chemical liquid heating under one-dimensional conditions.

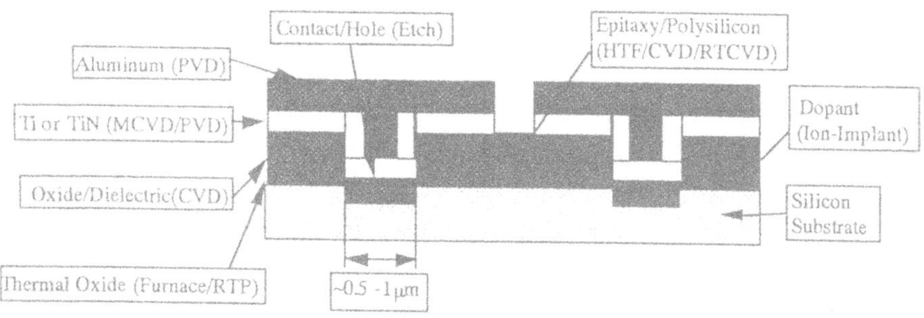

Figure 7. Schematic of the cross-section of an Integrated Circuit.

2.2. HEAT EXCHANGERS FOR IC MANUFACTURING

Each step in IC manufacturing is performed in a heat exchanger, commonly called a process chamber or reactor. (i) The growth of a thin SiO_2 layer on a silicon wafer is achieved in a device called a high-temperature film (HTF) furnace with an oxidizing environment. It is generally a very slow process. This type of furnace has been gradually replaced by the rapid thermal processing (RTP) furnace, wherein the thickness of these SiO_2 films can be precisely controlled. Figure 8-a is a schematic of RTP/HTF furnace. In the RTP furnace, very high temperature rises are rapidly achieved through controlled radiative heating, and the surface reaction between the flowing oxidant and the substrate takes place at rapid rates. This results in the occurrence of a well-controlled chemistry. (ii) An ion-implant chamber is where the doping of the masked substrate with appropriate ions takes place. Ion implantation is a process in which ionized atoms are accelerated towards to wafer surface to an energy level high enough to penetrate beyond the surface region and become implanted. (iii) In the etch chamber, Fig. 8-b, layers of unmasked dielectric or metal films are selectively removed from the substrate surface by wet or dry etching. The process, usually involves removing the dielectric molecules by Ar bombardment (called Ar plasma sputtering), or chemical removal with reacting gases in a plasma (reactive halogen plasma), or by a combination of the two (reactive ion plasma sputtering). (iv) Chemical Vapor Deposition (CVD) reactor involves chemical reaction between the substrate and the reacting gas or gases flowing over it in a high temperature (500 to 700°C) environment with or without plasma, (see Fig. 8-b). Dielectric layers or polysilicon films are deposited in a high temperature (500 to 600°C) CVD or higher temperature (about 1000°C) film (HTF) reactor. In this case, the wafer substrate reacts with the flowing gas to form a dielectric compound on the surface. Rapid thermal CVD (RTCVD) is an alternative means for this process. (v) Physical Vapor Deposition (PVD) chamber: The metallization process generally involves deposition of a thin layer of metal as liner (Ti) or a barrier (TiN). It can be done by either metal CVD, where the reactants are composed of the metallic compound to be deposited on the substrate, or by PVD also known as sputtering. Figure 8-c shows a PVD chamber where PVD is carried out under high vacuum with Ar plasma to remove metal atoms from a target and deposit them onto the substrates being heated to a temperature of 200 to 500°C. PVD is employed to deposit the aluminum metal layer to

560

completely fill the "contact". In the case of multilevel metallization (MLM) ICs, the process of dielectric layer deposition, photolithography, etching and metal deposition are repeated to obtain the second and/or higher levels of interconnect. The connections between these metals are called "vias".

The processes in IC chip manufacturing are carried out in heat exchangers called reactors or process chambers. Important physical phenomena involved in the equipment include chemically reacting flows, surface reactions, conjugate heat transfer, fluid mechanics, mass (species) transfer and plasma physics. For example, chemically reacting flows are important in CVD, ETCH, RTP and HFT, while radiation and conduction play major roles is RPT, HTF, high temperature PVD and CVD. It should be noted that semiconductor manufactory processes take place in a vacuum environment and involve radiative heat sources.

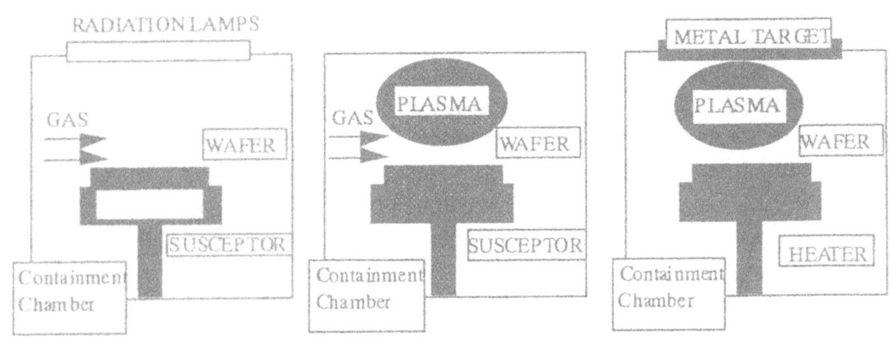

(a) RTP/HTF   (b) CVD/ETCH   (c) PVD

Figure 8.   Schematic of a few Process Chambers/Reactors for IC Manufacturing.

Figure 9 shows the process of semiconductor manufacture at Komatsu Electronics [1], Inc., Hiratsuka, Japan, and its related equipment. It begins with manufacturing the silicon wafer, followed by wafer cleaning using a chemical fluid. A radiative type heater (Fig. 5-b), with a capacity of 3 to 6kW, is employed for heating and temperature control of liquid chemicals. Meanwhile, a radiative-type in-line heater is used to produce hot di-water (ultra-pure-water) for the cleaning of silicon wafers and glass plates using in liquid crystals, replacing CFC (Freon) cleaning. The next process is photolitography consisting of three steps: coating, exposure and development. The coating process is performed in a cup of the spin coater under highly accurate temperature and humidity control in a precise air-conditioning unit, Fig. 10. An excimer laser is used in the exposure process, while a circulator is employed in the development process. Photolithography is followed by etching, doping, wire bounding, mould marking, inspection, and finally, packaging.

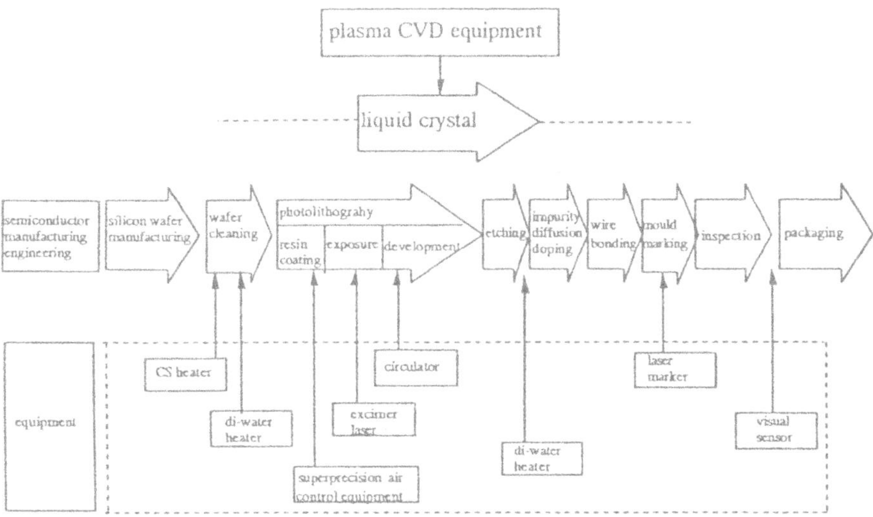

Figure 9. Semiconductor manufacturing and related equipment.

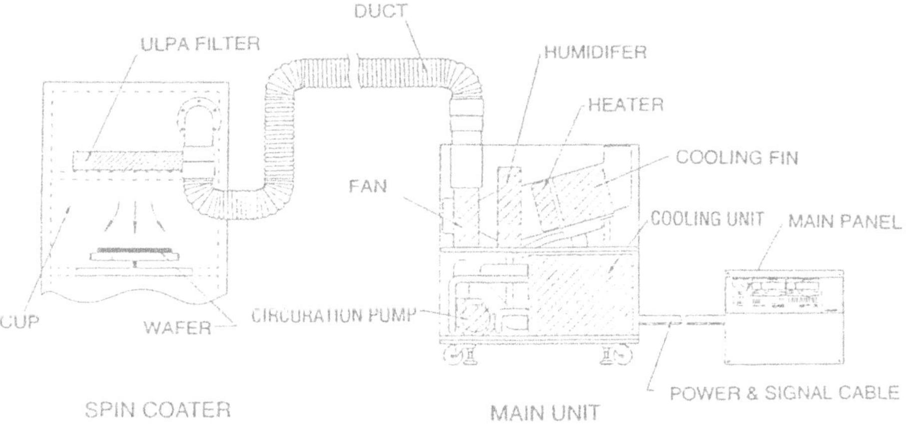

Figure 10. Air-conditioning unit.

3. Enhancement of Radiative Heat Exchangers

A radiative type heat exchanger with the heating capacity of 3 to 6 kW is most appropriate for heating liquid chemicals for semiconductor manufacturing. Based on this heating capacity, the mechanism and method of heat transfer performance enhancement are investigated here.

562

3.1.   HEAT TRANSFER MECHANISM

Figure 11 illustrates heat transfer mechanisms in a radiative-type, chemical liquid heater.
The principal heat transfer originates from radiative heat source, for example a lamp. It
is accompanied by both convective and conductive processes as shown by different
arrows. Conventionally, one would use a macroscopic, Stefan-Boltzmann equation as the
rate equation:

$$q = F_{1\text{-}2}\, A_1\, (E_{b1} - E_{b2}) \tag{1}$$

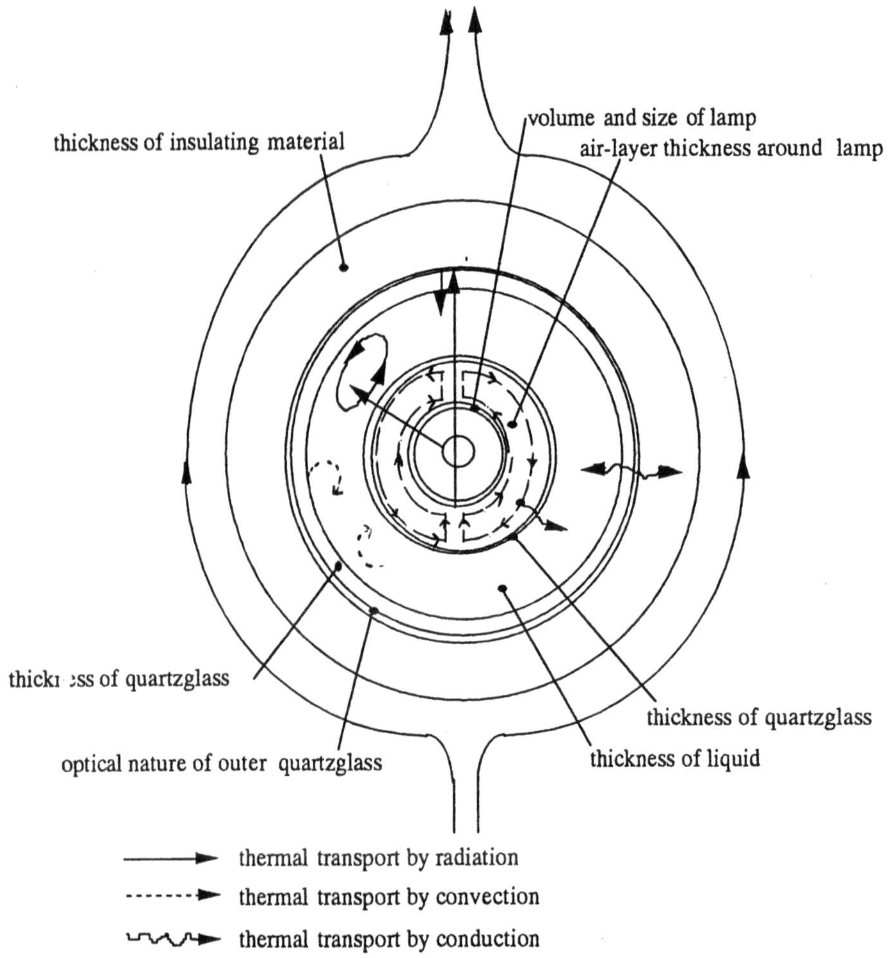

Figure 11.   Heat transfer mechanisms in a radiative-type, chemical liquid heater.

Here, q denotes the rate of radiative heat transfer, $F_{1\text{-}2}$, the shape factor viewed from
surfaces 1 to 2, A_1, area of surface 1, and E_b, black-body emissive power. $E = \sigma T^4$ where
σ is the Stefan-Boltzman constant equal to 5.67×10^{-8} [W/m^2 • K^4] and T is the surface

temperature in the absolute scale [K]. However, in order to enhance radiative heat transfer performance it is necessary to understand the microscopic aspect of the radiative heat transfer mechanism. The direct heating of chemical liquid by radiative heat transfer is induced by the collision of phonons having the same oscillating frequency as the molecular oscillations of chemical liquid with the molecules of chemical liquid, thus triggering resonant absorption that is converted into thermal energy. Hence, heat transfer performance is strongly dependent upon the radiative absorptivity of the chemical liquid α (λ) where λ denotes the wavelength. Since the main component of the chemical liquid is water, the absorptivity of water α_w (λ) is considered in the analysis. Figure 12 shows the τ_w -λ relationship as a function of the thickness of water layers, where τ_w denotes the transmissivity of water. It is observed that τ_w varies with δ. However, since δ=10 mm (or over) in practical applications, the water layer would completely absorb infrared radiation, with a wavelength exceeding about 1.3μm, while permitting most thermal radiation with shorter wavelengths to transmit through the layer. One may thus conclude that the ideal heat source is one that would emit the most thermal radiation having wavelengths of greater than 1.3μm and would not emit thermal radiation with the wavelengths of less than 1.3μm. Based on the Stefan-Boltzmann law prescribing an upper limit to the total emissive power of a black body, E_b, and the Wien displacement law relating to the wavelength for the maximum spectral emissive power, λ_{max}, it is practically impossible to find a heat source with such ideal characteristics.

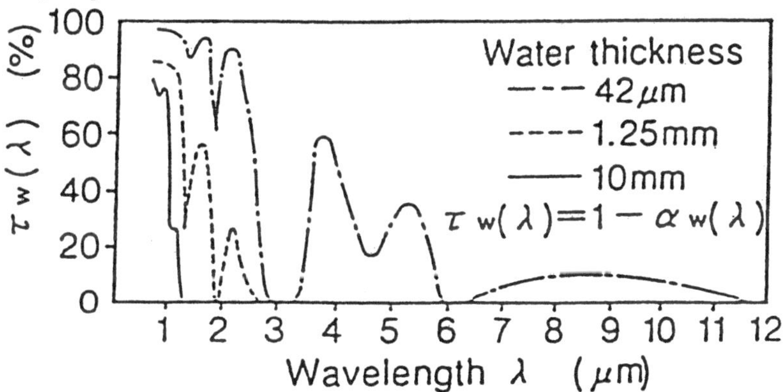

Figure 12. τ_w -λ relationship as a function of the thickness of water layers.

3.2. SELECTION OF HEAT SOURCE

Various heat sources were evaluated based on two criteria, wavelength distribution characteristics and total black body emissive power. The evaluation yielded the halogen lamp as the final choice. Figure 13 shows a schematic of a halogen-lamp of 3 kW heating capacity, together with the temperature ranges at different sections. One observes that the filament temperature ranges from 2,000 to 3,100°C. It is desirable to have this temperature range between the upper wavelength for maintaining high transmissivity through the inner quartz glass, 3μm, and the lower wavelength for maintaining high absorptivity of water (representing chemical liquids), 1.3μm. In other words, the filament temperature should be set in this range of wavelengths from 1.3μm to 3.0μm,

564

which corresponds to the near-infrared region. Figure 14 depicts the ratio of the monochromatic emissive power to the total emissive power of the blackbody. Here, the near-infrared region is indicated by the shaded area, from which the lowest temperature of 2,000°C is selected as the filament temperature. The spectral distribution of monochromatic emissive power density irradiated from this heat source, $E_o(\lambda)$, is presented in Fig. 15 for various filament temperatures together with that of transmitting of the quartz glass of 1mm thickness $\tau_g(\lambda)$. The total emissive power, E_o, can be expressed as

$$E_o = \int_o^\infty E_o(\lambda)d\lambda \qquad (2)$$

$E_o(\lambda)$-$E_o(\lambda)\tau_g(\lambda)$, represents the power density being lost by the quartz-glass.

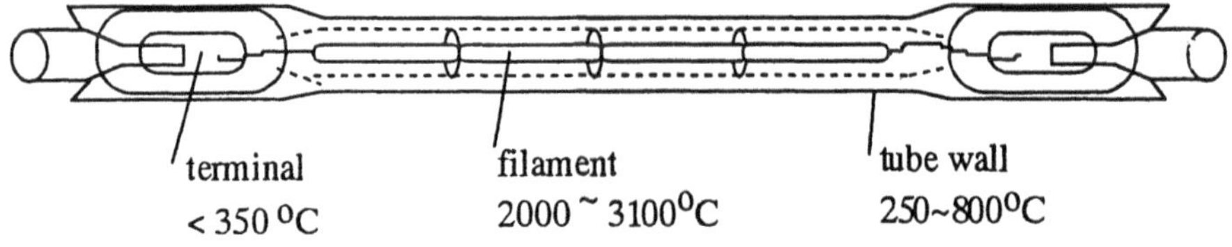

Figure 13. A schematic of a halogen-lamp of 3 kW heating capacity.

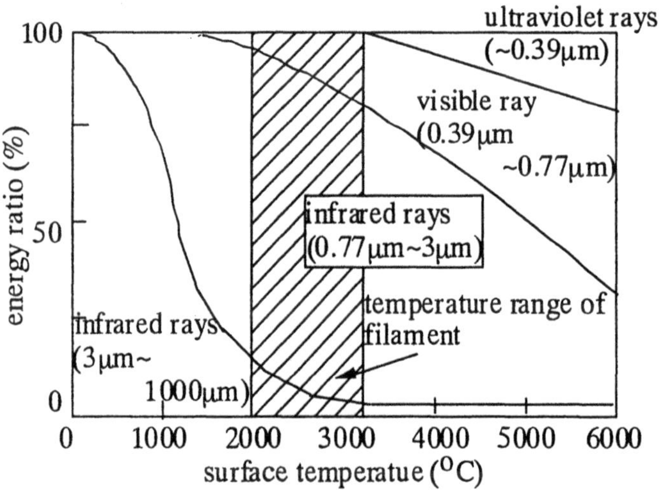

Figure 14. Ratio of the monochromatic emissive power to the total emissive power of the blackbody.

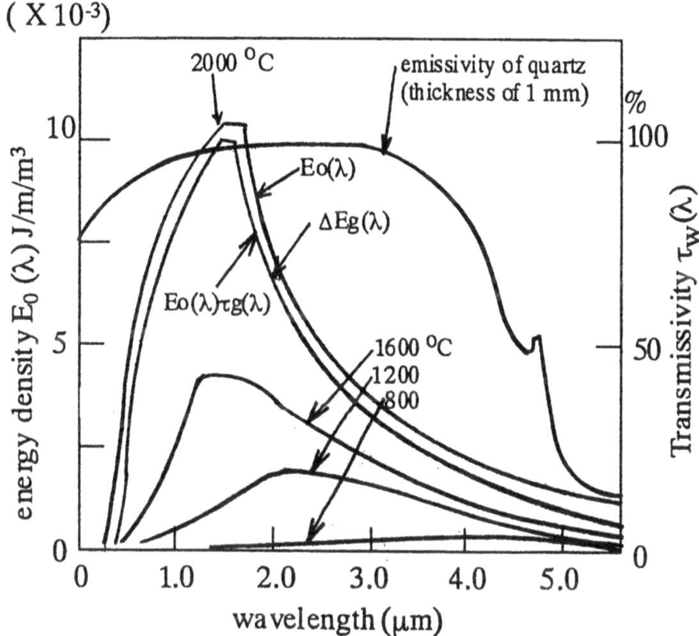

Figure 15. The spectral distribution of monochromatic emissive power density irradiated from heat source and transmissivity of quartz-glass.

3.3.   EFFECT OF THE AIR-LAYER THICKNESS

Let d_1 and d_2 be the diameters of the outer surface of the halogen lamp and the inner surface of the inner quartz-glass tube, respectively. T_1 and T_2 denote the wall temperatures of the halogen lamp and the inner quartz-glass tube, respectively (assume each wall has a uniform temperature). The thickness of the air layer enclosed in the annular space between the two walls, $(d_2-d_1)/2$, is determined taking into account T_1 which must not exceed 800°C, as indicated in Fig. 13. An example is presented here where the ambient temperature T_a is at 25°C, the heating capacity is 3kW, heat source temperature T_s is 2000°C and a halogen-lamp tube is in a horizontal position. Figure 16-a illustrates the procedure of determining the air-layer thickness. An infrared camera detects the lamp wall temperature as 415°C, while a theoretical computation predicts the rate of heat dissipation from the lamp wall by radiation and natural convection, q to be 135 W. Next, the halogen lamp tube is installed concentrically as the heat source within a heat exchanger for heating chemical liquids. The air-layer thickness is varied to theoretically determine q. It is assumed that T_1 remains unchanged at 415°C and that T_2 is at 100°C, that is, 20°C higher than the prescribed temperature of chemical liquids of 80°C. Results are presented in Fig. 16-b for q against δ and d_2/d_1. It is seen that the minimum value of q is 148W, which is higher than q for the halogen-lamp tube in the ambient space. Hence, one can predict that within a heat exchanger for chemical liquid heating, the lamp wall temperature is below a value of 415°C which is within the allowable temperature of 800°C. Furthermore, one finds that the air-layer thickness can be determined regardless of the lamp wall temperature.

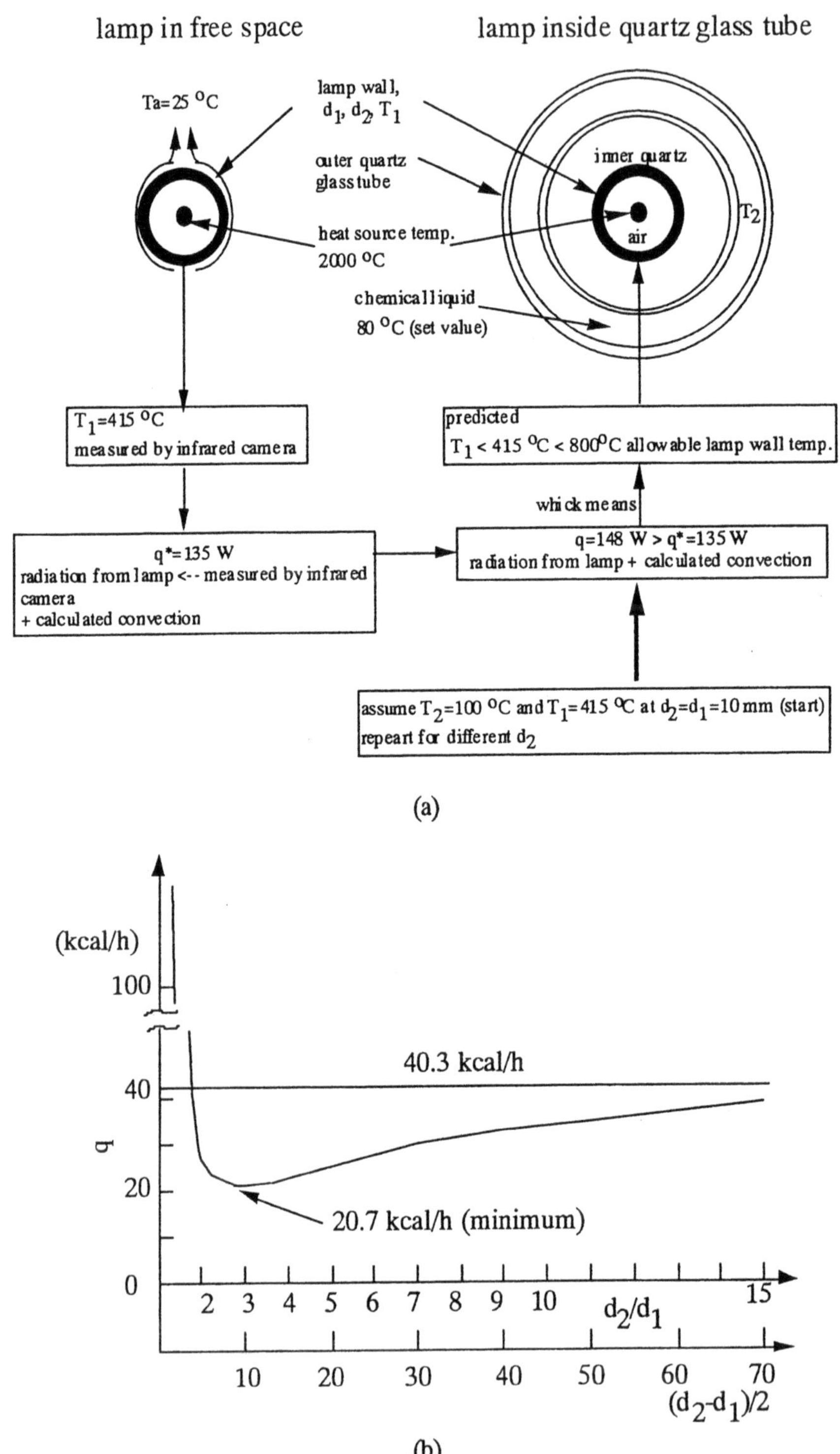

Figure 16. (a) Procedure to evaluate air-layer thickness and (b) q against δ and d₂/d₁.

3.4.   EFFECT OF INNER QUARTZ-GLASS TUBE

The diameter and thickness of the inner quartz-glass tube are determined, taking into consideration both the strength to withstand the pressure due to the chemical-liquid layer and the transmissivity of thermal radiation from the heat source. In view of size, it is preferred to have a more compact inner quartz-glass tube. However, considering the space for changing lamps, pressure concerns, transmissivity concerns, an inner diameter of 20mm and a thickness of 1mm were selected. The spectral transmissivity of a 1-mm thick, quartz-glass tube τ_g (λ) is superimposed on Fig. 15.   The monochromative emissive power density that passes through the inner quartz-glass tube is E_o (λ) τ_g (λ). Hence, the total transitive power E_o can be expressed as

$$E_o = \int_o^\infty E_o(\lambda)\tau_g\ (\lambda)d\lambda \tag{3}$$

Consequently, the total radiative heat lost in the quartz-glass tube, ΔE_g, can be written as

$$\Delta Eg = E_o - E_g = \int_o^\infty E_o(\lambda)\big[1 - \tau_g(\lambda)\big]d\lambda \tag{4}$$

It corresponds to the dark area in Fig. 15.   Note that the quantity $[(1-\tau_g(\lambda)]$ in Eq. (4) is the monochromatic absorptivity of the quartz-glass tube.

3.5.   THE EFFECT OF THICKNESS OF THE CHEMICAL-LIQUID (i.e. WATER) LAYER

The monochromatic transmissivity distribution curves in Fig. 12 indicate that thermal radiation is hindered with an increase in the thickness of the water layer. Beer's law can be described by the expression

$$\frac{I_o(\lambda)}{I_o(\lambda)} = \exp\left[-CL\varepsilon(\lambda)\right] \tag{5}$$

This can be applied to predict the radiant-heat absorption by a transparent solution. Here, $I_o(\lambda)$ denotes the intensity of incident monochromatic radiation; $I(\lambda)$, intensity of radiation exiting from the thickness of water layer L (length of optical path); ε (λ), monochromatic absorption coefficient, whose value depends on the pressure and temperature of the liquid; and C, concentration. The difference between the intensity of radiation entering the liquid and that leaving the liquid layer at L is

$$I_o(\lambda) - I(\lambda) = I_o(\lambda)\left\{1 - \exp\left[-CL\,\varepsilon(\lambda)\right]\right\} \tag{6}$$

568

or the amount of heat absorbed by the liquid. The quantity in the parentheses represents the absorptivity of the liquid α_λ at the wavelength λ. Equation (5) indicates a decrease in the amount of heat transmitted through the liquid, while Eq. (6) suggests an increase in the amount of heat absorbed by the liquid as its thickness (L) increases. For a liquid layer of 10-mm thickness, the monochromatic transmissivity $\tau_w(\lambda)$ is measured using a spectroradiometer. Results are presented in Fig. 17. It can be seen that the shape of the spectroradiometric curve agrees well with that for t = 10mm in Fig. 12, except at their maximum values.

The monochromatic emissive power absorbed by the liquid is $E_o(\lambda)\,\tau_g(\lambda)\,[1-\tau_w(\lambda)]$. The total absorptive power, E_w, can be expressed by

$$E_W = \int_o^\infty \tau_g(\lambda)\big[1 - \tau_w(\lambda)\big]d\lambda \tag{7}$$

E_w corresponds to the area enclosed by the curve and the abscissa in Fig. 17. The total amount of radiant heat transmitted through the liquid layer (but not being absorbed), ΔE_w can be expressed as

$$\Delta E_w = \int_o^\infty E_o(\lambda)\tau_g(\lambda)\,\tau_w(\lambda)d\lambda \tag{8}$$

It corresponds to the shaded area in Fig. 17. The fraction of the emissive power from the heat source being absorbed by the liquid layer, E_w/Eo, denotes the efficiency of directly heating the liquid by radiative heat transfer. A combination of Eqs. (2), (4), (7) and (8) yields

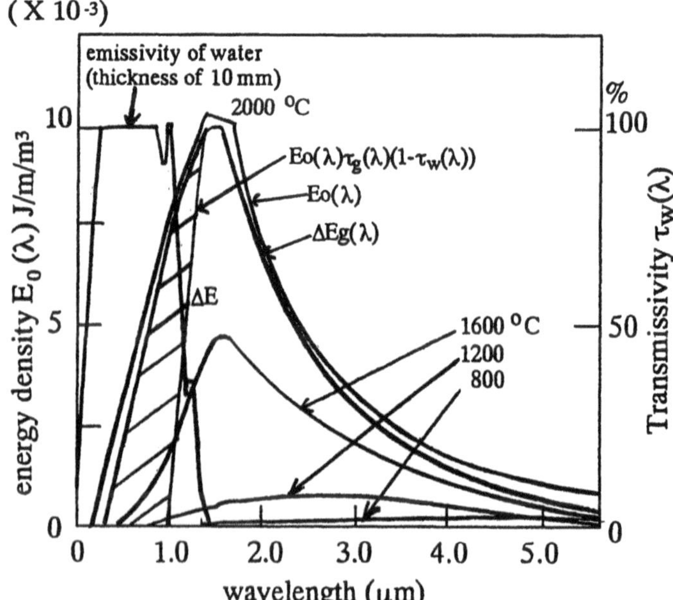

Figure 17. Spectral distribution of monochromatic emissive power density irradiated from heat source and transmissivity of water.

$$\frac{E_w}{E_o} = 1 - \frac{\left(\Delta E_q + \Delta E_w\right)}{E_o} \tag{9}$$

The value of E_w/E_o amounts to about 75% of that obtained from Fig. 17. In order to enhance the direct heating efficiency, it is necessary to reduce both the lost heat ΔE_w by about 10% and the transmitted heat by 15%. ΔE_q signifies either reflected or absorbed energy by the inner quartz-glass tube, which is eventually transferred to the chemical liquid by conduction or convection. It is expected that the direct heating efficiency may reach 85% if convection, conduction and radiation components are all included. Without some means to protect the outside surface of the outer quartz-glass tube, ΔE_w will be entirely lost to the ambient.

3.6. EFFECT OF OUTER QUARTZ-GLASS TUBE

In order to recover the transmitted last energy ΔE_w, appropriate measures must be taken at the outside surface of the outer quartz-glass tube, some of which are shown in Fig. 18. They include, for example, (a) direct reflection, (b) absorption/conduction, (c) wavelength-conversation reflection, and (d) absorption/convection. Tests and analyses are needed to evaluate the feasibility of each type.

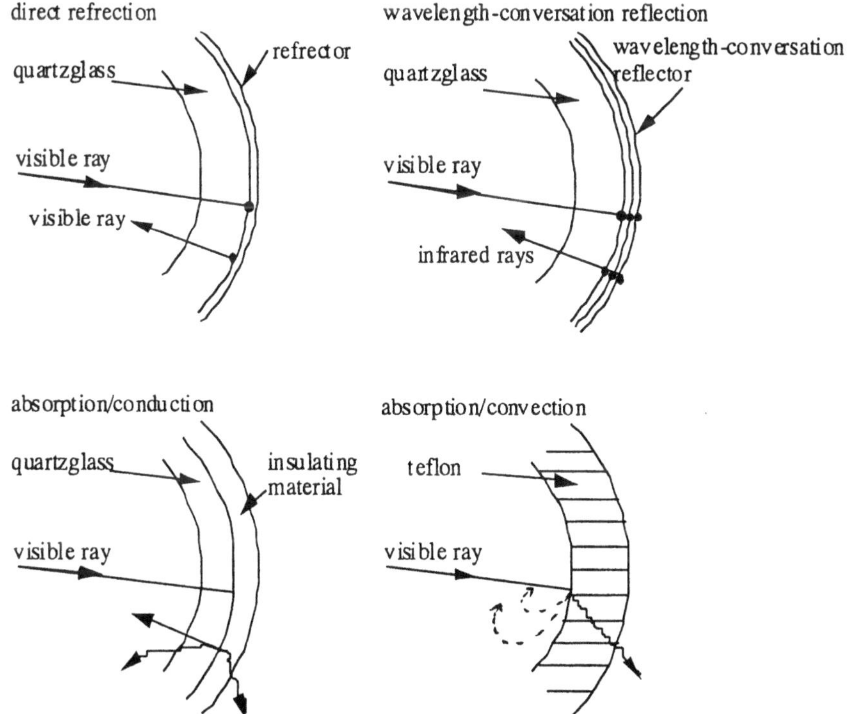

Figure 18.   Optical nature of the outside surface of the outer quartz-glass tube.

4. Heat Exchangers in Semiconductor Manufacturing Processes

4.1. CRYSTAL GROWTH

Both popular methods [2] for bulk crystal growth, namely Czochralski (CZ) and Bridgman methods, are external-heating type, as depicted in Figs. 19 and 20, respectively.

(a) Czochralski method

The CZ method is the most fundamental means of growing bulk crystals. Figure 19 shows lifting of a Si single bulk crystal. Figure 19 shows lifting of a Si single crystal. A heating coil is used to melt Si crystals in the quartz crucible placed with a graphite boat. The Si bulk crystal thus produced is also called CZ crystal. This method is also employed in the growth of bulk crystals of compound semiconductors.

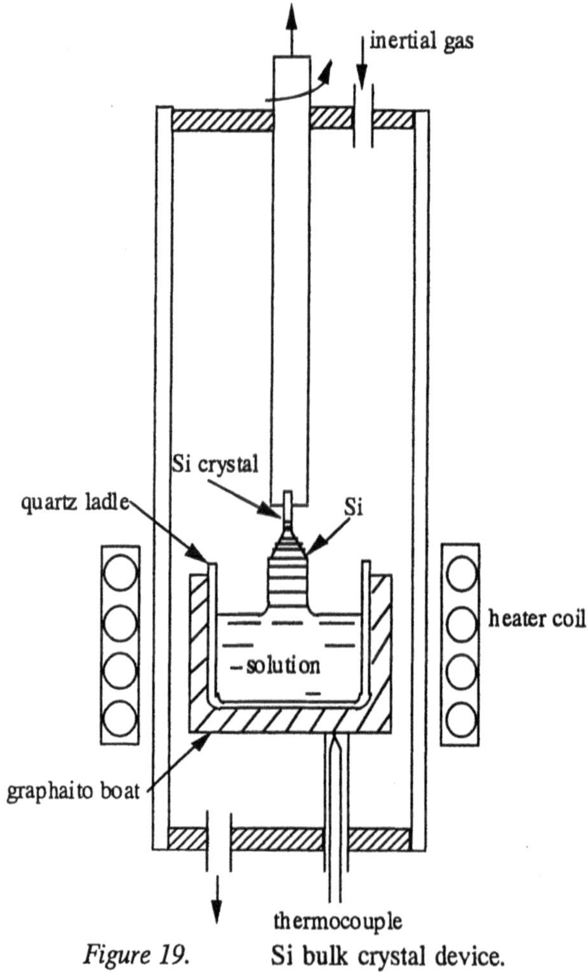

Figure 19.   Si bulk crystal device.

(b) Bridgman method

In this method, the material is melted inside a sealed quartz ampule, and the melt is gradually solidified from one end, where bulk crystallization begins with the growth nucleus acting as a seed. Figure 20 shows an example for growing the bulk crystal of GaAs inside a quartz boat due to volume expansion during solidification. This is called the horizontal Bridgman method. The quartz ampule is installed inside a quartz tube, which is heated externally by two moving low- and high-temperature furnaces.

4.2. GROWTH OF THIN SiO_2 LAYER ON A Si WAFER

Insulating films such as oxidized films and nitrified films are important materials for semiconductor chips, as mask materials in photolithography, as gate materials for MOS transistors, etc. Figure 21 is a schematic of a HTF furnace with an oxidizing environment. Note that it is an external heating type, radiant heat exchanger for oxidizing Si wafers by heating. This is called the thermal oxidization method. There are various thermal oxidation methods depending upon purposes, including

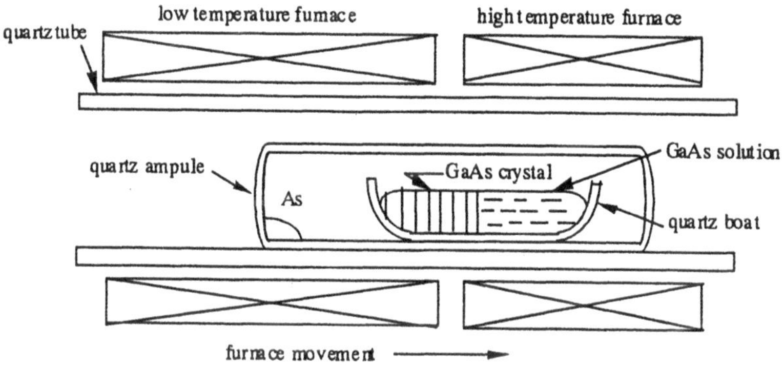

Figure 20.   Horizontal Bridgman device.

(a) dry oxidization method using dry high-purity oxygen,

(b) wet oxidization method by introducing wet oxygen,

(c) high-pressure oxidization method, and

(d) partial-pressure oxidization method

In any case, the basic principle is based on a process during which an oxidization agent (0) undergoes a chemical reaction with a Si atom on the surface of a Si crystal to form SiO_2. After forming the SiO_2, the oxidization agent diffuses through the film to reach the Si surface where new SiO_2 is formed. The gas flowing through the electric furnace in Fig. 21 is dry oxygen, wet oxygen, or other kinds.

In manufacturing insulating films, the reacting gas is supplied to the substrate surface for chemical reaction to form the film. It is called chemical vapor deposition (CVD): the thermal CVD method (under normal or reduced pressure) and plasma CVD method. A typical example of the plasma CVD method is illustrated in Fig. 22, which utilizes radio-frequency (rf) discharge to generate the plasma.

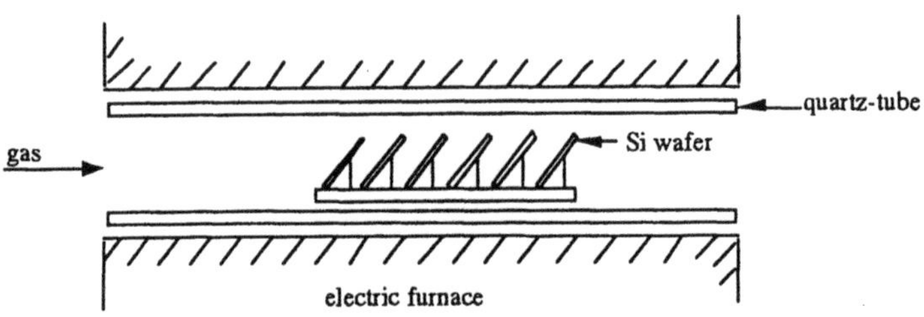

Figure 21.   HTF furnace with oxidizing environment.

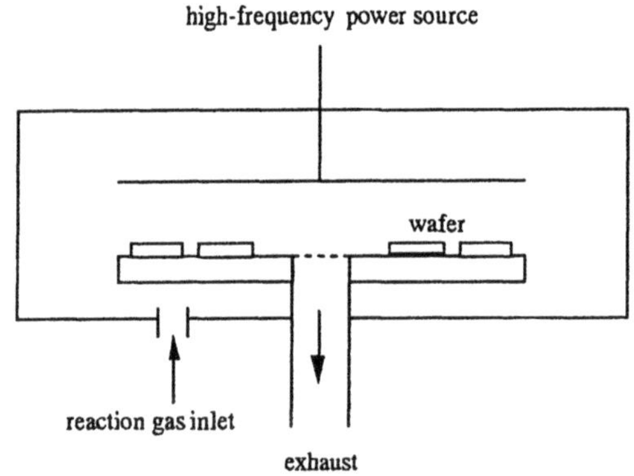

Figure 22.   Plasma CVD method.

4.3.   METALLIZATION

The metallization process generally involves deposition of a thin layer of metal as liner or barrier by either CVD or PVD known as sputtering. Figure 23 depicts (a) electron beam vapor deposition and (b) magnetron-sputtering. In the former, an electron beam illuminates the metal of a vaporization source being placed in a vacuum chamber to heat, vaporized and deposited on the substrate. This device is used for Al and other metal film deposition. The magnetron sputtering device has magnetic lines from a ring shape of magnets to surround the target surfaces (in circular form).

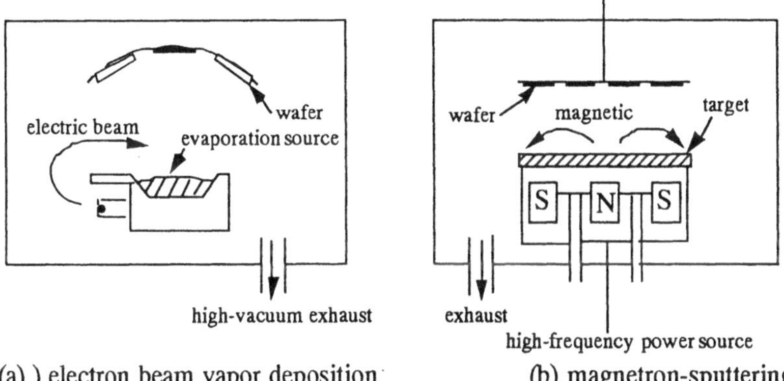

(a)) electron beam vapor deposition (b) magnetron-sputtering

Figure 23.   Electron beam vapor deposition and magnetron-sputtering.

4.4. ETCHING

IC's are manufactured according to design patterns via etching, doping and film growing processes. Figure 24 illustrates the steps of photolithography for etching a SiO_2 film, including (i) oxidation, (ii) photoresist coating, (iii) exposure, (iv) development, (v) etching, and (vi) resist removal. There are two etching methods: wet etching, Fig. 25, and dry etching, Fig. 26. In the dry etching case, ions are accelerated in the electrical field induced between the discharge plasma and the test surface and impacts on the test surface. Since etching reaction is accelerated and induced by the ion impact, etching progresses in the direction of incoming ions.

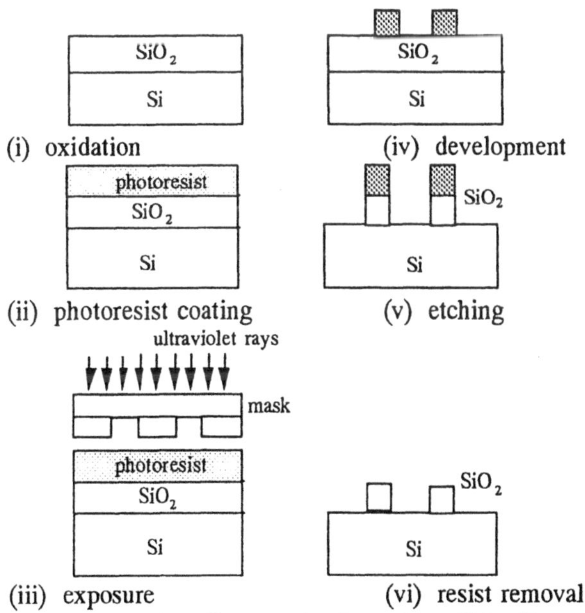

Figure 24.   Photolithography for etching SiO_2 film.

574

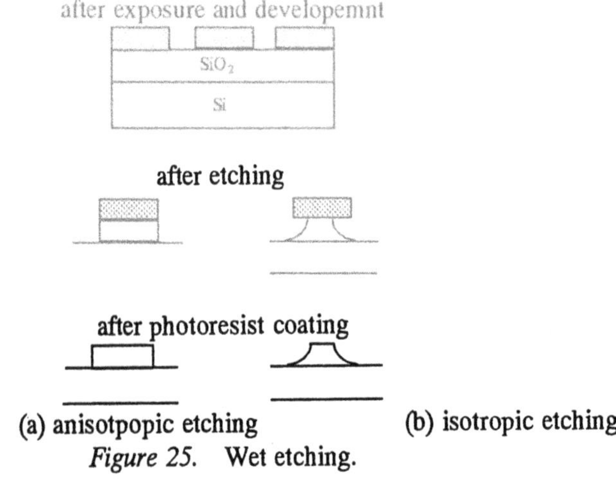

Figure 25. Wet etching.

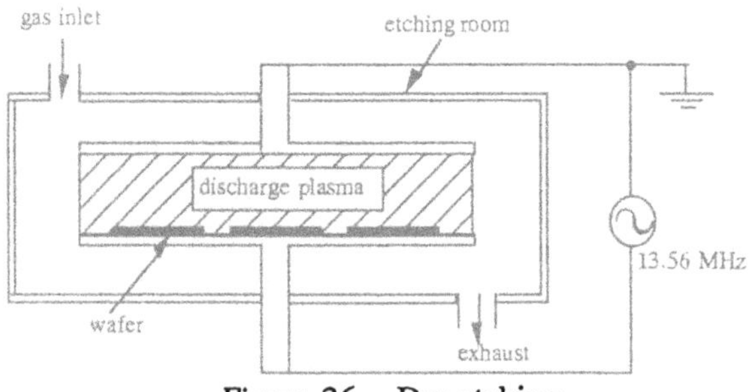

Figure 26. Dry etching.

4.5. EPITAXIAL GROWTH METHOD

With advances in semiconductor lasers and optoelectronic units, the epitaxial growth method has become important which is to grow semiconductor thin films of several μm on single-crystal substrates. Three epitaxial growth methods and apparatus are introduced here.

(a) Vapor-phase epitaxial growth method

This is thin-film single-crystal growth method from materials in a vapor phase. Figure 27 shows a vapor-phase-Si epitaxial growth apparatus, an external-heating type heat exchanger. By supplying a vapor mixture consisting of either $SiCl_4$ or $SiHCl_3$ and H_2 into a heat exchanger (namely, a furnace), Si single-crystal thin films will grow on the surface of Si substrate crystals heated at about 1200°C, through reduction reaction on the Si substrate. The crystal growth speed or film thickness can be controlled by the substrate temperature, molar ratio in vapor mixture, vapor flow speed, or other factors. If PH_3 or B_2H_6, together with the vapor mixture are supplied, then n-type or p-type epitaxial film can be produced, respectively. This method is called the hydrogen reduction method, in contrast to the thermal dissociation method. Like the apparatus used in the former method, the thermal dissociation method has a vapor mixture of SiH_4 and H_2 supplied to the heated Si substrate surface on which Si single-crystal thin film grows by means of thermal dissociation.

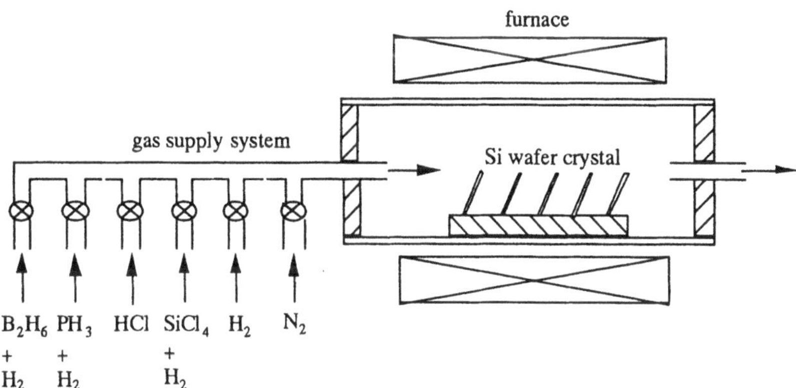

Figure 27. Vapor-phase-Si epitaxial growth apparatus.

(b) Metal organic vapor phase epitaxy (MOVPE)

It is also called metal organic chemical vapor deposition (MOCVD). The method is basically same as the vapor-phase epitaxial growth method except metal organic vapor, such as TMG [Ga $(CH_3)_3$] or TEG [GA $(C_2H_5)_3$] is used.

(c) Liquid-phase epitaxial growth method

The most commonly used slide-boat method is depicted in Fig. 28, another external-heating type. Multiple solution tanks are installed on a graphite boat. Materials with specified compositions are poured into each tank to prepare its solution. These solutions are made in contact with the substrate in series, and slowly cooled at constant temperature for crystal growth until a desired film thickness is achieved. By sliding the boat, as shown in the figure, thin-film crystals of different compositions are progressively grown on the substrate crystal. Thus, it enables growth of multiple-level thin-film crystals. An example is the application of this method in developing a Ga As/GaAl As double hetro-connect semiconductor laser.

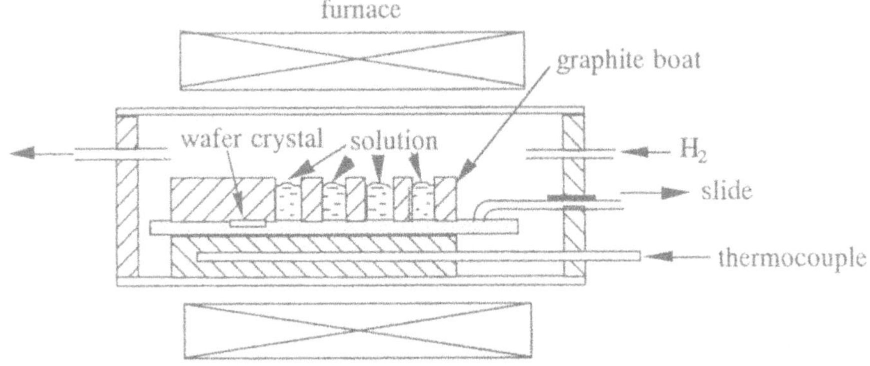

Figure 28. Slide-boat growth apparatus.

4.6. DOPING

Pure materials are rarely used in semiconductor IC's. Hence, it is necessary to dope a desired, small amount of impurity into semiconductor materials. For semiconductor wafers, doping can be conducted using zone leveling, diffusion, and ion-plantation methods. The zone leveling method is by external heating using coils, as depicted in Fig. 29. This is done by placing a seed crystal at one end of the refined semiconductor and moving the melting zone after the seed crystal has been in sufficient contact with the melt. When a small amount of appropriate dopant with small segregation coefficient is placed in contact with the seed crystal beforehand, a simultaneous action of single crystallization and uniform doping takes place. This method is simple in operation and can produce larger crystals with good dopant distribution, and thus is popularly utilized in manufacturing Ge and other semiconductor crystals.

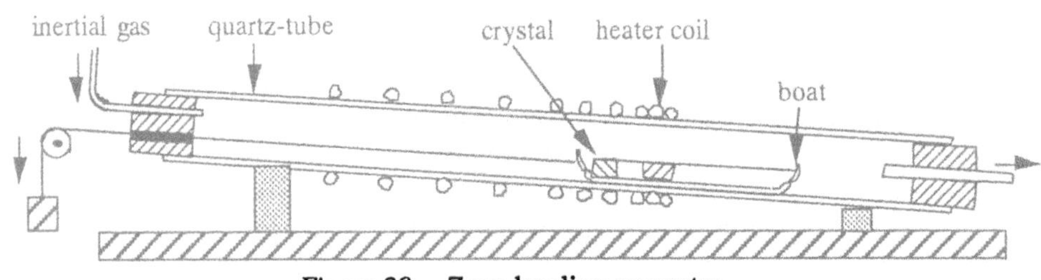

Figure 29. Zone leveling apparatus.

Nomenclature

A_1 area of surface 1

E emissive power, $E = \sigma T^4$

E_b black-body emissive power

E_w total absorptive power

F_{1-2} shape factor viewed from surfaces 1 to 2

I intensity of radiation
Io intensity of incident monochromatic radiation
L thickness of water layer
q rate of radiative heat transfer
T surface temperature
α radiative absorptivity
ε monochromatic absorption coefficient
σ Stefan-Boltzman constant
λ wavelength
τ_w transmissivity of water

References

1. Kadotani, K., Kubota, K. and Onishi, T. (1992) *Development of Radiative Type Heater for Heating of Chemical Fluid for Manufacturing Semi-Conductors*, Komatsu Technical Reports, Vol. 38, No. 130, pp. 1-13
2. Yamaguchi, J., Tanaka, T., Inuishi, Y. and Hamakawa, Y. (1995) *Semiconductor Engineering*, 3rd ed., Ohmsha, Tokyo, Japan.

EVAPORATION AND CONDENSATION HEAT TRANSFER ENHANCEMENT FOR ALTERNATIVE REFRIGERANTS USED IN AIR-CONDITIONING MACHINES

T. EBISU

Mechanical Engineering Laboratory, Daikin Industries, Ltd.
1304 Kanaoka-cho, Sakai, Osaka 591-8511, Japan

Abstract. The new advanced heat transfer tube called "Herringbone Heat Transfer Tube", having a unique inner surface geometry, has been experimentally investigated with the aim to enhance the heat transfer and improve the heat exchanger performance with R407C. Experimental examinations have been carried out to obtain the heat transfer coefficient and pressure drop for R407C flowing inside the horizontal herringbone tube, and the data have been compared with those for the existing inner-grooved tube. The heat transfer enhancement mechanism for the herringbone tube has been proposed in that the thin film layer spots that occur inside the tube could cause the heat transfer enhancement. The heat exchanger performances of the herringbone tube and the standard inner-grooved tube are compared toassess thire impact on energy conservation. The heat transfer coefficients for the herringbone tube are much larger than those for the standard tube, but the significant increase in heat transfer has been accomplished with an increase in pressure drop. The circumferential local heat transfer measurements have suggested that the heat transfer mechanism proposed by the authors is qualitatively verified, but the quantitative verification of the enhancement for the herringbone tube is not sufficiently clarified. The practical comparison of the heat exchanger performance showed that the evaporator performance is improved only at relatively high refrigerant flow rates, while the condenser using the herringbone tube seems to be better than that using the standard inner-grooved tube for all the refrigerant flow rates.

1. Introduction

The CFCs and HCFCs developed in 1930s have been widely used as working fluids in air-conditioning and refrigeration machines, since they possess excellent

S. Kakaç et al. (eds.), Heat Transfer Enhancement of Heat Exchangers, 579–600.

characteristics, such as thermal and chemical stability, non-toxicity, nonflammability and thermodynamic properties suitable for air-conditioning and refrigeration machines. However, they have a damaging effect on the ozone layer. In September 1987, the Montreal Protocol set restrictions on the production of specific CFCs. In a subsequent meeting in London (1990), the schedule for phase-out of CFCs was brought forward in comparison to the first regulation. In 1992 at the Copenhagen Conference, a new phase-out schedule for HCFCs by the year 2030 was agreed. At present, the production of CFCs is already phased-out in many countries, and production caps have been established for HCFCs.

Alternative refrigerants are forced to be environmentally friendly chlorine-free compounds that meet the requirement of zero ozone depletion potential (ODP). Moreover, the global warming issue is recently growing more and more important and affecting strongly the choice of HCFC alternatives as well as the ozone depletion issue. A low global warming potential (GWP) characteristic is required for the alternatives to attain energy conservation and to reduce CO_2 emissions.

With regard to the alternative refrigerant for R22, which plays a dominant role in residential, unitary and commercial air-conditioning applications all over the world, the environmentally friendly replacements with zero ODP are present, a binary zeotropic refrigerant mixture of R32/R134a, a ternary zeotropic refrigerant mixture of R32/R125/R134a (R407C) and an azeotropic mixture R32/R125 (R410A). For a comprehensive evaluation to employ such alternative refrigerants into existing air-conditioning machines currently running on R22, investigations, such as lubrication, thermal properties and flammability are required. Besides the basic characteristics mentioned above, the heat transfer and pressure drop characteristics are inquired by the designer of heat exchangers.

Torikoshi and Ebisu (1992, 1993, 1994, 1995) have carried out heat transfer experiments to study average evaporation and condensation for several alternative refrigerants flowing inside a horizontal tube. Among these alternatives, R407C introduces some complications for air-conditioning units, because it has the severe problem of heat transfer degradation during evaporation and condensation, attributed to the mass transfer resistance of the zeotropic characteristic. The results by Ebisu and Torikoshi (1994) demonstrated that the average evaporation and condensation heat transfer coefficients fell about 40% and 50%, respectively, below those for R22. Furthermore, the evaporation and condensation pressure drop for R407C was found to be even or larger than that for R22. Recently, Sundersan et al. (1994) and Uchida et al. (1995) have undertaken experimental investigations on evaporation and condensation heat transfer characteristics for R407C inside a horizontal tube. Their experimental results were similar to those of Ebisu and Torikoshi (1994) in that R407C has significantly lower heat transfer coefficients than those for R22. Additionally, Ebisu et al. (1996) conducted an experimental study concerning the air-cooled heat exchanger performance using

R407C to clarify the effect of the heat transfer degradation on air-cooled heat exchanger performance. Their results showed that the heat exchanger performance for R407C was about 5 to 10% lower than that for R22 in both cooling and heating modes, due to the degradation of the heat transfer coefficients of R407C. Hence, the heat transfer enhancement for R407C flowing inside a heat transfer tube is required to improve air-cooled heat exchanger performance.

The overall purpose of this work is to develop heat transfer enhancement techniques and to improve air-cooled heat exchanger performance for R407C. To fulfill this purpose, the new advanced heat transfer tube, called "herringbone heat transfer tube", has been developed in this study. The herringbone heat transfer tube has unique geometical surface inside a tube, made by welded-tube fabrication. This paper describes the characteristics of the heat transfer and flow friction of the herringbone heat transfer tube and the compares the overall heat transfer enhancement of the herringbone tube and the existing inner-grooved tube. The heat exchanger performance using the herringbone tube has been examined to clarify the effect of the heat transfer enhancement on the air-cooled heat exchanger performance. Additionally, the heat transfer mechanism for the herringbone heat transfer tube has been conjectured, and the qualitative verification of this presumption has been attempted experimentally.

2. Experiments

2.1. HEAT TRANSFER TUBE DEVELOPED

Photos and the detailed specifications of the herringbone heat transfer tube and the existing inner-grooved tube are shown in Figure 1 and Figure 2, respectively. The existing inner grooved tube is routinely used in air-conditioning applications, and is conventionally made by forming a set of grooves inside a seamless smooth tube. The new herringbone tube has a pair of groove patterns like a herringbone inside a tube. It is fabricated by embossing a flat strip of copper to form the desired herringbone geometry, thereafter the embossed strip is rolled into a round tube and axial seam is induction-welded. As the fin height for the herringbone tube is higher than the inner grooved tube, the increase in inside surface area compared to a smooth tube is 1.6 and 2.1 times for the inner grooved tube and the herringbone tube, respectively. The outer diameters, D_o ,of both the tubes are 7.0mm.

The heat transfer enhancement mechanism of the herringbone tube is supposed as shown in Figure 3. In addition to the surface area increase provided by the inner fins, the non-uniform liquid film layer forming inside the tube is considered to play a great role in the heat transfer enhancement for the herringbone tube.

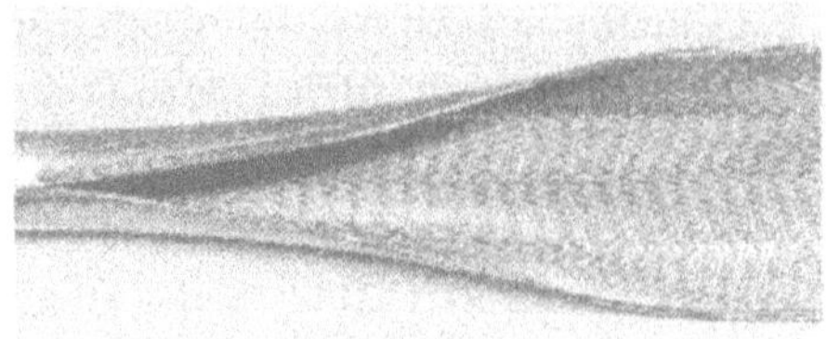

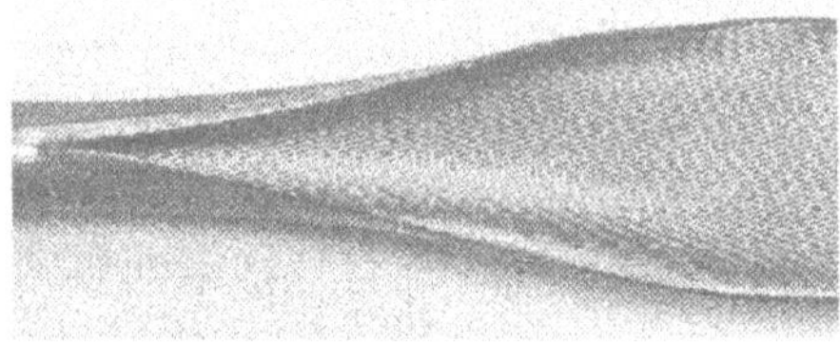

Figure 1. Photos of testes tubes

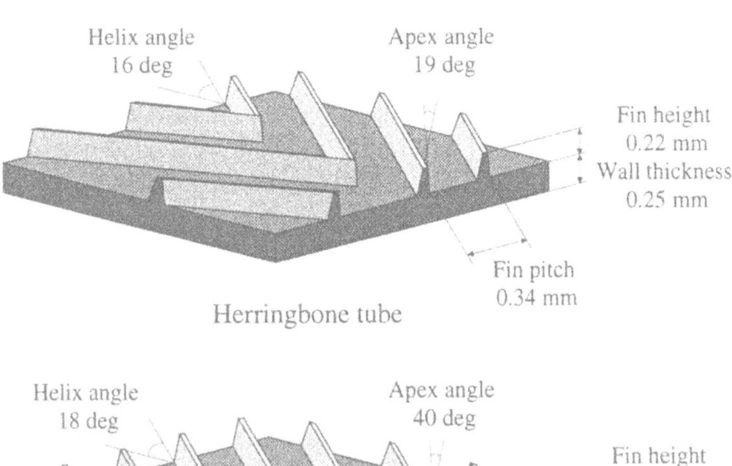

Figure 2. Geometries of tested tubes

For the inner-grooved tube, the heat transfer enhancement is probably caused by forming the uniform film layer around the entire tube wall, which the capillary wetting of the helical narrow grooves brings about. Contrariwise, the herringbone groove pattern is expected to carry liquid in two different flow directions by V-shaped grooves, thus, a circumferential non-uniform film layer is created inside the tube. The film layer can be thicked on the top and bottom. The film is supplied or removed continuously by the V shape grooves. The thin film on the side has a high heat transfer coefficient, thereby, contributing to the great heat transfer enhancement for the herringbone tube.

2.2. EXPERIMENTAL APPARATUS

A schematic drawing of the present experimental apparatus and the test section used to evaluate the heat transfer characteristics of the horizontal heat transfer tube is shown in Figure 4. This test rig, which consists mainly of a refrigerant loop and three independent water loops, has been designed to determine local

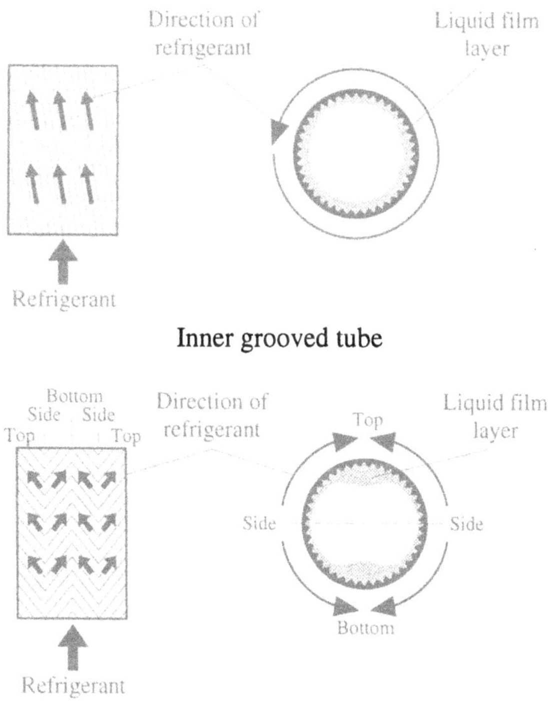

Inner grooved tube

Herringbone tube

Figure 3. Heat Transfer mechanism taking place inside a tube

584

evaporation and condensation heat transfer coefficients and pressure drops of the refrigerant flowing inside horizontal heat transfer tubes.

The refrigerant loop is connected to the heat transfer tube side. The main components are: a main heat exchanger, a subcooler, a circulating pump, a mass flow meter and three heat exchangers in-a-line. The large capacity of the main heat exchanger allows good control of the operating pressure of the refrigerant

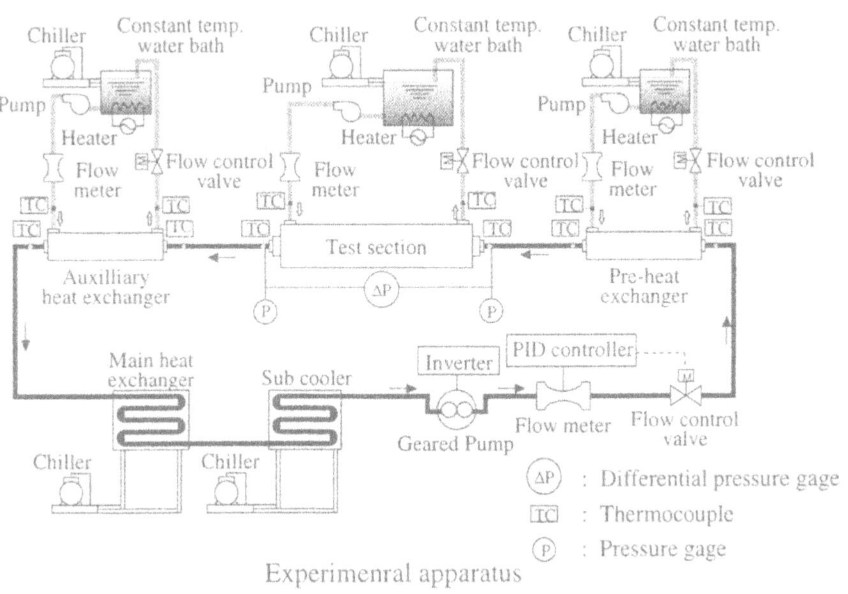

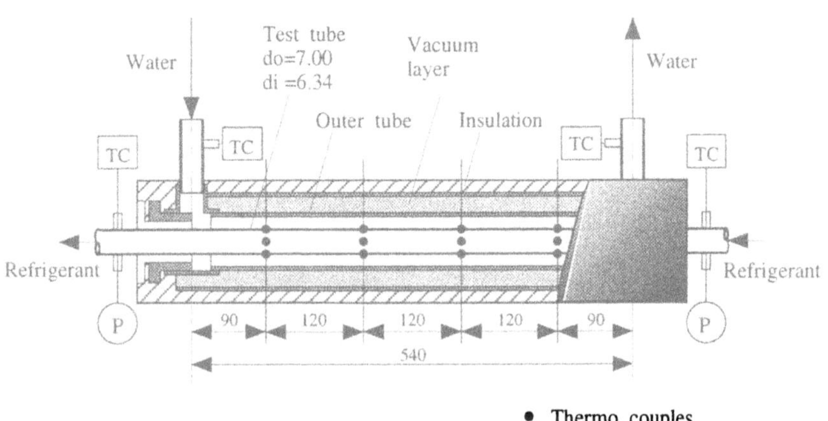

Figure 4. Experimental apparatus and test section (dimension is in mm)

loop, and the subcooler makes the refrigerant completely subcooled before entering the circulating pump. In this test rig, a gear pump is used instead of a compressor to circulate the refrigerant in order to measure the heat transfer characteristics of the refrigerant without lubricating oil. The refrigerant flow rate is adjusted by the flow control valve and the frequency of the pump, and it is measured with a mass flow meter. The heat exchangers, including a pre heat exchanger, a test section and an auxiliary heat exchanger, are all double-tube-type heat exchangers with refrigerant flowing in the inner tube and countercurrent water in the surrounding annulus. The pre heat exchanger adjusts the refrigerant vapor quality at the inlet of the test section, and the axillary heat exchanger make the refrigerant evaporate or condense completely. A couple of calibrated platinum resistance sensors with an accuracy of 0.02K and pressure gages with 1.0kPa accuracy are located at the inlet and outlet of each heat exchanger to measure the bulk temperatures and the pressures of the refrigerant, respectively.

The independent water loop is connected to each of the heat exchangers to supply heat-source or heat-sink water into the annulus side. Each loop has its own heating and chilling unit in order to obtain the desired temperature and flowrate of the water.

The test section has 700 mm overall length and 540 mm effective heat transfer length. The surface temperatures of the inner tube are measured with 8 thermocouples soldered at the top, both sides and bottom of the tube wall at two locations along the tube. The pressure drop of the refrigerant across the test section is measured with a differential pressure transducer. The test section has a vacuum layer outside the outer tube, and is thermally insulated using a 50mm thick piece of a plastic foam.

It is noted that the heat transfer performance for the herringbone tube varies with the location of the weld seam, because the axial seam may influence the liquid film layer and the flow pattern of the refrigerant. In order to eliminate the effect of the weld seam on the heat transfer performance, the herringbone tube has been always set up in the test section so that the axial seam line was located at the top of the tube.

2.3. EXPERIMENTAL CONDITIONS

The evaporation and condensation temperatures of the refrigerant at the test section were maintained at 278K and 323K, respectively. The refrigerant mass fluxes varied from $G=150kg/m^2s$ to $300kg/m^2s$.

The experiments provided the local heat transfer coefficients versus vapor qualities at a constant heat flux of $7.5kW/m^2$, which approximately corresponds to the practical condition of the air-cooled heat exchanger for a residential air-conditioning machine. The water temperature in the test section was controlled to attain the constant heat flux condition of the test section so that the vapor quality change (Δx) across the test section depends on the refrigerant mass flux, that is, Δx for G=300kg/m^{2}s and 150kg/m^{2}s are about 0.05 and 0.025, respectively.

2.4. DATA REDUCTION

The average refrigerant heat transfer coefficient , α_{ave}, was determined from the following equation:

$$\alpha_{ave} = \frac{q}{T_{w,ave} - T_r} \tag{1}$$

where q is the overall heat flux based on the outside area of the tube. The heat flux, q, was calculated from the water side enthalpy change and the test section dimension. $T_{w,ave}$ represents an average for eight thermocouples mounted on the tube wall surface. The temperature of the refrigerant, T_r, is defined as the equilibrium saturation temperature of the refrigerant corresponded to the pressure and quality in the test section. The local heat transfer coefficients at each location of the tube (top, sides and bottom) α_{local}, is additionally obtained from the temperature difference between ($T_{w,local}$ - T_r). REFPROP Ver. 5.0 by NIST was used as the database of the refrigerant in the present study.

To check the energy balance between water and refrigerant sides of the pre heat exchanger and the test section, single-phase refrigerant flow tests were preliminarily carried out. Comparisons between both heat quantities are made and found to be within 5% for all data runs. The overall heat loss or gain around the heat exchangers was kept less than 3% during evaporation and condensation. The experimental uncertainty in the present study is analyzed as follows. The mass flow meters to measure the flow rates of the refrigerant and water were calibrated with an experimental uncertainty of 2%. The errors of the thermocouple and platinum resistance sensor were found to be about $0.1K$ and $0.02K$, respectively. As a consequence, the experimental uncertainty in the heat transfer coefficient is estimated to be about 8% for all data runs.

3. Heat Transfer Performance of Herringbone Tube

3.1 BASIC HEAT TRANSFER CHARACTERISTICS OF R407C

First of all, typical heat transfer characteristics for R407C flowing inside a smooth tube are introduced to clarify the needs in developing the enhanced heat transfer tube. The evaporation heat transfer coefficient results for R407C are plotted in Figure 5 as a function of the refrigerant quality. Here, the quality means the average of inlet and outlet qualities of the test section. In this figure, the data for R22 are also displayed to compare with those for R407C. The figure shows that the heat transfer coefficients for both the refrigerants increase with the increase of quality, but the trend for R407C is more gentle than that for R22. When the data for R407C are compared with those for R22, the former shows 30% - 90% (50% average) lower heat transfer coefficients than the latter.

Figure 6 shows the heat transfer coefficients during condensation. The trend of the condensation heat transfer coefficients for both refrigerants is that the heat transfer coefficient decreases as the amount of liquid refrigerant (1-x) increases. Comparing to the results for both refrigerants, the heat transfer coefficients for R407C are significantly lower (40% average) than those for R22. As described above, it is found that the heat transfer enhancement for R407C is urgently required to improve the air-cooled heat exchanger performance using R407C.

3.2. PERFORMANCE OF HERRINGBONE TUBE

3.2.1. *Heat Transfer Enhancement and Mechanism*

In order to investigate the heat transfer characteristics and the heat transfer enhancement mechanism for the herringbone tube, the experimental data have been reduced to show the variations of the circumferential local heat transfer coefficient. Figure 7 shows the typical evaporation experimental results of the

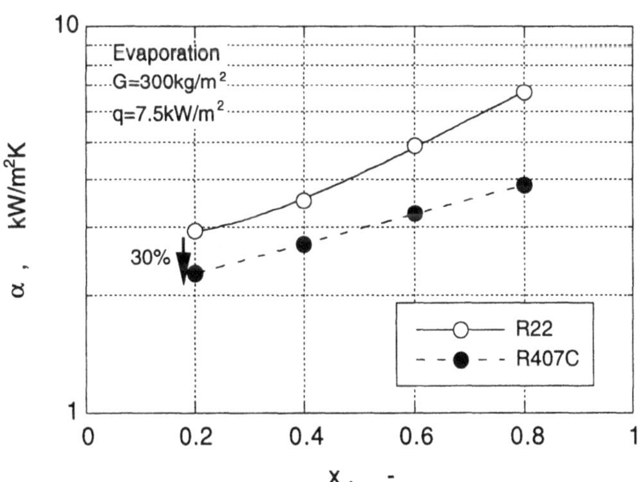

Figure 5. Experimental results of heat transfer coefficients during evaporation inside a smooth tube

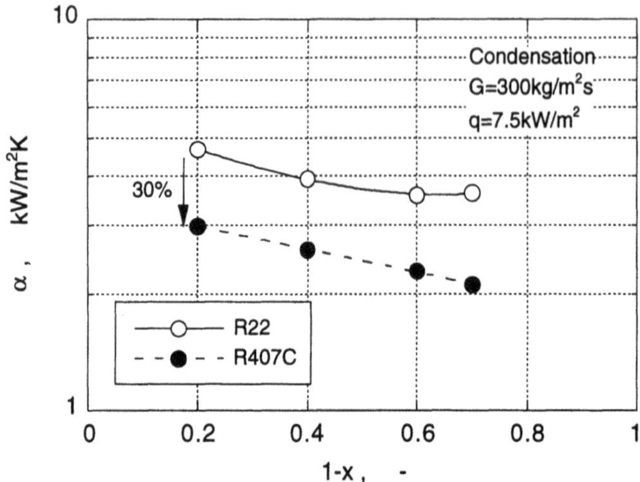

Figure 6. Experimental results of heat transfer coefficients during condensation inside a smooth tube

circumferential tube wall temperature difference, $T_{w,ave}$-$T_{w,local}$, and the local heat transfer coefficient, α_{local}, of the top, sides and bottom of the tube. The data are plotted as a function of the vapor quality, x, at G=300kg/m²s. Additionally, the data of α_{local} for the existing inner-grooved tube are displayed for comparison. With respect to the results for the herringbone tube, the temperature difference, $T_{w,ave}$-$T_{w,local}$, at both sides of the tube are higher by 0.1 to 0.2K than those at the top and bottom, and accordingly, the heat transfer coefficients are at both sides are approximately 10 to 15% higher than those at the others. On the contrary, there is no apparent difference of the circumferential heat transfer coefficients for the inner grooved tube. Quantitative comparisons of the average heat transfer coefficients between the herringbone and inner grooved tubes are shown in *Figure 8*. It implies the results that the evaporation heat transfer coefficient for R407C is enhanced as much as 90% by the herringbone tube. This significant performance improvement of the herringbone tube offers the possibility that the heat transfer coefficient for R407C degraded by the zeotropic characteristic can be recovered by the herringbone tube, and the heat transfer coefficient for R407C using the herringbone tube may equal to that for R22 using the existing inner-grooved tube.

Condensation results for the tube wall temperature difference and the heat transfer coefficient are shown in Figure 9. The results for the herringbone tube show that the temperature for both sides of the tube are lower by 0.1 to 0.3K, and the heat transfer coefficient for both sides are approximately 20 to 30% better than those for the top and bottom. In contradiction to the results for the herringbone tube, the circumferential distributions of the heat transfer coefficients for the inner grooved tube are within the limit of experimental error. Comparing the average heat transfer coefficients between the herringbone and inner grooved tubes, as shown in Figure 10, the former are as much as 200% higher than the latter, thereby, the heat transfer coefficients for R407C using the herringbone tube surpass those for R22 using the inner-grooved tube.

The noteworthy better heat transfer coefficients at both sides for the herringbone tube than those at the top and bottom can be explained by the expected heat transfer enhancement mechanism proposed by the authors. The film layer at both sides of the herringbone tube is thinned by the V shape grooves of the herringbone tube so that the heat transfer is enhanced better at the side rather than at the top and bottom. However, it cannot be explained well by the proposed

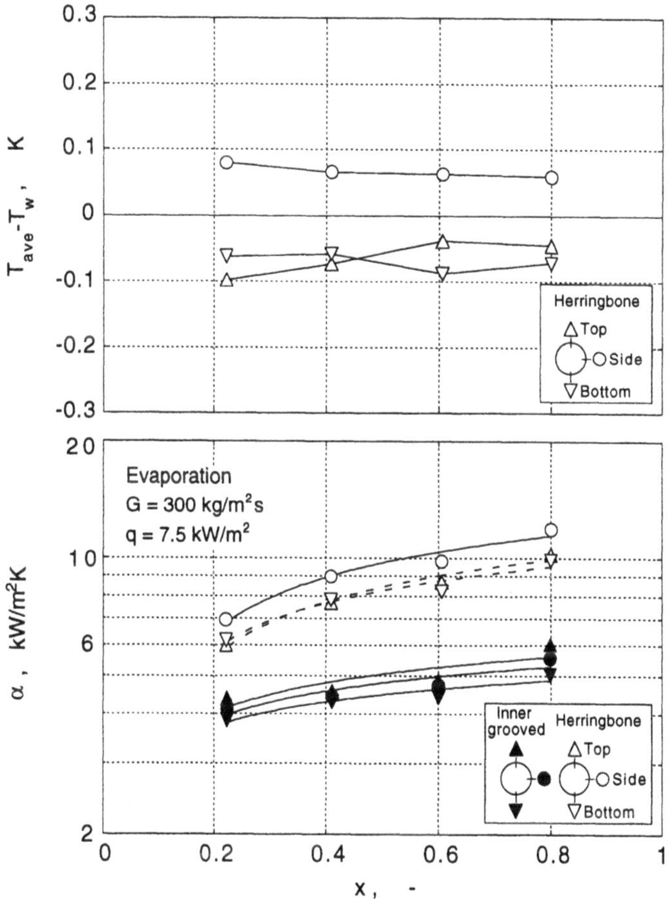

Figure 7. Circumferential distribution of tube wall temperature and heat transfer coefficient for R407C during evaporation

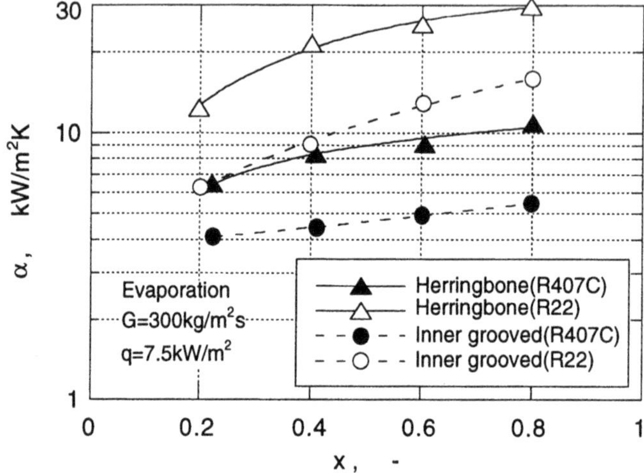

Figure 8. Comparisons of average heat transfer coefficients between the herringbone and inner grooved tubes during evaporation

heat transfer enhancement mechanism that the heat transfer coefficients at the top and bottom are higher than those for the inner grooved tube, because the film layers at the top and bottom of the herringbone tube are considered to be thicker than the uniform film layer of the inner grooved tube. One of the reasons of this result may be attributed to the sensitivity of tube wall measurements. As the copper heat transfer tube having high thermal conductivity is employed in the present measurements, the circumferential distribution of the tube wall temperature may be considerably moderated by the heat conduction through the tube wall. Therefore, sensitive measurements of the tube wall temperatures may not have been performed. And the above result may be due to the another possibility of the heat transfer enhancement mechanism. That is, the confluences of the liquid film flow at the top and bottom may have a good effect on the evaporation and condensation heat transfer.

3.2.2. *Pressure Drop*

A comparison of the evaporation pressure drop for the herringbone tube and the inner grooved tube is shown in Figure 11. The pressure drop increases with an increase of the vapor qualities for both heat transfer tubes. However, the pressure drop per unit length for the herringbone tube is about 60% higher than that for the inner grooved tube.

Figure 12 shows the results during condensation. A si,ilar qualitative trend of an increase of the pressure drop with the vapor quality is observed for both the herringbone tube and the inner-grooved tube. The pressure drop for the

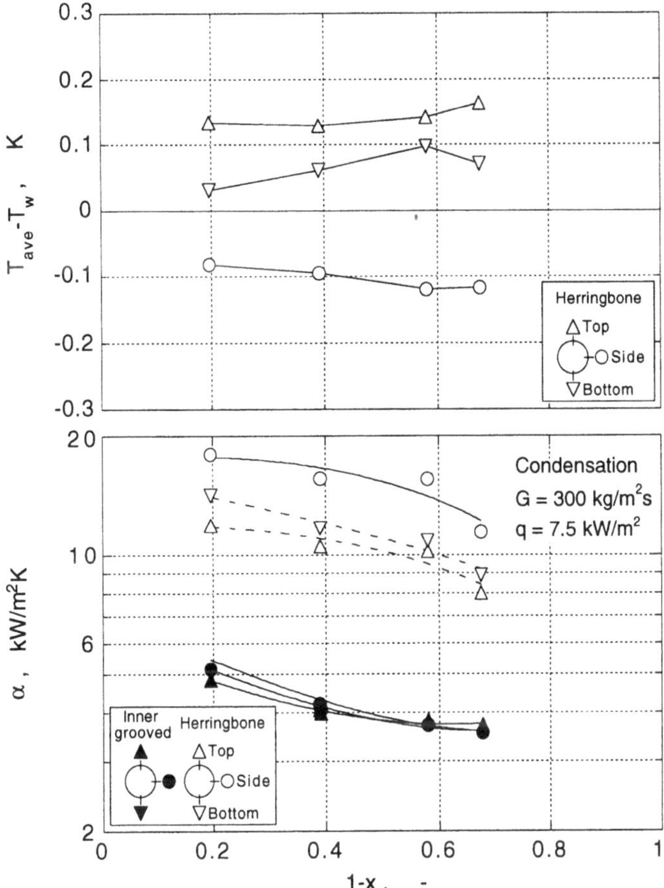

Figure 9. Circumferential distribution of tube wall temperature and heat transfer coefficient for R407C during condensation

herringbone tube is about 1.3 times as that for the inner-grooved tube. As the larger pressure drop of the herringbone tube during evaporation and condensation may be considered due to the higher groove and the two way helix angles of the herringbone tube, the optimization of the herringbone geometry, such as the height, helix angle, number of grooves and more, will diminish the pressure drop increase.

3.2.3. *Effect of Refrigerant Mass Flux*

The influence of the refrigerant mass flux on the heat transfer coefficient and the pressure drop has been additionally investigated in order to obtain the information for designing an air-cooled heat exchanger using the herringbone tube. Figure 13 shows the experimental results of the evaporation heat transfer

592

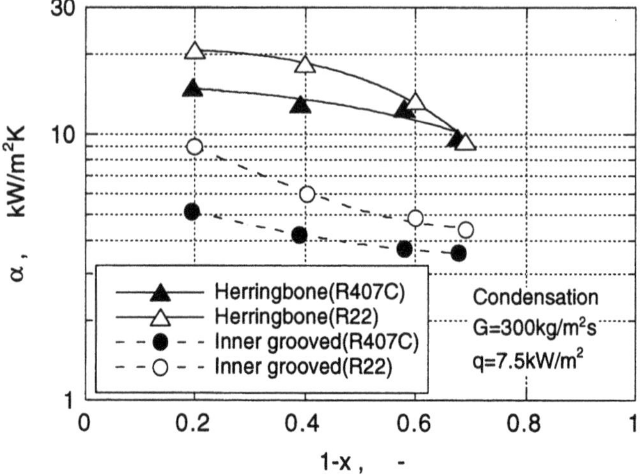

Figure 10. Comparison of average heat transfer coefficients between the herringbone and inner grooved tubes during condensation

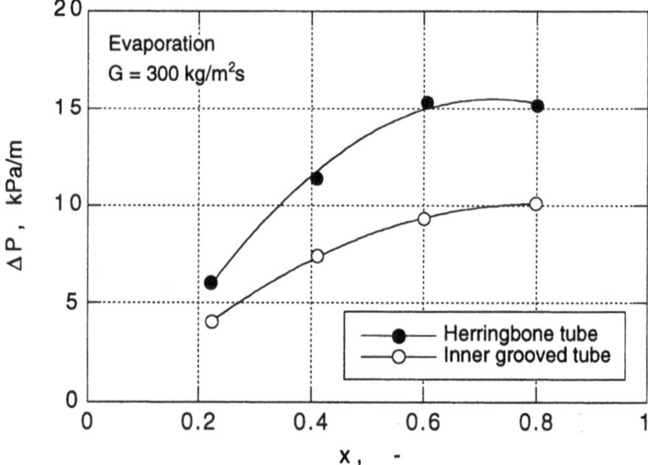

Figure 11. Comparisons of pressure drops between the herringbone and inner grooved tubes during evaporation

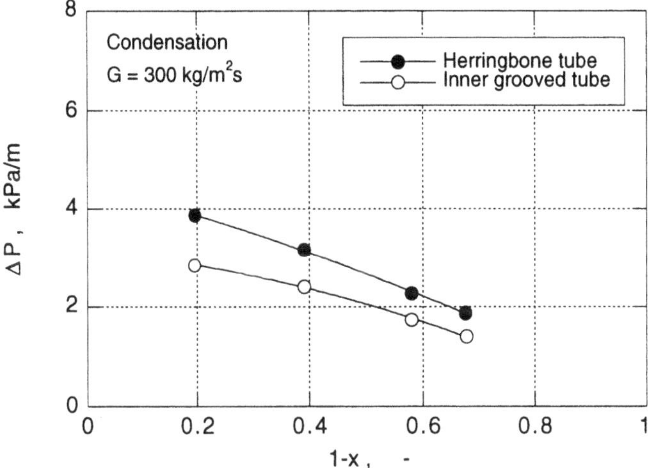

Figure 12. Comparisons of pressure drops between the herringbone and inner grooved tubes during condensation

coefficient. The data at $x=0.6$ with the variation of the refrigerant mass flux are typically plotted for both the herringbone and inner grooved tubes. The heat transfer coefficients for both tubes increase with increasing the refrigerant mass flux, but the trend for the herringbone tube is steeper than that for the inner grooved tube. Therefore, the herringbone tube can provide only a small positive effect on the evaporation heat transfer enhancement for mass flux lower than $G=200\text{kg/m}^2\text{s}$, while at larger the refrigerant mass fluxes, the more the herringbone tube contributes significantly to evaporation heat transfer enhancement.

Figure 14 shows the results of the condensation heat transfer coefficients. Similar to the results of the evaporation, the curve for the herringbone tube seems to be steeper than that for the inner grooved tube. The quantitative comparison of the heat transfer coefficient between both the tubes shows that the former achieves about 100% and 200% higher performance than the latter at $G=150\text{kg/m}^2\text{s}$ and $G=300\text{kg/m}^2\text{s}$, respectively. Therefore, the heat exchanger performances as a condenser using the herringbone tube can surpass those using the existing inner grooved tube for the entire range of the refrigerant mass fluxes.

The evaporation and condensation pressure drops are shown in Figure 15. The pressure drops for both tubes increase with an increase of the mass flux, and the herringbone tube shows about 50% and 30% higher pressure drops than the existing inner grooved tube during evaporation and condensation, respectively.

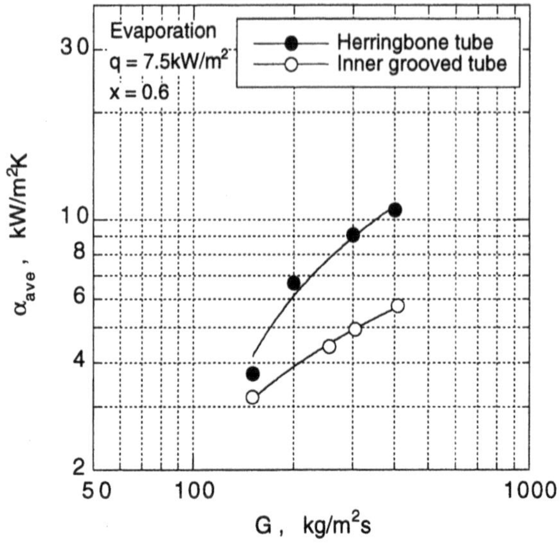

Figure 13. Effect of refrigerant mass flow rate on evaporation heat transfer coefficient

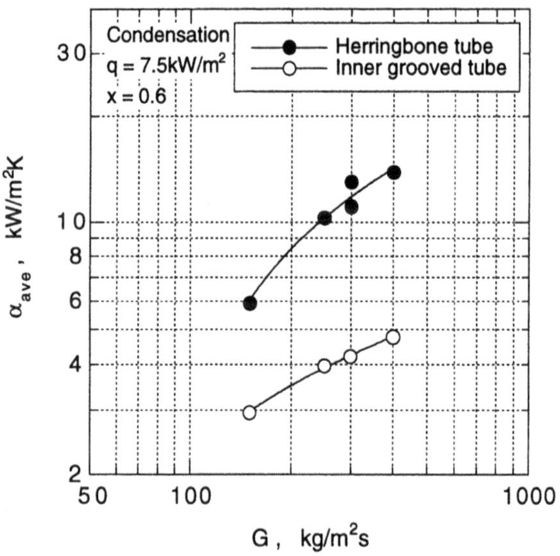

Figure 14. Effect of refrigerant mass flow rate on condensation heat transfer coefficient

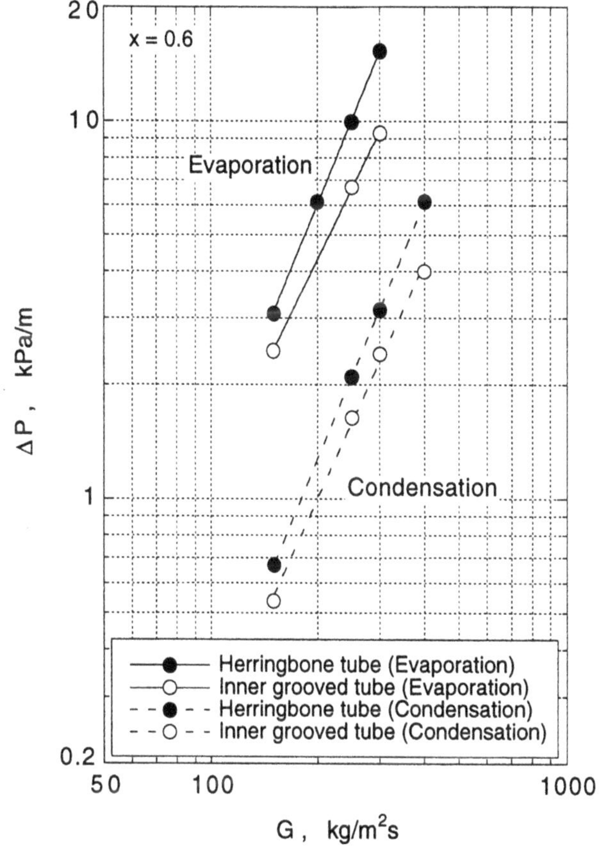

Figure 15. Effect of refrigerant mass flow rate on pressure drop

The higher pressure drop for the herringbone tube is due to the higher fin height and the greater surface area of the herringbone tube relative to the inner-grooved tube. The above-mentioned characteristics of the pressure drop of the herringbone tube may also influence the heat exchanger performance.

3.3. HEAT EXCHANGER PERFORMANCE USING HERRINGBONE TUBE

Practical implications for energy conservation and the size reduction of the heat exchanger through the heat transfer enhancement are obtained by comparing the heat exchanger performance between the developed and existing ones. In the present study, two of the same size heat exchangers, one for the herringbone tube and another for the existing inner grooved tube, have been employed for this

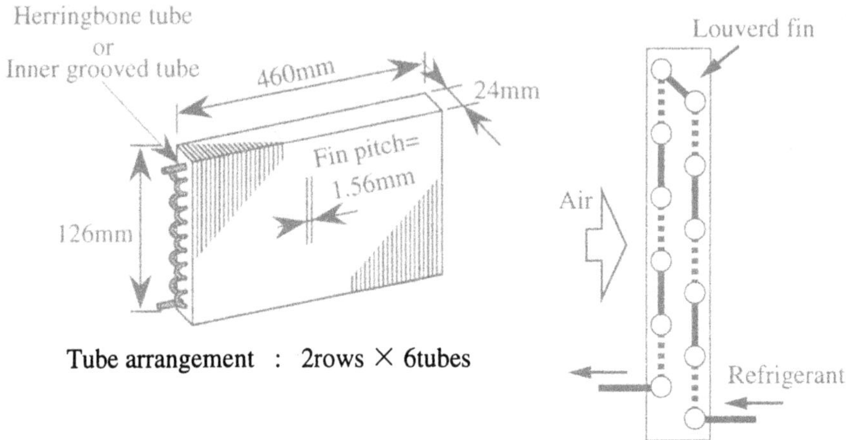

Tube arrangement : 2rows × 6tubes

Figure 16. Configurations of the air-cooled heat exchanger tested

			Evaporation	Condensation
Refrigerant	Evaporation pressure	[kPa]	588.4	—
	Inlet quality	[-]	0.27	—
	Outlet superheat	[K]	5	—
	Condensation pressure	[kPa]	—	2091
	Inlet superheat	[K]	—	30
	Outlet subcool	[K]	—	5
Air	Temperature (Dry bulb) [ ]		27	20
	Temperature (Wet bulb) [ ]		19	15
	Air velocity	[m/s]	0.9 ~1.8	1.0 ~2.0

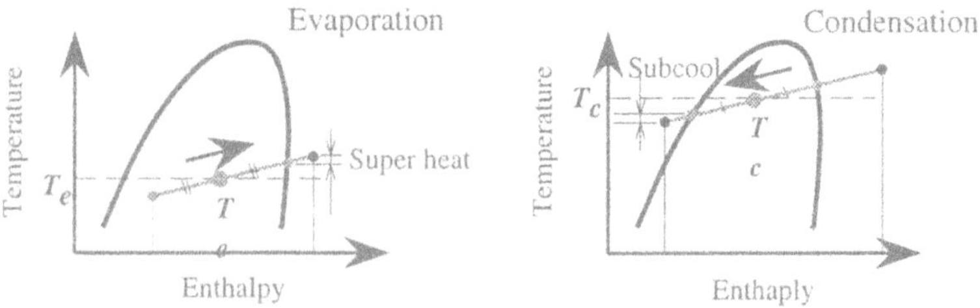

Definition of Eva./Cond. Temperature

Figure 17. Experimental conditions to evaluate
the heat exchanger performance

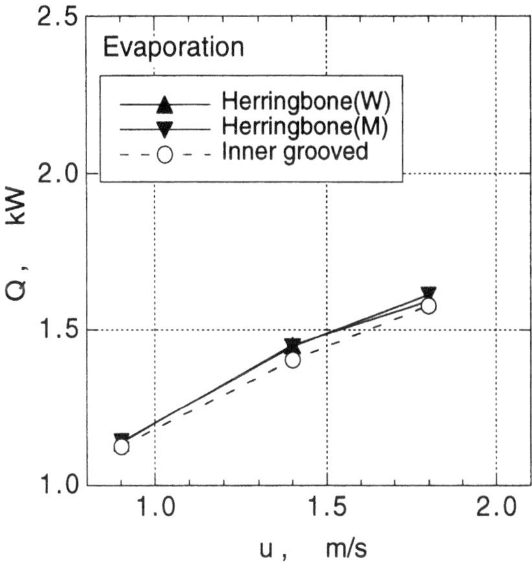

Figure 18.   Comparisons of evaporator performance between herringbone and inner grooved tubes

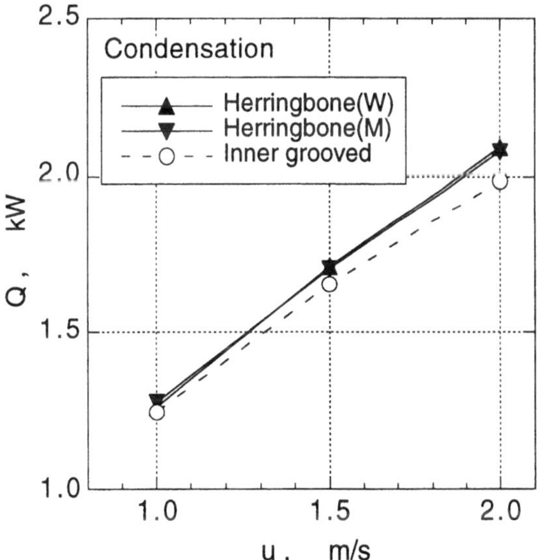

Figure 19.   Comparisons of condenser performance between herringbone and inner grooved tubes

comparison. Figure 16 and Figure 17 show the test heat exchanger configuration and the experimental conditions to investigate the heat exchanger performance. The experiments have been conducted with 2-row fin-and-tube heat exchangers which have cross-countercurrent flow configuration of the refrigerant and air. The working fluid employed was R407C.

Figure 18 shows the experimental results of the evaporator performance. The data are plotted as a function of u, which corresponds to the frontal air velocity of the test heat exchanger. The performance using the herringbone tube is about 3% better than that using the inner grooved tube at $u=1.8$m/s and 1.4m/s, while both the performance at $u=0.9$m/s are compatible. These results are due to the findings that the heat transfer coefficients for the herringbone tube are enhanced only at the relatively high refrigerant flow rate, and that the pressure drop is enlarged at all the refrigerant flow rates. Thus, more efforts to reduce the pressure drop and enhance the heat transfer coefficient at low refrigerant flow rate will be required to improve the evaporator performance in near future.

Condenser performances are compared between the herringbone tube and the existing inner grooved tube in Figure 19. The herringbone tube has improved the condenser performance about 3 to 5% compared to the inner grooved tube. It is noted that the air-side heat transfer is dominant in the air-cooled heat exchangers installed in air-conditioning machines. Hence, the significant heat transfer enhancement of the herringbone tube contributes a little to the improvement of the heat exchanger performance. However, even this small improvement results in a large energy conservation, because the total sales of residential and commercial air-conditioning machines is about 8 million in Japan alone.

4. Concluding Remarks

The new advanced heat transfer tube called, "Herringbone Heat Transfer Tube", having a unique inner surface geometry, has been experimentally investigated, with the aim to enhance the heat transfer for R407C and improve the heat exchanger performance using R407C. The local heat transfer coefficient and pressure drop for R407C flowing inside the horizontal herringbone tube have been measured, and the data have been compared with those for the existing inner-grooved tube. Additionally, the heat transfer enhancement mechanism for the herringbone tube has been proposed by the authors, and verified by the experiments. The heat exchanger performances for the herringbone tube and the inner-grooved tube have been compared for the energy conservation evaluation.

As the results of the experiments, the evaporation heat transfer coefficient for the herringbone tube is about 90% higher than those for the inner-grooved tube at high mass flux, while it provides a small positive effect on the evaporation performance under $G=150\text{kg/m}^2\text{s}$.   For condensation performance, the herringbone tube achieves about 100 to 200% higher heat transfer coefficient than the inner grooved tube for entire range of refrigerant mass fluxes.   However, the significant increase in heat transfer is accomplished with about 30 to 60% increase in pressure drop.

Although the circumferential local heat transfer data suggest that the heat transfer mechanism proposed by the authors is qualitatively verified, the quantitative verification of the heat transfer enhancement for the herringbone tube is not sufficiently clarified.   The practical comparison of the heat exchanger performance shows that the evaporator performance is improved only at relatively high refrigerant mass flux, while the condenser using the herringbone tube seems to be better than that using the inner-grooved tube.   In this study, the herringbone shape has been proven to be superior to the existing inner-grooved tube.   Future works should be done to optimize the herringbone shape to modify the heat transfer performance, and also to compare the performance with that of other inner-grooved tubes for the verification of the advantage of the herringbone tube.

Nomenclature

D	= diameter	(m)
G	= refrigerant mass flux	$(\text{kg/m}^2\text{s})$
q	= heat flux	(W/m^2)
T	= temperature	(K)
u	= air velocity based on frontal area	(m/s)
x	= local vapor quality	(-)

Greek Symbols

α	= heat transfer coefficient	$(\text{W/m}^2\text{K})$
ΔP	= pressure drop	(kPa)
Δx	= change in vapor quality	(-)

Subscripts

ave	= average
local	= local position on the tube surface
r	= refrigerant
w	= tube wall

600

References

1. Ebisu, T., Torikoshi, K. and Kawabata K. (1992) Heat transfer and pressure drop characteristics of HFC-134a in a horizontal heat transfer tube, *Proceedings of 1992 International Refrigerant Conference at Purdue*, **1**, 167-176.
2. Ebisu, T. and Torikoshi, K. (1993) Heat transfer and pressure drop characteristics of R134a, R32 and a mixture of R32/R134a inside a horizontal tube, *ASHRAE Transactions*, **99**-2, 90-96.
3. Ebisu, T. and Torikoshi,K., (1994) In-tube heat transfer characteristics of refrigerant mixtures of HFC-32/134a and HFC-32/125/134a, *Proceedings of 1994 International Refrigeration Conference at Purdue*, 293-298.
4. Ebisu, T. and Torikoshi,K., (1995) Experimental studies on crossflow heat exchanger performance using non-azeotropic refrigerant mixture, *Proceedings of 19th International Congress of Refrigeration*, **4a**, 163-170.
5. Sundarsan, S.G., Pate, M.B. and Doerr, T.M. (1994) A comparison of the effects of different lubricants on the in-tube evaporation of an HFC- blend refrigerant, *1994 International Refrigeration Conference at Purdue*, 323-328.
6. Uchida, M., Itoh, M. Shikazono, N. and Kudoh. M. (1996) Experimental study on the heat transfer performance of a zeotropic refrigerant mixture in horizontal tubes, *Proceedings of 1996 International Refrigeration Conference at Purdue*, 133-138.

DEVELOPMENT OF NEW CONCEPT AIR-COOLED HEAT EXCHANGER FOR ENERGY CONSERVATION OF AIR-CONDITIONING MACHINE

T. EBISU

Mechanical Engineering Laboratory, Daikin Industries, Ltd.
1304 Kanaoka-cho, Sakai, Osaka 591-8511, Japan

Abstract. A new concept air-cooled heat exchanger using mesh fins and tiny tubes, "mesh-finned heat exchanger", has been proposed with the aim of enhancing the air-side heat transfer. The basic characteristics of the air-side heat transfer and pressure drop characteristics for the mesh finned heat exchanger have been examined experimentally, and the information required to design the practical heat exchanger has been additionally investigated, with the variation of the fin and tube arrays. The comparisons of the heat exchanger performance between the proposed heat exchangers and the existing one show that the former are superior to the latter by approximately twice. Therefore, this new concept heat exchanger can help energy conservation through heat transfer enhancement.

1. Introduction

Requirements of enhancing energy efficiency and energy conservation continue to receive considerable attention in the design of more efficient air-conditioning units for residential and commercial use, because almost 30% of the electrical demand in Japan is spent for air-conditioning use. Moreover, the amendment to UNFCCC-COP3, which was held at Kyoto, Japan, in December 1997, has enforced energy conservation in air-conditioning equipment to reduce CO_2 emissions. In order to meet the requirement for developing more efficient air-conditioning units, improvements in the performance of air-cooled heat exchangers, which are one of the major components of air-conditioning units, are subjects of fundamental importance.

Up until now, considerable research and development relating to improvements of the air-cooled heat exchangers used in air-conditioning units have been conducted to enhance the air-side heat transfer coefficients, because heat transfer coefficients for the air side dominate the performance of air-cooled

S. Kakaç et al. (eds.), Heat Transfer Enhancement of Heat Exchangers, 601–620.

heat exchangers. The major technique that has been used practically to enhance the air-side heat transfer is fin surface interruption, for example, by using offset-strip or louvered fins. These fins provide enhancement of the average heat transfer of the entire fin surface due to repeated thermal boundary layer interruption and wake destruction. Consequently, the air-cooled heat exchangers employing these high performance fins contribute greatly to the volume reduction and energy conservation of air-conditioning units. With the aim of developing more efficient heat exchangers, when compared to the existing ones, much research is directed at improving the offset-strip fins and louvered fins: Joshi and Webb (1987), Kurosaki et al. (1988), Lee (1986), Nakayama and Xu (1983) and more.

Another possibility of further improving the air-side heat transfer coefficients is to control the velocity and thermal boundary layers occurring over each fin segment, and also yield turbulence and vortex in the air passage. In the present study, a new heat exchanger based on the above-mentioned concept has been developed in response to the requirement for further improvement of the air-side heat transfer. In order to fulfill this objective, a wire mesh and a small diameter heat transfer tube have been employed as the fin and tube, respectively. We call this new concept heat exchanger "mesh-finned heat exchanger".

This paper describes the basic information on the heat transfer and flow friction characteristics of the mesh-finned heat exchanger. At first, the geometrical parameters of the mesh finned heat exchanger element have been examined to provide the basic information on the heat transfer and flow friction characteristics. Then, an additional evaluation of the practical fin and tube arrangement has been performed to optimize the geometries of the fin array and the tube arrangement, and to enhance the performance of the mesh-finned heat exchanger. Furthermore, the overall performance of the mesh-finned heat exchanger has been compared with the conventional fin-and-tube heat exchanger to clarify the advantage of the proposed mesh-finned heat exchanger.

2. Experiments

2.1. HEAT EXCHANGER CONFIGURATION

The configuration of the mesh-finned heat exchanger developed in the present study is shown in Figure 1. In order to prevent fully developed flow and thereby to take advantage of enhanced heat transfer by the use of thermal leading edge effect, a wire mesh fin having a lattice shape is proposed as a fin. The detailed fin shape is displayed in the figure. Since each layer of the fin is fabricated by cutting and thereafter expanding a thin copper sheet, the wire mesh fin has a rhombic open area. With respect to the heat transfer tube, small diameter copper tubes are employed to reduce the air-side pressure drop of the air flowing through the element. In addition, the heat transfer tubes have the micro fins in their insides to enhance the heat transfer performance of the refrigerant flowing inside the tube.

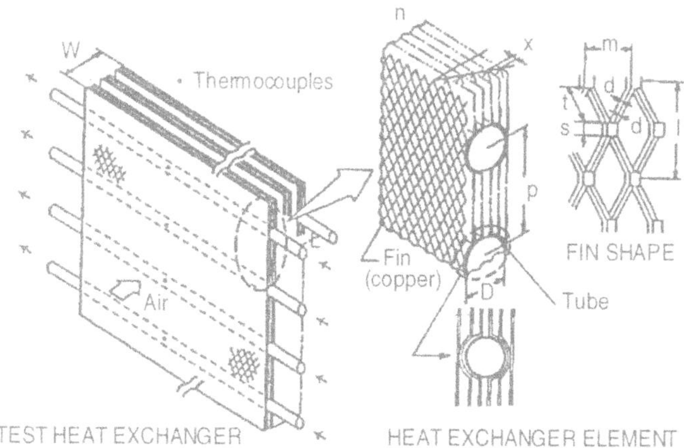

Figure 1. Test heat exchanger configuration, heat exchanger element and fin shape

Each fin is piled up in the streamwise direction, and heat transfer tubes are firmly sandwiched and soldered in parallel with multi-layers of the fins. Thus, each element of the mesh finned heat exchanger is made up of a row of the copper tubes and multi-layers of the fins. For initial experiments, the geometrical parameters (see Table 1) of this element are examined to provide the basic information on the heat transfer and flow friction characteristics of the mesh finned heat exchanger.

The tested heat exchanger consists of three or four piles of the above elements spaced at a given interval. Iinformation on the optimization of the fin array and tube displacement are required for the designer of heat exchangers, because there are possibilities to enhance the heat exchanger performance by the optimization of the fin array and the tube arrangement. The next examination in this study obtains the practical heat transfer and flow friction characteristics for parametric variations of the fin array and the tube displacement. The geometrical parameters are listed in Table 2.

2.2. EXPERIMENTAL APPARATUS AND PROCEDURE

A schematic diagram of the experimental apparatus is shown in Figure 2. It

TABLE 1.   Geometrical parameters for first examination

NO.	Symbol	d mm	l mm	m mm	D mm	p mm	n -	L -	W mm	s mm	t mm
1	○	0.2	2.0	1.1	3.2	12	2	4	20	0.60	0.65
2	◑	0.2	3.0	1.5	3.2	12	2	4	20	0.62	1.08
3	◑	0.2	5.0	2.0	3.2	12	2	4	20	0.77	1.95
4	⊖	0.2	5.0	2.0	3.2	16	4	4	20	0.77	1.95
5	⊖	0.2	5.0	2.0	3.2	20	4	4	20	0.77	1.95
6	⊕	0.2	5.0	2.0	3.2	20	4	3	15	0.77	1.95
7	⊖	0.2	5.0	2.3	2.0	20	6	3	19	0.68	2.09
8	●	0.2	5.0	2.3	2.0	24	6	3	19	0.68	2.09

TABLE 2.   Geometrical parameters for second examination

Symbol	d mm	l mm	m mm	n	x mm	Do mm	p mm	y mm	N
●	0.20	5.0	2.3	6	0.0	4.0	20.0	0.0	3
◑	0.20	5.0	2.3	6	0.19	4.0	20.0	0.0	3
◑	0.20	5.0	2.3	6	0.58	4.0	20.0	0.0	3
○	0.20	5.0	2.3	6	1.15	4.0	20.0	0.0	3
⊖	0.20	5.0	2.3	6	1.15	4.0	20.0	2.0	3
⊖	0.20	5.0	2.3	6	1.15	4.0	20.0	5.0	3
⊖	0.20	5.0	2.3	6	1.15	4.0	20.0	10.0	3
⊕	0.15	6.0	2.7	10	1.35	4.0	15.0	6.0	3

includes a wind tunnel having an 80 x 200mm rectangular cross section (height by width), flow control valves, a flow metering section, a blower and a hot water (instead of refrigerant) supply system.

The wind tunnel consists of a test section and a contraction section to homogenize the airflow. The test heat exchanger is mounted in the middle portion of the test section as shown in the figure. Pressure taps are located upstream and downstream of the test heat exchanger to determine the friction loss across it. Air from a temperature-controlled room is drawn through the contraction section into the wind tunnel by the blower, and a calibrated orifice flow meter is used in order to measure air flow rates passing through the heat exchanger. The average air temperatures entering and leaving the heat exchanger are measured by copper-constantan thermocouple grids installed at the inlet and outlet of the test section. The test section and the airflow duct are surrounded by a styrol insulation and multi-layers of glass wool insulation to minimize heat losses.

The hot water supply system consists of a constant temperature bath, a flow control valve and a flow rate meter. Hot water flows into each tube of the test heat exchanger through the flow meter so that the uniform temperature on the surface of the test heat exchanger can be achieved by water heating.

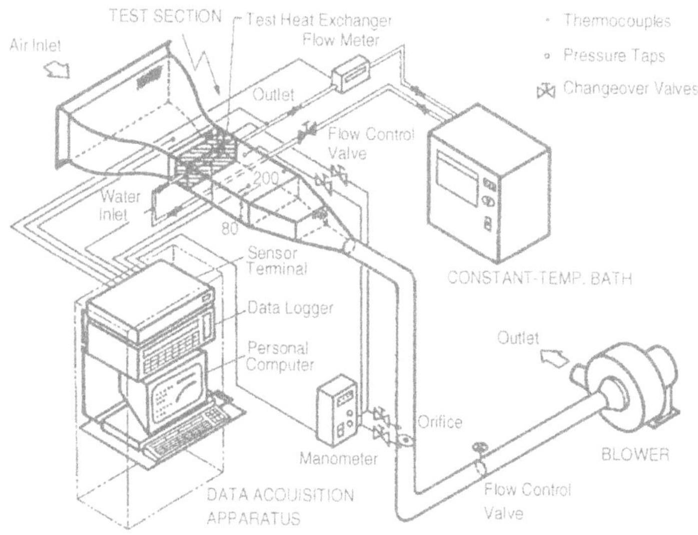

Figure 2. Schematic diagram of the experimental apparatus

The heat exchanger is affixed to both side walls of the test section to ensure proper positioning of each element. Two thermocouples, 180 deg. apart, are welded locations of mid-way along the length of the tube to measure the surface temperatures of each tube.

For all of the test runs, the flow rate of the hot water entering each tube is controlled to maintain the desired surface temperatures of all the tubes. After all of the temperatures are monitored to assure a steady-state condition, experimental data such as the air temperature downstream of the test heat exchanger, the surface temperatures of the tubes and hot water temperatures entering and leaving the tubes are recorded by a data logger.

3. Characteristics of Mesh-Finned Heat Exchanger

3.1. BASIC CHARACTERISTICS OF MESH FIN ELEMENT

In this section, the experimental results for the mesh-finned heat exchanger listed in Table 1 are described in order to obtain the basic characteristics of heat transfer and flow friction.

3.1.1. *Averaged Air-Side Heat Transfer Characteristics*
The experimental results of the average heat transfer coefficients are plotted

606

in Figure 3 as a function of air velocity, U_{max}, which corresponds to the air velocity passing through the open areas of the fin. h_o, the average air-side heat transfer coefficient of the mesh finned heat exchanger, is determined from Equation (1):

$$ho = \frac{Q}{A_t \cdot \Delta T_m}$$
(1)

where, Q and A_t are the heat flow rate of the air-side and the total surface area, including both the tube wall and the fin surface, respectively. ΔT_m is the logarithmic mean temperature difference based on the measured inlet and outlet air and tube wall temperatures.

The figure reveals a similar trend in the results of the ho-U_{max} relationship for all of the data, although there are noteworthy differences in the values of the ho among the geometrical parameters. That is, all of the average heat transfer coefficients increase with an increase in air velocity, while the geometrical

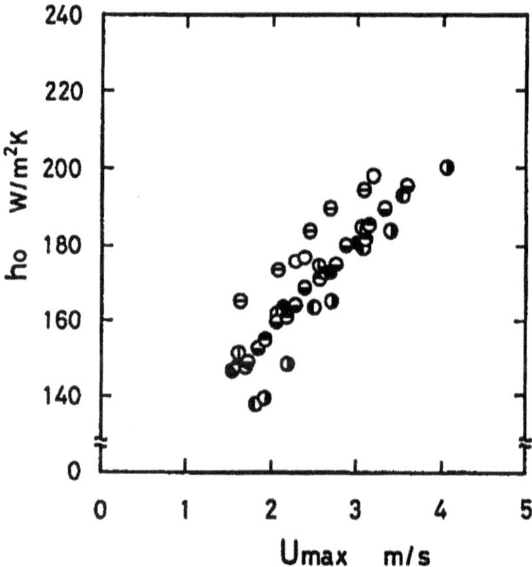

Figure 3. Average heat transfer coefficients of test heat exchangers

differences (fin size, tube pitch) play a role in the variations of the air-side heat transfer coefficients. In addition , the heat transfer coefficients of the mesh finned heat exchanger depend on the tube pitch in the element, the surface area of the fin and total surface area of the heat exchanger.

3.1.2. *Heat Transfer Characteristics of the Tube Bundle and Mesh Fins*

In order to clarify the basic heat transfer characteristics of the mesh finned heat exchanger, the heat transfer coefficients for the tube bundle and the mesh fins have been investigated individually.

In-Line Tube Bundle Heat Transfer. To provide quantitative information about the heat transfer coefficients for a bundle of the tubes, additional experiments have been performed with an in-line array of tubes without fins, but having the same diameter as the heat transfer tube employed in the test heat exchangers. For this experiment, a small size cartridge heater has been inserted inside each tube to serve a heat source, and these tubes have been arranged in an array of four rows, with four tubes per row arranged in the spanwise direction (4 × 4 arrangement). For all of the heat transfer measurement runs, the surface temperature of each tube maintained at a constant value by regulating the electric power to each cartridge heater.

Experimental data on heat transfer coefficients for a bundle of the tubes are shown in Figure 4, in the form Nu_D versus Re_D. The broken line in the figure is the curve from the empirical correlation for a single tube (Zukauskas, 1972). The coordinates appearing in the figure are defined as follows:

$$Nu_D = \frac{h_D D}{\lambda_a} \tag{2}$$

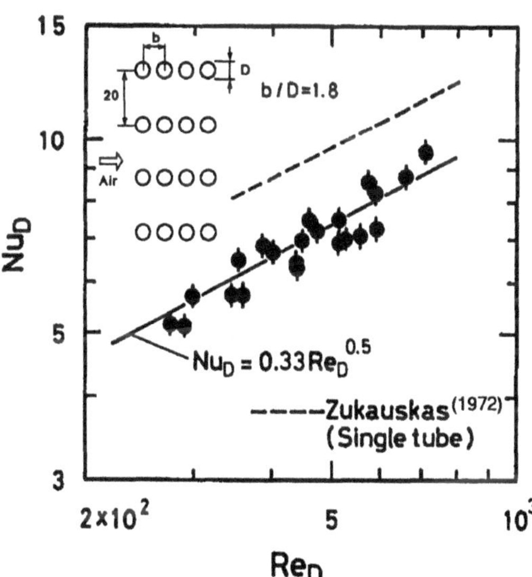

Figure 4. Average heat transfer coefficients of in-line tubes

$$Re_D = \frac{U_{max}\,D}{v_a} \tag{3}$$

h_D in Equation (2) is calculated using the mean heat flux and the logarithmic mean temperature difference based on the measured inlet and outlet air and tube wall temperatures .

The comparisons between the results for a bundle of the tubes and those for a single tube show that the in-line tube bundle provide about 30% lower Nu_D than the single tube. The reason that the heat transfer coefficients for the in-line tube bundle are less than those for a single tube is believed due to the flow bypass effect through the interval of tubes. A simple least squares fit of the present data has yielded the following empirical correlation of the heat transfer coefficients for a bundle of tubes,

$$Nu_D = 0.33\,Re_D^{\,0.5} \tag{4}$$

Mesh fin heat transfer. A common macroscopic analysis of the average air-side heat transfer coefficient for the entire heat transfer surface is in the form of separate contributions of the fin heat transfer coefficient and the tube heat transfer coefficient as:

$$ho = \left(1 - \frac{A_f}{A_t}\right)h_D + \eta\left(\frac{A_f}{A_t}\right)h_f \tag{5}$$

in which h_f is the heat transfer coefficient of fins, A_f/A_t is the ratio of the fin surface area to the total heat transfer surface area and η is the fin efficiency.

The fin efficiency is determined from the conventional model as

$$\eta = \frac{\tanh(M \cdot H)}{M \cdot H}$$

$$M = \left(\frac{4h_f}{d \cdot \lambda_f}\right)^{1/2} \quad , \quad H = \frac{(p - 0.87D)(s + t)}{l} \tag{6}$$

The h_f value is calculated from equations (4), (5) and (6) by iteration. The heat transfer coefficients thus determined, which are recast in dimensionless form as Nu_{df} ($=h_f{\cdot}4d/\lambda_a$), are plotted in Figure 5. Superimposed is the result (chained line) of experiments performed with woven-screens of circular wires by Hamaguchi et al. (1983). The present results of the heat transfer coefficients for

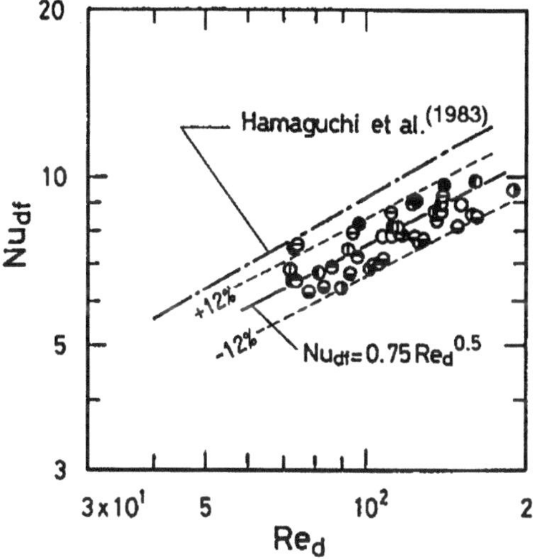

Figure 5.   Average Nusselt numbers of mesh fin

the mesh fin are correlated by the following equation:

$$Nu_{df} = 0.75 \, Re_d^{0.5} \tag{7}$$

It is possible to correlate the dependence of the Nu_{df} on Re_d within $\pm$ 12% by Equation (7).   A comparison of the heat transfer coefficients between the present study and Hamaguchi's correlation (shown by the chained line) shows that the multi-layers of the woven-wire screens give 20% more heat transfer relative to the present correlation (shown by the solid line).   One reason which affects the above comparison may be attributed to differences in fin geometries, and another is the wire screen stacking conditions.   The correlation curve in the previous study was obtained from experimental data for randomly stacked woven wire screens, while the data in the present study have been for a regular fin arrangement. It is reasonable to expect that a regular fin arrangement is less affected by enhancement due to wake turbulence and mixing than a random arrangement.

To demmonstrate the above consideration qualitatively, visualization by laser holographic interferometry has been conducted.   Figure 6 shows the experimental results of the thermal wake among the fin arrays.   The values of x/m are the experimental parameters, where x/m indicates the offset ratio of the each fin deviation to the regular portion.   The figure reveals that the more thermal wake occurs for the methodical fin array, $x/m=0$, compared to the other arrays.   This thermal wake indicates that the heat transfer for the regular array is less augmented

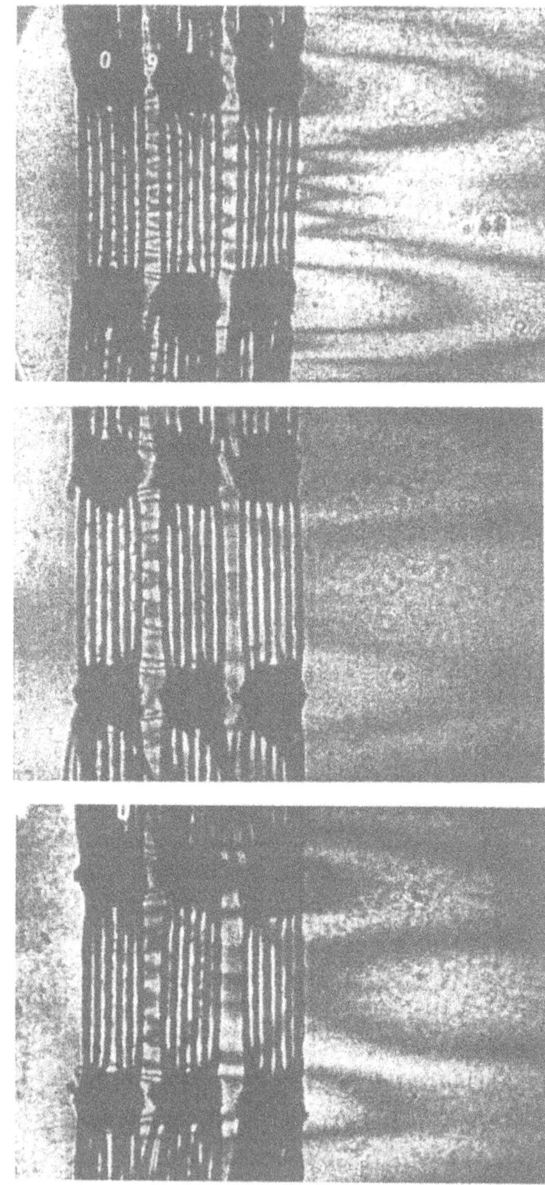

Figure6. Visualization of thermal wake by laser holographic interferometry with variation of fin deviation x/m (upper:x/m=0, middle:x/m=1/4, lower:x/m=1/2)

by the wake turbulence and mixing than those for the random arrays, due to the bypass flow of air flowing through each lattice. The quantitative comparison of the heat transfer performance among the fin arrays will be discussed in the next examination.

3.1.3. *Pressure Drop Characteristics*

The pressure drop results for the per-fin number of all the heat exchangers are plotted in Figure 7 as a function of U_{max} . The values of $(\Delta P / n \cdot L)$ are significantly affected by the fin surface area , rather than the tube pitch in the element. The results for $(\Delta P / n \cdot L)$ of each test heat exchanger are proportional to $U_{max}^{1.7}$.

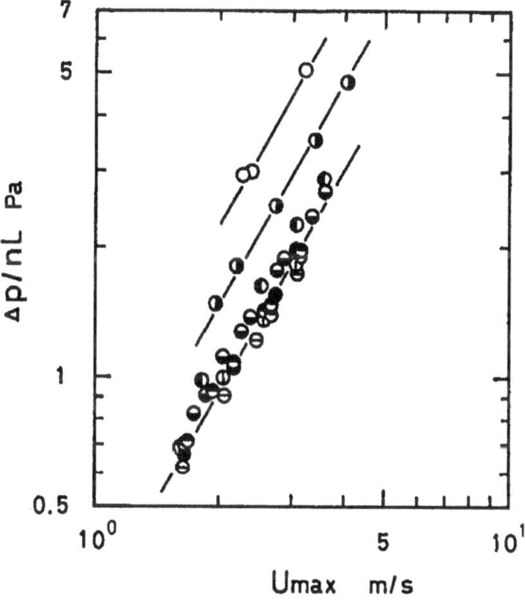

Figure 7. Pressure drop across test heat exchangers

3.2. EFFECT OF FIN ARRAY AND TUBE DISPLACEMENT

Here, the practical information of the mesh finned heat exchanger in Table 2 is introduced, in addition to the basic characteristics mentioned in the previous section. Examination of the effects of the fin array and the tube arrangement on

612

the heat transfer and flow friction performance gives very important information to the heat exchanger designers.

3.2.1. *Effects of Fin Array*

Here, the effects of the fin array offset from the regular in-line array on the heat transfer and pressure drop are investigated. The fin array offset is defined as the ratio of the each fin deviation from the regular position, x/m, displayed in Figure 1.

The experimental results of the average heat transfer coefficients for the present mesh finned heat exchangers are plotted in Figure 8 as a function of air velocity passing through the open area of the mesh fin. The curves are parametrized by the ratio x/m.

This figure reveals that the fin array has a decisive effect on the heat transfer, but the extent of the effect decreases with increasing x/m. For $x/m=1/2$, the heat transfer enhancement due to the fin displacement is about 134%, when compared to the results for the regular arrangement, $x/m=0$. However, there is no apparent difference between the results for $x/m=1/2$ and 1/4. The reason that the heat transfer for the regular in-line fin array ($x/m=0$, black circular data symbol) is smaller than those for the other arrays is believed to be related to the bypass flow through the lattice. This speculation is supported by the visualization results shown in Figure 6.

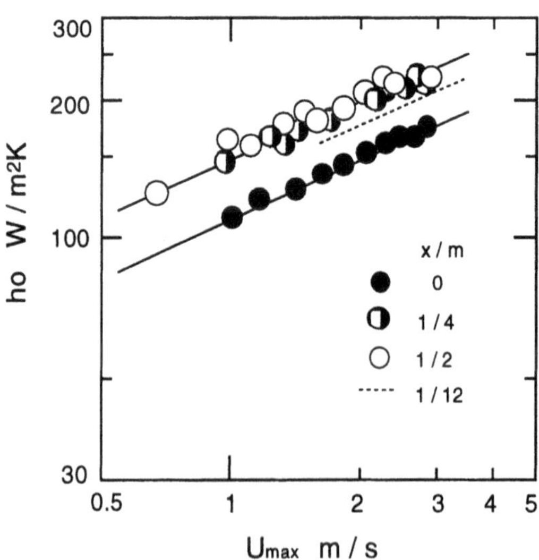

Figure 8.   Heat transfer coefficients for different fin arrays

The results for the per-number fin pressure drop for three different fin arrays are shown in Figure 9. The slope of the curve of $\Delta P / n{\cdot}N$ for $x/m{=}1/2$ and 1/4 is similar to that for $x/m{=}0$, while $\Delta P / n{\cdot}N$ for $x/m{=}1/2$ and 1/4 is about 40% larger than that for $x/m{=}0$.

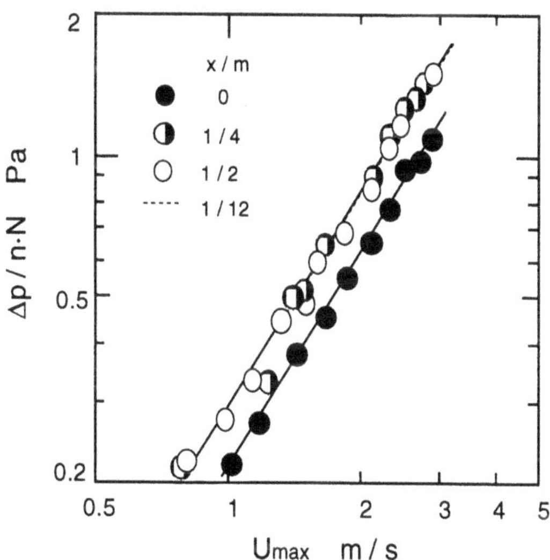

Figure 9. Pressure drops for three different fin arrays

3.2.2. *Effects of Tube Arrangement*

In the basic examination of the mesh-finned heat exchanger, the data have been obtained for an in-line tube array. However, there are possibilities of heat transfer enhancement due to the variation of the tube arrangements. The objective of this examination is demonstrate the effect of the tube arrangement on the heat transfer and flow friction. The tube arrangement is defined as y/p, which corresponds to the ratio of the tube deviation from the in-line position.

The heat transfer and pressure drop data for four different tube arrays are shown in Figure 10 and Figure 11, respectively. The data are plotted ho/ho_0 and $\Delta P/\Delta P_0$ versus y/p, where ho_0 and ΔP_0 correspond to the heat transfer coefficient and the pressure drop for the in-line tube array. These figures verify that both ho/ho_0 and $\Delta P/\Delta P_0$ are strongly affected by y/p, and the value of y/p where the heat transfer coefficient assumes the highest value appears in the range from 0.25 to 0.35. However, the pressure drop is also the largest in the same range of y/p.

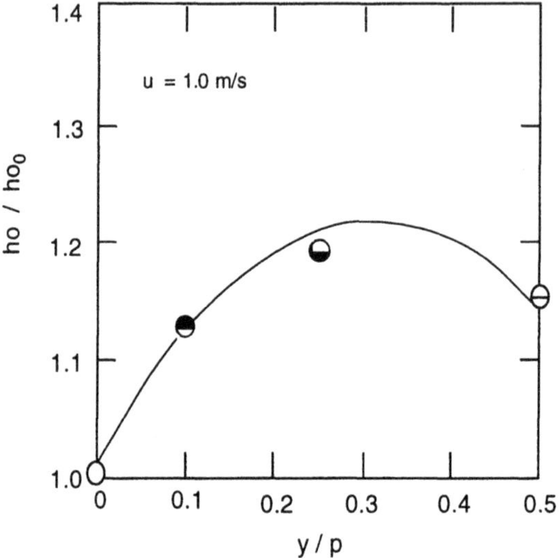

Figure 10. Heat transfer coefficients for four tube arrangements

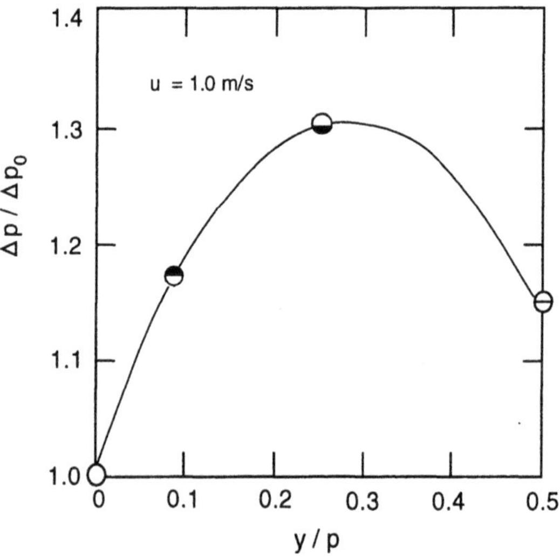

Figure 11. Pressure drops for four tube arrangements

To provide further perspective for the interesting effect of tube arrangement on the heat transfer, visual observation experiments were performed using the hydrogen bubble technique to obtain the flow patterns around the tubes.

Figure 12 shows the visualization results for four different tube arrangements. For $y/p \leqq 0.1$, the tubes situated at the second and subsequent rows in the tube bundle are surrounded by the separation flow taking place from the tubes upstream, while the tubes subsequent to the first row for $y/p=0.25$ and 0.5 (staggered tube array) tend to be insensitive to the separation flow from the first row. However, the tubes of the third row for $y/p=0.5$ are influenced by the wake of the first row. The significant difference in the flow patterns suggests that the different flow patterns caused by the different tube arrangements may give rise to the variation of the heat transfer results.

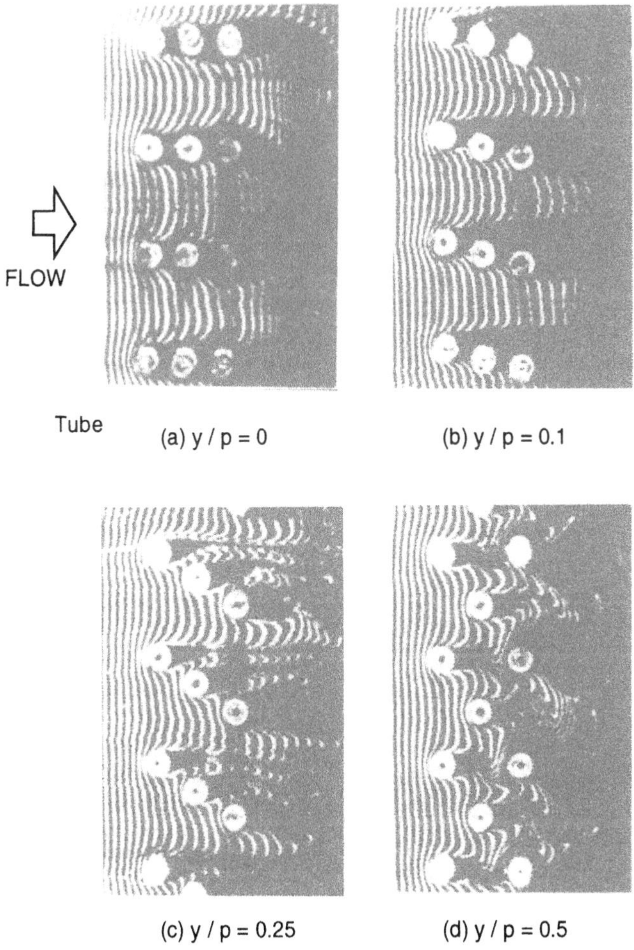

Figure 12. Flow patterns for different tube arrangements

Figure 13 shows the visualization results to examine the effects of the tube arrangements on the thermal wake occurring in the mesh fin elements. Visualization has been accomplished by laser holographic interferometry. In both figures, the thermal wakes from the heat transfer tubes can be seen. For $y/p=0.25$, the thermal wakes from the heat transfer tube have less influence on the subsequent rows, while the thermal wakes from the first row affect directly the third row for $y/p=0.5$. The heat transfer degradation for $y/p=0.5$ rather than $y/p=0.25$ shown in Figure 10 might be attributed to the effects of the thermal wakes on the subsequent rows.

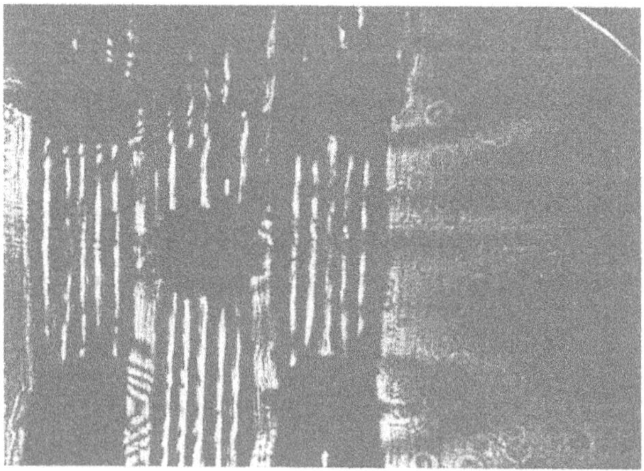

Figure 13. Thermal wake visualized by laser holographic interferometry
(upper : y/p = 0.25 , lower : y/p=0.5)

3.3. COMPARISON OF PERFORMANCE BETWEEN MESH FIN AND FIN-TUBE HEAT EXCHANGERS

In order to provide a reasonable guide for designing a more efficient heat exchanger, it is important to evaluate the performance of the proposed heat exchanger compared to the existing one. In the present study, the heat exchange quality per unit volume, E, and the pumping power per unit volume, P, presented by Soland et al. (1978) have been employed to fulfill the performance evaluation. These parameters are given by

$$E = \frac{ho \cdot Ao}{V} \quad , \quad P = \frac{Ga \cdot \Delta P}{V} \tag{8}$$

in which Ga is the air flow rate passing through the heat exchanger, and V is the heat exchanger volume.

The performance results based on such performance evaluation method are plotted in Figure 14. Data for the mesh-finned heat exchanger listed in Table 1 are indicated by the smaller circles, and data for Table 2 correspond to larger circles. The figure also includes data for both an existing heat exchanger and a plain-fin heat exchanger. The existing heat exchanger consists of louver fins and tubes having 9.5mm outside diameter equilaterally spaced on 25.4mm centers, while the plain fin tube heat exchanger consists of plain fins having the same fin thickness and tube diameter as those of the existing heat exchanger.

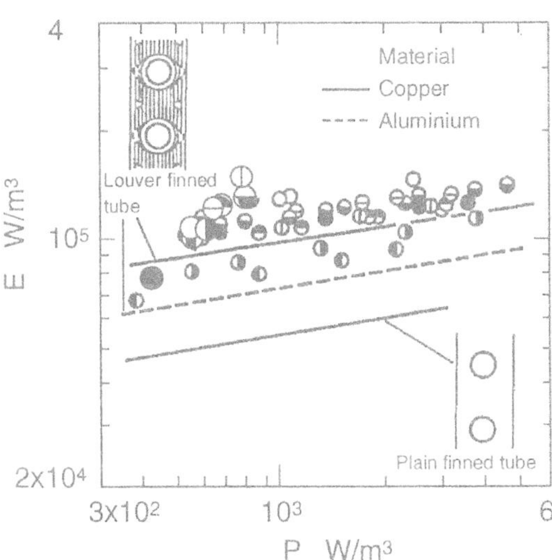

Figure 14. Performance data of test heat exchangers

A considerable difference in performance for E with P appears in the different geometries of the test heat exchangers. From the experimental findings, the highest E value in Table 1 is No. 1 and the value for No.1 is about 1.5 times higher than that for No. 3 at the same P value. In addition, the highest value of E in the list of Table 2 becomes about 30% superior to that of the highest in Table 1. The advanced heat exchanger performance in Table 2 is attributed to the modification of the fin array and the tube arrangement. Comparisons of the performance of the mesh-finned heat exchanger and the existing louvered heat exchanger show that the former is approximately double than the latter at the same P value. This finding corresponds to the significant energy conservation and also about 30% volumetric reduction of the actual heat exchanger.

4. Concluding Remarks

A new concept air-cooled heat exchanger using mesh fins and tiny tubes has been proposed with the aim of enhancing the air-side heat transfer. The basic characteristics of the air-side heat transfer and pressure drop characteristics for the mesh-finned heat exchanger have been examined experimentally, and the empirical correlations of the heat transfer coefficients for a tube bundle and mesh fins have been obtained individually.

In addition, the information required to design a practical heat exchanger has been obtained with variation of fin arrays and tube arrangements. The findings of the practical investigation suggest that the performance of the mesh-finned heat exchanger can be enhanced by optimization of the fin array and the tube arrangement.

The comparisons of the heat exchanger performance between the proposed heat exchangers and the existing ones show that the former are superior to the latter by about a factor of two. Therefore, this new concept heat exchanger, "mesh finned heat exchanger" can help energy conservation through heat transfer enhancement.

5. Nomenclature

A_f fin surface area, m^2
A_t total heat transfer surface area, m^2
a displacement in fin arrangement, mm
D diameter of tube, mm
d side length of fin, mm
E heat transfer quality per unit heat exchanger volume, W/m^3

G air flow rate, m^3/s

ho average air-side heat transfer coefficient, W/m^2K

ho_0 average air-side heat transfer coefficient for in-line tube arrangement, W/m^2K

h_D heat transfer coefficient of a tube bundle, W/m^2K

hf heat transfer coefficient of fin, W/m^2K

L number of elements

l height of lattice in mesh fin, mm

M parameter defined in Equation (6)

m width of lattice in mesh fin, mm

N number of element

Nu_D average Nusselt number of in-line tubes

Nu_{df} average Nusselt number of fins

n number of fins

P fluid pumping power per unit heat exchanger volume, W/m^3

p pitch between the tubes in an element, mm

ΔP pressure drop, Pa

ΔP_0 pressure drop for in-line tube arrangement, Pa

Q heat flow rate, W

Re_D Reynolds number based on tube diameter

Re_d Reynolds number based on a four times the side length of a fin

s side length of the crossing portion in a mesh fin, mm

t side length of lattice in a mesh fin, mm

ΔTm logarithmic mean temperature difference, K

u air velocity, m/s

U_{max} air velocity through the minimum flow passage, m/s

V volume of the test heat exchanger, m^3

W entire width of heat exchanger, mm

x deviation of fin from regular position

y deviation of tube from in-line position

Greeks

η fin efficiency

λ_a thermal conductivity of air, W/m K

λ_f thermal conductivity of fin, W/m K

ν_a kinematic viscosity of air, m^2/s

REFERENCES

1. Hamaguchi, K., Takahashi, S., and Miyabe, H. (1983) Heat Transfer Characteristics of Regenerator Matrix (Case of Packed Wire Gauzes), *Trans.*

JSME, Vol.49B, No.445, pp.2001-2009.

2. Joshi, H. M., and Webb, R. L. (1987) Heat Transfer and Friction in the Offset Strip-Fin Heat Exchanger, *Int. J. Heat Mass Transfer*, Vol.30, No.1, pp.69-84.

3. Kurosaki, Y., Kashiwagi, T., Kobayashi, H., Uzuhashi, H., and Tang, S.C. (1988) Experimental Study on Heat Transfer from Parallel Louvered Fins by Laser Holographic Interferometry, *Experimental Thermal and Fluid Science*, Vol.1, No.1, pp.59-67.

4. Lee, Y.N. (1986) Heat Transfer and Pressure Drop Characteristics of an Array of Plates Aligned at Angles to the Flow in a Rectangular Duct, *Int. J. Heat Mass Transfer*, Vol.29, No.10, pp.1553-1563.

5. Nakayama, W. and Xu, L. P. (1983) Enhanced Fins for Air-Cooled Heat Exchangers Heat Transfer and Friction Factor Correlations, Proc. *ASME-JSME Thermal Eng. Joint Conf., Honolulu*, Vol. 1, pp. 495-502.

6. Soland, J. G., Mack, Jr., and Rosenow, W. M. (1978) Performance Ranking of Plate-Fin Heat Exchanger Surfaces, *J. Heat Transfer*, Vol. 100, pp. 514-519.

7. Zukauskas, A. (1972) Heat Transfer from Tubes in Crossflow, *Advances in Heat Transfer*, Vol.8, Academic Press, New York, pp.93-160.

MULTI-HOLE COOLING EFFECTIVENESS ON COMBUSTION CHAMBER WALLS

B. LEGER and P. ANDRE
Laboratoire Aquitain de Recherche en Aérothermique (LARA)
Université de Pau et des Pays de l'Adour
c/o Turboméca
64511 Bordes cedex (France)

Abstract. Because turbo engines need to increase the combustion temperature to improve their global efficiency, there have been many different devices used to protect combustion chamber walls. Multihole cooling is one of them and it has been used for a long time but the thermal mechanism needs to be studied further. A research program has been carried out and is presented here. A special test bench has been built and the temperature measurement is made using an Infra-Red camera. The main results are detailed in terms of adiabatic effectiveness and in heat transfer coefficient. The dynamic effects are also analysed to help us to understand what happens in such a difficult heat transfer situation.

1. Introduction

During the present and past decades, aircraft or helicopter engine manufacturers have made many efforts to lower fuel consumption. This can be improved by the use of higher pressures and temperatures inside combustion chambers. Therefore, more efficient cooling techniques or devices for the chamber wall are needed to maintain its thermomechanical properties for a longer life time. Even though wall protection by using a reinforced insulation barrier, such as a ceramic wall coating is now widely used, it remains necessary to cool the wall by injecting a fresh air flows taken from the compressor outlet. Injection of cooling air and then wall heat transfer have been studied industrially. Among them, the emphasis is placed, in the present paper, upon a transpiration scheme often called "multihole cooling". In this technique, cooling air is admitted inside the combustion chamber through rows of discrete holes in the chamber wall. Whatever the cooling device is, the quantity of injected fresh air must be carefully controlled in order to avoid chemical pollutant reactions and thereby violate engine emission regulations. On the other hand, experimental results obtained in realistic

621

S. Kakaç et al. (eds.), Heat Transfer Enhancement of Heat Exchangers, 621–639.

environments are urgently needed by engine designers to check design computer codes. Surprisingly, little literature is available on experimental data for multiholed wall cooling with respect to thermal, dynamical and geometrical parameters close to those found in manufacturer engines. Most of these results are obtained with low temperature gas flows (100 to 250°C) which is obviously far away from the 1400K reached in common combustion chambers. [1,2,3]

In this paper, cooling devices for the combustion chamber wall are briefly presented and discussed. In order to demonstrate the effectiveness of multihole cooling, an experimental test bench working with realistic conditions of combustion pressure, temperature of hot gases and geometry of holes has been developed. Fuel combustion occurs in a tubular chamber and hot gases flow along a multiholed plate representing the wall chamber. Cool air tangentially flows on the opposite face of the wall, and is admitted on the hot side through inclined perforations. The two flows, i.e., incoming cooling air through the first row of holes and hot gases, mix together along the plate. The mixing induces a thermal protection for fresh air jets emitted from the downstream rows. It then results in the effect called film cooling which extends along the plate in the flow direction. Measurement methods such as I.R. thermography and laser doppler anemometry are also presented. Then, heat transfer modelling of the multiholed plate is discussed. By combining original experimental results and model, heat transfer coefficient distribution are obtained. These data are correlated to the main features of the dynamical behaviour of i) cooling air flows ii) the mixing zone between air and hot gases and iii) the cooling film. A discussion is finally engaged on heat transfer improvement on combustion chamber walls by multihole cooling.

2. Techniques for combustion wall cooling

2.1. PURE FILM COOLING

The simplest way for cooling a solid wall subjected on one side to heat fluxes from a hot gas stream is to remove heat by a flowing coolant on the opposite side of the wall (scheme 1, fig.1). In that case, no mass transfer and more particularly no rate of coolant is transferred from one side to the other one and heat transfer process remains purely convective on both side (with respect to conduction within the wall thickness). The earliest devices designed to improve purely convective situation used a certain amount of coolant flowing in a slot or a hole through the wall thickness (scheme 2, fig.1). This flow is ejected on the hot side and a mixing zone of hot and cold fluids is generated along the wall giving the wall protection to high heat fluxes from hot gases. This basic situation has been extensively evaluated since 1950. As shown in figure 1, the axial repetition of slots or holes oriented in the main flow direction can easily extend the coolant film length (schemes 3.1 to 3.3).

Increasing the cooling performance of a film cooling is mainly reliant upon two opposite processes, i) strictly parallel to wall film, ii) impinging jet effects. This later technique is illustrated by schemes 4.1 to 4.2 in figure 2. By using small discrete jets impinging on chamber double wall, it is shown out that heat transfer coefficients are 3 to 5 times higher than those obtained by forced convection parallel to the wall [4].

However, the impingement technique does not imply the cold surface as heat transfer surface. This was overcome by fabricating a porous skin from separate laminates incorporating separate passages (scheme 5) [5]. Surprisingly, with such device a slight decrease in cost and weight is reached if compared to one sheet wall situation as claimed out in [4]. Anyway, the cooling performance is inferior to that obtained from a transpiration device such as multiholes (scheme 6). Associated with ceramic wall coating before machining holes, multihole cooling is more frequently used in modern combustion chambers. This technique generally needs to beoptimized on account of the number of involved geometrical and fluid parameters.

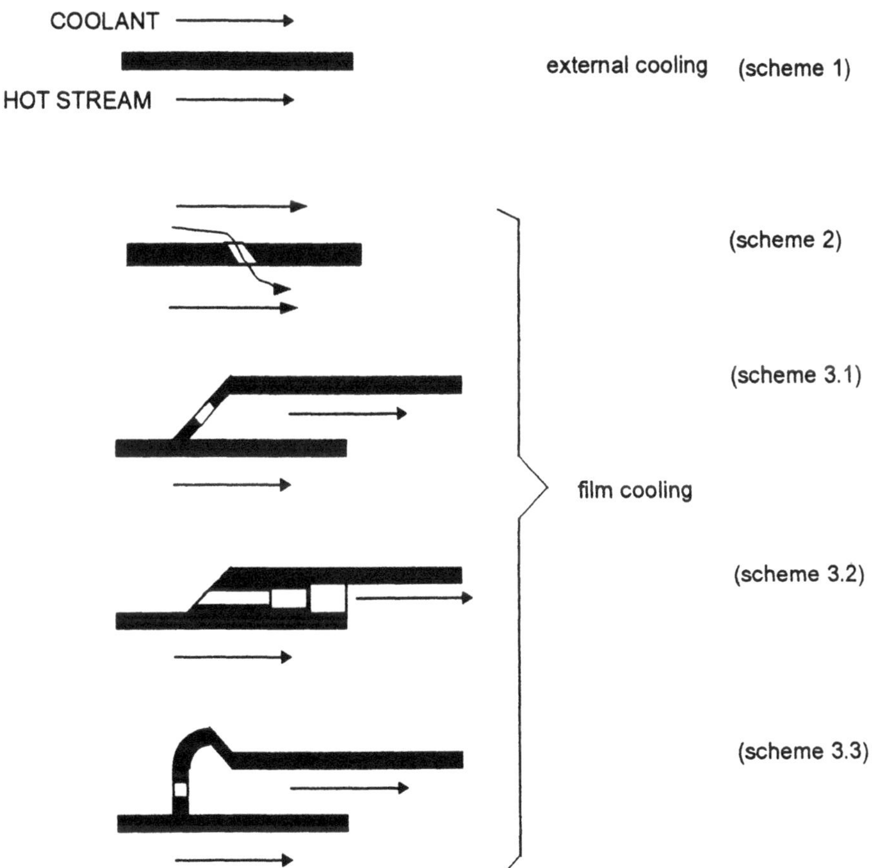

Figure 1: Film cooling devices

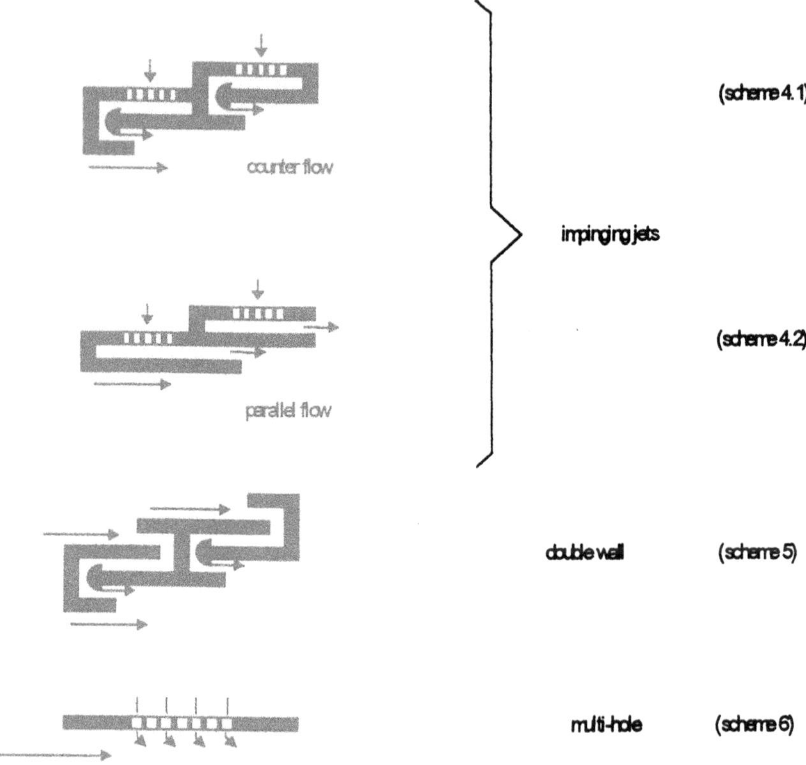

Figure 2: advanced film cooling devices

2.2. MULTIHOLE COOLING

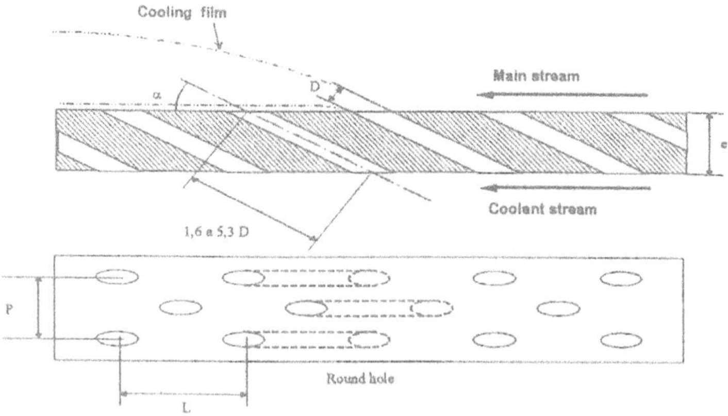

Figure 3. Geometry of multihole cooling

In this technique, it can be noticed that three convective heat transfer processes are competing (Fig. 3)
- a) convection along the cold side of the wall.
- b) convection inside each hole
- c) convection along the hot side.

If by increasing the number of holes an augmentation of heat transfer surface might be obtained, a mechanical limitation is obviously reached. A number of experimental results are available in literature [6,7,8,9,10]. Most of them are obtained with an experimental environment largely differing from the true engine conditions. As an example, injection of coolant in holes is oftent taken from a plenum such that no parallel flow of coolant exists along the wall. The advantage of such a device is to allow a precise measurement of the cold flow and consequently of the flow going through the holes. Moreover temperature ranges are limited to few hundred degrees Celsius. From the various available data, main conclusions given in terms of adiabatic effectiveness and heat transfer coefficient on hot side only are summarized as follows:

Geometry parameters:
- a) hole axis angle with respect to the wall plane: α. AFEJUKU et al. [11] have shown that inclined holes at $35°$ lead to a weak mixing between coolant jets and hot gases; jets are confined to the near wall region and result in a film cooling, whereas for an angle of $90°$ jets are strongly mixed and dispersed within the hot stream.
- b) Geometry of rows: by staggering holes, it is anticipated that hot spots between two neighbouring holes can be avoid; then, the adiabatic effectiveness is shown to be 0.2 to 0.35 in regions between staggered holes whereas with aligned holes, it reaches only 0.1 to 0.25 [11]
- c) Blowing rate η is defined as:

$$\eta = \frac{\rho_j U_j}{\rho_c U_c} \tag{1}$$

Blowing rate plays an important role through either the density fraction ρ_j/ρ_c or the density velocity ratio U_j/U_c. Whereas in true engine conditions, it is hardly feasible to change the density fraction, experiments have shown that by increasing this factor, adiabatic effectiveness is increased only in wall regions some distance from hole exit [12]. Somewhat similar conclusions are obtained by increasing the jet velocity U_j: within an axial distance $x/d<10$ from holes no significant effect is shown out on effectiveness; over $x/d=10$ effectiveness increases along the longitudinal axis and remains nearly constant transversally.

By increasing the velocity of hot stream U_c, results from [13] and [14] reveal a dependency of the related effectiveness augmentation to the nature of wall material. This has to be explained by a combined convection-conduction process.

Using realistic experimental conditions (pressure, temperatures, hole geometry...) J.L. Champion et al [15] have shown that with respect to the convective heat balance equation formulated on the two sides of the multiholed plate and within the holes: i) The heat transfer coefficient on the hot side is found independent of the cold temperature ii) the Nusselt number built on combine heat transfer on the cold side and within the holes varies with Re^n where Re is the Reynolds number of the cold flow and $n \approx 0.5$.

However, radiation fluxes are not fully considered in this study which could significantly alter the given conclusions.

3. Present experimental facilities

3.1. THE EXPERIMENTAL TEST BENCH

This test bench has been designed to reproduce geometrical, dynamic and thermal conditions really found in a turbo engine combustion chamber. Even the pressure loss can be automatically controlled, so the blowing rate can be chosen within the real values. As seen on figure 4, only a part of the cold air flows through the multiholed plate, in such a way that the three convective effects described above are fully reproduced.

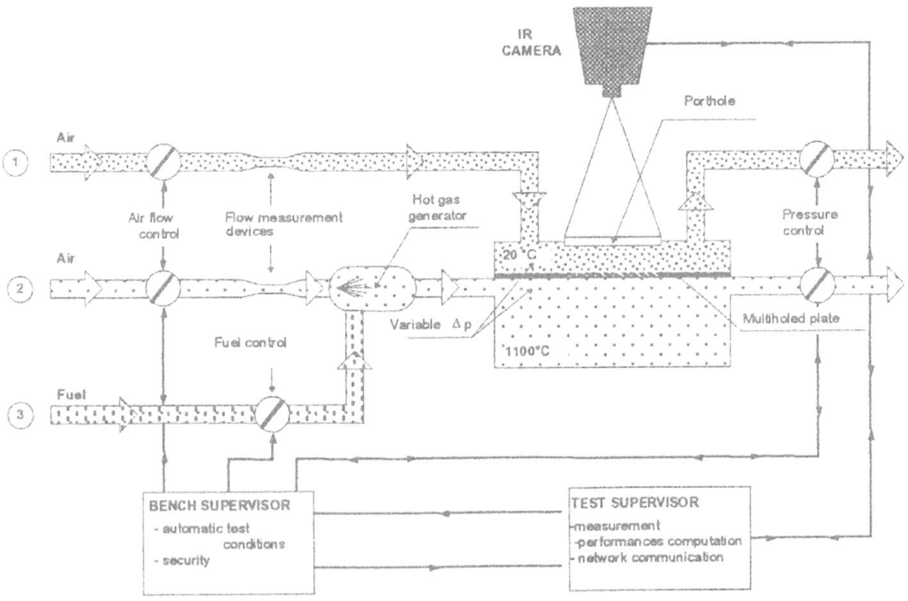

Figure 4: LARA test facilities

3.2.THE MEASUREMENT TECHNIQUES

3.2.1.Infra-red thermography

A large porthole has been designed in the test apparatus. This porthole in fluorine allows the measurement of the heat flux emitted by the heated multiholed plate with an I.R. camera. Fluorine has been chosen because of the very high transmittivity in the wave length range used by the I.R detector.

The plate temperature measurement is a very important and difficult step of the heat transfer coefficient measurement method and requires a considerable attention. The I.R. camera is designed to measure the heat flux emitted within the range of 2.5 to 5.6 µm. The temperature computation depends on different parameters relating Planck's law to these parameters is not sufficient. This is the reason for which we developed a specific algorithm. The specialized codes given with the acquisition system with the camera are, in some cases, not relevant. The following different steps are quoted:
- modelling of different fluxes in the test device
    - conduction
    - convection in holes and on plate on both sides
    - radiation
- noise filtering
- isochronic process of I.R. picture
- signal deconvolution
- spatial and temporal filtering

This calculation process allows us to determine the plate temperature with a maximum of accuracy.

3.2.2.Laser-doppler anemometry

The pure aerodynamic mechanism is very important in the physics of the multihole cooling. Convection in the holes, creation of film cooling downstream the multiperforated zone, and mixing of cold and hot streams must be dynamically studied.

To do so a special wind tunnel has been designed with Plexiglas walls. In order to separate the different effects, particularly to avoid three-dimensional effects, a transverse slot in a plate with two parallel air streams are studied (fig. 5). The main parameters are :
- Air velocity on both sides
- Pressure loss between the two flows

Of course, the temperature is ambient temperature, but we will deduce the flow structure in terms of velocity or stream function but also in terms of turbulence levels.

628

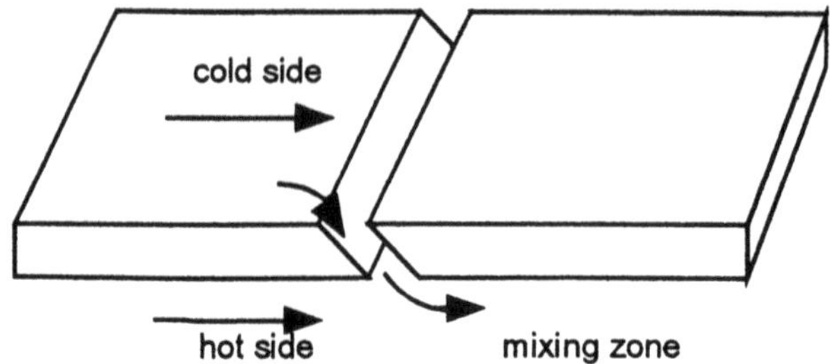

Figure 5: slot test facility for Laser Doppler Anemometry

3.2.3.Discharge coefficient measurement

A new method based on CO_2 mass flow rate measurement has been used here. We inject a measured quantity of CO_2 upstream on the cold side and by comparing the measurement downstream on both sides, it becomes easy to calculate the flow through the multiholed plate. As the pressure is measured on both sides of the test device, we obtain the pressure loss versus the air mass flow rate through the multiholed plate.

4. Results

This section provides examples of the results achieved within the present study. The quantities of interest are discharge coefficients, wall effectiveness, heat transfer coefficients and adiabatic wall effectiveness.
Different geometries of test plates have been designed to determine the influence of parameters D, α, P, L. (fig. 3)

4.1. EXPERIMENTAL PROCEDURE

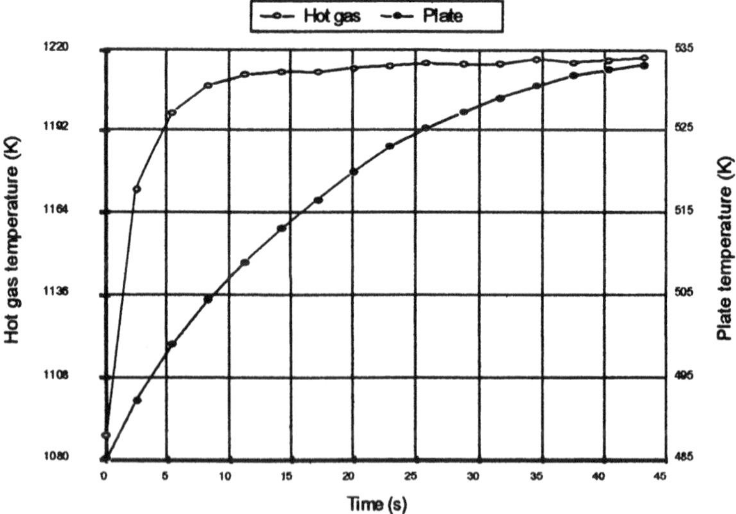

Figure 6: Experimental temperature warm-up

To calculate heat transfer coefficient and adiabatic temperature, we use a transient method which consists of a step of temperature of hot gases by increasing the fuel flow in the hot gas generator. A film of this evolution is recorded with the I.R equipment. Fig. 6 shows an example of this evolution and it may be noted that the step of temperature is only about 150°C which is big enough and but not too large to assure that the heat transfer coefficient on the hot side is a constant during the warm-up process.

4.1.1. Heat transfer equations

Figure 7, presents the thermal problem with conduction, convection on hot side, cold side and in hole, and also radiation which in our situation must be taken in account. This thermal scene must be carefully analyzed to be sure that all thermal fluxes will be described to obtain the true temperature with the I.R. camera.

<u>HOT SIDE</u>

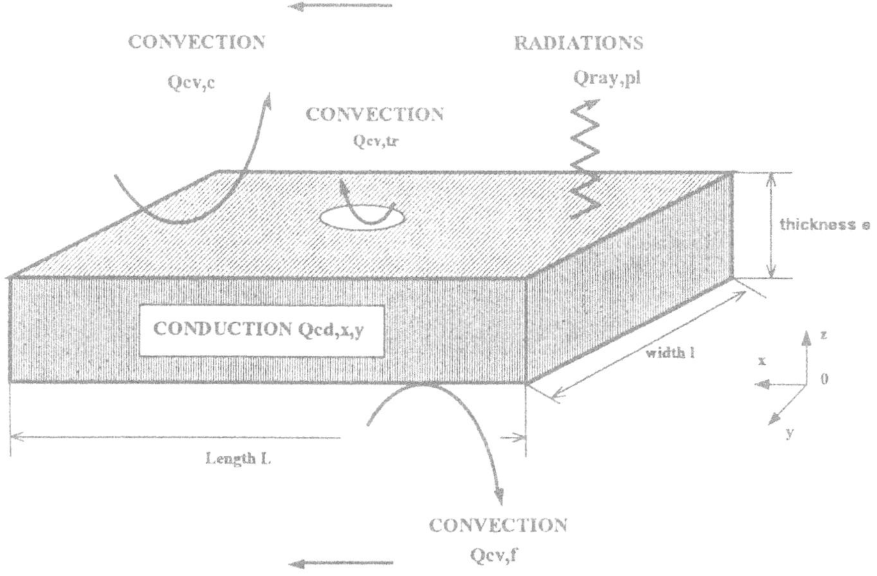

<u>COLD SIDE</u>

Figure 7: Heat fluxes on multiholed plate

Where:

Qcd,x = rate of conductive heat transfer in the flat plate material in the X-direction

Qcd,y = rate of conductive heat transfer in the flat plate material in the Y-direction

$Qray,c$ = rate of radiation heat transfer from hot wall box and hot gases to the flat plate

Qcv,c = rate of convection heat transfer from hot gases to the flat plate

$Qray,f$ = rate of radiation heat transfer from the plate to cold case and fluorine porthole

630

Qcv,f = rate of convection heat transfer from the flat plate to the cold gases
Qcv,tr = rate of convection heat transfer from the flat plate to multihole coolant gases

Then, the governing heat equation is:

$$\rho_{pl} C_{pl} \frac{\partial T_{pl}(x,y,z,t)}{\partial t} = \lambda_{pl} \frac{\partial^2 T_{pl}(x,y,z,t)}{\partial x^2} + \lambda_{pl} \frac{\partial^2 T_{pl}(x,y,z,t)}{\partial y^2} \qquad (2)$$

$$+ \lambda_{pl} \frac{\partial^2 T_{pl}(x,y,z,t)}{\partial z^2}$$

With the boundary conditions:

Hot gas side (z = e):

$$\lambda_{pl} \left(\frac{\partial T_{pl}(x,y,z,t)}{\partial z} \right)_{z=e} = \left(Q_{conv,c} + Q_{ray,c} \right)_{x,y,z=e,t} \qquad (3)$$

Cold gas side (z = 0):

$$\lambda_{pl} \left(\frac{\partial T_{pl}(x,y,z,t)}{\partial z} \right)_{z=0} = \left(Q_{conv,f} + Q_{ray,f} + Q_{conv,tr} \right)_{x,y,z=0,t} \qquad (4)$$

4.1.2. Thermal pictures processing

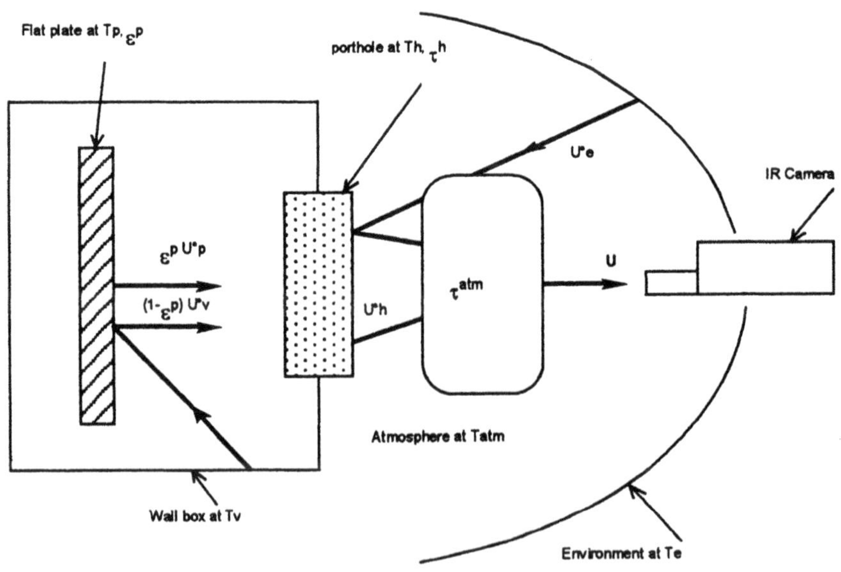

Figure 8.

As said above, measurements of these different values are carried out on a transient warm-up: after a short increase of hot gases temperature, the plate temperature is recorded by the infrared camera. This is the transient so called "warm-up" corresponding to the short time "t" equation (5) of the model. The steady state condition is then recorded to solve the long time equation (6):

$$\left[h_{conv,c}\left(T_{aw}-T_{pl}\right)\right]_t = \begin{bmatrix} \rho_{pl}C_{pl}e\dfrac{\partial T_{pl}}{\partial t}-Q_{cd,x}-Q_{cd,y}-Q_{ray,c} \\ -Q_{ray,f}-Q_{conv,f}-Q_{conv,tr} \end{bmatrix}_t \qquad (5)$$

$$\left[h_{conv,c}\left(T_{aw}-T_{pl}\right)\right]_\infty = \begin{bmatrix} -\dot{Q}_{cd,x}-\dot{Q}_{cd,y}-\dot{Q}_{ray,c} \\ -\dot{Q}_{ray,f}-\dot{Q}_{conv,f}-\dot{Q}_{conv,tr} \end{bmatrix}_\infty \qquad (6)$$

We can now calculate the h convection coefficients: if the distribution of the h coefficient and the T_{aw} adiabatic wall temperature are not dependent on the plate temperature T_{pl} over the studied range of temperature, we have:

$h_{cv,c}(t)= h_{cv,c}(t=\infty)$
$T_{aw}(t)= T_{aw}(t=\infty)$
Then, we can determine $h_{cv,c}$ and T_{aw} from the resolution of the two equations (5),(6).

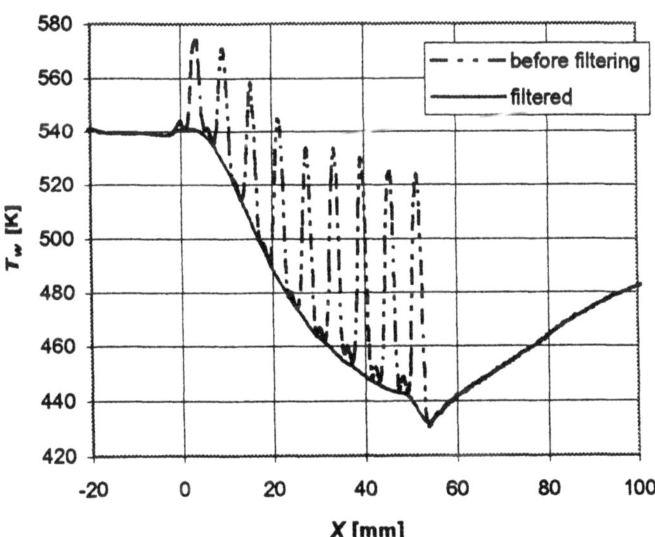

Figure 9: distribution of plate temperature before and after the filtering process

632

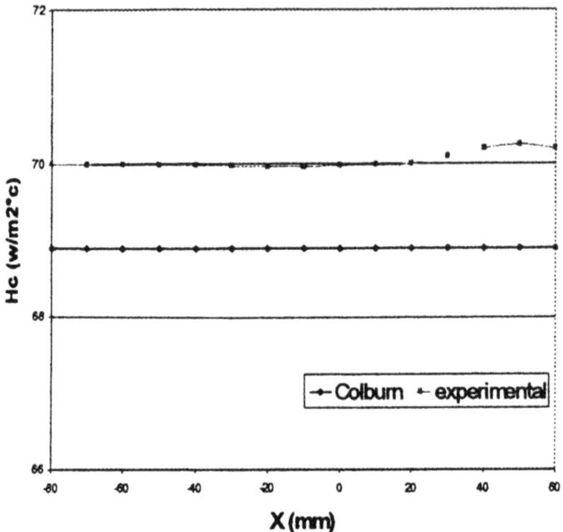

Figure 10: comparison of calculation and Colburn Nusselt correlation on an unholed flat plate

Figure 10 shows results of experiments on an unholed flat plate. This case is well known in literature and can be approached with the Colburn correlation. The comparison is very accurate and validates the developed process to calculate heat transfer coefficient and adiabatic temperature.

Investigations on several geometries and in different dynamic and temperature conditions lead us to determine Nusselt correlations based on Nu_0 which is the Nusselt number calculated with the classical Colburn correlation on a flat plate.
Nusselt correlations are given as follows:

$$\frac{Nu_x}{Nu_0} = 1 + \Omega l_{zmp}\sqrt{\frac{\psi}{\pi}}\;exp\left[-\psi\left(Xe - \frac{l_{zmp}}{2}\right)^2\right]exp\left[-\psi\left(X - \frac{l_{zmp}}{2}\right)^2\right] \quad (7)$$

with

$$\Omega = (\Delta P)^\alpha (Re)^\beta (Re)^\chi (T_c/T)^\delta (L/D)^\varepsilon (q)^\phi (P)^\gamma$$

Adiabatic wall effectiveness Eaw can also be deduced from the two heat transfer equations (5) and (6) introduced before. Figure 11 shows the longitudinal evolution of the calculated adiabatic wall effectiveness, along a multiholed plate, equation (7), compared to a typical experimental result.

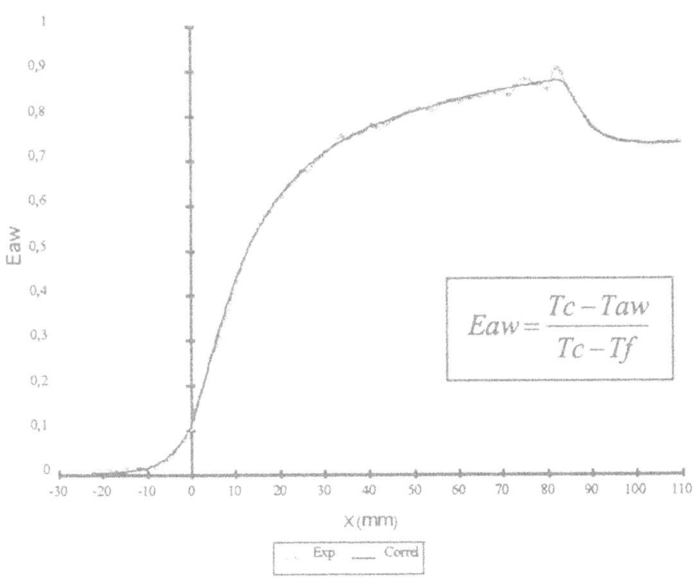

Figure 11 : measured and calculated adiabatic effectiveness

In the situation presented in Figure 11, the multiholed zone is extending between X=0 to X=85. Upstream this zone, the hot gas temperature is equal to the adiabatic temperature. As the cooling multiholed zone begins at X=0, conduction within the plate thickness explains the cooling effect between x = -10 to 0. We can observe the film cooling effect downstream the multiholed zone. This is an important conclusion, which shows the longitudinal extension of the thermal protection and therefore the possibility to reduce cold air flow rate. This effectiveness is still higher than 70%, 30mm after the end of the multiholed zone, and it can remain at high values depending on the mixing effect between the cold film close to the plate and the main hot stream. This shows the importance of the dynamic studies to point out the stability conditions of the cooling film. Then, we can easily imagine the possible devices to reduce the air consumption to protect combustion chamber walls.

Experimental and calculated longitudinal distributions of plate temperature are given in Figure 12 for a multiholed plate and show a good accuracy of the developed code.

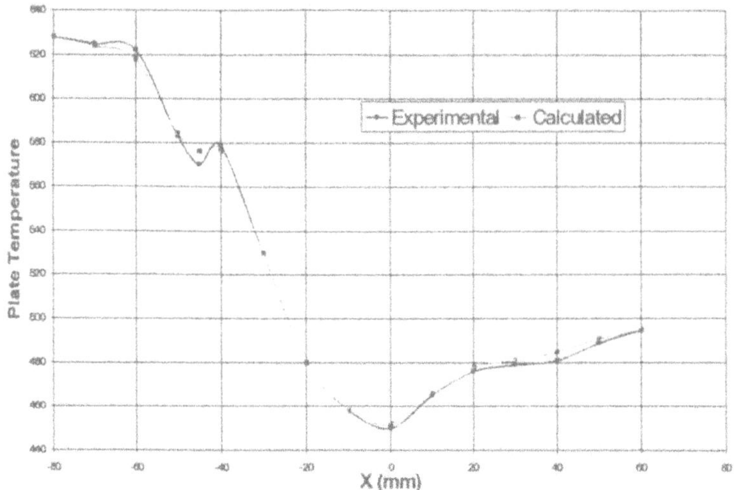

Figure 12 : Comparison between experimental and Calculated plate temperature for a multiholed flat plate

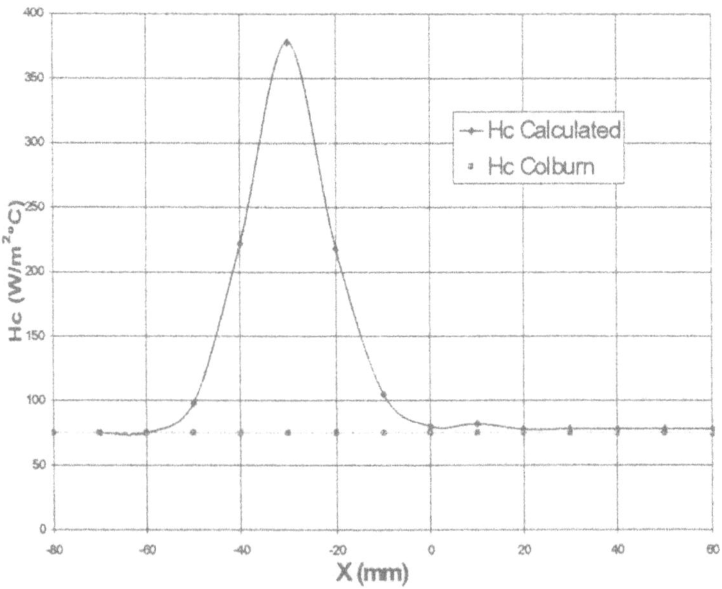

Figure 13 : evolution of the heat transfer coefficient on a multiholed plate and on a flat plate.

In Figure 13, a longitudinal evolution of the heat transfer coefficient on a multiholed plate is compared to heat transfer coefficient on a plain flat plate calculated with Colburn correlation. Heat transfer coefficients drastically increase in the multiholed zone and are nearly equal to the Colburn heat transfer coefficients downstream and upstream this zone. The film effect downstream themultiholed zone is not due to an augmentation of heat transfer coefficient but to a decrease of the adiabatic wall temperature. It is important to remember here that the wall temperature of a combustion chamber is a result of different heat transfer processes, and the reference temperature must be defined. We choose the adiabatic wall temperature which is the most convenient, in our opinion. Another temperature reference would lead to other conclusions.

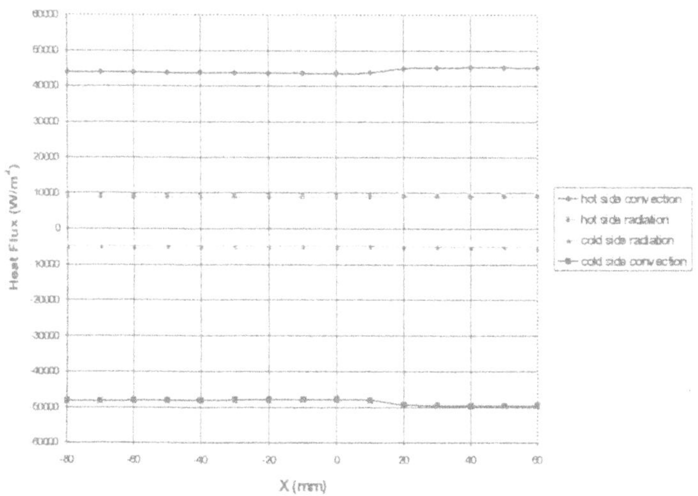

Figure 14 : Evolution of heat transfer rates for a standard flat plate

Figure 14 shows typical levels of the different heat fluxes rates encountered in our test module. These evaluations are done using classical computation to evaluate radiation an convective fluxes as explained before.

4.2. DISCHARGE COEFFICIENT

The discharge coefficient Cd is the ratio of the actual mass flow rate to the ideal mass flow rate through holes. The ideal mass flow rate is calculated assuming an isentropic one-dimensional expansion from the total pressure of the coolant stream. Figure 15 shows a typical result for a hole diameter of 0.5 mm and an angle of 30°. The discharge coefficient is plotted versus the pressure ratio across the hole at three different hot main stream Mach numbers.

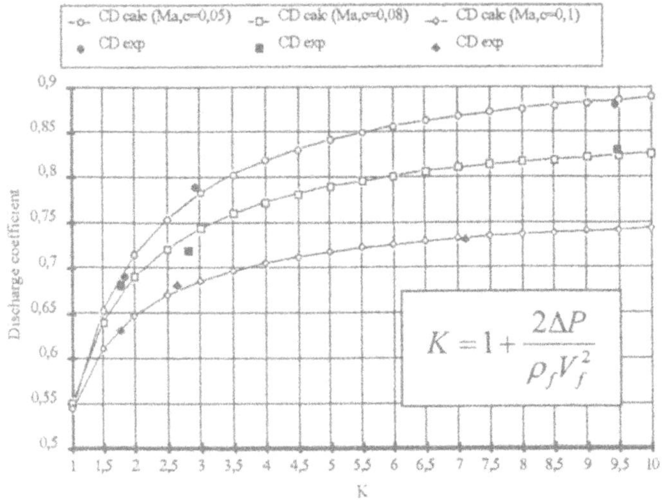

Figure 15: Discharge Coefficient plotted vs pressure ratio K

4.3. DYNAMICS

Figures 16 describes a different way to investigate the dynamic effects of a jet issuing normally to a main stream. The complexity of the behaviour of such jets is visible. Instability of vortex structure, three-dimensional effects, recirculating zones and mixing flows can be noticed. For a multiholed plate, all these effects are coexisting and depend on blowing rates, geometry and boundary layer structure. Three dimensional visualizations made with multiholed plates by Fric T.F. et al. [17] are presented for different jet Mach numbers; they clearly suggest added mixing effects between jets issuing from staggered holes.

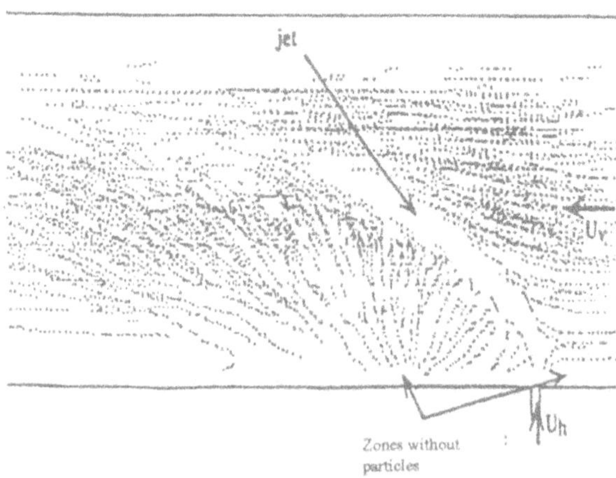

Figure 16 : Example of Particle tracking velocimetry of a perpendicular jet [16]

Figure 17 gives map velocity vector measurements done in our Plexiglas wind tunnel. Whereas heat transfer is not assessed in this test section, this map figures out the mixing between a cold air jet issuing from an inclined (60°) slot and hot gas stream flowing along to the plate. Different typical features can be noticed: a recirculating zone is existing downstream the slot exit. The vortex is stable with such an angle of injection, and delivers a weak thermal protection to the wall. The radiation heat flux captured in that zone is not transported by convection and the local temperature will increase. This effect can be observed on multiholed plate where between rows number 2 to 4, the temperature is increasing (fig. 14). This recirculating flow can be avoided by decreasing hole diameter and pitch between holes, or also, by decreasing the angle of hole or slot. Measurement done with slot inclination of 30° shows the complete disappearing of the recirculating flow.

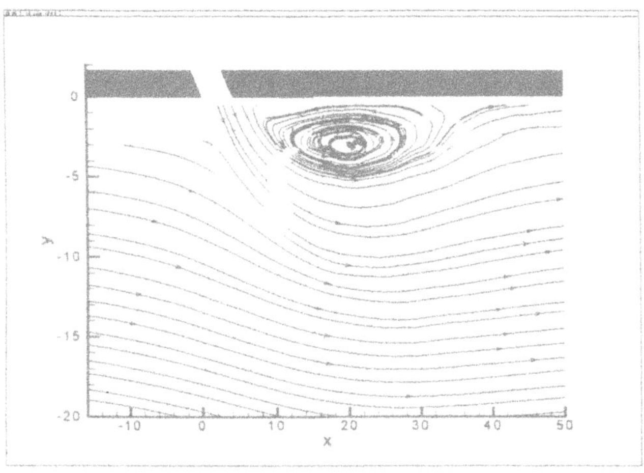

(Uh=Ub=3 m/s et ΔP=60 Pa)

Figure 17: measured velocity vectors and stream function

Even with the slot at 60°, the injected flow does not extend too much into the main stream but stays close to the wall downstream of the vortex with these dynamic conditions. Tests done with a slot at 30° show that this film mixes less with the main stream than at 60°. Therefore, we can imagine that the wall protection will be more efficient with low inclination of slot or holes.

The velocity measurement allows the computation of turbulence levels ($<u'^2>$, $<v'^2>$,$<u'v'>$). There are two zones where turbulence production rate is high . The first one is located on the interface between the upstream part of the jet and the main stream. This turbulence creates the mixing of the two streams and must be limited if possible. Film cooling efficiency decreases with increasing turbulence. The second turbulence zone is located in the jet itself. It is due to the important velocity found in the jet . The stability of jet depends on the level of turbulence created in this region.

As a concluding remark, it may be pointed out that a decreasing inclination from 60° to 30° will avoid recirculation and ensure stability of the film effect. The decrease of the observed turbulence levels increases the efficiency of the film cooling.

5. Conclusion

Multihole cooling was introduced to thermally protect wall combustion chambers were previous film cooling (Fig. 1) was not efficient enough. The cooling effect in the multiholed zone is very important and depends on the geometry of holes themselves (inclination, diameter and pitch), the pressure difference between the cold and hot streams, and the velocity ranges of the two flows. As a first consequence, decreasing of the injection angle of jet has increased the hole exchange surface and so the heat convective flux in holes. The most important result is the improvement of the film effect downstream of the multiholed zone. This subject is not actually completely understood and several laboratories are working on it with different calculation and experimental metods.

Several advantages lead combustion chamber designers to prefer multihole cooling than other cooling devices. The temperature of the wall can be decreased locally more than with other devices with less air consumption and a downstream film cooling effect is created. The effectiveness of this film and its stability (length, thickness) seem to be controlled by hole inclination and the blowing rate but also by the geometry of the multihole design (diameter and pitch).

6. Nomenclature:

C	Specific heat, J/kg.K	Index	
e	Plate thickness, m	pl	Multihole plate
h, H	Convective heat transfer coef-	g	Gas
ficient, w/m².K		ext	Outside
T	Temperature, °C	env	Sides
α	Hole angle, °	cv	Convection
D	Hole diameter, m	cd	Conduction
x, y, z	coordinates	ray	Radiation
P	transversal pitch, m	c	Hot gas side
Q	Heat flux, W/m²	f	Cold gas side
l_{zmp}	multiholed length, m	tr	Hole
		j	Hole flow
		aw	Adiabatic

Greek symbols

λ	Thermal conductivity, W/m.K
ε	Emissivity
τ	Transmission
ρ	Density, kg/m³
η	blowing rate
ψ	geometry factor

Adimensional numbers:

Nu   Nusselt number
Re   Reynolds number
Pr   Prandtl number

7. References

1. Mayle R.E., Camarata F.J. (1975) Multihole cooling film effectiveness and heat transfer. Journal of heat transfer, pp 534-538

2. Kodotani K., Golstein R.J. (1979) Effect of mainstream variables on jets issuing from a row of inclined round holes. Journal of engineering for power, vol. 101, pp 289-304

3. Le Grives E., Nicolas J-J, Genot J. (1979) Internal aerodynamics and heat transfer problems associated to film cooling of gas turbine ASME paper N° 79-GT-57

4. Wassel A.B., Banghu J.K., (1980) the development and application of improved combustor wall cooling techniques ASME paper N° 80-GT-66

5. Nealy D.A., Reidler S.B. (1979) Evaluation of laminated porous materials for combustor liner cooling ASME paper N° 79-GT-100

6. Sprrow E.M., Carranco Ortiz M. (1982) Heat transfer coefficients for the upstream face of a perforated plate positioned normal of an oncoming flow International journal of heat and mass transfer vol. 25,1, pp. 127-135

7. Mills A.F. (1962) Experimental investigation of turbulent heat transfer in a thermal entrance region of a circular conduit Journal of mechanical engineering and science vol. 4,1, pp 112-119

8. Andrews G.E., Alikhanzadeh M., Asere A.A., Hussain C.I., Khoshkbar azari M.S., Mkpadi M.C. (1986) Small diameter film cooling holes: Wall convective heat transfer ASME paper N° 86-GT-225

9. Giovannini A., Meurat A., Millan P., Servage J.L. (1991) Detailed heat transfer mappings around an isolated perforation hole. Eurotherm Seminar N° 25.

10. Kasagi N., Hirata M., Kumada M. (1981) Studies of full coverage film cooling ASME paper N° 81-GT-37, 38.

11 Afejuku W.O., Hay N., Lampard D. (1983) Measured coolant distributions downstream of single and double rows of film cooling holes Journal of engineering for power, vol. 105, pp. 172-177

12. Forster N.W., Lampard D. (1975) Effects of density and velocity ratio on discrete hole film cooling, AIAA Journal, vol. 13, N°8, pp 1112-1113

13. Ericksen V.L., Golstein R.J. (1974) Heat transfer and film cooling following injection through inclined circular tubes. Journal of heat transfer, pp. 239-245

14. Crowford M.E., Kays W.M., Moffat R.J. (1980) Full-coverage film cooling –part I: Comparison of heat transfer data for three injection angles. ASME paper N° 80-GT-43

15. Champion J.L., Dorignac E., Deshaies B. (1995) Etude experimentale du processus de refroidissement d'une paroi multiperforée. Colloque Société Française des Thermiciens

16. Blanchard J.N., Brunet Y. Merlen A. (1997) Etude par visualisation d'une interaction d'un jet avec un écoulement transversal perpendiculaire, 7ème colloque national de visualisation et de traitement d'images en mécanique des fluides, pp. 209-213.

17. Fric T.F., Campbell R.P., Rettig M.G. (1997) Quantitative visualisation of full-coverage discrete–hole film cooling, ASME Orlando, Florida, June.

EXPERIMENTAL STUDIES ON INFLUENCE OF PROCESS VARIABLES TO THE EXERGY LOSSES AT THE DOUBLE TUBE HEAT EXCHANGER

AHMET CAN and DOĞAN ERYENER
Faculty of Engineering and Architecture, Trakya University, 22030 ,Edirne ,Turkey

ERTAN BUYRUK
Faculty of Engineering, Cumhuriyet University,58140 , Sivas , Turkey

Abstract. Heat Exchangers are widely used at different applications related with heat science and techniques. Today at the investigation of heat exchanger projection, First Low of Thermodynamics is not anymore sufficient. The reason of this more effective use of heating sources and systems, not the quantity of energy but quality of energy must be taken into consideration. Energy transformation in the heat exchangers, in the scope of Second Law of Thermodynamics can be listed in two groups depending on chosen variables for analysis. These are a) entropy based work b) exergy based work. In this study, it is aimed to analysed theoretically and investigate experimentally effect of inlet and outlet temperatures of temperatures differences and flow directions for same path parallel flow and counter path parallel flow by using double tube heat exchanger. According to the obtained results, it is aimed to determine necessary information for projection and to bring down to the most economical value of the heat exchanger cost and of energy consumption.

1. Introduction

Concept of exergy was first time proposed by Rudolf Planck and instead of technical work ability, it was introduced as a new word by Zoran Rant [1]. In the different written sources, instead of this term maximum work and in the Anglo-Saxon sources, "availability" are used. Exergy is defined a maximum work performance by a system at the present environment and known situation. Maximum work can be obtained only in the reversible systems, while in reversible situations, with every entropy rise there is a loss in work.

Work can not be produced by the heat exchanger, if liquids or gases are heated or cooled inside them. If the amount of transferred heat is taken into consideration under ΔT temperature difference, taking place in the heat exchanger as irreversibility then exergy loss is formed. Furthermore friction depending on ΔP pressure loss forming in the tubes is source of irreversibility.

It is possible to set up a system with maximum effectiveness and minimum exergy loss in the heat exchanger. There are many theoretical investigations for this purpose [3-6]. This study is the experimental application of previous theoretical works of [7] and [8]. With respect to exergy is presenting the actual engineering value then as with the all reversible energy systems analysis of the heat transfer in the heat exchanger, exergy loss must also be calculated. However, according to this, during the heat transfer, process variables and projections of equipment can be determined in the way to decrease exergy loss.

2. Theoretical Analysis of Exergy Loss in the Double Tube Heat Exchanger

Hot and cold fluids were taken as incompressible fluids and pressure losses were ignored when fluids flow in the heat exchanger.

General exergy loss expression for open system, when applied to the heat exchangers;

641

S. Kakaç et al. (eds.), Heat Transfer Enhancement of Heat Exchangers, 641–648.

642

$$E = \dot{m}_c\left(h_{c,i} - h_{c,o}\right) + \dot{m}_h\left(h_{h,i} - h_{h,o}\right) + T_o\left[\dot{m}_c\left(s_{c,o} - s_{c,i}\right) + \dot{m}_h\left(s_{h,o} - s_{h,i}\right)\right] \qquad (1)$$

is obtained. Supposing, heat that is given by the hot liquid, is taken by cold liquid then Eq. (1) can be written as;

$$E = T_o\left[\dot{m}_c\left(s_{c,o} - s_{c,i}\right) + \dot{m}_h\left(s_{h,o} - s_{h,i}\right)\right] \qquad (2)$$

If the changes in entropy for cold and hot fluids is written with specific heat expressions, in the case of constant pressure then exergy loss equation can be obtained as;

$$E = T_o\left[\dot{m}_h c_{p_h} Ln\left(\frac{T_{h,o}}{T_{h,i}}\right) + \dot{m}_c c_{p_c} Ln\left(\frac{T_{c,o}}{T_{c,i}}\right)\right] \qquad (3)$$

With values of $C_{min} = m_h \cdot c_{ph}$ and $C_{max} = m_c \cdot c_{pc}$ for the case of when hot fluid is the minimum fluid then

$$E = T_o\left[C_{min} Ln\left(\frac{T_{h,o}}{T_{h,i}}\right) + C_{max} Ln\left(\frac{T_{c,o}}{T_{c,i}}\right)\right] \qquad (4)$$

is written. Heat transfer effectiveness, ε, heat capacity ratio of the actual rate of heat transfer in the exchanger to the maximum possible rate of heat transfer

$$\varepsilon = \frac{\left(T_{c,o} - T_{c,i}\right)}{\left(T_{h,i} - T_{c,i}\right)} \qquad (5)$$

can be defined. By using the effectiveness , ε, heat capacity ratio, $C_r = C_{min}/C_{max}$ and temperature ratio $T_r = T_{hot}/T_{cold}$ then

$$E = T_o\left[C_{min} Ln\left[1 - \varepsilon\left(1 - \frac{1}{T_r}\right)\right] + C_{max} Ln\left[1 + \varepsilon C_r\left(T_r - 1\right)\right]\right] \qquad (6)$$

is found. If this expression is divided to T_o, Cmin then, dimensionless exergy loss can be written as;

$$e = Ln\left[1 - \varepsilon\left(1 - \frac{1}{T_r}\right)\right] + \frac{1}{C_r} Ln\left[1 + \varepsilon C_r\left(T_r - 1\right)\right] \qquad (7)$$

From this equation, it can be seen that dimensionless exergy loss depends on three variables, $e = f(\varepsilon, T_r, C_r)$. This expression can also be used for double tube heat exchanger when minimum fluid is hot fluid and for same path parallel flow or counter path parallel flow.

Amount of heat transfer in heat exchanger with heat transfer effectiveness, Eg.(8) can be written as;

$$Q = \varepsilon \cdot C_{min}\left(T_{h,i} - T_{c,i}\right) \qquad (8)$$

If both side of this expression is divided with $(C_{min} \cdot T_{c,i})$ then dimensionless heat flux is obtained as;

$$q = \varepsilon\left(T_r - 1\right) \qquad (9)$$

When dimensionless exergy loss and dimensionless heat flux are rationed then

$$\frac{e}{q} = Ln\left[1-\varepsilon\left(1-\frac{1}{T_r}\right)\right]+\frac{1}{C_r}Ln\left[1+\varepsilon\,C_r\left(T_r-1\right)\right] \qquad (10)$$

is found. If similar process is made for the case of cold fluid is the minimum fluid then dimensionless exergy loss can be written as;

$$e = Ln\left[1+\varepsilon\left(T_r-1\right)\right]+\frac{1}{C_r}Ln\left[1-\varepsilon\,C_r\left(1-\frac{1}{T_r}\right)\right] \qquad (11)$$

In this case the ratio of the dimensionless exergy loss and dimensionless heat flux is obtained with Eg.(12);

$$\frac{e}{q} = \frac{Ln\left[1+\varepsilon\left(T_r-1\right)\right]+\dfrac{1}{C_r}Ln\left[1-\varepsilon\,C_r\left(1-\dfrac{1}{T_r}\right)\right]}{\varepsilon\left(T_r-1\right)} \qquad (12)$$

3. Computer Program Developed for Exergy Loss Calculation

Computer program, which was developed for this study, was written with QBasic version 7.1. Exergy loss and other expressions were calculated by using experimental data finding from the double tube heat exchanger. Algorithm of the program can be seen at Figure 1.

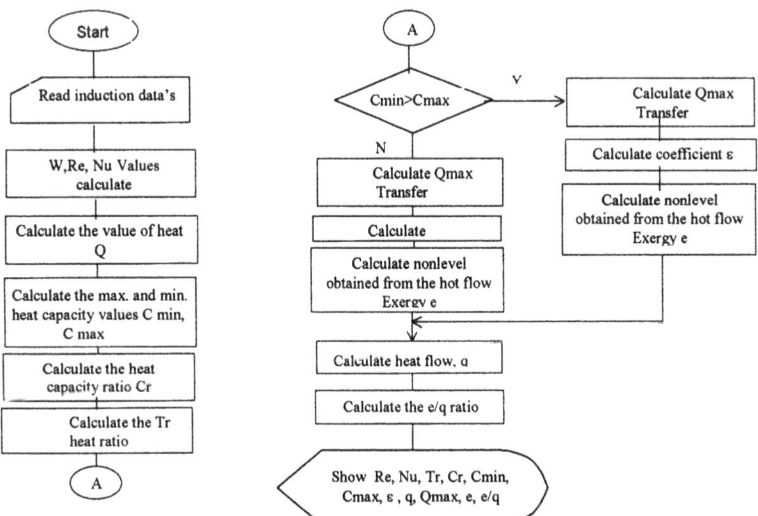

Figure 1 : Algorithm of the program.

Calculations were carried out three stages are

Stage 1: Heat exchanger total heat transfer coefficient was determined as;

$$K = \frac{1}{\dfrac{1}{\alpha_{in}} + \dfrac{1}{\alpha_{out}}} \tag{13}$$

Total heat transfer is determined as;

$$Q = \dot{m}.c_p.\Delta T \tag{14}$$

Stage 2: With comparing the heat capacity values for both fluids, Cmin and Cmax values are determined. After that, in the case of when hot fluid is minimum fluid then maximum amount of heat transfer is calculated as;

$$Q = C_{min}\left(T_{h,i} - T_{c,i}\right) \tag{15}$$

Heat transfer effectiveness ε;

$$\varepsilon = \frac{Q}{Q_{max}} \tag{16}$$

Stage 3: Dimensionless exergy loss and dimensionless heat flux was calculated according to the minimum fluid. At the end of the calculation, ratio of those two values is determined (e/q).

4. Presentation of Experimental Bench and Experimental Work

Experimental Bench is schematically illustrated in Figure 2. This is of the double pipe with hot water flowing through the central tube while cooling water flows through the annular space [9]. Thermocouples sense the entry and exit temperatures of the hot and cold water streams end of the tube metal at either end of the heat transfer surface. A digital thermometer with a selector switch, displays the temperature sensed by water and metal temperature thermocouples. Resolution of the digital temperature indicator is 0,1°C. Hot water provided by an electric resistance type water heater (3,1 kW) is fed into the upper end of the central tube (d_0 = 9,5 mm, d_i = 7,5 mm) of the heat exchanger . The water cools as it flows downward through the heat exchanger and on leaving the bottom passes through a pump which provides the circulation head. The water then flows through either a high or a low flowmeter (according to the flow rate) and then back to the water heater where it is reheated.

Mains cold water passes through a flow control valve and flow meter to quick release union on the face of the panel. A similar quick release union is connected to the water drainage point. Flexible hoses connecting with the annulus at the either end of the heat exchanger can be connected either of these unions.

Values are provided to control the flow rate of the hot and cold streams of water. A trial control determines the power input to the water heater and thermostat, sensing to the temperature in the water heater limits the water temperature to approx. 80°C .

During the experimental work, cold water flow rate, mc, hot flow rate, m_h, hot water inlet temperature, T_h, were adjusted. The measured values were hot and cold water streams inlet and outlet temperatures and of the tube metal at either end of the heat transfer surfaces. Measurements were repeated at least 4 times by changing one parameter for known flow condition and situation and their averages were taken as final. Total heat transfer, Q, occurred in heat exchanger, minimum heat capacity, maximum heat capacity, heat capacity ratio, C_r, heat transfer effectiveness, ε, dimensionless exergy loss, e, dimensionless heat flux, q, maximum amount of heat transfer, Qmax, and ratio of e/q values were calculated.

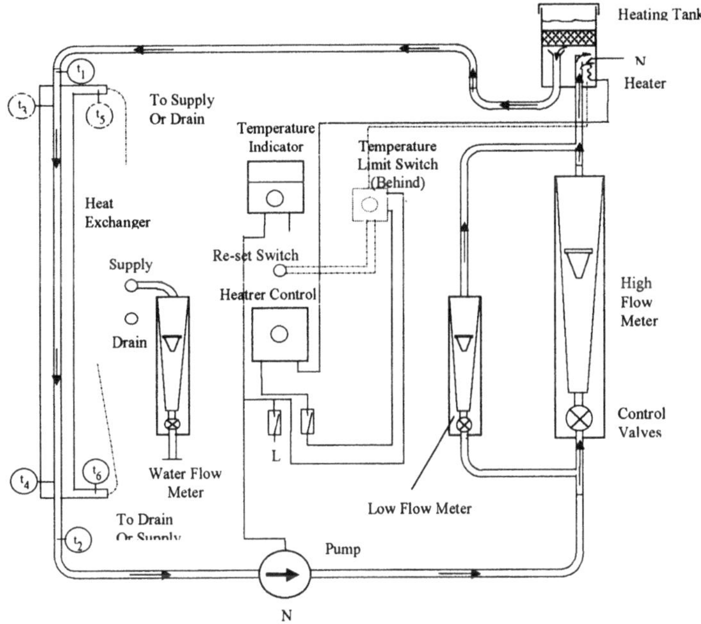

Figure 2: Water-water parallel flow double tube heat exchanger experimental bench.

5. Results and Discussions

By using experimental results from the double tube heat exchanger, process variable's effects to the exergy loss is determined. Figure 3 shows the change on the dimensionless exergy loss depending on hot water flow rate.

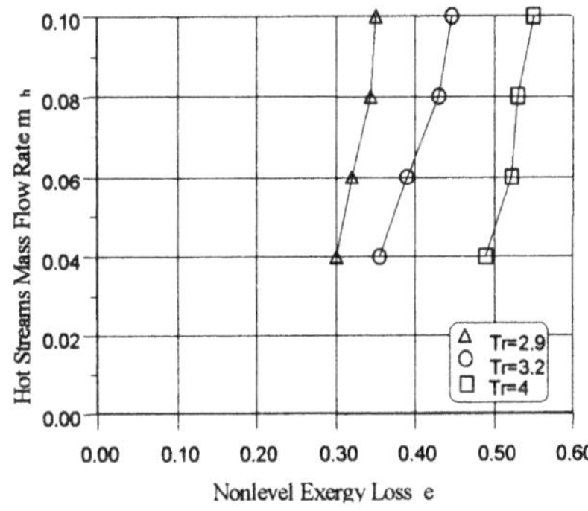

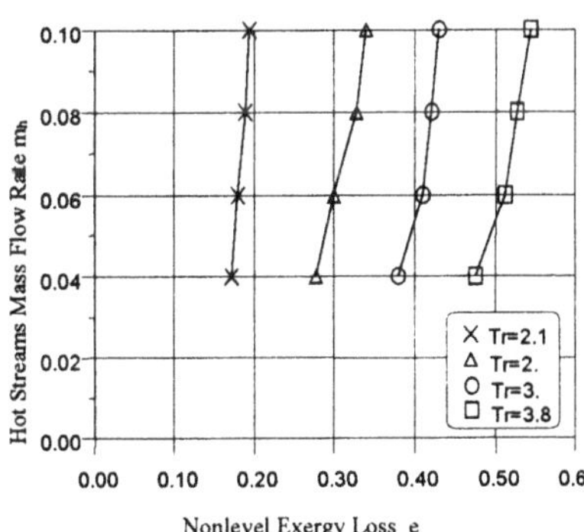

a) Same path parallel flow

b) Counter path parallel flow

Figure 3: Dependence of Nonlevel exergy loss to hot water mass flow rate

According to the graph increasing hot water flow rate causes the exergy loss to increase for both path at constant temperature ratio. This pattern can also be seen for different temperature ratios.

Figure 4 shows change in the dimensionless exergy loss depending on entrance temperature ratio.

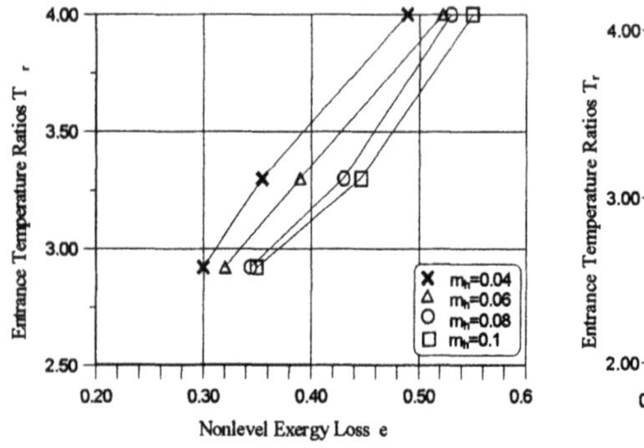

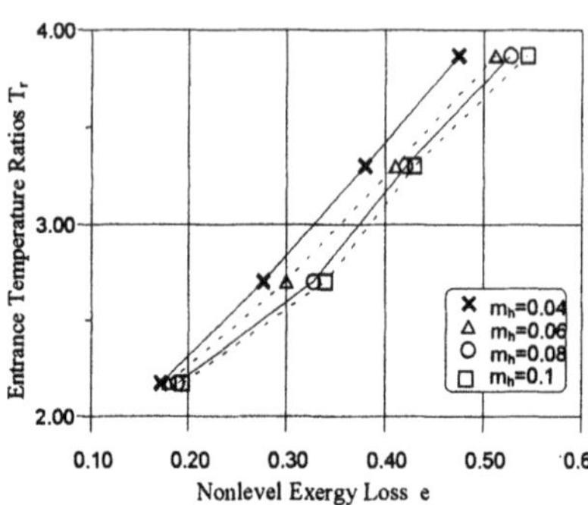

a) Same path parallel flow   b) Counter path parallel flow

Figure 4: Dependence of Nonlevel exergy loss to the entrance temperature ratio.

As it can be seen from the figure 4 increasing entrance temperature ratio causes the exergue loss to increase with keeping the hot water flow rate constant. Same patterns are seen for different hot water flow rate.

Figure 5 shows the change on the dimensionless exergy loss depending on dimensionless heat flux. Again if the dimensionless heat flux is increased then it causes the exergy loss to increase with constant heat capacity ratio. Same patterns of this relation are seen for different heat capacity ratio.

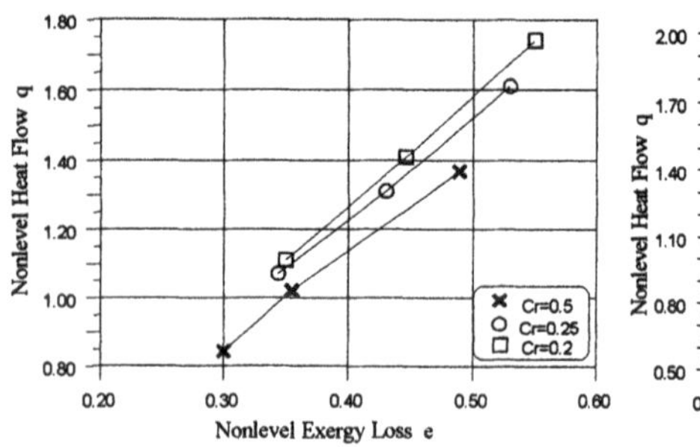

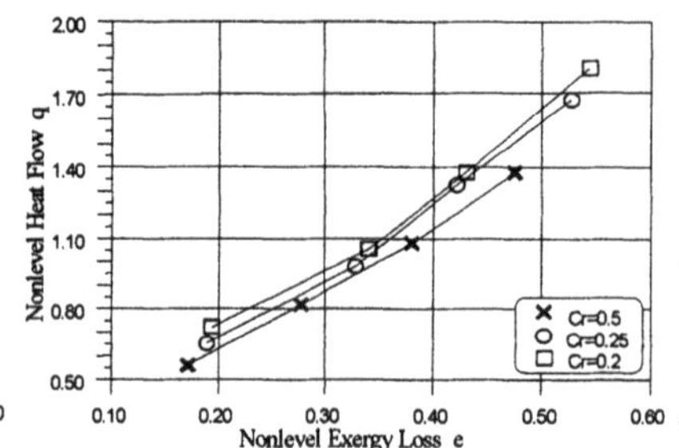

a) Same path parallel flow   b) Counter path parallel flow

Figure 5: Dependence of Nonlevel exergy loss to nonlevel heat flow.

Figure 6 shows the change on the dimensionless exergy loss depending on effectiveness. Increasing the effectiveness causes the exergy loss to increase for both paths at constant temperature ratio.

From the obtaining results, in the case of known temperature ratio $T_r = 3,3$ and known effectiveness $\varepsilon = 0,55$, dimensionless exergy loss values are e=0,43 (same path parallel flow), e=0,41 (counter path parallel flow). The case counter path parallel flow situation according to the First Law of Thermodynamics, it is necessary to have a small heat transfer surface because of the high ΔT_{log} According to the Second Law of Thermodynamics it was also found that the counter path parallel flow is more advantageous.

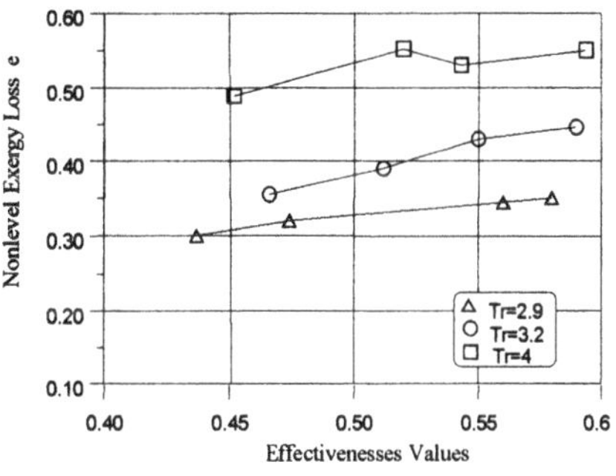

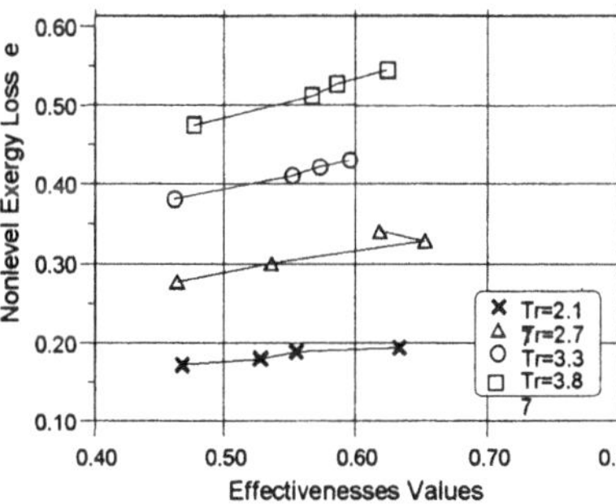

a) Same path parallel flow   b) Counter path parallel flow

Figure 6: Dependence of exergy losses at the constant temperature ratio with coefficient's values.

Nomenclatures

E	Exergy loss, kW
$\dot{m}$	Mass flow rate, kg/s
s	Specific entropy, kJ/kgK
h	Specific enthalpy, kJ/kg
T_o	Absolute temperature of the Environment K
c_p	Specific heat capacity at constant pressure, kJ/kgK
C	Heat capacity, kJ/s.K
ε	Effectiveness
C_r	Heat capacity ratio
T_r	Temperature ratio
q	Nonlevel heat flow
K	Total heat transfer coefficient

Subscripts

h	Hot stream
c	Cold stream
i	Entering
o	Leaving
min	minimum
max.	maximum

References

1. Rant, Z. (1956) *Forschung Ingeniuerwesen* , 22, 36.

2. Hahne, E (1993) *Technische Thermodynamik*, 2. überarbeitete Auflage, Addision - Weslwey Publishing Company.GmbH Deutschland.

3. Paykoç, E (1986) Çift Borulu Eşanjörlerde Tersinmezlik, *Türk Isı Bilimi ve Tekniği Dergisi*, Cilt 9, Sayı 3, s. 35 -41, Ekim.

4. Bejan, A (1977) *The Concept of Irreversibility in Heat Exchangers Design Counterflow Heat Exchangers For Gas – to Gas Applications*, Transactions of ASME, Vol. 99, August.

5. Bejan, A. *For Entropy Generation Trough Heat and Fluid Flow*, Department of Mechanical Engineering, University of Colarado Boulder.

6. Kays, M.W. and London, A.L. (1984) *Compact Heat Exchangers*, McGraw - Hill Inc., New York

7. Can, A and Senyücel, N (1995) Termodinamiğin İkinci Yasasına Dayanan Bir Yönteme Göre Isı Aktarıcılarının Bilgisayar Programı İle Tasarımı ,*III. Balıkesir Mühendislik Sempozyumu, Bildiriler Kitabı*, 10 - 11 Nisan, Balıkesir.

8. Can, A and Senyücel, N (1995) Isı Değiştirgeçlerinin Ekserji Ekonomik Analizi ve Optimizasyonu, *Gazi Üniversitesi ve TIBTD (ODTÜ) işbirliği ile düzenlenen ULIBTK' 95 " Uluslararası Katılımla 10. Ulusal Isı Bilimi ve Tekniği Kongresi " Bildiriler Kitabı*. 2. cilt, s. 1393 – 1403, 6 - 8 Eylül, ANKARA.

9. P.H. Hilton LTD, *Water - Water Turbulent Flow Heat Transfer Unit*, H950M/E/5/047, APR 93, Hampshire, SO20 6 PX, England.

ENHANCEMENT OF DIRECT-CONTACT HEAT TRANSFER IN CONCENTRIC ANNULI

TÜLAY A. ÖZBELGE AND MANİJEH K. SHAHİDİ
Department of Chemical Engineering
Middle East Technical University
Ankara 06531, Turkey

Abstract. Compared to conventional heat exchangers, direct contact heat exchangers have many advantages especially in an annular geometry; because, the unequal shear stresses exerted on the liquid drops located in the annulus by the inner and outer pipe surfaces cause the deformation of drops and the break-up of drops into droplets, thus the enhancement of heat transfer is favored. This paper is a study of direct contact heat exchange between two immiscible liquids (water and oil), without phase change, in co-current turbulent flow through a horizontal concentric annulus. A correlation exists among the Nusselt number, dispersed phase volume fraction, continuous phase Weber number, annular gap and the ratio of inner radius to outer radius, considering the effects of the physical properties of the phases, the operating conditions and the geometry of the system on the overall volumetric heat transfer coefficient. A discussion of the experimental parameters and a parametric study with a calculation scheme for estimating drop size, shear stress between a drop and the continuous phase, volumetric heat transfer coefficient and the difference between wall shear stresses exerted on a drop are presented in the paper.

1. Introduction

Direct contact heat transfer between liquids has the advantage of eliminating problems associated with heat transfer surfaces such as corrosion and fouling. Since the principle in direct contact heat exchangers is to provide intimate contact between a hot fluid and a cold fluid, the mixing technique determines the type of heat exchanger: "layer heat exchanger" or "spray heat exchanger"[1]. In a layer heat exchanger, one of the fluids is stagnant while the other flows on top. In the spray type, one is dispersed in the other. The latter type has been studied in sea-water desalination [2,3]. As reported by Sideman and Taitel [4], Simpson et al [5] and Sambi [6], a dispersed liquid droplet changes its shape from spherical through ellipsoidal to a cap-shaped bubble during the course of evaporation in a continuous immiscible liquid medium; the mechanism of heat transfer and the dynamics of the system are greatly affected by the transformation in the shape of the dispersed phase during phase change [7]. In the desalination of seawater by direct contact heat transfer, a two-phase system consisting of a mineral oil (Energol WM-2) and water has been evaluated for its workability and heat transfer properties under boiling conditions by Özmen [8].He recommended that his results be used in the design of heat exchangers, evaporators and pumps used for conducting or recirculating the immiscible liquid mixtures. Recently, there are very valuable contributions in this field, one of them being an extensive experimental and numerical study by Tadrist et al [9].

In previous direct contact heat transfer studies without phase change, Grover and Knudsen [10] determined the rate of heat transfer between a petroleum solvent and water in a circular pipe used as a co-current heat exchanger; better dispersion gave higher volumetric heat transfer coefficients. Porter et al . [11] and Yalvaç [12] investigated the direct contact heat transfer between oil and water in turbulent horizontal pipe flow using a semi-empirical method of correlating the data. Köksal [13] performed a similar study with oil and salty water in turbulent pipe flow. Takahashi et al.[14] studied the thermo-hydrodynamic characteristics of a co-current stratified flow of liquid metal and water in a horizontal rectangular channel. Leib et al. [15] investigated the annular laminar flow of two immiscible liquids (kerosene-water) in a vertical pipe. The outer phase was kerosene around the inner liquid phase, which was water, and it acted as a thermally insulating layer between the electrically heated wall and water flowing in the core region of the pipe. The lubrication of a non-Newtonian liquid by a Newtonian fluid

S. Kakaç et al. (eds.), Heat Transfer Enhancement of Heat Exchangers, 649–663.

(oil-water system) in the concentrically stratified laminar core-annular flow of two immiscible liquids in a horizontal tube has been investigated by Bakhtiyarov and Siginer [16], the results of which are useful in high wax content crude oil transportation problems and in the design of lubricated pipelines. The fluid flowing in the core region was oil in that study.

Most of the studies reported in the literature are concerned with circular pipes; relatively little research, of either experimental or theoretical nature, on liquid-liquid flows in non-circular channels exists. In considering the factors which may increase the drop deformation and break-up, thus provide a greater heat transfer surface and better contact between the phases in a given volume, the interfacial shear is an important one among many others [1]; therefore, choosing the annulus geometry to study direct contact heat exchange between two immiscible liquids, firstly without phase change seems useful.Thus, this subject was investigated in horizontal concentric annuli with different annular gaps and aspect ratios, κ, between a hot oil (BP-Transcal-N) and cold tap water phases at different operating conditions of co-current flows by the present authors [17,18]. The annular geometry was found to be effective in enhancing the heat transfer rate and the obtained volumetric heat transfer coefficients were higher than those of previous work performed by Yalvaç [12] using the same liquid-liquid system in cocurrent pipe flow. The objective of this paper is to present additional correlations that may be useful in designing similar systems and to give a parametric study with some theoretical considerations.

2. Experimental Work

The immiscible liquids used in this study were water and a heating oil manufactured by British Petroleum with a trade name of BP-Transcal-N. A schematic diagram of the experimental set-up is given in Figure 1.

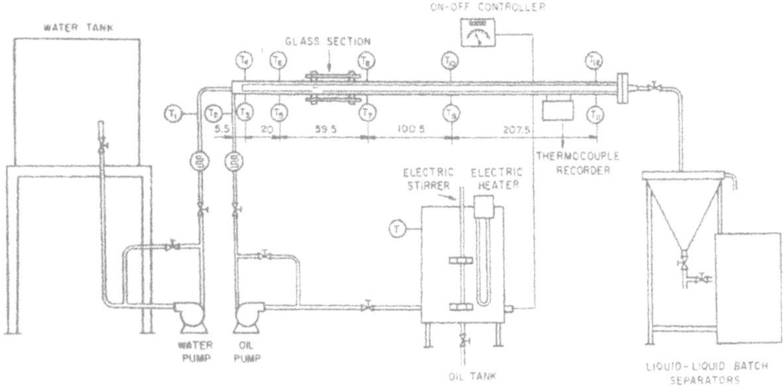

Figure 1.Schematics of the experimental set-up

It consisted of a large water tank, a horizontal annulus, a liquid-liquid separator tank, oil and water pumps, an oil tank equipped with a 3 kW electrical heater, an electrical stirrer and a resistance thermometer connected to a temperature controller to keep the inlet oil temperature constant at the desired value. The test section was a horizontal concentric annulus made of galvanised iron, approximately 4 m in length. A short glass pipe section of 25 cm in length, close to the entrance region and   having the same diameter as that of the outer pipe,was used for visual and photographic observations of flow patterns, formation of drops and size of drops of the dispersed phase. The oil and water streams were introduced into a very short pipe section of 2 cm in length where they hit each other at an angle, so that a good dispersion and a proper mixing of two liquids before entering the annulus was provided. Experimental data for direct contact heat transfer were obtained in a variety of flow conditions to study the effects of oil volume fraction, total liquid velocity, inlet oil temperature and aspect ratio by the present authors and reported elsewhere with further details of the experimental set-up and the experimental procedure [17,18].

3. Discussion of Experimental Results

The volumetric heat transfer coefficients U_{vo} calculated from experimental data at different oil volume fractions Φ_o keeping the total liquid velocity $V_T=(\dot{V}_o+\dot{V}_w)/A_f$, aspect ratio κ and the inlet oil temperature T_{oi} constant are given in Figure 2.

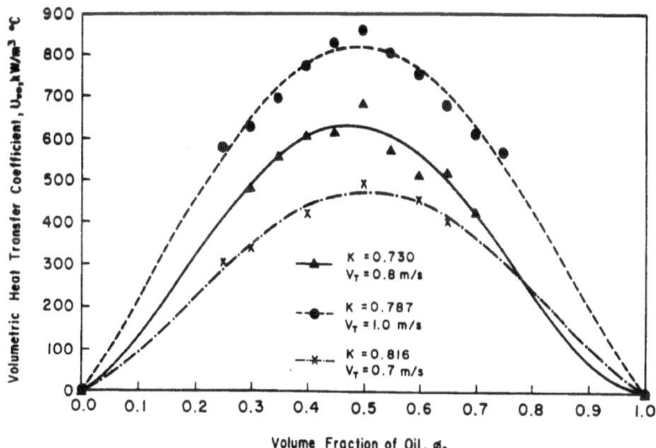

Figure 2. Plot of U_{vo} vs, Φ_o in the horizontal annuli with different aspect ratios
(From [18])

It was found that in each concentric pipe-set, at any chosen value of the total liquid velocity, U_{vo} values (calculated over the total length of the annular heat exchanger, $L = 3.93$ m, including the 2 cm long mixing section just before the annulus) increased with increasing dispersed phase volume fraction Φ_d and reached a maximum at almost $\Phi_d \cong 0.5$. When Φ_o was less than 0.5, water was the continuous phase, and when Φ_o was equal to or greater than 0.5, oil was the continuous phase. Therefore $\Phi_d = 0.5$ can also be written as $\Phi_o =\Phi_w = 0.5$. If this result is compared with those of previous studies [11-13] in pipe flow, it is observed in Figure 3 that the volumetric heat transfer coefficient increases with the dispersed phase concentration up to $\Phi_o=\Phi_w = 0.25$; then it remains constant between 0.25 and 0.75 oil volume fraction. In this region of constant heat transfer coefficient, the rate of increase in the interfacial area due to the increasing concentration of the dispersed phase was claimed to be counterbalanced by the increasing rate of coalescence. Porter et al. [11] reported that the transition point from a water- continuous phase to an oil- continuous phase was approximately determined at $\Phi_w = 0.3$ in pipe flow, though this transition was not clear-cut.

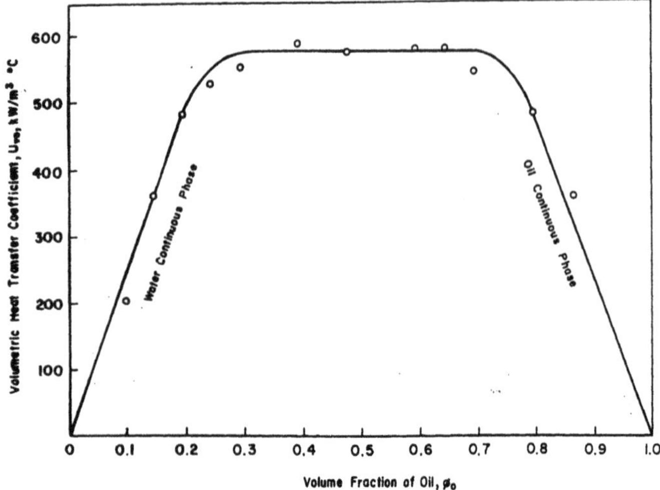

Figure.3 Plot of U_{vo} vs , Φ_o in a pipe (From [12])

However, in the present work, this point was clearly observed at around $\Phi_o = \Phi_w = 0.5$, where a maximum value of U_{vo} was expected, since U_{vo} increased with the dispersed phase concentration. This can be qualitatively explained by the unequal shear stresses exerted on the drops located in the annulus by the inner and the outer pipe surfaces [19, 20], which cause the deformation of spherical liquid drops to pear-shaped drops and the break-up of drops into droplets. Thus, the rate of coalescence decreases, and the interfacial area of the dispersed phase per unit volume of the test section continues to increase with the increasing dispersed phase volume fraction. Therefore, U_{vo} values do not remain constant, and increase to their maximum values at $\Phi_o = \Phi_w = 0.5$. This effect becomes more significant as the κ value decreases; the difference between the shear stresses exerted on the two wall-side peripheries of the drop in the annulus increases. This changes the shape and the rotational behaviour of the drop, and in turn enhances the heat transfer rate between the dispersed and the continuous phases [19, 20].

When the behaviours of dispersed and continuous phases were visually observed through the transparent section of the system, the following points were noted: at each constant liquid velocity and each oil volume fraction, the highest U_{vo} values were obtained in pipe-set II with the smallest aspect ratio of 0.73. Apparently, this pipe-set created the most favorable flow patterns for the enhancement of direct contact heat transfer between immiscible liquids. In most of the runs, a good dispersion was observed in each phase. At lower velocity ranges, partial stratification was observed, and the upper phase consisted of dispersed water in oil while the lower phase consisted of dispersed oil in water. A significant deformation and elongation of dispersed droplets in the flow direction was observed due to the shear stresses. For the smallest flow area and medium κ value of 0.787, at high velocities fully dispersed flows and at low velocities stratified-dispersed flows were observed. Elongation of droplets was not significant and the shape of the droplets was spherical. For the largest κ value of 0.816 and the largest flow area, at low flow velocities due to low shear, dispersion between the phases was low; the two phases flowed in a stratified manner [17,18].

4. Correlation of Data With Theoretical Considerations

Two important factors to be considered in this study are: the dependence of the drop size (hence, the interfacial area) on the physical properties of the phases, operating conditions, geometry of the system and the rate of heat transfer to or from an individual drop. Using the statistical theory of turbulence [20], Porter et al . [11] obtained the characteristic drop size *"d"* as follows:

$$d/D \sim We^{3/5} \tag{1}$$

where D is the pipe diameter and We is the Weber number. They also reported the ratio of the relative velocity *"u"* for two fluid elements to the mixture velocity V_T as follows:

$$u/V_T \sim (d/D)^{1/3} \tag{2}$$

From simple volumetric considerations, the interfacial area A in the test section of volume V is,

$$A \sim \Phi_d V/d \tag{3}$$

where Φ_d is the volume fraction of the dispersed phase. After considering internal and external heat transfer coefficients in their analysis, the volumetric heat transfer coefficient for liquid-liquid pipe flow was given [11] as:

$$\frac{U A D^2}{V k_d} \sim \Phi_d We^{-6/5} \tag{4}$$

In the present study, the same approach is used in obtaining correlations from experimental data [18]. The major resistance to heat transfer is considered to be in the oil phase due to its low thermal conductivity; an additional dimensionless group, the aspect ratio κ is introduced into Eq. (4) to account for the annulus geometry. Besides, the pipe diameter in Eq. (4) is replaced by the hydraulic diameter of the annulus. The unknown regression coefficient and the powers of dimensionless groups of the non-linear correlations derived from the modification of Eq. (4) are obtained by Marquardt's algorithm [21] which yields the following correlations in Eqs. (5), (8) and (10) giving the best fit to the present experimental data [18] within ± 20 % accuracy:

$$Nu = 40.219 \ \Phi_o^{\ 0.54} \ We_w^{\ 0.33} \ \kappa^{-3.86} \qquad (0< \Phi_o <0.5) \tag{5}$$

In Eq. (5), the Nusselt number based on the overall volumetric heat transfer coefficient ($U_{vo} = UA/V$) is expressed in terms of U_{vo}, the hydraulic diameter D_h , and the thermal conductivity of oil k_o , as follows:

$$Nu_{vo} = U_{vo} D_h^{\ 2} / k_o \tag{6}$$

Eq.(5) is valid in the region where oil is the dispersed phase and the Weber number has been calculated with the density of the continuous phase, namely the density of water ρ_w , as it is given in Eq. (7):

$$We_w = \rho_w (V_T)^2 D_h / \sigma \tag{7}$$

The other correlation obtained, in the region where water is the dispersed phase, is given as follows:

$$Nu_{vo} = 41.741 \ \Phi_w^{\ 0.70} \ We_o^{\ 0.42} \ \kappa^{-2.95} \qquad (0< \Phi_w <0.50) \tag{8}$$

654

where Nu_{vo} is defined in Eq. (6), the Weber number has been calculated with the density of oil ρ_o, oil being the continuous phase, as it is given in Eq. (9):

$$We_o = \rho_o \, (V_T)^2 \, D_h / \sigma \tag{9}$$

Comparing the powers of κ in Eq. (5) and Eq. (8) it is obvious that the effect of κ is greater in the region where oil is the dispersed phase (Eq. 5). The increasing shear stresses with the decreasing κ value can more easily deform oil droplets rather than water droplets. This can be explained with the functional relationship by Hinze [20] between the Weber number $We_o = \rho_c \, (V_T)^2 D_h / \sigma$, the viscosity ratio of the immiscible liquid phases (μ_c/μ_d), and the viscosity group defined by $\mu_d / (\sqrt{(\rho_d \, \sigma \, d)})$ where μ, ρ, and σ, are the viscosity, density and the interfacial tension, respectively; the subscripts c and d indicate the continuous and the dispersed phases, respectively. D_h is the pipe diameter or the hydraulic diameter depending on the geometry of the system and d is the diameter of the liquid globule or drop. The Weber number generally refers to the dynamic pressure set up in the continuous phase around the globule and the critical Weber number is reached when the deformation or the break-up of the globule occurs. Further, that the critical Weber number occurs at lower values of the viscosity group as the viscosity of the continuous phase decreases. From this explanation it can be deduced that as the external force on the liquid drops increases with decreasing κ, oil drops in the water-continuous phase can be more easily deformed than the water drops in the oil-continuous phase. This is because, in the latter case, the occurrence of the critical Weber number for the deformation of the water drops corresponds to a higher value of the viscosity group due to the higher viscosity of the oil-continuous phase. In other words, the external force required to deform water drops in the oil phase should be greater to overcome the viscous forces.

Eq. (5) and Eq. (8) can be represented by a single correlation, obtained by using all the experimental data of this study, as follows:

$$Nu_{vo} = 42.278 \; \Phi_d^{0.67} \; We_c^{0.38} \; \kappa^{-3.37} \tag{10}$$

where Φ_d is the dispersed phase volume fraction, and We_c is defined for the continuous phase. It can be summarized that the volumetric Nusselt number in direct contact heat transfer of two immiscible liquids in horizontal concentric annuli is directly proportional to the (2/3) power of a dimensionless group, namely $(\Phi_d \sqrt{We_c} / \kappa^5)$, as shown in Eq. (11):

$$Nu_{vo} \sim [\Phi_d \sqrt{We_c} / \kappa^5]^{2/3} \tag{11}$$

Eqs. (5), (8) and (10) are $\pm$ 20 % accurate for the following ranges of the variables: oil volume fraction Φ_o , $0.25\text{-}0.75$: mixture velocity, V_T, 0.4-1.1 m/s; aspect ratio, κ, 0.730-0.816; inlet water temperature, T_{wi}, 15-22°C; inlet oil temperature, T_{0i}, 50-75°C; hydraulic diameter, D_h $7.6\text{x}10^{-3}$ -$9.6\text{x}10^{-3}$ m; flow area, A_f $3.80\text{x}10^{-4}$ - $6.98\text{x}10^{-4}$ m^2. The ranges of the Weber number in Eqs. (5), (8) and (10) are $(37 < We < 190)$, $(30 < We < 150)$, and $(30 < We < 190)$, respectively. Nu_{vo} values predicted by Eq. (10) are plotted against the experimental Nu_{vo} values, and the scattering of points around the diagonal is shown in Figure 4.

The accuracy of $\pm$ 20 % in the above-given correlations can be justified by considering the occurrence of different flow patterns for the oil and the water phases at different operating conditions. Depending on the mixture velocity, the phase concentration, and the κ value, different patterns of flow such as stratified flow, dispersed-stratified flow and fully dispersed flow could be visually observed through the transparent section of the flow system.

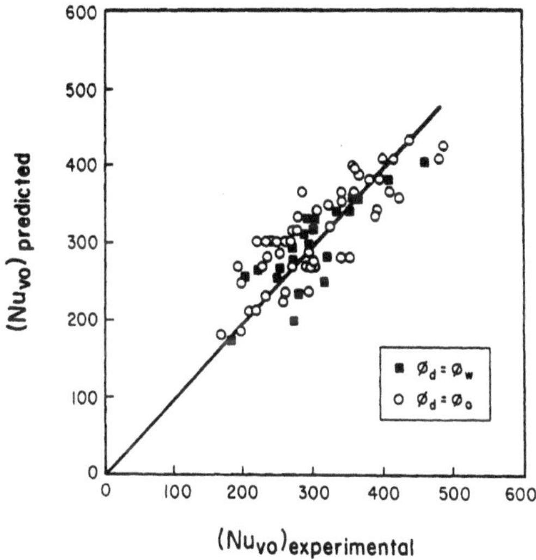

Figure 4. Plot of $(Nu_{vo})_{pred.}$ vs , $(Nu_{vo})_{exptl.}$

5. A Parametric Study and Further Considerations

In this section, a parametric study is aimed to check the effects of main parameters and especially the ones related to the geometry of the system (e.g., shearing action in the annular gap) on the heat transfer rate. In this analysis, considering the complexity of flow, the following assumptions are made:

(a) spherical, uniform size, rigid drops exist in the flow section,

(b) volume fraction of dispersed phase is uniform in axial and radial directions,

(c) the flow is steady and one-dimensional,

(d) relative velocity between the phases is proportional to V_T defined in this study [11,20] ,

(e) coalescence and break-up of drops do not occur ,

(f) all drops travel with the same velocity of u_d, the velocity of dispersed phase.

The number of drops passing a cross-section per unit time $(\dot{N}$ = drop number flow rate) is a function of the relative velocity or V_T (see assumption-d), shear stress between a drop and the continuous phase τ_{dc} ,the viscosity, density and the size of the drop. Assuming that some of the variables remain almost constant, the dimensioual analysis has been carried out among the following variables:

$$\dot{N}= func \ (V_T, \ \tau_{dc} , \ \mu_d) \tag{12}$$

656

and the result is:

$$\dot{N} = K_1 \left(\tau_{dc} / \mu_d \right) \tag{13}$$

where K_1 is a constant to be determined experimentally. Obviously, $\dot{N}$ can be written in terms of the mass flow rate of the dispersed phase $\dot{m}_d$ divided by the mass of one drop m_{dr},

$$\dot{N} = \dot{m}_d / m_{dr} \qquad \text{or} \qquad 6 \dot{V}_d / \pi d^3 \tag{14}$$

where $\dot{V}_d$ is the volumetric flow rate of the dispersed phase and d is the uniform drop diameter. From Eqs. (13) and (14), the drop size can be estimated as follows:

$$d = (6/\pi K_1)^{1/3} \left(\dot{V}_d \, \mu_d / \tau_{dc} \right)^{1/3} \qquad \text{where } K_2 = (6/\pi K_1)^{1/3} \text{ is a constant,}$$

$$\text{then} \qquad d = K_2 \left(\dot{V}_d \, \mu_d / \tau_{dc} \right)^{1/3} \tag{15}$$

The local heat transfer coefficient h_{loc} between a drop and the continuous phase can also be predicted by dimensional analysis, considering the most important variables, as follows:

$$h_{loc} = func \left(d, \mu_d, Cp_d, V_T, \sigma \right) \tag{16}$$

$$h_{loc} = K_3 \left(Cp_d \, \sigma / d \, V_T \right) \left(\mu_d \, V_T / \sigma \right)^b \tag{17}$$

Since σ is a constant for the specific oil-water system chosen in this study, the power "b" of the dimensionless group in Eq. (17) can be equated to one, then Eq. (18) follows:

$$h_{loc} = K_3 (Cp_d \, \mu_d / d) \tag{18}$$

where K_3 is another constant. The rate of heat transferred from the dispersed phase to the continuous phase between the inlet and the chosen cross-section at a certain length of the concentric pipe-set is equal to

$$Q = \dot{m}_d Cp_d (T_{di} - T_d) \tag{19}$$

where T_{di} and T_d are the inlet temperature and the local temperature of the dispersed phase at the chosen exit length L_e, respectively. The arrival time of the drops to the chosen specific area at a distance L_e from the entrance is $t = L_e / u_d$, the assumed uniform velocity of drops being u_d. Therefore, the time average rate of heat exchange between the dispersed phase and the continuous phase can be written as follows:

$$q = Q / (L_e / u_d) \tag{20}$$

The local q value can also be obtained at a distance L_e from Eq. (21) ;

$$q = h_{loc} \left(\pi d^2 \dot{N} \right) \left(T_d - T_c \right) \tag{21}$$

where $\dot{N}$ is the number of drops passing that cross-section per unit time; T_d and T_c are the local temperatures of the dispersed phase and the continuous phase at a length L_e from the entrance, respectively. From Eqs. (14), (15), (18 - 21), the following relationship is obtained:

$$(T_{di} - T_d) / (T_d - T_c) = K_4 (\mu_d \, \tau_{dc}^2 / \dot{V}_d^2)^{1/3} (L_e / u_d \rho_d) \tag{22}$$

where $K_4 = \pi K_1 K_2 K_3$ is a lumped constant. Since,

$$\dot{m}_d = \dot{V}_d \rho_d \text{ and } \dot{V}_d = u_d A_f \Phi_d \tag{23}$$

then Eq. (22) has the following form:

$$(T_{di} - T_d)/(T_d - T_c) = K_4 (\mu_d \rho_d^2 \tau_{dc}^2 / \dot{m}_d^5)^{1/3} (V_e \Phi_d) \tag{24}$$

where, the volume of the test section is $V_e = A_f L_e$. From the experimental data, all the parameters in Eq. (24) are known for each run, except τ_{dc} which is a function of drop diameter as it is given in Eq. (15).

The heat transfer rate Q, is also obtained by using a mean heat transfer coefficient h_m, the total surface area of all the drops that travel along the total length of the heat exchanger and the logarithmic mean temperature difference, ΔT_{lm} . In this case,

$$Q = h_m (\pi d^2 \dot{N}) (L / u_d) \Delta T_{lm} \tag{25}$$

where L is the total length of the heat exchanger. From Eqs. (14), (15), (19), (23) and (25), h_m can be found as follows:

$$h_m = K_5 (\dot{m}_d^4 \mu_d / \tau_{dc} \rho_d)^{1/3} (Cp_d \Delta T / V \Phi_d \Delta T_{lm}) \tag{26}$$

where $K_5 = 1 / (\pi K_1 K_2^2)$; $\Delta T = T_{di} - T_{do}$

and $\Delta T_{lm} = [(T_{di} - T_{ci}) - (T_{do} - T_{co})] / ln [(T_{di} - T_{ci}) / (T_{do} - T_{co})]$

All parameters in Eq. (26) have been measured except τ_{dc} ; therefore, τ_{dc} needs to be estimated, for which the following can be written [22]:

$$\tau_{dc} = C_D (\rho_c u_r^2 / 2) \tag{27}$$

where u_r is the relative velocity between the drop and the continuous phase; C_D is the drag coefficient, which is a function of Reynolds number of the drop of diameter d ; Re_d is written in Eq. (28) ;

$$Re_d = (d u_r \rho_c / \mu_c) \tag{28}$$

Since the drop size is not known, Eq. (15) can be substituted in Eq. (28):

$$Re_d = K_2 (\dot{V}_d \mu_d / \tau_{dc})^{1/3} (u_r \rho_c / \mu_c) \tag{29}$$

where u_r is given by Eq. (30), and ρ_c and μ_c are the density and the viscosity of the continuous phase, respectively.

$$u_r = u_d - u_c = \dot{V}_d / A_f \Phi_d - \dot{V}_c / A_f \Phi_c \tag{30}$$

5.1. PROPOSED CALCULATION SCHEME

From this parametric study using the experimental data [18], the following calculation scheme can be proposed:

1. A drop diameter can be assumed.
2. From Eq. (29), Re_d is calculated using Eq. (30) for u_r .
3. From an available correlation [23] for liquid-liquid systems

658

between C_D and Re_d, C_D can be obtained; thus τ_{dc} is calculated using Eq. (27),

4. From Eq. (15), the constant K_2 is determined, and thus $K_1=6/(\pi K_2^3)$ is found.
5. Using the available data [18] and a τ_{dc} value corresponding to the assumed drop size, the constant K_4 is obtained, using Eq.(24); hence $K_3=K_4/(\pi K_1 K_2)$ is calculated.
6. From known K_1 and K_2 values , the K_5 constant of Eq. (26) can be found, since K_5 is equal to $[1/(\pi K_1 K_2^2)]$.
7. Using the available experimental data and the known K_5 value, h_m can be calculated from Eq. (26).
8. Calculated h_m values are multiplied with the interfacial area between the drops and the continuous phase in the test volume V to obtain calculated U_{vo} values , which have also been determined experimentally.
9. These calculated $U_{vo}=h_m(\pi d^2 \dot{N} L/u_d V)$ can be written by using $\dot{N}=K_1(\tau_{dc}/\mu_d)$, $V=LA_f$ and $u_d=\dot{m}_d/(\rho_d A_f \Phi_d)$ to obtain the following equality, $U_{vo}=h_m(\pi d^2 K_1\tau_{dc}\rho_d \Phi_d/\mu_d \dot{m}_d)$.
10. Calculated U_{vo} values are compared with the experimentally determined U_{vo} values and the iteration procedure is continued.

5.2. EFFECT OF ANNULUS GEOMETRY ON HEAT TRANSFER

In the previous work by the present authors [18] , it was observed that as the aspect ratio of the annulus decreases, the volumetric heat transfer coefficients increase. This may be evaluated by studying the effect of the difference between the shear stresses exerted by the outer and inner walls on the drops.

As shown in Figure 5, when a drop is located in the annulus where the two wall side peripheries A and B experience different shear stresses, a rotation of the drop has to occur. As a result, the drop must move to a region where the energy required for rotation is minimum. This location should be the region where the shear stress is zero (point-C).Zero-shear points are at a radial distance of $r=(\lambda-\kappa) D_o/2$ from the horizontal central axis of the concentric annulus , where r is the radial distance and λ is given [19] for an annulus as follows:

$$2\lambda^2 =(1-\kappa^2)/ln(1/\kappa) \tag{31}$$

If it is assumed that the outer wall and inner wall shear stresses, τ_{oi} and τ_{io} , respectively, act on one-half surface area of a spherical drop with a diameter $"d"$, the following equalities can be written:

$$F_{oi}=\tau_{oi}(\pi d^2/2) \tag{32}$$

$$F_{io}=\tau_{io}(\pi d^2/2) \tag{33}$$

A torque Ω exerted on the spherical rigid drop about its center due to the difference of these forces is defined [24] as the cross-product of the moment arm and the net force:

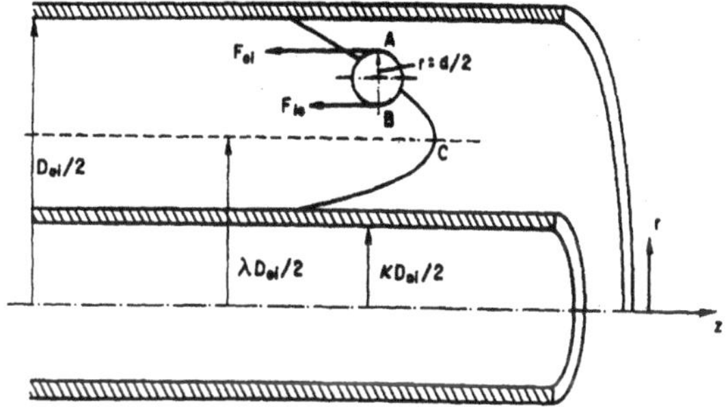

Figure 5. Forces on a drop in an annulus

$$\Omega = r \times (F_{oi} - F_{io}) \tag{34}$$

The direction of Ω is normal to the plane formed by r and , $(F_{oi} - F_{io})$ and its magnitude is given [24] in Eq. (35)

$$\Omega = (F_{oi} - F_{io})\, r\, \sin\theta \tag{35}$$

Considering that the angle θ between each force and its corresponding moment arm is 90°, and the moment arm of each force is equal to $(d/2)$ of the drop, from Eqs. (32), (33) and (35), the scalar magnitude of the torque can be written as follows:

$$\Omega = (\tau_{oi} - \tau_{io})\, \pi d^{3} / 4 \tag{36}$$

From basic physics [24], Eq. (37) can be written for the rotational motion about the center of mass of the drop:

$$\Omega = I\,\alpha \tag{37}$$

where I is the rotational inertia or moment of inertia of the body about a particular axis of rotation and α is the angular acceleration. For a rigid sphere, $I = (2/5)\, m_{dr}\, (d/2)^{2}$ and the relationship between angular acceleration of a point on the drop surface and the translational acceleration, $"a"$ of the center of mass of the drop given [24] as $a = \alpha\,(d/2)$ are substituted into Eq. (37), then Eq. (36) is equated to Eq. (37) to obtain the following equation :

$$\tau_{oi} - \tau_{io} = 4\, m_{dr}\, a \,/\, (\,5\, \pi d^{2}\,) \tag{38}$$

Then, substituting $a = (2\, u_{r}^{2} / d)$ and $m_{dr} = (\pi d^{3} \rho_{d} /6\,)$ into Eq. (38), Eq. (39) is obtained:

$$\Delta\tau = \tau_{oi} - \tau_{io} = 4\, u_{r}^{2}\, \rho_{d} /15 \tag{39}$$

The difference in wall shear stresses $\Delta\tau$ can also be related to the overall volumetric heat transfer coefficient U_{vo} by dimensional analysis considering the variables such as hydraulic diameter D_{h} which accounts for the effect of annular gap, drop number flow rate $\dot{N}$, and the heat capacity of drops Cp_{d}

660

besides $\Delta\tau$:

$$U_{vo} = func\ (D_h,\ \dot{N},\ Cp_d,\ \Delta\tau) \tag{40}$$

the following relationship has been obtained:

$$U_{vo} = K_6\ (\Delta\tau)\ Cp_d\ /\ D_h^2\ \dot{N} \tag{41}$$

Substituting $\dot{N}$ from Eq. (13) into Eq. (41), the result is found in Eq. (42):

$$U_{vo} = (K_6\ /\ K_1)\ (\Delta\tau\ Cp_d\ \mu_d\ /\ D_h^2\ \tau_{dc}) \tag{42}$$

The calculation scheme proposed in this paper for the estimation of the drop size and the shear stress between a drop and the continuous phase can be continued here to check the $\Delta\tau$ values obtained using Eq. (39). After determining the approximate drop size from the proposed iteration procedure, which yields the calculated U_{vo} values close to the experimental U_{vo} values within a reasonable error margin, these U_{vo} values can be plotted against the group of variables $(\Delta\tau\ Cp_d\ \mu_d\ /D_h^2\ \tau_{dc})$ on the right hand side of Eq. (42). Since $(K_6\ /K_1) = (K_7)$ is a constant, in this plot a straight line relationship is expected. Deviations from the straight line may occur, mainly due to the assumption of considering the dispersed phase as of rigid, spherical and uniform-size drops. In order for the moving drop to be spherical, the forces due to surface tension at its boundary must exceed the forces due to presssure differences, which tend to make the drop nonspherical [25].

6. Conclusions

In direct contact heat transfer of immiscible liquids in annuli, the experimental data show that volumetric heat transfer coefficient increases with increasing dispersed phase volume fraction and reaches a maximum at around $\Phi_o = \Phi_w \cong 0.5$, keeping all other parameters constant. The magnitude of this maximum coefficient at $\Phi = 0.5$ and at a constant velocity increases with increasing total liquid velocity of oil-water mixture and with decreasing aspect ratio of annulus. The Nusselt numbers based on the overall volumetric heat transfer coefficient in direct contact heat transfer between two immiscible liquids flowing co-currently and turbulently through the horizontal concentric annuli were predicted with an accuracy of $\pm$ 20 %. The obtained correlations for the overall volumetric heat transfer coefficients consist of easy-to-measure parameters; therefore, they are quite practical to be used in the design of such systems.

In addition to the above work, a parametric analysis of the problem considering the most important variables, eliminating the ones remaining almost constant in this study, with its calculation procedure has been presented to estimate a uniform drop size, the shear stress between a drop and the continuous phase, and the effect of the difference in the outer and inner wall shear stresses on the volumetric heat transfer coefficient. This procedure needs to be checked with the available experimental data [18].

Acknowledgement
Inspiring discussions on the subject matter of this paper with a late Professor of Chemical Engineering Department, M.E.T.U., Prof. Dr. Tarık G. Somer is highly appreciated.

Nomenclature
a	Translational acceleration, m/ s^2
A	Total interfacial area in the test volume, m^2
A_f	Cross-sectional area of annulus, m^2

C_D	Drag coefficient
Cp	Heat capacity, kJ / kg .K
d	Drop diameter, mm
D	Pipe diameter, mm
D_h	Hydraulic diameter of annulus, $D_h = D_{oi} - D_{io}$, mm
D_{io}	Outside diameter of the inner pipe, mm
D_{oi}	Inside diameter of the outer pipe, mm
F_{io}	Force applied on a drop by the inner wall, N
F_{oi}	Force applied on a drop by the outer wall, N
h_{loc}	Local heat transfer coefficient, kW / m^2. K
h_m	Mean heat tansfer coefficient, kW / m^2. K
I	Moment of inertia of a spherical drop, kg m^2
K_i	Constants, $i = 1\text{-}7$
L_e	Length of test section upto a specified cross-section of exchanger, m
L	Total length of heat exchanger, m
m_{dr}	Mass of a drop, kg
$\dot{m}_d$	Mass flow rate of dispersed phase, kg/s
$\dot{N}$	Number of drops passing a cross-section, per unit time, or drop number flow rate, #/s
Nu_{vo}	Nusselt number based on U_{vo}
q	Time-average heat transfer rate, kW/ s
Q	Heat transfer rate, kW
Re_d	Reynolds Number of a drop
t	Time, s
T	Temperature, ^{0}C or K.
u_r	Relative velocity between continuous and dispersed phases, m/s
u_d	Velocity of a drop or of the dispersed phase, m/s
u_c	Velocity of the continuous phase, m/s
U_{vo}	Volumetric heat transfer coefficient, kW/m^3 $^\circ$C
V_e	Volume of test section up to a specific length L_e, m^3
V	Total volume of heat exchanger, m^3
$\dot{V}$	Volumetric flow rate , m^3/s
V_T	Total liquid velocity or mixture velocity, m/s
We	Weber Number based on continuous phase

Greek Symbols

α	Angular acceleration, rev/s^2
ΔT	Temperature difference, $^\circ$C
ΔT_{lm}	Logarithmic mean temperature difference, ^{0}C
θ	Angle between a force and its moment arm
κ	Aspect ratio of annulus, D_{io} / D_{oi}
λ	A parameter defined in Eq.(31)
μ	Dynamic viscosity, kg/ m . s
ρ	Density, kg/m^3
σ	Interfacial tension between two liquids, N/m
τ_{dc}	Shear stress between a drop and the continuous phase, N/m^2
τ_{io}	Shear stress applied by inner wall on a drop, N/m^2
τ_{oi}	Shear stress applied by outer wall on a drop, N/m^2
Φ	Volume fraction
Ω	Torque, N m

662

<u>Subscripts</u>

c	Continuous phase
ci	Continuous phase at the inlet
co	Continuous phase at the outlet
d	Dispersed phase
di	Dispersed phase at the inlet
do	Dispersed phase at the outlet
in	Inlet
out	Outlet
o	Oil
oi	Oil inlet
oo	Oil outlet
w	Water
wi	Water inlet
wo	Water outlet

References

1. Tadrist, L., Seguin, P., Santini,R., Pantaloni,J. (1985) *Experimental and Numerical Study of Direct Contact Heat Exchangers*, Int. J.Heat & Mass Transfer, **28**, pp. 1215-1227.
2. Letan, R., Kehat, E. (1968) *The Mechanism of Heat Transfer in a Spray Column Heat Exchanger*, A.I.Ch.E.J ,**14**, pp. 398-405.
3. Spiegler, K.S. (1966) *Principles of Desalination*, Academic Press, New York,N.Y.
4. Sideman ,S., Taitel,Y. (1964) *Direct Contact Heat Transfer with Change of Phase: Evaporation of Drops in an Immiscible Liquid Medium*, Int. J. Heat & Mass Transfer, 7, pp. 1273-1289.
5. Simpson,H.C., Beggs,G.C.,Nazir, M. (1973) *Evaporation of Butane Drops in Brine*, Proceedings of the 4th International Symposium on Fresh Water from the Sea, **3**, pp.409-420.
6. Sambi,S.S. (1981) *Direct Contact Heat Transfer to Vaporizing Drops in Immiscible Liquids*, Ph.D. Thesis, I.I.T., Delhi.
7. Raina,G.K., Wanchoo,R.K. (1968) *Direct Contact Heat Transfer with Phase Change: Bubble Growth and Collapse*, The Can. J. of Chem.Eng., **64** , pp.393-398.
8. Özmen,S. (1971) *Heat Transfer and Frictional Properties of a Two-Phase Liquid System*, M.Sc. Thesis in Chem.Eng., M.E.T.U., Ankara, Turkey.
9. Tadrist,L.,Sun,J., Santini,R., Pantaloni,J. (1991) *Heat Transfer with Vaporization of a Liquid by Direct Contact in Another Immiscible Liquid: Experimental and Numerical Study*, Trans. ASME, J. Heat Transfer , **113**, pp. 705-713.
10. Grover,S.S., Knudsen,J.G. (1955) Chem.Eng.Prog. Symp. Ser., **51**, p.71.
11. Porter, J.W., Goren,S.L., Wilke,C.R. (1968) *Direct Contact Heat Transfer Between Immiscible Liquids in Turbulent Pipe Flow*, A.I.Ch.E. J., **14**, pp. 155-158.
12. Yalvaç,S. (1974) *Direct Contact Heat Transfer Between Two Immiscible Liquids in a Pipe*, M.Sc. Thesis in Chem.Eng., M.E.T.U., Ankara, Turkey
13. Köksal,D. (1980) *Direct Contact Heat Transfer Between Immiscible Liquids in Turbulent Pipe Flow*, M.Sc. Thesis in Chem.Eng., M.E.T.U., Ankara, Turkey
14. Takahashi, M., Inoue, A.,Aoki, S., Aritomi,M., Endo,H. (1981) *Interface Heat Transfer of Horizontal Co-current Liquid-Liquid Stratified Flow*, Bull. JSME, **24** , pp.1002-1010.
15. Leib, T.M., Fink,M., Hasson,D. (1977) *Heat Transfer in Vertical Annular Laminar Flow of Two Immiscible Liquids*, Int.J.Multiphase Flow, **3**, pp. 533-549.
16. Bakhtiyarov,S.I., Siginer, D.A. (1998) A *Note on the Laminar Core-Annular Flow of Two Immiscible Fluids in a Horizontal Tube*, Proceedings of International Symposium on Liquid-Liquid Two-Phase Flow and Transport Phenomena, pp.107-112, Begell House Inc.,ISBN:1-56700-111-4,New York,N.Y.
17. Shahidi,M.K., Özbelge, T.A. (1995) *Direct Contact Heat Transfer Between Two Immiscible Liquids Flowing in a Horizontal Concentric Annulus*, Int. J. Multiphase Flow, **21**, No.6, pp1025-1036.
18. Shahidi,M.K. (1994) *Direct Contact Heat Transfer Between Two Immiscible Liquids Flowing in*

a Horizontal Concentric Annulus, M.Sc. Thesis, Chem. Eng. , M.E.T.U., Ankara, Turkey

19. Bird,R.B.,Stewart,W.E., Lightfoot,E.N. (1960) *Transport Phenomena,* Wiley & Sons, Inc. New York, N.Y.

20. Hinze, J.O. (1955) *Fundamentals of the Hydrodynamic Mechanism   of Splitting in Dispersion Processes,* A.I.Ch.E. J., **1**, pp.289-295.

21. Marquardt, D.W. (1963), J. Soc. Ind. Appl. Math, **11** , p.431

22. McCabe, W.L., Smith, J.C., Harriott, P. (1993), *Unit Operations of Chemical Engineering* Fifth Edn. McGraw-Hill, Inc., New York, N.Y.

23. Clift,R., Grace, J.R., Weker, M.E. (1978) *Bubbles, Drops and Particles,* Academic Press, New York, N.Y.

24. Resnick,R., Halliday, D. (1962) *Physics,* Part 1, pp.216-241, John Wiley & Sons,Inc., New York, N.Y.

25. Landau,L.D., Lifshitz, E.M. (1959) *Fluid Mechanics*, Pergamon Press, Addison-Wesley Inc., p.70.

UNRESOLVED PROBLEMS AND SUGGESTIONS FOR FUTURE RESEARCH

MAZHAR ÜNSAL
Department of Mechanical Engineering
University of Gaziantep
Gaziantep 27310, Türkiye

Abstract. This paper is an edited summary of the short presentations made by the contributing Lecturers and Participants during the "Panel Discussions" session of the NATO Advanced Study Institute on "Energy Conservation through Heat Transfer Enhancement of Heat Exchangers". The "Panel Discussions" session was organised during the Advanced Study Institute to identify unresolved problems and suggestions for future research.

1. Introduction

The NATO Advanced Study Institute on "Energy Conservation Through Heat Transfer Enhancement of Heat Exchangers" was held in Çesme, Türkiye from May 25,1998 to June 5, 1998. The "Panel Discussions" session chaired by the present author during the NATO Advanced Study Institute was scheduled by Dr. Sadik Kakaç for identification of unresolved problems and topics for future research and development. All of the participants of the NATO Advanced Study Institute were invited by the present author for contribution to the "Panel Discussions" session prior to this session. The following Distinguished Lecturers and Participants have then contributed to the "Panel Discussions" session (given in alphabetical order):

Dr. Arthur E. Bergles, Renssalear Polytechnic Institute, USA
Dr. Takeshi Ebisu, Daikin Industries Ltd, Japan
Dr. Martin Fiebig, Ruhr-Universitat Bochum, Germany
Dr. Oleg A. Kabov, Russian Academy of Sciences, Russia
Dr. Tuncer Kuzay, Argonne National Laboratory, USA
Dr. Ralph L. Webb, Pennysylvania State University, USA
Dr. Wen-Jei Yang, University of Michigan, USA

Dr. Arthur Bergles had to leave the meeting before the "Panel Discussions" session and was not able to attend the session. Instead, he submitted his suggestions and recommendations in advance, which were then presented during the session by the present author. All the other contributors participated in the "Panel Discussions" session by means of 10-15 minute short presentations. This paper consists of a condensed version of the suggestions and recommendations for future research as presented by these Distinguished Experts. Their suggestions and recommendations are included in this paper with a minimum amount of editing in order to render the contents of these contributions as close as possible to the original presentations given during the "Panel Discussions" session of the NATO Advanced Study Institute.

S. Kakaç et al. (eds.), Heat Transfer Enhancement of Heat Exchangers, 665–673.

The first presentation was given by the present author and it is basically a review of the suggestions and recommendations given by Dr. Bergles. In the next presentation, Dr. Ebisu describes unresolved problems important in the design and development of air cooled heat exchangers and emphasises the importance of manufacturing methods and materials and also the importance of third generation heat transfer enhancement technology. In the following short presentation, Dr. Fiebig points to the research needs for the development of efficient heat transfer enhancement devices or generators and to the importance of materials and combined enhancement effects. Dr. Oleg A. Kabov brings to our attention three unresolved problems, one problem related to the prediction of forced flow condensation inside longitudinally finned tubes, a second problem related to condensation on three dimensionally finned and sawtooth shaped surfaces and a third problem related to the horizontal standing wave formation in two phase flows. In the next presentation, Dr. Tuncer Kuzay gives information on new and emerging fields such as Microscale Energy Transport, which are open for potential research to mechanical engineers and the thermal scientist. Dr. Webb points to the importance of sophisticated experimental tools, CFD, predictive models based on fundamental understanding of the enhancement mechanisms, materials and the tendency toward smaller sized structures in design. Finally, Dr. Wen-Jei Yang discusses emerging new research areas in thermal sciences and possible future changes in the undergraduate curriculum.

2. Summaries of the Short Presentations made by the Contributors

2.1 SUMMARY OF THE SHORT PRESENTATION BY Dr. MAZHAR ÜNSAL

Dr. Arthur Bergles had to leave early and is not with us in this session today. I would like to thank Professor Arthur Bergles, in his absence, for giving me a written list of his suggestions and recommendations in advance. What I would like to do now is to try to quickly go over the suggestions and recommendations Professor Bergles would like to convey to the audience in this "Panel Discussions "session.

(1) The first point he would like to make is to remind us of the following statement by Santayana: "Those who cannot remember the past are condemned to repeat it". What Professor Bergles is trying to emphasise here is that one should learn the existing literature very well and know what has been accomplished in the past before starting with a new research in a given technical field. This way you will not repeat something done in the past and which is already available to you.

(2) Secondly, he would like to point to the importance of "technology transfer". Professor Bergles is trying to emphasize here that the mechanism by which research results go into practical applications must also be enhanced and that we should be putting more effort on this matter. I can give an example backing up his comment. Here in this resort they did not have an air-conditioning system in the past. Two years ago they installed a new air-conditioning system. The technical manager of the resort told me today that they do not have a regenerator for energy recovery on this air-conditioning system. As you know, in air-conditioning systems, we have outdoor air intake and exhaust air outlet and in between these air streams we can use a regenerator for energy recovery. And if we include a regenerator in the design phase of an air conditioning system, the heat load of the air-conditioning system is reduced. This reduces the size of the air conditioning system to be installed and also there is a reduction in the running costs. In Turkey, most air-conditioning systems except a few do not use these regenerators. In this NATO Advanced Study Institute we are discussing enhancement of regenerators, considering the presentation by Dinglreiter et.al. which is titled "Enhancement of Combined Heat and Mass Transfer in Rotary Exchangers", but we see that even ordinary regenerators without any enhancement are not utilized in practical applications. This is an example for the phase lag between research and utilization of this research in applications. Therefore it is important that these new techniques should make a rapid transition from the academic or industrial research laboratories into engineering design practice.

Other suggestions and recommendations by Professor Bergles are given in condensed form as follows:

(3) Enhanced heat transfer will be more important in the future when energy prices rise again.

(4) Manufacturing methods and materials requirements may be overriding considerations in enhancement technologies and it is important to do research on anti-fouling surface development.

(5) Enhancement technology is still largely experimental, although great strides are being made in analytical/numerical descriptions of the various technologies. Accordingly it is imperative that experimentation is a necessary tool for the study of enhancement which must be kept viable. Experimentation is still a vital art needed for direct resolution of transport phenomena in complex enhanced geometries as well as benchmarking computer codes. As such, experimental skills should continue to be taught and conventional laboratories should be maintained. Heat transfer itself must be thought in the Mechanical Engineering Curriculum and enhancement should also be part of the courses on heat transfer.

It is also worthwhile to emphasize the importance of third generation enhancement techniques, which have potential for new research projects. In the "Concluding Remarks" section of his paper titled "The Imperative to Enhance Heat Transfer" which was presented in this NATO advanced Study Institute, Professor Bergles have already mentioned what he means by third generation enhancement technology.

Finally, I would like to bring to your attention the potential of compound enhancement as another emerging area in enhancement research. What Professor Bergles means by compound enhancement is combined utilization of more than one enhancement mechanism in a given heat transfer enhancement problem.

2.2 SUMMARY OF THE SHORT PRESENTATION BY Dr.TAKESHI EBISU

In this discussion session, I would like to mention some unsolved problems in relation with the design and development of air-cooled heat exchangers. I would like to introduce unsolved problems for both the airside and the refrigerant side, which are important in the design of air-cooled heat exchangers.

The most important problem on the airside is how to design a heat exchanger in wet conditions. Condensate water on the outer surface has an effect on heat transfer coefficient and we do not have a calculation method which considers the effect of the condensate water. We do not have design information on this subject, which requires more attention.

Another unsolved and I think the most severe problem for the airside is "the development of a next generation heat transfer enhancement technique for enhancement of the airside heat transfer". I showed in this morning our idea of enhancement for the airside heat transfer coefficient, which is the utilization of meshed type fins. But, is there any other better idea to enhance the air side coefficient significantly. One possible solution may be the aluminum brazed heat exchanger, which is being investigated by Prof. Webb. I think the aluminum brazed heat exchanger is a possible solution but we really need a more effective significant heat transfer enhancement method for this application.

Up until now we have used the inner grooved tube for the tube side heat transfer enhancement. We now need a next generation heat transfer technology to replace the inner grooved tube because we have used the inner grooved for over 20 years and I think this inner grooved tube is now considered as old technology. We need a next generation heat transfer enhancement idea. One possible solution is the Herringbone tube. This Herringbone tube is based on a new heat transfer enhancement mechanism. Conventional inner grooved tube is based on a uniform inner liquid layer inside the tube. However the heat transfer mechanism for our Herringbone tube is based on a non-uniform liquid film thickness inside the tube. This is just our idea but we really need a better next generation heat transfer enhancement idea to replace this inner grooved tube.

Another existing unsolved problem is the refrigerant flow distribution problem. The current air-cooled heat exchangers using small diameter tubes result in many channels for refrigerant flow. Achieving a balanced distribution of the two-phase flow refrigerant in these channels is a severe problem. We don't have a good method for an acceptable distribution of the refrigerant.

I agree with all of the suggestions and recommendations given by Prof. Bergles. Especially I agree with that on "the importance of manufacturing methods and materials". Because our new idea Herringbone tube is based on a new fabricating method. We hope for a fusion between the new heat transfer enhancement methods and the new manufacturing methods in the next generation heat transfer enhancement technology. What we now need is the third generation advanced enhancement technology.

2.3 SUMMARY OF THE SHORT PRESENTATİON BY Dr. MARTIN FIEBIG

I believe for single-phase passive heat transfer enhancement only three modes exist: (1) developing boundary layers, (2) swirl (longitudinal vortices) and (3) self-excited oscillations (flow destabilization). The first two modes may be generated at all Reynolds numbers, but they become more effective at higher Reynolds numbers. The third mode needs a minimal critical Reynolds number.

Here heat transfer enhancement means the enhancement of wall temperature gradient. The definition of the Nusselt number, the dimensionless wall temperature gradient, shows that the heat transfer coefficient is proportional to the product of thermal conductivity times Nusselt number divided by the hydraulic diameter.

High heat transfer coefficients can be achieved by small hydraulic diameters. This has been extensively discussed in this ASI. But a small hydraulic diameter implies higher manufacturing costs and increased fouling problems. These are the limits for the minimal hydraulic diameter.

For the minimal hydraulic diameter, enhancing the Nusselt number can further increase the heat transfer coefficient. There are only the mentioned three modes to increase the Nusselt number. Self-excited fluid oscillations also reduce fouling. So they are very useful in two ways. To generate self-excited flow oscillations at the minimum possible Reynolds number is important for an efficient enhancement device. So I see a need in research for the development of efficient heat transfer enhancement devices or generators.

With regard to the enhancement devices, the following statement can be made:
For generation of swirl, wing type structures are very efficient. For roughly hundred years research has been conducted on wings to generate high lift which means high swirl at minimum drag, which implies minimum pressure loss. Wing type devices generate also developing boundary layers and induce self-excited oscillations. They can be easily incorporated into heat transfer surfaces by punching, embossing or attaching. But the best heat transfer enhancement device depends on the specific heat exchanger. For finned heat exchangers, you can alter the fins by adding, punching or embossing.

I agree with Prof. Bergles and Dr. Ebisu regarding the importance of materials and combined enhancement effects. Certainly the three passive enhancement modes may be combined with active modes. For instance an active frequency source when tuned to the natural frequency of the flow oscillation will further add to the heat transfer enhancement.

2.4 SUMMARY OF THE SHORT PRESENTATİON BY Dr. OLEG A. KABOV

I would like to bring to your attention three unsolved problems. The first problem is condensation in the horizontal longitudinally finned tube. Prof. Web and Prof. Bergles gave us information about the condensation inside the microfin tubes. In the microfin tube, the height of the fin is about 0.3 mm and the fins have a spiral or helical angle to the axis of the tube. What I describe here is condensation in a different tube with longitudinal fins. Height of these fins is 1 mm or a little more than that. This tube has applications in space technology. This tube is produced from an aluminum alloy by extrusion. The inside tube diameter ranges from 4 to 8 mm usually. There is no theoretical model in the literature for prediction of heat transfer enhancement in such a tube. The flow pattern map for this process does not exist in the literature. We have complete vapor condensation in such a tube and we don't know if we can use equations valid for condensation inside smooth tubes.

The second problem is related to the problem mentioned by Prof. Bergles in his suggestions and recommendations. This is vapor condensation on three-dimensional finned or sawtooth shaped surfaces. These surfaces are very promising for enhancement of condensation heat transfer and now many companies use these tubes. One problem I think is that the heat transfer coefficient for this third generation heat transfer enhancement technique is only 40% to 60% higher than the heat transfer coefficient for the second generation heat transfer enhancement technology according to the book by Prof. Webb and the recent results obtained in the Moscow Power Engineering Institute. The third generation enhancement techniques provide a high area ratio coefficient than integral finned tubes and a better condition for condensate film thinning. It is well known now that the main enhancement of heat transfer takes place on the top part of the integral fin tubes. Why don't we have a large enhancement for this third generation heat transfer enhancement technique? We need valid theoretical models for condensation on these third generation heat transfer surfaces and formulation of theoretical models requires systematic experimental investigations in this field.

The third unsolved problem is the horizontal standing wave formation in two-phase flows, which I have described in my first lecture. In a series of studies during 1984-1994, we discovered in our Institute a new phenomena in two-phase flows. Local heating of a slowly moving liquid film generates a horizontal standing wave near the upper age of the heater. And below this horizontal wave, the fluid moves in the form of rivulets with thin liquid films between these rivulets. When we change the dimensions of the heaters in the horizontal direction about ten times and in vertical direction several times, we still obtain the same physical phenomena, which is very stable for different heaters and liquids. We measured the temperature gradient and found that this phenomenon has a thermocapillary nature. A strong temperature gradient appears at certain heat fluxes close to 13 Kelvin per mm near the upper edge of the heater. Why do we need to study this problem? First of all because this phenomena can take place in two-phase flow systems and instability can take place. Prof. Kakaç yesterday showed us the importance of instability in two-phase flows. These phenomena can also appear in space applications under micro gravity conditions. In space applications under micro gravity conditions, we usually have low mass flow rates and annular flow appears at a superficial gas velocity of about 1 cm per second. The second reason for studying these phenomena is that we can try to use it to enhance heat transfer in liquid films. We suggested this enhancement technique based on a nonuniform heat flux distribution on surfaces. Such a surface contains two sections; a heated and another unheated part. With the horizontal standing wave, we can have heat transfer enhancement. We need theoretical models for this horizontal standing wave structure. At this time, we do not even have an acceptable physical modelling of this problem.

2.5 SUMMARY OF THE SHORT PRESENTATION BY Dr. TUNCER KUZAY

In the beginning of the workshop I believe Dr. Bergles made a statement indicating that certain schools in America are planning to take the heat transfer and fluid mechanics out of their curriculum or that some have already have done so. That statement created some excitement and I did respond to that in some fashion and I did make some suggestions. Then Dr. Bergles said, "well maybe these matters should be included in a separate discussion period" which is this. And subsequently chair gave me time to present my thoughts on this. Because I am working at the Advance Photon Source in the Argonnne National Laboratory, I have the chance of witnessing what is being done in the high tech area today with the most advanced X-rays and this gives me some sort of an advantage for telling you what I see in relation to what is being done; the direction, the future and done by whom. So, therefore I am taking it to a different avenue and really my presentation is more on the educational side visevis the engineering education and what our role should be at the universities teaching the next student generation coming in and all those researchers to be trained for the demands of the market tomorrow.

Heat transfer and fluid mechanics are mature sciences with at least 100 years of history and so what follows next in curriculum in the universities and the role of mechanical engineers in the age of high tech? Now, what are the indicators the telling science come from different directions in the US. It may be coming from different directions in Germany, Europe, Russia, etc. too but from my perspective in the US first of all ASME recently as you know included a new section which is called MEMS-"micro electromechanical systems" is a very strong and very rapidly growing section at ASME. And micro scale energy transport is a session that you now routinely see in ASME conferences. That has become a very important part in the ASME's program. You may also know that in America the National Academy of Sciences predicts future for the funding purpose mostly for the government. And national academy of sciences have three branches; national research council, and the national academy of engineering, and the institute of medicine. These three entities have boards made of learned people and they try to read the future and then make advisory suggestions to the government. So, looking at these one can see where we are heading to. So high tech is only on the recommendation of these entities. What we see today is that high tech is only possible if you go down in length and in time. That means length has to be smaller and smaller to really discover the nature and the time scale has to be shorter because many events take in almost non-equilibrium state in the high tech industries. So therefore we are heading to direction of micro scaling and then miniaturization engineering. So that is the direction. That goes along with the Dr. Fiebig's suggestion that we can reduce the hydraulic diameter to enhance the heat transfer. That is the direction at a practical application. So what are the emerging fields? We are more and more looking to solid state devices, material physics, human genome, DNA, microbiology, structural biology. micro machining, micro sensors and photonics. Those are the real emerging fields. Engineers have a lot to play these emerging fields. Today, nobody knows how to prepare the genome samples in crystalline form to look with X-rays. And then only heat transfer people can handle these things. Now sources are lasers, X-rays, and laser X-rays. Optical tools are being used more and more in fact I complement Dr. Mayinger for being in the forefront of setting up a very nice optical laboratory at his University. This is the direction because these tools are able to dissect the dimensions down to the level that we need for microelectronic and microelectromechanical systems. Now who are doing these things today? About 99% physicists, there is hardly any engineering people in all these fields today. That is where the deficiency is which should be remedied. Because not necessary in entirety but in parts these subjects are also subjects that the engineers can handle and should handle. So that takes me to what Prof. Naim Afgan said that we should give engineers more physics indeed. We had to know more physics to handle these issues in the mechanical engineering curricula.

So, therefore I am suggesting to you that in order to be in the high tech area engineering curriculum have to include a very good basis on micro scale energy transport and micro electromechanical systems. Now who is doing that today? In America these subjects are thought today in a very small number of the leading Universities. And in fact, these people have already put a textbook, Microscale Energy Transport. This is a book that probably engineers should learn as a minimum. And this just came out last year. The ASME MEMS proceedings are available for the past few years. You can look at them and then, there is a routine international workshop on micro scale energy transport and two volumes of the proceedings are out already. There is a new journal now on micro scale energy transport so these areas are becoming very important for the future.

2.6 SUMMARY OF THE SHORT PRESENTATION BY Dr. RALPH WEBB

One of the comments made by Prof. Bergles (as presented by Dr. Unsal) was he thought enhanced heat transfer was still mostly an empirical experimental science. However, over the 35 years that I have worked in this field, I have seen significant advances in work to develop semi-empirical or mechanistically based models that provide understanding of the enhancement mechanism and the effect of the geometry and fluid variables. A key purpose of my book "Principles of Enhanced Heat Transfer" (John Wiley & Sons, 1994) was to document these advances. Much of the subject relates to heat transfer from complex geometrical structures. Much of the work to develop predictive models of these geometries has been based on curve fitting or multiple regression type techniques. These are strictly empirical techniques that do not give us mechanistically based understanding. Such empirical techniques should be used only when one does not know how to develop a predictive model based on fundamental understanding.

The tools we have today are much more sophisticated than were available 20 or 30 years ago. Now that CFD is a commonly used analysis tool, I believe that it will make an important contribution to more advanced detailed analysis of heat transfer and friction for complex surface geometries typical of enhanced heat transfer surfaces. An example of where CFD should be useful is for condensation and evaporation in the microfin tube. I presented a paper on Monday that said the condensation coefficient is directly linked to the single-phase heat transfer coefficient. I don't know how to analytically predict the single-phase heat transfer coefficient and friction in microfin tubes, which is complicated by flow separation caused by the helix angle. CFD should be a very useful tool to predict the single-phase heat transfer and friction characteristics. Another example is analysis of the effect of vortex generators. Prof. Fiebig has done excellent CFD analysis on vortex generators.

Another point is that, in addition to prediction, we need to understand the mechanism by which these things work. Today, I talked about understanding the mechanism of nucleate boiling on surfaces having subsurface structures. If we understand the mechanism, we are then set to determine how to make things better, or to do better modelling. Sophisticated experimental tools available today include use of liquid crystals, or holographic interferometry such as done by Prof. Cila Herman. Such tools will be important in advancing understanding the mechanism of enhanced heat transfer. So I think we are moving into a period, in which these sophisticated analysis and experimental tools should be used replace the old approach of just "testing something and measuring its performance."

I have heard references to MEMS (Micro Electro Mechanical Systems) several times in this conference. I definitely believe that we are on a march moving to smaller size structures. I think one of the things important in enhanced heat transfer is that we should not see it as a discipline in itself. Many applications for enhanced heat transfer are in multi-disciplinary problems. We should not approach enhanced heat transfer with the attitude "I am going to make a better tube and it is someone else's job to learn how to use it". I think we have to understand the multi-disciplinary technology that is involved and say "how can we use our enhanced heat transfer skills and integrate them with the multi-disciplinary objective". One example is in the area of micro-electronic technology, which encompasses the making of computer chips, and in the their cooling. Heat transfer considerations should be integral to the chip design, and to the design of the composite chip/heat sink and computer in which the chips are used. We have to join on this team and bring our skills into this technology.

I would also like to comment on changes that I see in materials used in heat exchangers, and competition among materials. In the past, for example, we have seen steam condenser tubes principally made of copper alloys. Today, corrosion considerations are causing copper alloys to be replaced by titanium or stainless steel tubes. Although titanium is highly corrosion resistant, it has a very low thermal conductivity. Use of fins on titanium steam condenser tubes is not very practical, because of its low thermal conductivity. As an example, one may consider a bi-metal (copper/titanium) condenser tube. Such a tube would use a titanium liner on the waterside for good corrosion resistance and an enhanced finned surface on the steam side. Although this makes "heat transfer sense", we must learn how to make this type of a bi-metal structure. We are moving into an area in future, in which coatings and composite structures (e.g., bi-metal structures) are going to be very important. Coatings can also be integrated with enhancement technology. Today, hydrophilic coatings are used to provide improved wetting and condensate drainage in air-cooled evaporators. In dropwise condensation, we want a non-wetting coating. We need to bring enhancement technology into these multi-disciplinary areas and include the materials people in seeking the overall program objectives.

2.7 SUMMARY OF THE SHORT PRESENTATION BY Dr. WEN-JEI YANG

I would like to share with you my concerns regarding heat transfer studies for the future. By coincidence, my point of view is shared by Dr. Kuzay. My remarks will address two different topics. The first is changes emerging in our research and educational priorities. The evolution of the study of thermo-physics began with thermodynamics. This was very long ago, but around the time of the Second World War a new field was established called "heat transfer" which is essentially non-equilibrium thermodynamics. Over the past half-century, as some of my colleagues have already indicated, the study of heat-transfer has become a classical field. But, we must ask, what is to become of the field now? Will it reach a dead end? Industry, of course, does not care because it is product-oriented, concerned only with making money. But academia is different. In academia, we have a mission to seek out new knowledge, an obligation to extend the scope of our research and educate the new generation of graduates. Our product will be these graduates who we will send out into both industry and society. You may have heard of something called "curriculum 2000". The term refers to expectations that, by the year 2000, engineering will have undergone fundamental revisions, revisions that may seriously affect heat transfer studies. Although the heat transfer field is now well established, as long ago as the 1980s ASME tried to eliminate the heat transfer division. Indeed, they had already agreed to rename it the "thermal engineering division", but at the last minute decided against the change. In ASME's classification, heat transfer and fluid mechanics are now included within basic engineering.

What does curriculum 2000 entail in this regard? At the University of Michigan graduate program, to take one example, we have three courses in heat transfer. In the undergraduate program there is also one heat transfer course which includes the three mechanisms of conduction, convection, and radiation. Right now, we offer three three-credit courses each in fluid mechanics, heat transfer, and thermodynamics. However, the department is currently in the process of breaking up the undergraduate heat transfer course, to be distributed between thermodynamics and fluid mechanics. Instead of being divided between the three subjects, conduction and radiation will be included within thermodynamics, and convection heat transfer will be included within fluid dynamics. The result will be two courses, thermodynamics and fluid mechanics, each of four credits each. At this historical moment, the college of engineering, and society more generally, feel it is more appropriate to emphasize design and manufacturing, which are more application-oriented, rather than heat transfer, fluid mechanics, or thermodynamics, which are seen to be too basic. By the turn of the century this perspective will have generated a significant change in American engineering education.

The second thing I would like to address is some specific directions for future research in heat transfer. Let me highlight three of them. One concerns the scale of physical systems: we should direct our interests toward micro, nano and mini scales. I have noticed in this institute the use of terms such as "microfins" and "microchannels". But in fact what these refer to are not micro in scale at all; one cannot make a "micro heat exchanger" merely by reducing its size. The resulting difference will be a mere terminological one. Dr. Kuzay has mentioned MEMS and non-Fourier conduction, and I am currently undertaking research on both of these.

A second direction of research is an emphasis on specific applications, for example heat transfer with applications in manufacturing and materials processing, such as the semiconductor manufacturing that I will lecture about tomorrow. Other applications include those within electronic packaging and biomedicine. There is particular promise for applications in biomedicine. For example, heat shock is known to affect DNA. So far, only biologists and a few medical doctors are working on this topic, and they know little about heat transfer. We need to get involved.

The third direction is research on heat transfer in composite systems, for example, porous media and ceramics in food processing. Thermal conductivity can be increased in composite materials such as graphite in single crystal form embedded in copper or aluminum. I have used this material for boiling. Other materials include C-C (Carbon-Carbon) composites, and structured media such as Dr. Webb mentioned in his lecture. These are my personal perspectives on extending the heat transfer discipline into the future.

3. Concluding Remarks

According to the classification given by Dr. Bergles, we have 14 enhancement methods, six heat convection modes and the radiation heat transfer mechanism. We also have various enhancement techniques, limitless different geometric orientations, different applications with thousands of previous papers published on heat transfer enhancement. Even if we have intended to concentrate only on enhancement, not all of the enhancement methods could have been covered in the limited number of presentations and discussions which took place in the "Panel Discussions" session of this NATO Advanced Study Institute. Consequently, It should be noted that this "Panel Discussions" session has been limited mostly to the areas of expertise of our lecturers and participants.

The statement made by Dr. Bergles about enhancement still being mostly an experimental art indicates that a grater part of future research on enhancement is still going to be experimental in nature. The statement made by Dr. Webb about the importance of predictive models also indicates need for more experimental work on the subject which may lead to a better understanding of the enhancement mechanisms and contribute to the modelling and prediction research in this field. Theoretical research based on computation is also important and will contribute to improved design of enhanced heat transfer equipment.